Studies in Logic
Volume 62

Argumentation and Reasoned Action
Proceedings of the 1st European
Conference on Argumentation,
Lisbon 2015

Volume I

Volume 52
Inconsistency Robustness
Carl Hewitt and John Woods, eds.

Volume 53
Aristotle's Earlier Logic
John Woods

Volume 54
Proof Theory of N4-related Paraconsistent Logics
Norihiro Kamide and Heinrich Wansing

Volume 55
All about Proofs, Proofs for All
Bruno Woltzenlogel Paleo and David Delahaye, eds

Volume 56
Dualities for Structures of Applied Logics
Ewa Orłowska, Anna Maria Radzikowska and Ingrid Rewitzky

Volume 57
Proof-theoretic Semantics
Nissim Francez

Volume 58
Handbook of Mathematical Fuzzy Logic, Volume 3
Petr Cintula, Petr Hajek and Carles Noguera, eds.

Volume 59
The Psychology of Argument. Cognitive Approaches to Argumentation and Persuasion
Fabio Paglieri, Laura Bonelli and Silvia Felletti, eds

Volume 60
Absract Algebraic Logic. An Introductory Textbook
Josep Maria Font

Volume 61
Philosophical Applications of Modal Logic
Lloyd Humberstone

Volume 62
Argumentation and Reasoned Action. Proceedings of the 1st European Conference on Argumentation, Lisbon 2015. Volume I
Dima Mohammed and Marcin Lewiński, eds

Volume 63
Argumentation and Reasoned Action. Proceedings of the 1st European Conference on Argumentation, Lisbon 2015. Volume II
Dima Mohammed and Marcin Lewiński, eds

Studies in Logic Series Editor
Dov Gabbay   dov.gabbay@kcl.ac.uk

Argumentation and Reasoned Action
Proceedings of the 1st European Conference on Argumentation, Lisbon 2015

Volume I

Edited by
Dima Mohammed
and
Marcin Lewiński

© Individual author and College Publications 2016
All rights reserved.

ISBN 978-1-84890-211-4

College Publications
Scientific Director: Dov Gabbay
Managing Director: Jane Spurr

http://www.collegepublications.co.uk

Original cover design by Orchid Creative   www.orchidcreative.co.uk
Printed by Lightning Source, Milton Keynes, UK

All rights reserved. No part of this publication may be reproduced, stored in a retrieval system or transmitted in any form, or by any means, electronic, mechanical, photocopying, recording or otherwise without prior permission, in writing, from the publisher.

Table of contents

1. Deontic authority in legal argumentation: A case study — 1
By Michał Araszkiewicz and Marcin Koszowy
Commentary by Luís Duarte d'Almeida — 21

2. Investigating the impact of moral relativism and objectivism on practical reasonableness — 25
By Michael D. Baumtrog
Commentary by David Hitchcock — 47

3. Advocacy vs. inquiry in small-group deliberations — 53
By J. Anthony Blair
Commentary by Floriana Grasso — 69

4. The bearable ambiguity of the constitutional text: Arguing, bargaining and persuading in the Italian Constituent Assembly — 75
By Giovanni Damele

5. Willingness to trust as a virtue in argumentative discussions — 91
By José Ángel Gascón
Commentary by G.C. Goddu — 109

6. Cooperation in legal discourse — 113
By Stefan Goltzberg
Commentary by Maurizio Manzin — 129

7. A Descriptive and comparative analysis of arguing in Portugal — 135
By Dale Hample, Marcin Lewiński, João Sàágua and Dima Mohammed
Commentary by Paula Castro — 159

8. "All things considered" — 165
By David Hitchcock
Commentary by Erich Rast — 181

9. Collaborative reframing: How to use argument mapping to cope with "wicked problems" and conflicts — 187
By Michael H. G. Hoffmann
Commentary by Sally Jackson — 217

10. Canons of legal interpretation and the argument from authority 221
By Michael Hoppmann
Commentary by Marcin Koszowy & Michał Araszkiewicz 239

11. Modeling argumentative activity in mediation with Inference Anchoring Theory: The case of impasse 245
By Mathilde Janier, Mark Aakhus, Katarzyna Budzyńska and Chris Reed
Commentary by Sara Greco 265

12. A new approach to argumentation and reasoning based on mathematical practice 269
By Andrzej Kisielewicz
Commentary by Andrew Aberdein 287

13. Symbolic condensation in visual and multimodal argumentation 293
By Jens Kjeldsen
Commentary by Michael Gilbert 315

14. Fairness, definition and the legislator's intent: Arguments from *Epieikeia* in Aristotle's *Rhetoric* 319
By Miklós Könczöl
Commentary by Serena Tomasi 337

15. Fair and unfair strategies in public controversies: The case of induced earthquakes 343
By Jan Albert Van Laar and Erik C. W. Krabbe
Commentary by Chris Tindale 363

16. Working with Open Argument Corpora 367
By John Lawrence, Mathilde Janier and Chris Reed

17. Towards an online social debating system 381
By João Leite, João G. Martins and Sinan Egilmez
Commentary by Michael Hoffmann 399

18. How to conclude practical argument in a multi-party debate: A speech act analysis
By Marcin Lewiński — 403

Commentary by Steve Oswald — 421

19. Journalists' emotionally colored standpoints: A path leading to foster existing stereotypes in the audience?
By Margherita Luciani — 429

Commentary by Henrike Jansen — 453

20. Malleability and predictability of source credibility: American election candidates as a case study
By Jens Koed Madsen — 457

Commentary by Dale Hample — 477

21. An epistemological theory of argumentation for adversarial legal proceedings
By Danny Marrero — 481

22. Not just rational, but also reasonable: Critical testing in the service of external uses of public political arguments
By Dima Mohammed — 499

Commentary by Jean Goodwin — 515

23. Prosodic constraints on argumentation, from individual utterances to argumentative exchanges
By François Nemo, Camille Létang and Mélanie Petit — 521

Commentary by Andrea Rocci — 541

24. Repetition as a context selection constraint: A study in the cognitive underpinnings of persuasion — 547
By Davis Ozols, Didier Maillat and Steve Oswald

25. Practical argumentation and multiple audience in policy proposals — 567
By Rudi Palmieri and Sabrina Mazzali-Lurati

Commentary by Jean Goodwin — 589

26. Does public deliberation really need normative constraints? Recovering the Aristotelian rhetorical theory 593
By Salvatore Di Piazza, Francesca Piazza and Mauro Serra
Commentary by Amnon Knoll 609

27. Argumentatively evil storytelling 615
By Gilbert Plumer
Commentary by Paula Olmos 631

28. "Rationality as use": On the nature of rationality in argumentation 635
By Menashe Schwed
Commentary by Nuno Venturinha 659

29. Dialogue grammar induction 665
By Mark Snaith and Chris Reed
Commentary by João Leite 683

30. On cognitive environments 687
By Christopher W. Tindale
Commentary by Andrea Rocci 705

31. HLA Hart on logic and interpretation 711
By Cosmin-Marian Vaduva
Commentary by António Marques 727

32. Speech acts in a dialogue game for critical discussion 731
By Jacky Visser
Commentary by Alice Toniolo 751

33. Speech acts and burden of proof in computational models of deliberation dialogue 757
By Douglas Walton, Alice Toniolo and Timothy J. Norman
Commentary by Jan Albert van Laar 777

34. Using argumentation within sustainable transport communication 781
By Simon Wells and Kate Pangbourne
Commentary by Mark Aakhus 803

35. Giving reasons *pro et contra* as a debiasing technique in legal decision making 807
By Frank Zenker, Christian Dahlma, Rasmus Bååth and Farhan Sarwar
Commentary by Fabrizio Macagno 821

36. Against visual argumentation: Multimodality as composite meaning and composite utterances 829
By Igor Ž. Žagar

Preface

What we are offering you here are the proceedings of the 1st European Conference on Argumentation: Argumentation and Reasoned Action, held at the ArgLab, Nova Institute of Philosophy, Universidade Nova de Lisboa, 9-12 June 2015, Lisbon, Portugal. The European Conference on Argumentation (ECA) is a new pan-European initiative launched in 2013 aiming to consolidate and advance various streaks of research into argumentation and reasoning. ECA's chief goal is to organise, on a biannual basis, a major conference that provides an opportunity for exchanging research results and networking in all areas related to the study of argumentation: philosophy, communication, linguistics, discourse analysis, computer science, psychology, cognitive studies, legal theory, etc. While based in Europe, ECA involves and further encourages participation from argumentation scholars all over the world.

The proceedings comprise what we think is the most complete statement of the state of the art of argumentation studies today. From ancient rhetoric to Artificial Intelligence, and from analytic philosophy to detailed empirical research, the contributors have submitted argumentation theory and practice to a thorough examination. It's gratifying to have all this research collected in one source and realise the breadth and depth of the lively debates in argumentation studies.

The proceedings are divided into two printed volumes. *Volume I* includes 36 long papers along with the commentaries on them delivered during the conference. *Volume II* comprises 10 regular papers presented within three thematic panels, as well as 71 regular papers from the general conference programme. All these papers were subject to a double-blind peer review of the Scientific Panel of more than 50 experts in the field, who evaluated extended abstracts for long papers and short abstracts for regular papers before the conference. As a result, out of 286 submissions nearly 200 were accepted for presentation. Most of them are collected here. We have excluded poster presentations and invited keynote addresses by John Searle, Isabela and Norman Fairclough, and Simon Parsons.

As a collaborative effort, these proceedings cannot be published without duly acknowledging, first, João Sàágua, Fabrizio Macagno and Giovanni Damele, ArgLab members who were part of the ECA Lisbon 2015 Organising Committee. Also, the entire 12-person ECA Steering Committee (see www.ecargument.org) was indispensable in shaping the conference. We would also like to thank Nuno Mora, Marisa Campos and Rita Luís for their crucial assistance with the conference and the proceedings. We wouldn't be able to make it without you!

With this publication, we are happy to conclude ECA Lisbon 2015. Yet, ECA is an on-going project, and for years to come. As we write this, the 2nd European Conference on Argumentation: Argumentation and Inference is being prepared by our colleagues at the University of Fribourg, Switzerland, where it will take place 20-23 June 2017. This will be another chance to present, discuss and publish the most up-to-date research on argumentation. You're welcome to be part of it.

Dima Mohammed
Marcin Lewiński

Lisbon, May 2016

1

Deontic Authority in Legal Argumentation: A Case Study

MICHAŁ ARASZKIEWICZ
Jagiellonian University, Poland
michal.araszkiewicz@uj.edu.pl

MARCIN KOSZOWY
University of Białystok, Poland
koszowy@uwb.edu.pl

The complexity of arguments from deontic authority about what should be done requires employing specific distinctions that would capture those phenomena which are not directly recognized in the existing argument studies. In order to justify this view we elaborate a case study of a legal controversy between the Constitutional Tribunal and the Supreme Court in Poland which is helpful for proposing the way in which existing taxonomies of arguments from authority may be refined.

KEYWORDS: arguments from deontic authority, deontic authority, legal arguments, non-deontic authority

1. INTRODUCTION

The variety of argumentative appeals to authority encompasses a rich repertoire of techniques which establish, employ or attack authorities who have (or are claimed to have) normative powers to tell people what they should do (e.g. Walton & Koszowy, forthcoming). The source of these powers may be of a different sort, e.g.: command, expertise and dignity (see e.g. Goodwin, 1998). Despite the diversity of ways in which these arguments are structured and the important connection with recent models of practical reasoning about what should be done (Atkinson et al., 2006), we may observe the lack of thorough analyses of deontic or normative aspects of using authorities in argumentation. In this paper we employ the distinction between cognitive (*de facto*) authority, i.e. the expert authority in a given field and deontic (*de iure*) authority, i.e. the authority which indicates what should be done (Bocheński, 1974; Walton, 1997; Walton & Koszowy, 2015).

Apart from the thorough study of arguments from expert authority in the philosophy of argument, quite recently, some authors pointed out the need of exploring, in a systematic way, also some deontic components of arguments from expert opinion (Walton, 2014; Walton & Koszowy, 2015) or the need of reconstructing appeals to authority by means of argumentation schemes for reasoning about trust (Parsons et al., 2014). These accounts may constitute a point of departure for a further systematic inquiry into the variety of argumentative techniques employing deontic authority. However, although these works offer some useful notions and distinctions which allow us to grasp some key aspects of this type of arguments, we may observe that the main approach present in the current study of deontic authority in argumentation is rather sketchy and does not provide neither a thorough systematic analysis nor a thorough bottom-up method of the study of arguments based on deontic authority. This view may be illustrated by the fact that there are some cases which show that the existing accounts do not grasp the diversity of argumentative appeals to deontic authority. For instance, in the context of judicial interpretation of law we may observe the difference between the genuine and the apparent deontic authority of the interpretive statements present in the judgments of high courts. In this paper, we defend the claim that the complexity of phenomena related to arguments based on authority requires an introduction of more specific distinctions which would allow to capture those features which are not directly recognized by the existing accounts. In order to justify this view, we elaborate a case study of a legal controversy between the Constitutional Tribunal and the Supreme Court in Poland. The analysis of this case will allow us to propose the taxonomy of appeals to deontic authorities.

We do not claim, however, that our approach provides a comprehensive bottom-up method of dealing with appeals to deontic authority in (legal) argumentation. Our task is more limited: by elaborating our case study we attempt at justifying the legitimacy of extending the existing set of distinctions in order to grasp the variety and the complexity of arguments from deontic authority. We believe that this type of inquiry should constitute an indispensable step towards elaborating a proper method of mining argument structures which are related to deontic type of authority.

The paper is not a contribution to theory of legal argumentation, but rather to the general theory of argument schemes. It is shown how fine-grained distinctions developed and used in legal theory (and implicitly present in judicial practice) may deepen our understanding of the concept of deontic authority and its use in the process of argumentation.

2. THE TREATMENT OF EPISTEMIC AND DEONTIC AUTHORITY IN THE PHILOSOPHY OF ARGUMENT

In this section, we will briefly report on the current state of the art in the study of arguments from authority in the field of philosophy of argument. This short exposition is aimed at showing that although argumentation theory has elaborated useful concepts and distinctions which allow us to analyse and evaluate arguments from authority, its repertoire of devices should be extended (and made more specific) in order to grasp some particular complex phenomena related especially to the normative or deontic dimension of appeals to authority.

In the contemporary philosophy of argument, apart from arguments from authority of "someone who knows" – such as appeals to expert opinion or arguments from position to know (see e.g. Walton, 1997; Walton, Reed & Macagno, 2008), there is a great diversity of arguments based upon the authority of someone who is socially authorized to formulate directives (i.e. orders, commands, requests etc.). The need for a systematic inquiry into this kind of arguments stems at least from the fact that some typical social interactions based on authority involve not only cognition, but also institutional (deontic) powers. For example, the professor who is the authority for his student in a given field of knowledge, may at the same time have "institutional" authority concerning procedures governing the operations of a laboratory (Bocheński, 1974). The authority of the first kind is called "cognitive", "epistemic" or "*de facto*" (Walton, 1997). The second kind may be associated with labels such as "administrative", "deontic" or "*de iure*" – it is commonly characterized as the authority of someone who is socially predestinated to formulate directives (Bocheński, 1974; Wilson, 1983; Woods & Walton, 1974; Walton, 1997).[1]

The general intuition underlying the distinction between the two kinds of authority is that they are different types of relationships:

> The cognitive (epistemic, de facto) type of authority is a relationship between two individuals where one is an expert in a field of knowledge in such a manner that his pronouncements in this field carry a special weight of

---

[1] A similar intuition is present in using some other labels for two kinds of authority, such as "theoretical" and "practical" (Fox, 1972) or "epistemic" and "executive" (De George, 1985) (see Mizrahi, 2013, p. 59).

presumption for the other individual that is greater than the say-so of a layperson in that field. The cognitive type of authority, when used or appealed to in argument, is essentially an appeal to expertise, or to expert opinion. By contrast, the administrative (deontic, de jure) type of authority is a right to exercise command or influence, especially concerning what should be done in certain types of situations, based on an invested office, or an official or recognized position of power (Walton, 1997, pp. 77-78).

Whereas arguments based on the relationship of being for someone an expert in a given field are the subject-matter of thorough studies in the philosophy of argument (see, e.g. Walton, 1997; Walton, 2006; Walton, Reed & Macagno, 1998; Mizrahi, 2013), scant and patchy attention has been paid to arguments which involve the relationship of commanding or influencing someone's views, attitudes or actions by means of appealing to deontic powers. In other words, the diversity of case studies of appeals to cognitive authorities (see e.g. Woods & Walton, 1974; Walton, 2013) contrast with a much less systematic focus on detailed analyses of uses of arguments involving a deontic type of authority. Recent works devoted to the appeals to deontic authority have rather the form of outlines and proposals or explications of typical rhetorical and cognitive mechanisms than the form of coherent and systematic research programs. Here we may observe the gap between the important social role of appeals to deontic authorities in the decision-making processes and the insufficiency of research tools which are incapable of grasping the specific character of appeals to deontic authority.

A starting point of an inquiry into the structure of appeals to deontic or administrative authority may be found in (Budzynska, 2010) who points to the fact that since above argumentation schemes do not include appeals to deontic authority, there is a need of proposing such schemes which would encompass the following circumstance of performing speech acts:

Premise1. i performs F(A)
Premise 2. i is authorized to perform F(A)
Conclusion. A

The key feature of this account would be to precisely grasp the difference between the two types of authority (Walton & Koszowy, 2015). This difference can only be brought about precisely by interpreting how each of them is used as the speech act in a dialectical exchange between two parties. Budzynska has shown the basis for this distinction by describing speech acts appropriate for the use of argument from administrative authority.

According to Walton and Koszowy (2015), the way of defining administrative authority proposed by Budzynska refers to a relationship between two agents in a dialogue where one has a right to exercise command or influence over the other. Typical speech acts for this type of communication are those of command, require and forbid. The dialectical framework for appeals to cognitive or epistemic authority involves following speech acts.: (i) the proponent brings forward an argument based on expert opinion; (ii) the respondent is the party to whom the argument was addressed, and (iii) the expert he proponent starts putting forward an argument (an assertion taken to be supported by reasons) and the respondent has the role of asking critical questions at the next move (or putting forward counter-arguments) (see Walton & Koszowy, 2015).

Administrative authority is more difficult to specify with precision than the authority of expert opinion. For instance, the pronouncement of a judge who has arrived at a decision on the outcome of a trial is final in some ways. For example, in a criminal trial the same defendant cannot be tried for the same crime twice, what is the case of double jeopardy. Even so, the finding of a criminal trial is subject to review in some cases, and a retrial can be ordered, for example if it was found that certain evidence was overlooked in the first trial that might have made a significant difference to its outcome (Walton & Koszowy, 2015).

On this basis, the following argumentation scheme has been advanced to represent a form of argument from administrative or deontic authority (Walton & Koszowy, 2015):

> Premise 1. δ is an administrative (deontic) authority in institution Ω.
> Premise 2. According to δ, I should do A.
> Conclusion. Therefore I should do A.

Matching the scheme is a set of basic critical questions that are devices to raise doubts about whether the argument holds a given instance.[2] They also provide guidance to a person who evaluates arguments from deontic authority:

---

[2] For extended list of critical questions for evaluating arguments from deontic authority see (Walton & Koszowy, 2015).

CQ1: Do I come under the authority of institution Ω?
CQ2: Does what δ says apply to my present circumstances C?
CQ3: Has what δ says been interpreted correctly?

These and other tools may play an instructive role in analysing and evaluating arguments from deontic authority. But the question arises whether they allow to capture the diversity of deontic types of authority? This question is justified because the problem of exploring and representing the structure of attacking arguments from deontic authority depends on the kind of authority which is employed by a given argument. For instance, we may ask whether present distinctions allow us to analyse arguments referring to such types of authority as (1) resulting in establishing an obligation, prohibition or permission, (2) aiming at establishing certain inconclusive, yet normatively relevant results, (3) based on unconditional norms, (4) based on conditional norms designed for attaining certain objectives, (5) based on certain legal norms, and (6) related to social roles that are not regulated by the state.

Our working hypothesis is that some distinctions regarding the notion of authority employed by legal theorists may constitute a promising point of departure for developing the study of arguments from deontic authority in argumentation theory. In the next section we will pave the way for exploring this hypothesis by giving an overview of the study of arguments from authority from the legal perspective.

3. ARGUMENTS FROM AUTHORITY IN LAW

The law is perhaps the most natural area to study arguments based on deontic authority. One of the most important functions of law is to guide behavior by classifying certain acts as obligatory, prohibited or permitted. The source of normative power of the law is a subject of continuous and vivid debate, but in this context it is worth mentioning that there is an influential conception of law according to which the defining feature of law is its authority: law is authoritative in the sense that its commands should be obeyed simply because the law says so; there is no need to appeal to any extra-legal reasons to justify a conclusion based on the law (Raz, 2009).

The scheme of an argument from deontic authority based on a legal rule may be reconstructed as follows:

> Premise 1. According to the law, if the conditions C1, ...Cn are fulfilled, legal consequence LC should follow.
> Premise 2. The conditions C1, ...Cn are fulfilled.

Conclusion. Legal consequence LC should follow.

Obviously, the scheme presented above may be seen as a specific case of an argument scheme from administrative authority (as defined in Section 2), provided that the legal consequence LC is about A that should be done. There is an enthymematic premise in this scheme concerning the obligation to follow the rules of law, where law is classified as the institution Ω and the legal rule in question is an instance of administrative authority δ. The question concerning the normativity of law is very meaningful and it led to development of many sophisticated legal-theoretical conceptions (Araszkiewicz, Banaś, Gizbert-Studnicki & Płeszka, (Eds.), 2015). Scientific investigations notwithstanding, it is a fact that in contemporary democratic countries the argument from deontic authority based on legal rule is a very powerful one. However, one should take into account that legal rules are subject to interpretation (MacCormick & Summers, 1991). Different subject may have different opinions with respect to proper interpretation of a legal rule. For example, the district court and the appellate court may disagree with regard to the right interpretation of a rule constituting a legal basis for the decision in a given case; legal scholars notoriously argue about the abstract interpretation of legal rules and concepts (Araszkiewicz, 2014b). Legal rules may contain ambiguous terms: in this context, legal interpretation will aim at the choice of one of the competing meanings. However, the terms used there may also be vague, open textured or otherwise indeterminate (Endicott, 2001). In this context, sophisticated argumentative methods may be required to define the scope of the term in question (Araszkiewicz, 2014a).

Some entities may be authorized to give binding decisions concerning legal interpretation. The structure of an argument scheme from deontic authority concerning legal interpretation may be presented as follows.

> Premise: According to the opinion of entity E, legal rule R should be interpreted in accordance with the interpretation I.
> Conclusion: Legal rule R should be interpreted in accordance with the interpretation I.

However, some problems may arise when the source of the enthymematic premise of this scheme (according to this premise legal rules should be interpreted in accordance with opinions of entity E) is a legal rule itself, because this legal rule is also subject to interpretation.

The case study we intend to discuss deals precisely with such a situation.

4. CASE STUDY: THE BINDING POWER OF INTERPRETIVE JUDGMENTS ISSUED BY THE CONSTITUTIONAL TRIBUNAL

In order to explore our main hypothesis (which is that some legal-theoretic distinctions regarding the notion of authority may be helpful in developing the study of arguments from deontic authority), in this section we propose to discuss the case related to the legal controversy between the Constitutional Tribunal and the Supreme Court in Poland. This important controversy has become the subject of investigation in Polish jurisprudence, in particular in the context of the phenomenon of the so-called multi-centric character of contemporary legal systems (*locus classicus*: Łętowska, 2005). Here, we are intending to look at this problem from a more abstract perspective, concerned with the structure of the arguments used and not with the actual normative consequences for the Polish legal system. However, a brief introduction to the content of relevant norms of the Polish law is necessary in this place.

The Constitutional Tribunal is authorized to decide on constitutionality of legal provisions which may be found in different normative acts. According to the Polish Constitution[3], the judgments of the Constitutional Tribunal are universally binding (art. 190. 1 of the Constitution). The Supreme Court, in turn, supervises the common courts and military with regard to the development of their case law (which is, however, not formally binding in Poland, which belongs to the culture of civil law)[4]. In case of doubts concerning proper interpretation of the law, it is authorized to enact the so-called resolutions[5]. Their purpose is to unify the case law in order to avoid discrepancies in judiciary interpretation of legal provisions. The resolutions in principle not formally binding on Polish courts, but practically they are followed due to very high status of the Supreme Court.

At certain point of development of the Polish legal culture, the Constitutional Tribunal have begun to issue not only judgments concerning constitutionality or unconstitutionality of legal provisions,

---

[3] Constitution of the Republic of Poland of 2 April 1997, Journal of Laws 1997, No. 78, position 483, with amendments.

[4] See art. 183. 1 of the Constitution and the Supreme Court Act of 23 November 2002, Journal of Laws 2013, position 499, cons. version with amendments.

[5] According to the art. 61 of the Supreme Court Act.

but also, among others the so-called interpretive judgments (which rulings are of the form "Legal rule R, interpreted in accordance with interpretation I, is (in)consistent with the Constitution"). The question arose whether judgments of this type can be classified as formally binding on the Polish courts. The main context for this issue has been provided by the institution of retrial in civil cases. According to the art. 4011 of the Code of Civil Procedure, it is possible to apply for a retrial, if the Constitutional Tribunal rules on the unconstitutionality of a provision on which the issued judgment was based on. A specific legal issue has been identified whether the issuance of an interpretive judgment by the Constitutional Court also creates a basis for a successful application for a retrial. An affirmative answer to this question would lead to the conclusion that common courts are in certain sense bound by the interpretation of law contained in the judgments of the Constitutional Tribunal.

Let us explain the reasoning leading to this conclusion in a more detailed manner: assume that a common court decides a case of the basis of a rule R in accordance with interpretation I. If afterwards the Constitutional Tribunal issues and interpretative judgment according to which the rule R, if read in accordance with I, is unconstitutional, then the application for a retrial of the case decided by the common court would be successful. The common court would be obligated to decide the case once again, adopting another interpretation of the legal rule R. Such conclusion seems to be grounded in the art. 190.1 of the Constitution, according to which the judgments of the Constitutional Tribunal are universally binding. Let us add that the art. 190.4 of the Constitution explicitly states that the judgments of the Constitutional Tribunal concerning unconstitutionality of legal rules provides a basis for a retrial in cases where decisions were based on those unconstitutional rules.

The issue led to disagreement among the Polish judiciary. The Supreme Court initially represented a rather ambiguous attitude. It issued a few decisions according to which the interpretive judgments of the Constitutional Tribunal provided grounds for retrials of civil cases. One of this decisions had a status of a resolution (of 9 June 2009, II PZP 6/09). The Constitutional Tribunal itself, as a matter of course, supported a thesis according to which its interpretive judgments create a basis for a retrial.

However, eventually, the Supreme Court adopted a negative stance and, in an enlarged panel of 7 judges, issued a resolution of 17 December 2009, III PZP 2/09. This resolution has been classified as the so-called Legal Principle, which means that it is formally binding for the Supreme Court. According to the resolution, the interpretive judgments of the Constitutional Tribunal can provide no basis for a retrial of a civil case. Hence, if a case has been decided on the basis of legal rule R and its interpretation I, assessed an unconstitutional as the Constitutional Tribunal, there is no right to retrial of this case. The Supreme Court justified its position by means of, inter alia, the following arguments:

- before the Constitution of 1997 entered into force, the Constitutional Tribunal was explicitly vested with a power to establish the so-called universally binding interpretation of statutes. However, the Constitution does not provide for such a solution; therefore, actually there is no normative basis for issuing of interpretive judgments by the Constitutional Tribunal; hence, there is no basis for either formally binding character of the interpretation of law promoted by the Constitutional Tribunal on the Supreme Court and the common courts, or the other way round;

- the resolutions issued by the Supreme Court, although they have an explicit statutory basis, are generally not formally binding on the common courts (they are binding only in the case in which they have been issued and additionally resolutions classified as Legal Principles are binding on the proper panels of the Supreme Court);

- the binding character of the interpretive judgments issued by the Constitutional Tribunal would impair the independence of the judiciary, who are subordinate to the Constitution, the international law and the statutes only;

- the conception of interpretive judgments is also at odds with the constitutional principles of separation of powers, because the said judgments introduce normative novelty to the existing regulation, while only the legislator is authorized to introduce such novelties.

The resolution of the Supreme Court outlined above offers multiple possibilities for analysis of different types of arguments based on authority. As observed above, any argument based on a legal rule may be reconstructed as appealing to authority due to the authoritative component included in the concept of law. For similar reasons, of a conclusion of an argument is the content of a judicial decision, then by definition this conclusion is supported by the appeal to authority of the court on the one hand, and of the legal basis of this judgment on the other hand. Let us observe that the arguments based on authority may have a varying scope of binding power. While typically arguments appealing to the authority of law, if properly built, lead to the conclusions that are binding on everyone, the arguments based on judicial decisions, concretizing the abstract and general legal norms, are formally binding on the parties to the dispute only. However, certain judgments have more extensive formal binding power, and the judgments issued by the Constitutional Tribunal concerning the (un)constitutionality of legal rules have are a good and in a way extreme examples of those: they are universally binding similarly to legal rules. The issue of the present case study is whether the interpretive judgments of the Constitutional Tribunal also have the same feature.

The fact that legal rules are subject to interpretation adds complexity to the discussion of arguments based on authority in legal contexts. If an interpretation of a rule is contestable, then the persuasive force of an argument based on this rule is seriously limited. The case study outlined above is particularly interesting, because it pertains not only to the issue of contestable interpretations of legal rules, but it introduces a multi-level context where interpretations of legal rules are the subject of judgments the scope of binding power of which is contestable, because the interpretation of rules on the basis of which the judgments are issued is in itself contestable.

5. ANALYSING THE CASE BY MEANS OF ARGUMENTATION SCHEMES

Let us now attempt to systematize the structure of argumentation present in the case study by means of the argumentation schemes presented in the section 3 of this paper. Assume that the legal rule R, interpreted in accordance with an interpretation I has been assessed as unconstitutional in an interpretive judgment of the Constitutional Tribunal. The same legal rule R interpreted in the manner I has been a basis of an earlier judgment by the court C and a motion for retrial of civil proceedings is filed to the court C (the circumstances referred to in the previous two sentences should be jointly referred to as *conditions*). The court C asks itself: am I obligated and authorized to initiate the retrial, or should I reject the motion?

Below, we present a series of arguments from authority employed to support the former conclusion (Arguments I-IV) and the latter one (Arguments A-D).

The first argument from authority applicable to the situation is based on the article 401[1] of the CCP:

Argument I.
Premise 1. According to the article 401[1] of the CCP, if *conditions* are fulfilled, the court should initiate a retrial of civil proceedings.
Premise 2. Conditions are fulfilled.
Conclusion. The court should initiate a retrial of civil proceedings.

Now, this argument may be attacked by pointing out that it is contestable that the *conditions* are actually within the scope of application of the rule expressed in the article 401^1 of the CCP. Concretely, it may be contested whether interpretive judgments of the Constitutional Tribunal are within the scope of application of the article 401^1 of the CCP. As we know, the Constitutional Tribunal promotes an interpretation of this provision according to which its interpretive judgments are within the scope of application of the said provision. This argument from authority based on this opinion of the Constitutional Tribunal may be represented in the following manner:

> *Argument II.*
> *Premise.* According to the opinion of the Constitutional Tribunal, the article 401^1 of the CCP should be interpreted in such way that the *conditions* are within the scope of the article 401^1 of the CCP.
> *Conclusion.* The article 401^1 of the CCP should be interpreted in such way that the *conditions* are within the scope of the article 401^1 of the CCP.

Let us note that the conclusion of Argument II is equivalent with regard to its meaning with the premise 1 of Argument I. However, the argument II may be attacked by showing that the opinions of the Constitutional Tribunal concerning the interpretation of legal rules are not binding on the courts. This kind of attack would be an undercutting attack, directed against the relation between the premise and conclusion of Argument II. However, such attack may be responded by another argument from authority based on the art. 190.1 of the Constitution.

> *Argument III.*
> *Premise.* According to the article 190.1 of the Polish Constitution, the judgments of the Constitutional Tribunal are universally binding.
> *Conclusion.* The judgments of the Constitutional Tribunal are universally binding.

Let us note that although the Argument III is based on a legal rule, it is not necessary to discuss the problem of conditions of application of this rule, because the said rule expresses an unconditional feature of judgments of the Constitutional Tribunal. The conclusion of the Argument III reinforces the link between the premise and the conclusion of the Argument II, basing on the true assumption that the interpretive opinion expressed in the premise of Argument II was a part of a judgment of the Constitutional Tribunal. The Argument III may be attacked, too, by pointing out that the interpretive judgments of the Constitutional Tribunal, contrary to the literal formulation of the provision of the Constitution, are excluded from the scope of the expression "the judgments of the Constitutional Tribunal" used in the formulation of art. 190.1 of the Constitution.

The final argument for the favor of the conclusion according to which the interpretive judgments of the Constitutional Tribunal are a basis for a retrial of civil proceedings is based on the article 190.4 of the Constitution:

> *Argument IV.*
> *Premise.* According to the article 190.4 of the Polish Constitution, the judgments of the Constitutional Tribunal concerning the unconstitutionality of legal provisions are a basis for retrial of civil proceedings.
> *Conclusion.* The judgments of the Constitutional Tribunal concerning the unconstitutionality of legal provisions are a basis for retrial of civil proceedings.

Again, this argument, directly supporting the premise 1 of the Argument I, may be attacked by pointing out that interpretive judgments of the Constitutional Tribunal cannot count as "the judgments of the Constitutional Tribunal concerning the unconstitutionality of legal provisions".

Let us now present arguments from authority in favor of the second of the deliberated conclusions, namely, that a motion for a retrial of a civil case, where the remaining *conditions* are satisfied, should be dismissed. The first of these arguments is a peculiar type based on legal rule. Its peculiarity stems from the fact that it is a negative argument: the lack of legal rule authorizing the Constitutional Tribunal to issue binding interpretive judgments leads to the conclusion that the interpretive judgments of this tribunal cannot bind the courts, and therefore cannot be a basis for a retrial of civil proceedings:

> *Argument A.*
> *Premise 1.* Only if there is an explicit legal rule authorizing the Constitutional Tribunal to issue binding interpretive judgments, the interpretation of law presented in its judgments is binding on the courts.
> *Premise 2.* There is no such rule in the legal system.
> *Conclusion.* Interpretation of law contained in interpretive judgments of the Constitutional Tribunal is not binding on the courts.

The Argument A is based on the appealing assumption according to which there should be an explicit legal basis for any competence of any public authority, especially if this competence leads to issuance of binding decisions, directed towards other authorities and/or subjects of law. The controversial part of this argument is, of course, its second premise, which in fact is a conclusion of assumedly successful attacks on the Arguments I-IV, which aimed to show that there are legal grounds for binding character of the Constitutional Tribunal's interpretive judgments. The question is, then, how the attacks on Arguments I-IV could be justified. The answer is that they are justified by arguments based on the authority of the Supreme Court and its resolution discussed above. The conclusions of the arguments B, C and D are propositions attacking the Arguments I-IV:

Argument B.
Premise. According to the opinion of the Supreme Court, the article 401^1 of the CCP should be interpreted in such way that the *conditions* are not within the scope of the article 401^1 of the CCP.
Conclusion. The article 401^1 of the CCP should be interpreted in such way that the *conditions* are not within the scope of the article 401^1 of the CCP.

Argument C.
Premise. According to the opinion of the Supreme Court, the article 190.1 of the Constitution should be interpreted in such way that the interpretive judgments of the Constitutional Tribunal are not universally binding.
Conclusion. The article 190.1 of the Constitution should be interpreted in such way that the interpretive judgments of the Constitutional Tribunal are not universally binding.

Argument D.
Premise. According to the opinion of the Supreme Court, the article 190.4 of the Constitution should be interpreted in such way that the interpretive judgments of the Constitutional Tribunal are not "judgments concerning unconstitutionality of legal provisions".
Conclusion. The article 190.4 of the Constitution should be interpreted in such way that the interpretive judgments of the Constitutional Tribunal are not "judgments concerning unconstitutionality of legal provisions".

Note that the Argument B is a "mirror image" and a direct rebutting attack on the conclusion of Argument I. The Argument C undercuts the

Argument II and attacks the premise of the Argument III (to recall: the Argument III justified the link between the premise and the conclusion of Argument II). Finally, the Argument D attacks the premise of Argument IV.

The main critical question which could be asked in connection with the arguments B-D concerns the binding character of the opinions issued by the Supreme Court. Why should these interpretive opinions be followed by the courts? As indicated above, the interpretation of law advocated by the Supreme Court is not formally binding on any subject, except for the resolutions classified as the Principles of Law, which are formally binding on the panels of the Supreme Court itself, although they may be rebutted by a subsequent resolution of its larger panels. As a consequence, the premises of arguments B, C and D might be contested on the basis of lack of actual binding authority of opinions of the Supreme Court concerning the interpretation of law. However, although they are not formally binding – they are not precedents in the strict sense of the word – the resolutions of the Supreme Court are practically followed by the courts and it is expected from the court that they should follow the resolutions. The acceptance of this type of weaker authority of the resolutions of the Supreme Court reinforces the arguments B-D, thus leading to the successful undercutting of the Argument II and excluding the interpretive judgments of the Constitutional Tribunal from the scope of premises of Argument I and Arguments III and IV.

Obviously, the analysis of the case study presented above does not exhaust the full scope of legal problems and relations between arguments based on authority and present in the case. However, it provides a basis for certain distinctions in the general category of administrative authority as defined in section 2 above.

6. TOWARDS A BROADER VIEW

This case clearly shows that current accounts of arguments from authority about what should be done which distinguish only a couple of general categories (such as cognitive or *de facto* and administrative or *de iure* authority), are not always capable of grasping the variety of argumentative appeals to authority. However, in our opinion, the contemporary research landscape concerning different normative concepts enables us to introduce more refined distinctions into the theory of arguments based on authority. The most general term available in this context is normative authority (see the point (ii) below where we distinguish between deontic and non-deontic authority). We are able to form arguments based on authority focused on desired behavior of the addressee of the argument irrespective of:

> (i) whether this authority is based on any type of support by the State (formal authority) or on other sources (informal authority),
>
> (ii) whether according to this authority certain behavior is defined as obligatory or permitted and this qualification is binding on the addressee of the argument (deontic authority) or there are only weak, although normative, guidelines to follow this pattern of behavior (non-deontic authority),
>
> (iii) whether the authority requires certain type of behavior be fulfilled unconditionally, in any circumstances (categorical or unconditional authority) or only upon fulfillment of certain conditions (hypothetical or conditional authority).

As for the first distinction (between formal and informal authority), the case study encompassed only the instances of formal authority. Informal authority may be found in social contexts where direct influence of the State is absent: for instance, in the families, societies and clubs, sports etc. The legal context implies the presence of formal authority. However, the case study presented above clearly shows that there are different sources of arguments based on formal authority: the text of statutory law and different types of judicial decisions.

The second distinction (between deontic and non-deontic authority) may be assessed as the most interesting and important one. The discussed case study clearly shows that conclusions stemming from an argument based on authority may consist in imposition of a strict obligation or prohibition, or be otherwise formally binding on the addressees. However, this is not always the case. The arguments based on authority may yield conclusions which are normative, although in a weaker sense. Such feature is characteristic for the judgments of the Supreme Court. Although the common courts are not formally obligated to follow the interpretation of law expressed in the judgments of the Supreme Court, it is expected from them to adopt this interpretation. The justification of such behaviour of common courts is not purely instrumental (the risk of overturning of a decision going against the opinion of the Supreme Court), but it is normative (adoption of another interpretation would be assessed as infringement of a norm valid in the community of the judiciary (although this is not a legal norm). According to the opinion of the Supreme Court, also the interpretive judgments of the Constitutional Tribunal can have only this type of weak normativity, but they cannot be formally binding on courts. The

Constitutional Tribunal is of a different opinion, stating that its judgments have the qualification of deontic authority.

As for the third distinction (categorical – hypothetical authority), certain conclusions stemming from arguments based on authority are binding in any circumstances, and some of them only if certain conditions are fulfilled. Arguments based on authority may fit into one of these categories depending on the type of rule on which the argument is based. While certain rules are unconditional (it is necessary to follow them in any circumstances), the scope of application of majority of them is restricted to satisfaction of certain conditions. The rule expressed in the article 190.1 of the Constitution is an example of an unconditional rule, while the remaining legal rules quoted in the case study are conditional rules.

The criteria of division outlined above may be used simultaneously to define more specific types of normative authority. These distinctions may be useful in analysis of actual judicial argumentation, especially in cases of disagreement between the opinions of higher courts, as in the case study discussed above.

7. CONCLUSION

Our proposal for the contribution to the field was twofold. First, we elaborated a real-life case study to show how diverse actual types of arguments from normative authority in law are; we also investigated certain types of conflicts between these arguments. Second, this case study allowed us to propose the way in which existing taxonomies of arguments from authority may be refined. The conceptual framework proposed in this paper may constitute a motivation for a further systematic inquiry which would have two directions: top-down and bottom-up. The top-down approach would consist of the proposed taxonomy of appeals to deontic authority as an instructive tool for classifying appeals to authority.

However, we should note that the proposed case study is only a very limited sample and it is just aimed at justifying the legitimacy of elaborating a thorough method of classifying and analyzing argumentative appeals to deontic authorities. Hence, in order to check whether or not this taxonomy is close to the actual argumentative practice, we should also employ the bottom-up approach which would consist of the analyses those communication contexts in which the notion of authority plays a key role. Among such contexts there are not only cases of appealing to the authority of institutions in legal cases but also cases of quoting institutional, administrative and moral authorities in the media.

Furthermore, the proposed approach focused on combining two disciplines which study arguments from deontic authority, i.e. the philosophy of argument and legal theory might be mutually beneficial for both of them. On the one hand, the philosophy of argument may benefit from legal theory by making existing taxonomies of arguments from deontic authority more specific. On the other hand, legal theory may employ some philosophical distinctions which reflect some key intuitions that might be employed in mining arguments from deontic authority in legal texts.

ACKNOWLEDGEMENTS: We gratefully acknowledge the support of the Polish National Science Center for Marcin Koszowy under grant 2011/03/B/HS1/04559.

REFERENCES

Araszkiewicz, M. (2014a). Legal Interpretation: Intensional and extensional dimensions of statutory terms. In E. Schweighofer, M. Handstanger, H. Hoffmann, F. Kummer, E. Primosch, G. Schefbeck & G. Withalm (Eds.), *Zeichen und Zauber des Rechts. Festschrift für Friedrich Lachmayer* (pp. 469-492). Bern: Editions Weblaw.
Araszkiewicz, M. (2014b). Scientia Juris: A Missing Link in the Modelling of Statutory Reasoning. In R. Hoekstra (Ed.), *JURIX 2014, The Twenty-Seventh Annual Conference*, (pp. 1-10). Amsterdam: IOS Press.
Araszkiewicz, M., Banaś, P., Gizbert-Studnicki, T. & Płeszka, K. (Eds.). (2015). *Problems of Normativity, Rules and Rule-Following*. Cham: Springer.
Atkinson, K., Bench-Capon, T., & McBurney, P. (2006). Computational representation of practical argument. *Synthese, 152*, 191-240, doi: 10.1007/s11229-005-3488-2.
Bocheński, J. M. (1974). An analysis of authority. In F.J. Adelman (Ed.), *Authority* (pp. 58-65). The Hague: Martinus Nijhoff.
Budzynska, K. (2010). Argument analysis: Components of interpersonal argumentation. In P. Baroni et al. (Eds.), *Frontiers in artificial intelligence and applications. Proceedings of 3rd International Conference on Computational Models of Argument (COMMA 2010)* (pp. 135–146). Amsterdam: IOS Press.
De George, R.T. (1985). *The Nature and Limits of Authority*. Lawrence: University Press of Kansas.
Endicott, T. (2001). *Vagueness in Law*. Oxford: Oxford University Press.
Fox, R. (1972). Two kinds of authority. *Philosophy in Context, 1*, 32-35.
Goodwin, J. (1998). Forms of authority and the real ad verecundiam. *Argumentation, 12*, 267-280.

Łętowska, E. (2005). Multicentryczność współczesnego system prawa i jej konsekwencje (The Multi-centric Character of the Contemporary Legal System and the Consequences Thereof). *Państwo i Prawo (Law and State)*, *4*, 3-10.

MacCormick, N., & Summers, R. (Eds.). (1991). *Interpreting Statutes: A Comparative Study*. Darthmouth: Darthmouth Publishing Co Ltd.

Mizrahi, M. (2013). Why arguments from expert opinion are weak arguments. *Informal Logic*, *33*, 57-79.

Parsons, S., Atkinson, K., Li, Z., McBurney, P., Sklar, E., Singh, M., Haigh, K., Levitt, K., & Rowe, J. (2014). Argument schemes for reasoning about trust. *Argument & Computation*, *5*(2-3), 160-190, doi: 10.1080/19462166.2014.913075.

Raz, J. (2009). *The Authority of Law*. 2nd edition. Oxford: Oxford University Press.

Walton, D. (1997). *Appeal to Expert Opinion*. University Park, PA: Pennsylvania State University Press.

Walton, D. (2013). An argumentation model of forensic evidence in fine art attribution. *Artificial Intelligence & Society*, *28*(4), 509-530, doi: 10.1007/s00146-013-0447-1.

Walton, D. (2014). On a razor's edge: Evaluating arguments from expert opinion. *Argument and Computation*, *5*(2-3), 139-159.

Walton, D., Koszowy, M. (2015). In B. Garssen, D. Godden, G. Mitchell & A. F. Snoeck Henkemans (Eds.), *Proceedings of the eight international conference of the International Society for the Study of Argumentation* (pp. 1483-1492). Amsterdam: Sic Sat (CD ROM).

Walton, D., Koszowy, M. (forthcoming). Explaining the fallacy of argument from authority (*argumentum ad verecundiam*), under review.

Walton, D., Reed, C., & Macagno, F. (2008). *Argumentation Schemes*. Cambridge etc.: Cambridge University Press.

Wilson, P. (1983). *Second-hand Knowledge: An Inquiry into Cognitive Authority*. Westport: Greenwood Press.

Woods, J., Walton, D. (1974). Argumentum ad verecundiam. *Philosophy and Rhetoric*, *7*, 135-153.

Legal Authority: Three Distinctions?
Commentary on Araszkiewicz's and Koszowy's Deontic Authority in Legal Argumentation

LUÍS DUARTE D'ALMEIDA
University of Edinburgh, UK
luis.duarte.almeida@ed.ac.uk

Araszkiewicz and Koszowy claim that there are three important distinctions that existing accounts of arguments from authority fail to recognise. I want to challenge them on that claim.

Let me start with the basic notions of theoretical and practical authority, which Araszkiewicz and Koszowy refer to—unhelpfully, I think (for reasons that I will explain)—as "cognitive" and "deontic" authority, respectively. According to the standard characterisation, in an argument from theoretical authority we rely on the pronouncements of an expert on a certain topic in order to conclude that there is reason to believe that a certain proposition is true. In an argument from practical authority, by contrast, we rely on the pronouncements of the holder of a normative power in order to conclude that there is reason for someone to act in a certain way.

Araszkiewicz and Koszowy are concerned with arguments from practical authority in law. Such arguments, they say, instantiate the following scheme (which I reproduce with minor notational changes):

Araszkiewicz's and Koszowy's scheme
(1) According to the law, if conditions C1, . . . Cn are fulfilled, legal consequence LC should follow.
(2) Conditions C1, . . . Cn are fulfilled.
Therefore,
(C) Legal consequence LC should follow.

They add that "[t]here is an enthymematic premise in this scheme, concerning the obligation to follow the rules of law".

I don't think, however, that this is an adequate reconstruction of the structure of arguments from practical authority in law. I see no reason that the pattern should feature a pair of premises like Araszkiewicz's and Koszowy's premises (1) and (2). One suggestion would be to revise their scheme as follows:

A first revision of Araszkiewicz's and Koszowy's scheme
(1') According to the law, legal consequence LC should follow.
Therefore,
(C) Legal consequence LC should follow.

A premise like (1') is what lawyers call a proposition of law. It is a statement of what the law is on a given matter in a given legal system. And to my mind that is all that Araszkiewicz and Koszowy need by way of an explicit premise in the scheme of arguments from practical authority in law. They need a premise stating a legal requirement—a premise reporting what it is that should be done according to the law.

But notice two things in Araszkiewicz's and Koszowy's original scheme. First, their premise (1) is also a proposition of law. Second, their premises (1) and (2), taken together, seem to deductively entail the claim that

(3) According to the law, legal consequence LC should follow.

Now, this claim, too, is a proposition of law. As such, it is all that is required in Araszkiewicz's and Koszowy's scheme. And it is precisely this claim that appears—under the guise of premise (1')—in my proposed revision of their scheme. The only difference between the claim in (3) and Araszkiewicz's and Koszowy's premise (1) is that while the latter claim might be directly established on the basis of, let us say, a valid statutory provision, the claim in (3) would be established on the basis of a valid statutory provision *together* with the fact that the antecedent of the conditional is satisfied. But that has nothing to do with the structure of the argument. It has to do only with the reasons that may support the adoption of the premise that reports the content of the relevant authoritative pronouncement.

I should also say that I don't find Araszkiewicz's and Koszowy's language particularly clear. I would rephrase premise (1') more plainly in terms of actions that should be performed, rather than in terms of "consequences" that should "follow". We could rewrite it like this:

(1'') According to the law, agent A should perform action φ.

In fact this could be further refined. The following would capture even more clearly, I think, the fact that we are dealing with arguments from authority:

(1''') The law requires that agent A perform action φ.

So here is how Araszkiewicz's and Koszowy's scheme could once again be revised:

> *A second revision of Araszkiewicz's and Koszowy's scheme*
> (1''') The law requires that agent A perform action φ.
> Therefore,
> (C') Agent A should perform action φ.

The enthymematic premise that they mention—a necessary premise for the argument to run—would accordingly read as follows:

> (E) We should behave as the law requires.

These are minor points of criticism. But another, more substantial worry is that Araszkiewicz's and Koszowy's three distinctions—the distinctions which existing accounts of arguments from authority in law supposedly fail to recognise—appear to add little, if anything, to our understanding of arguments from practical authority.

Araszkiewicz's and Koszowy's first distinction is between what they call "formal" sources of authority, like statutory texts or judicial decisions, and "informal" sources of authority, like the prescriptions issued to children by their parents or the rules of clubs or sports. I am not sure that I fully grasped the difference; I don't know that the rules of a sport or a club are any less "formal" than the ones we find in statutes. But the main objection is that it is far from clear that by being alert to this distinction we will have made any progress in our understanding of arguments from authority. If someone's requirements are indeed authoritative, then they afford the opportunity for arguments from authority that instantiate the scheme regardless of whether their authority is formal or informal.

The third distinction that Araszkiewicz and Koszowy draw—I will come to their second distinction in a moment—is a distinction between what they call "categorical" and "hypothetical" authority. This turns on whether the action required by the authority is prescribed only under certain conditions. But again this seems to me to have little bearing on our understanding of arguments from authority. It is simply a matter of making it clear, when describing the content of the relevant requirement, that it may involve the verification of certain conditions:

> (1'''') The law requires that (agent A perform action φ if *p*).
> Therefore,
> (C'') Agent A should perform action φ if *p*.

As to Araszkiewicz's and Koszowy's second distinction, they call it "the most interesting and important" of the three, and I agree. It is the distinction between what they term "deontic" and "non-deontic" authority. Here I have one quibble and one substantive suggestion. The quibble concerns Araszkiewicz's and Koszowy's terminology. Their distinction is meant to track a difference between two types of practical authority. In their paper, however, as I have already noted, Araszkiewicz and Koszowy also use the phrase "deontic authority" to refer to practical as opposed to theoretical authority. So what we now have is, it seems, a distinction between *deontic* deontic authority, and *non-deontic* deontic authority. That can be confusing.

My substantive suggestion regards the content of their distinction. Araszkiewicz and Koszowy present it as a distinction between cases in which the person subject to authority has an *obligation* to perform the relevant action, and cases in which that person's normative position is the weaker one of having a reason—but not necessarily an obligation—to act in a certain way. My suggestion is that what I have offered above as a revised version of Araszkiewicz's and Koszowy's scheme provides us with a satisfactory way of representing this distinction. We just need to disambiguate the word "should", which occurs both in the enthymematic premise and in the conclusion of the argument. Depending on context, such claims might be read in a number of ways:

(E') We have reason to behave as the law requires.
(E'') We have strong reason to behave as the law requires.
(E''') We have a duty to behave as the law requires.

The bottom line is that the distinction between reasons of different strength can easily be accommodated. Contrary to Araszkiewicz's and Koszowy's suggestion, it does not seem to upset the standard way of accounting for arguments from authority in law.

2

Investigating the Impact of Moral Relativism and Objectivism on Practical Reasonableness

MICHAEL D. BAUMTROG
ArgLab, Universidade Nova de Lisboa, Portugal
baumtrog@fcsh.unl.pt

The first part of this paper outlines the model for practical reasoning and argumentation developed by Sàágua and Baumtrog (forthcoming) while highlighting the role of value considerations. The second part applies the model to an example of practical reasoning from Anders Behring Breivik, first, using moral relativist views and second, using moral objectivist views. The third stage discusses the impacts of the results.

KEYWORDS: Anders Behring Breivik, objectivism, practical reasoning, reasonableness, relativism

1. INTRODUCTION

While the philosophical notions of a reason and reasoning have often been a major component of moral theory (Scanlon, 1998; Dancy, 2004; Parfit, 2011; Gert, 2005), only a small amount of recent work has drawn out the importance of considering values in practical reasoning and argumentation construction and evaluation from a reasoning and argumentation perspective (Walton, 2007; Bench-Capon & Atkinson, 2008; Fairclough & Fairclough, 2012; Baumtrog, 2013). In these works, questions about the moral values included in practical reasoning have come to light and there is general agreement that a full evaluation of reasonableness needs to consider the positive character (goodness) of the values employed therein. However, while these frameworks have much to say about the place in the structure of reasoning where value considerations occur, they say little about how the standards used to determine positive value can impact evaluations of reasonableness. In this paper I aim to begin to explore the connections between moral standards and standards of reasonableness by investigating the implications of using differing moral perspectives when evaluating the reasonableness of practical reasoning. Accordingly, this exploratory paper has two main aims. The first is to discuss the place of moral considerations in practical reasoning and argumentation. The second is to highlight the importance that the selection of a moral perspective bears on coming to and evaluating a decision to act.

To achieve these aims, the paper proceeds in three steps. First, a model of practical reasoning is outlined. Second, the model is applied to an instance of practical reasoning conducted by mass murder Anders Behring Breivik. Zooming in on the part of the scheme where moral considerations arise most forcefully, the third part of the paper contemplates the impact that differing moral views, divided broadly between moral relativist and moral objectivist positions, can have on Breivik's practical reasoning.

2. AN INTEGRATED MODEL OF PRACTICAL REASONING AND ARGUMENTATION

Fruitful suggestions and models for evaluating practical argumentation have recently been forwarded by Walton (2007), Atkinson and Bench-Capon (2008), Hitchcock (2011), and Fairclough and Fairclough (2012). Using insights from all of these scholars, but overcoming what we saw to be some limitations, Sàágua and Baumtrog (forthcoming) have developed a new model of integrated practical reasoning and argumentation. In this section I will provide a brief overview of the model before zeroing in on where moral considerations arise.

Generally speaking, practical reasoning is an activity of the human mind aiming at forming an intention to complete the actions required to alter the state of the world. Practical argumentation, however, is a dialectical/dialogical (or polylogical) situation through which human agents support or criticise a given line of practical reasoning, or a step of that reasoning. Practical reasoning is a mental and individual process. It is an activity of the mind through which an individual, starting from certain mental states – propositional attitudes – and following a rational process according to rules, leads his mind into a new mental state that concludes the process (Broome, 2002). Practical argumentation begins when one or more parts of the process of practical reasoning comes into conflict. This can happen with another individual if practical reasoning is externalized, or can happen within an individual who "argues with himself" or imagines himself arguing with an Other.

Accordingly, our model has been designed to be an *integrative, descriptive, and normative model*. It is aimed at integrating the structure of both practical reasoning and practical argumentation, including the variants usually differentiated in both – i.e., instrumental, normative, and value. It is descriptive in the sense that following the model corresponds to the real practice of reasoners and arguers. It is normative in the sense that it prescribes a chain of inferences (for reasoning) or a chain of primitive argumentative schemes (for argumentation) that should occur, and in a certain order, to promote maximally plausible formulations, conclusions, and decisions to act. The model is presented here:[1]

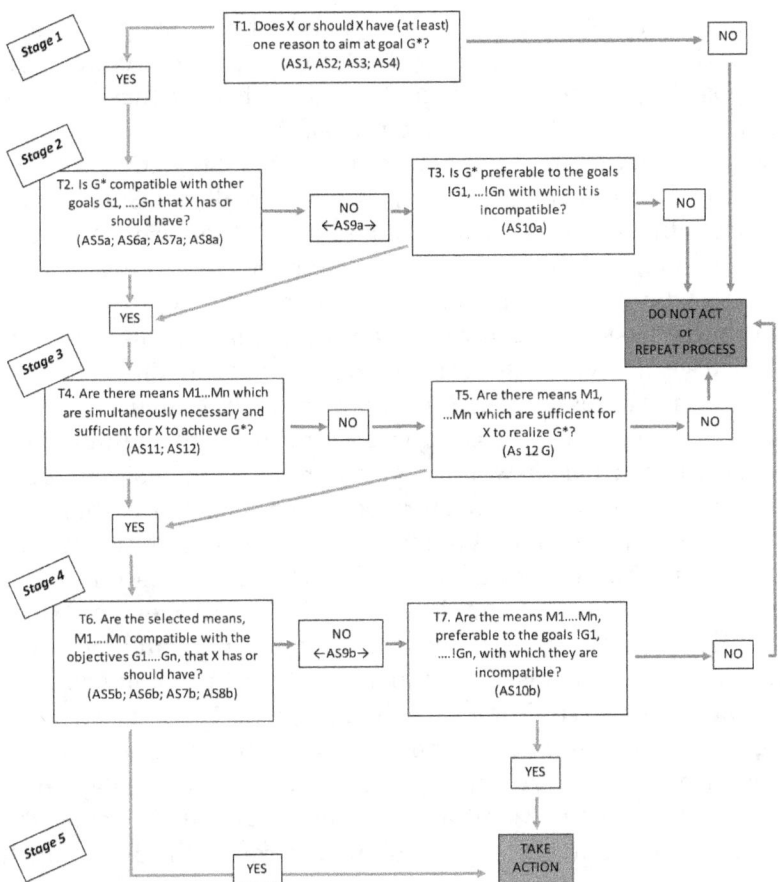

Figure 1 – Integrated Model of Practical Reasoning/Argumentation

Imagining a human agent in any given circumstance, the model begins by asking if the agent has a reason to alter the current state of the world.

---

[1] The complete list of corresponding argumentation schemes is included in the Sàágua and Baumtrog forthcoming publication. The schemes important for present purposes are provided below, but the complete list is not crucial here and has not been provided out of respect for spacial consideration.

While such wording might sound drastic, "altering the state of the world" is most often meant on a local rather than global scale - to mean "change the circumstances the agent is in". For example, if in the current state of the world my cats don't have food, I can reason toward feeding the cats by asking if I have a reason to do so. We envisage determining the motivations for aiming at a goal as the first step in explicit practical reasoning and argumentation and thus label it "Stage 1".

The complete model is composed of five stages: Stage one addresses the agent's motivation for action; Stage two is concerned with the proposed goal and other goals; Stage three concerns available means for achieving the proposed goal; Stage four deals with the relation between the means and other goals; and Stage five is simply the decision to act. If the agent progresses through all five Stages, they will have reasonable grounds for deciding to act. If they are stopped at any stage, they will then have reasonable grounds for not acting.

In order to licence moving from one Stage to the next, the reasoner must answer one or two yes or no Topic questions. In any case, an affirmative answer results in a "green light" to move to the next Stage. In some cases, a negative answer or "red light" will lead to another Topic and thus a second chance to move to the next Stage. In other cases, a negative answer leads straight to a conclusion not to act.

Each Topic questions an aspect of the general theme of the Stage. For example, given that Stage 2 is concerned with the agent's goal and other goals, Topic two questions if the goal is compatible with other goals the agent has or should have. If the answer is "Yes, it is compatible," then the agent can follow the green path and move to Stage 3. If the answer is "No, it is not compatible," then the agent should follow the red path and answer Topic 3 (still in Stage 2). Topic 3, asks the agent if the goal in question is preferable to the goal it conflicts with. If the agent determines that "Yes, it is preferable," then he or she can follow the green path to Stage 3 and address the means. If, however, it is not preferable, then the agent should follow the red path to the conclusion not to act.

Answering "Yes" or "No" to the Topics is not, however, based merely on the free thinking or intuitions of the agent. In order to reasonably answer the Topic questions, the agent must have reasons supporting their answer. Those reasons can be specified using an appropriate argument scheme. The model indicates what we consider to be the basic necessary schemes to justify an answer to each one of the Topics, though, in practice an agent may of course use schemes over and above the provided list.

One last, though important note about the model before moving on to see how it works in action. The reasons which emerge from the schemes are to be considered *pro tanto*, or contributory reasons, in the way that Jonathan Dancy (2004) has characterized them. This consideration is important because of two major implications it carries through the reasoning. First, it means a reason on one side is not, by itself, enough to licence moving to a conclusion to act or not act. The questions and schemes are set up in oppositional fashion so that contributing reasons from both sides can be weighed. For example, an agent using the schemes associated with Topic 1 could come up with four reasons to pursue the goal. Rather than jumping straight to a conclusion to pursue it, however, Topic 2 is aimed at finding reasons not to pursue it. Only after both reasons for and reasons against have been addressed is the agent free to look for means.

Second, reasons being contributory also means that one reason may outweigh all opposing reasons. In other words, the number of reasons and weight provided to one side of the Yes or No answer are not in a strict relationship. Thus, even though there may be four reasons for accepting the proposed means and one reason against, that one reason may outweigh the other four.

With this overview in mind, we now turn to an application of the model.

3. APPLYING THE MODEL

In this section I will apply the model just outlined to the reasoning of Anders Behring Breivik.[2] For some background before we begin, Breivik is the man who in Oslo on the 22 of July, 2011 detonated a bomb outside the office of the Prime Minister of Norway killing 8 people before proceeding to the island of Utøya, where killed a further 69 people, most of whom were under the age of 18 attending a labour party youth camp.

The underlying assumption in choosing this example is an intuitive one - that Breivik's practical reasoning was unreasonable. While such an assumption puts "the cart before the horse" by assuming the reasoning to be unreasonable before testing it, it does not do so needlessly. To take as a default that the reasoning leading to mass murder is reasonable, thus placing the burden of proof on theories of reasoning to show the counter, empties humanity of any non-technical

---

[2] For present purposes there is no need (and indeed not enough space) to discuss all of Breivik's answers to the critical questions. Accordingly, only his reasons are presented here. Below, however, we will focus on one critical question for AS10 which is crucial for our investigation.

notion of reasonableness. While there is perhaps room for this level of philosophical discussion elsewhere, I ask my readers for charity in presuming that Breivik's case works as a paradigm example of unreasonableness leaving the burden of proof on others to demonstrate the opposite.

It is also worth noting that since Breivik did not discuss his plans with anyone prior to his actions,[3] the model will proceed accordingly. In other words, the assumption in applying the model is that it is used on Breivik's reasoning and argumentation within himself rather than with any physical Other.

Without further ado, in a number of places Breivik makes it clear that his goal is to spread his compendium. Although it might seem as though his goal is to blow up the Prime Minister's office and conduct his mass killing on Utøya Island, he explicitly denies this in his first psychological interview: "*Spreading the compendium is the goal of the operations*, he says, *and the operation's success is measured by the spreading of the compendium*" (Husby & Sørheim, 2011, sec. 5.8). In that light, for the following application G* will represent "*spreading the compendium*".[4]

TOPIC 1: Do I or should I have (at least) one reason to aim at *spreading the compendium*?

Reason 1 - AS1. Assumption of Objectives by Teleology
Premise 1: I have expressing my love for my own people and country and getting rid of the evil in the country as a finality.
Premise 2: Spreading the compendium belongs to expressing my love for my own people and country and getting rid of the evil in the country.
Therefore, plausibly
Conclusion: There is a teleological reason for me to assume *spreading the compendium*.

---

[3] See Berwick (2011, p. 1381)

[4] In their report, Husby and Sørheim use italics to indicate direct quotes from Breivik (Husby & Sørheim, 2011, sec. 1.3) Taking inspiration from them I have done the same for this modelling. All regular font is based on, but is not a direct quote of, Breivik's own words. My forthcoming dissertation will contain an appendix providing the exact sources of all of the quotes used to reconstruct Breivik's reasoning. That appendix is long and is not crucial for our purpose here. It can, however, be provided upon request.

Reason 2 - AS2. Argument from Positive Values
Premise 1: *Cultural preservation* is a positive value.
Premise 2: Cultural preservation values spreading the compendium.
Therefore, plausibly
Conclusion: There is a second reason for me to assume *spreading the compendium*.

Reason 3 - AS2. Argument from Positive Values
Premise 1: *"Logic" and rationalist thought* is a positive value.
Premise 2: "Logic" and rationalist thought value spreading the compendium.
Therefore, plausibly
Conclusion: There is a third reason for me to assume *spreading the compendium*.

Reason 4 - AS3. Argument from Positive Consequences

Premise 1: If *spreading the compendium* is realized by me, then more people will join our cause.
Premise 2: More people joining our cause is to be valued positively.
Therefore, plausibly
Conclusion: There is a fourth reason for me to assume *spreading the compendium*.

Having provided four reasons to support continuing reasoning toward his goal, Breivik can now address Topic 2 – reasons against pursing his goal.

TOPIC 2: Is *spreading the compendium* compatible other goals I have or should have?

Reason 1 - AS5a. Argument from Negative Values
Premise 1: Fascism is a negative value.
Premise 2: Fascism negatively values *spreading the compendium*.
Therefore, plausibly
Conclusion: There is a first reason for me not assuming *spreading the compendium*.

Reason 2 - AS6a. Argument Contradicting Positive Values
Premise 1: Democracy is a positive value.
Premise 2: *Spreading the compendium* contradicts democracy.
Therefore, plausibly
Conclusion: There is a second reason for me not assuming *spreading the compendium*.

Reason 3 – AS6a. Argument Contradicting Positive Values
Premise 1: Having children and a good job is a positive value.
Premise 2: *Spreading the compendium* inhibits having children and a good job.
Therefore, plausibly
Conclusion: There is a third reason for me not assuming *spreading the compendium*.

Reason 4 - AS7a. Argument from Negative Consequences
Premise 1: If spreading the compendium is realized by me, I will have to *leave my old life behind*.
Premise 2: *Leaving my old life behind* is to be negatively valued.
Therefore, plausibly
Conclusion: There is fourth reason for me not assuming *spreading the compendium*.

With four reasons against pursuing his goal and thus a conflict between Topics 1 and 2, Breivik must address the question posed in Topic 3 – choosing the proposed goal or the alternatives.

TOPIC 3: Is spreading the compendium preferable to the alternatives with which it is incompatible?

Reason 1 - AS10a.1. Argument Based on Rational Preference
Premise 1: The alternatives and spreading the compendium are contradictory.
Premise 2: Expressing my love for my own people and country and getting rid of the evil in the country, promoting cultural preservation, promoting "Logic" and rationalist thought, and more people joining our cause count for spreading the compendium.
Premise 3: Not promoting fascism, not contradicting democracy, not contradicting the value of getting a good job and starting a family, and not *leaving my old life behind* count for alternatives.
Premise 4: The reasons for spreading the compendium are preferable to the alternatives.
Therefore, plausibly
Conclusion: I should spread the compendium and abandon the alternatives.

Having reasoned through acceptance of the goal, Breivik can now address his proposed means. He does so through addressing Topic 4. Due to the fact that he sees his means as necessary and sufficient, he answers Topic 4 with a "Yes".

Topic 4: Are there means which are simultaneously necessary and sufficient for me to spread the compendium?

Reason 1 - AS11. Necessary Condition Argument
Premise 1: I have the objective of *spreading the compendium.*
Premise 2: Performing *a deadly shock attack* is necessary for me to spread the compendium.
Therefore, plausibly
Conclusion: I have a reason to perform a *deadly shock attack.*

Reason 2 - AS12. Sufficient Condition Argument
Premise 1: I have the objective of *spreading the compendium.*
Premise 2: If I carry out the shock attack it will be widely reported in the media which will spread the compendium.
Therefore, plausibly
Conclusion: I have a reason to carry out *a deadly shock attack.*

Since he deems the means to be simultaneously necessary[5] and sufficient, Breivik can skip Topic 5 and proceed to check the compatibility of the means with the goals he has or should have, which is done through providing an answer to Topic 6.

TOPIC 6: T6. Is performing a *deadly shock attack* compatible with the alternatives that I have or should have?

Reason 1 - AS5b. Argument from Negative Values
Premise 1: Terrorism is a negative value.
Premise 2: Terrorism negatively values carrying out a *deadly shock attack.*
Therefore, plausibly
Conclusion: There is a first reason for me not to carry out a *deadly shock attack.*

Reason 2 - AS7b. Argument from Negative Consequences
Premise 1: If I carry out a shock attack then innocent civilians will die.
Premise 2: Innocent civilians dying is to be negatively valued.
Therefore, plausibly

---

[5] Breivik also describes his plan for distributing (rather "spreading" the compendium), by which he means physically distributing it. This aspect can also be included as a necessary means for him, but I have omitted it in the above to avoid unnecessarily confusing the terms here.

Conclusion: There is a second reason for me not to carry out a *deadly shock attack*.

Reason 3 - AS7b. Argument from Negative Consequences
Premise 1: If I carry out a shock attack I will die or live in a hellish situation.
Premise 2: Me dying or living in hell is to be negatively valued.
Therefore, plausibly
Conclusion: There is a second negative consequence reason for me not to carry out a *deadly shock attack*.

Faced with an incompatibility between his proposed means and other goals he should have, Breivik now has to determine if his proposed means and goal are preferable to the other goals with which they are in conflict. He does this by providing a "Yes" answer to Topic 7.

TOPIC 7: T7. Is performing a *deadly shock attack* preferable to the alternatives with which it is incompatible?

Reason 1 - AS10b.1. Argument Based on Rational Preference
Premise 1: The alternatives and performing a *deadly shock* attack are contradictory.
Premise 2: Being necessary and sufficient for the goal [in conjunction with the reasons supporting the goal from above] count for performing deadly shock attack.
Premise 3: Not conducting terrorism, not killing innocent civilians, not dying myself or living in a hellish situation count for the alternatives.
Premise 4: The reasons for performing a *deadly shock attack* are preferable to the alternatives.
Therefore, plausibly
Conclusion: I should perform a *deadly shock attack* (and abandon the alternatives).

4. THE PLACE OF MORALS

Moral considerations appear in this structure of individual argumentative reasoning in schemes 2, 5-8, and 10. Since, however, by design, scheme 10 takes all of the schemes used before it into account by calling for an "all things considered" comparison, it is the scheme most important for practical reasoning overall (and the most difficult to formulate/capture!). In order to rationally justify his choice of whether pursuing his goal and means is more preferable than not, Breivik must consider if the goal (AS10a) and means (AS10b) are 1) better than and

2) more probable to be accomplished than the alternatives. The AS10 critical questions scrutinize these two basic factors for deciding preference - the first question scrutinizes how he decided which option is better and the second scrutinizes how he decided which is more probable. For our purposes, the first question is most important. It asks, "What makes the standard used for the valuation of the reasons associated with the goal/means, the best standard for this situation? Why might the standard used for the valuation of the reason associated with the goal/means not be the best for this situation?"

Other "value based" models, have pointed to the role of values in supporting the goal, but none have yet (to my knowledge) pointed out how to consider the role of values associated with the means (cf. Walton 2007; Fairclough & Fairclough 2012). In this light, given that value considerations appear throughout practical reasoning, the model used here might be more appropriately described as a model for "value laden" practical reasoning rather than "value based". The title "value laden" also captures the idea that the values associated with the goal do not have to be the same as the values associated with the means. In the present case, this is crucially important. There may be nothing wrong with valuing cultural preservation - the value supporting the goal of spreading the compendium. The value allowing murder which underlies the means for achieving cultural preservation, is however, in the present case, devastatingly problematic.

For both the goal and means, since there has yet to be any universal agreement on a singular authoritative moral philosophy, the question concerning the betterness of the alternatives is phrased so as to allow for any standard to be used. The idea is that when the choice of pursuing the goal/mean(s) or alternative(s) was made, the options were measured against some "covering value" (Chang, 1998 p.5). This covering value functions as a standard against which to compare the betterness of the proposal vs. its alternatives. The idea is that if there is no clear moral authority to be used in all cases, the best a reasonable person can do is argue for the selection of the standard they have employed.

Arguing for the selection of the standard is important because the expectation is that differing moral theories will yield differing answers to whether or not to pursue the goal or means. Accordingly, the content of the applied moral standard becomes crucial for making reasonable decisions. If my initial intuition in choosing Breivik's case as an example of the unreasonable is correct, then any standard which would allow for the performance of his murders, should be excluded. This is not to say that the standards excluded here should never be used in any other situation. That may or may not be the case and would require developed argumentation in a different work. The point here is that if there is a risk that certain moral content can allow an unreasonable decision to pass a test of reasonableness, then it is crucially important to investigate the content of the standard used and to put it through scrutiny if we are to adequately declare an action as reasonable.

For clarity, in what follows I will focus on Breivik's use of AS10b – his choice to perform a deadly shock attack - to shed light on the following questions: Can Breivik's practical reasoning, all things being equal, be considered reasonable when it employs standards from moral relativism and moral objectivism? In the best case scenario, Breivik's reasoning will be deemed unreasonable in both instances. If, however, using one or the other moral perspective returns a result of "reasonable" we will have to either adjust the model for practical reasoning, abandon the moral view point for use in the model, or seriously revaluate our intuitive notions of reasonableness to allow for Breivik's example to be included under the umbrella title of reasonable.

4.1 Considering differing moral perspectives in practical reasoning

A leading proponent of moral relativism, Gilbert Harman, describes moral relativism as

> a claim about reality. It is a version of moral realism. It is the that (sic) there are many moralities or moral frames of reference and whether something is morally right or wrong, good or bad, just or unjust, virtuous or not is always a relative matter. Something can be right or good or just only in relation to one moral framework and wrong or bad or unjust in relation to another. Nothing is simply right or good or unjust or virtuous (Harman, 2014).

In addition to the descriptive claim that many moralities exist, moral relativists typically also share the view that fundamental moral disagreements between these moralities cannot be rationally resolved (Gowans, 2012, p. 4). Thus, for moral relativists, different moral frameworks may carry equal moral authority.

A relativist approach can appear quite threating for a number of reasons, three of which have been succinctly pointed out by Thomas Scanlon. The first is that it could stop people from taking basic moral rules, like not committing frivolous murder, as authoritative (1998, pp. 331-334). The "second reason is grounded in the confidence we have or would like to have in our judgment that certain actions are wrong" (p. 334). If morals are relative, it might undermine any force in a moral condemnation. The third reason is that it can make it seem as though there are no genuine moral disagreements – that the disagreements "disappear" (p. 335; Gowans, 2012, p. 20).

As serious as these general objections to moral relativism may be, some inspirations for developing moral relativism stem from equally as serious objections against moral objectivity. Moral objectivity can be understood as the claim that "moral judgments are ordinarily true or false in an absolute or universal sense, that some of them are true, and that people sometimes are justified in accepting true moral judgments (and rejecting false ones) on the basis of evidence available to any reasonable and well-informed person." (Gowans, 2012, p. 6). Using dichotomous terms like true and false in regards to morality raises a great number of questions (such as how we can access moral truth) and places a huge amount of pressure on the moral agent. Moral objectivity pressures the agent because it seems to restrict human freedom, under the risk of moral condemnation, to live and organize our lives and societies in what are commonly thought of as differing but equally valuable ways. Further, it seems to lead to an intolerance of those individuals and societies who maintain different values and promote different conceptions of the good. If objectivity is true, it might be that there is only one right way to live (and it's mine!). This absolute universal conception of morality goes against the common notions of cultural tolerance and continuous learning and spurs an impulse away from it for many.

Given their fundamental differences, if used as evaluative standards in reasoning toward a decision to act, these two perspectives are hypothesised to provide quite different recommendations – in other words, quite different *pro tanto* reasons.

4.2 Moral Relativism

Like other broad philosophical topics, moral relativism comes in different flavours. Three of the most predominant are subjective, social/cultural, and naturalistic.

4.2.1 Subjective moral relativism

Subjective moral relativism is the idea that every individual has her or his own viewpoint (Häyry, 2005, p. 9) and that, as noted above, disagreements between viewpoints are irresolvable. Although it is another step to claim that each individual's moral view ought to be authoritative over others, which might better be called "absolute subjectivism" (ibid), if moral disagreements between individuals cannot be resolved then there is nothing preventing one from using their subjective moral view as authoritative.

Accordingly, if in his reasoning Breivik took only his own moral view as authoritative, and that view allowed committing murder, he could quite easily conclude that according to the standard of subjective moral relativism he is licenced to perform his shock attack. The moral component of his AS10b could include implicit reasoning along the lines, "Performing deadly shock attacks is preferable to the alternatives because it is better (or at least no worse) than the alternatives according to my standard of subjective moral relativism."

When the critical question is asked, summarized as "What makes this standard the best standard for the situation and why might it not be?" Breivik could not help but conclude that it is the best standard for this and every situation because it is his and cannot be shown to be better or worse than any other. Thus, because this standard fails our test in that it could advise Breivik to commit his murders, and since the critical question cannot safeguard against its use, it does not appear to be an acceptable standard to use for reasonable practical reasoning.[6]

4.2.2 Social/cultural moral relativism

Rather than focusing on the self, a social or cultural moral relativist could appeal to societal/cultural norms and values as the authoritative moral force. Indeed, cultural relativism is often proposed as the standard moral relativism (Gowans, 2012, pp. 4ff.). According to it, different societies may have different but equally authoritative moral guidelines and it would not be possible to clearly demonstrate the moral superiority of any one to another. This account is appealing for a number of reasons, a major one being that it seems to include the idea of tolerance for different cultures (Lukes, 2008, pp. 38-44), but protects against the dangers of subjective moral relativism by appealing to popularity.

In one way, applying the standard of social moral relativism might have prevented Breivik from being able to reason to his means. If he thought of Norwegian law as the definition of social moral authority, for example, he could have concluded that murder was not the preferable means– declared unequivocally wrong. In another way, however, because the definitions of "society" and "culture" are notoriously problematic (Lukes, 2008, pp. 112-122), the culture and society of which Breivik was a part – what he would call the "pan-

---

[6] At least in this instance.

European and national resistance movement" (Berwick, 2011, p. 1352) – could be seen to endorse his means.

Using a social/cultural relativist standard, AS10b could then include implicit reasoning along the lines, of either "Performing a deadly shock attack is not preferable to the alternatives because according to the standard of social moral relativism, as indicated by national law, it is not the better choice" or "Performing a deadly shock attack is preferable to the alternatives because according to the standard of social moral relativism, as indicated by the pan-European national resistance movement, it is the better choice."

After applying the standard of social/cultural relativism, when asking the critical question about what makes it the best standard for the situation, Breivik could again appeal to the irresolvability between fundamental moral differences inherent in the idea of moral relativism. This would make his choice of the standard the best (or at least no worse than) alternative cultural standards. Thus, using this standard carries a chance of recommending that Breivik conclude not to perform what we have *prima facie* declared his unreasonable acts, but it seems to be a slim chance. It seems more likely that he would stick to the momentum in his proposal of means, thus choosing his own relative culture for justification, rather than countering the inertia of his means proposal by identifying with an opposing cultural authority.

4.2.3 Naturalistic moral relativism

One of the most developed self-proclaimed relativist theories comes from David Wong who takes a naturalistic approach to defend what he calls "pluralistic relativism" (Wong, 2006, p. XV). For Wong, while there is no one single true morality there are constraints on what can be considered an adequate morality, which, he claims, are derived "from the functions of morality, human psychology, and the nature of human cooperation" (Wong, 2006, p. 65). The constraints include:

> requiring human beings to seek only that which they have some propensity to seek; inclusion of norms of reciprocity in light of strong self-interest; in specification of norms and reasons, balancing self- and other-concern in ways that include putting less pressure on other-concern through provision of some "payoff " in terms of self-interest; justifiability of norms and reasons to the governed in terms of their interests when presented without falsification; and finally the value of accommodation of moral disagreement. (ibid)

These constraints set limits on what can count as an adequate morality but do not say anything about the way moralities can take shape within the constraints. Accordingly, different people and societies can share basic values but prioritize them differently. These different orderings allow for more than one adequate moral framework which all fulfil the basic functions of morality such as facilitating social cooperation and articulating "character ideals and conceptions of the good life specifying what is worthwhile for the individual to become and to pursue" (Wong, 2006, p. 43).

Wong's pluralistic relativism seems closest to the moral perspective included in Fairclough and Fairclough's articulation of practical reasoning. Because they say the most, at least from the side of argumentation, about substantive value issues in practical reasoning, a brief comment of comparison is worthwhile here. In their articulation of the structure of practical reasoning and their suggestions for how to evaluate it, Fairclough and Fairclough want to allow for a value pluralist approach but specifically deny being moral relativists. They argue,

> Some value differences are unreasonable and cannot withstand critical examination. For instance, some values are indefensible from a purely instrumental point of view, because they contradict the agent's goals: valuing a life of leisure is not reasonable if your goal is to get high grades. But some value differences are unreasonable in a deeper, non-instrumental sense: a racist conception cannot remain indefinitely in play alongside one which rejects racism. Disagreement over this issue is unreasonable and a reasonable resolution can be legitimately expected. Sometimes, however, people disagree in a reasonable way and the disagreement is also irresolvable. Such disagreements often depend on the way people rank the values and goals that matter to them. Reasonable disagreement, we suggest, is generated by conflicting but reasonable values and goals or by different rankings of the same values and goals. (2012, p. 60)

To define what count as reasonable values which can be disagreed upon, Fairclough and Fairclough pick up the capabilities approach advocated by Martha Nussbaum and support a notion of "values that closely approximate a list of universal human rights, or duties/obligations that we have towards our fellow beings" (ibid).

Thus, like Wong, Fairclough and Fairclough see legitimate value disagreements stemming from the way different individuals or groups rank their value preferences within the confines of a conception of universal human rights and human flourishing, or in Wong's terms, an adequate conception of morality.

Nonetheless, performing a deadly shock attack falls outside the constraints of an adequate morality in a number of ways, some of the most obvious being that it denies reciprocity and fails to balance others' concerns. Using a naturalistic relativist standard, AS10b could then include implicit reasoning along the lines, "Performing deadly shock attacks is not preferable to the alternatives because it is outside of the constraints of an adequate system of morality."[7]

When asking the critical question about what makes Naturalistic Relativism the best standard for the situation, Breivik could site Wong's reasons supporting his characterization of the function of morality. To argue for what might not make it the best standard, Breivik would have to show the weakness or falsity of the argued for constraints.

So much for our brief discussion of the application of moral relativism to practical reasoning. We now move to the other side of the coin, to objective moral theory, to discuss some of the ways it can have an impact.

4.3 Moral Objectivism

4.3.1 Consequentialism

There are many importantly different consequentialist theories. The basic theory, known as act consequentialism can be summarized with the familiar phrase "the greatest happiness for the greatest number" (Sinnott-Armstrong, 2014, p. 2). Somewhat clearly, this basic form of consequentialism cannot function as a moral standard for our case. Indeed, it can be argued that Breivik used just this standard in his actual reasoning, as shown by his admittance that "Innocent people will die, in the thousands. But it is still better than the alternative; millions of dead Europeans, which is the worst case phase 3 scenario." (Berwick, 2011, p. 1360).

Eliminating basic consequentialism as the overall moral standard to be used in deciding preference is not to say that the positive or negative consequences of the goal or means are unimportant. Rather, consequences are better understood as one of the *pro tanto*

---

[7] One might ask about the extent to which Wong's approach ought to be grouped under "relativism" at all? If the backbone of the theory posits universal, objective (derived from nature or otherwise) guidelines for morality, then all of the possible moralities can only be judged immoral against the objective guidelines, which are not relative.

considerations for which an overall moral standard can rule. This is why in the model outlined above, consequences are specifically addressed in AS3 and AS7 and are not considered the *pro toto* consideration of value.

There is not space here to review each available consequentialist theory, though such a review would be worthwhile. The point I wish to highlight with this very short discussion of consequentialism is just that considerations beyond the consequences of the proposed goal and means need to be taken into account for an adequate assessment of value and that, though objective, there is a possibility for Breivik to conclude to perform his shock attack.

4.3.2 Deontology

Deontological theories argue that what is right or wrong depends on the kind of action under consideration. Deontologists point to rules meant to hold across contexts and apply to all (reasonable or rational) people. The most famous deontological rule is Kant's Categorical Imperative: "Act only according to that maxim whereby you can, at the same time, will that it become a universal law" (Kant, 1993. p. 30).

Many theories inspired by Kant have since been developed. For instance, Thomas Scanlon has argued that judgments of right and wrong

> are judgments about what would be permitted by principles that could not reasonably be rejected, by people who were moved to find principles for the general regulation of behavior that others, similarly motivated, could not reasonably reject. In particular, an act is wrong if and only if any principle that permitted it would be one that could reasonably be rejected by people with the motivation just described (or, equivalently, if and only if it would be disallowed by any principle that such people could not reasonably reject). (Scanlon, 1998, p.4)

Using Scanlon's rule, there is no way Breivik could have concluded to conduct his deadly shock attack. This is because any principle permitting it could be reasonably rejected by people motivated to find principles for the general regulation of behaviour. There is not space in this paper to see if all, or even the bulk of deontological moral theories would prohibit Breivik from conducting his means, but to my knowledge no such theory allows for mass murder.

When addressing the critical question regarding why this standard is the best, Breivik could use Scanlon's arguments for the primacy of the notion of a reason and its use in justification for guiding moral action. He could argue against the standard on consequentialist grounds – that despite being unjustifiable to those killed, his ends will justify his means.

5. CONCLUSION

In this paper I have attempted to highlight the importance of paying attention to the selection of moral content used in practical reasoning and argumentation. If moral goodness is important for assessing reasonableness, but the same action can be deemed good and bad depending on the moral theory employed, then research into how differing moral theories ought to relate to reasonableness is an important undertaking. This exploratory paper has only attempted to make first steps in this pursuit. To do so I reviewed an instance of unreasonable practical reasoning through the lens of different moral theories to see which ones might licence the decision to act.

The idea was that in the best case scenario, Breivik's reasoning would be deemed unreasonable when employing standards from both moral relativism and objectivism. However, this was not the case. From our brief overview it seems that in both cases there is a chance it could be deemed reasonable – with subjective relativism and act consequentialism being the most likely candidates to allow it to pass. Thus, it seems as though we will have to adjust the model for practical reasoning, abandon the differing moral view points for use in the model, or seriously revaluate our intuitive notions of reasonableness to allow for Breivik's example to be included under the umbrella title of reasonable.

I do not think we need to adjust our intuitive notion that reasoning to mass murder is unreasonable (and I do not think this claim needs argumentation). Accordingly, shall we adjust the model or abandon the viewpoint? "Both" might be the correct answer. At least in this situation, subjective moral relativism and act consequentialism ought to be abandoned as potential moral standards. Whether or not they are appropriate standards in any other situation is a worthwhile question for future investigation.

Attempting to capture the processes of justifying a rational preference between alternatives, the function of AS10, is notoriously difficult. Thus, I am confident that further refinements to the model, and especially this scheme, are inevitable and warranted. One prospective avenue for research is an investigation into what is involved in a meta-procedure for rationally justifying the selection of a moral theory. In other words, faced with a variety of possible moral authorities, which principles should guide the way we go about choosing between them?

With such a procedure, however, the possibility of an infinite regress looms large.

For now, it is important to recall that the recommendations of each of these moral views is *pro tanto* and operates in the mix of all the other reasons as well. Accordingly, on this version of practical reasoning, moral theory cannot tell us *pro toto* what is reasonable, but it is indispensable as a *pro tanto* consideration when appropriately scrutinized by asking what makes the selection used the best possible.

REFERENCES

Atkinson, K., & Bench-Capon, T. (2008). Addressing moral problems through practical reasoning. *Journal of Applied Logic*, 135-151.

Baumtrog, M. D. (2013). Considering the role of values in practical reasoning argumentation evaluation. In D. Mohammed & M. Lewiński (Eds.), Virtues of Argumentation. Proceedings of the 10th International Conference of the Ontario Society for the Study of Argumentation (OSSA), 22-26 May 2013. Windsor, ON: OSSA.

Berwick, A. [. (2011). *2083 - A European Declaration of Independence*.

Broome, J. (2002). Practical Reasoning. In J. Bermùdez, & A. Millar (Eds.), *Reason and Nature: Essays in the Theory of Rationality* (pp. 85-111). Oxford: Oxford University Press.

Chang, R. (1997). Introduction. In *Incommensurability, Incomparability, and Practical Reason* (pp. 1-34). Cambridge: Harvard University Press.

Dancy, J. (2004). *Ethics without Principles.* Oxford: Oxford University Press.

Fairclough, I., & Fairclough, N. (2012). *Political Discourse Analysis: A Method for Advanced Students.* London: Routledge.

Gert, B. (2005). *Morality: Its Nature and Justification.* New York: Oxford University Press.

Gowans, C. (2012). Moral Relativism. (E. N. Zalta, Ed.) *The Stanford Encyclopedia of Philsophy.* Retrieved from http://plato.stanford.edu/archives/spr2012/entries/moral-relativism/

Harman, G. (2014, February 22). Moral relativism is moral realism. *Philosophical Studies.* doi:10.1007/s11098-014-0298-8

Häyry, M. (2004). A defense of ethical relativism. *Cambridge Quarterly of Healthcare Ethics, 14*, 7-12.

Hitchcock, D. (2011). Instrumental rationality. In P. McBurney, I. Rahwan, & S. Parsons (Eds.), *Argumentation in Multi-Agent Systems: 7th International Workshop, ArgMAS 2010 Toronto, ON, Canada, May 10, 2010 Revised, Selected and Invited Papers* (pp. 1-11). Berlin, Heidelberg: Springer.

Husby, T., & Sørheim, S. (2011). *Court Psychiatric Report (First Report).* Retrieved from https://sites.google.com/site/breivikreport/documents/anders-breivik-psychiatric-report-of-2011

Kant, I. ([1785} 1993). *Grounding for the Metaphysics of Morals* (3 ed.). (J. W. Ellington, Trans.) Indianapolis/Cambridge: Hackett.
Lukes, S. (2008). *Moral Relativism.* New York: Picador.
Parfit, D. (2011). *On What Matters* (Vol. 1). Oxford: Oxford University Press.
Sàágua, J., & Baumtrog, M.D. (Forthcoming). Integrating Practical Reasoning and Argumentation.
Scanlon, T. M. (1998). *What We Owe to Each Other.* Cambridge: Harvard University Press.
Sinnot-Armstrong, W. (2014). Consequentialism. (E. N. Zalta, Ed.) *The Stanford Encyclopedia of Philosophy.* Retrieved from http://plato.stanford.edu/archives/spr2014/entries/consequentialism/
Walton, D. (2007). Evaluating practical reasoning. *Synthese: An International Journal for Epistemology, Logic and Philosophy of Science, 157,* 197-240.
Wong, D. B. (2006). *Natural Moralities: A Defense of Pluralistic Relativism.* New York: Oxford University Press.

Commentary on Baumtrog's Investigating the Impact of Moral Relativism and Objectivism on Practical Reasonableness

DAVID HITCHCOCK
Department of Philosophy, McMaster University, Hamilton, Canada
hitchckd@mcmaster.ca

1. INTRODUCTION

Michael Baumtrog uses Anders Breivik's mass killings in Norway to test both his (Baumtrog's) model of practical reasoning and various perspectives on morality. He takes it as evident that Breivik's homicides are practically unreasonable. He shows that Breivik's well-documented reasoning leads from his goal through all the steps of Baumtrog's model to a decision to slaughter dozens of Breivik's fellow Norwegians—a decision that is practically unreasonable only if moral considerations outweigh Breivik's reasons. Subjective moral relativism and act consequentialism, Baumtrog argues, would not block a decision to proceed with the killing, and so ought to be abandoned, at least for Breivik's reasoning. Baumtrog concludes that he needs to do more work on how one selects a moral theory when one is engaging in practical reasoning.

2. ARE MASS KILLINGS ALWAYS UNREASONABLE?

It is not self-evident that mass killings like Breivik's are practically unreasonable. Depending on the circumstances, mass killings may be well calculated to achieve a defensible goal, even if one takes into account the suffering of those killed and injured and of their families and friends, as well as side-effects and possible alternative means for accomplishing the goal. For example, dropping the atomic bomb on Hiroshima and Nagasaki in 1945, which intentionally killed thousands more innocent people than Anders Breivik did, has been defended as the best means in the circumstances of bringing the war with Japan to an end, with minimum casualties. We may be convinced that the intentional killing of large numbers of innocent human beings is grossly immoral, in fact an outrage. But it does not follow automatically that killing of this sort is practically unreasonable.

3. BAUMTROG'S MODEL OF PRACTICAL REASONING

In Baumtrog's model of practical reasoning, it is reasonable for an agent to implement a set of means if and only if this set of means is sufficient for achieving a goal that the agent has a reason for pursuing and both the goal and the means for achieving it are preferable to any incompatible goals that the agent has or should have. Baumtrog understands the concepts of goal and means in a broad sense that for example counts doing an honest deed as a means to achieving the goal of being honest. Even allowing for this broad sense, some decisions are reasonable even though they do not meet Baumtrog's conditions. For example, in playing bridge it is often reasonable for declarers to adopt a strategy that is not sufficient to achieve their goal of making the contract and that may in fact fail to achieve it, since declarers must make choices on the basis of probabilities and even of remote possibilities.

4. APPLICATION OF THE MODEL TO ANDERS BREIVIK'S REASONING

Baumtrog takes at face value Breivik's statement to the first pair of investigating psychiatrists (Husby & Sørheim, 2011, sec. 5.8) that his goal was spreading his so-called "compendium", a work of more than 1,500 single-spaced pages in which he makes the case, using his own writings and various extensively quoted sources, that political Islam is trying to take over the world and that European cultural Marxists are abetting this goal by their weak response to European Islamization. Baumtrog then fits into his model the reasoning that Breivik used in his compendium and reported later (to the police and the psychiatrists) to get to the conclusion that he should go on his killing spree. A crucial part of this reasoning is the claim that the mass killing is a necessary means for achieving his goal of spreading the compendium. Breivik did in fact claim that his attack was necessary, both for revenge and to get people to "open their eyes to what is happening in Norway and Europe" (Husby & Sørheim, 2011, sec. 5.8). But Baumtrog simply registers the claim of necessity, without any attempt to evaluate it. If we think about means of raising awareness of a supposed worldwide Islamic conspiracy abetted by supine European cultural Marxists, a lot of alternatives come to mind, not mutually exclusive. One can get one's views published and discussed in a variety of forums, by various means. Massacre of some of those deemed responsible for abetting European Islamization will certainly get attention, but its main effect will be revulsion at the ideas that motivated the killings. Thus Breivik's killing spree was neither a necessary nor a sufficient means for achieving his goal of making people aware of the ideas in his compendium. Thus the fact that Baumtrog's model fits Breivik's stated reasoning does not show that, aside from moral considerations, Breivik's reasoning was practically reasonable.

5. MORALITY AND PRACTICAL REASONABLENESS

Breivik's reasoning is however simply a hook by which Baumtrog raises the question of whether moral considerations can make it unreasonable to do something that is grossly immoral but otherwise justified within his model of practical reasonableness. So we might imagine some other act that seems to be grossly immoral, say the intentional mass killing of many innocent people, but that is a sufficient (and even necessary) means for achieving a goal that the decision-maker has a reason to achieve, where both the goal and the means are preferable to any non-moral goals incompatible with them that the decision-maker has or should have. We can stipulate, for the sake of argument, that dropping the atomic bomb on Hiroshima and Nagasaki was a sufficient (and even necessary) means of achieving a quick unconditional surrender by Japan, that there was a reason for pursuing this goal, and that both the goal and the means were preferable to non-moral "goals" incompatible with them. Baumtrog's question is what accounts of morality would show that under these stipulated conditions the decision to drop the atomic bomb was practically unreasonable.

In undertaking such an exercise, we should be more careful than Baumtrog to distinguish meta-ethical theories from normative ethical theories. It is the normative ethical theories that have implications for a judgment of the moral status of dropping the atomic bomb on Hiroshima and Nagasaki, not the meta-ethical theories. Whether one is a relativist of some sort about the status of moral judgments or an objectivist of some sort has no implications for one's position on a substantive moral question. A cultural relativist for example will take a moral judgment to be binding for a culture whose morality endorses that judgment, but the culture in question may subscribe to hedonistic utilitarianism or the Ten Commandments or Kant's categorical imperative or an unsystematic collection of rules with vaguely specified exceptions. In assessing the morality of dropping the atomic bomb on Hiroshima and Nagasaki, what does the work is not the meta-ethical position of cultural relativism but the specific content of the morality of the culture of the decision-maker, who in this case was Harry Truman. The assessment would be the same for an objectivist who takes the morality of Harry Truman's culture to be the correct morality. What counts is the content of morality, not the meaning and justification of that content.

As for the dropping of the atomic bomb, under the stipulated conditions any form of consequentialism would probably endorse it as morally permissible and even morally obligatory. Not only does the act of dropping the atomic bomb have better consequences under the stipulated conditions than its alternatives, but so too does adoption of any rule that generalizes Truman's thinking. On the other hand, a deontological moral theory with an absolute prohibition on intentionally killing innocent human beings, such as traditional Roman Catholic moral theology, would condemn the dropping of the atomic bomb as immoral. Even a deontological moral theory that permits exceptions, such as Bernard Gert's account of what he takes to be the common morality of all humanity (Gert, 2005), might condemn the dropping of the atomic bomb as immoral, depending on the theory's procedure for justifying violations of the moral rules. Gert proposes a two-step procedure, the first step being a determination of the morally relevant features of the act, and the second a comparison of the consequences of its being publicly known that acts with those morally relevant features are morally permitted and the consequences of its being publicly known that acts with those morally relevant features are morally prohibited. It is not clear to me whether applying Gert's procedure to Harry Truman's decision would imply that it is strongly morally justified, weakly morally justified, or morally unjustified. My uncertainty is due to ignorance of the likely consequences of the two alternatives.

To sum up, consequentialist moral theories would assess as morally justified dropping the atomic bomb under the stipulated conditions, whereas some but not all deontological theories would condemn it as immoral. But so what? The divergence gives us no reason to reject consequentialist or weak deontological moral theories for Truman's reasoning, since there is no agreement on whether Truman's decision, given the stipulated conditions, was practically reasonable. Consequentialists and weak deontologists will be inclined to think that it was practically reasonable, whereas strong deontologists will take it to be morally wrong, even though it chose an effective means for achieving Truman's goal. It is not even clear that those who assess Truman's decision as morally unjustified would treat it as practically unreasonable; it is a coherent position that a decision is practically reasonable but immoral.

This result suggests that a model of practical reasonableness cannot be used to test moral theories. If some moral theories count an otherwise practically reasonable decision as morally unjustified and other theories imply that it is morally justified, it is likely that people's untutored moral judgments of that decision will vary in the same way.

6. CONCLUSION

Anders Breivik's unusually comprehensive and articulate descriptions of the reasoning that led him to slaughter dozens of his fellow Norwegians provide a basis for testing models of practical reasoning, practical rationality and practical reasonableness. Michael Baumtrog's model takes a decision to be practically reasonable if and only if it adopts a set of means that are sufficient for realizing a goal that the decision-maker has a reason to pursue, provided that both the goal and the means are preferable to any incompatible goals that the decision-maker has or should have. Breivik's reasoning fits Baumtrog's model, but his homicides are neither necessary nor sufficient for achieving his stated goal of making people aware of his manifesto. Thus his decision was not practically reasonable, even aside from its gross immorality.

Other mass killings might however be practically reasonable on Baumtrog's model, if one abstracts from moral considerations. We can stipulate that U.S. President Harry Truman's decision in 1945 to drop the atomic bomb on Hiroshima and Nagasaki was sufficient to achieve his goal of an immediate unconditional surrender by Japan and that both the goal and the means were preferable to other goals that he had or should have had. In assessing the impact of different moral perspectives on the all-things-considered reasonableness of his decision, we should pay attention to substantive ethical theories, not to meta-ethical positions. If we do, we discover that his decision is morally justifiable on consequentialist or weak deontological theories, but unjustifiable on strong deontological theories. But this result does not show that a strong deontological theory was the appropriate moral theory for Truman to use, since even people who take moral considerations into account will disagree about the reasonableness of Truman's decision. Thus it is unlikely that a model of practical reasoning can be used to determine which moral theory is appropriate in a given decision-making situation.

REFERENCES

Gert, B. (2005). *Morality: Its nature and justification*, revised edition. Oxford: Oxford University Press.

Husby, T., & Sørheim, S. (2011). *Anders Behring Breivik psychiatric report 2011-11-29*. Retrived from: https://sites.google.com/site/breivikreport/documents/anders-breivik-psychiatric-report-of-2011; accessed 2015 09 19.

3

Advocacy vs. Inquiry in Small-Group Deliberations

J. ANTHONY BLAIR
CRRAR, University of Windsor, Canada
<u>tblair@uwindsor.ca</u>

The paper compares using arguments for advocacy and using them for inquiry as methods of decision-making in small groups. In both cases, arguments can be used to support and to critique a position. But significantly, the advocate's commitments differ from the inquirer's. Ideally, inquiry precedes advocacy, yet common procedural rules enforce decision-making by advocacy. The paper suggests alternative procedures small groups might follow to permit the use arguments to inquire in their decision-making.

KEYWORDS: advocacy, argument, argument functions, argument roles, argumentation, decision-making, deliberation, inquiry, small group decision making

1. PREFACE

This paper is motivated by a couple of personal experiences. One is over 40 years of working with my colleague Ralph Johnson in which we planned a course together, produced four editions of a textbook and several joint papers, co-organized three conferences and co-edited a journal, almost always using arguments to inquire as well as to advocate. The other is a couple of terms as a department head chairing meetings when members' advocacy argumentation produced or reinforced divisions in the group and blocked opportunities for open-minded investigations of alternatives.

2. INTRODUCTION

If people think about using arguments to decide what to do, they tend to have in mind arguing for and against some proposal about what action should be taken. Such activity consists of using arguments to advocate acceptance or rejection of the proposal. Indeed, one popular approach, Pragma-dialectics, proposes that this is precisely what argumentation consists of, namely a complex of various kinds of communication act in which one party, having adopted a position, uses arguments to try to get a disagreeing party to give up its doubt about it, and the other party resists by critiquing those arguments or else by using arguments to try to get the first party to accept some contrary alternative position that *it* has adopted (van Eemeren & Grootendorst, 2004). In such activity, both parties use arguments to advocate something, that is, in support of the acceptance or rejection of some position.

There might be no doubt that argumentation so understood is one use of arguments, but some of us (Walton e.g., 1998; Blair, 2004) have held that it is not the only one. For instance, it has seemed to me that arguments can be and are used to come to a decision about—among other things—what to do, without anyone's having at the outset adopted a position on the matter requiring action (or forebearance). Such a use of arguments cannot, by hypothesis, be to advocate a position, because the people using the arguments have not adopted a position; they are using arguments to figure out what position they ought to adopt or commit to, or at least that they are entitled to adopt or commit to. They are in a pre-advocacy state.

I have called the use of arguments to figure out what attitude or decision to take *inquiry*, because I have thought of it as the activity of inquiring into what to do. But this is an unfortunate choice of terminology for a couple of reasons. For one thing, Douglas Walton (e.g., 1998) has already appropriated the term as the name for scientific inquiries and public inquiries such as air disaster investigations, and that is not what I have in mind. The term *investigation* has the virtue, like *inquiry*, of having the connotation that the person so using arguments has not yet made up his or her mind about what position to take, but it also has the connotation of looking for an explanation.

Of the candidates I have so far been able to think of, the term that comes the closest to describing the use of arguments I have in mind is *exploration*. We explore the unknown; when exploring we make discoveries; we are not sure what we will find when we set out to explore something. These are all characteristics of the use of arguments that I am talking about. Even so, the fit isn't perfect. For the goal of deliberation, as Aristotle noted long ago, is choice (*Nicomachean Ethics* III.3), and we don't exactly explore in order to make choices or decisions.

Hans Hansen (in conversation) has suggested that the contrast I have in mind is analogous to the distinction made in the philosophy of science between the logics of *discovery* and of *justification*. The use of arguments in deliberation is to discover the best and so choice-worthy alternative, and the use of arguments to advocate approval of a particular alternative is to offer a justification of the recommended choice. So perhaps my title should have been, "Advocacy vs. discovery in small group deliberations." I like the metaphors of exploration and discovery.

3. THE USE OF ARGUMENTS IN ADVOCACY

I need to spell out in more detail what I mean by using arguments to advocate and using them to inquire or discover in small-group decision making. Their use in advocacy is well understood because we tend to be familiar with their use in the decision-making of large groups, such as the legislative or governing assemblies of states, provinces, municipalities, school boards, universities, clubs and other kinds of organizations. The decision-making discussions in such groups consist of debating proposals about what to do, and are regulated by adherence to adopted rules of order. There are many such rules of order.[1] I will select one for illustrative purposes—the one that is in widespread use in Canada and the United States: *Robert's Rules of Order* (Robert, 1986).

Robert's Rules have the effect of prescribing the use of arguments to advocate in decision-making. For one thing, no discussion is allowed until someone proposes a particular course of action, so there is already an advocated proposal for the decision about what to do put before the body.[2] Second, the proposer of that recommendation gets to speak first with arguments in its favor, which is to advocate it.[3] Third, there is a relevance rule requiring members of the body to speak to the motion.[4] The result is that the deliberations of the body consist principally of arguments advocating acceptance of the proposal and arguments advocating its rejection.

Let me draw attention to some of the features of this advocacy. One feature follows from the fact that in moving a motion, the mover is recommending a particular choice of action. The rule entitling the

---

[1] *Bourinot's Rules of Order, Beauchesne's Parliamentary Rules and Forms*, Maleau and Montpetit's *House of Commons Procedure and Practice* are just three used in Canada.

[2] "Before any subject is open to debate it is necessary, first, that a motion be made by a member who has the floor . . ." (*Robert's Rules of Order* 1986, p. 32: Article I, §3).

[3] "By parliamentary courtesy, a member upon whose motion a subject is brought before the assembly is first entitled to the floor, . . ." (*ibid.*, p 95: Article V, §34).

[4] "All remarks must be ... confined to the question before the assembly" (*ibid.*, p. 203: Article XIII, §64).

mover to speak first in favour of the motion simply recognizes the mover's burden of proof. Burden of proof tends to be a conservative principle. One does not have an onus to defend the status quo, unless and until someone has argued that it be changed. A motion before a legislative body either calls for a change by way of implementing a new policy or it calls for resistance to pressure to change from a challenge to the status quo. Either way, the *onus probandi* falls on the person moving the proposal, and the procedural rules acknowledge this fact.

Another feature is that the mover is expected to argue for his or her motion. Thus the mover normally expects the assembly to contain some resistances to the motion, and even when the assembly is expected to approve it unanimously, it still has to be supported *pro forma*. This is precisely what is meant by "advocacy", namely arguing in support of a claim of some kind.

The mover of the motion is not permitted to argue against it. Thus his or her advocacy of the decision it embodies entails a fixed commitment to it and to his or her arguments for it as long as it is on the floor. However, a mover may, with the permission of the assembly, withdraw his or her motion. Doing so has the effect of the motion's never having been moved.

A third feature of these decision-making discussions in legislative bodies is that others besides the mover have an opportunity to debate the proposal. That is, others can advocate against accepting it, or produce new arguments in its favour, or produce arguments against the arguments in its favour, or produce arguments against the arguments opposing it, and so on. In each case the asserter of a new claim has the burden of proof, and so is expected to argue for it. The objective is to give the proposal a thorough examination, bringing both its merits and its defects to light, so that the members of the body can make informed decision when they vote on it.

Yet another feature of procedural rules like *Robert's Rules*, is that they permit a measure of flexibility in the body's decision-making. If it emerges during the debate that the proposal has a flaw that can be repaired by a modification, it is possible to amend the motion. If the arguments used in the debate expose a generally-agreed fatal flaw in the motion, and at the same time suggest a better alternative, it is possible to defeat the original motion and move a new one.

There are procedural rules that prevent one person or a small group from dominating the discussion and that permit those who want to speak for or against the motion to do so, rules that permit the body to call an end to the debate, and rules that invoke the operation of a decision-procedure: some form of voting.

These features of decision-making by the exchange of arguments advocating for and against a proposed choice, which are important in themselves, have a consequence of some importance as well. Because it is focused on the motion that initiates it, the discussion is thereby focused by the frame that the motion presupposes.[5] Any proposed decision is a particular solution to a question about what to do under a particular description. Take, for example, the recent question in the United States whether to approve the building of the Keystone pipeline to transport oil from the Alberta tar sands to the Texas refineries on the Gulf of Mexico. Some framed it as a straight economic question: did the pipeline make business sense? Others framed it as an environmental policy question: how did the pipeline affect global climate change? Different ways of framing the decision yield different arguments and different choices. Thus, by putting a motion before a decision-making body, the mover frames the issue, proposing an answer to a question that assumes a particular problem or opportunity, which might preclude other ways to think about the issue. The ensuing debate can then become preoccupied by that focus and so, overlooking the other possible ways to frame the issue, can miss an opportunity to find better solutions. Decision-making by using arguments to advocate following procedural rules like *Robert's Rules* has to rely on the capacity of the participants to recognize and draw attention to salient alternatives of the way the motion on the floor frames the issue to be decided. It takes deliberate effort and skill to do that, capacities that are often absent from the decision-making body.

4. THE ROLES AND FUNCTIONS OF ARGUING PRO AND CON

In order to compare advocacy and inquiry (or discovery), I need to set out the mechanics—the roles and functions—of the arguing involved in advocating for (and against) a choice in the debate in such bodies as those that rely on rules of procedure to keep order and ensure fair debate. The person who puts the motion on the floor and those who argue in its favour are its *proponents*; those who argue against its adoption by the body are its *opponents*. These are roles. The person who moves the motion is—at least under *Robert's Rules*—stuck in the role of proponent, but others in the body are free to switch roles. For instance, someone who initially speaks against the motion, thereby occupying the role of opponent, can be won over by the proponents' arguments and, in a later turn, functioning as a proponent, add arguments in its support. It

---

[5] The *locus classicus* on the concept of framing is Goffman (1974).

takes a good measure of self-assurance to change one's mind publicly, so such role-reversals tend to be rare in the cut and thrust of debate.

The kinds of argument move available are relatively few and can be spelled out. Arguments in favour of a position can offer either *direct support* for it or *indirect support* for it. A directly-supporting (a.k.a "ground-level"[6]) argument will have the position being advocated (or its approval) as its conclusion, and will consist of reasons for such support. A reason for such a claim will be expressed as a set of one or more premises, the acceptance of which, the advocate contends, supplies grounds for accepting the claim. There are four locations for indirectly supporting arguments. Each of them is an answer to a critique of one of the different elements of a directly supporting argument. These critiques can be voiced by opponents during the debate, or they can be anticipated by a proponent and attributed by him or her to the opponents.

I take a single argument to consist of a reason, i.e., a set of premises, that its author asserts as true or plausible or probable—in general, as worthy of acceptance by the members of the body—and a conclusion that its author invites his or her interlocutors to accept or endorse on the basis of having accepted those premises. An argument so conceived is in principle vulnerable to critique at four points. (1) The critic can challenge the adequacy of the premises in various ways, such as: of one or more of them, argue that it is false, or improbable, or implausible, or that it begs the question (e.g., by being acceptable only if the conclusion is already accepted), or that it is problematic (i.e., it is controversial and so itself requires support), or that it is inconsistent with another premise. (2) The critic can challenge the force or the bearing on the conclusion of one or more of the premises, in various ways. He or she can argue that the conclusion is a non-sequitur, or that a premise is beside the point, or that its evidentiary weight is negligible. (3) The critic can argue that the amount, the quality or the kind of ground the premises supply is inadequate to support the conclusion— e.g., that more evidence is needed (e.g., more studies), or more robust evidence is needed (e.g., double-blind studies), or different evidence is needed (e.g., systematic, not anecdotal, evidence). (4) Last, the critic can ignore the support the argument supplies and argue that its conclusion is false or is a bad idea or won't work, etc., in other words, that

---

[6] Directly-supporting arguments are what Finocchiaro (2013) calls "ground-level" arguments.

regardless of the proponents' reasons offered in its support, the conclusion should be rejected.[7]

In each case, the critic needs to support the critique with reasons, i.e., with arguments, because the critic here has the burden of proof. By hypothesis, the proponent has discharged his or her burden of proof by offering ground-level arguments in support of the motion. Responding by simply disagreeing or simply asserting that such support is flawed identifies the critic's opinion, but gives the proponent no reason to accept that opinion and acknowledge the flaw(s). So the critic owes arguments in support of his or her critique. Having supplied the needed arguments, the critic puts the proponent's directly-supporting arguments in jeopardy; the critic has provided reasons that raise questions about them. Hence, by responding to those arguments in turn, and in particular by showing that they are themselves flawed, the proponent restores the legitimacy of the initial argument and thereby provides indirect support for the initial conclusion. As mentioned above, the proponent can either wait for a critique before responding to it, or can anticipate it and respond to it as part of his or her initial support for the motion.[8]

To be sure, the response to a criticism need not be an argument refuting it. The proponent can acknowledge well-founded criticisms and respond to them by repairing the argument, either by dropping the offending part, or by adding new or altered premises, or by qualifying the conclusion, or by altering the substance of the conclusion—or else by conceding that the argument is beyond repair and abandoning it. Friendly criticisms are offered with a view to improving the position being advocated or improving the arguments used to support it; hostile criticisms aim to discredit the position being advocated or destroy its supporting arguments.

---

[7] The reader might have noticed the similarity between these kinds of critique and Pollock's (e.g., 2008, pp. 452-453) so-called "defeaters." The second and third are akin to his "undercutters" and the fourth to his "rebuttals." Pinto (e.g., 2001, pp. 103-104) independently identified "underminers" and "overriders." But I am persuaded by Johnson (2013) that there are differences between my account and theirs. Some readers will recognize in the first three kinds of critique the footprint of the criteria of acceptability, relevance and sufficiency that Johnson and I introduced in 1977. Their latest incarnation is Johnson and Blair (2006).

[8] The critic's argument in support of his critique, being an argument about an argument, is in Finocchiaro's (2013) terms, a meta-argument. The proponent's answering argument would thus be a meta-meta-argument. The consideration of, and response to, meta-argument critiques of the original ground-level argument are what Johnson (2000) calls the dialectical tier.

The above-described classification of uses of arguments to support and to critique proposed decisions, and the arguments offered in their support, constitutes a tool for the analysis of the argumentation in actual decision-making debates. It is doubtful that most participants are aware of these functions of their arguments, nor need they be. However, making them explicit provides one dimension for a comparison between decision-making by using arguments to advocate and using them to inquire or explore, and it is to an account of the latter that I now turn.

5. THE USE OF ARGUMENTS IN INQUIRY

I have been describing decision-making by debate in large groups that consist of dozens, scores, or even hundreds of people. That activity requires procedural rules to ensure order and fairness. I have argued that the procedures prescribed by the likes of *Robert's Rules of Order* have the result that the arguments employed are used to advocate particular proposals. However, at least in Canada and the United States, procedural rules like *Robert's Rules* are so well known and widely implemented that they often become the default for *any* decision-making body. For instance, my university has adopted them officially as the default rules of order for all university committees, from the university senate and faculty councils down to small departments and departmental committees. Another example: in a small fishing club I belong to, at our annual meetings a discussion of trout stocking will barely have begun when someone calls out, "I move that ..."; or the boats manager will end his report with a motion to purchase a new outboard motor, and the pro and con advocacy begins. Organizations of all kinds and sizes run their meetings along lines like those prescribed in *Robert's Rules*. Decision making by advocacy arguing can seem to be the only model available.

Nevertheless, there is at least one different model for the use of arguments in making decisions. It derives from the use of arguments to inquire or explore. One example of the use of arguments to inquire or explore is a method designed for students to follow when writing argumentative essays for their university courses, described by Jack Meiland in his handbook for students, *College Thinking* (1981, Ch. 4). Here is Meiland's advice (much abbreviated and paraphrased). Start by spending several days mulling and over (and writing notes about) the question or problem with which the paper will deal; then state it as clearly and precisely as possible, and explain why it matters and why it is problematic. Formulate a serious answer to the question. Produce a plausible argument that supports that answer. Then turn around and critique that argument, using the strongest objections to it you can think of. Next, switch back and see if you can think of replies that defend the argument against those objections. Acknowledge it when an objection seems sound. Repeat this argument-objections-replies cycle for all the plausible arguments you can think of or have time and space to deal with. The finding of the essay will be either that the hypothesized answer to the question or problem that motivated it is a good one because there are good arguments for it that stand up to criticism, or that the answer is a poor one because the best arguments for it can be shown not to be very good—or its merit lies somewhere in between. Meiland calls such an essay a "Dialectical argumentative paper" (1981, p. 67).

Now, suppose that instead of each student writing the essay alone, a dozen and a half students are invited by the instructor to sit down together to compose such an essay jointly. Each person in the group is at liberty to suggest formulations or reformulations of the initial problem, reasons why it is important and reasons why it is a problem—until all are satisfied. They then propose plausible solutions to the problem. Focusing on the candidate solutions one at time, they produce the most compelling reasons they can think of for each candidate, expressing those reasons as arguments that support it. They then all switch roles and try to find flaws or weaknesses in those arguments, producing arguments to support their critical judgements. Next they switch roles again and think about how to respond to the arguments that are critical of the reasons that seemed to justify choosing the proposed candidate solution. In the case of criticisms they find cogent, some might be answered by modifying the argument: changing its wording, qualifying it, deleting parts of it, adding to it, and so on; other criticisms might justify abandoning the argument.

It is a short hop to generalize the example of the class of students to any small decision-making group. The group is presented with a problem needing a solution, a question needing an answer, a situation needing a decision. The first thing they do is examine the assumptions of the task they have before them. Is there really a problem and if so is it best understood in the way it was presented to them? Is that the right question to ask? Is that the right or best way to understand the situation? In short, are there other, better ways to frame the issue than in the terms in which it was originally framed? Having identified what seems to them to be the best way to understand the issue, the members of the group next compile a list of what seem on the face of it to be plausible candidates for solutions, answers or choices. Each candidate is treated as having the standing of a scientific hypothesis. Like an hypothesis, a candidate has to be tested. The testing is rigorous for all hypotheses, and especially so for favoured hypotheses (to guard against bias). The members of the group test the hypotheses or candidate using arguments. They formulate the strongest arguments they can think of or find that support each hypothesis; they subject those arguments to challenging criticisms backed up by their own supporting arguments; and they decide by assessing their supporting arguments whether those criticisms themselves stand up to scrutiny. In these ways they discover which hypotheses are well supported, which have to be rejected or abandoned because they have a disqualifying flaw or because the reasons that can be found to support are too weak, and they discover which hypotheses, upon critical scrutiny of their supporting reasons, are found to have some merit. The group produces a ranking of the hypotheses based on the strength of their support. The best-supported hypothesis is the solution to the problem, the answer to the question or the choice to be made.

6. COMPARING ARGUMENTS' USE IN ADVOCACY AND IN INQUIRY

Several features of such a procedure mark ways it differs from using arguments to advocate in decision-making. First, in such inquiry, unlike in advocacy, the position is not a given at the outset and then defended; the preferred position results from the inquiry, it doesn't initiate it. Second, whereas in advocacy the initiator of the proposal is obliged to occupy the role of its proponent throughout, in inquiry no one has a constant burden of proof for some position. Everyone in the inquiry group is free to switch roles from proponent presenting arguments in support of a position to opponent presenting critiques of those arguments, and is indeed bidden to do so. The ideal is open-mindedness towards what position is best. Third, in advocacy the proponent's commitment to the position and to the arguments he or she uses to support it usually reflects the proponent's own beliefs. One normally doesn't advocate a position that one does not believe in. (There are exceptions. The college debater is assigned to the affirmative or the negative, and the lawyer taking on a case is obliged to advocate the client's best interests.) In inquiries the arguer is committed to the argument he or she offers for consideration just until an unrecognized flaw in it is pointed out by a fellow inquirer. Fourth, inquirers seek the best arguments as measured by their efforts to be rigorous and comprehensive—unlike advocates, who seek the arguments most likely to convince the majority. The inquirer's objective is the best position ascertainable, the advocate's objective is the best position that is winnable. Fifth, the advocate makes a tactical decision about which objections to his or her position to discuss and attempt to defuse or discredit, whereas the inquirer seeks out the most telling objections in order to put the position being explored to the most demanding tests. Sixth, in a well-regulated inquiry, attention is devoted to the way the issue that initiated it is framed, and alternative ways of framing the issue are considered. The question of appropriate framing is foregrounded. Advocacy, in contrast, begins with the frame behind the proposal being advocated already assumed. In advocacy, the appropriateness of the framing is backgrounded. Seventh, advocacy narrowly focuses the group's deliberation whereas inquiry spreads it widely; thus inquiry is time inefficient as compared to advocacy.

While these differences in the use of arguments in advocacy and in inquiry are significant, notice that the argument roles and functions are the same in both. A person in the role of proponent of a position produces arguments supporting it; a person in the role of opponent of that position produces critical arguments against the supporting arguments or against the position; a person in the role of proponent produces arguments that challenge the critical arguments. The iterations can go on indefinitely in theory, but in practice time, patience, ingenuity eventually run out and the need to act or decide brings the process to a close.

7. THE NEED FOR GUIDELINES FOR INQUIRIES

It seems on the face of it that the course of wisdom is to make decisions with an open mind based on a thorough assessment of the merits and disadvantages of a broad range of alternatives. Of course, the nature of the decision-making situation makes a difference. I have been told that experienced chairs of company boards of directors like to run meetings following something like the inquiry approach: a free-wheeling discussion of the alternatives and their respective pros and cons until a consensus emerges, at which point the chair calls for a motion. I have also been told that in Canadian federal government cabinet meetings, where ministers have been well-briefed on the relevant pros and cons and there is a need for issues to be decided by the end of the meeting, inquiry-like discussions are precluded and advocacy interventions are required.

Nevertheless, when institutional practice calls for small groups such as committees to make decisions, or when there is an opportunity to benefit from the contributions of several different minds with a variety of knowledge sets and value perspectives, the use of arguments to explore the merits of alternatives seems more closely to approximate the ideal than the use of arguments to advocate for and against one option. And in institutional settings in which advocacy is ingrained or unavoidable, it seems desirable that any proposal put before the group be the product of a prior inquiry of the kind described above, with the arguments used to advocate the proposal being selected from among those that stood up well to critical scrutiny in that inquiry.

However, even when there is ideally a need and in fact an opportunity for a small group to conduct an explorative argument-based inquiry, there is to my knowledge no well-known procedure to call upon, analogous to available guidelines such as *Robert's Rules* designed for argument-based advocacy. All too often, in a group ideally positioned to conduct an inquiry, the discussion will begin by someone making a motion, and advocacy will take over, for advocacy trumps inquiry. "There is a motion on the floor" renders "let's first consider all the alternatives" out of order. With a view to correcting that deficiency, I offer here a draft of that might be called "Lisbon Rules of Order" as guidelines for argument-base decision-making inquiries. I am sure it can be improved, but it is offered as a start.

Any such set of guidelines ought to satisfy the following desiderata. The guidelines should be *clear and simple*. The objective of widespread use means that they should be readily understandable by non-specialists and that no great effort should be required to master them. In order that imagination and creativity can nurture the inquiry, they should be as *permissive or uninhibiting* as possible. At the same time, they should *block* an expected tendency to revert to *advocacy argument*. They should *serve as a guide* through the stages of an orderly inquiry, so that the participants can know where they are heading and what to expect next at any point. They should *encourage open-mindedness*. And of course the rules should be *consistent* with one another and *economical* in the sense that no particular guideline is redundant.

The following draft of guidelines consists of four strict rules plus advice for carrying out three basic tasks.

Lisbon Rules of Order for Inquiries (draft)

Rule 1: No substantive motions are permitted until the completion of Task 3.

Rule 2: Participants have the right, and an obligation, to express arguments both for and arguments against each hypothesis, and to express judgements of the merits of those arguments. Thus they ought switch back and forth between the roles of proponent of arguments pro and proponent of arguments con and between critic trying to refute arguments and defender trying to refute objections or repair arguments.

Rule 3: Each member of the group has the right, and must have the opportunity, to contribute suggestions and arguments.

Rule 4: The following three tasks are to be completed in sequence. It is undesirable to proceed to the next task before substantially completing the present one. However, it is permissible to revisit and revise elements of previously-completed tasks should it become desirable to do so.

Task 1: Identification of options

1.1 Identify different ways of framing the issue before the group so as to avoid unduly limiting the range of possible decisions. If there is disagreement about how to frame the issue, make how to frame the problem itself a topic for a preliminary inquiry using these guidelines.

1.2 List what seem to be the plausible positions that might be taken. Any possible decision that any member of the group thinks is worth considering is to be included.

1.3 If research is needed to expand the range of possible decisions, assign research duties and recess for a reasonable time for research. (May combine with 2.3.)

1.4 Once the list of viable options seems complete, go to Task 2.

Task 2: Investigation using arguments

2.1 Treat each option as an hypothesis to be tested by seeing how good a case can be made for it by weighing the merits of the arguments in its favour vs. the arguments against it.

2.2 In particular, for each hypothesis: Formulate the apparently best reasons for it and those against it. Look hard for flaws or weaknesses in those reasons. Judge whether the alleged flaws are indeed flaws, and whether the admitted flaws can be corrected, either by modifying the reasons or by revising the hypothesis. Decide how strong is the net case for the hypothesis.

2.3 It is permissible (and desirable) to modify hypotheses in the light of weaknesses or flaws exposed by arguments criticizing them that are acknowledged to make good points.

2.4 If research is needed to find possible reasons for or against the hypotheses, recess in order to provide time for such research before proceeding. (May combine with 1.3.)

2.5 Abandon consideration of an hypothesis if clearly disqualifying reasons against it exist or the case for it is obviously comparatively weak.

2.6 Once the various hypotheses have been examined and judged as thoroughly as time and resources permit, go to Task 3.

Task 3: Summation and application

3.1 Determine which one is the best-supported hypothesis. This may be done in any systematic way—e.g., pairwise comparisons, ranking them and dropping off the weakest, using new arguments for and against particular rankings to break ties. The best-supported hypothesis is the decision to take on the issue that initiated the inquiry.

3.2 Produce a case for the proposed decision. A case consists of a review of the arguments for the hypothesis, together with a mention of the principal objections and why they are mistaken or can be overcome, and a review of the arguments against the hypothesis, together with a mention of why they are mistaken or can be overcome.

8. CLOSING COMMENTS

I close with four comments.

First, the end of 2.2 ("Decide how strong is the net case for the hypothesis.") and the beginning of 3.1 ("Determine which one is the best-supported hypothesis.") are placeholders. How to come to on-balance judgements that weigh pros and cons of different kinds and weights is a topic for evaluation theory. So is how to identify the best-supported hypothesis from among a group of similarly qualified alternatives. Such judgements are necessary and are made all the time, so the fact that these guidelines require them is not a defect of the guidelines; but the fact that no guidance for making them is supplied is a gap in the guidelines.

Second, only Rule 3 speaks to group members' interactional needs. An effective inquiry envisaged by such guidelines requires group members to be tolerant, helpful, patient, encouraging, and constructive. It will require participants to take on process roles in addition to the task roles of arguing as proponent and opponent, repair builder and so on. Participants will need to monitor the discussion, interject when a point has been made and it is time to move on, keep order, make sure some don't dominate and everyone has a chance to contribute, and so on. I am well aware that there is an enormous literature on group dynamics and effective group interaction. Whether useful advice from that literature can be added to the guidelines without making them too cumbersome is a question that needs to be answered.

Third, indeed, whether groups with more than three or four members can effectively carry out an inquiry along these lines is an empirical question on which I have only anecdotal evidence.

Finally—speaking of unanswered empirical questions—whether decision-making by small groups that follow this inquiry procedure produce better decisions that those that follow the advocacy model is a matter that needs to be tested. Here I appeal to the knowledge of social scientists, who, I assume, know how to design experiments to test such assumptions.

ACKNOWLEGEMENTS: My thanks to colleagues at the Centre for Research on Reasoning, Argumentation and Rhetoric at the University of Windsor—Pierre Boulos, Hans Hansen, Marcello Guarini, Ralph Johnson, Suzanne Mcmurphy, Christopher Tindale and Douglas Walton—for helpful constructive comments on an earlier draft. A conversation with Ivo Krupka, principal of Public Policy and Management consultancy and former senior executive in the public services of Canada, and Peter Dey, most recently Chairman of Paradigm Capital and member of several corporate boards, helped me to clarify and sharpen some points.

REFERENCES

Aristotle. (1984). *Nicomachean ethics.* In J. Barnes (Ed.), *The complete works of Aristotle*, Vol. 2. Princeton: Princeton University Press.
Blair, J. A. (2004). Argument and its uses. *Informal Logic, 24*(2), 137-151. Reprinted in J. A. Blair. (2012). *Groundwork in the theory of argumentation* (Ch. 14, pp. 185-195). Dordrecht: Springer.
Eemeren, F. H. van & Grootendorst, R. (2004). *A systematic theory of argumentation: The pragma-dialectical approach.* Cambridge: Cambridge University Press.
Finocchiaro, M. A. (2013). *Meta-argumentation: An approach to logic and argumentation theory.* London: College Publications.

Goffman, E. (1974). *Frame analysis: An essay on the organization of experience.* New York: Harper and Row.

Johnson, R. H. (2000). *Manifest rationality: A pragmatic theory of argument.* Mahwah, NJ: Lawrence Erlbaum Associates.

Johnson, R. H. (2013). Defeasibility from the perspective of informal logic. In D. Mohammed & M. Lewiński (Eds.), *Virtues of Argumentation. Proceedings of the 10th International Conference of the Ontario Society for the Study of Argumentation (OSSA)* (pp. 1-12). Windsor, ON: OSSA.

Johnson, R. H., & Blair, J. A. (2006). *Logical Self-Defense.* New York: IDEA Press.

Meiland, J. W. (1981). *College thinking: How to get the best out of college.* New York: New American Library, Mentor Books.

Pinto, R. C. (2001). Argument schemes and the evaluation of reasoning. In R. C. Pinto, *Argument, Inference and Dialectic* (Ch. 10, pp. 98-104). Dordrecht: Kluwer.

Pollock, J. L. (2008). Defeasible reasoning. In J. E. Adler & L. J. Rips (Eds.), *Reasoning* (Ch. 23, pp. 451-470). Cambridge: Cambridge University Press.

Robert, H. M. (1986). *Robert's rules of order: The standard guide to parliamentary procedure.* New York: Bantam Books.

Walton, D. N. (1998). *The new dialectic: Conversational contexts of argument.* Toronto: University of Toronto Press.

The Deliberation and the Advocacy Rooms Commentary on Blair's Advocacy vs. Inquiry in Small-Group Deliberations

FLORIANA GRASSO

Department of Computer Science, University of Liverpool, UK
floriana@liverpool.ac.uk

1. INTRODUCTION

Blair offers us a convincing case on the need to spell out the differences between "the use of arguments in deliberation", as a way to "discover the best and so choiceworthy alternative", and "the use of argument to advocate approval of a particular alternative". Whilst advocacy is not only well understood but also regulated and common place in many formal and informal situations, deliberation, or inquiry, Blair observes, does not share the same recognition.

Blair lists a number of features that set the two argumentative situations aside, which emphasise in some way the contrast between a somewhat restrictive mindset of the advocates, who are deeply concerned about defending their own positions, and the open-mindedness of the deliberators, who set themselves the objective of explore as many alternatives as possible, and analyse each of them deeply and without prejudice. Blair concludes by proposing the *Lisbon Rules of Order for Inquiries*, or a collection of guidelines that can regulate a process of deliberation, with a view, generally speaking, of preserving exactly this characteristic of openness and lack of prejudgement and bias.

2. IS THERE A DIFFERENCE IN THE ESSENCE OR THE MANIFESTATION?

A question worth asking ourselves, however, is whether the difference between the two styles of debate lies in effect on the way in which the participants argue, or rather on the way in which they are attached to their own propositions. In other words, from the point of view of the single individuals participating to the two types of debate, one may conclude that their attitude towards the claims they are putting forward, or the positions they are supporting or simply enquiring about, is the key. Not only the propositional attitude towards the claim, but also the entrenchment to it, and most crucially, the attitude towards any value associated with the *outcome* of a particular piece of argumentation they are engaged into. Hence, we have a more "relaxed" attitude in inquiry ("let's explore this claim, let's see where it takes us") and a more "tense" one in advocacy ("what if I lose my position? What does this entail? Losing face? Losing a debate? Losing an election? I shall not allow this to happen.").

However, the robustness with which each of the arguers intends to participate to the discussion regulates the depth and breadth of the options explored in an inquiry too: Blair observes that in advocacy "the mover is expected to argue for his or her motion. Thus the mover normally expects the assembly to contain some resistance to the motion [...]". But a thorough deliberator also will have the same expectation from the participants to the inquiry: he or she will expect and hope each option to be scrutinised deeply, and every attempt to defend it, to argue for it, to be put forward, not so much to demonstrate attachment to the option, but with the aim to elicit the most varied set of criticisms, and hence either reinforce the position, if all criticisms can be addressed, or informatively dismiss it, if they cannot.

The question is then: are the two processes really so different once we take out of the equation the level of personal attachment to a position by the participants? Can eliminating personal attachment blur the boundaries enough to make the two processes equivalent? And what does "equivalent" mean in this context?

2. THINKING IN MODELS

It is not perhaps difficult to imagine that if we decouple participant attitudes toward a proposition, from the act of putting forward the possible actual list of arguments pro and con the proposition, we can indeed minimise the difference between the two types of debate, as long as we have enough participation to the two processes to guarantee that, in advocacy, all the counterarguments to a motion will be explored, and in inquiry, all criticisms to the various options will be thoroughly and exhaustively put forward.

It helps think in terms of models: what can be generalizable? Is there a pattern? What does the behaviour of a "system" look like, be it an individual or a collection of individuals? Can this be modelled? Can the behaviour be reproduced?

This does not have to be necessarily, or solely, driven by an interest in artificial models, their implementation or simulation, but rather in the deep understanding of a complex behaviour, and its features, and the way it can be deconstructed in components.

So for instance, suppose one can enumerate exhaustively all possible arguments and claims that can be put forward on a certain topic, and suppose also that a mechanism is put in place to ensure that all of them are, in due course, put forward before the end of the debate and a conclusion is reached. Suppose also that one manages to elaborate an appropriate model for the behaviour of the various participants, and regulate it to avoid situation that can compromise the outcome of the process. Then, what about this outcome? And what about the process? Can we model the behaviour of the system of individuals, regarded as a whole entity?

Could both inquiry and advocacy be seen as substantially two equivalent ways to reach a "good" conclusion, whatever we want to associate with the notion of "goodness" of a conclusion?

More precisely, can we assume that, given an appropriate selection of participants to the two processes (and what an appropriate selection is, is still to be decided, but we will come back to that later), would both processes reach the same conclusion?

An idealised version of both group processes then, both discovery and advocacy, would require a comprehensive and exhaustive exploration of all possibilities: all options are considered in the discovery process, all voices are heard in the advocacy process. If this is done systematically, then is it plausible to assume that the outcome of the process, and hence the features of the systems when seen as a whole, are almost indistinguishable, from the outside?

3. IDEALISED PROCESSES

It remains to be established what a "good selection" of participants might look like. Blair's description of group members participating to an effective inquiry is most interesting: they have to be "tolerant", "helpful", "patient", "encouraging", "constructive".

While these are not features which are easy to pinpoint or model, one can think of different roles that different agents can take in a debate so that the actual outcome of the process is to avoid intolerance, impatience, destructiveness. One can think of very specialised agents, which are assigned a specific task within a debate, and one can associate a different set of rules to each different task or role, as laid out in Blair's paper.

The agents operating in the advocacy environment have a precise remit in the process, those in the discovery environment can have different roles with respect to a claim.

In such an "idealised" debate, each agent will simply be "programmed" to perform its part, and in the end, each process will terminate with a collectively "agreed" solution to the initial problem.

The model of the process becomes more interesting and challenging when "values" are put in the equation. If agents have different attitudes towards a proposition and towards the outcome of the whole process ("what will it happen if I lose?") then this ideal process is jeopardised, arguments can be hidden, on purpose, or ignored, and objections can be dismissed with no further analysis; motivated belief formation and dissonance would guide some reasoning processes, and one cannot guarantee anymore that the ideal solution, ideal insofar as all participants agree to it, can be reached.

In analogy with the Artificial Intelligence metaphor of Intelligent Autonomous Agents, this can perhaps be addressed by having a "large enough" pool of agents to engage in the process, one for each of the complete set of claims that can be put forward on a particular issue, supposing for one moment one can in fact come up with such a list. So, one does not need to be concerned anymore that one particular argument or motion or objection will not be put forward, because the agent associated with it will be "activated" at a given point in time.

Let us amuse ourselves pondering on how this can happen, let us think of it as a game.

4. A PARTY GAME: THE DELIBERATION AND THE ADVOCACY ROOM

Suppose a party is organised, with an unlimited number of participants, perhaps an online party when people can join and leave anytime. At some point, a debate game starts: participants are randomly assigned and asked to direct themselves to one of two different rooms, the Deliberation Room, and the Advocacy Room.

By the entrance of each room, there are two enormous baskets, one for each room. Before entering the room assigned to them, participants are asked to pick randomly a card from the basket outside their room: this card contains the statement which is assigned to them, with also some information on the context in which this claim can be put forward. The assumption is that each basket enumerates **all** possible claims that are relevant and pertinent to the topic under consideration. The two baskets at the entrance of each room contain therefore exactly the same set of claims, as the two rooms will debate on the same topic.

In each room, participants are given the set of rules for the game. These are the rules governing the engagement to the debate: Blair's *Lisbon Rules of Order for Inquiries* to the participants in the

Inquiry room, and, say, *Robert's Rules of Order* to those in the Advocacy Room.

The final piece of information each participant receives is the role they will play in the game, the attitude they need to show towards their claim (perhaps a penalty associated to their claim being discarded at the end of the process? Or a prize associated to the way in which their claim has contributed to the final decision? Or a measure of the esteem they acquire by engaging in the process? One can think of many different variations of this mechanism).

After this preparatory phase, the two games will start. What happens in the rooms is not observed by those outside: it is assumed the two rooms will make sure their internal rules are followed, and all participants have the chance to explore their claim to the full.

In the advocacy room, claims are put forward as motions, and participants are encouraged to argue for the motions, or resist to them, in turn putting forward new motions and claims, by following Robert's rules of order.

In the inquiry room, claims are put forward as proposals, and participant are encouraged to deeply scrutinise the proposals, finding reasons to corroborate them, or discard them, but following Blair's rules of order.

At the end of the two debates, each room will produce a claim, or perhaps a set of claims, a "theory", that has collectively been agreed to stand by those in the room. This is the only outcome of each room: all the other dynamics that took place in the two rooms, the voices of the various participants, the strength of their positions, the prizes or penalties they collected in the room, all stay in the room.

The question is therefore:
1. provided the idealised preparatory phase was completed to satisfaction, and all possible claims enumerated and placed in the baskets;
2. provided the information on the card for each participant was sufficient enough to let them figure out when to intervene in response to which other claim or motion;
3. provided the rules of order were precise enough to guarantee that the two debates were conducted fairly and in line with their spirit, even if the participants joined in with no prejudgements on the topic and the claim assigned to them; and
4. provided sufficient time was given to the two rooms so that all options could be extensively explored in detail;

will the two rooms arrive to the same conclusion?

4

The Bearable Ambiguity of the Constitutional Text. Arguing, Bargaining and Persuading in The Italian Constituent Assembly

GIOVANNI DAMELE
ArgLab, Universidade Nova de Lisboa, Portugal
<u>giovanni.damele@fcsh.unl.pt</u>

Jon Elster suggested that the constitution-making enterprise can be understood resorting to two types of "speech acts": "arguing" and "bargaining". He also makes reference to "rhetorical statements". Thus, the model seems to be triadic: arguing, bargaining and persuading. The analysis of constitution-making debates could be improved by developing this triadic model. In order to test this model, I will analyze the debate that led to the promulgation of article 29 of the Italian Constitution.

KEYWORDS: comparative constitutional law, constituent assembly of Italy, constitution-making process, Jon Elster, political argumentation, political rhetoric

1. ARGUING AND BARGAINING

In his article *Arguing and bargaining in two constituent assemblies*, Jon Elster suggested that the constitution-making enterprise can be understood more generally by resorting to two types of "speech acts": "arguing" and "bargaining". These two types are used by Elster with the aim of exploring two distinct issues: "the role of the rational argument" and that of "threat-based bargaining" ("bargaining on the basis of extra-parliamentary resources"). This theoretical model is eventually tested by resorting to the two examples of the American and French post-revolutionary constituent assemblies. The reason for privileging constituent assemblies, according to Elster, is that they are "often more polarized than ordinary law-making bodies", mainly because "the matters that have to be decided are far removed from petty, self-interested, routine politics" (Elster, 2000, p. 347). Thus, constituent assemblies seem to "exhibit both arguing and bargaining in their most striking forms". They oscillate between "higher law-making" (in Bruce Ackerman's words (Ackerman, 1991)) and "sheer appeal to force" (Elster, 2000, p. 348). More generally and in other words, in this and in others articles (1995; 2012a; 2012b) Elster seems to develop a kind of "general theory of constituent process" based on an (Habermasian) argumentation theory – even if such a "general" theory should comprehend not only "bottom-up" processes, such as the Philadelphia Convention and the French revolutionary "Assemblée constituante", but also "top-down" and even informal constituent processes.

Taking into consideration the debates developed in the two constituent processes, Elster insists mainly on the role of rational argument in constituent decision-making, claiming that even the actors with "purely self-interested" concerns may be "forced or induced to substitute the language of impartial argument for the language of self-interest". A kind of "substitution" that Elster attributes to what he calls "the civilizing force of hypocrisy". However, Elster also makes reference (without further explanation) to a "third type of speech acts", namely "rhetorical statements aiming at persuasion", defining them as an "appeal to the passions of [the] audience, rather than to their reason or self-interest" (Elster, 2000, p. 371 n. 116). For this reason, Elster's seems to suggest a triadic, rather than dyadic, model: arguing ("reason speaks to reason"), bargaining ("interest to interest"), and persuading ("passion to passion").[1]

In this paper, I will simply try to verify if Elster's model is applicable to an example of debate taken from the *travaux preparatoires* of the Italian Constituent Assembly. The analysis will be developed following the *dyadic* model explicitly presented by Elster and taking into consideration the relevance of the third component: the rhetorical dimension.

1.1 The model and its components: arguing, bargaining and persuading

Despite the reference to "rhetorical statements", Elster focuses his analysis on the French and American constituent assemblies starting from the double dimension of argumentation (a word that Elster seems

---

[1] "I distinguish among reason, passion and interest as motives of speakers in constitutional or legislative assemblies. Applying the same distinction to the motives imputed by the speakers to their audience, rhetoric may perhaps be defined by the feature that its practitioners appeal to the passions of their audience rather than to their reason or self-interest. In some debates, reason speaks to reason: in others, interest to interest; in still others, passion to passion" (Elster, 2000, p. 371, n. 116).

to use as a short-hand for "rational argumentation") and negotiation. According to him: "rational argumentation on the one hand, threats and promises on the other, are the main vehicles by which the parties seek to reach agreement" (Elster, 2000, p. 372). However, the neglected component, rhetoric, can be characterized as a "common dimension" shared by both the argumentation and negotiation stages. Indeed, as I will try to suggest, both the strategic use of arguments (fundamental in Elster's model) and the strategic use of threat and warning (on which the negotiation stage is based) are developed through rhetorical moves.

1.1.1 Arguing

As I mentioned before, in Elster's model negotiation is subject to criteria of credibility, while "rational argumentation" is subject to criteria of validity. And "validity" should be understood according to the three "validity claims" (propositional truth, normative rightness and truthfulness or sincerity) to which, from the point of view of Habermas' theory of communicative action, a speaker who aims at achieving understanding rather than success is committed.

Considering the "type" of arguments used in constitutional debates, Elster makes broad reference to two general categories: consequentialist or deontological arguments. In the first case, the framers "appeal to overall efficiency", while in the second "to individual rights". This second type, that of "right-based arguments", seems to be more "impartial" because "the rights are assigned to everybody" [Elster, 2000, p. 379]. Elster also notices that "consequentialist" arguments can imply a certain amount of impartiality, however the use of deontological arguments seems to be much more effective: "Framers can go to great lengths to make it appear that a measure whose real justification is obviously utilitarian can also be defended in terms of rights" (Elster, 2000, p. 391).

This latter consideration leads Elster to conclude that the two original Habermasian commitments to truth and impartiality can coexist with a *strategic* use of impartiality and sincerity. Even if the framers are not genuinely committed to these values, "they may find it in their interest to appear to be so committed". In this case, "they engage [...] in strategic uses of purportedly non-strategic argument". In other words, self-interested actors may try to "ground their claims in principle" when "their self-interest tells them to appeal to an impartial equivalent of self-interest". This strategic function of "citing a general reason" has, according to Elster, an obviously persuasive goal, that is increasing the ability of the speaker to persuade others, and, particularly, to persuade "the neutrals to agree" with him (Elster, 2000, pp. 408-409).

Interestingly enough, this passage, which takes us from a sincere commitment to truth to a *strategic* sincerity, partly coincides with the Aristotelian definition of rhetorical discourse. In rhetoric, the possibility of the discourse being true is taken into consideration only as a means of persuasion: independently of the truthfulness of the premises or of the sincere commitment of the speaker to truth. What matters is the *appareance* of truth: the *persuasion* of the audience about truth (Viano, 1955, p. 284). In other words, the *civilizing force of hypocrisy* is, in its essence, based on a rhetorical move.

1.1.2 Bargaining

If, as according to Elster, rational discussion is supposed to be based on "the power of the better argument", then constitutional bargaining "rests on resources that can be used to make threats (and promises) credible". In his paper, Elster focuses mainly on extrapolitical resources, that is sources of bargaining power which "exist independently of the political system" - such as money, manpower, and foreign allies. However, Elster does not ignore that "constitutional bargaining may be based on resources created in the assembly itself" (Elster, 2000, p. 392). One important example in the constitutional context is that of vote-trading or logrolling, that is, the practice of exchanging favors by reciprocal voting for each other's proposed legislation. In this case, the practice of bargaining is based on resources that are not extrapolitical, but arise within the assembly itself. In particular, Elster notices that when a constitutional assembly does not set its own rules, the parties can resort to the pre-existing institutional framework, using it for strategic purposes (Elster, 2000, p. 405).

Finally, a kind of "substitution", analogous to that established in an argumentative context by the "civilizing force of hypocrisy", can be also found in a bargaining context. In the same way in which they may substitute an impartial argument for a direct statement of their interest, strategic actors "may also find it useful to substitute truth claims for credibility claims". In this case, "instead of making a threat whose efficacy depends on its perceived credibility, they may utter a warning that serves the same purpose and avoids the difficulties associated with threats". Indeed, according to Elster "warnings" are "factual equivalents" of a threat. Threats, "are statements about what the speaker will do", while warnings are statements "about what will (or may) happen, independently of any actions taken by the speaker" (Elster, 2000, pp. 414-415). Thus, this substitution seems to be motivated by a strategic preference for (more persuasive) claims based on truth instead of (possibly less persuasive) claims based on credibility. Also in this case, the "substitution" seems to be based on the rhetorical dimension of verisimilitude. A reference to (supposedly true) factual consequence is used in order to increase the persuasiveness of a mere appeal to (subjective) credibility.

1.1.3 Persuading?

The strong accent on the argumentative dimension makes Elster's theory particularly attractive when studying constitutional debates. Nevertheless, the analysis of real constitution-making debates could be improved by developing the triadic version of the model, suggested, but not elaborated by Elster. Such an improvement can be achieved by introducing the third "type of speech acts" (in Elster's words).

Elster's theory seems to be based on two pure models of arguing and bargaining, i.e. "rational argumentation" and negotiations based on threats, and two "impure" models, i.e. the strategic use of impartial arguments and negotiations based on warnings. In both cases, what happens is a "substitution" with strategic purposes: impartial arguments are substituted for "direct statements of their interest" and "truth claims" for "credibility claims". Again in both case, this kind of "substitution" seems to have an eminently persuasive function. They seem to be, in other words, purely rhetorical.

Oddly enough, Elster also seems to attribute a negative judgement to what he calls a "rhetoric statement". For example, he makes a distinction between, on the one hand, "many of the debates at the Federal Convention", described as "remarkably free from cant and remarkably grounded in rational argument" and, on the other hand, discussions in the *Assemblée Constituante*, which are "heavily tainted by rhetoric, demagoguery and overbidding" (Elster, 2000, p. 411). When noticing Mirabeau's enormous influence in the French Assembly, in spite of his inconsistencies and his "erratic behavior", Elster warns that "rhetoric, as a distinct from argument, may have been at work here" (Elster, 2000, pp. 377-378). Due to this negative judgement, Elster fails to see the rhetorical character of the strategic use of impartial arguments and claims based on warnings. This constitutes an underestimation of rhetoric that seems rather inadequate for the purpose of gaining insights into the discursive processes of constitutional decision-making, especially considering the relevance of those strategic uses for Elster's theory of the "civilizing force of hypocrisy". Indeed, because of this relevance the rhetorical goal of persuading seems to have a kind of "logical priority". In other words, the three "types of speech acts" - arguing, bargaining and persuading - are hierarchically interconnected in a model in which "arguing" and "bargaining" are both submitted to the rhetorical purpose of persuasion.

In order to test this "triadic" model, I will analyze the constitutional debate that led to the promulgation of article 29 of the Italian constitution: the article on family and marriage. There are many reasons for this choice. First of all, the debate was characterized by use of both consequentialist and deontological reasons. In other words, both "questions of principle" and "utilitarian" justifications had been invoked on both sides of the debate. Second, it represents a kind of "temporary breakdown of cooperation", which is important in this context because Elster considers the ability to specify what will happen in such cases as "the most important requirement" of a "bargaining theory". Finally, it is also important because the participants to the debates tried to find a way through the impasse using strategic arguments, together with "impartial" arguments and threats, within the pre-existent institutional framework.

2. THE DEBATE ON THE ARTICLE 29 IN THE ITALIAN CONSTITUENT ASSEMBLY

2.1. Context

The Italian Constituent Assembly (*Assemblea Costituente*) was elected in 1946 (using proportional representation) with the first free and full elections in Italian history. In the same election, through an institutional referendum, the monarchy was removed and a Republic was instituted.[2] The result of the election set up a balance between the main antifascists parties, the centrist Christian Democrats on one side (who won 35 percent of the vote) and the two leftist main parties on the other (the socialists of the PSI, with 21 percent of the votes, and the communists of

---

[2] In 1944 the Lieutenant-general of the realm, the crown prince Humbert, under pressure from the reformed democratic parties in the South, issued a decree-law which stated that "after the liberation of the national territories the institutional forms would be chosen by the Italian people" and that after direct elections a "Constituent Assembly would be elected to deliberate the new constitution". This decree-law constitutes part of the, in Elster's terms, "upstream legitimacy" of the Italian constitution. In 1944, the heir apparent was acting as king, since the king was "suspended" from exercising his duties due to the fact that the anti-fascists parties looked at him as the main responsible of the rise of the fascist regime, given his decision to appoint Mussolini as Prime Minister in 1922.

the PCI, with 19 per cent).³ A third political party, the liberals, was also very influential even if numerically less significant (less than 10 per cent).

The work of the Italian Constituent Assembly lasted 18 months, from 1946 to 1947 (with over 170 sittings). In order to organize the work in the Assembly, the task of drafting constitutional provisions was delegated to a "Constitutional Commission" of 75 deputies (also known as "the Commission of the 75"), divided into three sub-commissions, each one chaired by a deputy of one of the three mains parties. The first sub-commission, tasked with drafting constitutional articles on "Rights and Obligations of the Citizens", was chaired by the Christian democrat deputy Umberto Tupini. The second and the third, on "Constitutional Organization of the State" and "Economical and Social Relationships", were chaired by the communist Umberto Terracini and the socialist Gustavo Ghidini, respectively. Finally, a Committee of 18 deputies (the "Committee of the 18") was given the task of writing an overall draft of the constitution, in accordance with the work of the three sub-commissions.

During this period, the work of the Assembly was threatened by external and internal circumstances. Internally, there were the obvious tensions inherent in the antifascist parties belonging to different political traditions: the Christian democrats, the social-communists, and the liberals. Externally, there was the risk of a political crisis leading to civil war and, eventually, to foreign intervention. However, the main external influences affecting the work of the Assembly were the relations between political parties in view of the political elections of 1948. In particular, in May 1947 the Prime Minister Alcide De Gasperi, general secretary of the Christian Democratic Party, drove the communists out of the government with the aim of forming a new government without the extreme left and obtaining the full support of the government of the United States. As John Foot noted, Italy could have followed the path of Greece in 1946-1947, with a bloody civil war and foreign intervention (Foot, 2003). However, De Gasperi and the general secretary of the Communist Party, Palmiro Togliatti, continued to collaborate even after May 1947 to draw up the Constitution. Further, the fact that the Constituent Assembly had a special "commission" of deputies concentrated only on the drafting of the Constitution allowed for collaboration and helped to produce a "high-level" debate, separated from other more "political" spheres.

The result of this process has been a democratic constitution, which is the obvious result of a set of political compromise. The main

---

³ Only 21 of the 556 elected constituent members were women.

example of such a compromise is the Article 7, which established a privileged status for the Church in the new republic, including (with the approval of the communists) the Concordat between Church and the fascist State of 1929 (the so-called "Lateran Traties"). Other compromises ended with the independence of the Constitutional Court – a defeat for the Communists – or the Christian Democratic retreat from a corporativist and regionalist model, which can be viewed as a victory of the liberals (Einaudi, 1948, p. 662). According to Mario Einaudi, one of the main Christian Democratic defeats came when the final constitutional text no longer proclaimed the indissolubility of marriage (Einaudi, 1948, p. 663).

2.2 Positions

As a matter of fact, the Italian Constitution declares under article 29 that the Republic recognizes the rights of the family as a "natural society"[4] founded on marriage. Further, the second paragraph of the Article asserts the principle of the moral equality of the married couple, focused in particular on the position of the woman to whom it wishes to guarantee equal dignity and rights within the marital relationship.[5] This final version of the article was arrived at after bitter debates.[6]

---

[4] This expression is generally interpreted as a way to stress that the contemplated concept of "family" had original rights which pre-existed the State, and which the latter was obliged to recognize. Recently (Decision No. 138 of 2010), the Italian Constitutional Court gave a narrow interpretation of the article, affirming that the absence of explicit references to homosexual unions in the *travaux preparatoires* of the Constituent Assembly does not allow embracing, through interpretation, situations and problems that were not considered at all when the article was enacted, such as homosexual marriage (Damele G., Sulle motivazioni della recente sentenza della Corte Costituzionale italiana in materia di matrimonio omosessuale. Con un confronto con la giurisprudenza del *Tribunal Constitucional* portoghese, *Diritto e Questioni Pubbliche*, 1/11 (2011), 631-661).

[5] Art. 29 of the Italian Constitution:

"The Republic recognizes the rights of the family as a natural society founded on matrimony"

"Matrimony is based on the moral and legal equality of the spouses within the limits established by law to guarantee the unity of the family "

[6] The minutes – in Italian – are available in the online archive of the Italian Chamber of Deputies (www.camera.it). The debates about Article 29 can be found in the 1946 dates: 26 July, 13 September, 30 October, 5-7, 12-13 and 15

Besides the Concordat with the Catholic Church, the nature and the role of the family were central to the debates in the Constituent Assembly and were a main source of division among the political parties. Being related with "rights and obligations of the citizens", this issue fell under the competence of the "First Subcommission". Here, two rapporteurs, the communist Nilde Iotti and the Christian democrat Camillo Corsanego, where charged with drafting a first proposal. This actually lead to two proposals, in which five fundamental questions were raised regarding: 1) the definition of family, 2) equality between spouses and the representation of women's roles, 3) the status of children born out of wedlock, 4) the indissolubility of marriage, and 5) the degree of autonomy of the family vis-à-vis the state (Naldini, 2003, pp. 55-56). During the debate, the members of the commission began an active effort to achieve a political compromise. At one point, an agreement seemed possible on all of the issues, except the indissolubility of marriage.

To be sure, the Christian democrats strongly supported the inclusion of the indissolubility of marriage, mainly because they wished to avoid any introduction of divorce in the Civil Code. For many of them, including the rapporteur of the draft Corsanego and the young but influential deputy Giuseppe Dossetti, the question at stake was of enormous importance. Dossetti (who later became a catholic monk) very clearly stated: "for us, the issue we are debating [indissolubility of marriage] is the most fundamental issue of the Constitution". Conversely, the social-communists opposed the inclusion of the "indissolubility" in the constitutional text. However, there were nuances. While the socialists' position was basically justified by their traditional anticlericalism, shared by the majority of their electorate, the communists where in a rather uncomfortable position. On one hand, they were seeking to avoid the inclusion of the "indissolubility" of marriage in the Constitution, something that would have made the introduction of divorce through ordinary (not constitutional) legislation impossible. On the other hand, they were trying to present themselves as, if not "moderate", at least "not so radical", with the aim to extend the attraction of a part of the Catholic electorate to the Communist Party. Togliatti repeatedly affirmed that neither he nor the party endorsed the introduction of divorce in the Italian Civil Code. However, he was also strongly against the introduction of the indissolubility of marriage in the constitution. This exposed him to the charge of inconsistency.

November (First Subcommission). As well as 4-8, 10-11 and 17 March, 15, 17-19, 21-24 April of 1947 (Assembly).

In other words, the Christian democrats were not inclined to withdraw from a fundamental position of their political platform, centered on the defense of the Concordat and of catholic schools and on the opposition to divorce. It is worth noting that this political platform was "fundamental" not only for ethical reasons, but also for its capacity to mobilize the catholic electorate. Conversely, for social-communists and (many) liberals it was impossible to accept the introduction in the constitution of the indissolubility of marriage. In the background, there was the same reason. The Christian democrats feared (and the social-communist hoped) major reforms in the Italian family law.

2.3 Strategies

Confronted with this situation, Christian democrats deputies seemed divided, even if this division could have been a part of their strategy. Some deputies, such as Aldo Moro, were willing to reach a compromise, other, such as Giorgio La Pira or Dossetti were trying to maximize the result, even considering the risks pursuant to breaking off cooperation. In order to resolve the impasse, Togliatti and Moro (together with Corsanego, Dossetti and Iotti) were given the task of reformulating the text. They finally cut the Gordian knot by splitting the draft into two articles, one on family and the other on marriage. In the former (the provisional "article 1"), they included a definition of family as a "natural society", in the latter (the provisional "article 2") they dropped any reference to the "indissolubility" of marriage. Thus, the agreement was reached on the basis of a (strategically) ambiguous reformulation of the text. The expression "natural society" was intended to stress that the "family" has original rights which pre-exist the State and which the latter is obliged to recognize. Whether the "indissolubility" of marriage was or was not included in this concept of "natural society" was matter of interpretation. Moro perhaps thought that it was included; Togliatti certainly thought that it was not included.

When the text was submitted in the Commission of 75 for approval, Togliatti, following the previous agreement, gave instructions to the communists deputies about how to vote. However, an amendment presented by La Pira and reintroducing the indissolubility unexpectedly passed in the commission. Now, the definition of family as "natural society" and the indissolubility of marriage were both included in the Constitution. It was a complete victory for Christian democrats and a complete defeat for communists, which, according to the logrolling strategy, allocated their votes with the aim to approve the expression "natural society". At this point, the discussion ended in a second deadlock between those insisting on the inclusion of the "indissolubility" (mainly Christian democrats) and those (communists, socialists, and also liberals) contrary to that inclusion. The socialist Lelio Basso, in particular, threatened to withdraw from the commission and cause secession.

To make the things even more complicated, the Committee of 18 merged the two articles again into one. In this article (provisional article 23), the expression "indissolubility of marriage" was included, together with "natural society", in the definition of "family": "The Republic recognizes the rights of the family as a natural society founded on the indissolubility of marriage". At this point, the last passage was the vote of the amendment of the socialist deputy Grilli, proposing to eliminate the term "indissolubility". Before voting, twenty deputies called for a secret ballot. This was a quite unusual demand, based, however, on the rules of the pre-fascist Chamber of Deputies, which the Constituent Assembly decided to adopt. Thanks to this strategic use of a pre-existing institutional framework, which reduced the influence of the parties (and also of the public opinion) over the deputies, the inclusion of the indissolubility of marriage, as a principle declared in the Constitution, failed by only two votes (out of 384).

2.4 Arguing, bargaining and persuading

Analyzing the debate in the commissions, we find many characteristics included in Elster's model. First of all, we find a bargain conducted resorting to internal and external resources. Among the internal resources, the use of logrolling was, ultimately, ineffective. Besides that, at least in one case, the direct threat to withdraw from the commission was used. However, the final restoration of the compromise was due to a strategic use of the rules of the Assembly.

Considering the argument used, both parties tried to use "impartial" arguments, even if not of the same type. We know that both sides had strongly "egoistic" reasons, mainly related to the imminent electoral campaign. However, they substituted the language of impartial argument for the language of self-interest, though with some differences. Those that were against the inclusion of the "indissolubility of marriage" were actually worried about the consequent impossibility of introducing divorce into the Italian legislation without a constitutional amendment. They cannot, however, use this argument - in some cases for electoral reasons, but also because it was a major source of (further) division. Both liberals and socialists were also against the introduction of the expression "natural society", even if for different reasons. The socialists, because they feared that it would have been possible to interpret it as a "synonym" of "indissolubility of marriage"; the liberals, because they were against a solution that seemed to introduce a jusnaturalistic element into the constitution. Neither the socialist nor the liberals (with few individual exceptions) explicitly endorsed the introduction of divorce. Instead, they used "historical" or "philosophical" arguments based on the impossibility of giving a univocal definition of an institution like the family who is, ultimately, historically and culturally grounded.

From this point of view, the position of the Christian democrats was easier and much more coherent. However, even if they were convinced that this position was not simply related to catholic doctrine, but was ultimately based on the very "nature" of marriage, they also felt that it should have been justified on the basis of "universalizable" arguments. For this reason, they made it clear that they were referring to marriage not as a sacrament (which was already recognized by the church as indissoluble), but as a legal institution. Thus, they presented philosophical, consequentialist (based on "the unity of family"), and even "scientific" reasons for defending its indissolubility. La Pira used a kind of "*tu quoque*" argument, suggesting that Togliatti's opposition to the indissolubility of marriage was ultimately hypocritical, considering that divorce was practically banned in the Soviet Union. This argument was probably used with the aim of implicitly attacking Togliatti with an "*ad personam*" argument, since everybody knew that Nilde Jotti, one of the two rapporteurs of the initial draft, was the partner of Togliatti, who was at that time married to another communist deputy, Rita Montagnana. Nevertheless, that type of argument was somewhat common in the Assembly at the time. It was also used in its version of a kind of "impartial" use of the argument from authority, based on the appeal to someone who is mainly considered as a political authority by the adversary. As Mario Einaudi noticed, "the Communists called in the authority of George Washington and relied on Benjamin Franklin to weaken the argument for an upper chamber, while the Christian Democrats quoted at length the authority of Stalin to support the thesis that the two chambers had to be of equal power" (Einaudi, 1948, p. 662).

3. CONCLUSIONS

The example of the debates about Article 29 of the Constitution could help to reinterpret Elster's model from a rhetoric-centered point of view. Moreover, it represents an interesting case to analyze the possibility of an efficient agreement, "an outcome on the Pareto-frontier" in Elster's words, after a "temporary breakdown of cooperation" (Elster, 2000, p. 399). In this case the cooperation collapsed twice, leading to the crystallization of two mutually irreconcilable positions. Due to the impossibility of resolving the impasse through argumentative means, only a strategic use of the institutional framework was ultimately effective in reestablishing the initial compromise - a kind of strategic use of the rules of the assembly which is not, itself, cooperative (like logrolling).

As mentioned above, the deputies resorted to deontological and consequentialist arguments in order to justify their position. From this point of view, three main strategies can be identified: 1) the "objective strategy", centered on the use of *teleological arguments*, based on the necessity to protect the unity and avoid the "disintegration" of family (Corsanego, among others), together with *arguments from general principles* (La Pira, Dossetti), *appeal to the authority* (both political and "scientific" authorities – La Pira made a generic reference to "scientific researches" supporting the social dangerousness of divorce), *reference to the historical and cultural context* (Lelio Basso, among others, disapproving the introduction of the expression "natural society"), *appeal to popular opinion* (Corsanego, speaking about the "clear opinion" of the "authentic Italian people") and *to popular practices* (the communist Umberto Nobile, *a contrario*, in opposition to the divorce and with reference to the "disintegration" of families in the United States, where the divorce was allowed); 2) a second (ancillary) strategy, based on *ad hominem* attacks, allusively based on *tu quoque* and *ad personam* arguments: and 3) a strategy based on a strategic use of an ambiguous reformulation of the original draft, as a result of an (argumentative) bargain.

Neither the first nor the second strategy were ultimately effective. The first strategy was certainly based on a strategic substitution of more "partial" and less persuasive arguments with others more impartial and (supposedly) persuasive. This kind of substitution, on which Elster bases his theory of the "civilizing force of hypocrisy", was quite obvious in this context. However, the point here is not the "strategic use" of arguments, but, more in general, the strategic argumentation as a discursive technique whose purpose is the persuasion of the audience. This is especially so if we consider that the deputies were addressing at least two audiences: one composed of their colleagues, directly, and the other by their electors, indirectly. Again, this strategy is purely rhetorical, and it couldn't be otherwise.

From the point of view of the (argumentative) negotiation, however, an interesting result was reached with the third strategy. And it is precisely the result that was reestablished with the final vote. In order to resolve the first impasse, Togliatti and Moro agreed on a new version of the text based, on the one hand, on the substitution of the word "indissolubility" with the more vague term "unity", on the other hand, on the introduction of the ambiguous expression "natural society". This process seems to be an example of what Cass Sunstein would call an "incompletely theorized agreement", something that originates from conflicting opinions on the common good, but which nevertheless concludes with participants agreeing on a single outcome, but for different reasons (Mansbridge et al., 2010). This peculiar method of statute-making, aiming at reduce the potential for conflict, is based, according to Sunstein, on an agreement on abstractions, to which an agreement on the particular meaning of those abstractions does not correspond.[7]

Thus, ambiguity is used strategically to foster agreement on abstractions without limiting specific interpretations (Eisenberg, 1984). This technique, which according to Eisenberg corresponds to a "more rhetorical view of communicator as strategist", is particularly used by collective agents like legislatures. In this context, people "confront multiple situational requirements, develop multiple and often conflicting goals, and respond with communicative strategies which do not always minimize ambiguity, but may nonetheless be effective" (Eisenberg, 1984, pp. 227-238). The strategic use of ambiguity allows for multiple interpretations, preserving future opinions. It is, in Eisenberg's words, a way not for minimizing, but for managing disagreement and idiosyncrasy. This technique presents some interesting (for the purpose of this research) characteristics and a peculiar consequence. The former have to do with the "upstream legitimacy" of the constitution, the latter with the "downstream legitimacy".

First of all, the example of Article 29 shows that in some cases this strategic use of ambiguity is obtained through what we can call a kind of "rhetorical definition". Definitions (or apparent definitions) in which the *definiens* is, in fact, ambiguous and not univocal and in some cases, such as in the case of the (apparent) definition of "family" as a

---

[7] "Well-functioning constitutional orders try to solve problems through incompletely theorized agreements. Sometimes these agreements involve abstractions, accepted as such amidst severe disagreements on particular cases. Thus people who disagree on incitement to violence and hate speech can accept a general free speech principle, and those who argue about same-sex relationships can accept an abstract antidiscrimination principle. This is an important phenomenon in constitutional law and politics; it makes constitution-making possible. Constitution-makers can agree on abstractions without agreeing on the particular meaning of those abstractions". (Sunstein, 2007).

"natural society", reflects its ambiguity to the *definiendum* itself. Secondly, as Eisenberg pointed out, strategic ambiguity allows for managing (even if not minimizing) disagreement and idiosyncrasy. This is surely a crucial aspect in constitution making, and above all was crucial in the Italian constituent process. However, the deputy and renowned legal scholar Piero Calamandrei criticized the frequent practice to resort to ambiguous terms and verbal *escamotages* without resolving the conflict.

Finally, this technique has a relevant consequence concerning constitutional interpretation. This consequence can easily be identified resorting to the theoretical distinction between "normative text" and "norm" *sensu stricto*. The former is a set of normative statements, the latter is the interpretation of the normative text, made by an authorized interpreter (in this case, by an authorized interpreter of the Constitution: in the Italian case, the Constitutional Court). The Italian legal scholar Giovanni Tarello noticed that this kind of agreement is, actually, an agreement *on the normative text*, and not on the norm itself. Such an agreement can be ultimately regarded as a delegation of powers to the bodies in charge of interpreting the statutes. In this sense, constitution-makers can move strategically in order to find a (provisional) solution on a sufficiently ambiguous normative text, which can be interpreted in different ways and according to different political approaches. However, the main consequence of this move will be an increase in the discretionary power of the interpreters (Tarello, 1980, p. 365).

In conclusion, the constitutional debate ended with a compromise solution based on a certain amount of ambiguity. This kind of strategic agreement can surely be explained as a combination of negotiation and, in Elster's words, strategic uses of "impartial" or "purportedly non-strategic" arguments. A more realistic approach, however, suggests that this combination of negotiation and argumentation has been developed through rhetorical moves, accordingly to the rhetorical goal of persuading a given audience. Interestingly, but not surprisingly in an institutional context such as that of the Italian Constituent Assembly, the final solution was reached only through a strategic use of the pre-existent institutional framework.

ACKNOWLEDGEMENTS: This research was supported by the FCT (*Fundação para a Ciência e a Tecnologia*) grant SFRH/BPD/68305/2010. The author would like to acknowledge Marcin Lewiński, Fabrizio Macagno, Francesco Pallante, Anna Pintore and Persio Tincani for their valuable comments and suggestions.

REFERENCES

Ackerman, B. (1991). *We the People. Vol. 1: Foundations.* Harvard: Harvard U.P.
Einaudi, M. (1948). The constitution of the Italian Republic. *The American Political Science Review, XLII*(4), 661-676.
Eisenberg, E.M. (1984). Ambiguity as strategy in organizational communication. *Communication Monographs, 51*, 227-242.
Elster, J. (1995). Forces and mechanisms in the constitution-making process. *Duke Law Journal, 45*, 364-396.
Elster, J. (2000). Arguing and bargaining in two constituent assemblies. *Journal of Constitutional Law, 2*(2), 345-419.
Elster, J. (2012a). Clearing and strengthening the channels of constitution making. In T. Ginsburg (Ed.), *Comparative Constitutional Design* (pp. 15-30). New York: Cambridge University Press.
Elster, J. (2012b). The optimal design of a constituent assembly. In H. Landemore & J. Elster (Eds.), *Collective Wisdom. Principles and Mechanisms* (pp. 148-172). Cambridge: Cambridge University Press.
Foot, J. (2003). *Modern Italy.* Hampshire-New York: Palgrave MacMillan.
Mansbridge, J. et al. (2010). The place of self-interest and the role of power in deliberative democracy. *Journal of Political Philosophy, 18*(1), 64-100.
Naldini, M. (2003). *The Family in the Mediterranean Welfare States.* London-Portland, OR: Frank Cass.
Sunstein, C.R. (2007). Incompletely theorized agreements in constitutional law. *Social Research, 74*(1), 1-24.
Tarello, G. (1980). *L'interpretazione della legge.* Milano: Giuffrè.
Viano, C.A. (1955). *La logica di Aristotele.* Torino: Taylor.

5

Willingness to Trust as a Virtue in Argumentative Discussions

JOSÉ ÁNGEL GASCÓN
Universidad Nacional de Educación a Distancia (UNED), Spain
jagascon@bec.uned.es

The virtue of critical thinking has been widely emphasised, especially the habit of calling into question any standpoint. While that is important, argumentative practice is not possible unless the participants display a willingness to trust. Otherwise, continuous questioning by one party leads to an infinite regress. Trust is necessary in order to allow for testimony and expert opinion, but also to exclude unwarranted suspicions that could damage the quality of an argumentative discussion.

KEYWORDS: authority, deliberation, expert opinion, testimony, trust

1. INTRODUCTION

The capacity to scrutinise arguments and to call claims into question is doubtless a fundamental quality for a virtuous arguer. Argumentation theory and critical thinking—the word 'critical' is symptomatic here— have correctly emphasised the importance of that skill. Moreover, the extent to which criticism and doubt are allowed in an argumentative discussion is an indication of the quality of the process. For this reason, for example, one of the rules of the pragma-dialectical model of critical discussion states that (van Eemeren & Grootendorst, 2004, p. 144):

> Rule 6:
> b. The antagonist may always attack a standpoint by calling into question the propositional content or the justificatory or refutatory force of the argumentation.

However, van Eemeren and Grootendorst explain that, although rule 6 gives the antagonist the *right* to call into question any standpoint, the antagonist is not *obliged* to do so (2004, p. 151). Indeed, such an obligation would easily lead to a dead end in the discussion. If the opponent calls into question every reason that the proponent puts forward, both arguers will be unable to make any progress in the discussion. In order to avoid this problem, of course, argumentation theorists consider the notion of *shared premises* or *common ground*, a "zone of agreement" on the basis of which it could be possible to "conduct a fruitful discussion" (p. 60). Van Eemeren and Grootendorst recognise that (p. 139): "A critical discussion is impossible without certain shared premises and without shared discussion rules."

Nevertheless, even if the arguers do not share enough common ground, this fact only does not prevent an argumentative discussion from being possible and fruitful. The common ground may often be sufficiently broad to allow engagement in successful discussions, but sometimes it is not. In those cases, other resources can make the discussion possible. For example, arguers frequently present testimonies and arguments from authority as reasons in support of their standpoints. Such reasons are *not* part of the common ground, but they frequently pave the way towards agreement. The effectiveness of testimonies and appeals to authorities depends on a fundamental component of argumentation: trust.

Even though the actual practice of argumentation largely relies on trust—and trust is given great value in studies on mediation—this component is not frequently present in philosophical accounts of argumentation. Furthermore, Daniel Cohen (2013) argues that argumentation theory is biased toward scepticism. According to Cohen, argumentation theory, by having as a fundamental principle that everything is arguable, and by promoting a set of skills that can be easily abused, might make it too easy for the sceptic to reject knowledge. A virtue approach to argumentation, suggests Cohen, with "its focus on *how* arguers argue, its distinction between *skills* and *virtues*, and its embrace of the difference between *rational* and *reasonable* arguing," (pp. 10-11) can help us understand these biases and learn from them. I believe Cohen is right and I will present one of the virtues that, in my view, could make arguers more reasonable: *willingness to trust*.

In this paper I intend to show why the presence or absence of trust is crucial in every discussion, how it influences the course of the discussion, and why it is so important that arguers be willing to trust each other. Obviously, trust is not the same as credulity, and being willing to trust does not mean being open to believe anything and anyone. Therefore, an explanation of the virtue of willingness to trust must address the question of when it is wise to trust and when it is not. In the following sections, I attempt to cast some light on those issues.

2. WHAT IS TRUST?

Trust is a more widespread attitude than we might think, even though sometimes we are willing to trust when we should not, or are not willing to trust when there is no reason for suspicion. We not only trust friends, with whom we have a very close relationship and share past experiences, to tell us the truth; we also trust our doctor, whom we might barely know, to be genuinely concerned about our health and to have the necessary knowledge to treat us. When we ask for directions to a complete stranger in the street, we trust him or her to be sincere. We only worry about trust when our expectations are not fulfilled and someone disappoints us, but usually the presence of trust is not noticed when everything goes as expected.

People's views on trust are enormously varied, and unfortunately there is also a large variety of academic views on trust—views from philosophy, psychology and sociology. However, the good news is that here we do not need a general account of trust, but rather an explanation of the presence and importance of trust in argumentative discussions. For this reason, I will use only those theoretical concepts that are relatively uncontroversial and can help us understand why willingness to trust is an argumentative virtue.

What most conceptions of trust have in common is that they characterise it as an *expectation*, that is, a belief or attitude (Asen, 2013, p. 4; Govier, 1997, p. 32). The psychologist Julian Rotter defined interpersonal trust as "a generalized expectancy held by an individual that the word, promise, oral or written statement of another individual or group can be relied on" (1980, p. 1). This might be a useful characterisation of trust for argumentation theory, which suits better our present needs than other definitions that make reference to beliefs about the general goodness of people or to optimism about the future—even though those definitions might be in general preferable because they capture the open-ended character of trust (Govier, 1997, p. 13).

It is also commonly accepted that trust involves beliefs about the other person's *competence* and *motivation* (Fricker, 2007, p. 45; Govier, 1998, p. 6; Hardin, 2006, p. 36). When we trust someone, we believe that he or she is competent enough to do what we expect him or her to do, and that he or she has the appropriate motivations—that, for example, he or she is not acting *entirely* in his or her own interests and this benefits us by chance (Hardin, 2006, p. 67).

It is also useful, regardless of the account of trust one adopts, to think of trust in terms of *commitment* (Hardin, 2002, p. 5). Suppose I expect a friend to meet me at the airport tomorrow morning, but I have not told him so and he is not aware of my expectations. Or suppose I have the unrealistic expectation that my friend—a nurse—will cure my chronic illness, even though he has repeatedly told me that he cannot do that. In both cases, my friend cannot be said to have disappointed me if he does not do what I expect him to do. The reason is very simple: *he has not committed himself* to do that. This is also the reason why people can be trustworthy in areas where their knowledge is limited, so long as they know their limits and do not commit themselves to do what they cannot do.

Finally, virtually all theorists agree that trust involves a *risk*. Even though this condition leaves out some uses of "trust"—such as when somebody, probably a poet or a philosopher, says "I trust the sun to rise tomorrow"—it seems that trust entails uncertainty. As Hardin says (2002, p. 12): "More generally, one might say trust is embedded in the capacity or even need for choice on the part of the trusted." It does not make sense, for instance, to say that I trust my sister not to spend all my money on a ridiculously expensive car if she does not have access to my bank account. Trusting involves being vulnerable to some extent (Hardin, 2002, p. 46): "If I trust you to act on my behalf, I set myself up for the possibility of disappointment, even severe loss."

Trust, then, is an attitude based on beliefs about a person. It is therefore a cognitive concept (Hardin, 2002, p. 10). This has important implications: since we cannot freely decide to believe or not a proposition, it follows that we cannot choose whether or not to trust. Of course, I can decide to cooperate with someone I do not trust, or to *pretend* that I trust him or her, but that does not make trust more real. Thus Hardin argues (2002, p. 59):

> I just do or do not trust to some degree, depending on the evidence I have. I do not, in an immediate instance, choose to trust, I do not take any risk in trusting. Only actions are chosen—for example, to act as I would if I did in fact trust or to take a chance on your being trustworthy beyond any evidence I have that you will be trustworthy.

While Trudy Govier includes not only beliefs but also feelings in her characterisation of trust, she also claims that we cannot choose to trust (1997, p. 45):

> Trust is based on beliefs and feelings that, though sometimes alterable after critical reflection and deliberation, cannot be created or abolished at will.

Actually, this fact can be seen as a reason in support of a virtue-based normative account of trust. We cannot choose to trust someone to do something *in an immediate instance*, as Hardin says. However, our trusting or not largely depends on our character—apart, of course, from the other person's trustworthiness—so we *can* cultivate a character that make us trust the right people in the right situations. We *can* become sensible to what the other person's knowledge and motivations are, of his or her commitments, and of the risks involved. Klemens Kappel, who also rejects the idea that we can decide to trust, acknowledges that (2014, p. 2026): "I can, of course, decide to *cultivate* epistemic trust in you, or at least I could decide to *try* to cultivate a certain pattern of epistemic trust."

Moreover, even if we could choose to trust in a particular situation, there are just too many factors to be taken into account by general rules or principles. If willingness to trust is to be studied from a normative perspective, a sensibility to the specificity of every situation seems more appropriate—the kind of sensibility that is entailed by virtue. In addition, if—as we have seen—trust is based on beliefs, then we can benefit from the insights provided by virtue epistemology.

3. IS IT WISE TO TRUST?

In the last section it was pointed out that we cannot choose to trust or not in a particular situation. There is an additional limitation regarding trust: in the real world it is impossible for any of us *never* to trust *anybody*. As Trudy Govier says (1997, p. 62): "There is no real alternative to trusting other people for the truth." From the moment we are born, trust is a precondition of knowledge and even of our having any experience at all. Govier says (p. 61):

> Such trust can be argued to be a priori because there is a sense in which it is logically prior to experience itself. It is prior because it is a *condition* of experience.

Without trust, we could not even be sure of information as basic as our birthday or our real name, for we do not have direct evidence of that— we must trust our parents, our doctor, the institution that issued our ID card, or what have you. Children are predisposed to unquestioningly trust their parents and other people, and that makes them grow and learn (Govier, 1998, p. 68). Govier even places trust at the foundation of meaningful communication (p. 8): "we must believe that the other says what he means and means what he says."

Of course, the fact that we cannot dispense with trust altogether does not imply that we must childishly believe everybody. As Hardin notes (2002, p. 71), "infant trust would be stupid in an adult." As we grow up, we learn to question some—perhaps many—of the beliefs that we have acquired. We develop the capacity of reasoning and of asking ourselves whether someone is *trustworthy*, and by asking questions about the people's trustworthiness we obtain knowledge that determines our *degree* of trust (Hardin, 2002, p. 71). As we grow up, then, our unquestioning trust becomes a more nuanced and reasonable capacity for trust.

Apart from the degree of trust, we must also take into account *what* we trust the other person to do. Trust not only involves a truster and a trusted, it also takes place in a particular situation or action. Nobody trusts anybody without restriction. We might, for example, trust a friend to take care of our car, but do not trust her to give us back two thousand dollars if we lend the money to her. Therefore, trust can be considered as a three-part relation: a person trusts someone *to do X* (Hardin, 2002, p. 9).

When we have a virtuous willingness to trust, we are sufficiently sensitive to know who we can trust, to what degree, and to do what. Here I will focus on the kind of trust that several argumentative settings require. As we will see in the next sections, this includes believing the claims of trustworthy experts and witnesses, accepting the arguments of trustworthy arguers, and being willing to cooperate with trustworthy partners in a deliberation.

Trusting in this sense will be wise if it involves a prudential assessment of the components that we saw in the previous section, especially the "sort of person the other is, with regard to motivations and to competence" (Govier, 1997, p. 4), and the risks involved in the particular situation. Our past experience with the other person is, of course, useful as well; for example, we will not continue to trust someone who repeatedly disappoints us (Hardin, 2002, p. 72). Sometimes, however, we will have no past experience with the other person, as when we deal with complete strangers. The most obvious example is asking someone for directions in the street. In those cases, we tend to believe the information that the strangers give us because the risks involved are very low—the worst-case scenario would be for us to get lost. If, however, we are the editors of a journal, we would not accept a stranger's paper in the street, for the risks are higher—our reputation is at stake.

As Trudy Govier puts it, "trust makes a leap" (1997, p. 47). Whether or not we are willing to make that leap depends on the elements mentioned above, but in any case we will be vulnerable to some extent. For this reason, our willingness to trust also depends on our character, on whether we are "self-confident and secure enough to cope with disappointments and adapt to changing circumstances" (Govier, 1997, p. 29). The question, then, is not whether or not it is wise to trust, but when and to what extent.

4. TRUST IN ARGUMENTATION

4.1. Appeals to expert opinion and testimony

Are arguments from expert opinion legitimate? Is it wise to trust experts? Recently, Moti Mizrahi (2013) argued that arguments from expert opinion are *all* weak, in the sense that their premises provide little or no support for their conclusion. He cites several studies that show—among other things—that, statistically, experts' predictions are only slightly more accurate than mere chance, and that experts' findings are likely to be refuted after a few years (p. 64). Therefore, given that the fact that an expert holds a claim p does not make p significantly more likely to be true, all arguments from expert opinion must be weak.

Mizrahi's article was followed by a response from Markus Seidel (2014). Seidel points out that our dependence on expert opinion is so strong that arguments in support of the absolute rejection of appeals to expert opinions, like Mizrahi's, are self-undermining. He argues that, even in order to support the conclusion that arguments from expert opinion are weak, we need to resort to some kind of argument from expert opinion (p. 213):

> Mizrahi is relying on the expertise of others in conducting empirical studies on expertise in order to come to his claim that there is empirical evidence for the conclusion that arguments from expert opinion are weak arguments.

I believe Seidel is right. As Trudy Govier points out (1997, p. 54): "We can check some claims and reports made by other people, but only by relying on the claims and reports of still other people." But Mizrahi's contention is actually a little more complex and interesting than that. Mizrahi makes clear that, according to him, arguments from expert opinion are those which do not rely on empirical evidence or even agreement among experts *at all* (2013, p. 71):

> In other words, once we take into account considerations of evidence for p and whether or not p is consistent with common knowledge in a field, then an argument from expert opinion is no longer just an appeal to expert opinion. Rather, it

is an appeal to expertise, evidence, and agreement among experts.

Hence, it seems that, according to Mizrahi, an argument from expert opinion relies *solely* on the expert's claiming that *p*, and taking into account any other consideration would entail adding premises to the argument and therefore rendering it a different type of argument. This, however, is a rather limited conception of the argument from expert opinion. It seems to lead us away from reasonable trust and closer to blind faith. Moreover, I believe it is misleading in two respects. Firstly, it overlooks the fact that, even if the expert opinion is based on empirical evidence, *some degree of trust is still required* for the argument to be convincing. Thus, for example, Mizrahi did not personally conduct the studies he cites in support of his position (2013, p. 76): "Granted, I did not conduct any experimental studies on expertise. Luckily, I don't have to. Others have done the hard work already." How do we know that the research was properly conducted? And that the results are not forged? Results often admit of several interpretations, why should we accept the author's interpretation as the best? Responses to this questions always depend partly on our degree of trust in the expert. Of course, trusting does not mean blindly believing anything any expert says; but, as Seidel holds, "reasonably scrutinizing authorities should not lead us to a rampant scepticism about expertise" (2014, pp. 192-193).

Secondly, the fact that issues about empirical evidence and agreement among experts are taken into account does not mean that the argument put forward is not a genuine appeal to expert opinion. Those components can be an intrinsic part of the *strength* of the argument from authority, even though they are not premises of the argument. For example, Douglas Walton proposes the following scheme (1997, p. 210):

> E is an expert in domain D.
> E asserts that A is known to be true.
> A is within D.
> Therefore, A may (plausibly) be taken to be true.

Walton takes into account further information in the critical questions he proposes for the evaluation of the strength of the argument from authority (p. 223):

> *Expertise question*: How credible is E as an expert source?
> *Field question*: Is E an expert in the field that A is in?
> *Opinion question*: What did E assert that implies A?
> *Trustworthiness question*: Is E personally reliable as a source?

> *Consistency question*: Is A consistent with what other experts assert?
> *Backup evidence question*: Is A's assertion based on evidence?

Critical questions, then, are not part of the argument scheme; they are not premises. Instead, they are part of the dialectical framework for the evaluation of arguments from expert opinion (p. 158). This shows how we can consider empirical evidence and agreement among experts as relevant components of the strength of the argument from expert opinion, without necessarily incorporating them into the argument as premises and, contrary to what Mizrahi claims, without turning it into a different type of argument. By doing this, we can better understand that appeals to expert opinion involve both reasonable scrutiny and trust.

Similar considerations support the legitimacy of arguments from testimony. Govier (1993, p. 93) defines testimonial claims as "those which describe or purport to describe a particular person's observations, experience and related memories." The epistemologist John Hardwig (1991, p. 698) argued that beliefs based on testimony might be not only unavoidable but also *epistemically superior* to beliefs based on empirical evidence. The reason is that, individually, we cannot gather all the necessary empirical evidence in support of every one of our beliefs. Therefore, if only first-hand empirical evidence should be taken into account as reasons in support of our beliefs, most of our reasons would be very poor. However, we all have very good evidence for at least some of our beliefs—especially if we have witnessed an event or are experts in some domain—that constitutes our reasons. If we take into account testimonial evidence, that means that we take into account other people's reasons, including the experts' and witnesses', so we will have much better reasons that justify our beliefs. Thus, Hardwig states his *principle of testimony* (p. 697):

> If A has good reasons to believe that B has good reasons to believe p, then A has good reasons to believe p.

A will not believe that B's testimony gives him or her good reasons to believe p, Hardwig adds (1991, p. 700), unless A trusts B.[1] But, actually, Hardwig was not referring to testimonies of common people, or to laymen's trust in experts, but to the very scientific enterprise (p. 706):

---

[1] Kappel (2014) proposes a reliabilist interpretation of Hardwig's ideas, according to which epistemic trust implies the existence of a reliable belief-forming process that is discriminating and defeater-sensitive. However, he is concerned with the conditions for justification and the definition of knowledge, and here I focus on internal traits that make an individual virtuous.

"Often, then, a scientific community has no alternative to trust, including trust in the character of its members."

The question, then, is whether or not to trust a person that presents his or her testimony in a particular situation. An argument based on testimony belongs to the kind of arguments that Douglas Walton names *arguments from position to know*, and the critical questions that he proposes—where *a* stands for the other person and *A* stands for what he or she claims—are (2006, p. 86):

> Is *a* in a position to know whether *A* is true (false)?
> Is *a* an honest (trustworthy, reliable) source?
> Did *a* assert that *A* is true (false)?

Hence, Walton's critical questions for arguments from authority as well as for arguments from position to know provide helpful guidelines for deciding whether to trust someone in a particular situation. Note, though, that critical questions are neither clear-cut rules nor an algorithm that yields a unique answer. They are very useful as a guide, and they are questions that the respondent can ask to the proponent, but they cannot remove the need for practical wisdom and sensibility to particular situations. Willingness to trust is, after all, a virtue.

4.2. Arguments that rely on trust

The acceptability of the conclusion of certain arguments, then, is a matter of trust—and, I would add, this also happens sometimes with some premises in *any* kind of argument. But in some cases the inference relies on trust as well. In his response to Bowell and Kingsbury (2013), who argued against the legitimacy of a virtue approach to argumentation, Andrew Aberdein (2014) claims that facts about the arguer are sometimes relevant to the evaluation of his or her argument. Bowell and Kingsbury themselves provide a compelling example (p. 27):

> Suppose someone tries to convince me that Tom is not fluent in German, on the grounds that Tom is a New Zealander and only 2% of New Zealanders are fluent in German. This looks like a good enough inductive argument. However, there could be information that I lack which would undermine the argument without falsifying the premises; for example, the information that Tom is the New Zealand ambassador to Germany. Given this, facts about the arguer might matter. [...] Is the arguer the sort of person who would tell me if he knew that Tom was the New Zealand ambassador to Germany, or is he the sort of person that would delight in tricking me into

thinking that the New Zealand ambassador to Germany doesn't speak German?

Bowell and Kingsbury argue that either the argument put forward is inductively strong regardless of whether information is being hidden, or it actually contains the unstated premise "There is nothing unusual about Tom that bears on the likelihood of his speaking German". However, when discussing arguments from expert opinion, we saw how the strength of the argument can be assessed without including every criterion as a premise in the argument. This case is very similar in this regard. In particular, here the strength of the argument depends in part on our trusting the arguer not to hide information from us.

The great majority of arguments we normally use are defeasible, that is, their conclusion is *plausibly* true and the inference may lose its strength if *new evidence* appears. For this reason, virtually any defeasible argument will be more convincing if it is put forward by someone whom we trust to share *all* the information he or she has with us, even if that information could undermine his or her own position. One and the same argument might be more convincing if presented by a trustworthy arguer than if presented by someone untrustworthy—and for good reasons.

Consider another, probably more realistic example. A petroleum company intends to extract crude oil in a populated region, and after some empirical research it publishes a report supporting the conclusion that there will be no undesirable consequences for the population or the environment. Some of the inhabitants read—and understand—the report, and although the data is consistent with other, impartial reports and they have no other information about the possible environmental impact of the extractions, they distrust the company's arguments. They do not accept them because of the company's obvious interests and because that company has omitted relevant information from its reports in the past. Perhaps they are not convinced that there will be undesirable consequences either, but they suspend judgement instead of accepting the conclusions of the report. The inhabitants do not accept the arguments because they do not *trust* the company—and, in this case, surely it is not their fault.

4.3. Deliberations and trust

Deliberations are a kind of argumentative dialogue that is intended to resolve on a course of action or a normative issue. Robert Asen (2013, p. 5) defines deliberation as "an encounter among interlocutors who engage in a process of considering and weighing various perspectives and proposals for what they regard as issues of common concern." I will discuss certain dimensions of trust that are probably more crucial in deliberations than in other kinds of argumentations.

Trust is doubtless an essential basis of successful and satisfactory deliberations. However, trust should not be considered as a necessary condition for deliberations—they benefit from trust, but they can also take place in the absence of trust and subsequently *foster* it. As Asen holds (p. 15): "People need not wait for trust to deliberate. Instead, deliberation itself may serve as a means by which we come to trust others, and our trust may become stronger with practice." Several factors, circumstances, or behaviours promote trust. In this section, I will present them as *signals*, that could warrant our trusting someone in the context of a deliberation, and to which we should be sensitive.

Whether or not to trust our partners in a deliberation is not so much a matter of the outcomes of the process as of the process itself. If the process of deliberation is conducted in a way that makes everyone involved feel included, recognised, and respected, the deliberation will very likely be satisfactory to all. Thus (Asen, 2013, p. 8):

> Relations of trust may enable affirmative answers to questions that participants regularly confront in deliberation: Can I believe what other people say? How shall I evaluate their evidence? Are they listening to me? Will the other people involved in the deliberation heed our decision? Trust strengthens deliberation not by ensuring an outcome, but by committing participants to the process of producing a deliberative outcome, namely, a judgment.

Asen (2013, p. 9) proposes four attitudes that help build trust in deliberation: flexibility, forthrightness, engagement, and heedfulness. Firstly, it seems intuitively correct that a participant in a deliberation who is *flexible* about his or her beliefs and proposals conveys a sense that he or she is genuinely concerned about reaching an agreement. Flexibility allows the arguer to acknowledge the others, to form his or her views in collaboration with them, and to recognise different positions as reasonable and justified (p. 10). Secondly, *forthrightness* means that the arguer is honest, that he or she means what he or she says, makes plain his or her motives and goals, offers reasons in support of his or her position, and does not deceive or hide information. This is perhaps the quality that is most directly relevant to trustworthiness, for it is the arguer's honesty what is often called into question in deliberations—for example, by accusing him or her of having a hidden agenda. Thirdly, trustworthy arguers try to *engage* each other's perspectives, learning about (p. 12):

> different perspectives, including understanding why people hold their beliefs, how these beliefs may be different from and similar to one's own, how people may take a different route to a shared judgment, and how similar starting points may lead to different interpretations and judgments.

And, finally, when an arguer displays *heedfulness* by truly paying attention to what the others have to say, he or she shows that deliberation matters and that, for example, it is not just a means of trying to "provide political cover for a decision that already has been made", and that he or she will not "conduct their future actions without any reference to relevant deliberations" (pp. 13-14).

5. THE DANGERS OF UNGROUNDED DISTRUST

Why speak of the dangers of distrust? Is it not more dangerous and more frequent to overly trust people? Gullibility is no doubt a vice in argumentation, but perhaps it is only indicative of the absence of the virtue of critical thinking. When discussing the virtue of willingness to trust, the related vice, I believe, is not gullibility but ungrounded distrust or suspicion. Rotter explains (1980, p. 4):

> If trust is simply believing communications, then high trust must be equated with gullibility. However, if we redefine trust as believing communications in the absence of clear or strong reasons for not believing (i.e. in ambiguous situations) and gullibility as believing when most people of the same social group would consider belief naïve and foolish, then trust can be independent of gullibility.

Actually, there are reasons to define trust as independent of gullibility. Rotter continues:

> In fact, anecdotal evidence suggests that it is the low truster who is taken in by the disarming dishonesty of the con artist and is the frequent victim of con games.

Thus, surprisingly, it seems that one of the dangers of distrust is that it could lead to gullibility. Actually, there might be a very good explanation for that. As has been argued, absolute distrust is not a real alternative; we all need to trust in some people in order to have not only knowledge, but also most of our beliefs and experiences. For this reason, low trusters cannot distrust *everybody*; instead, they do not trust in most people. They trust in a small number of people only, and that makes them dependent on fewer sources of knowledge and therefore they cannot check the reliability of many of those sources (Govier, 1997, p. 130). Low trusters are, then, more uncritical and more prone to error.

Consider the case of people who do not trust scientists. When arguing with someone who does not consider scientific opinions as expert opinions, he or she will not accept any appeal to those authorities. In reality, however, those people are bound to trust other—alleged—authorities. Complete distrustfulness, as has been argued, is impossible. So what usually happens is that those people—so-called low trusters—put their trust instead in homoeopaths, astrologists, religious authorities, or the like. This means that they put their trust in *fewer* people, becoming more dependent on them than high trusters are on the *more numerous* people they trust. This path is even more manifest in the case of people who believe in conspiracy theories. Ironically, an initial attitude of low trust leads to gullibility.

A second danger of ungrounded distrust does not directly affect the arguer himself or herself—as happens when it causes gullibility—but the others. The problem arises when we distrust certain people due to prejudice and stereotypes. When this distrust is widespread, those people's voices are silenced in practice and there is a real risk that their experiences, beliefs, and proposals are not taken into account. Feminist authors have drawn our attention to this problem, which has been called *rhetorical disadvantage* (Govier, 1993) or *testimonial injustice* (Fricker, 2007).

We frequently deal with strangers or hear their testimonies and opinions—for example, on television and in newspapers—and we have to decide whether or nor not to trust them without much evidence. In order to make a judgement in such circumstances, we commonly resort to stereotypes, which function as heuristics and are not necessarily bad (Fricker, 2007, p. 32). Some of those stereotypes, however, are unreliable, are maintained in the face of counter-evidence, and undermine the speaker's credibility. They are *prejudiced judgements* that distort the hearer's perception of the speaker (p. 36). According to Miranda Fricker, the testimonial injustice that results from prejudiced judgements "excludes the subject from trustful conversation" (p. 53), but unfortunately it is "a normal feature of our testimonial practices" (p. 43).

Trudy Govier (1993) explains that prejudice and stereotypes can act in any of the four different levels of assessment of testimonies. First, one can dismiss a testimony because it is assumed that the speaker is not *serious*—he or she is just joking or being ironic. Presumably the prejudice here involves our own ways of communication (p. 97): "Standards of rationality, seriousness, and maturity incorporate norms that are not neutral as regards age, gender, race, class, culture, and style." In the second place, assuming the speaker is serious, he or she may be considered *not to be truthful* on the basis of stereotypes regarding the social group to which he or she belongs. Thirdly, some stereotypes can similarly attribute a lack of *competence* to that social group, and hence to the speaker in question. And, finally, even assuming that the speaker is serious, truthful and competent, one can ultimately decide *not to accept* his or her testimony because it contradicts some of our beliefs. While it is normally a good practice to question beliefs that somehow contradict our own—and, of course, to question our own beliefs at the same time—Govier explains that this norm can make us reject the testimonies of people who have different experiences (p. 98):

> To the extent that A is a person different from B in experience, social standing, gender and so on, B is likely to have established beliefs and preconceptions different from those of A. Ironically the very features that make A's testimony necessary, intellectually interesting, and important to B may also serve to render it unbelievable.

If general principles and norms might cause those problems, what can be done? Prejudices need not be conscious beliefs from which we *infer* that certain speaker is not trustworthy. Instead, they are often a sort of "background theory" that affects our *perception* of people's credibility (Fricker, 2007, p. 71). Fricker argues that the "model for judgement" in the testimonial sphere "is perceptual, and so non-inferential" (p. 72). For this reason, Fricker advocates a virtue approach to epistemic testimony, which does not rely on sets of rules but on "a sensitivity to epistemically salient features of the situation and the speaker's performance" (*Ibid.*). Rules and norms might, of course, be useful as general guidelines, but they are not the whole story. For example, among other things, virtue involves feeling the appropriate kind of emotions. I conclude with Fricker's own words (p. 80):

> When it comes to epistemic trust, as with purely moral trust, it can be good advice to listen to one's emotions, for a virtuous hearer's emotional responses to different speakers in different contexts are trained and honed by experience. The feeling of trust in the virtuous hearer is a sophisticated emotional radar for detecting trustworthiness in speakers.

6. CONCLUSION

I have emphasised the importance of being willing to trust and the different dimensions of trust in argumentative situations. Yet, of course, trusting *anyone*, trustworthy or not, is not wise. In fact, it could be dangerous or even morally wrong. Trusting a complete stranger to take care of a baby would be not only foolish but also reprehensible. When referring to the problem of declining trust, discussed by many scholars, Hardin says (2002, p. 30): "Commonly, the best device for creating trust is to establish and support trustworthiness."

It is also very difficult—and probably a bad idea—to have even slight trust in some especially dangerous circumstances. The degree of trust that one can afford, as has been said, depends on the risks involved, and some particular situations might dramatically raise the risks of even small degrees of trust. Trudy Govier presents a particularly extreme example (1997, p. 134):

> If relatives simply disappear, if one is starved, beaten, and tortured, if friends and colleagues may be spies for a brutal regime, people are unlikely to be high trusters, and a recommendation to trust more makes little sense.

Therefore, in the absence of trustworthiness or in risky situations, willingness to trust *the trustworthy person* will not make any difference. Why not focus on trustworthiness then? The reason is that trustworthiness alone is not sufficient; a *sensibility* to trustworthiness, or disposition to believe the trustworthy person, is also necessary. As has been explained, ungrounded distrust can seriously harm the course of a discussion, and trustworthiness cannot solve this problem.

REFERENCES

Aberdein, A. (2014). In defence of virtue: The legitimacy of agent-based argument appraisal. *Informal Logic, 34*(1), 77-93.

Asen, R. (2013). Deliberation and trust. *Argumentation and Advocacy, 50*(1), 2-17.

Bowell, T., & Kingsbury, J. (2013). Virtue and argument: Taking character into account. *Informal Logic, 33*(1), 22-32.

Cohen, D. H. (2013). Skepticism and argumentative virtues. *Cogency, 5*(1), 9-31.

van Eemeren, F. H., & Grootendorst, R. (2004). *A systematic theory of argumentation*. New York: Cambridge University Press.

Fricker, M. (2007). *Epistemic injustice: Power and the ethics of knowing*. New York: Oxford University Press.

Govier, T. (1993). When logic meets politics: Testimony, distrust, and rhetorical disadvantage. *Informal Logic, 15*(2), 93-104.
Govier, T. (1997). *Social trust and human communities*. Montreal: McGill-Queen's University Press.
Govier, T. (1998). *Dilemmas of trust*. Montreal: McGill-Queen's University Press.
Hardin, R. (2002). *Trust and trustworthiness*. New York: Russell Sage Foundation.
Hardin, R. (2006). *Trust*. Malden, MA: Polity Press.
Hardwig, J. (1991). The role of trust in knowledge. *Journal of Philosophy, 88*(12), 693-708.
Kappel, K. (2014). Believing on trust. *Synthese, 191*, 2009-2028.
Mizrahi, M. (2013). Why arguments from expert opinion are weak arguments. *Informal Logic, 33*(1), 57-79.
Rotter, J. B. (1980). Interpersonal trust, trustworthiness, and gullibility. *American Psychologist, 35*(1), 1-7.
Seidel, M. (2014). Throwing the baby out with the water: From reasonably scrutinizing authorities to rampant scepticism about expertise. *Informal Logic, 34*(2), 192-218.
Walton, D. (1997). *Appeal to expert opinion: Arguments from authority*. University Park, PA: Pennsylvania State University Press.
Walton, D. N. (2006). *Fundamentals of critical argumentation*. New York: Cambridge University Press.

Commentary on Gascón's Willingness to Trust as a Virtue in Argumentative Discussions

GEOFF GODDU
University of Richmond, USA
ggoddu@richmond.edu

1. INTRODUCTION

We try to inculcate in our students, our children, our readers, our colleagues even, the proper balance between credulity and skepticism. No one can, on pain of inconsistency, believe everything. Nor can anyone function in the world or communicate with others if one doubts everything, so we must find the right balance. It should also be uncontested that the proper balance is situationally sensitive. We are rightly less skeptical of a literal claim such as "I saw Cristiano Ronaldo walk across the street today" and more skeptical of a literal claim such as "I saw Cristiano Ronaldo walk on water today" (regardless of what Real Madrid fans, or Portuguese fans, or Ronaldo himself might think). Another example: as the risk associated with believing falsely goes up, so does our tendency to double check sources that we might otherwise leave undoubted.

Since functioning in the world or communicating requires a (situationally influenced) balance between excessive credulity and excessive skepticism and arguing is a functioning in the world and communicating, arguing requires a (situationally influenced) balance between excessive credulity and excessive skepticism. None of this strikes me as controversial (or exciting or interesting), so what is the relevance to argumentation theory?

2. RELEVANCE?

Gascón says that "In this paper I intend to show why the presence or absence of trust is crucial in every discussion, how it influences the course of the discussion, and why it is so important that arguers be willing to trust each other." But, none of these things, with perhaps the exception of the last, happens—what does happen is a description of the uncontroversial core of trust, a Govier inspired defense of the claim that complete distrust is impossible, some examples of trust in argumentation (which since it is inescapable, of course it appears in argumentation), and a final section on the dangers of ungrounded distrust which might be support for why it is important for arguers to be willing to trust each other.

Perhaps Gascón takes his discussion to be a counterbalance to the perception that some theorists (or practitioners) put too much weight on skepticism with the result that the theory (or practice) is skewed too far towards skepticism. Is it true that theory or practice has improperly skewed the balance towards skepticism? I grant that argumentation allows almost anything to be challenged (though not all at once). I grant that the attitude of challenging can certainly be carried too far to make arguings or discussions unfruitful—but is it true that argumentation theory or practice in fact promotes a too skeptical attitude? (See, for example, Cohen, 2013) I don't know. I don't know what the appropriate level of trust/distrust ought to be generally or particularly. I suspect that "critical thinking" teachers believe the general populous is too credulous. But do they overcompensate in the skeptical direction or not go far enough? Perhaps we should trust others (and ourselves) less than we do, even after a critical thinking course, (at least if we are interested in attaining the truth) given how much of our perception of the world, say of the color combination of a particular dress, or our own judgments about our own objectivity are mental interpolations and fabrications. We impute causation to mere correlations, we make explanatory patterns out of noise, we let irrelevancies anchor important judgments and decisions (see, for example, Kahneman, 2011).

Perhaps Gascón is trying to convince us that this balancing of trust/distrust is best understood as an argumentative virtue. I certainly grant that such balancing is consistent with being a virtue, though the virtue in question is probably not willingness to trust, but trusting appropriately. After all, Aristotelean virtues are supposed to have not one (the ungrounded distrust Gascón suggests), but two corresponding vices. Courage, for example, is the balance between recklessness and cowardice. Trusting appropriately would presumably be the balance between credulousness and excessive doubt. I certainly do not doubt that the ideal arguer has the property of trusting appropriately (given the evidence, the context, the goals of the parties, the risks involved as a result of cognitive error, etc.) But whether this property needs to be accounted for or explained as a virtue, I have no idea. I grant we can talk the virtue talk, but I have no idea whether we ought to talk this way in relation to the 'trusting appropriately' norm. For example a "maximize true belief, minimize false belief" advocate can argue that adhering to the trusting appropriately norm will maximize true belief while minimizing false beliefs without ever talking in terms of virtues and vices.

3. CONCLUSION

Whether we talk in terms of virtues or not, there certainly are the problems of determining (i) what the appropriate level of trust is in a given situation, (and in particular a given argumentative situation), (ii) whether cultivating the appropriate level of trust is something that can be taught or not (and if so, how), and (iii) whether trust plays a special role in some arguments.

Regarding this last, Gascón makes some provocative claims about "arguments that rely on trust". He writes of a particular induction example from Bowell and Kingsbury (2013):

> (1) "Here the strength of the argument depends in part on our trusting the arguer not to hide information from us."

and

> (2) "One and the same argument might be more convincing if presented by a trustworthy arguer than if presented by someone untrustworthy."

Taken literally, it appears that (1) is making a claim about trust making arguments themselves stronger (or weaker). My worry here is not to conflate *strength of arguments* with either the *convincingness of arguments or arguings*. Consider the sentence, "Cristiano Ronaldo walked across the street yesterday." Uttered by person A the sentence is either true or false. Uttered by person B the sentence is either true or false. The epistemic credence I put on the sentence will likely be influenced by my judgments of the trustworthiness of A and B. How convinced I am that the sentence is true depends upon how much I trust A and B to report truthfully. But how convinced I am is independent of the actual truth/falsity of the sentence. Similarly for arguments. A might utter the same argument as B and regardless of the actual truth/falsity of the premises or how much support the premises actually provide the conclusion, my judgments about the epistemic credence I place in the conclusion are likely to be influenced by my judgments about the reliability of A and B in uttering the truth or in providing sufficient reasons. But again, how convinced I am is independent of how strong the argument actually is. So while (2) is plausible, (1) is much less so unless it is really talking about the convincingness of the argument rather than the strength of the argument. And for someone who thinks convincingness is a property of arguings (as in "he argued convincingly" and not arguments, then (2) will be false, since one and the same arguing cannot be done by two separate people.) But, unlike a claim about trust influencing the strength of arguments, there is nothing special about trust playing a role in the convincingness or epistemic credence placed in arguments or their conclusions.

REFERENCES

Bowell, T., & Kingsbury, J. (2013). Virtue and argument: Taking character into account. *Informal Logic, 33*(1), 22-32.
Cohen, D. H. (2013). Skepticism and argumentative virtues. *Cogency, 5*(1), 9-31.
Kahneman, D. (2011). *Thinking, fast and slow.* New York: Farrar, Straus and Giroux.

6

Cooperation in Legal Discourse

STEFAN GOLTZBERG
Perelman Centre for Legal Philosophy, ULB, Belgium
stefan.goltzberg@ulb.ac.be

Legal lines of reasoning sometimes appear absurd. One non-trivial way to explain this absurdity is to characterize legal discourse as *intentionally non-cooperative*, i.e., as a mode of discourse in which people deliberately mislead and evade. But is legal discourse really non-cooperative in this way? Is it non-cooperative at all, or less cooperative than other discourses? Is legal discourse a *strategic* discourse in the sense that it fails to cooperate?

KEYWORDS: cooperation, cooperative principle, flouting, Grice, legal reasoning, Marmor, pragmatics, strategic attitude

1. INTRODUCTION

Legal lines of reasoning sometimes appear absurd. One non-trivial way[1] to explain this absurdity is to characterize legal discourse as *intentionally non-cooperative*, i.e., as a mode of discourse in which people deliberately mislead and evade. But is legal discourse really non-cooperative in this way? Is it non-cooperative at all, or less cooperative than other discourses? Is legal discourse a *strategic* discourse in the sense that it fails to cooperate?

When a judicial decision seems absurd, laypersons (and some lawyers) may indeed conclude that those behind it are misleading or introducing irrelevant arguments. But this occurs in ordinary discourse too. The main difference between an absurd statement in legal discourse and in other discourses is that, in non-legal discourse, an absurd statement must typically be normalized after it is revealed, i.e., an absurdity cannot stand in an ordinary functional conversation,

---

[1] This phenomenon is not reducible to the trivial fact that in every context *some* lines of reasoning may happen to be absurd.

whereas in law, a (seemingly) absurd conclusion is occasionally allowed.[2]

Here we are beginning to achieve a more nuanced account of absurdity in legal discourse. We can go further by entering the field of pragmatics. Specifically, it is useful to employ Paul Grice's influential theory of linguistic cooperation to decide whether and why absurd statements are knowingly made in legal discourse. Importantly, Grice defines non-cooperation as a failure to fulfil certain *conversational maxims*, rather than just intentionally misleading. He stresses that *failing* to cooperate is not the same as *choosing* not to cooperate.[3]

Let us take an example of apparent linguistic non-cooperation in a judicial decision.

In Smith v. United States, a person was charged with numerous firearm and drug trafficking offenses for having offered to trade an automatic weapon to an undercover officer for cocaine. The courts had to decide whether specific penalties had to be imposed on Smith because, under Title 18 U.S.C. 924(c)(1), he, "during and in relation to ... [a] drug trafficking crime use[d] ... a firearm." In this case, Smith "used" the firearm as a means of exchange, not as a weapon.

The legal question was: what is the scope of *use* in the phrase "use of a weapon"? Could the use of a weapon in a "non-weaponly" way fall within the statute? In affirming Smith's conviction and sentence, the Court of Appeals held that 924(c)(1)'s plain language imposes no requirement that a firearm be used *as a weapon*, but applies to *any* use of a gun that facilitates in any manner the commission of a drug offense.

---

[2] Indeed, the phenomenon in law sometimes occurs at the highest level of the judicial system and cannot be considered as an accidental result. On the contrary, some decisions are deliberately taken even though they would be (and are) perceived as absurd.

[3] Another useful way to account for absurdity in legal discourse is to employ the so-called "new formalism" — illustrated in particular by the views of Frederick Schauer. New formalism sees absurd statements simply as an unfortunate output of the application of legal rules; the law is a suboptimal system that allows for rules to be applied to some extent automatically. This, Schauer claims, is the price of a legal system. A judge may have a good reason to suspend the application of a law and yet may apply it even though she knows that some results will be wrong. Indeed, if the judge were entitled to suspend the application of the rule each time she considered the rule to be wrong, she would simply be replacing the legislator. This formalist approach is extremely interesting, not least because it is compatible with an *ex ante* approach to judicial decisions (explained in Farnworth 2007).

In other words, the law does not specify that use must be "weaponly" use. The Supreme Court then confirmed this interpretation.

Now there is something strange if not absurd about saying that Smith did *use* the firearm – it seems counterintuitive or to run against common sense.

In a dissenting opinion, Justice Scalia found this interpretation so absurd that he suggested the following analogy[4]:

> To use an instrumentality ordinarily means to use it for its intended purpose. When someone asks, "Do you use a cane?" he is not inquiring whether you have your grandfather's silver-handled walking stick on display in the hall; he wants to know whether you walk with a cane. Similarly, to speak of "using a firearm" is to speak of using it for its distinctive purpose, i.e., as a weapon. To be sure, "one can use a firearm in a number of ways," [...] including as an article of exchange, just as one can "use" a cane as a hall decoration - but that is not the ordinary meaning of "using" the one or the other.[5]

To anticipate our discussion of Grice's theory of cooperation below, we could explain the absurdity of this Supreme Court decision as a violation of one of Grice's conversational maxims: the *maxim of relevance* or the *maxim of quantity*: the court did not take into account the fact that the legislator was relevant and was as informative as required. Surely the broader meaning of "use of a firearm" is irrelevant in the sense that one practically never uses it in ordinary discourse.

But if we explore Grice's maxims further, we find that the Supreme Court fulfilled another maxim, the requirement not to read anything into what someone (or the law) says beyond what she (or it) actually says (*What isn't said, isn't the case*, to use Levinson's wording). In other words, the Supreme Court in this case did not mislead; rather, it resolved a contradiction of linguistic cooperation in a surprising or unexpected way. It turns out that this may be what distinguishes legal from non-legal discourse.

2. GRICE'S FRAMEWORK FOR UNDERSTANDING COOPERATION

Grice formulates the principle of cooperation in the following way:

---

[4] Ironically, Justice Scalia is usually known for his textualist and originalist approach, i.e., he usually suggests that judges have a duty to apply a law exactly as worded, rather than reading meanings into it.

[5] *Smith vs U.S.* 508 U.S. 223 (1993)

> Make your conversational contribution such as is required, at the stage at which it occurs, by the accepted purpose or direction of the talk exchange in which you are engaged. (Grice, 1989, p. 26)

Grice then immediately proceeds to implement the cooperative principle with specified conversational maxims,[6] which he structures along four main categories (echoing Kant): Quantity, Quality, Relation and Manner. Here are Grice's four main conversational maxims:

> Quantity:
> 1. Make your contribution as informative as is required (for the current purpose of the exchange).
> 2. Do not make your contribution more informative than is required.
>
> Quality:
> 1. Do not say what you believe to be false
> 2. Do not say that for which you lack adequate evidence.
>
> Relation:
> Be relevant
>
> Manner:
> 1. Avoid obscurity of expression.
> 2. Avoid ambiguity.
> 3. Be brief (avoid unnecessary prolixity).
> 4. Be orderly.

It comes as no surprise that the two maxims of Quality are usually considered crucial: to be insufficiently orderly (even on purpose) is less frowned upon than to deliberately tell untruths or to fail to guard against telling untruths.

---

[6] *Conversational* implicatures must be distinguished from another type of inference: *conventional* implicatures. These do not strongly depend on the content of a given sentence but apply as default rules that are semantically encoded. For example, the word "but" conventionally suggests (i) that there is a contrast between what occurs to the left of *but* and what occurs to the right; and (ii) that the segment following the "but" is presented as stronger (more decisive?) than what comes before. "That restaurant is good but expensive," all things considered, typically speaks against the choice of this very restaurant. By contrast, "That restaurant is expensive but good" emphasises the quality in spite of the price.

For our purposes, however, the two maxims of Quantity (1 and 2) are more important. To flesh out the meaning of these two maxims, we can employ a Neo-Gricean approach that builds on them to distinguish two apparently complementary maxims (Carston, 2011, p. 13):

>   1. Q-principle: Say as much as you (truthfully) can.
>   2. I-principle: Say no more than you must.

Each maxim is associated with a heuristic:

>   1. Interpretive heuristic (Q): What isn't said, isn't the case.
>   2. Interpretive heuristic (I): What is simply or briefly described is the stereotypical or normal (default) instance.

Expressing Grice's maxims of Quantity as interpretive heuristics allows us to relate them directly to concepts used in legal argument, each related to a typical line of reasoning:

>   1. *a contrario*.
>   2. analogy.

In addition, both arguments are couched as Latin phrases or canons:

>   1. *Expressio unius est exclusio alterius*
>   2. *Eiusdem generis*

Llewellyn (1950) further devised a series of principles of legal interpretation that also allow for a reformulation of the distinction between these two legal principles:

>   1. Thrust: A statute cannot go beyond its text.
>   2. Parry: To effect its purpose, a statute may be implemented beyond its text.

Now we see more clearly how Grice's theory of cooperation, and specifically his two conversational maxims of Quantity, can be applied to *Smith v. United States*. The Supreme Court was required to fulfil both these maxims of Quantity, and discovered them to be in conflict.

All the above expressions of the maxims of Quantity are represented in table 1 [*A contrario* and analogy]:

Grice	Neo-Gricean pragmatics		Argument	Canon	Llewellyn
	2 principles	Interpretive principle			
First Maxim of Quantity (be as informative as required)	Q-principle: Say as much as you (truthfully) can.	(Q): What isn't said, isn't the case.	A contrario	*Expressio unius est exclusio alterius*	Thrust: A statute cannot go beyond its text.
Second Maxim of Quantity and of relevance	I-principle: Say no more than you must.	(I): What is simply/briefly described is the stereotypical or normal (default) instance.	Analogy	*Eiusdem generis*	Parry: To effect its purpose, a statute may be implemented beyond its text.

Now that Grice's framework of linguistic cooperation has been briefly described, the question is: what is linguistic non-cooperation (legal or otherwise)? But before getting into how to fail to fulfil Grice's maxims, it is important to carefully define Grice's framework and respond to some critiques.

3. IS GRICE'S PRINCIPLE OF COOPERATION NORMATIVE?

Grice's framework would be less useful if it was a norm to be followed rather than a tool to describe actual discursive communication. Jenny Thomas draws attention to the fact that many readers have understood that, because Grice's cooperative principle is in the imperative form – "Make your conversational contribution such as is required" – it is in fact normative (a prescription). And if the principle is a norm, where does this principle take its authority from?

But it seems that the normative reading of the cooperative principle is not necessary. Indeed, Grice himself wrote that:

> It is just a well-recognized empirical fact that people do behave in these ways; they learned to do so in childhood and have not lost the habit of doing so. (Grice, 1989, p. 29)

One can thus interpret the imperative wording of Grice's principle not as a prescription, but rather as a description of the assumption that a certain set of (learned) rules is in operation (Thomas, 1995, p. 62).

But Grice next writes that he is not content with mere facts but wants to unravel the structure underlying this crude fact:

> I am, however, enough of a rationalist to want to find a basis that underlies these facts, undeniable though they may be; I would like to be able to think of the standard type of conversational practice not merely as something that all or most do *in fact* follow but as something that it is *reasonable* for us to follow, that we *should not* abandon. (Grice, 1989, p. 29)

This sounds much more like a normative principle—"that we *should not* abandon."

4. IS GRICE SPEAKING ABOUT LINGUISTIC COOPERATION AT ALL?

Some scholars have concluded that Grice is not really focussed on discourse. With his conversational imperative – "Make your conversational contribution such as is required" – he is promoting a sort of general benevolence. It is cooperation *tout court*, not only cooperation within the context of linguistic exchange. If this is true, Grice's framework wouldn't shed much light on legal discourse per se.

Pavlidou stresses this distinction between substantial (or social) cooperation and formal (or linguistic) cooperation (Pavlidou, 1991). While Grice may have had formal cooperation in mind when he devised his cooperative principle, using ordinary conversation to plan activities is substantial cooperation. When you fulfil Grice's maxims in order to coordinate your shopping with your neighbour, this can hardly be described as linguistic cooperation.

However, as Lindblom writes, Grice's theory rather subtly distinguishes between benevolence and linguistic cooperation. Grice's point is that linguistic communication demands a certain type of cooperation that is limited to discourse, i.e., that you have to learn certain rules of conversational cooperation to be able to talk at all. This is separate from any considerations of benevolence:

> The crucial subtlety of Grice's theory is this: interlocutors do not necessarily cooperate with each other; they cooperate with a set of conventions that allows each interlocutor to produce approximate enough meanings for communication to work. This form of cooperation is not necessarily benevolent

at all; even the bitterest of verbal fights require linguistic cooperation to work. (Lindblom, 2009, p. 154)

Thus, many authors (but not Grice) incorrectly conflate the two types of cooperation.

Lumsden also offers a way to relate substantial and linguistic cooperation without conflating them. For Lumsden, the distinction between the two kinds of cooperation is "only defensible if it is accepted that linguistic cooperation can be determined by an extra-linguistic goal" (Lumsden, 2008, p. 1896). And linguistic cooperation *may* be determined by extra-linguistic cooperation if the latter is shared:[7] "When an extra-linguistic goal determines the linguistic goal it is on the presumption that the goal is shared" (Lumsden, 2008, p. 1901). In other words, substantial cooperation can set the goal for linguistic cooperation, but linguistic cooperation still proceeds following its own rules. If I try to help someone (substantial cooperation) who tells me on the phone "I cut my finger", I will therefore understand the implicature that he is talking about *his* own finger (formal cooperation).

5. IS GRICE'S PRINCIPLE OF COOPERATION COHERENT?

Returning to the question of whether Grice's principle is normative or descriptive, it is amenable to reservations. As Lumsden rightly suggests, we can separate two claims that Grice makes about conversations:

1). The audience needs to *assume* the speaker is cooperating.
2). Conversations *are* usually cooperative.

It is possible to attack claim (2) while maintaining claim (1) (Lumsden, 2008, p. 1900). It is not necessarily the case that speakers in fact cooperate with each other — note that Grice never said that speakers *always* cooperate. Lumsden's point is that we can separate both claims: whether you agree or not with the factual statement according to which conversations *are* usually cooperative, you may find good grounds to substantiate the claim that there is good reason to *assume* the speaker is cooperative.

Indeed, Grice's pragmatics become more useful once we accept that people often do not cooperate. The whole point of Grice's pragmatics is to account for contributions that either *fail* or *seem* to fail

---

[7] Again, "the cooperative principle does indeed only apply to linguistic cooperation but in some cases the extra-linguistic goal determines linguistic cooperation." (Lumsden, 2008, p. 1901)

to fulfil the cooperative principle and the conversational maxims. While Grice's cooperative principle suggests that the speaker should respect the maxims, the principle is mentioned especially (if not only) in order to *repair* the failure in communication or at least the failure in respecting the maxims.

6. HOW TO FAIL TO FULFIL GRICE'S MAXIMS?

However you fail to fulfil a maxim, the cooperative principle is called for in order to explain away the failure. Accordingly, we first of all assume that the speaker is cooperating (trying to make the conversational contribution that is required). If something goes wrong (at least according to an observer), a participant in a talk exchange has – at least apparently – failed to fulfil a maxim in one of various ways:

> 1. He may quietly and unostentatiously *violate* a maxim; if so, in some cases he will be liable to mislead.
>
> 2. He may *opt out* from the operation both of the maxim and of the Cooperative Principle; he may say or allow it to become plain that he is unwilling to cooperate in the way the maxim requires. He may say *I cannot say more; my lips are sealed*.
>
> 3. He may be faced by a *clash*: he may be unable, for example, to fulfil the first maxim of Quantity (*Be as informative as is required*) without violating the second maxim of Quality (*Have adequate evidence for what you say*).
>
> 4. He may *flout* a maxim; that is, he may *blatantly* fail to fulfil it. On the assumption that the speaker
>   a. is able to fulfil the maxim without violating another maxim (because of a clash);
>   b. is not opting out;
>   c. is not, in view of the blatancy of his performance, trying to mislead,

the hearer is invited to consider that the speaker is being only apparently uncooperative, and therefore a *conversational implicature* is generated. If there is an *implicature*, then the maxim is said to be *exploited.* (based on Grice, 1989, p. 30)

Although Grice presents the failures in this order, intentional violation could also be described as the *last resort* explanation: indeed, in the event of a failed conversation, you will assume that the speaker was cooperating, but opted out (if this was explicit enough), faced a clash, or flouted a maxim (again if this was manifest, blatant). Only if none of these explanations work will you consider that the speaker intentionally misled you. Indeed, deliberate violation of a maxim seems to have to be covert, since an overt failure to fulfil (flouting) generates an implicature.

Grice distinguishes three main groups of cases (which are not meant to be exhaustive) that generate conversational implicatures.

> **GROUP A:** *Examples in which no maxim is violated, or at least in which it is not clear that any maxim is violated.* (Grice, 1989, p. 32)
>
> (A) I am out of petrol
> (B) There is a garage round the corner

(B) is assumed to respect the conversational maxim of relevance. Therefore (B) suggests that the garage is open or at least he has no reason to believe otherwise. (B) utters this sentence having in mind that somehow (A) will be able to purchase some petrol at that garage. Interestingly, (B)'s contribution is described by the literature as being an implicature even though this is a case where no maxim is violated.

> **GROUP B:** *Examples in which a maxim is violated[8], but its violation is to be explained by the supposition of a clash with another maxim.* (Grice, 1989, p. 32)
>
> (A). Where does he live?
> (B). Somewhere in the South of France

This example illustrates a clash between the maxim of truthfulness and the maxim of relevance. (B) supposedly wants to tell the truth – maybe this is all (B) knows is true – but on the other hand the indication is vague and possibly not relevant anymore as the maxim of quantity (Be as informative as required) is not fulfilled.

> **GROUP C:** *Examples that involve exploitation, that is, a procedure by which a maxim is flouted for the purpose of getting in a conversational implicature by means of something of the nature of a figure of speech.* (Grice, 1989, p. 33)

---

[8] Here Grice uses the verb "violate" in its wide sense. In this paper the strict sense is usually used.

(A) *War is war*

This statement sounds trivial — it is the illustration of the principle of identity 1=1. The speaker is thus blatantly failing to fulfil the requirement to say something informative. But what is meant is not necessarily trivial and could possibly be worded as: you cannot expect an army not to injure people. So this apparently trivial statement could be a nontrivial comment justifying or at least explaining the number of civilian casualties in a conflict.

(A) [in a letter of reference] *His command of English is excellent*

The case of (B) is different. This statement is supposed to be one of the few sentences written by a professor in a letter of reference. Upon reading this, one would conclude that the professor has a bad opinion of the assistant considered – otherwise the professor would have written something more positive. In Shakespearean language, the professor is "damning with faint praise." In Grice's framework, the professor is blatantly exploiting the first maxim of Quantity (*Make your contribution as informative as required*). While having an excellent command of English is in itself a valuable asset, it is not a particularly attractive quality in an assistant (especially if it is the only quality).[9]

7. NO COOPERATION IN LEGAL DISCOURSE?

Andrei Marmor claims that legal discourse is less cooperative than ordinary discourse: while in ordinary discourse speakers engage in a cooperative exchange of information, legal discourse displays a non-cooperative form of communication:

> Unlike regular conversational contexts, where the parties to the conversation aim at a cooperative exchange of information, a non-cooperative form of communication is present in the legislative context. (Marmor, 2008, p. 438)

Marmor calls this form of communication a *strategic* behavior.[10] Recently, Ohlendorf (2013) has mounted a very interesting attack on

---

[9] Grice lumps many other conversational phenomena into Group C (where an implicature is generated). To name a few: irony, metaphor, meiosis (understatement) and obscurity.

[10] One remark: not only does Marmor describe ordinary discourse as enabling or promoting a cooperative exchange of cooperation, he even writes that it

Marmor's view. Ohlendorf claims that legal discourse is in fact cooperative in that the judge does or at least should take into account the implicatures of the legal texts.

Marmor's claim is too wide but captures some insights that I will try to make more explicit. Let us confront Marmor's claim with Grice's framework, specifically the four ways whereby a speaker fails to fulfil a conversational maxim and therefore fails to respect the cooperative principle:

> (1.) violation of the maxim, which is misleading;
> (2.) opting out;
> (3.) failing to fulfil one maxim because fulfilling it clashes with another maxim;
> (4.) flouting, i.e. exploiting the maxim.

First, let's consider the first way not to cooperate: *violating* maxims (which could produce lies). It is true that in some specific legal or judicial contexts, a speaker is expected to not *fully* cooperate. For example, the witness who is cross-examined is asked questions that make sense only if one assumes that he is being less than cooperative[11].

But if Marmor's notion of strategic behavior means that in legal discourse (this notion of legal discourse should of course be narrowed down) speakers *habitually* violate the maxims, that would mean that legal discourse habitually chooses to *mislead*. This would be "no holds barred": lying would not be reprehensible in a legal context. It is hardly reasonable to consider seriously that lying is acceptable — even in a legal context! The contrary may be true: while some people would not mind too much lying, there comes a time when under oath the perspective of being sentenced for lying would deter the liar from lying.

Let's move on to the next type of failure to cooperate, *opting out*. I cannot think of many illustrations of opting out in legal discourse apart from possible responses to the Miranda warning, with which the arrested individual is reminded of his rights:

> You have the right to remain silent. Anything you say or do can and will be held against you in a court of law. You have the right to speak to an attorney. If you cannot afford an attorney, one will be appointed for you. Do you understand these rights as they have been read to you?

---

aims at a cooperative exchange. I would like to qualify this claim: ordinary discourse could hardly be described as *aiming* at a cooperative exchange.

[11] Mosegaard Hansen (2008).

Here, the person under arrest may very well answer that he is not willing to speak until his lawyer is present. This is opting out. But this is an extremely specific context.

This leaves us with the final two varieties of failure to cooperate: inability to fulfil a maxim due to a *clash* with another maxim; and blatant *flouting* of a maxim. Let me address the latter first. When you *flout* a maxim ("War is war" or "Well, his command of English is excellent"), you make it very clear that you do as if you were not cooperating, generating an implicature that is supposed to be conveyed appropriately. Again, remember that the cooperative principle is precisely what enables the hearer to move from what is said to what is meant. In the case of flouting, both the speaker and the hearer understand that there is a gap between what is said and what is meant, and the hearer is *invited* to fill it, under the reasonable and charitable assumption that the speaker is rational and does not speak nonsense.

But then the question arises: is legal discourse *more* prone to flouting? Do lawyers exploit maxims *more* than the other people? As far as I know, nothing indicates that there would be more flouting in law than in non-legal discourse. Indeed, legislation, at least modern legislation, is supposed to be as explicit as possible. That many statements in statutes need be ascertained and explained has nothing to do with the phenomenon of flouting, which is an *intentional* move whereby you deliberately say something while meaning something else (remember that this is not the same as deliberately misleading). It also happens that the legislator remains vague — for example, in several countries the law forbids abandoning a child but does not say exactly when leaving the room qualifies as abandoning the child. This abandonment is a vague notion but being vague intentionally cannot be equated with *exploiting* a maxim.

This leaves us with an inability to fulfil a maxim due to a *clash* between two maxims. For example, you cannot be precise (enough to be relevant) because you want to be truthful: *he lives in the South of France*. Or vice versa: you cannot be accurate because you want to be relevant (when you say: *there are 200 miles between London and Manchester*).
Now we can return to *Smith v. United States*, and see that the question of "use" involves a clash between two maxims.

First, the legislator imposes some specified penalties if the defendant, during and in relation to a drug trafficking crime "uses a firearm." Now the judge — or for that matter whichever lawyer is asked to determine what the law says/means — is supposed to follow two maxims and is supposed to assume that the legislator followed two maxims:

1. The maxim according to which when the text of the law is clear (*as informative as required*) it should be given effect, and so "use" should be interpreted to include a large number of non-technical meanings; and

2. The maxim according to which "uses a firearm" should be understood according to its normal (default) meaning of *using a firearm as a weapon*.

Now a case comes along in which the petitioner used the firearm as barter. The Supreme Court faced a clash between these two maxims and finally confirmed the comprehensive understanding of "use" (*what isn't said, isn't the case*), thereby respecting the first maxim to the detriment of the second.

8. CONCLUSION

To sum up, lawyers neither fully cooperate just like laymen nor completely fail to cooperate. It looks as if lawyers sometimes do not cooperate (at least in the expected manner) because they are faced with a clash of principles, and they must fail to fulfil one of them. But this is not yet a specific trait of legal discourse; indeed most people from time to time are forced to choose between two maxims that clash with each other. If *pace* Ohlendorf (2013) legal discourse is to be described as strategic — as opposed to cooperative — I think it means that the legal agent, facing a clash between two maxims, will at times choose to fulfil the maxim that was *less expected* to be fulfilled.

Let us add a last example to that of *Smith v. United States*. Imagine a piece of legal advice — or more interestingly a judicial decision — that interprets the letter of a tax law so as to make tax avoidance possible. Such a memo or ruling would often work against the spirit of the law, which obviously *means* the opposite and *aims* at the opposite. When the interlocutor within a legal conversation fails to cooperate, he is not — at least not necessarily — *violating* a maxim. Rather, he is emphasizing *another* maxim that clashes with the one that turns out to be unfulfilled.

ACKNOWLEDGEMENTS: I would like to thank Philippe De Brabanter, Maurizio Manzin and Izabela Skoczeń for their comments on previous versions of this paper.

REFERENCES

Carston, R. (2011). Legal Texts and Canons of Construction: A View from Current Pragmatic Theory. In M. Freeman & F. Smith (Eds.), *Law and Language. Current Legal Issues* (pp. 8-33). Oxford: Oxford University Press.

Chapman, S. (2007). *Paul Grice, Philosopher and Linguist.* London: Palgrave Macmillan.

Cosenza, G. (2001). *Paul Grice's Heritage.* Turnhout: Brepols.

Davies, B. L. (2007). Grice's Cooperative Principle: Meaning and Rationality. *Journal of Pragmatics, 39,* 2308-2331.

Davis, W. (1998). *Implicature: Intention, Convention, and Principle in the Failure of Gricean Theory.* Cambridge: Cambridge University Press.

Farnworth W. (2007). *The Legal Analyst. A Toolkit for Thinking about the Law.* Chicago and London: University of Chicago Press.

Grice, P. (1989). *Studies in the Way of Words.* Cambridge MA: Harvard University Press.

Ingram, P. (1998). Implicature in Legal Language. *International Journal for the Semiotics of Law, 1,* 51-70.

Kasher, A. (1976). Conversational maxims and rationality. In A. Kasher (Ed.), *Language in Focus: Foundations, Methods and Systems* (pp. 197-216). Dordrecht: Reidel.

Levinson, S. (2000). *Presumptive Meaning. The Theory of Generalized Conversational Implicature.* Cambridge, MA: MIT Press.

Lindblom, K. (2001). Cooperating with Grice: A cross-disciplinary metaperspective on uses of Grice's cooperative principle. *Journal of Pragmatics, 33,* 1601-1623.

Lindblom, K. (2009). Cooperative Principle. In J. L. Mey (Ed.), *Concise Encyclopedia of Pragmatics,* 2nd Edition, (pp. 151-158). Amsterdam: Elsevier.

Llewellyn, K. N. (1950). Remarks on the theory of appellate decisions and the rules of canons about how statutes are to be construed. *Vanderbilt Law Review, 3,* 395-405.

Lumsden, D. (2008). Kinds of conversational cooperation. *Journal of Pragmatics, 40,* 1896-1908.

Marmor, A. (2008). The Pragmatics of Legal Language. *Ratio Juris, 4,* 423-452.

Mosegaard Hansen, M.-B. (2008). On the availability of 'literal' meaning: Evidence from courtroom interaction. *Journal of Pragmatics, 40*(8), 1392–1410.

Neale, S. (1992). Paul Grice and the philosophy of language. *Linguistics and Philosophy, 15,* 509–559.

Neale, S. (2008). Textualism with Intent. Excerpt from manuscript for discussion at Oxford University, Law Faculty, November 18.

Ohlendorf, J. D. (2013). Textualism and Obstacle Preemption. *47 Georgia Law Review 369,* 369-443.

Pavlidou, T. (1991). Cooperation and the choice of linguistic means: Some evidence from the use of the subjunctive in Modern Greek. *Journal of Pragmatics, 15*, 11-42.

Perry, J. (2013). Textualism and the Discovery of Rights. In A. Marmor & S. Soames (Eds.), *Philosophical Foundations of Language in the Law* (pp. 105-129). Oxford: Oxford University Press.

Poggi, F. (2011). Law and Conversational Implicatures. *International Journal for the Semiotics of Law, 24*, 21-40.

Scalia, A., & Garner, B. A. (2012). *Reading Law. The Interpretation of Legal Texts.* West.

Schauer, F. (2009). *Thinking like a lawyer. A New Introduction to Legal Reasoning.* Cambridge, MA & London: Harvard University Press.

Searle, J. (2007). Grice on meaning: 50 years later. *Teorema, 26*, 9–18.

Sperber, D., & Wilson, D. (2012). *Meaning and Relevance.* New York: Cambridge University Press.

Ziff, P. (1967). On H. P. Grice's Account of Meaning. *Analysis, 28*, 1–8.

Is the Distinction Between "Cooperative" and "Strategic" Crucial for Jurisprudence and Argumentative Theory? Commentary on Goltzberg's Cooperation in Legal Discourse

MAURIZIO MANZIN
Faculty of Law / CERMEG / University of Trento, Italy
maurizio.manzin@unitn.it

1. INTRODUCTION

As far as I can understand, the basic points in Goltzberg's paper are the following: 1. (in general) to what extent legal discourses can be said "cooperative" 2. (in particular) if legal discourses are "cooperative" even when they conclude with seemingly *absurd* statements – and if not: are they non-cooperative in Andrei Marmor's sense of "strategic"? (Marmor, 2008, 2014).

In my view, all of these questions imply a set of further clarifications about (at least): the meaning of "absurd" when concerning a statement in a *single* communicative context, compared with its meanings in *different* communicative contexts; the philosophical substance (and practical usefulness) of Marmor's distinction between "cooperative" and "strategic", in the light of Grice's principle of cooperation and his well-known Four Maxims (Grice, 1991); the peculiar kind of reasonableness involved in legal argumentation (recently: Van Eemeren, 2011).

2. AN EXAMPLE FROM ITALIAN CASE LAW

Let me start with an example: a legal case recently decided by the Italian Court of Cassation (Cass. pen. Sez. I, 24/02/2015, n. 8163/15).

In 2011 Mrs Melania Rea, a 29-year-old mother, was murdered during a countryside trip she went on together with her husband Salvatore Parolisi and their 18-months daughter. Melania was killed with 35 knife wounds, and a syringe was stuck into her body to simulate a drug crime. At the end of the trial (in the district court as well as, one year later, in the Court of Appeal), Mr Parolisi was convicted of homicide and charged of 30 years in prison. In February, 2015 the Supreme Court of Cassation ordered the Court of Appeal in Perugia to reduce the punishment, because in its view it was not the case for the application of the aggravating circumstance of cruelty as established in Italian Penal Code 61.4.

To most of the people in Italy it seemed very strange, and to some extent *absurd*, to consider the murder as "not cruel", having the husband killed his young wife with 35 strokes of knife (in presence of their little daughter), and having abandoned her in agony in a wood. But is the idea of "cruelty" as commonly conceived by public opinion compatible with the one fixed in the body of penal laws and in authoritative legal precedents? Clearly, it is not. Killing a young woman in the way Mr Parolisi would have done to his wife, is no doubt "cruel" according to the common sense, but it is not necessarily the same in the legal context, where the definition of "cruelty" implies (according to both Italian case law and legal doctrine) the evidences that the murderer acted with the precise intention of inflicting a strong and prolonged pain to the victim.

The opinion unanimously shared by all the three Courts which decided the case was that Mr Parolisi killed his wife in a burst of anger ("explosion of rage") and without any premeditation, and that he didn't want to make her suffer during the murder. He "only" wanted to kill her. So, *in the terms of legal discourse*, the homicide was not "cruel". For this reason, the Supreme Court decided that the measure of the punishment had to be changed, so the Court of Appeal turned it later into 20 years in prison instead of 30.

3. ABSURDITY, COOPERATION AND LEGAL ARGUMENTATION

My example on cruelty aims at showing that (as questioned in Introduction, point 1) we should distinguish between what can be intended as *not* "absurd" in a *single* communicative context and what could even seem crazy in *another* and *different* one. To judge "not compatible with cruelty" the above mentioned crime is, in the (Italian) legal context and from the inside of its practices, not absurd at all, whereas in that of public opinion, which simply ignores legal interpretation, the Supreme Court's decision could be (and actually was) perceived as an additional and somehow ironic wound inflicted to the victim.

This point has to do with the question 3 of my Introduction: the existence of different kinds of reasonableness. In my example, the *legal* reasonableness and the *ordinary* one. Both are governed by rules which are explicitly or implicitly shared by the arguers, although such rules could change from one context to another. Then, when it happens that some legal statements seem absurd to us, it is not necessarily the case of a clash among law and reason, but rather a result of the enforcement of legal reasonableness (which has its own rules) in an appropriate context. Finding, evaluation and coherent application of such rules are parts of what we call "legal argumentation".

Turning back now to Introduction, point 2 (Andrei Marmor's distinction between "cooperative" and "strategic", and its relevance for legal reasoning), I would say that, apart from some critical aspects of Grice's account about the principle of cooperation (e.g. is it obtained by induction? is it normative?), it seems to me that in the legal context a number of Grice's maxims is *always* involved – I mean, a discrete quantity of "cooperation" is exploited, in line with many procedural statutes concerning the judicial process (for the Italian legal system: Tomasi, 2012, 2015. It is obviously impossible to deepen here this point).

In conclusion, supposed that some maxims of Grice's always run in a legal context (in which all actors are expected to fulfill them under the very rule of the statute law regulating the judicial process), I should be favorable to accept a "cooperational" account of legal discourses rather than a fully "strategic" one. Of course, I realize that, notwithstanding the "cooperational" character of some phases in the judicial procedure (basically due to the statutes), it should seem hard to retain that the arguers in this peculiar sort of communicative situation don't act "strategically" – it is the case, for instance, for lawyers and prosecutors, who interpret the law and describe the facts in order to win the case. But the parties' strategies do not imply a complete lack of cooperation between the opponents, just as in whatever critical discussion.

4. COOPERATION AND LEGAL REASONABLENESS

These last concerns lead to a reformulation of the question under point 2 of my Introduction. In fact, after assuming that some apparent "absurdities" in legal discourses have to do with the difference between the contexts of communication, and that their strategic character does not mean that any cooperation among the parties is broken, it remains to clarify in what sense we should take Marmor's divide between "cooperative" and "strategic".

There are three possible scenarios:
1. legal discourses are *always strategic*,
2. legal discourses are *always cooperative*,
3. legal discourses are *in some circumstances strategic (or cooperative)* and in some others not.

Evidently, such reformulation ends up by bringing into question the idea of a strict divide between "cooperative" and "strategic" in the judicial context, suggesting instead a *weaker* account of Marmor's theory.

Does such possible *weakness* of Marmor's divide have something in common with Goltzberg's conclusion about the clash –and the consequent choice– between cooperational maxims in legal reasoning (which should be, according to Goltzberg, the cause of some apparent "absurdities" in legal discourses)?

To answer such question let me refer to D'Agostini, 2010. In her work, D'Agostini addresses the "dialogue" and the "dispute" as the two basic modalities of any critical discussion. In the first case ("dialogue"), opponents A and B are engaged in showing the acceptability of one's own claim (*p* alone, or *p* vs *non-p*). In the second case ("dispute"), A and B want to demonstrate that only *p* (or only *non-p*) is acceptable. In a "controversy" (which is a species of the genre "dispute"), and namely in a legal one, both A and B aim at persuading a "third" –the judge(s) or the jury– instead of one another (Manzin, 2014). But whatever the disagreement of A and B about *p*, either in a "dialogue" or in a "dispute" the opponents always act in line with some rules that can be referred to an underlying scheme of cooperation (like Grice's maxims, Van Eemeren & Grootendorst's "ten commandments", Govier's "ARG-conditions", Apel's or Habermas' rule for argumentation, *et al.*) (D'Agostini, 2010).

In other words, the strategies for victory are applied by each actor –lawyer(s), prosecutor etc. depending on the specific judicial field– in a way not very different from that of a chess player, who "struggles" against her rival within the frame of fixed and accepted rules. According to such rules, some moves will be intended as regular while some others should be considered absurd. It is obvious that from the perspective of a non-chess context (e.g. the one of draughts) the absurd moves could seem not absurd and vice versa.

5. CONCLUSION

The just mentioned example of chess can be connected to what I assumed in the beginning of my Commentary about the question under point 3, regarding the peculiar kind of reasonableness involved in judicial discourses. As I have repeatedly stressed, cooperational rules work well with legal disputes (and therefore such disputes are *reasonable* in the terms of a critical discussion). The peculiarity of legal reasonableness deals with the argumentative context, which is an "institutionalized" one – that is to say, a one having its own procedural (and cooperational) rules to set up the discussion and the final judgment; and, too, having some specific points (statute and case law, prudential maxims, legal arguments etc.) which are supposed to be shared by everyone acting in the practice.

Taking now once again into account the decision of the Italian Court of Cassation about the "not-cruelty" of the wife-murderer's behavior –and the way it was perceived as "absurd" by public opinion– we should admit that, being coherent with all the provided legal-institutional rules and points, the Supreme Court's decision was nothing but reasonable!

So, to some extent, I agree with Goltzberg's assumption on the so-to-say 'mixed nature' of legal discourses (and of some others) – in his own words: «lawyers neither fully cooperate just like laymen nor completely fail to cooperate» (above: 126). What I finally wander is: are we sure that Marmor's distinction between "cooperative" and "strategic" is really crucial for jurisprudence and argumentative theory?

ACKNOWLEDGEMENTS: Special thanks to Serena Tomasi, Phd. for discussing with me about the enforcement of cooperational rules in the judicial process. Her work in progress on argumentation and procedural law, together with her concrete experience in legal practice, helped me a lot in the comprehension of the nature of "institutionalization" in legal context.

REFERENCES

D'Agostini, F. (2010). *Verità avvelenata.* Torino: Bollati Boringhieri.
Eemeren, F.H. van (2011). *In alle redelijkheid. In reasonableness.* Amsterdam: Rozenberg.
Grice, P. (1991). *Studies in the way of word.* Cambridge, MA: Harvard University Press.
Manzin, M. (2014). *Argomentazione giuridica e retorica forense.* Torino: Giappichelli.
Marmor, A. (2008). The pragmatic of legal language. *Ratio Iuris, 21*(4), 423-452.
Marmor, A. (2014). *The language of law.* Oxford: Oxford University Press.
Tomasi, S. (2015). La struttura argomentativa dell'eccezione. In S. Bonini, L. Busatta & I. Marchi (Eds.), *L'eccezione nel diritto.* Trento: Università degli Studi di Trento.
Tomasi, S. (2012). *Teorie dell'argomentazione e processo penale. Un'analisi comparata delle principali teorie argomentative contemporanee con profili applicativi al processo penale.* Tesi di Dottorato. Trento: Università degli Studi di Trento.

7

A Descriptive and Comparative Analysis of Arguing in Portugal

DALE HAMPLE
University of Maryland, USA
dhample@umd.edu

MARCIN LEWIŃSKI
ArgLab, Universidade Nova de Lisboa, Portugal
m.lewinski@fcsh.unl.pt

JOÃO SÀÁGUA
ArgLab, Universidade Nova de Lisboa, Portugal
jsaagua@fcsh.unl.pt

DIMA MOHAMMED
ArgLab, Universidade Nova de Lisboa, Portugal
d.mohammed@fcsh.unl.pt

This empirical project reports data on Portuguese understandings of, and orientations to, interpersonal arguing, based on a survey conducted in Portugal (N=252). We report information on levels of argumentativeness, verbal aggressiveness, personalization of conflict, and argument frames. We compare results between Portuguese men and women, and between US and Portuguese respondents. Our results reveal significant differences between the American and Portuguese orientations to argumentation, which this paper further investigates and explains.

KEYWORDS: arguing motivations, argumentation, argument frames, conflict, interpersonal arguing, Portugal, verbal aggressiveness

1. INTRODUCTION

Portugal is a nation of more than 10 million people located in the Iberian Peninsula in Western Europe. It has a long documented history:

first inhabited by the Celts, Galicians and Lusitanians, it was subsequently integrated into the Roman Empire, then colonized by Germanic tribes, and later conquered by the incoming Arabs. In 1139 it was founded as an independent monarchy. It was one of the European nations that started the "Age of Discovery" in the 15th and 16th century and established new connections with Africa, Asia, and South America. This paved the way for the long period of colonization which spread Portuguese language and culture around the world to places like Brazil, Angola, Mozambique, and Macau. As a result, Portuguese is today the sixth most spoken language in the world with some 220 million native speakers (190 million of them in Brazil). There is also a sizable Portuguese diaspora in the ex-colonies as well as developed countries, especially France, the United States, and Switzerland. In the 20th century, after four decades of right-wing dictatorship, Portugal established itself as a parliamentary democracy in the 1970s and became a member of the European Community in 1986. In all these senses, Portugal shares much of its history with its larger Iberian neighbor – Spain – although important cultural differences remain.

Today, Portugal is a parliamentary republic, a member of the Eurozone, with a developed economy with a per capita income of about €16,200. Together with other Western European countries, it has a legally protected and factually existing open public sphere, with a multi-party political system. Article 37 (sec. 1) of the constitution adopted in 1976 assures the right of free speech: "Everyone shall possess the right to freely express and publicise his thoughts in words, images or by any other means, as well as the right to inform others, inform himself and be informed without hindrance or discrimination." Portugal has a dozen state-run universities, sixteen public polytechnics, ten private universities, five private polytechnics, and a number of other public or private institutes that provide higher education.

Portugal also has a long, but relatively scarce, tradition of argumentation studies. Peter of Spain (Petrus Hispanus) – the medieval author of the *Tractatus*, a standard textbook on logic across European universities until the 17th century – is believed to be Pedro Julião (c. 1215 – 1277), a Portuguese physician, and later a Roman Catholic Pope, John XXI. (This, at least, is the Portuguese version; in Spain, Petrus Hispanus is instead identified as a Spanish friar.) Later on, a 16th century philosopher, Francisco Sanches, followed that tradition, although his work, *Quod nihil scitur (That Nothing is Known)*, was focused on endorsing a skeptical view about attaining genuine knowledge through syllogistic reasoning. In the 20th century, argumentation studies in Portugal were pursued mostly in the context of Law Studies, Rhetoric in Literature Departments, Linguistics, and Logic (mostly syllogistic logic and modern formal logic) in Philosophy Departments. Portuguese argumentation scholarship belonged largely to the francophone "sphere of influence," with the chief references being Perelman's new rhetoric and French linguists such as Ducrot or Plantin, whose main works were translated and published in Portugal (see van Eemeren et al., 2014, pp. 753-761, for a synthetic overview of argumentation studies in Portugal and other "Portuguese-speaking areas," mostly Brazil). Only in the 21st century can we see in Portugal an institutionally organized interest in argumentation studies, as they are currently conducted across the world (e.g., Ribeiro, 2009). As part of this development, the Argumentation Lab of the Nova Institute of Philosophy (IFILNOVA), housed in the Universidade Nova de Lisboa, was founded in 2009 with a mission to carry out the philosophical and linguistic investigation of argumentation (van Eemeren et al., 2014, pp. 757-758). All that said, the social scientific study of interpersonal argumentation does not appear to have been a topic of published investigation (whether in English, Portuguese, or French) for any researchers studying Portugal and its people, and we believe that the present study is the first of that sort.

To that end, we have decided to explore basic orientations and understandings of interpersonal arguing in Portugal. This research – both the theory and the methods – originated in the US, and a basic question is whether American theories, concepts, and findings can be exported to Portugal. Studies parallel to the present one have focused on China (Xie, Hample, & Wang, 2015), India (Hample & Anagondahalli, 2015), the United Arab Emirates (Rapanta & Hample, 2015), Malaysia (Waheed & Hample, 2016), and Chile (Santibáñez & Hample, 2015). In each of those investigations, nationals of various countries produced some results quite similar to those from the US, but some interesting differences were also noticeable. We will follow the general strategy of those prior studies by comparing new Portuguese data to findings already generated in the US.

2. UNDERSTANDINGS OF INTERPERSONAL ARGUING

Our aim is to create a platform for further study of Portuguese arguing practices, with the eventual goal of developing a Portuguese-centric description and theory of argumentation, should one be justified by empirical results. To that end, we have selected some fundamental topics of investigation. These are well researched in the US, and book-length summaries of the work are available (Hample, 2005; Rancer & Avtgis, 2014). The particular measures we implement here have many known covariates and causal connections (in the US), and consequently our results should serve as immediately convenient clues to what research should be done next. The variables are in three groups: arguing motivations, argument frames, and personalization of conflict.

2.1 Motivations

Arguing motivations are captured here with two variables, argumentativeness (Infante & Rancer, 1982) and verbal aggressiveness (Infante & Wigley, 1986). Both are sorts of aggressiveness and they are conceptualized as personality traits. Each is actually composed of two subscales, one addressing approach motivations and one measuring avoidance impulses. These measures have been applied in several nations, but not in Portugal (Rancer & Avtgis, 2014, ch. 7).

Argumentativeness refers to the motivation to attack another person's position, reasoning, evidence, or case. Verbal aggressiveness is the motivation to attack another person's character, background, habits, or nature. Research consistently shows that argumentativeness is a constructive trait with many positive life consequences, and verbal aggressiveness is corrosive to personal, educational, and professional experiences (Rancer & Avtgis, 2014). Commonly the avoidance subscale's total is subtracted from the approach subscale's total, to obtain a cumulative measure of either argumentativeness or verbal aggressiveness. In the US, the subscale pairs have substantial negative correlations and this is reasonable statistical practice. However, this strong negative relationship has not consistently appeared in other nations (Hample & Anagondahalli, 2015; Rapanta & Hample, 2015; Santibáñez & Hample, 2015; Xie, Hample, & Wang, 2015). Even in the US, research sometimes finds that one of the subscales will have important connections to some other measure, but the opposite subscale does not have parallel negative connections.

Consequently, we will report results for each subscale separately. For argumentativeness, the subscales are argument-approach and argument-avoid, and for verbal aggressiveness the two subordinate measures are verbal aggressiveness-antisocial and verbal aggressiveness-prosocial. As the names suggest, the two argumentativeness measures indicate a person's general willingness to engage or dodge another person's arguments. The verbal aggressiveness measures represent antisocial tendencies (sometimes suggested as a sufficient measure of verbal aggressiveness) and prosocial inclinations. Collected together, these measures will yield rough summaries of Portuguese willingness to argue and attack.

2.2 Frames

Argument frames is a battery of self-report measures that were developed to answer the question, "What do ordinary people think they are doing when they argue?" (Hample, 2003; 2005). The frames are organized into three groups, each of which gives answers of a somewhat different flavor. This line of work also originated in the US.

The first group of frames captures people's primary goals for arguing. Four such goals have been studied: utility (to obtain or preserve some benefit), identity display (to show off some valued aspect of self), dominance (to assert one's superiority over the other person), and play (to argue for enjoyment). Notice that all of these goals implicitly comprehend the other arguer as no more than a means to the arguer's end: as the person who yields benefits, who applauds an identity disclosure, who subordinates self in response to an assertive display, or who enjoys a playful episode.

The second group of frames assesses the respects in which the arguer takes the other person into account in a more genuine way. These measures are blurting, cooperation, and civility. Blurters say what they think without any authentic acknowledgement of the other person, whereas not blurting means that arguments are edited and adapted to the other person, often in an effort to soften the argument's character and make it more polite. Cooperation is the opposite of competition. Both require that the other person's goals be perceived and understood, but competition uses those goals as targets to be confronted or manipulated, whereas cooperation involves trying to accommodate to what the other person wants. Civility refers to whether arguments are understood as proceeding politely, constructively, and with mutual respect.

The final group of frames has only one measure, called professional contrast. A review of literature revealed that on at least seven points, ordinary arguers often have understandings that are opposite to those of argumentation scholars (Hample, 2003). For instance, many people believe that arguments are incitements to physical fighting but scholars believe that arguing is an alternative to violence. Another example is that ordinary actors often believe that arguments are damaging to relationships but professionals understand that arguments can often improve an interpersonal relationship by putting it on a better footing. This subscale is scored so that high scores mean that the respondent is in agreement with argumentation professionals.

These measures have been in development for some years but have now reached stable form. Several of the frames have been singled out for pointed study (Hample, Han, & Payne, 2010; Hample & Irions, 2015; Hample, Richards, & Skubisz, 2013). These have also been involved in studies of other countries (Hample & Anagondahalli, 2015; Santibáñez & Hample, 2015; Xie, Hample, & Wang, 2015).

2.3 Taking conflict personally

The last set of measures assesses the degree to which people personalize conflicts (Hample & Cionea, 2010; Hample & Dallinger, 1995). When engaged in a face-to-face disagreement, people can orient to the substantive topic or can react as though they had been personally assaulted. In fact, the latter is possible in some conflicts, and every person has something that we might call a conflict biography that is summarized in the reflexive impulse to regard a disagreement as substantive or as a personal attack.

This battery of measures has six subscales. The first, direct personalization, most immediately expresses the idea of taking conflict personally. The next is persecution feelings, which reflects the sense that the other people are participating in the conflict in order to victimize the respondent. Stress reactions is third, and this collects physical and psychological feelings of stress and anxiety due to participation in conflicts. The next two scales are a pair, positive and negative relational effects. People may believe that conflicts can damage relationships, improve them, both things, or neither. The final measure is valence, which is scored to indicate how much a person enjoys participating in conflicts. People who take conflict personally have high scores on direct personalization, persecution feelings, stress reactions, and negative relational effects; these same people also have low scores for positive relational effects and valence. People who do not personalize have the opposite profile. Personalization of conflict has been measured in several nations besides the US (Avtgis & Rancer, 2004; Kim, Yamaguchi, Kim, & Miyahara, 2015; Rapanta & Hample, 2015; Waheed & Hample, 2016; Xie, Hample, & Wang, 2015), but not in Portugal.

2.4 Argument concepts summary

Collected together, these instruments should generate a good description of Portuguese orientations to interpersonal arguing, and their understandings of it. We will see the degree to which they approach arguments constructively (argumentativeness), destructively (verbal aggressiveness), or defensively (taking conflict personally). We should get estimates of why they argue (first set of frames), how they interact with others (second set of frames), and their degree of sophistication about the activity (professional contrast). All of these measures have been substantively related to many other traits, causes, and outcomes. The profile of Portuguese orientations that this study generates ought to give a first indication of the degree to which US theories are exportable to Portugal, and may suggest where further work should be aimed.

3. ARGUMENT-RELEVANT CONSIDERATIONS IN PORTUGAL

3.1 The public sphere

Argumentative practices in Portugal, as perhaps everywhere else, cannot be disconnected from some specific traits of people's history, life, and culture. Since it is not possible to present here all that matters in these areas over the long Portuguese history, we will briefly focus on the 20th and 21st centuries with regards to public "argumentative habits" in relation to the political situation.

From 1910 to 1928, in the First Republic (created after the regicide of the last king Carlos) argumentation freely entered the public space. A number of political parties were established and managed to get their deputies elected to the Portuguese Parliament. Newspapers at a national and regional level expressed different and competing views about what society, education, and economy were going to spring from the Republican Revolution. Plazas, cafés, and other public venues were filled with people having open debates.

The short period of the First Republic ended in chaos and was succeeded by the right-wing dictatorship of Salazar and Caetano (1928-1974). A political security agency (PIDE / DGS) was created, and together with it came two important limitations of argumentative practices: censorship at all levels (the media, theater, public discussions) and prohibition of public gatherings. It was effectively forbidden to publicly discuss politics or even social issues. Since public argumentative practices were greatly repressed, they migrated into the homes of the Portuguese, and occurred largely within the families and between close friends. In the public sphere, courageous humor and innuendo, sometimes of a sophisticated kind, tended to replace articulated argumentative discussions.

After the Carnation Revolution of the 25th April of 1974, the open public sphere experienced during the First Republic re-emerged, aided by the by-then popular broadcast media. For instance, TV debates between Mário Soares (the leader of the Socialist Party) and Álvaro Cunhal (the leader of the Communist Party) are famous till today. They have been widely discussed in the public space and have had a lasting impact on Portuguese public opinion (de Sena, 2002). As a whole, the media and the public spaces were populated with a wide range of different points of view, strongly argued for, by all the players involved in the Carnation Revolution and its outcomes. This period gradually morphed into the current shape of the public sphere characteristic of many other Western European nations: where political maturity signifies bitterness between the 'Left' and the 'Right' and citizens who freely exercise their right to criticize the laws of the country and the practices of public officials, with a certain dose of skepticism of whether such criticism will significantly change the policies.

3.2 "Argument" in Portuguese

Before briefly discussing the scarce research on how Portuguese communicate in interpersonal contexts, we need to mention some peculiarities in the Portuguese vocabulary for arguing. First of all, "argumentation" ("argumentação") is a somewhat academic term which is not commonly used in everyday language to denote argumentative practices. "Conversation" ("conversa") or "discussion" ("discussão") are, by far, much more frequently used. Second, "an argument" ("um argumento") is always a product (a reason or a set of reasons) and never a process (a verbal activity); in O'Keefe's (1977) terms, it refers only to $argument_1$ and never $argument_2$. In Portuguese one cannot say "I had an argument with João" ("Eu tive um argumento com o João"); instead, one would say "I had a discussion with João" ("Eu tive uma discussão com o João"). But by saying this, one conveys a certain negative connotation related to quarrelling; if that connotation is not intended, one should say "I had a conversation with João" ("Eu tive uma conversa com o João"). This leaves the Portuguese with a certain dilemma – at least when looking from the perspective of the capacious English notion of "argument". In the first case ("uma discussão"), they risk injecting possibly unintended aggressiveness into the description of the arguing process. In the second case ("uma conversa"), they risk dissipating the argumentative character of the communicative encounter. Interestingly, the verbs "argumentar" ("to argue") and "discutir" ("to discuss") are largely interchangeable and clearly associated with a verb "conversar" ("to have a conversation"). Portuguese people tend to tacitly associate the action conveyed by the verb as something friendly and cooperative: a conversation where reasons are going to be exchanged and assessed – it is not primarily about me or you, it is about reasons. This is strongly contrasted with the nominalization of "discutir" preceded by an indefinite article – "uma discussão" ("a discussion") – that has an aggressive connotation of a verbal fight not present in the verb. The activity described by a noun tends to be perceived as the result of something that I already have in my mind and want to confront you with – it is a more personal situation of "me against you".

3.3 Portuguese culture and communication

While empirical studies on argumentation in Portugal are conspicuously wanting, a few pointers towards how Portuguese communicate in argumentative situations can be gleaned from a more general literature on communication styles, where Portugal is typically part of a cross-national or inter-cultural comparison. Tixier, who studied management and communication styles in Europe in early 1990s based on interviews with managers in European corporations, concludes that "Portugal presents non-Latin characteristics in southern Europe" (1994, p. 25). Most notably:

> Among the Latins themselves the attitude towards conflict can be markedly different. The Portuguese, for example, claim to be more peaceable and less conflictual in their relationships than the Spanish. For proof they cite the fact that they are fond of bullfights on horseback in which the bull is not put to death. In meetings in Portugal managers avoid direct conflicts as much as possible. (Tixier, 1994, p. 16)

Moreover, "temperaments in certain European countries – for example Portugal, Luxembourg, Belgium and Sweden – are clearly less extroverted, causing people to weigh their words, to distrust flowery language, and not to express themselves at great length" (Tixier, 1994, p. 19). This is markedly different from other southern European Latin countries such as Spain or Italy, characterized by a clear tendency for a more confrontational, extroverted, verbose, and "flowery" communication style. Moreover, Tixier stresses the role of personal relationship and trust in business communication among the Portuguese, who often prioritize them over efficiently setting and achieving goals.

In general in cross-cultural studies, such as those of Hofstede, Hofstede, and Minkov (2010), Portuguese culture is characterized by a relatively high power distance, low individualism (i.e., high collectivism), and low masculinity (i.e., high femininity). Interestingly, of the 76 countries and regions studied, it has the second highest score of uncertainty avoidance, right behind Greece. *Uncertainty avoidance* is defined as "*the extent to which the members of a culture feel threatened by ambiguous or unknown situations.* This feeling is, among other manifestations, expressed through nervous stress and in a need for predictability: a need for written and unwritten rules" (Hofstede, Hofstede, & Minkov, 2010, p. 191; italics original).

Such general results are important to our study for two reasons: a methodological and a substantial one. First, it has been argued that the above-mentioned cultural features can impact the very way participants in a given country respond to social scientific surveys (Harzing, 2006). For instance, members of cultures with high power distance, that is, with high reverence for authority, would respond with more acquiescing answers: they will more likely produce the "agree" or "strongly agree" responses on the standard Likert-scales, presumably to satisfy the researcher imbued with scientific authority. Moreover, respondents from collectivistic cultures would typically provide responses somewhere in the middle of the scale, rather than on its extremes (say, 3, rather than 1 or 5 on a 5-point scale), to better fit in the average. If these hypotheses were correct, then Portuguese respondents would provide responses different from the US respondents in virtue of the very way they treat surveys – rather than in virtue of the content of their attitudes to the questions. (Note that the US is basically a low power distance, highly individualistic, masculine, and low uncertainty avoidance culture – all these in contrast to Portugal; see Hofstede, Hofstede, & Minkov, 2010.) However, as Harzing's (2006) study of survey response style differences across 26 countries shows, the US and Portuguese respondents basically display a very similar response pattern, with no significant differences. If these results are generalizable, then we can assume that what we are testing in our study are Portuguese-American differences in substantial attitudes to interpersonal argumentation, rather than different ways of responding to a survey.

Second, high reverence for authority, collectivism, and avoidance of uncertainty would all significantly impact the role argumentation has to play in daily encounters. In asymmetric situations (employee-employer, student-professor) Portuguese would tend to be less inclined to attack the superior's position or reasoning. Given the collectivism, collaborative and civil forms of argumentation and reasoning would predominate. Finally, playful and personal forms of argument would give way to standardized and formal ways of managing disagreement through argument – otherwise strong stress reactions would ensue. All these are, however, but mere speculative suppositions based on some general studies – all of them in need of solid empirical investigation.

4. RESEARCH QUESTIONS

Because we have found no prior research using our instruments in Portugal, we lack the literature to justify hypotheses. We do not regard prior results from the US or other nations as sufficiently on point to expect those outcomes to occur in Portugal as well. In fact, part of the motivation for this study is the suspicion that every national culture may need to be thought about separately. Therefore we pose research questions. Some of these solely involve Portugal and others require comparisons to US data. These comparisons are made to test the exportability of US theory, which would be convenient (but only if justifiable) because of the bulk of work that has been done there.

Our first research question concerns the possibility of sex differences in Portugal. These are commonly reported in the US and other nations. Although the same detailed patterns do not always appear in every country, one can often discern a pattern of males having higher scores on various aggression-related measures and women displaying more anxiety-related orientations to argument and conflict. Biological sex is a fundamental social category. It is often the first thing noticed about another person, and a person's sex is immediately associated with a variety of social affordances and responsibilities, for good or ill (Buss, 1995). Therefore we inquire:

> RQ1: Do Portuguese men and women differ in their orientations and understandings of interpersonal arguing, as measured by verbal aggressiveness, argumentativeness, argument frames, and personalization of conflict?

Our second research question begins a direct evaluation of the pertinence of US results to Portugal. A first matter bearing on this issue is whether respondents from the two nations have parallel scores on the various instruments used in this study.

> RQ2: Do Portuguese and US respondents have different mean scores on the instruments assessing orientations and understandings of interpersonal arguing?

A related but more delicate question is whether our measures have the same dynamics in both nations. Does variable X have a similar relationship to variable Y in both countries? Examination of these correlations is a more important theoretical step than merely examining the means scores from the two nations. Even if one nation has a lower mean on some variable than the other nation, if that variable is associated with the other measures in the same way, we would have the same theoretical relationship, just at a lower range of scores. Correlations are fundamental hints about causality, and so they are a critical first step in describing the outline of a theory, and the possible need to break off from American theorizing. Therefore:

> RQ3: What are the correlational systems among the variables in Portugal, and are they comparable to those in the US?

5. METHOD

5.1 Participants and procedures

The study was conducted at a major Portuguese state university located in Lisbon, the capital city. While most respondents can be assumed to be students, faculty, and staff of that university, other Portuguese citizens were also invited to participate through social networks (Facebook) and e-mail messages, including those sent to e-mail lists of professional organizations such as Associação Portuguesa de Linguística (APL: Portuguese Linguistics Association). Data were collected online, with the survey being hosted by SurveyMonkey. The entire survey took about half an hour to complete. Respondents were not compensated or given course credit for their participation.

A total of 252 complete survey responses were received. Many respondents were employed, likely at a university (104 full-time and 24 part-time). Students accounted for 39% of the sample, with 97 respondents. Another 17 people said merely that they were unemployed (and were possibly students), and 10 preferred not to answer the question. Women comprised 64% of the sample. About 29% of the sample was married, with another 8% divorced or widowed. A plurality (32%) described themselves as single, but another 26% said they were in a relationship. Only 2% said their highest educational experience was high school; 19% had a PhD, 25% a master's degree, and another 12% had some graduate work. University graduates were 16% of the sample and another 26% said they were enrolled at the university at the time of the survey. The sample's mean age was 35.2 years ($SD = 13.6$).

5.2 Instruments

All of the instruments originated in the US. Bilingual speakers translated the measures. First, the items were translated into Portuguese by the third author, a native speaker of the European Portuguese. Then, the Portuguese materials were translated back into English by another bilingual speaker who had not seen the original English versions and who, while having an MA degree, is not an academic accustomed to a scientific jargon of argumentation theory. The chief aim was to have the survey read well for ordinary Portuguese speakers. The few discrepancies between the English original and back-translation were discussed among the authors of the paper (three of whom are bilingual), and further with their Portuguese colleagues. As a result, the Portuguese materials were adjusted accordingly. (The Portuguese translations are available from the authors.)

Importantly, given the complex semantics of "argumentation" in Portuguese (see above), we added the following explanation in the introduction to the online survey:

> No que se segue usa-se o termo "argumentação" no sentido habitual de uma troca de argumentos entre duas ou mais pessoas. Preferiu-se este termo aos termos "discussão" ou "debate", eventualmente mais correntes, para poder abarcar com um só termo 'mais neutro' uma argumentação que tanto pode ser calma e ponderada, e por isso mais conotada com o termo "debate", como animada ou mesmo conflituosa, e por isso mais conotada com o termo "discussão".
>
> In what follows we use the term "argumentation" in the usual sense of an exchange of arguments between two or more people. This term is preferred over the more common terms "discussion" or "debate" in order to cover with one 'more neutral' term the practice of argumentation, which can be both calm and deliberated, and therefore close in its meaning to "debate", as well as animated or even conflictual, and therefore associated with the term "discussion".

In accordance with this, whenever possible, we used consistently throughout the survey the terms "argumentação" and "argumento." When the English survey explicitly disambiguated "argument" to a "conflict discussion," we accordingly translated it as "discussão conflictuosa."

Descriptive statistics for all measures, including internal reliabilities, are in Table 1. All self-report measures used a 1-10 metric for Likert responses, ranging from strongly disagree (1) to strongly agree (10). Scale scores were calculated so that higher scores represent more of the named characteristic.

Argument Motivations. Argumentativeness was assessed with the (translated) Infante and Rancer (1982) items. The standard verbal aggressiveness (Infante & Wigley, 1986) instrument was used. These produce four measures: argument-avoid, argument-approach, verbal aggressiveness (antisocial), and verbal aggressiveness (prosocial).

Argument Frames. The argument frames instruments have undergone revision. The earliest complete version of the scales was reported in Hample, Warner, and Norton (2006). The blurting scale was later finalized in Hample, Richards, and Skubisz (2013), and the utility scale was then revised in Hample and Irions (2015). These subscales afford measures of the utility, identity display, dominance assertion, and play goals; the sensitivity to blurting, cooperation, and civility possibilities; and the reflectiveness implied in the professional contrast measure.

Personalization of Conflict. The Taking Conflict Personally (TCP) scales originated with Hample and Dallinger (1995), and considerable psychometric data was cumulated for the analyses in Hample and Cionea (2010). These measures include direct personalization, persecution feelings, stress reactions, positive relational effects, negative relational effects, and conflict valence.

5.3 US Comparison Data

Hample and Irions (2015) collected data on all these measures in the US for another study. Their respondents were all undergraduates at a large public university in the Eastern portion of the US. A total of 461 respondents provided data in exchange for minor class credit. About 70% of these were women. Most (53%) described themselves as Euro-American, followed by 11% African-American, and 7% Asian-American; the remainder were scattered among other ethnicities or nationalities. Their average age was 20.3 years ($SD = 1.7$). Additional detail on this sample is available in the original report.

6. RESULTS

Table 1 displays descriptive statistics regarding the measures. Readers should notice that 5.5 is the theoretical midpoint of the 1 – 10 scales, and this will give a rough guide to the degree to which Portuguese respondents found our measures to be salient considerations in regard to interpersonal arguing. Another note of interest is the satisfactory Cronbach's alphas for the measures. These were generally comparable to those obtained in the US. If these instruments had seemed odd or incoherent to Portuguese respondents, reliabilities might have been quite low. The adequacy of these tests of internal consistency suggests that the instruments resonated reasonably well with Portuguese participants.

6.1 Research Question 1

The first research question was, "Do Portuguese men and women differ in their orientations and understandings of interpersonal arguing?" Results bearing on this matter are also in Table 1. For the most part Portuguese men and women had closely comparable scores on our measures. The only exceptions were that men were willing to be more personally hostile and antisocial; men were more open to arguing for play; and women had more pointed stress reactions to conflict. These few significant outcomes were consistent with the research tradition's general finding that men are more aggressive than women on these and related instruments. However, that pattern was missing in regard to quite a few other opportunities for men to be more aggressive and women to be more anxious and nurturing (e.g., prosociality, argumentativeness, the other primary arguing goals, civility, and most of the personalization measures).

Since much of the prior research showing sex differences on these measures involved only undergraduate respondents, we repeated these analyses with data restricted to the 97 students (32 men and 65 women). The following comparisons were statistically significant: play (t (df = 95) = 2.63, $p < .01$; men = 6.69, women = 5.61), stress reactions (t (df = 95) = 3.75, $p < .001$; men = 4.56, women = 6.04), positive relational effects (t (df = 95) = 2.53, $p < .05$; men = 5.98, women = 5.11), and conflict valence (t (df = 95) = 2.99, $p < .01$; men = 6.03, women = 5.04). The verbal aggressiveness (antisocial) result present in the whole sample was not quite significant in the student sample (t (df = 95) = 1.77, $p = .08$; men = 3.44, women = 2.93). The play and stress results reproduced those of the whole sample (Table 1), and the positive relational effects and valence outcomes were also similar to those in the whole sample except that for students the differences achieved statistical significance. These post hoc analyses showed the same weak pattern as for the whole sample (men are more aggressive, women are more anxious and nurturing), but again with many opportunities to express the pattern showing null results. The contrast between the results for students and the whole sample suggests that some sex differences may slightly dissipate with age and life experience.

So our answer to the first research question is that sex differences were observed for verbal aggressiveness (antisocial), the play goal, and stress reactions. These differences were consistent with the idea that men are more aggressive and women more anxious. However, the measures in the study afforded many more opportunities for those inclinations to express themselves, and those other results were null.

6.2 Research Question 2

Our second concern was to answer the question, "Do Portuguese and US respondents have different mean scores on the instruments assessing orientations and understandings of interpersonal arguing?" The pertinent results are in Table 2.

As the Table immediately indicates, Portuguese and US residents differed significantly on nearly all the measures. US respondents were more verbally aggressive (antisocial) and more argument avoidant, whereas the Portuguese were readier to engage the other person's arguments. Portuguese were more interested in arguing in order to display identity or to play, but less oriented to dominance displays. The Portuguese were also less likely to blurt and more pointedly interested in arguing cooperatively and civilly. They were more sophisticated in their reflections about arguing, as indicated by their professional contrast scores. People from the US took conflict more personally, when the direct personalization, persecution, and stress measures are considered. Portuguese were more optimistic about arguing's effects on relationships, whether the positive or negative measures are examined. And Portuguese respondents had more positive valence for conflicts.

These specific results formed a fairly clear pattern. The Portuguese understood arguing as less hostile, more pleasant, and less personal (and so more substantive), and (perhaps consequently) were less likely to avoid arguments and more embracing of the opportunity to engage in conflict. If we assume that arguing in Portugal is more pleasant, more constructive, and better mannered than in the US, as the data imply, all the results in the table are accounted for. The higher Portuguese score for professional contrast also suggests that they have a more productive experience with interpersonal arguing than Americans do.

6.3 Research Question 3

The final research question was, "What are the correlational systems among the variables in Portugal, and are they comparable to those in the US?" Here the results are in Table 3. To simplify the correlational report, Table 3 is divided into the instruments' categories and therefore omits cross-category correlations (e.g., between argument-avoid and professional contrast). Full tables of correlations are available from the authors.

First, consider the motivational patterns. The prosocial and antisocial subscales of verbal aggressiveness are understood (in the US) as polar opposites, as are argument-avoid and argument-approach. In

Portugal these subscale correlations were negative, as theory predicts. The US correlation between the verbal aggressiveness subscales was noticeably more extreme than the Portuguese result, but the argumentativeness correlation was comparable. Antisocial impulses in the US clearly implied approaching arguments, but not in Portugal. But in Portugal prosocial inclinations did predict argument approach although they did not in the US. Prosociality was unrelated to argument avoidance in Portugal but strongly associated with it in the US. These two sorts of aggressive motivations (argumentativeness and verbal aggressiveness) played different roles in the two countries.

The frames results show patterns that are recognizable from one nation to the other. Blurting and the first order frames (utility, identification, etc.) were all positively correlated in both nations. Cooperation, civility, and professional contrast also displayed a pattern of clear positive correlations in both nations. In fact, the whole correlation matrices for both countries were quite similar, suggesting that argument frames were working in similar fashion for both national samples.

The Taking Conflict Personally scales also showed marked similarity in both countries. The group of subscales that in the US are regarded as directly measuring personalization – direct personalization, persecution, stress, and negative relational effects – were all strongly and positively correlated in both samples. The two relational effects measures were correlated negatively in Portugal but not in the US. This particular US result is atypical, and the two scales normally have a significant negative correlation with American samples (Hample & Cionea, 2010). In both countries, valence was depressed by the personalization measures and had a positive connection to positive relational effects. All these results conform to the originating theory and American experience with the measures.

Examination of the dynamic associations among the variables (represented by their correlations) showed clear correspondence between Portugal and the US for argument frames and taking conflict personally. Although the argumentativeness and verbal aggressiveness subscales did not produce completely alien results in the two countries, some differences in how those inclinations operate were evident.

7. DISCUSSION

Several of our results were prefigured in our very general description of Portuguese culture. We found several differences between Portuguese men and women, but in the context of what has been reported in other nations, those differences in Portugal were relatively rare. This is consistent with Portugal's status as a high feminine nation (Hofstede, Hofstede, & Minkov, 2010). The pattern we found was the usual one (men aggressive, women mild) but it did not assert itself very strongly. We interpret this as indicating that Portuguese men have an unusually pronounced share of what are women's virtues in other nations – cooperativeness, unaggressiveness, and concern for others.

Mothering takes on certain distinguishable features in Portugal. During the times of dictatorship mothers, who typically stayed at home, tended to do the arguing that was necessary to educate their children much more than working fathers. And they were, and still are, quite good in so doing. Here we are talking mainly about practical and even moral argumentation since that kind of argumentation is the one that drives the educational processes. Now that Portuguese women have fully entered the workplace, their argumentative skills, long developed through generations, fit in nicely.

Similarly, our review suggested that, in comparison with other national cultures, Portuguese have an unusual aversion to uncertainty (Hofstede, Hofstede, & Minkov, 2010). This might well reduce risk taking in interpersonal arguments, damping out tendencies to take chances by exposing one's views immediately, clearly, and aggressively. We found that the Portuguese were less impulsive in arguing, as indicated by their lower propensity to blurt (compared to Americans), and were more careful to cooperate and aim themselves toward civil interactions. Portuguese were not averse to arguing, and in fact were somewhat more eager to engage. Compared to US respondents, Portuguese were quite willing to argue to display identities, but only pleasant ones because they were less willing to argue in order to assert dominance. These patterns of Portugal-US difference are also consistent with Tixier's (1994) description of the Portuguese as careful and non-confrontational in their interactions.

At least some of the differences in orientations regarding RQ2 and RQ3 where the US and Portugal are compared can be explained through the very way the concept of "argument" is perceived in both cultures. Basically, argument tends to be positively valued in Portugal (just like a "good reason" is), so Portuguese have little reason to avoid it, take it too personally, or be worried about its interpersonal effects. Our understanding of these issues led us to certain translation decisions. The Portuguese items in the argumentativeness scales used the verb "argumentar" and the noun "argumentação," neither of which carry the negative connotations of English "arguing." This might partly explain the non-standard results in the motivational patterns, viz. the fact that in Portugal prosocial inclinations predicted argument approach, while in the US they predicted argument avoidance. Portuguese can well manifest their prosocial concerns by arguing – presumably understood as engaging in a cooperative conversation where reasons are jointly produced, exchanged, and assessed. By contrast, Americans are prosocial by avoiding arguments, i.e., by avoiding verbal fights. While we have done our best to limit these linguistic differences – through the introductory note and careful context-sensitive translation – we cannot claim to have eradicated deeply held understandings of argument, discussion, and conversation in ordinary Portuguese. We have taken effort to reflect them accurately, yet these understandings are different from the standard (if not hegemonic) English connotation. Van Eemeren et al. (2014, pp. 3-4) stress that the meaning of the Latin-based counterparts of the English "argument" and "argumentation" in many other European languages – including Portuguese "argumentação" – are devoid of the sense of conflict or quarrel. "Argumentation" signifies the process of producing and exchanging reasons, and its products.

In sum, if now we try to come up with some common traits of Portuguese argumentative practices that the locals might agree upon, we would stress the following matters. First of all, there is a deep rooted Portuguese trait of character, or tacit belief, that offending someone is a terrible thing to do. So, people are inclined to be very careful with, e.g., *ad hominem* arguments (compare our verbal aggressiveness results). For the same reason, competition is likely to be more sportive than real (compare our play frame results): points of view are more likely to be possible topoi we are prepared to argue about, than real positions that you, or I, are going to assume and attack – more so, if the issue under discussion is known to be of a sensitive nature. It is not uncommon to start a discussion by asking "What do you think can be argued pro This and That? And against?", instead of saying "I believe that This and That on account of such and such reasons."

A related matter is that as a rule, (almost) nobody will take you seriously if you brag in one way or another, even via showing off your argumentative skills. Portuguese people "have a nose" to detect when, instead of having a real discussion for "honest" purposes, you are just showing off and, in the latter case, you may end up a target of mockery. Thus, the Portuguese were willing to show off certain features of their identities by arguing, but were not willing to attempt domination in this way. In a sense people feel slightly offended if you pretend to be better than them. Showing off is not seen as something Portuguese applaud, at least traditionally. Eusébio (the football player) and Amália (the Fado singer), both known worldwide, were locally perceived as humble non-bragging people. Even today, Portuguese national heroes tend to be perceived in this way.

ACKNOWLEDGEMENTS: We would like to acknowledge the invaluable help of our colleagues at the Universidade Nova de Lisboa, especially João Costa and Nuno Mora, in preparing and distributing the Portuguese survey. Their nuanced comments on the aspects of arguing in Portugal significantly improved the study. Rita Azevedo back-translated the first Portuguese version of the survey into English and provided important feedback regarding final translation decisions.

REFERENCES

Avtgis, T. A., & Rancer, A. S. (2004). Personalization of conflict across cultures: A comparison among the United States, New Zealand, and Australia. *Journal of Intercultural Communication Research, 33*, 109-118.

Buss, D. M. (1995). Evolutionary psychology: A new paradigm. *Psychological Inquiry, 6*, 1-30.

Eemeren, F. H. van, Garssen, B., Krabbe, E.C.W., Snoeck Henkemans, A.F., Verheij, B., & Wagemans, J.H.M. (2014). *Handbook of argumentation theory*. Dordrecht: Springer.

Hample, D. (2003). Arguing skill. In J. O. Greene & B. R. Burleson (Eds.), *Handbook of communication and social interaction skills* (pp. 439-478). Mahwah, NJ: Lawrence Erlbaum.

Hample, D. (2005). *Arguing: Exchanging reasons face to face*. Mahwah, NJ: Lawrence Erlbaum.

Hample, D., & Anagondahalli, D. (2015). Understandings of arguing in India and the United States: Argument frames, personalization of conflict, argumentativeness, and verbal aggressiveness. *Journal of Intercultural Communication Research, 44*, 1-26.

Hample, D., & Cionea, I. A. (2010). Taking conflict personally and its connections with aggressiveness. In T. A. Avtgis & A. S. Rancer (Eds.), *Arguments, aggression, and conflict: New directions in theory and research* (pp. 372-387). New York, NY: Routledge, Taylor, and Francis.

Hample, D., & Dallinger, J. M. (1995). A Lewinian perspective on taking conflict personally: Revision, refinement, and validation of the instrument. *Communication Quarterly, 43*, 297-319. doi: 10.1080/01463379509369978.

Hample, D., Han, B., & Payne, D. (2010). The aggressiveness of playful arguments. *Argumentation, 24*, 405-421. DOI 10.1007/s10503-009-9173-8.

Hample, D., & Irions, A. (2015). Arguing to display identity. *Argumentation, 29*(4), 389-416. doi: 10.1007/s10503-015-9351-9.

Hample, D., Richards, A. S., & Skubisz, C. (2013). Blurting. *Communication Monographs, 80*, 503-532. doi:10.1080%2F03637751.2013.830316.

Hample, D., Warner, B., & Norton, H. (2006). The effects of arguing expectations and predispositions on perceptions of argument quality and playfulness. *Argumentation and Advocacy, 43*, 1-13.

Harzing, A.-W. (2006). Response styles in cross-national survey research: A 26-country study. *International Journal of Cross Cultural Management, 6*, 243-266. doi: 10.1177/1470595806066332.

Hofstede, G., Hofstede, G. J., & Minkov, M. (2010). *Cultures and Organizations: Software of the Mind*. 3rd ed. New York: McGraw Hill.

Infante, D. A., & Rancer, A. S. (1982). A conceptualization and measure of argumentativeness. *Journal of Personality Assessment, 46*, 72-80. doi: 10.1207/s15327752jpa4601_13

Infante, D. A., & Wigley, C. J. (1986). Verbal aggressiveness: An interpersonal model and measure. *Communication Monographs, 53*, 61-69. doi: 10.1080/03637758609376126

Kim, E. J., Yamaguchi, A., Kim, M.-S., & Miyahara, A. (2015). Effects of taking conflict personally on conflict management styles across cultures. *Personality and Individual Differences, 7*, 143-149.

O'Keefe, D. J. (1977). Two concepts of argument. *Journal of the American Forensic Association, 13*(3), 121-128.

Rancer, A. S., & Avtgis, T. A. (2014). *Argumentative and aggressive communication: Theory, research, and application*, 2nd. ed. New York: Peter Lang.

Rapanta, C., & Hample, D. (2015). Orientations to interpersonal arguing in the United Arab Emirates, with comparisons to the United States, China, and India. *Journal of Intercultural Communication Research, 44*, 263-287. doi: 10.1080/17475759.2015.1081392.

Ribeiro, H. J. (Ed.) (2009). *Rhetoric and argumentation in the beginning of the XXIst century*. Coimbra: Coimbra University Press.

Santibáñez, C., & Hample, D. (2015). Orientations toward interpersonal arguing in Chile. *Pragmatics, 25*(3), 453–476. doi: 10.1075/prag.25.3.06san.

Sena, N. M. de. (2002). *A interpretação política do debate televisivo 1974-1999*. Lisboa: Instituto Superior de Ciências Sociais e Políticas.

Tixier, M. (1994). Management and communication styles in Europe: Can they be compared and matched. *Employee Relations, 16*, 8-26. doi: 10.1108/01425459410054899.

Waheed, M., & Hample, D. (2016, June). *Argumentation in Malaysia and how it compares to the U.S., India, and China*. Paper presented at the annual conference of the International Communication Association, Fukuoka, Japan.

Xie, Y., Hample, D., & Wang, X. (2015). A cross-cultural analysis of argument predispositions in China: Argumentativeness, verbal aggressiveness, argument frames, and personalization of conflict. *Argumentation, 29*(3), 265-284. doi: 10.1007/s10503-015-9352-8.

Table 1: Descriptive Statistics and Male/Female Comparisons for Portugal

	Items	Cronbach's Alpha	N	Mean	SD	Male Mean	Female Mean	t
VA Antisocial	10	.82	252	3.04	1.34	3.45	2.82	3.69***
VA Prosocial	10	.79	252	6.40	1.49	6.32	6.45	0.63
ArgAvoid	10	.83	252	3.74	1.50	3.65	3.80	0.83
ArgApproach	10	.86	252	6.78	1.52	6.66	6.84	0.93
Utility	8	.80	252	5.14	1.52	5.21	5.11	0.48
Identification	8	.74	252	6.90	1.33	6.97	6.87	0.61
Play	4	.82	252	5.52	2.09	5.88	5.32	2.03*
Dominance	6	.80	252	3.92	1.76	4.10	3.82	1.22
Blurting	10	.84	252	4.20	1.49	4.29	4.14	0.76
Cooperation	7	.74	252	7.76	1.30	7.55	7.88	1.93†
Civility	10	.79	252	7.82	1.15	7.66	7.91	1.62
Prof Contrast	7	.85	252	8.09	1.49	8.00	8.14	0.71
Direct Persnl	7	.81	252	4.78	1.72	4.53	4.92	1.78
Persecution	6	.78	252	4.15	1.72	4.00	4.23	1.02
Stress	5	.69	252	5.24	1.83	4.51	5.64	4.92***
Pos Relatnl	7	.80	252	5.11	1.55	5.28	5.01	1.33
Neg Relatnl	5	.83	252	6.78	1.84	6.42	6.98	2.35*
Valence	7	.70	252	5.28	1.52	5.52	5.14	1.93†

Note. For Cooperation, the final item in the standard ordering was dropped to improve reliability. This item is also often dropped in U.S. research. All variables were measured on a 1-10 metric. For the tests of sex differences, 90 respondents were male and 162 were female. t-tests were adjusted for unequal group variance when necessary. Significance tests were two-tailed.
† $p < .06$ * $p < .05$ ** $p < .01$ *** $p < .001$

Table 2: Comparisons of Means from Portugal and United States

	Portugal			United States			t
	N	Mean	SD	N	Mean	SD	
VA Antisocial	252	3.04	1.34	441	4.58	1.53	-13.25***
VA Prosocial	252	6.40	1.49	441	6.36	1.23	0.37
ArgAvoid	252	3.74	1.50	441	5.77	1.40	-17.88***
ArgApproach	252	6.78	1.52	441	5.76	1.34	8.86***
Utility	252	5.14	1.52	461	5.32	1.26	-1.58
Identification	252	6.90	1.33	461	6.57	1.40	3.13**
Play	252	5.52	2.09	461	4.44	2.17	6.84***
Dominance	252	3.92	4.44	461	4.44	1.86	-3.60***
Blurting	252	4.20	1.49	461	5.11	1.58	-7.49***
Cooperation	252	7.76	1.30	461	6.83	1.50	8.62***
Civility	252	7.82	1.15	461	6.26	1.16	17.18***
Prof Contrast	252	8.09	1.45	461	6.29	1.86	14.29***
Direct Persnl	252	4.78	1.72	461	5.82	1.70	-7.74***
Persecution	252	4.15	1.72	461	4.59	1.58	-3.39***
Stress	252	5.24	1.83	461	5.70	1.61	-3.36***
Pos Relatnl	252	5.11	1.55	461	6.02	1.45	-7.84***
Neg Relatnl	252	6.78	1.84	461	6.06	1.40	5.42***
Valence	252	5.28	1.52	461	4.12	1.76	9.19***

Note. U.S. data are from Hample & Irions (2015). t-tests were corrected for unequal group variances when necessary. Significance tests were two-tailed. All variables were measured on a 1-10 scale. Negative signs for the t-tests indicate that the U.S. mean was higher, and no sign means that the Portuguese mean was higher.
** $p < .01$ *** $p < .001$

Table 3: Correlations Among Variables in Portugal and the US

	1	2	3	4	5	6	7
Portugal							
1 VA Antisocial							
2 VA Prosocial	-.21***						
3 Argument Avoid	.18**	.03					
4 Argument Approach	.12	.29***	-.34***				
United States							
1 VA Antisocial							
2 VA Prosocial	-.31***						
3 Argument Avoid	.01	.30***					
4 Argument Approach	.32***	.06	-.35***				
Portugal							
1 Utility							
2 Identification	.31***						
3 Dominance	.39***	.36***					
4 Play	.34***	.41***	.33***				
5 Blurting	.26***	.11	.28***	.29***			
6 Cooperation	.14*	.32***	-.12	.04	-.03		
7 Civility	-.08	.08	-.35***	-.10	-.33***	.38***	
8 Professional Contrast	-.03	.05	-.26***	-.02	-.20**	.29**	.51***
United States							
1 Utility							
2 Identification	.36***						
3 Dominance	.46***	.22***					
4 Play	.33***	.41***	.48***				
5 Blurting	.39***	.19***	.46***	.26***			
6 Cooperation	.12*	.35***	-.21***	-.09	-.04		
7 Civility	-.13**	.13**	-.38***	-.06	-.37***	.31***	
8 Professional Contrast	-.03	.14**	.22***	.02	-.09*	.20***	.48***
Portugal							
1 Direct Personalization							
2 Persecution Feelings	.59***						
3 Stress Reactions	.47***	.45***					
4 Positive Relational Effects	.07	.12	-.05				
5 Negative Relational Effects	.30***	.39***	.44***	-.24***			
6 Conflict Valence	-.41***	-.30***	-.49***	.29***	-.44***		
United States							
1 Direct Personalization							
2 Persecution Feelings	.57***						
3 Stress Reactions	.63***	.46***					
4 Positive Relational Effects	-.09	-.11*	-.15**				
5 Negative Relational Effects	.40***	.39***	.39***	-.02			
6 Conflict Valence	-.46***	-.12*	-.59***	.19***	-.39***		

Note. US data are from Hample & Irions (2015).
* $p < .05$ ** $p < .01$ *** $p < .001$

Commentary on Hample, Lewiński, Sàágua and Mohammed's A Descriptive and Comparative Analysis of Arguing in Portugal

PAULA CASTRO
Instituto Universitário de Lisboa (ISCTE-IUL) and CIS-IUL, Portugal
paula.castro@iscte.pt

1. INTRODUCTION

I am very pleased to be commenting on this paper and want first of all to congratulate the authors on undertaking the long neglected task of studying argumentation in the Portuguese context. I further wish to underline how important it is that they do this as part of a systematic program of research. The first author in particular, Dale Hample, has been developing this program for a number of years now and extending it to several national contexts (e.g. Hample & Anagondahalli, 2015). Thus the paper receiving commentary is not the result of a single idea quickly tested and put to rest to make room for the next one. It instead forms part of a systematic endeavor that has produced a stream of associated investigations, partnerships and publications. These qualities cannot be commended enough. Moreover, the clarity of the paper bears testimony to the clarity of the authors' goals and made this an easy reading.

As the paper examines argumentation in Portugal with a type of analysis traditionally used in psychology, a Portuguese psychologist seemed to the authors to be a suitable invitee for a commentary; this is how I – a Portuguese psychologist – find myself in a position from which to thank them for this invitation and to make the above congratulations.

In a commentary, congratulations are usually followed by the "however" – i.e. the questions and/or objections – often of a theoretical or methodological nature. My first "however" will nevertheless be one of perspective. I need to confess from the outset that although invited as a Portuguese psychologist, this was not the identity that came forward while I was reading the paper. It was my identity as a European academic interested in communication, discourse and argumentation that came forward for organizing this commentary.

Consequently, my goals will be twofold. The first will be to clarify what in the text of Hample and colleagues brought forward the

European in me and what perspective that afforded me for unpacking and questioning some of the implicit theoretical assumptions of the paper. The second goal will be to offer some suggestions for further analysis of the data used in the paper, based on the assumptions I put forward when commenting on their theory.

2. US-EUROPE SCIENTIFIC RELATIONS

Starting, then, with the reasons that brought forward the European in me, the following quotation was especially important:

> This research – both the theory and the methods – originated in the US, and a basic question is whether American theories, concepts, and findings can be *exported* to Portugal. (p. 137, this volume, my emphasis)

The export–import language of this paragraph reminded me of how, in the Post War years, the American inclination towards a *nomothetic project for the social sciences* had important repercussions for the European social sciences (Wallerstein et al., 1996). The nomothetic goal of devising general laws requires the export of methods and theories for testing whether these in fact work similarly in other parts of the world. This necessitates partners in other parts of the world, a requirement that stimulated, during the Post War years, our American colleagues to sponsor the re-birth of social sciences in Europe. The case of social psychology is particularly well documented (Farr, 1996; Castro, 2002; Moscovici & Markova, 2006). It is well known that in the early 1960s an American – J. Lanzetta – *based "in the American Embassy in London"* (Farr, 1996, p. 9) and a funding institution – the *US Office for Naval Research* – played a crucial role in the re-birth of European social psychology. They contacted and brought together in two meetings (1963 & 1964) the (up until then) scattered European social psychologists, funded the meetings and sponsored the formation of the European association of social psychology.

Heated debates on both sides of the Atlantic ensued from these efforts, revolving around two main topics (Moscovici & Markova, 2006; Israel & Tajfel, 1972). One regarded the best way for envisaging the future of US-Europe scientific relations: some voices defended the path of export and replication, while others called for funding autonomous research agendas, hoping for local creativity to later feed back to the US (Moscovici & Markova, 2006). The second topic concerned the underlying epistemological and ontological assumptions of the research being exported.

Regarding these assumptions, a group of European social psychologists claimed from the 1970s onwards (see Israel & Tajfel, 1972) that US export efforts focused too narrowly on intra-individual aspects, assuming stable characteristics of subjects (such as traits, motivations and information processing mechanisms) to be the source of meaning and action (Bruner, 1990). This also meant that the focus was mostly on simple Subject-Object relations. However, claimed the European authors, this neglected the constitutive role of relations in meaning and action, i.e., how meaning and action emerged from triadic Subject-**Other**-Object relations (Moscovici, 1972). Adopting the triadic model meant that the role of culture, context and interaction had to be taken into account and social psychology had to look at communication (Moscovici, 1972). This helped open a space for psychology to pay increasingly more attention to language, discourse, argumentation and rhetoric.

3. SOME QUESTIONS ABOUT ASSUMPTIONS

Let us now try to discern what theory and concepts the paper in question uses, in order to understand where it is situated in this debate. According to the authors, their research uses sub-scales for operationalizing three Components: (1) arguing motivations, (2) argument frames, and (3) personalization of conflict. Let's first look at how Component 1 is defined:

> Arguing *motivations* are captured here with two variables, argumentativeness (...) and verbal aggressiveness (Infante & Wigley, 1986). Both are sorts of aggressiveness and they are conceptualized as *personality traits*. Each is actually composed of two subscales, one addressing approach *motivations* and one measuring avoidance *impulses*. (p.138, this volume)

The key concepts are underlined: motivations, personality traits, impulses. Their implicit assumptions are clear: what matters for understanding argumentation happens within the subject, expressing stable characteristics, and does not so much happen between Subject and Object (e.g. the topic being argued about), or between Subjects jointly relating to an object (e.g. arguing about a topic).

Looking now at the implicit assumptions underlying Component 2, we can see that, according to the authors, these scales were:

> developed to answer the question, "What do ordinary people *think they are doing* when they argue?" (...) (so) what we are testing (...) are Portuguese-American differences in substantial *attitudes* to interpersonal argumentation. (p. 139, this volume)

Here the concepts used are also clear: beliefs about argumentation, attitudes. However, the implicit assumptions are less clear. This is because there is a long and ongoing debate about the nature of beliefs or attitudes – two of the more central concepts of social psychology. Some see them as (again) stable characteristics of individuals, developed from direct, unmediated, subject-object relations; others see them as emerging from culture and responsive to context and interaction (see Howarth, 2006).

I am not sure that it is clear in the text to which tradition the authors subscribe regarding Component 2: are they assuming that attitudes about argumentation are stable traits of Subjects, and that what matters for understanding them happens in direct Subject-Object relations? Or are they, instead, assuming attitudes and beliefs to emerge from triadic Subject-Other-Object relations, being contingent upon culture and the ongoing interaction?

Answering this would also help clarify the results of this comparative study. One example:

> US respondents were more verbally aggressive (antisocial) and more argument-avoidant, whereas the Portuguese were readier to engage the other person's arguments. Portuguese were more interested in arguing in order to display identity or to play, but less oriented to dominance displays. (p.150, this volume)

If the authors assume that these (and similar) results reflect motivations, personality traits and beliefs as stable dispositions of individuals (not contingencies of culture, context and interaction), i.e., if it is assumed that these concepts and results reflect essentialist aspects, several consequences follow. One is that the authors need to clarify what *theory of the subject* they are using. Do they indeed work with a theory of the subject potent enough for explaining how and why US and PT respondents are essentially different regarding argumentation? Is there in fact such a theory, able to account for how essential characteristics of the subject can be linked to nationality?

Or might it be the case that – and this is a crucial question – the authors are in fact using subject and essentializing concepts to tackle what are Subject-Other, i.e., interactional, relational and cultural (i.e. non-essential) aspects? I would argue that this is the case. I would further argue that the authors seem to be trying to describe regularities of action (how people argue/say they argue); however - and because they resort to essentializing concepts -, they end up presenting them as regularities of being (how people are) and taking nationality to be an explanatory principle for the production of those regularities.

4. SUGGESTIONS

I now reach the moment of offering some suggestions. The first would be to explore the potentialities of a re-conceptualization. I argue that Component 1 (and the scales that measure it) can be re-conceptualized *in a non-essentialist* way. Component 1 can be seen as tackling and describing *what people* **do** *when arguing*, instead of what people **are**.

Some respondents report more frequently using avoidance or approach strategies. Does this make them necessarily, in essence, avoidants or approachers? In my view this is an unwarranted leap from a description to (what is taken to be) an explanation. The scales used in the paper can thus be seen in a (alternative) descriptive way as a measure of the extent to which respondents are **doing** avoidance, or using avoidance strategies, instead of a measure of how much they **are** avoidant by nature, trait or motivation. This would have several consequences: it would open up new questions about how culture, context and topic stimulate avoidance strategies, and also suggest new ways for re-analyzing the present data.

One such way would be to focus more on internal **relations** among the sub-scales within each component and between components, and less on simple one-to-one descriptions and comparisons. This will help to develop thicker descriptions of the processes at work, and is anyway a good option for non-representative samples such as the present one. It can mean, for example, exploring inter-correlations amongst the sub-scales of Components 1 and 2. Or it can even mean going a step further and using regression analyses to reverse the direction of what an intra-subject reading would suggest: instead of using sub-scales of Component 1 (traits) to predict Component 2 (beliefs about argumentation), I suggest testing the extent to which Component 2 helps predict (re-conceptualized) Component 1 (now seen as what people do when arguing, or how they argue).

5. CONCLUSION

In sum, my argument has been that using a non-essentialist position for studying argumentation has advantages. From this position, we are not prevented from trying to find regularities and patterns within and across cultures, and achieving some level of generalization. But, by avoiding the assumption that meaning and action emerge from a Subject furnished of stable traits and alone facing an Object, we more easily avoid the fallacy of taking as explanatory what is descriptive, and remain vigilantly suspicious of the over-generalizations that essentializing concepts invite. This makes research more resistant to succumbing prematurely to the nomothetic temptation. And by remaining un-convinced that we already have general explanations (the Portuguese are... Americans are), we can keep trying to understand how culture, context and interaction work together to affect individual argumentation.

To finalize, I want to express again my pleasure in doing this commentary and my hope to again be invited as a Portuguese and a psychologist, despite my difficulties in viewing these identities as uni-dimensional.

REFERENCES

Bruner, J. (1990). *Acts of meaning*. Cambridge, MA: Harvard University Press.
Castro, P. (2003). Dialogues in social psychology, or how new are new ideas. In J. Lazlo & W. Wagner (Eds.), *Theories and controversies in societal psychology* (pp. 32-54). Budapest: New Mandate.
Farr, R. (1996). *The roots of modern social psychology: 1872–1954*. Malden: Blackwell Publishing.
Hample, D., & Anagondahalli, D. (2015). Understandings of arguing in India and the United States: Argument frames, personalization of conflict, argumentativeness, and verbal aggressiveness. *Journal of Intercultural Communication Research, 44*, 1-26.
Howarth, C. (2006). How social representations of attitudes have informed attitude theories: the consensual and the reified. *Theory and Psychology, 16*, 691-714.
Israel, J. & Tajfel, H. (Eds.) (1972). *The context of social psychology*. London: Academic Press.
Moscovici, S. (1972). Society and theory in social psychology. In J. Israel & H. Tajfel (Eds.). *The context of social psychology*. London: Academic Press.
Moscovici, S. & Markova, I. (2006). *The making of modern social psychology: the hidden story of how an international social science was created*. Cambridge, UK: Polity Press.
Wallerstein, I., Juma, C., Keller, E.F., Kocka, J., Lecourt, D., Mudimbe, V, Mushakoji, K., Prigogine, I., Taylor, P., & Trouillot, M-R. (1996). *Para abrir as ciências sociais? Relatório da Comissão Gulbenkian sobre a reestruturação das ciências sociais*. Lisboa: Publicações Europa-América.

8

"All Things Considered"

DAVID HITCHCOCK
Department of Philosophy, McMaster University, Hamilton, Canada
hitchckd@mcmaster.ca

Diverse considerations may be relevant to deciding what to do, and people may disagree about their importance or even their relevance. Reasonable ways of taking such diversity into account include comprehensive listing of considerations, assessment of the acceptability and relevance of each consideration, reframing, adjusting the option space, debiasing, estimations of importance, and allocating the burden of proof.

KEYWORDS: Benjamin Franklin, considerations, debiasing, decision-making, framing, Harald Wohlrapp, pros and cons, Stephen Thomas, Trudy Govier, weighing

1. INTRODUCTION

Practical reasoning in the most general sense is reasoning about what policy to adopt. A policy decision is a decision to do or permit or require or forbid a certain kind of action or complex of actions in a certain kind of situation. For example, it is my policy to arrive a couple of days early for conferences in Europe, in order to get over the jet lag. A plan can be regarded as a policy limited to one occasion. It can be syntactically complex, with nested Boolean operators, and can be more or less completely specified. For example, two conference attendees might plan to have dinner together at an Indian restaurant on a particular evening if they can find one with a good reputation but otherwise to join any group of conference goers who are going to a restaurant together. An action is a limiting case of a plan, an immediately implemented plan with just one component, such as presenting a paper at a conference. In what follows, I use the phrase 'deciding what to do' in a broad sense that includes adopting policies and plans as well as forming an intention to perform a particular action.

In deciding what to do, in any of these senses, diverse considerations are often relevant. These considerations may point in different directions. Further, if the decision is a group decision, members of the group may differ from one another on which considerations are relevant, as well as on the absolute or relative importance of a given consideration. In this paper, I suggest reasonable ways of taking such diversity into account.

2. KINDS OF CONSIDERATION

Considerations may be of various kinds (McBurney, Hitchcock & Parsons, 2007).

A consideration may be a definite or probable or possible contribution to or frustration of a goal. For example, the declarer in a game of contract bridge has the goal of making the contract, and in most suit contracts preventing the opponents from trumping in contributes to that goal; thus in most suit contracts preventing the opponents from trumping in is a relevant consideration in deciding how to play the hand. In counter-insurgency warfare, it might be a goal to secure the loyalty of the general population that is not part of the insurgency, a goal that would be frustrated by indiscriminate attacks on populated areas held by the insurgents; hence avoiding civilian casualties in populated areas held by the insurgents is a relevant consideration in counter-insurgency warfare.

A consideration may also be a definite or probable or possible beneficial or adverse consequence. For example, a possible beneficial consequence of moving to a permissive legal framework on abortion is a reduction in crime rates starting 15 to 20 years after abortions become legal and available, because of the relatively high crime rate among men born to women who would have aborted their pregnancy if they had been able. A possible adverse consequence of legalizing assisted suicide or voluntary active euthanasia is social and moral pressure on aged and infirm people to make a "responsible" decision to ease the burden on themselves, their loved ones and society by deciding to end their life.

A consideration may be a prescription or prohibition, whether absolute or defeasible, by an authoritative norm. For example, the electricity safety rules of the University of St. Andrews in Scotland state: "There is an absolute duty to ensure that no electrical equipment is put into use where its strength and capability can be exceeded in such a way as may give rise to danger." (*https://www.st-andrews.ac.uk/staff/policy/healthandsafety/publications/electricalsafety/*; accessed 2015 04 27) The charter of the United Nations forbids its member states to initiate a war without authorization by the Security Council. This prohibition is defeasible, with the only permitted exception being self-defence against an armed attack (Article 51). An example of a defeasible prescription is the requirement in Quebec's Charter of Rights and Freedoms to come to the aid of anyone whose life is in peril. The law specifies the defeaters of this obligation: "unless it involves danger to himself or a third person, or he has another valid reason" (C-12, I.I.2, online at http://www2.publicationsduquebec.gouv.qc.ca/dynamicSearch/telecharge.php?type=2&file=/C_12/C12_A.htm; accessed 2015 04 27). Prescriptions and prohibitions can be treated as constraints on decision-making, but it is not irrational to override them for reasons other than those officially recognized. Almost all of us break the law on occasion, for reasons that the law does not recognize as legitimate; for example, we exceed the speed limit when driving our car or ride our bicycle on a pedestrian crosswalk. Moral constraints are even more elastic, given the lack of codification of morality and the lack of any formal adjudication of charges of immoral behaviour.

Finally, a consideration may be possession by a contemplated policy of an intrinsically desirable or undesirable feature. For example, vigorous physical exercise adopted as a means to being healthy and fit might feel good, because of the dopamine flooding into one's brain. Or a medication taken orally may have an unpleasant taste.

3. POLICY QUESTION AND OPTIONS

Surveying relevant considerations of all these types makes sense only against a background of a policy question and a range of options. For example, in thinking about what to do about human-caused climate change, there are two over-arching policy questions. First, how should humanity minimize future disruption of the Earth's climate by human activity? Second, how should humanity prepare for the anticipated effects of present and foreseeable climate disruption? It is important in formulating such questions to do so in a way that does not foreclose options by building part of the answer into the question. If for example the question of how to minimize future disruption of the Earth's climate by human activity were formulated as the question how to get to a global carbon-neutral economy, that would foreclose consideration of geo-engineering options that are compatible with a carbon-positive economy or of the more ambitious goal of a carbon-negative economy. Whatever care is taken in the initial formulation of a policy question, decision-makers need to be prepared to revise that formulation in the light of evidence-gathering, reasoning and discussion. Further, if the decision is to be made by a group, there will need to be resolution of any disagreements about how the policy question is to be formulated.

The options under consideration should be mutually exclusive, or confusion will result. In considering policies for reducing emissions of carbon dioxide from burning fossil fuels, for example, it would be a mistake to take the options to be regulation or carbon pricing, since there are options that combine both, such as cap-and-trade systems. The options need not be exhaustive, either logically or practically, but they should include the possibilities that seem initially most attractive. If the options are to be surveyed mentally rather than through an externalized process, they should be limited to at most three or four, which is as much as a purely mental consideration can manage. Constraints can be used to limit the initial option space. But flexibility may be needed as deliberation proceeds, if for example each of the options under consideration has serious drawbacks or is not feasible. This flexibility may take various forms: changing constraints, adding options, removing options, making options more specific, recasting options according to a different principle of division. And, as with the formulation of the policy question, in group decision-making there will need to be resolution of any disagreements about the option space.

4. LISTING CONSIDERATIONS

If the decision is important enough for thorough reflection, it makes sense to list all the considerations that anybody in the decision-making group thinks relevant. This was part of the approach of Benjamin Franklin, whose letter to the chemist Joseph Priestley of 19 September, 1772 is the earliest extant document describing an approach to taking diverse considerations into account:

> When these difficult Cases occur, they are difficult chiefly because while we have them under Consideration all the Reasons pro and con are not present to the Mind at the same time; but sometimes one Set present themselves, and at other times another, the first being out of Sight. Hence the various Purposes or Inclinations that alternately prevail, and the Uncertainty that perplexes us.
> To get over this, my Way is, to divide half a Sheet of Paper by a Line into two Columns, writing over the one Pro, and over the other Con. Then during three or four Days Consideration I put down under the different Heads short Hints of the different Motives that at different Times occur to me for or against the Measure. (Franklin, 1990/1772)

Franklin goes on to describe his method of estimating the respective "weights" of the listed considerations, a method which I will discuss later. For now, it should be noted that Franklin's method of listing the pros and cons is over-simplified in at least three respects.

First, it assumes a simple decision of whether or not to adopt a specific "measure". More typically, there are a number of positively specifiable options. For example, if a government wishes to reduce greenhouse gas emissions in its jurisdiction, it can do so by direct regulation alone, by merely taxing fuels that emit greenhouse gases, or by solely introducing a cap-and-trade system that gives or sells to emitters tradable permits to emit specified quantities. One cannot reasonably list the pros and cons of more than two options by drawing a line down the middle of a page and putting the pros on one side and the cons on another. In fact, once cannot reasonably do so even when there are just two options. In his treatment of decision-making in the textbook *Practical Reasoning in Natural Language*, Stephen Thomas repeatedly warns the student against assuming that a reason against one option is automatically a reason for its rivals; as an example, he notes that being on a diet that bans desserts made with sugar is a reason against choosing lemon pie but not thereby a reason for choosing chocolate cake instead (Thomas, 1997, p. 390). Similarly, a reason for one option is not automatically a reason against another; for example, a couple deciding whether to go out to see a movie or stay home and watch a documentary might count enjoyment as a reason for seeing the movie but note that watching the documentary would also be enjoyable. In general, then, for each identified option the pros and cons should be listed independently of the listings for the other options; thought is required to see if a pro or con for one option is respectively a con or pro for another.

Second, Franklin envisages listing of reasons pro and con by only one individual. If a group is making a decision, each member of the group should have the power to propose any reason that seems to them to be relevant, pro or con, to any option. One of the strengths of group decision-making, in fact, is the ability of a group to marshal more considerations than would have occurred to any one member of the group.

Third, Franklin omits any reasoning that may be involved in establishing the reality of a proposed consideration. In deciding how much saturated fat to include in one's diet, a relevant consideration might be the contribution of eating saturated fats to the level of low-density lipoprotein (LDL, or "bad cholesterol") in one's blood, which is a contributor to heart attacks and strokes. It is not self-evident that eating more saturated fats raises the level of LDL in one's blood, and in fact a recent systematic review of studies of the relation between dietary fats and heart disease found a more complicated picture:

> All lines of evidence indicate that specific dietary fatty acids play important roles in the cause and the prevention of CHD [coronary heart disease–DH], but total fat as a per cent of energy is unimportant. Trans fatty acids from partially hydrogenated vegetable oils have clear adverse effects and should be eliminated. Modest reductions in CHD rates by further decreases in saturated fat are possible if saturated fat is replaced by a combination of poly- and mono-unsaturated fat, and the benefits of polyunsaturated fat appear strongest. However, little or no benefit is likely if saturated fat is replaced by carbohydrate, but this will in part depend on the form of carbohydrate. Because both N-6 and N-3 polyunsaturated fatty acids are essential and reduce risk of heart disease, the ratio of N-6 to N-3 is not useful and can be misleading. In practice, reducing red meat and dairy products in a food supply and increasing intakes of nuts, fish, soy products and nonhydrogenated vegetable oils will improve the mix of fatty acids and have a markedly beneficial effect on rates of CHD. (Willett, 2012, p. 13)

A rather complicated line of reasoning is required to get from this authoritative review to a set of dietary recommendations in a country's food guide or to a personal policy on what to eat. In general, reasoning or appeal to authority will lie behind any proposed consideration, and should be made explicit in a thorough consideration of what to do. Argument mapping tools are a useful means of doing so.

5. EVALUATING CONSIDERATIONS

Franklin's method omits not only the reasoning supporting the reality of a listed consideration but also critical assessment of each consideration, whether or not its reality is supported by argument. Something that seems to be a "motive" for or against a "measure" might be merely apparent or irrelevant. Hence there needs to be a twofold evaluation of each identified consideration prior to any attempt to take it into account.

In the first place, it must be asked whether the consideration really obtains, and if so to what extent or in what form. How much would we really enjoy the documentary, in comparison to the enjoyment we would get from going out to a movie? What reduction in the risk of heart disease can be expected from replacing butter in my diet with polyunsaturated fats, red meat with fish, and whole milk with skim milk? If the reality of a listed consideration is supported by a line of reasoning, the reasoning needs to be assessed for its adequacy, with respect to both its ultimate assumptions and its inferential links, and with attention to alternative positions and the supporting evidence for them. The results of such assessments might typically be framed in terms of a probability distribution among possible values of a variable of interest. Perhaps the documentary would most likely be only moderately enjoyable, with a remote chance of being highly enjoyable and a bigger chance of being so boring that we will stop watching it midway through.

In the second place, each proposed consideration should be assessed for its relevance. Does the factor in the situation as we have assessed it really count for the option, as we have assumed? Perhaps it actually counts against it, or is irrelevant. Similarly for factors that have been listed as counting against an option. Harald Wohlrapp recommends that for each pro one consider how an opponent of the option might reply, and similarly for each con how a proponent of the option might reply. This procedure is part of what he calls "completing the discussion", the other parts being the questioning of each argument and counter-argument and seeking out the frames in which the decision-makers see the issue and state of affairs (Wohlrapp, 2014, p. 261). Supplying a counter-argument may result in changing a pro to a con, or vice versa. Wohlrapp gives some striking examples of such shifts in relevance status, in his discussion of a list in a textbook example (Govier, 1997, p. 393) of reasons for and against legalizing voluntary active euthanasia for people with a terminal illness. One reason listed as supporting such legalization is its sparing family members the agony of watching a loved one die a horrible and unworthy death. An opponent of legalization could reply that awareness of such suffering can deepen one's appreciation of the fragility of life, in a way that Bertrand Russell reports as his personal experience (Wohlrapp, 2014, p. 257). Sparing people the agony of watching others suffer might encourage a superficial depersonalized hedonism of the sort portrayed in Aldous Huxley's *Brave New World*. Another reason listed as supporting legalization of euthanasia is that responsible adults should be allowed to choose whether to live or die, a consideration whose postulated positive relevance depends on assigning a positive value to individual self-determination. An opponent could claim that legalizing active euthanasia would put social and moral pressure on people with a terminal illness to do the "responsible" thing and have their life ended in order to spare their family and friends the distress of seeing them fade away and society the cost of providing for their care. According to this response, the increase in self-determination for the few terminally ill people whose excruciating suffering can only be relieved if someone else kills them is more than balanced by the decrease in self-determination for terminally ill people whose situation is not so dire (Wohlrapp, 2014, p. 264). Thus the principle that people should be allowed to choose whether to live or die might be a reason against legalizing voluntary active euthanasia rather than for doing so.

These assessments of each proposed consideration include identification of possible objections to the judgments of its factual correctness and of its relevance, of possible replies to those objections, and so on potentially *ad infinitum*, a process that Thomas in particular emphasizes. For example, a claim that replacing red meat with fish in one's diet will reduce a person's risk of heart disease might conflict with a finding that people who eat more red meat and less fish are no more likely to develop heart disease than people who eat less red meat and more fish. A defender of the claim might reply by explaining away the anomalous finding.

Such thoroughness makes sense for important decisions where the decision-makers have the time and other resources to go through the process, but not for minor decisions.

6. REFRAMING

It may become appropriate to reframe the issue under discussion, for example as a way of introducing commensurability into what was previously a stand-off of incommensurable perspectives. Fed Kauffeld (2011) finds such a reframing in the debates in 1787 and 1788 over the ratification of the United States constitution. Anti-federalists objected that each of the threats to liberty in the proposed constitution was an overriding consideration against it. Federalists responded that the constitution's merits outweighed its defects. Thus there was a standoff, with the anti-federalists regarding the federalists as ignoring the overriding negative considerations that they had pointed out. *The Federalist Papers*, a series of essays written by three of the leading federalists under the pseudonym Publius, strove to reframe the issues. With respect to any power granted to the national government by the proposed constitution, Publius argued, the question was not whether the power was dangerous but whether it was necessary and if so whether adequate safeguards had been provided to protect against its abuse. Publius's reframing of the issues implied an allocation of the burden of proof that made back-and-forth discussion possible. First it was necessary for a defender of the constitution to show that a given power was necessary. Then the proponent had to show that there were adequate safeguards against the abuse of that power. Having accomplished these two tasks to "his" satisfaction, Publius declared that it was now up to the anti-federalist opponents of the constitution to justify continuation of the debate by addressing with reason and evidence the considerations as configured in the Federalist Papers.

Generalizing from this case study, we can conclude that sometimes a disagreement about the relative ranking or absolute degree of importance of one or more considerations can reasonably be finessed by recasting the issues under discussion. The epistemic success of any such reframing depends on whether it incorporates in a satisfactory way each party's perspective on the considerations in question. In the debates over the ratification of the United States constitution, for example, the reframing incorporated the federalists' focus on the merits of the proposed constitution in the arguments for having a national government and for the necessity that it have each of the powers deemed dangerous by the anti-federalists. It incorporated the anti-federalists' concern about personal liberty in the issue of whether for each dangerous power there were enough safeguards against its abuse.

Other sorts of reframing may be necessary. Harald Wohlrapp has drawn our attention in his recent book *The Concept of Argument* (Wohlrapp, 2014) to the ubiquity of frames in every person's encounter with the world and with other people. A frame as Wohlrapp understands it is a way of seeing an issue or situation. One sees B as A. For example, the anti-federalists saw a powerful national government as a potential threat to liberty. If two people frame a state of affairs differently, they will tend to talk past each other and to be incapable of appreciating the other person's perspective. This sort of mutual incomprehension is particularly common when neither party is aware that they are framing the state of affairs in a certain way. It needs to be dealt with by identifying the frames and attempting to transcend their differences.

Wohlrapp takes integration of divergent frames to be the second (and final) objective in a discussion of identified pros and cons (Wohlrapp, 2014, p. 261). He describes four ways of overcoming frame differences. *Frame criticism* directly attacks a frame as inappropriate. It implies that the critic has gone beyond the frame and can see it as a whole, so to speak from outside it. An example might be a rejection of framing homosexual relations as perverse and unnatural. *Frame hierarchization* makes competing frames explicit as aspects and puts those aspects into a hierarchy. For example, one can see a car that one is about to buy as a status symbol or as a mode of transportation. These frames might compete, for instance if nervousness about damaging the status symbol makes one reluctant to drive the car. Then one can externalize these frames as aspects of the situation and put psychological comfort in driving the newly purchased car above its status. *Frame harmonization* finds a way to reconcile two competing frames. The reframing of the issues in the debate on ratifying the United States constitution harmonized the federalists' framing of the constitution as a way to provide the national government with powers that it needed with the anti-federalists' framing of it as opening the way to dangerous abuse of those powers. *Frame synthesis* preserves two competing frames in a "higher" frame that accommodates them both in a Hegelian sublation. Wohlrapp gives an example of a frame synthesis from a speech by Saint-Just in the French parliamentary debate of 1792 over what to do with the deposed King Louis XVI. Previous speakers had argued that from a moral point of view the king deserved to be punished for his behaviour before he was deposed (e.g. trying to flee the country, conspiracy with foreign powers to have them restore him as an absolute monarch) but that from a legal point of view there was no way to try him for what he did while he was still king and therefore above the law. Thus there was a stand-off between incompatible frames. Saint-Just argued that both frames assumed falsely that the king was a member of society. On the contrary, from the perspective of those who are founding a Republic, the king is a usurper who could not have been part of an original social contract. Having attacked the Republic, he should be judged as an enemy alien and treated according to the practices of war.

7. DEBIASING

Throughout the processes described to this point, it makes sense to try to remove any distorting effect of one's initial biases, meaning by a bias "a disposition, implicit or explicit, to reach a particular kind of conclusion or outcome, or to remain in one" (Kenyon & Beaulac, 2014, p. 344). Removing distorting effects of one's biases is a difficult task, since a bias may take the form of a latent frame of which one is not even aware. Further, awareness may not be enough to prompt the appropriate correction. Bias in a decision-making situation may in this respect resemble confirmation bias, which has been shown to operate even when there is a very weak initial commitment to a hypothesis and to persist even after explicit recognition that one starts with an inclination to believe a certain explanation of some phenomenon (Nickerson, 1998). Recognition of bias is just a first step. It is however a step that needs to be taken, and it does not automatically occur. With respect to general cognitive and affective biases that we inherit from our species' evolutionary history as part of our intuitive "system 1" thinking processes, recognition can come from learning about such things as the availability heuristic and the representativeness heuristic

(Kahneman, 2011). It can also come from diagnosis of the causal pathway that led one to make a serious cognitive or affective mistake (Croskerry, Singhal & Mamede, 2013). With respect to biases specific to the particular decision-making situation, recognition is a matter of noting one's initially preferred option and one's initially privileged considerations.

Once recognized, biases need to be taken into account appropriately. Croskerry, Singhal & Mamede (2013) distinguish three types of strategies for overcoming bias: educational strategies designed to enhance future ability to debias, workplace strategies designed to be implemented at the time of dealing with a problem, and forcing functions designed to nudge a decision-maker towards a better outcome. They caution that these strategies are not mutually exclusive but lie on a spectrum, and that there is uneven evidence for their effectiveness. Some of the strategies they mention are specific to the context of medical diagnosis about which they are writing, but others are more generally applicable. Among the educational strategies, they cite evidence for limited effectiveness of a "consider-the-opposite" procedure and of teaching rules of statistical inference. Among the workplace strategies, they mention identifying more aspects of a problem, meta-cognitive reflection on the thought processes that have led one to a certain conclusion, group decision-making, being required to justify one's decision and be accountable for it, avoiding cognitive overload and fatigue, making an initial judgment on one's own before attending to what others have concluded, and using decision support systems. Two of their forcing functions that have general application are seeking evidence that supports a decision opposite to one's initial impression and considering whether there was a control group in a study of a suspected causal relationship.

Kenyon and Beaulac (2014) cite experimental evidence that teaching and warning people about kinds of bias and situations where they arise is ineffective in mitigating biased reasoning. One can be aware of a certain kind of bias but fail to recognize that one is oneself exhibiting that bias. In fact, monitoring oneself for bias in the process of thinking about some problem has been found to make matters worse, in that people think falsely that they have eliminated any possible bias by paying attention to the possibility that it is operative. The strategy that has been found in experimental studies to be most effective, and most generally effective, is to consider explicitly a range of alternative perspectives or counterfactual outcomes, and what would have had to happen in order for those outcomes to occur. Kenyon and Beaulac produce a useful taxonomy of levels of debiasing, ranging from the most individual and least effective to the most contextual and most effective:

> *Level 1 debiasing* (prior elimination of bias): General education, environment, habituation and training lead an individual to have no disposition to produce a particular sort of biased judgment. The bias does not arise.
> *Level 2 debiasing* (self-generated correction of bias): In a judgment situation, an individual uses previously learned behavioural or cognitive strategies to revise an occurrent or incipient biased judgment.
> *Level 3 debiasing* (environment-generated correction of bias): In a judgment situation, situational nudges prevent bias that would otherwise have affected the judgment.
> *Level 4 debiasing* (environment-generated overriding of bias): Biased judgment occurs, but situational constraints prevent the action or outcome from being biased. (Kenyon & Beaulac, 2014, pp. 350-352, paraphrased)

One example of level 4 debiasing is anonymous grading of students' work, which automatically removes any effect of the grader's preconceptions of the expected quality of work from the students that they have come to know personally–an effect that has been repeatedly demonstrated in educational research. Another example is having candidates for an orchestra position perform behind a screen, a procedure that has been shown to reduce dramatically a widespread bias against women in such hiring decisions. Level 4 debiasing strategies in decision-making contexts will be specific to the kind of context and kind of decision involved. They will have in common that a type of bias irrelevant to making the decision is identified in advance and the decision-making context is structured so that the bias can have no influence on the ultimate decision.

8. WEIGHING

At some stage in the decision-making process described so far, there will ideally emerge an agreed policy question with an agreed list of options, for each of which there is an agreed set of positively relevant considerations and an agreed set of negatively relevant considerations–the agreements resulting in each case from comprehensive reasoned, dynamic discussion. There then arises the issue of how to take this diversity into account in a reasonable way, an issue that becomes even more pressing if there is residual disagreement about either the policy question or the option space or the considerations relevant to one or more options. Wohlrapp insists that a listing of pros and cons reflects an interim state of the discussion. If the members of the decision-making group complete the discussion and integrate the frames as he recommends, he claims, either a single option will emerge as a "valid" decision or the parties will become aware of the basis for their residual disagreement (Wohlrapp, 2014, p. 263). If the parties resort prematurely to some sort of "weighing" of the pros and cons, he maintains, then they have shifted from seeking an argumentatively valid decision to seeking a balancing of interests (Wohlrapp, 2014, p. 261). Whether or not his claim is correct, it seems utopian to expect individual or group decision-making to reach a stage in all cases where there are no longer competing pros and cons for each of two or more options. If such competing pros and cons remain, something analogous to weighing seems inevitable.

Franklin describes his method of weighing pros and cons as follows:

> When I have thus got them all together in one View, I endeavour to estimate their respective Weights; and where I find two, one on each side, that seem equal, I strike them both out: If I find a Reason pro equal to some two Reasons con, I strike out the three. If I judge some two Reasons con equal to some three Reasons pro, I strike out the five; and thus proceeding I find at length where the Ballance lies; and if after a Day or two of farther Consideration nothing new that is of Importance occurs on either side, I come to a Determination accordingly.
> And tho' the Weight of Reasons cannot be taken with the Precision of Algebraic Quantities, yet when each is thus considered separately and comparatively, and the whole lies before me, I think I can judge better, and am less likely to take a rash Step; and in fact I have found great Advantage from this kind of Equation, in what may be called Moral or Prudential Algebra. (Franklin, 1990/1772)

Franklin's "moral or prudential algebra" leaves it quite unclear how one is to assign weights to the reasons pro and con some option. Trudy Govier has proposed a method for doing so in the fourth (Govier, 1997) and subsequent editions of her textbook. The weight or strength of a consideration, she claims, is inversely proportional to the number of exceptions to the principle in virtue of which the consideration counts for or against an option. Thus she takes it to be a weak consideration that legalizing voluntary active euthanasia for terminally ill patients would cut social costs, since there are many exceptions to the principle that a practice that cuts social costs should be legalized. It is a stronger consideration that legalizing voluntary active euthanasia would save many patients from great pain, she claims, since there are comparatively few exceptions to the principle that any practice that would save people from great pain should be legalized. Wohlrapp has objected (Wohlrapp, 2014, pp. 256-258) that it is possible to give only an intuitive estimate rather than an exact count of the number of exceptions to a general principle and that it would show a lack of understanding to calculate arithmetically the scores of the pros and the scores of the cons on the basis of a principle's estimated number of exceptions. Others expressed similar scepticism at a 2010 symposium on so-called "conductive arguments" (Blair & Johnson, 2011); in her subsequent "overview of the symposium" (Govier, 2011), however, Govier did not respond to their doubts.

Even if one could somehow develop a way of measuring or estimating the weight of a consideration, it would seem ludicrous to baptize (say, as a "graviton") a unit of weight for considerations, to assign so many such units with a plus or minus sign to each consideration, to calculate the total weight of the pros and cons for each option, and to choose the option with the highest resulting score. If in the lead-up to such a numerical calculation, members of a decision-making group disagree about the weight to be assigned some consideration, discussion will ensue about the reasons for giving it more or less weight than some other competing consideration. This discussion is the real core of the balancing of competing considerations; assignment of weights on some invented cardinal scale is a confusing epiphenomenon.

Fred Kauffeld has usefully classified and listed (Kauffeld, 2011, p. 160) the descriptive adjectives that people use to indicate the importance, relative or absolute, of a consideration. In relation to other considerations, a consideration can be on the one hand decisive, overriding or paramount, or on the other hand trivial or insignificant. With respect to the response it merits, a consideration can be noteworthy, weighty, sobering, tiresome, serious, compelling, powerful, persuasive, disturbing, reassuring or interesting. Such descriptors can provide guidance in the final stages of taking the relevant considerations into account. They are of course subject to challenge, and responses to such challenges are possible. Such challenges and responses can be supported by appeal to underlying values, which may themselves become subject to discussion.

In group decision-making that takes into account a variety of considerations, it can be helpful to agree on where the burden of proof lies (Bailin & Battersby, 2010). There is a presumption in favour of existing policy, whether of a family or a voluntary organization or a state. The force of the presumption is to put a burden on an advocate of change to show that the change would on balance have better results.

The presumption is weaker or even non-existent if the existing policy has obvious major disadvantages or has been adopted with a weak or absent evidential basis. It is stronger if the existing policy has been adopted as a result of careful deliberation and the relevant circumstances have not changed significantly.

9. SOUNDNESS AND COMPLETENESS

It seems difficult to conceptualize the processes described above in a way that makes possible the sort of proofs of soundness and completeness that we find in the meta-theory of various logical systems, in particular because the processes are in part dynamic and creative. However, a decision-making process that fails to take into account a major relevant consideration can reasonably be called incomplete, and one that results in a decision that flies in the face of the preponderance of relevant considerations can reasonably be called unsound. It is worth exploring how far we can get in making more precise and comprehensive such informal criteria for soundness and completeness.

10. PROSPECTS FOR QUANTIFICATION

It seems unlikely that the complexities involved in taking diverse considerations into account can be treated by a Bayesian or similar quantitative approach. Proponents of such a quantitative approach (Hahn & Hornikx, forthcoming; Selinger, 2014) could try to apply it to a complex case of practical decision-making about which the relevant facts are known, in the way that the Bayesian approach was applied in a book-length study to the Sacco-Venzetti trial (Kadane & Schum, 1996).

REFERENCES

Bailin, S., & Battersby, M. (2010). *Reason in the balance: An inquiry approach to critical thinking.* McGraw-Hill Ryerson.

Blair, J. A., & Johnson, R. H. (Eds.). (2011). *Conductive argument: An overlooked type of defeasible reasoning.* London: College Publications.

Croskerry, P., Singhal, G., & Mamede, S. (2013). Cognitive debiasing 2: Impediments to and strategies for change. *BMJ quality & safety, 22*(Suppl 2): ii65–ii72.

Franklin, B. (1990/1772). Letter to Joseph Priestley, 19 September 1772. In B. Franklin, *London, 1757-1775* (pp. 248-249). Raleigh, NC: Generic NL Freebook Publisher.

Govier, T. (1999). Reasoning with pros and cons: conductive arguments revisited. In T. Govier, *The philosophy of argument* (pp. 155-180). Newport News, VA: Vale Press.

Govier, T. (1997). *A practical study of argument*, 4th edition. Belmont, CA: Wadsworth.

Govier, T. (2011). Conductive arguments: overview of the symposium. In Blair and Johnson (pp. 262-276).

Hahn, U., & Hornikx, J. (2015). A normative framework for argument quality: Argumentation schemes with a Bayesian foundation. Synthese. Advance online publication. doi:1007/s11229-015-0815-0.

Kadane, J. B., & Schum, D. A. (1996). *A probabilistic analysis of the Sacco and Vanzetti evidence*. New York: John Wiley & Sons.

Kahneman, D. (2011). *Thinking, fast and slow*. New York: Farrar, Straus & Giroux.

Kauffeld, F. (2011). Ranking considerations and aligning probative obligations. In Blair and Johnson (pp. 158-166).

Kenyon, T., & Beaulac, G. (2014). Critical thinking education and debiasing. *Informal Logic, 34*(4), 341-363.

McBurney, P., Hitchcock, D., & Parsons, S. (2007). The eightfold way of deliberation dialogue. *International Journal of Intelligent Systems, 22*(1), 95-132.

Nickerson, R. S. (1998). Confirmation bias: A ubiquitous phenomenon in many guises. *Review of General Psychology, 2*(2), 175-220.

Selinger, M. (2014). Towards formal representation and evaluation of arguments. *Argumentation, 28*(3), 379-393.

Thomas, S. N. (1997). *Practical reasoning in natural language*, 4th edition. Englewood Cliffs, NJ: Prentice-Hall.

Willett, W. C. (2012). Dietary fats and coronary heart disease. *Journal of Internal Medicine, 272*(1), 13-24.

Wohlrapp, H. R. (2014). *The concept of argument: A philosophical foundation*. Dordrecht: Springer.

Commentary on Hitchcock's "All Things Considered"

ERICH H. RAST
ArgLab, Universidade Nova de Lisboa, Portugal
erich@snafu.de

1. INTRODUCTION

In "All things considered", David Hitchcock addresses some of most important and persistent problems of decision-making and particularly of *group decision-making* on which he focuses. How can people with varying backgrounds and with possibly different value systems come to a rational agreement on policy issues? How to deal with conflicting opinions in group decisions? As I believe correctly, Hitchcock emphasizes the need for distinguishing factual matters from more value related matters, and the importance of debiasing and frame harmonization techniques in order to resolve conflict prior to a weighing of options. I mostly agree with the way Hitchcock lays out the decision-making process and would therefore like to focus on some issues that are not explicitly mentioned but always lurk in the background of a discussion of policy making.

2. ON THE IMPORTANCE OF ARGUMENTATION IN GROUP DECISION-MAKING

To start with the most important issue: Why is it so important to develop an argumentative approach of group decision-making that eases conflict resolution using reframing and debiasing techniques? After all, it is very common to introduce a fixed scoring system and then combine the scores in some way, for example by averaging. What is wrong with that?

There are two answers to this question, an obvious and a less obvious one. The obvious answer is this: As long as it is possible to reduce conflicts of opinion among policy makers in a rational way and thereby come to an agreed upon policy recommendation, this seems more desirable than an automated method, for a decision based on less conflict between decision makers seems to better reflect the decision makers' opinions. A less obvious answer is given by Arrow's Theorem and a number of related impossibility theorems that have been explored in Social Choice.[1] From Arrow's theorem it follows, roughly speaking, that if three or more complete preference relations of three or more decision makers are to be combined into one preference relation over various options, they are allowed to have any kind of preferences, options are not pre-ordered in some fixed way, any preference between two options only depends on these two options and not on other options under consideration, and the choice process is unanimous (if all experts agree that A is better than B, then A is better than B in the outcome), then the aggregation method will be dictatorial: There is one decision maker whose preferences determine the outcome.

There are various ways to get around this theorem, for example by restricting the preferences to single-peaked ones, by making use of additional cardinal utility information or by using a rank-based aggregation method like in Kemeny (1959). However, at least in a purely qualitative setting there are still many highly restricting variants of Arrow's Theorem that need to be taken into account. I do not want to dwell upon these technical issues further and instead give the example of a Condorcet case[2] that illustrates the gist of the problem. Suppose there are three experts in a panel who have to evaluate three options A, B, C. (As Hitchcock rightly remarks, agreeing on such options is already a major undertaking, but let us grant that much for the sake of the argument.) Among the 3! = 6 possible strict linear orderings they might come up with are A>B>C, B>C>A and C>A>B. When combined naively, for example using a simple majority principle, the resulting preferences become circular, a phenomenon known as collective irrationality in Social Choice. Working around this, a simple additive utility model with equal weight for the voice of every decision maker would result in the recommendation A~B~C, but it is obvious that this outcome does not resolve any conflict of opinion and, depending on the circumstances, might overall not be desirable. Only through deliberation can such obstacles be overcome by shaping the preferences in a way that allows a fair determination of one or more winning options.

This point has recently been emphasized by List (2001, 2002) and by Fishkin et al. (2012) who show on the basis of preference polls before and after deliberation has taken place ("deliberative polls") that deliberation often shapes the preferences of a group in a positive way,

---

[1] See Arrow (1951). The impact of this theorem on qualitative decision-making is studied in Mas-Clolell & Sonnenschein (1972), Weymark (1984), Dubois (2002), Pini (2009) among others.

[2] See Condorcet (1785).

by making them closer to being single-peaked, which allows for a fair decision without falling prey to Arrow's theorem.

3. OBSERVATIONS AND CRITICAL REMARKS

Putting these important but slightly technical issues aside, I wish to make a few additional remarks about some points mentioned by Hitchcock that in my opinion deserve more attention.

First, Hitchcock lays out that considerations may contribute to or frustrate a goal and sometimes also prescribe or prohibit a certain goal. However, he believes that such constraints should generally be soft, defeasible or, as he calls them, "elastic". I believe that this is an extremely important point and wish to add that the elasticity of constraints, of moral constraints in particular, ought to be in a reciprocal relationship to the stakes. The higher the stakes, the lower the admissible elasticity, and the lower the stakes, the higher the admissible elasticity of a moral constraint. This connection does not seem to have been explored very thoroughly so far and it would be interesting to see it worked out in detail.

Another, more critical remark concerns the way in which Hitchcock talks about eliciting the relevance of a consideration in contrast to a weighing of the pros and cons. He seems to suggest that these two processes can be separated, one occurring at an early stage and the other at the last stage of policy making. If this is his view, then it seems implausible to me. It is hard for me to see what an "absolute degree of importance" might be, for example. The question of whether a certain plan has desirable or undesirable consequences is always judged on the basis of mixed considerations that involve subjective assessments of the importance of the respective aspect of the consequences and an evaluation of the goodness or badness of the consequences. In my opinion, "relevance" or "importance" are just other aspects of value-based reasoning and there is no direct path to them.

A second critical remark concerns Wohlrapp (2014) more than Hitchcock's contribution to this volume. As Hitchcock paraphrases, "[i]f the parties resort prematurely to some sort of 'weighing' of the pros and cons, then they have shifted from seeking an argumentatively valid decision to seeking a balancing of interests." I believe this to be a misleading, if not dangerous way of putting things. In reality, all arguments except strictly deductive ones come with a certain strength only, and there is no agreed upon methodology for assigning that strength in a group decision context, except, perhaps, for purely factual matters when all decision makers readily agree on some authority. In other words, it seems to me that assessing argumentative validity in a real-world policy making context is mostly weighing the pros and cons. Of course, there are fallacies that may rule out certain arguments, but identifying these does not suffice for determining the strength of the remaining arguments. Vice versa, you rationally seek a balancing of interests by arguing for or against certain considerations. In a rational approach, this is so even when a bargaining solution is available, because the attractiveness of a bargaining trade-off needs to be justified.[3] So overall I am not convinced that these two issues can be separated from each other, as the above passage seems to suggest.

Finally, I would like to make a more general remark about the role of fairness which may be taken as one more reason not to confound individual practical reasoning with group decision-making. (I am not claiming that Hitchcock mixes up the two processes in any way, so this is just another addition.) As I submit, considerations of fairness do not play a major role in individual decision-making, whereas they are substantial and not neglectable in group decision-making. To strengthen this thesis let me sketch two arguments why fairness is an important ingredient of group decisions that do not readily apply to individual decision-making. The first one is based on the claim that fairness has epistemic benefits in the long run. At first sight, this does not seem to be the case, though. Consider a factual component in a group decision such as the question of how much a new airport will cost.[4] If there were good reasons to believe that the contributions of one of the decision makers were not truth-conducive, for example if an expert turned out to lack the necessary credentials at a closer inspection, then that decision maker ought not have been part of the evaluation panel in the first place. At the other side of the scale is the case when one decision maker already knows the truth; then the others seem to be unnecessary. Or so, one might think. However, these cases are unrealistic because in reality it must, under normal circumstances, be assumed that all of the members of the group are competent and make errors with roughly the same probability, and that the group decision method should be epistemically reliable in repeated applications. If that is so, then it will be possible to justify fairness considerations in group decision-making by establishing

---

[3] Part of this justification is, of course, the fact that bargaining balances the wins and losses in the most optimal way, but this is still a rational consideration.

[4] A similar case can be made for value judgments, but it is more complicated to argue this way, because it is less clear what the 'right' value judgment is. In any case, real-world decision-making always concerns factual and value-laden questions at the same time.

their truth-conduciveness. This virtue-epistemological argument for fairness has no counterpart in individual decision-making.

The second argument concerns the social role of fairness, and may also involve the claim that fairness has a value of its own. Especially in policy making group decisions often take place in a setting in which democratic principles need to be respected, and fairness is one of them. So it could be argued that fairness is needed as a regulatory principle for properly distributing power and responsibility in group decision-making even if it leads to suboptimal decisions when being compared to other methods, for a similar reason as a less efficient democracy is generally desirable over the efficiency of a benevolent dictatorship. Of course, you could claim that an individual reasoner should give all reasons a fair amount of thought or other such truisms, but this use of "fair" is different from fairness in a social setting. Fairness in the social sense seems to play no role in individual decision-making.

Another, more mundane reason to keep practical reasoning of an individual reasoner and policy making carefully apart is that many examples of individual practical reasoning concern decisions of minor importance such as planning a holiday trip, but one might argue that for such decisions no recommendation theory of decision-making is needed. (One might even criticize certain branches of practical reasoning for lack of normative guidance and an unhealthy focus on unimportant low stake scenarios.) In policy making, on the other hand, decisions tend to be important in terms of money and resources spent and the stakes are often high. For example, estimates of statistical deaths and injuries play a role in practically all decisions in the health care and security fields. This calls for careful deliberation and much more exhaustive evidence gathering, such as a proper risk assessment, prior to deciding on a course of action, and I believe that Hitchcock has pointed out many of the complexities of this process in his article.

ACKNOWLEDGEMENTS: I am grateful to the Portuguese Foundation for Science and Technology for the grant SFRH/BPD/84612/2012 under which this work was conducted.

REFERENCES

Arrow, K. (1951). *Social choice and individual values.* New Haven, CT: Yale University Press.
Condorcet, M. de. (1785). Essai sur l'application de l'analyse à la probabilité des décisions rendues à la pluralité des voix. Paris: L'Imprimerie Royale.

Dubois, D., Fargier, H., & Perny, P. (2002). On the limitations of ordinal approaches to decision making. In D. Fensel, F. Guinchiglia, M.A. Williams, & D. McGuinness (Eds.), *Knowledge Representation 2002 – Proceedings of the 8th International Conference (KR'02)* (pp. 133-146). San Francisco, CA: Morgan Kaufmann.

Fishkin, J., McLean, I., List, C., & Luskin, R. (2012). Deliberation, single-peakedness, and the possibility of meaningful democracy: Evidence from deliberative polls. *The Journal of Politics, 75*(1), 80-95.

Kemeny, J. (1959). Mathematics without numbers. *Daedalus, 88*(4), 577-591.

List, C. (2001). *Mission Impossible? The Problem of Democratic Aggregation in the Face of Arrow's Theorem*. Dphil thesis, University of Oxford.

List, C. (2002). Two concepts of agreement. *The Good Society, 11*(1), 72-79.

Mac-Colell, A., & Sonnenschein, H. (1972). General possibility theorems for group decisions. *The Review of Economic Studies, 39*(2), 185-192.

Pini, M. S., Rossi, F., Venable, K. B., & Walsh, T. (2009). Aggregating partially ordered preferences. *Journal of Logic and Computation, 19*(3), 475-502.

Weymark, J. A. (1984). Arrow's Theorem with social quasi-orderings. *Public Choice, 42*, 235-246.

Wohlrapp, H. R. (2014). *The concept of argument*. Dordrecht: Springer.

9

Collaborative and Adversarial Reframing: How to Use Argument Mapping to Cope with "Wicked Problems" and Intractable Conflicts

MICHAEL H. G. HOFFMANN

School of Public Policy, Georgia Institute of Technology, USA
m.hoffmann@gatech.edu

"Wicked problems" and intractable conflicts require to look at something from different perspectives that often are determined by conflicting belief-value systems. They should be approached by inclusive deliberation. But how can deliberation lead to agreement when this requires changes in belief-value systems? This paper proposes a strategy of how to address this challenge.

KEYWORDS: computer-supported argument visualization, critical thinking, diagrammatic reasoning, reflection, reflective argumentation, reframing, self-regulation, self-correction, semiotics, social interaction

1. INTRODUCTION

When at the end of the 1980s a team of 46 experts in critical thinking engaged, over two years, in a consensus-oriented process to define the notion of "critical thinking" (using the "Delphi Method," Linstone & Turoff, 1975), their final "consensus statement regarding critical thinking and the ideal critical thinker" states in its first sentence: "We understand critical thinking to be purposeful, self-regulatory judgment ..." (Facione, 1990, p. 2). Accordingly, they listed the ability of "self-regulation" as the sixth critical thinking skill after "interpretation," "analysis," "evaluation," "inference," and "explanation" in their well-known "Delphi Report." To describe more precisely what they meant by "self-regulation," they further differentiated it into "self-examination" and "self-correction" and highlighted abilities such as "self-consciously to monitor one's cognitive activities, ... correcting either one's reasoning or one's results, ... to reflect on one's own reasoning and verify both the results produced and the correct application and execution of the cognitive skills involved" (Facione, 1990, p. 10-11).

It could be argued that self-regulation is indeed the most important of all critical thinking skills because it refers to a person's ability to improve all the other skills. In spite of its relevance, however, self-regulation is not one of those skills that were assessed in any of the many iterations of the California Critical Thinking Skills Test which may be the most widely known outcome of the "Delphi Report" on critical thinking, and to my knowledge there are only a few systematic efforts to actually teach self-regulation, self-examination, and self-correction. One of them is called "Teaching through critique." It intends to apply what is known about the role of critique in visual arts education to education in general (Inman, 2015). Another one is my own effort to facilitate reflection on one's own reasoning by specific design decisions that were realized in the collaborative argument visualization software AGORA-net (see http://agora.gatech.edu/learn/goal).

I find it very surprising that there is not more research on self-examination and self-correction in argumentation theory. There are many studies on the quality of arguments, on evaluation criteria, and on how arguments can or should be used to persuade an audience and to find consensus or resolve differences of opinion. But the point is: If we want somebody to be persuaded by our arguments and if we want that people resolve differences of opinion, we should acknowledge that both processes presuppose that people are able to change their positions, and that they actually do it from time to time. This again presupposes obviously that they critically examine the arguments for their positions, that they recognize certain weaknesses or gaps in these arguments, and that they change their mind as a consequence.

The need to change one's mind is particularly pressing in two types of situations: first when we start from a position of ignorance or knowledge that is insufficient given the complexity of problems we are facing, and secondly in situations where we know exactly what we know but others don't agree with us. We experience the first type of situation in learning processes and when groups try to cope with what Horst Rittel and Melvin Webber called "wicked problems" (Rittel & Webber, 1973). And we see the second type in conflicts that are deemed to be "intractable," as they are called in conflict research, or in situations of "deep disagreement," as they are frequently discussed in argumentation theory and in philosophy.

In this contribution I will theoretically explore how collaborative argument mapping can be used to cope with wicked problems and intractable conflicts. My starting point is the thesis, first formulated by Rittel and Webber with regard to wicked problems, that both these problems should be approached by "an argumentative process in the course of which an image of the problem and of the solution emerges gradually among the participants, as a product of incessant judgment, subjected to critical argument" (Rittel & Webber, 1973, p. 162). My goal in this contribution can be described as nothing more than just understanding what this can mean.

After briefly exploring wicked problems and intractable conflicts in the following section, I will present two theoretical tools that can help to understand how collaborative and adversarial argument mapping can support the kind of self-regulation, self-examination, and self-correction that seems to be necessary to move forward in the "argumentative process" described by Rittel and Webber. The first tool is provided by studies on "framing" and "reframing" in conflict research. I understand framing here primarily as a cognitive process of sense-making that is determined by given beliefs and values. World views, ideologies, cultural and other norms, but also needs and interests limit what makes sense to someone.

The second theoretical tool that I will use is semiotics, the theory of signs and representations, and here in particular Charles S. Peirce's notion of diagrammatic reasoning. Peirce's central idea was that the external representation of our reasoning in diagrams, and experimenting with those diagrams, allows us to learn by exploring options of reasoning. I think this idea can be crucial for self-regulation. On the one hand, diagrammatic reasoning seems to be just playing with a diagram, but on the other it can open up new horizons for our own reasoning.

Combining the notions of framing and diagrammatic reasoning, the core of the approach that I am proposing here is to use argument mapping to stimulate and organize reframing. If people are enabled to change their perceptions by means of collaborative or adversarial argument mapping, coping with "wicked problems" and the solution of intractable conflicts might be possible.

Next I will show how these ideas on framing and reframing by means of diagrammatic reasoning influenced core design decisions on which the AGORA software is based. As already mentioned, AGORA's main cognitive purpose is to stimulate reflection. However, since my own experience of using the software in the classroom indicates that people have a strong tendency to stick to the first solution they come up with, I argue in the final section that argument mapping needs to be supplemented by social interaction and critique. But there are also cases in which reframing happens as a reaction to deeply troubling problems and confusion. My conclusion is, thus, that both collaborative and adversarial argument mapping might be promising approaches to stimulate the kind of reflection and self-correction that is necessary for reframing and coping with wicked problems and intractable conflicts.

2. WICKED PROBLEMS AND INTRACTABLE CONFLICTS

Rittel and Webber characterized "wicked" problems in contrast to "tame" problems by ten criteria:
1. "There is no definitive formulation of a wicked problem" (see below).
2. "Wicked problems have no stopping rules," because any "solution" can still be improved.
3. "Solutions to wicked problems are not true-or-false, but good-or-bad."
4. "There is no immediate and no ultimate test of a solution to a wicked problem."
5. "Every solution to a wicked problem is a 'one-shot operation'; because there is no opportunity to learn by trial-and-error, every attempt counts significantly."
6. "Wicked problems do not have an enumerable (or an exhaustively describable) set of potential solutions, nor is there a well-described set of permissible operations that may be incorporated into the plan."
7. "Every wicked problem is essentially unique."
8. "Every wicked problem can be considered to be a symptom of another problem."
9. "The existence of a discrepancy representing a wicked problem can be explained in numerous ways. The choice of explanation determines the nature of the problem's resolution." This means that no theory that underlies a certain problem solution can simply be falsified by counter-evidence.
10. "The planner has no right to be wrong," because planners "are liable for the consequences of the actions they generate; the effects can matter a great deal to those people that are touched by those actions." (Rittel & Webber, 1973, pp. 161-167)

The first of these characteristics might be the most important feature of "wicked problems." The reason why there is "no definitive formulation of a wicked problem" is that there is a multitude of possible ways to describe the problem or to formulate what the problem is. Every "specification of the problem is a specification of the direction in which a treatment is considered." Rittel and Webber illustrate what they conceive as the identity of problem formulation and envisioning a solution by referring to poverty as a wicked problem:

> Does poverty mean low income? Yes, in part. But what are the determinants of low income? Is it deficiency of the national and regional economies, or is it deficiencies of cognitive and occupational skills within the labor force? If the latter, the problem statement and the problem 'solution' must encompass the educational processes. But, then, where within the educational system does the real problem lie? What then might it mean to "improve the educational system"? Or does the poverty problem reside in deficient physical and mental health? If so, we must add those etiologies to our information package, and search inside the health services for a plausible cause. Does it include cultural deprivation? spatial dislocation? problems of ego identity? deficient political and social skills? – and so on. If we can formulate the problem by tracing it to some sorts of sources – such that we can say, "Aha! That's the locus of the difficulty," i.e. those are the root causes of the differences between the "is" and the "ought to be" conditions – then we have thereby also formulated a solution. To find the problem is thus the same thing as finding the solution; the problem can't be defined until the solution has been found.
> (Rittel & Webber, 1973, p. 161)

Since wicked problems allow a multitude of different approaches that might contribute to their solution, the identity of problem formulation and conceiving a solution implies that there will be a multitude of possible problem formulations. Any sufficiently detailed description of what the problem "is" is already predetermined by a certain vision of its solution—a vision that is often biased by diverse values and interests. This results from the fact that in pluralist societies, in which a multitude of world views and values compete, the determination and formulation of a problem as well as the assessment of its "solution" can in itself be controversial and open to discussion.

When Rittel and Webber recommend that wicked problems should be approached "based on a model of planning as an argumentative process in the course of which an image of the problem and of the solution emerges gradually among the participants, as a product of incessant judgment, subjected to critical argument" (Rittel & Webber, 1973, p. 162), they seem to assume that something like a Habermasian model of public deliberation that is based on an open and free exchange of arguments should be the best strategy to address wicked problems. In any case, Rittel and Webber's approach stimulated the development of computer-supported argument visualization (CSAV), as can be seen most prominently in *Visualizing Argumentation: Software Tools for Collaborative and Educational Sense-making*

(Kirschner, Buckingham Shum, & Carr, 2003). As Simon Buckingham Shum explains there, CSAV tools promise to support students' interactions when they learn how to engage in "an open-ended, dialectical process of collaboratively defining and debating issues" by means of arguments (Buckingham Shum, 2003, pp. 12-13).

The main challenge posed by wicked problems is that individual problem perceptions are always limited. It is unlikely—and impossible to know beforehand—that one person is able to capture the multiperspectivity of a wicked problem. For intractable conflicts, by contrast, the main challenge is not limited knowledge but radical disagreement on what the conflict is about, what its root causes are, and how it should be resolved. This way, mutual understanding—or at least the acknowledgment of understanding—is almost impossible. Conflicts about controversial issues such as abortion or the conflict between Israelis and Palestinians are well-known examples. Daniel Bar-Tal defines intractable conflicts as follows:

> Intractable conflicts are existential from the point of view of the participating parties. They are perceived as being about essential and basic goals, needs, and/or values that are regarded as indispensable for the society's existence and/or survival. In addition, they are often of a multifaceted nature, involving various spheres such as territory, self-determination, statehood, economy, religion, or culture. (Bar-Tal, 2007, p. 1433)

Not all conflicts are intractable, of course, but elements of intractability can be found in many conflicts. There are often points of disagreement where a counter-position is experienced as a threat to one's very existence or identity. Conflicts can be serious, and we should take them seriously. For intractable conflicts, or intractable elements in conflicts, we need more than what, for example, the pragma-dialectical approach to argumentation can provide. Pragma-dialectics requires as a precondition for resolving differences of opinion that conflicting parties agree on just those things that are controversial in intractable conflicts: a consistent list of "shared premises"; "how they will decide together on the acceptability of other propositions"; agreement which rules apply in the "argumentation stage" and other procedural rules (van Eemeren & Grootendorst, 2004, pp. 137, 143, 145-150). Taking conflicts seriously means to develop tools and approaches that can be used when no agreement about anything controversial is easy to achieve.

3. FRAMING AND REFRAMING

In research on conflicts and controversies, as well as in communication theory, the concept of "framing" has proven useful for discussing possibilities and limits of mutual understanding.[1] In an attempt to organize the various ways in which the concept has been used and defined in the literature, I suggested in earlier work a basic distinction between "semiotic framing" and "cognitive framing" (Hoffmann, 2011a). Whereas semiotic framing can be defined as the process of producing signs with the deliberate or unconscious intent of setting the boundaries around an issue in a certain way and to construct the meaning of what is within those boundaries, I proposed to identify "cognitive framing" with "sensemaking":

> Sensemaking is the process of interpreting data in a way that they fit into a belief-value-attitude system. "Data" can be externally observable signs, people, things, events, etc., but also ideas or thoughts. A "belief-value-attitude system" is a web of beliefs, values, and attitudes that is consistent from its bearer's point of view. A "belief" is defined here as that cognitive state we are in whenever we take something to be the case or regard it as true, implicitly or explicitly; thus, a belief is representable in the form of a factual or conditional statement. "Values" are defined as behavior guiding beliefs that are based on principles, needs, interests, or preferences. "Attitudes" are emotions someone feels with regard to certain data. The "fit" of data into a belief-value-attitude system can be achieved in three different ways: (a) by constraining the data (neglecting what is incomprehensible, or interpreting it in a way that it fits); (b) by changing the system; or (c) by a mixture of (a) and (b). (Hoffmann, 2011a, p. 149)

Cognitive framing, according to this conceptualization, happens all the time when we are "making sense" of what is going on around us, that is, when we interpret signs, events, actions, other people, and so on. This way, any interpretation whatsoever is assumed to be determined by what I call a "belief-value-attitude system" in the definition above. When sensemaking is defined as "the process of interpreting data in a way that they fit into a belief-value-attitude system," then it is clear that the limits of these systems define the limits of what we are able to see and to understand.

---

[1] Cf. Schön & Rein, 1994; Donohue, Rogan, & Kaufman, 2011; Gray, 2006, 2007; Lakoff, 2004; Lewicki, Gray, & Elliott, 2003; Putnam & Holmer, 1992; Shmueli, Elliott, & Kaufman, 2006; Tannen, 1993.

Colleagues in conflict research who use the concept of framing to describe problems of mutual understanding often assume that the resolution of intractable conflicts is only possible if the parties to the conflict are able to "reframe" the way they perceive the conflict. This would presuppose—according to my understanding of cognitive framing—that people are able to overcome the limits of a given belief and value system, and that they are able to change their mind on positions they perceive as essential for their very existence and identity. In the case of wicked problems—to which I propose to apply the same model of framing and reframing—a similar change of mind and the development of belief-value systems would be required, even though this would less likely be experienced as a threat to one's identity.

4. SEMIOTICS AND DIAGRAMMATIC REASONING

The main challenge for coping with wicked problems and intractable conflicts is, thus, to develop and change our systems of beliefs, values, and attitudes. The idea to use argument mapping to stimulate such a change of mind can theoretically be supported by Peirce's work on "diagrammatic reasoning." Diagrammatic reasoning is reasoning by means of representations which visualize, in particular, structures and relations. A central idea of diagrammatic reasoning is that an external visualization of what we think about an issue allows us to identify problems and gaps in our own thinking, leads to the identification of relations we were not aware of before, and stimulates thus creativity and learning (Stjernfelt, 2007; Hoffmann, 2004, 2011c).

A "diagram," for Peirce, is a representation whose main function is to represent a certain group of relations, namely those relations in which objects are "rationally related," that is, "intelligible" relations or relations that are "rationally comprehensible" (Peirce, NEM IV 316). Relations are rationally comprehensible, I think, if they can be represented by means of a "consistent system of representation" (Peirce, CP 4.418).

The rational character of diagrams is crucial for diagrammatic reasoning. Peirce defines diagrammatic reasoning as reasoning with diagrams that are constructed by the means provided by a certain system of representation—be it a logical system, or an axiomatic system as in geometry, or simply the vocabulary and grammar of a language. Such reasoning with diagrams is realized when we experiment with a diagram according to the rules of the chosen representational system. Experimenting with diagrams, transforming them according to the rules of the system, and observing what happens is crucial for scientific discoveries and the development of new knowledge (Peirce, NEM IV 47 f.).

There are two ways in which diagrammatic reasoning can support the development of knowledge. First, experimenting with diagrams can lead to the discovery of regularities and of "relations between elements which before seemed to have no necessary connection" (Peirce, CP 1.383), which again can lead to the creation of new concepts and theories. Examples in the history of science are the discovery of incommensurability in geometry and irrational numbers in arithmetic (Hoffmann & Plöger, 2000), the formulation of Desargues' theorem in projective geometry (Hoffmann, 2011e), and Maxwell's development of the electromagnetic field concept (Nersessian, 2008). In all these cases it is important that the discovery of something new is conditioned on the rules of the chosen representational system. If there were no rules according to which experiments are performed in diagrammatic reasoning, it would never be possible to distinguish arbitrary observations from those that can be used to create new knowledge. The same is true for the development of new knowledge in the more mundane cases of ordinary learning by means of diagrammatic reasoning.

Secondly, diagrammatic reasoning can help to structure and organize one's reasoning, and it can support the evaluation and improvement of reasoning. As I showed elsewhere, diagrammatic reasoning can help students to reflect on their reasoning without being constrained by the limits of their working memory; analyze a problem more thoroughly and systematically; clarify and coordinate confused ideas about a problem; clarify implicit assumptions; identify background knowledge that might be inadequate; structure a problem space; change perspectives; identify unexpected implications; play with interpretations; discover contradictions; and distinguish the essential from the peripheral (Hoffmann, 2011c). All this is important for coping with wicked problems.

However, a precondition for such a reflection on one's own reasoning by means of diagrammatic reasoning is that students have clear standards at their disposal to structure their reasoning and to evaluate its quality. There need to be structures according to which reasoning can be organized and criteria with respect to which it can be assessed. Structures are usually part of the ontology of representational systems (the ontology provides the elements that can be represented) and criteria are given in the rules of these systems. Logic, for example, provides an ontology and rules that can be used to structure arguments and to perform self-controlled reasoning: to assess the inferences that we draw in reasoning. We all, as Peirce wrote, "have in our minds certain norms, or general patterns of right reasoning, and we can compare the inference with one of those and ask ourselves whether it satisfies that rule" (Peirce, EP II 250 = CP 1.606). Usually, of course, these norms need to be acquired and learned before they are "in our minds."

Logic, however, is only one possible normative standard for self-controlled reasoning. The "critical questions" that Douglas Walton developed for each of a multitude of different argument schemes fulfil the same function (Walton, Reed, & Macagno, 2008). They provide a standard that students can use to structure their reasoning and to assess—and to improve, if necessary—the arguments they encounter or create.

Whatever the standard of good reasoning that is realized in a system of representation, it is paramount that some standard is available for structuring reasoning and evaluating its quality, and to improve this quality if it is questionable. Diagrammatic reasoning, as defined by Peirce, can thus support self-regulation, self-examination, and self-correction.

What we can learn from these considerations about the potential of diagrammatic reasoning for acquiring new knowledge on one hand and for providing standards for structuring and assessing the quality of reasoning on the other is that a necessary condition for both is knowing and accepting the rules of a chosen system of representation; a system by means of which diagrams can be constructed, experiments be performed, and reasoning be organized and evaluated. Such a system constrains reasoning in a way that our cognitive energy gets focused on points that are crucial for reflection on one's own reasoning—just like a fireman's jet of water will be the more focused the more it is constrained. Without any constraints there would be no direction for our reasoning.

5. REFRAMING BY MEANS OF ARGUMENT MAPPING IN AGORA-NET

A major opportunity that computer-supported argument mapping provides regarding the potential of diagrammatic reasoning is that the rules of representational systems can be implemented in the user interface of software. This has been done in the collaborative argument visualization software AGORA-net (*http://agora.gatech.edu/*). As already mentioned, AGORA-net is designed to stimulate reflection. Elsewhere I coined the term "reflective argumentation" for those approaches to argument which see a major purpose of the construction of arguments in stimulating one or more of the following: reflection on the quality of one's own arguments; reflection on the quality of one's own reasoning (the process on one hand and beliefs, values, and attitudes on the other); the motivation to improve the quality of one's arguments and reasoning; creativity how to do it; or a change of perspective on an issue (Hoffmann, 2015).

In order to provide, on the one hand, an example of how computer-supported argument visualization (CSAV) tools can be designed to support reframing and, on the other, to discuss some of the problems that I encountered in trying it, I will briefly argue for a unique design feature of the AGORA software. The main characteristic of AGORA-net is that propositional logic is the system of representation that is built into its user interface. Four argument schemes from propositional logic—*modus ponens*, *modus tollens*, disjunctive syllogism, and not-all syllogism—provide the standard that guides the structuring of arguments. There is no need for the user to know anything about logic or deductive validity because the process in which arguments are constructed guarantees that at the end all resulting arguments are logically valid. After entering a claim and one or more reasons, the user needs to select one of the four argument schemes and the system creates automatically a further premise which transforms the given argument into a deductive form (see Figure 1 for a completed argument).

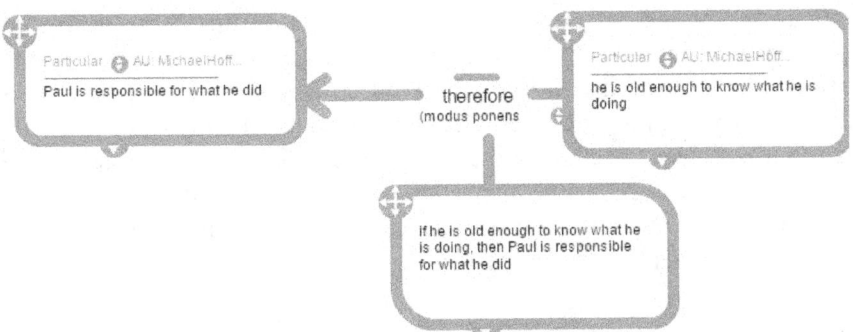

Figure 1 – A logically valid argument in AGORA-net. The premise underneath the "therefore"—called the "enabler" because this premise guarantees the truth or acceptability of the conclusion if all premises are true or acceptable—is automatically created by the software based on the conclusion, reason, and the user's selection of *modus ponens* as argument scheme and "if-then" as language form

It is important to note that I am not interested in deductive thinking. I do not think that everybody should learn how to construct deductive inferences, or that logical arguments are important in everyday life—they are obviously much less important than other forms of argument. I do not think that "all good arguments are deductively valid," a position that Leo Groarke (1992) defended, and I do not think that "natural language arguments should be understood as attempts to formulate deductive arguments" (defended by Groarke, 1999). Neither do I think that "reasoning is inherently deductive in character," as Ralph Johnson (2007, p. 73) characterized so-called "deductivism." But I do think that confronting the user of argument mapping software with the standards of propositional logic is *useful* for two purposes: first, to *support* the user's assessment of his or her own arguments, whatever they are, and, second, to *stimulate* critical reflection on the quality of these arguments. It is like using a ruler to draw a line. It helps, but the ruler itself is not interesting. Of course, one might ask whether we should have straight lines. But that, obviously, depends simply on the purpose. Sometimes we need straight lines, sometimes we don't.

Figure 2 presents my argument for the assumption that it is useful for the first purpose—to support argument assessment—to present arguments in logical form. Of course, this argument does not exclude the possibility that "clearly defined standards for the structure of good arguments" are also provided by non-logical approaches to argument. Any system that allows to distinguish clearly between reasons and conclusions supports the structuring of reasoning to a certain degree. More sophisticated systems, such as Rationale, Araucaria, and OVA, allow the use of large numbers of non-logical argument schemes that provide more guidance than simply distinguishing reasons and conclusions, and it would be possible—for example—to implement even the sets of critical questions that Doug Walton developed for individual argument schemes into CSAV tools. Also the Toulmin-model of argument with its six-entity ontology could easily be provided in the form of an argument template. All this would help to structure reasoning and arguments.

Collaborative reframing

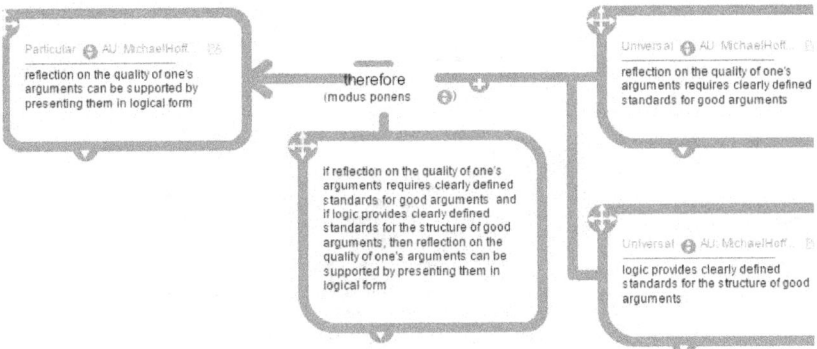

Figure 2 – An argument, constructed in AGORA-net, for the use of logical argument schemes in AGORA-net

This could be the place for a long discussion about the question whether logical or non-logical representations of arguments are more effective to support argument assessment, but at the end this is an empirical question that should be decided by well-designed experiments. The same will be true for my argument for the second assumption mentioned above: that presenting arguments in the four forms of propositional logic that are available in AGORA is useful to stimulate reflection on the quality of one's own arguments. But in this case, I think, I have a stronger theoretical point, as can be seen in Figure 3 which represents my argument for this assumption.

According to my own experience of constructing hundreds of argument maps in AGORA and evaluating even more student maps, reading carefully the enabler of such an argument and asking whether it is convincing—interpreting it as something like a universal law—can lead to five possible insights. Each of them describes a way how any given argument can be improved.
1. The reasons provided do not justify the conclusion, they might even talk about something completely different (for examples see Figures 4 and 5), or a particular reason is not needed.
2. Something is missing to justify the conclusion. Thus, another reason needs to be added (see Figure 6).

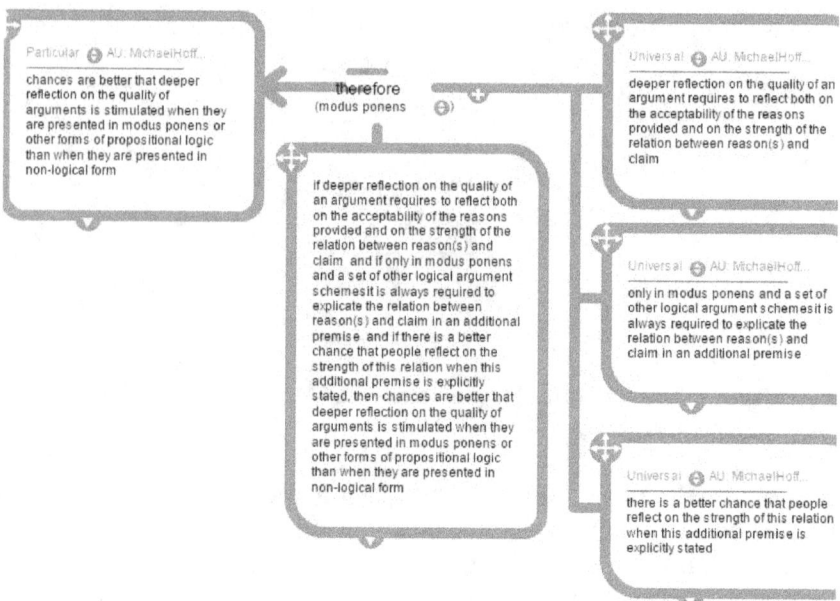

Figure 3 – This argument is based on the assumption that the explicit formulation of the connection between reasons and conclusion in an additional premise stimulates reflection on this connection. In AGORA, this additional premise is automatically created by the software based on user input. To keep it visually apart from what I call the argument's "reasons," it is always presented underneath the "therefore." – Leo Groarke makes a similar point. He stresses that in everyday arguments the relation between reasons and conclusions is usually not explicitly stated. But by treating it as an "unexpressed premise" and explicitly stating it we are able to "expose assumptions which need to be a focus of discussion when we decide whether an argument should be accepted." That is, the strategy to model arguments as deductions "can help identify the issues that need to be addressed in dialectical exchange. ... By recognizing these assumptions as unexpressed premises, a deductivist approach furthers the dialectical exchange which is the key to resolving differences of opinion" (Groarke, 1999, p. 9).

3. A particular conclusion cannot be justified at all. It is either plainly wrong or needs to be modified, for example by adding qualifiers such as "probably," "there is reason to assume that," or by specifying certain circumstances or conditions under which it is acceptable. (For an example, see the development between Figures 13 and 14. Adding qualifiers, by the way, is a strategy to

represent argument forms that are usually classified as non-deductive, such as induction and abduction, in deductive form.)

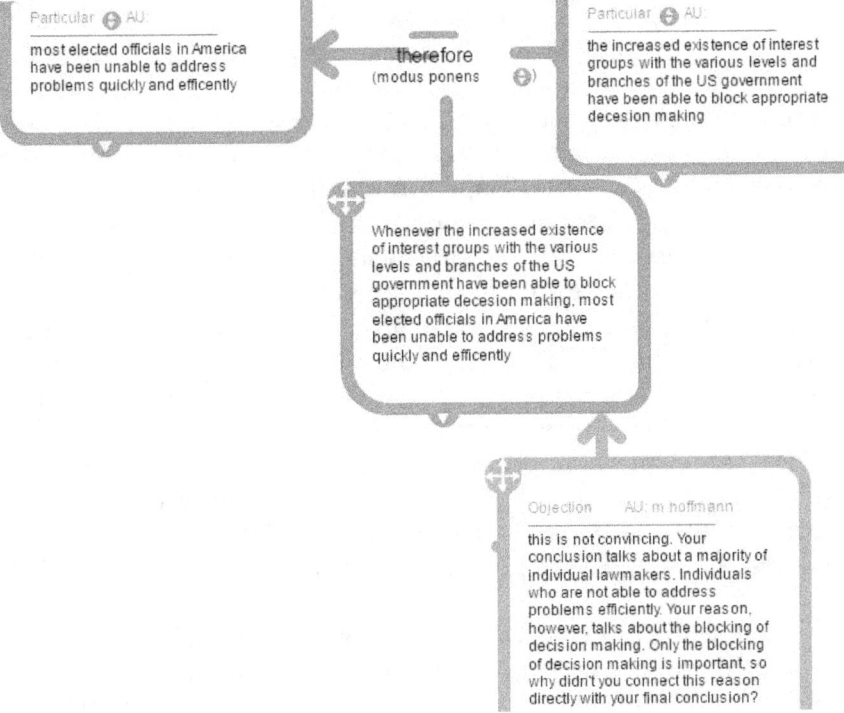

Figure 4 – Part of a student map with objection by the instructor. (This and the following argument maps are reproduced here with permission of the students who created them)

4. Particular reasons need to be modified. They might not fit grammatically into the enabler (in which they show up in exactly the form they were entered as reasons), or they do not provide what is needed to justify the conclusion.
5. The overall structure of the argument is not optimal:
    a. Reasons might include more than one proposition and should be divided. (Because it might be the case that each proposition needs to be justified by further arguments. Usually these arguments can only justify one of them, not all together.)
    b. Reasons might be too complicated and should be divided into chains of arguments.
    c. Reasons might be able to support the conclusion only in combination with certain other reasons, or they might

support it independently from other reasons. (In the former case, reasons are combined in one enabler, as in Figures 2 and 3, in the latter they show up in different arguments for the same conclusion, each with its own enabler. The problem is illustrated in Figure 6).

Even though the AGORA software has been designed to stimulate critical reflection on the quality of arguments—especially of one's own arguments—there are two major problems that turned up when the software was used in the "real world" of the college classroom. The first one is that I could observe far less critical assessment of arguments than I expected, and the second problem is that being able to improve reasoning, for example by restructuring an argument, refining the conclusion, or adding reasons so that the argument becomes stronger and more convincing, is still a long way from engaging in "reframing." In order to be able to cope with wicked problems and intractable conflicts it is not sufficient to improve the arguments that are used in the "argumentative process" described by Rittel and Webber; what is required is to engage in a process that ultimately may lead to a substantial change of one's position and its underlying system of beliefs and values.

One might argue that reframing could be possible as the result of gradually improving one's own arguments. This would amount to a change of beliefs and values that is due to changing one's position based on acknowledging certain weaknesses of the arguments that are supposed to legitimize this position. However, if not even the kind of self-correction is easily achievable that is required for improving one's position, how could then argument mapping be sufficient to stimulate reframing?

5. ADVERSARIAL OR COLLABORATIVE ARGUMENT MAPPING? THE ROLE OF SOCIAL INTERACTION, CRITIQUE, AND CONFUSION

In order to develop a better understanding of what makes it so difficult to improve the quality of one's arguments, let me provide a few examples of argument maps that small teams of two or three students developed in the context of a class project. The same class project was done, over a few weeks in the spring of 2015, in two philosophy classes for public policy students, one an undergraduate class, the other a graduate class. The project focused on the arguments that Francis Fukuyama formulated about the decay of democratic institutions in America. (We used Fukuyama, 2014a, which is an excerpt from Fukuyama, 2014c, as a starting point, but in the phase of the project work described here, the teams developed projects on their own.) The class curriculum realized a certain form of problem-based learning (PBL) in which parts of the role of a facilitator, who usually supports student teams in their learning process, were replaced by the AGORA software (see Hoffmann & Borenstein, 2014). All the examples provided here are parts of larger argument maps. I obtained permission from the students who created them to use them for presentations and publications.

Figure 5 shows part of an argument map that has been created by two undergraduates as an "intermediate" proposal about ten days before their final class presentation. It is one of those examples where the intended stimulation of reflection obviously failed.

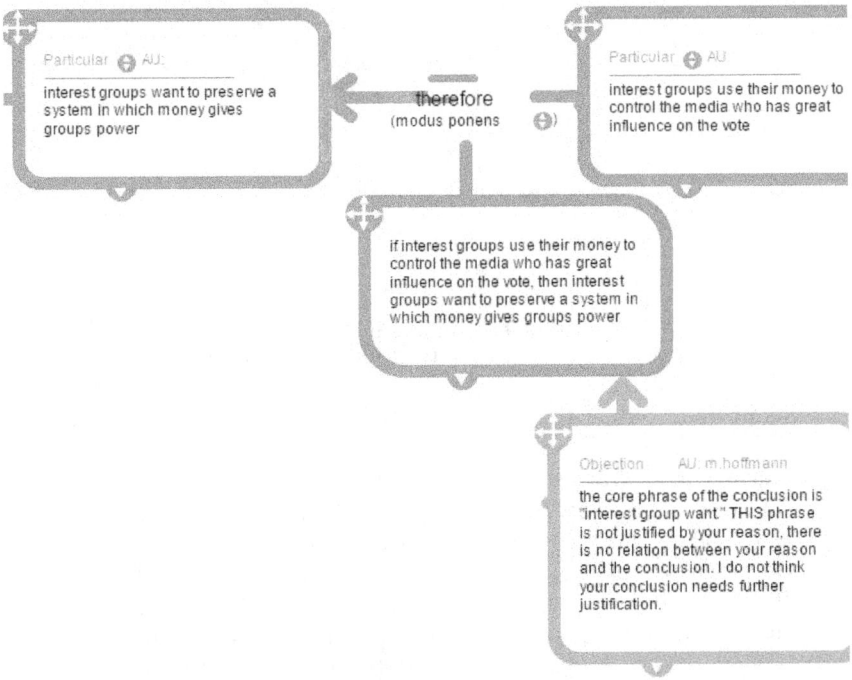

Figure 5 – Part of an argument about interest groups, with objection from the instructor

One only needs to read the argument's enabler to see that it simply does not make much sense: "if interest groups use their money to control the media who has great influence on the vote, then interest groups want to preserve a system in which money gives groups power." It does not make sense to infer the "wanting" from something they do.

Another group of three graduate students worked on racial differences regarding childhood obesity. One argument of their intermediate submission is shown in Figure 6. Again, if one reads the enabler and tries to assess its acceptability, there are many problems that should jump into one's eyes. Some of them are discussed in my objection to this enabler in Figure 6.

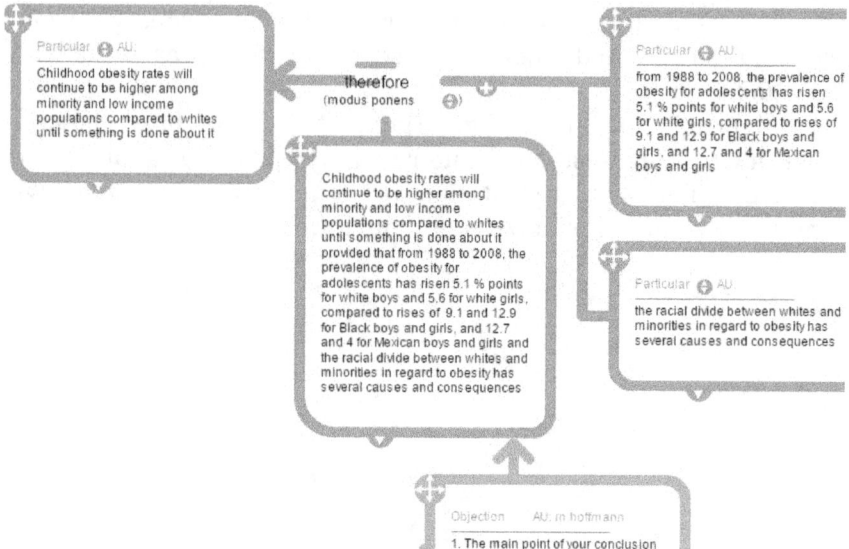

Figure 6 – The main argument of an argument map on childhood obesity. The complete objections at the bottom reads as follows: "1. The main point of your conclusion is "will continue." But you do not provide a reason for the continuation. 2. Your second reason is about cause for the racial divide. But the cause in itself does not tell you anything about continuation. 3. Why do you talk about "consequences" in your second reason? What are these consequences? Are they in relevant for your conclusion? I cannot see this. 4. Your first reason and the second are independent. They are independent because the first one provides evidence while the second one talks about cause. Both are completely different things. Since you can defend your conclusion either by evidence (if you add as a co-dependent reason something like "there is nothing that could cause a change of this previous trend") or by reference to cause (if you add to your second reason another co-dependent reason saying something like "there is no indication that any of these causes is no longer present"), I would recommend that you divide your 2 co-dependent reasons into 2 independent arguments, each with two co-dependent reasons"

Why doesn't the reflection work as intended? Didn't they read their enablers or didn't they see the problems when they read them? I don't have any idea. An analysis of all the argument maps that were submitted in both classes as an intermediate result shows that on average about 26% of all enablers on the four argument maps submitted by the graduate students were unacceptable and even 45% of the five undergraduate maps. These argument maps were created without interference from the instructor or feedback from students outside of the teams that developed these maps.

However, the results were better in the students' final submissions after they presented and discussed their intermediate argument maps in class and received detailed critique by the instructor in form of objections that were added to specific enablers. The number of acceptable enablers rose, on average, from about 74% to 83% on the maps created by the graduate students and from about 55% to 95% on those by the undergraduates (see Figure 7). The discrepancy regarding only slight improvements with the graduates and substantial improvements of the quality of the undergraduates' argument maps can probably be explained by my observation that the undergraduates were much more eager to talk to me to clarify my objections and to get suggestions on how an argument map could be improved.

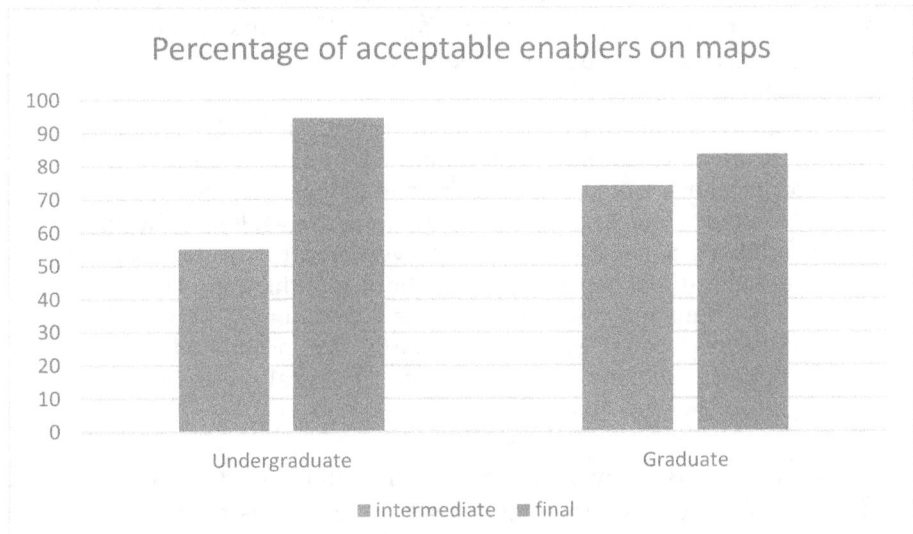

Figure 7 – Improved quality of argument maps between an intermediate and final sub-mission. Paired Student's T-test gave a significance of p=0.025 for the undergraduate maps. Results for the graduate maps are not significant with a p-value of 0.165.

This observation, but also the numbers presented in Figure 7, indicate that a crucial condition for the improvement of arguments is critical feedback from people outside of the group that produced a specific map. Neither the design of AGORA-net, nor the social interaction within the groups were obviously sufficient to stimulate the sort of critical reflection that would be required to produce better arguments.

However, it should be added that I do not have any information about what happened within the groups before they presented their intermediate map. It may be that 26% and 45% unacceptable enablers, respectively, is already an improvement compared to where they started; an improvement that resulted either from critical reflection of individual students when they drafted their arguments or from debates within the groups. The numbers above only indicate that the more adversarial style of my objections to particular enablers had an effect on the students' final revisions of their maps.[2]

It might be helpful to illustrate this effect with a few comparisons between intermediate and final versions of argument maps. Figure 8 shows the main argument of an intermediate argument map that a group of three undergraduates presented and discussed in class, and then submitted it as a graded assignment. This is an argument about so-called "Super PACs," a new kind of "political action committee" (PAC) created in July 2010 following the outcome of a federal court case known as *SpeechNow.org v. Federal Election Commission*. As one of the sources writes that the students used:

> Technically known as independent expenditure-only committees, Super PACs may raise unlimited sums of money from corporations, unions, associations and individuals, then spend unlimited sums to overtly advocate for or against political candidates. Super PACs must, however, report their donors to the Federal Election Commission on a monthly or quarterly basis -- the Super PAC's choice -- as a traditional PAC would. Unlike traditional PACs, Super PACs are prohibited from donating money directly to political candidates. (*https://www.opensecrets.org/pacs/superpacs.php*)

Whereas Figure 8 shows the main argument of the intermediate submission, Figure 9 shows what happened with this main argument in the final presentation and submission. The conclusion has been refined

---

[2] I added objections also to particular reasons on the maps, but the effects of these interventions are outside of the focus of this contribution. It suffices to say that all reasons (those that were not further justified by arguments) of all the final maps were acceptable.

and the overall structure of the argument has been revised so that the new reason that is depicted on the right of Figure 9 summarizes a more detailed argumentation that justifies this reason on the right. A part of this argumentation is depicted in Figure 10.

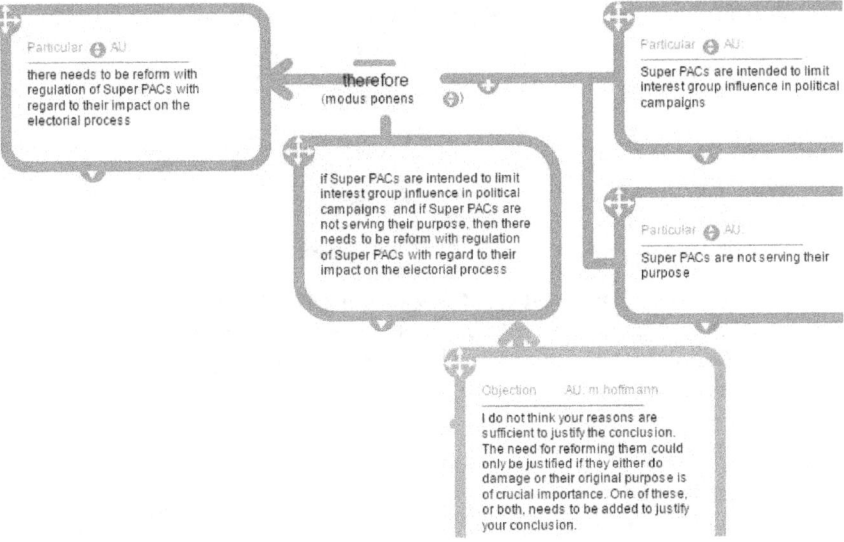

Figure 8 – The main conclusion and argument of the group working on Super PACs in their intermediate submission, with an objection by the instructor. On the students' map, the second reason is justified by further arguments

A comparison of the arguments depicted in Figures 11 and 12 provides another example for a revision that seem to be motivated by a critical objection.

The group that worked on the arguments that are represented in Figures 11 and 12 is of particular interest because these students were able to reframe their problem perception. They changed their point of view between their intermediate and final argument map. This, at least, can be inferred when we compare the main conclusions and main arguments of these two maps and interpret the development between them based on a comment one of the students wrote when she submitted the final argument map. She points out that their starting point was the following quote from Francis Fukuyama:

> most citizens have neither the time, nor the background, nor the inclination to grapple with complex public policy issues; expanding participation has simply paved the way for well-

organized groups of activists to gain more power. (Fukuyama, 2014a)

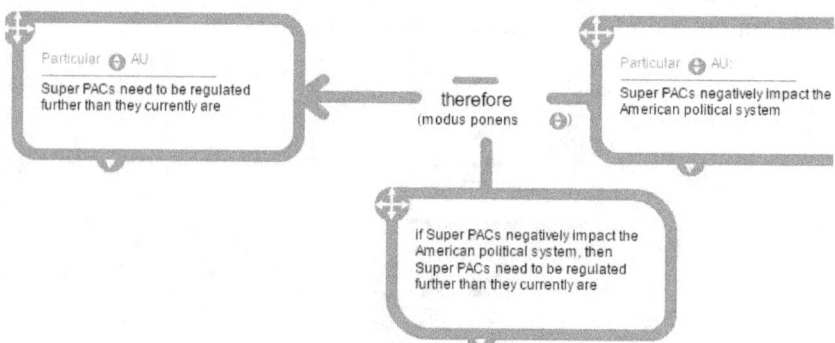

Figure 9 – The main conclusion and argument of the final version of an argument map that resulted from revising the map depicted in Figure 8. The reason on the right is further justified

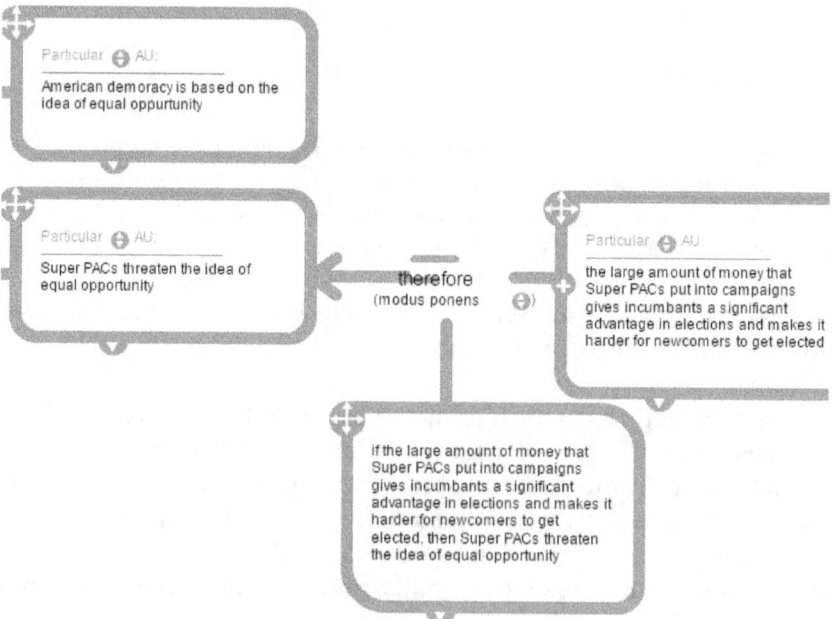

Figure 10 – Another part of the argumentation on Super PACs whose main conclusion is depicted in Figure 9. The reason on the right is again justified by another argument

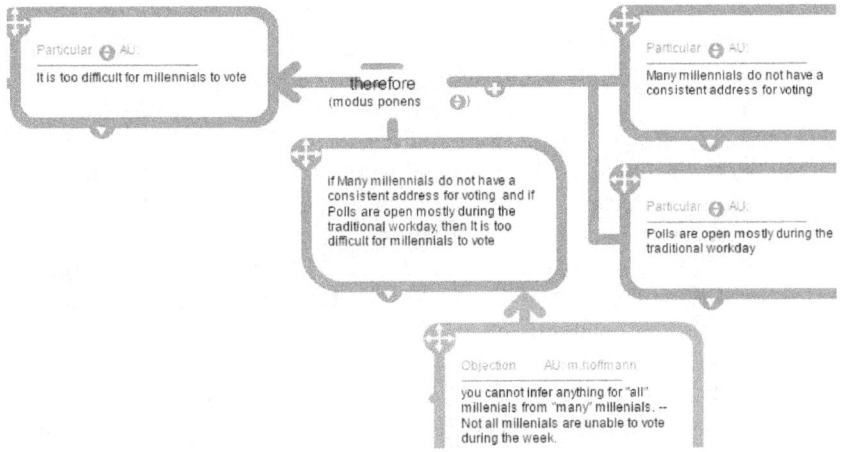

Figure 11 – Part of an argument about the voting behavior of "millenials" (defined by the authors of this argument map as those born between 1981 and 1997)

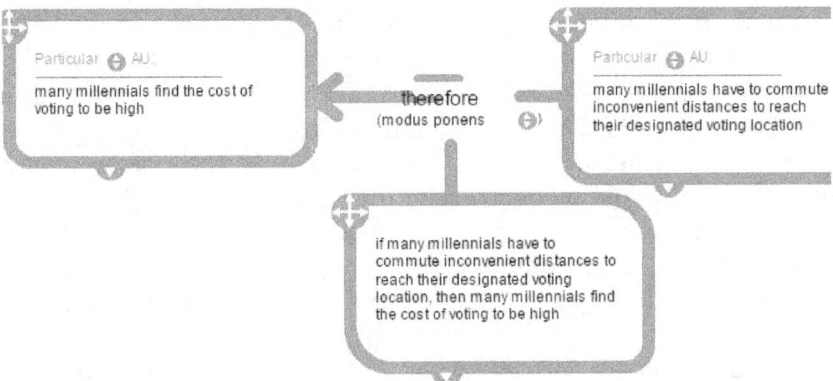

Figure 12 – Revision of the argument depicted in Figure 11. The reason on the right is then justified by two independent arguments. The entire map has been published by the students in AGORA-net. Go to http://agora.gatech.edu and search for map #11046

Then she continues:

> We questioned why this was the case, leading us to various subtopics we could pursue. We finally narrowed in on one topic: how this statement particularly applies to millennials and why they don't vote. We came up with several reasons, but they seemed unorganized. That's when I remembered a formula called the Calculus of Voting that we had learned

about in a contemporary issues class with Dr. Barke: R = PB - C +D, meaning that the likelihood that someone will vote is based on the the probability that their vote will count, the benefit they would receive from their candidate being elected, the cost of voting, and the good will feeling that voting brings based on our sense of duty (Riker). Discounting P because it varied by state and disregarding D because it's extremely difficult to measure led us to our final equation, R = B - C, and our hypothesis: if B is low and C is high, R will be low. Using this function as a guide, we were better able to organize our final map (Attached)[3]

It is important to note that the focus on the voting behavior of millennials determined already the intermediate argument map from which Figure 11 is taken. The genuine creative step of reframing the problem perception must have been stimulated by two things: a sense that the reasons they came up with in previous work to explain the assumed withdrawal of millennials from the political process "seemed unorganized," and then the idea that the "calculus of voting" (Riker & Ordeshook, 1968) might help to organize their material.

It is indeed fascinating to see how Riker's calculus of voting can be translated into an argument about the voting behavior of a specific part of the population. All the elements of his equation are presented in the argument, either as reasons—B shows up in the reason on top of Figure 13 and C in the second reason that is cut off in this picture—or as qualifiers in the argument's main conclusion (P and D; abbreviations are explained in the yellow comment box in Figure 13). Using the calculus of voting really contributed to a substantial improvement of the quality of this argument.

What does all this mean with regard to the question whether collaborative or adversarial argument mapping is more effective to stimulate reflection, self-regulation, and reframing? As I pointed out in the beginning of this section, I was rather disappointed that working with the AGORA software did not lead to as much critical reflection as I expected. With regard to these examples it seems to be indispensable to use the more adversarial style of adding objections to particular elements on an argument map to motivate cognitive change. However, the last example indicates that substantial changes that might be characterized as "reframing" are at least possible. If we ask what the conditions might be that support this kind of more substantial reflection and reframing, I think the students' experience of the situation they

---

[3] I am thankful to Laura-Margaret Burbach for allowing me to use this quote.

found themselves in as "unorganized" is significant. Even if it might be a stretch to use this analogy, this experience is similar to what Thomas S. Kuhn described in *The Structure of Scientific Revolutions* as the situation of a "crisis" in science before a major "paradigm shift" helps the scientific community to substantially reorganize knowledge.

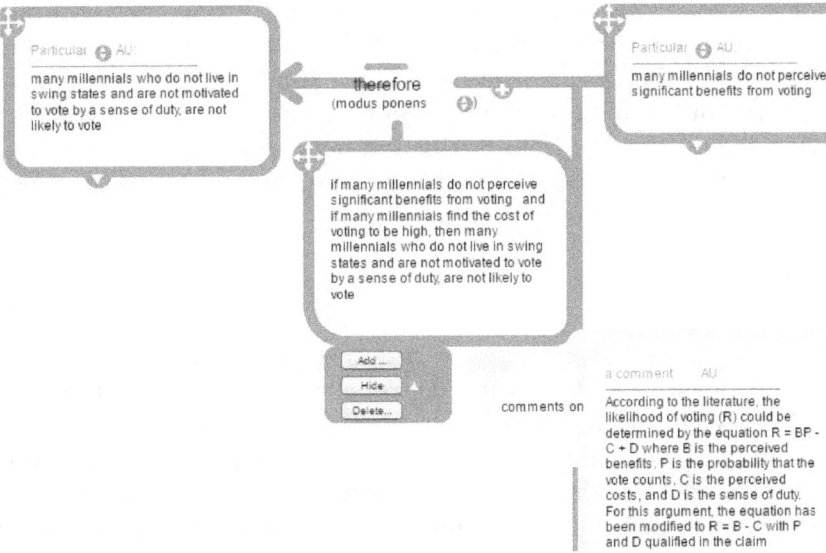

Figure 13 – The main argument from the final map from which also Figure 11 was created. The second co-dependent reason is cut off, but its content is represented in the argument's "enabler" (underneath the "therefore") after the "and if." The comment has been provided by the students

Kuhn illustrates this sense of crisis by citing Copernicus's complaint that "in his day astronomers were so 'inconsistent in these [astronomical] investigations' ...," that they would take well-formed parts "from diverse" models without realizing that these parts do not fit into "a single body" of knowledge. He cites Einstein's impression that the situation in physics in the beginning of the 20th century was "as if the ground had been pulled out from under one, with no firm foundation to be seen anywhere, upon which one could have built," and then Wolfgang Pauli's observation in the months before Heisenberg laid the foundations of the new quantum theory:

> At the moment physics is again terribly confused. In any case, it is too difficult for me, and I wish I had been a movie comedian or something of the sort and had never heard of physics. (Kuhn, 1996, pp. 83-84)

Plato already said that confusion and astonishment are "the beginning of philosophy." What he meant was that we have to take problems seriously to develop the kind of energy that is necessary to create new ideas and knowledge.

This kind of seriously felt confusion might provide the most important stimulation for reflection, self-correction, and reframing. Going back to what we learned from Peirce's concept of diagrammatic reasoning and the crucial role of clearly defined systems of representations, I would think that argument visualization software should provide a level of rule-based clarity and firmness that is rigid enough to produce just this kind of creative confusion—even if it requires additionally critique from other people.

6. CONCLUSION

My goal in this contribution was to assess the potential of argument mapping for coping with "wicked problems" and intractable conflicts. The underlying problem in both cases can be described by the notions of framing and reframing. People's perception of a problem or conflict is always limited, they "frame" what they see based on certain systems of beliefs, values, and attitudes. Thus, the main challenge is to stimulate and support "reframing," that is, a substantial reorganization and change of given knowledge, beliefs, or values. As a starting point to address the problem of reframing, I used Rittel and Webber's idea of an "argumentative process in the course of which an image of the problem and of the solution emerges gradually among the participants, as a product of incessant judgment, subjected to critical argument" (Rittel & Webber, 1973, p. 162). Such an "argumentative process" of critical exchange, I tried to show, can be supported and stimulated by argument mapping software that realizes, in its user interface, Peirce's ideas on "diagrammatic reasoning."

However, an analysis of argument maps that students submitted as part of a class project provided some indication that using rule-based systems to represent arguments is often not sufficient to stimulate and support reflection, self-regulation, and reframing. What is needed in addition in these situations is critique from the outside of established group dynamics to mobilize the energy that is required to change reasoning. In this sense, adversarial argument mapping—that is, argument mapping in which the creator of a map is confronted with objections to specific parts of his or her argument—seems often superior to collaborative argument mapping in which teams work on their own.

However, we saw one example in which a quite substantial reframing process could be observed. In this case, it might indeed have been the clear structure of the arguments that the students created that showed them that their thinking about the problem in question was still too "unorganized" and confusing. Thus, it seems reasonable to conclude that both adversarial and collaborative argument mapping can play a role in dealing with wicked problems, and maybe also with intractable conflicts.

ACKNOWLEDGEMENTS: This research has been supported by a grant from the Fund for the Improvement of Postsecondary Education (FIPSE), U.S. Department of Education (Grant P116S100006). I thank the following students for allowing me to reproduce parts of their argument maps: Lee Ayes, Len Berg, Laura-Margaret Burbach, Daniel D'Arcy, Kate de Give, Megan Haley, Janelle Johnson, Namrata Kolla, Joshua Lord, Mary Francis McDaniel, Amanda Nabors, and Nicole Qin. One student preferred to remain anonymous. Thanks also to Tobias Hoffmann who helped me to get the numbers of the empirical analysis in the section on "reframing" right.

REFERENCES

Bar-Tal, D. (2007). Sociopsychological Foundations of Intractable Conflicts. *American Behavioral Scientist, 50*, 1430-1453.

Buckingham Shum, S. (2003). The Roots of Computer-Supported Argument Visualization. In P. A. Kirschner, S. J. Buckingham Shum & C. S. Carr (Eds.), *Visualizing Argumentation: Software Tools for Collaborative and Educational Sense-making* (pp. 3-24). London: Springer.

Donohue, W. A., Rogan, R. G., & Kaufman, S. (Eds.). (2011). *Framing Matters: Perspectives on Negotiation Research and Practice in Communication.* New York, NY: Peter Lang.

Facione, P. A. (1990). *Critical Thinking: A Statement of Expert Consensus for Purposes of Educational Assessment and Instruction. The Delphi Report. Executive Summary.* Retrieved from http://assessment.aas.duke.edu/documents/Delphi_Report.pdf

Fukuyama, F. (2014a). America in Decay. *Foreign Affairs, 93*(5), 3-26.

Fukuyama, F. (2014c). *Political order and political decay: from the industrial revolution to the globalization of democracy.* New York: Farrar, Straus and Giroux.

Gray, B. (2006). Mediation as Framing and Framing Within Mediation. In M. S. Herrman (Ed.), *The Blackwell Handbook of Mediation: Bridging Theory, Research, and Practice* (pp. 193-216). Malden, MA; Oxford, U.K.; Carlton, AU: Blackwell.

Gray, B. (2007). Frame-Based Interventions for Promoting Understanding in Multiparty Conflicts. In T. Gössling, L. Oerlemans & R. Jansen (Eds.), *Inside Networks. A Process View on Multi-Organisational Partnerships, Alliances and Networks*. Cheltenham, UK; Northampton, MA: Edward Elgar Publishing.

Groarke, L. (1992). In defense of deductivism: Replying to Govier. In F. H. v. Eemeren, R. Grootendorst, J. A. Blair & C. Willard (Eds.), *Argumentation illuminated* (pp. 113-121). Amsterdam: Sic-Sat.

Groarke, L. (1999). Deductivism Within Pragma-Dialectics. *Argumentation, 13*(1), 1-16.

Hoffmann, M. H. G. (2004). How to Get It. Diagrammatic Reasoning as a Tool of Knowledge Development and its Pragmatic Dimension. *Foundations of Science, 9*(3), 285-305.

Hoffmann, M. H. G. (2011a). Analyzing Framing Processes in Conflicts and Communication by Means of Logical Argument Mapping. In W. A. Donohue, R. G. Rogan & S. Kaufman (Eds.), *Framing Matters: Perspectives on Negotiation Research and Practice in Communication* (pp. 136-164). New York, NY: Peter Lang (pre-print available at http://works.bepress.com/michael_hoffmann/37/).

Hoffmann, M. H. G. (2011c). Cognitive conditions of diagrammatic reasoning. *Semiotica, 186*(1/4), 189-212.

Hoffmann, M. H. G. (2011e). "Theoric Transformations" and a New Classification of Abductive Inferences. *Transactions of the Charles S Peirce Society, 46*(4), 570-590.

Hoffmann, M. H. G. (2015). Reflective argumentation: A cognitive function of arguing. *Argumentation*, 1-33. doi: 10.1007/s10503-015-9388-9

Hoffmann, M. H. G., & Borenstein, J. (2014). Understanding Ill-Structured Engineering Ethics Problems Through a Collaborative Learning and Argument Visualization Approach. *Science and Engineering Ethics, 20*(1), 261-276. doi: 10.1007/s11948-013-9430-y

Hoffmann, M. H. G., & Plöger, M. (2000). Mathematik als Prozess der Verallgemeinerung von Zeichen: Eine exemplarische Unterrichtseinheit zur Entdeckung der Inkommensurabilität. *Zeitschrift für Semiotik, 22*(1), 81-114.

Inman, J. (2015). Teaching through Critique: An Extra-Disciplinary Approach. *The National Teaching & Learning Forum, 24*(2), 6-8. doi: 10.1002/ntlf.30017

Johnson, R. H. (2007). Informal Logic and Epistemology. *Anthropology & Philosophy, 8*(1-2), 69-88.

Kirschner, P. A., Buckingham Shum, S. J., & Carr, C. S. (Eds.). (2003). *Visualizing Argumentation: Software Tools for Collaborative and Educational Sense-making*. London: Springer.

Kuhn, T. S. (1996). *The Structure of Scientific Revolutions* (3rd ed.). Chicago: The University of Chicago Press.

Lakoff, G. (2004). *Don't Think of an Elephant: Know Your Values and Frame the Debate—The Essential Guide for Progressives*. White River Junction, VT: Chelsea Green Publishing Company.

Lewicki, R. J., Gray, B., & Elliott, M. (Eds.). (2003). *Making Sense of Intractable Environmental Conflicts. Concepts and Cases.* Washington - Covelo - London: Island Press.
Linstone, H. A., & Turoff, M. (Eds.). (1975). *The Delphi Method: Techniques and Applications.* Reading, MA: Addison-Wesley (available online: http://www.is.njit.edu/pubs/delphibook/).
Nersessian, N. J. (2008). *Creating scientific concepts.* Cambridge, MA: MIT Press.
Peirce. (CP). *Collected Papers of Charles Sanders Peirce.* Cambridge, MA: Harvard UP.
Peirce. (EP). *The Essential Peirce. Selected Philosophical Writings.* Vol. 1 (1867–1893), Vol. 2 (1893–1913). Bloomington and Indianapolis 1992 +1998: Indiana University Press.
Peirce. (NEM). *The New Elements of Mathematics by Charles S. Peirce* (Vol. I-IV). The Hague-Paris/Atlantic Highlands, N.J., 1976: Mouton/Humanities Press.
Putnam, L. L., & Holmer, M. (1992). Framing, Reframing, and Issue Development. In L. L. Putnam & M. E. Roloff (Eds.), *Communication and Negotiation* (Vol. 20, pp. 128-155). Newbury Park, CA: Sage.
Riker, W. H., & Ordeshook, P. C. (1968). Theory of Calculus of Voting. *American Political Science Review, 62*(1), 25-42. doi: 10.2307/1953324
Rittel, H. W. J., & Webber, M. M. (1973). Dilemmas in a general theory of planning. *Policy Sciences, 4*, 155-169.
Schön, D. A., & Rein, M. (1994). *Frame reflection. Toward the resolution of intractable policy controversies.* New York: Basic Books.
Shmueli, D., Elliott, M., & Kaufman, S. (2006). Frame changes and the management of intractable conflicts. *Conflict Resolution Quarterly, 24*(2), 207-218.
Stjernfelt, F. (2007). *Diagrammatology: An Investigation on the Borderlines of Phenomenology, Ontology, and Semiotics.* Dordrecht, NL: Springer.
Tannen, D. (Ed.). (1993). *Framing in Discourse.* New York: Oxford University Press.
van Eemeren, F. H., & Grootendorst, R. (2004). *A Systematic Theory of Argumentation. The pragma-dialectical approach.* Cambridge: Cambridge University Press.
Walton, D. N., Reed, C., & Macagno, F. (2008). *Argumentation schemes.* Cambridge; New York: Cambridge University Press.

Commentary on Hoffmann's Collaborative and Adversarial Reframing

SALLY JACKSON
University of Illinois at Urbana-Champaign, USA
sallyj@illinois.edu

1. INTRODUCTION

I want to begin by thanking Professor Hoffmann for the excellent presentation, and for the great work that has gone into developing and deploying AGORA. I have followed this work with great interest, and since I too have spent some time thinking about how to use interactive online tools to improve critical thinking, I am deeply appreciative of how difficult this is. Hopefully, my comments will convey this appreciation.

The most significant feature of this work is the idea of preparing people to solve wicked problems and resolve intractable disagreements. If we could fulfil this ambition with AGORA or with some similar system, we would accomplish something of genuine worth. I want to begin by speculating about how AGORA might be more effective in training the kind of skills needed to fulfil this ambition, and then turn to what, besides skills, might be cultivated through experience with an online system.

2. A SECOND DESIGN PRINCIPLE FOR AGORA?

I want to raise a practical question that leads both to a theoretical point and to a possible second design principle. The question is why AGORA's intervention does not seem to be powerful enough to get people to criticize their own arguments. Professor Hoffmann reported some discouraging news: that students are really not very good at using AGORA's automated feedback to improve their reasoning. But he also reported some encouraging news: that when an instructor or other human adversary intervenes, students show much more ability to produce good reasons in support of their claims or to modify their claims to better reflect what they actually have support for. This process of responding to challenge, over and over and over, is very likely the key to developing a capacity for self-regulation, an ability to criticize and correct one's own reasoning. The hope is that AGORA can eventually do this without the intervention of an instructor. We aren't there yet, but perhaps we can get there.

Let's consider two possibilities: first, that interacting with software is never going to be the same as interacting with a human adversary; and second, that software might simulate a human adversary quite well if it behaves more like a human adversary.

On the first possibility, I see no reason in principle why we cannot design software that will train minds into considering other perspectives, elaborating their own arguments more fully, and revising their own thinking as they encounter challenges they cannot meet with elaboration alone. I strongly suspect, though, that this will not be through diagramming (AGORA's main design principle), but through interventions that are substantively challenging, the way interaction with a human adversary is. The reason I suspect this is that making connections explicit only goes so far; people need to meet opposition in order to motivate a deeper search for why they believe what they do. The way I have always thought about this is as a process of expansion around disagreement. If we want interactions with software to allow the same kind of experience as interactions with a human adversary, we have to train the software to disagree early and often. In one of my own efforts to use this principle, I designed a system that would disagree with students' responses to online lessons, whether or not their answers were correct. This did not immediately lead to correct beliefs, but it did lead to more thought, as evidenced by how much students wrote in response to disagreement as compared with how much they wrote in response to agreement. (See Jackson & Wolski, 2001, for details of methods and results.)

What about the second possibility, that AGORA might be more effective if more human-like in its response? As presently developed, AGORA is quite standoffish. When a user begins constructing an argument, AGORA prompts for a claim and a reason, then helps out by supplying an enabler and offering some coaching on how to self-criticize. But AGORA never disagrees, never actually gives any hint that anything is wrong with what the user proposed, as a human interlocutor does. At least in the case of modus ponens, the enabler contributed by AGORA is a nearly exact repetition of the user's own words. There is nothing here to force the user to think any harder than before. So I think Professor Hoffmann is right, that some sort of adversary is needed to trigger a real search for whether an argument makes sense—but there is no reason in principle why that adversary cannot be a suitably designed software system.

Whether a fully automated response will ever have the same effect as a human critic is unanswerable. That depends not only on how clever we are in designing an automated adversary, but also on how inventive human critics themselves become as they reflect on what can and can't be simulated through software. It may be that people will never really be motivated enough by a simulated opponent, but I doubt that, given the limitless motivation people have to overcome challenges built into computer games.

Trying to imagine how to create algorithms that would challenge an arguer in just the right ways brings us back to what is probably the single most important thing to know about argumentation: that it expands around disagreement (Jackson & Jacobs, 1980). We need to encounter disagreement with particular points in order to know what needs elaboration. And we need to have this experience many times before we start to anticipate it, and do it spontaneously. It takes repeated practice for people to develop a habit of self-criticism, and even then, they usually have to remind themselves to turn on their inner critic.

So disagreement, even the hint of disagreement, even the simulation of disagreement, is valuable. The more we expose our students to disagreement, the stronger their own inner critic will become. The original design idea behind AGORA is that diagramming exposes weaknesses that need correction; but it doesn't throw up any *particular* red flags. AGORA could build in the sort of substantive disagreements that help a student see every piece of an argument as having an associated disagreement space that might need to be explored. Indeed, this idea is already partly realized in AGORA's design, in the presentation of critical questions associated with particular argumentation schemes. A full realization would pose questions not in generic form but as real challenges to the arguer's reasoning.

3. WICKED PROBLEMS AND INTRACTABLE DISAGREEMENTS

It is not completely obvious that being better at argument criticism makes a person more effective in solving wicked problems or resolving intractable disagreements. I do not think it is guaranteed that training in argumentation, even really effective training, will equip a person to solve wicked problems. It is possible, after all, that people can be trained into hyper-competitiveness and even a sort of shallow gamesmanship.

But I like the idea of setting our aspirations this high—aspiring to equip people to do the things that are truly difficult to do in argumentation. How might an automated system like AGORA contribute to the kind of person who can respond well to wicked problems and to intractable disagreements? One thing we might do with automated systems that will not be done with human interaction alone is to guarantee that every individual, even the strongest, has a lot of experience with failure. I believe it would be all to the good if everyone were to come to difficult interactions with a healthy amount of humility. To cultivate that, we would want to design training to lead to a recognition that no argument is ever invulnerable, and that even a very poor argument can become a strong one under sustained critique.

REFERENCES

Jackson, S., & Jacobs, S. (1980). Structure of conversational argument: Pragmatic bases for the enthymeme. *Quarterly Journal of Speech, 66*(3), 251-265.

Jackson, S., & Wolski, S. (2001). Identification of and adaptation to students' preinstructional beliefs in introductory communication research methods: Contributions of interactive web technology. *Communication Education, 50*, 189-205.

10

Canons of Legal Interpretation and the Argument from Authority

MICHAEL J HOPPMANN
Northeastern University, USA
m.hoppmann@neu.edu

The paper argues that Legal Scholars and Argumentation Theorists could mutually benefit from each others' scholarship by understanding the legal "canons of interpretation" (first developed by von Savigny in the 19th century and significantly refined since) as a complex form of an argument from authority, thus merging a core piece of jurisprudence (norm interpretation) and argumentation theory (critical questions).

KEYWORDS: argument from authority, argument schemes, canons of interpretation, critical questions, norm interpretation

1. AIM AND PROBLEM

One of the most puzzling and at the same time very pressing problems in legal argumentation theory concerns the reasonable interpretation of laws and similar texts. While previously held beliefs that a law should be clear enough to "speak for itself" are – like the medieval legal syllogism – certainly a thing of the past, the best way of legal interpretation and reconstruction is still hotly contested. This is all the more significant because this question is not a mere academic puzzle, but has an important impact on crucial decision in courts, politics, and society at large. How we interpret a law or legal text matters.

Among the most promising attempts at solving this problem are the so-called "canons of interpretation" of legal texts. Originally developed by Friedrich Carl von Savigny in the early 19th century the model has since been widely debated and further refined by legal theorists mostly in Europe. In its original form, published in *System des heutigen Römischen Rechts* [The System of Contemporary Roman Law] in 1840, Savigny proposes that the interpreter of a law should distinguish between four independent levels or focal points of interpretation, the *grammatical*, the *logical*, the *historical*, and the *systematic* element. Distinguishing between these levels will produce four more precise interpretations, which can then in a second step be reasonably weighed against each other.

The canons of interpretation thus present a significant advantage over free holistic or intuitive approaches to legal interpretation, but they do not offer very precise internal criteria for judging the strengths and reasonableness of each of the interpretative levels, let alone a way to arbitrate between contradictory results on different levels. The second of these problems, concerning inter-element arbitration has attracted a certain amount of interest from legal scholars. Kriele (1976), Larenz (1991), Engisch (2005), and Alexy (2010) among others have attempted to organize the canons in a hierarchical order, but (with the exception of the first canon) these attempts have not been successful.

It is the first problem however that is the focal point of this paper. I will argue that it is possible to use relatively well-established criteria for judging the reasonableness of the interpretation within each individual canon by merging the results of legal theory (the canons of interpretation) with the results of argumentation theory (critical questions, especially for the argument from authority or expertize). Understanding each canon of interpretation as a specific argument scheme allows the critic to utilize the set of critical questions associated with this scheme.[1] These critical questions are a considerably more sophisticated tool compared to the status quo in legal theory and could thus help improve the reliability and reasonableness of legal interpretation with the help of the canons.

From this aim follows the basic structure of this paper: I will first provide a brief review of the two independent strains of thought in legal theory (i. e. norm interpretation) and argumentation theory (i. e. critical questions with an emphasis on the argument from authority), and then propose a possible solution in form of an interdisciplinary merger between the two strains. Wherever appropriate I will use a norm from the U.S. Uniform Code of Military Conduct (§904: Aiding the

---

[1] When I originally started thinking about this topic and proposed this paper I assumed that each of the six canons in Alexy's system could indeed be reconstructed as an argument from authority or expertise (I use these terms interchangeably here, see Hoppmann, 2011) – hence the title of this paper. I no longer hold this belief. As will be discussed in more detail below, some canons (and probably even the most interesting ones) can indeed be reconstructed as arguments from authority, but others are better fitted into different argument schemes. I decided to stick with the title for now, but hope that the argument of the paper changed for the better.

Enemy) as an example for challenges and solutions in argumentative norm interpretation.

It is a curious coincidence that both strains of thought go through a similar historical development: Recognized as worthy fields of inquiry in Greek antiquity, they encountered a brief period of naïve denial in the early modern period before coming to full prominence in modern legal and argumentation theory respectively. To provide some context I will accordingly include a very brief historical review of norm interpretation and the argument from authority.

2. INTERPRETING NORMS

The need to interpret legal norms was well recognized in ancient Greek rhetoric. Hermagoras of Temnos dedicated half of his stasis model to the so-called *zetemaka nomika* or legal questions (Hoppmann, 2007, pp.1337f.; Matthes, 1958, pp. 183ff.; Thiele, 1893, pp. 83f.). He distinguished between four aspects that could arise in practical rhetoric in a court of law, especially when interpreting the meaning of a legal document: 1) *rheton kai dianoia* – the disagreement between the parties could hinge on the difference between literal and intended meaning of the document or law in question; 2) *antinomia* – two or more laws might be applicable to the given case and lead to contradicting results; 3) *amphibolia* – the key document or law could be ambiguously phrased, thus leading to contradictory literal interpretations; and 4) *syllogismos* – there could be a (perceived) legal gap that might be filled by using parallel norms analogously. This collection of aspects of legal interpretation is of course far from being complete or systematic yet, but its prominent position in one of the key models of ancient rhetoric (i.e. stasis theory) points to its importance in early legal reasoning.

This awareness of the importance (and inevitability) of norm interpretation was lost in a later – in some regard more naïve – period of legal reasoning. Under the influence of Montesquieu's doctrine of the division of powers 19th century legal theorists turned to classical formal logic as a methodological role model in designing the so-called "legal syllogism" (Feteris, 1999, pp. 5ff.; Hoppmann, 2008, pp. 16ff.; Perelman, 1979, pp. 39ff.). In this construction the law was seen as the major premise, the deed of the accused as the minor premise, and the punishment as the conclusion. The only task of the judge was to test the validity of the syllogism by comparing the middle term: did the deed of the accused match the prohibition in the law? Under no circumstance should the judge independently interpret the law, as this would be seen as a violation of the division of powers: the legislator makes the law, and the judges apply it. This is of course only possible if legal texts are seen as self-explanatory and unambiguous in all cases – an assumption that is very hard to uphold under the influence of day-to-day jurisdiction. One might even argue that it is a logical impossibility to ever apply syllogistic reasoning to legal cases, since the norm by definition addresses general, abstract classes whereas the deed (equally by definition) is composed of individual, concrete cases. Whether a given case falls under an abstract class will necessarily require a certain amount of interpretation. Unsurprisingly then, the legal syllogism was relatively short lived as a paradigmatic form of legal reasoning and soon gave way to more modern forms of norm interpretation.

In 1840 Friedrich Carl von Savigny publishes his *System des heutigen Römischen Rechts* and introduces the so-called "canons of interpretation" (Savigny, 1840, pp. 212ff.; Alexy, 2010, pp. 234ff.). These canons[2] provide a novel approach to norm interpretation that significantly shaped modern continental legal theory. Rather than offering a method by which the interpreter could find the one true meaning of a legal text, Savigny divided possible interpretations into four distinct levels: the *grammatical*, the *logical*, the *historical*, and the *systematic*. The interpreter (e.g. a judge or attorney) would then in a first step try to determine the meaning of the law for each distinct level and in a second step weigh the four interpretations against each other. If all four interpretations concur, by either agreeing on judging the concrete case to be an instance of the abstract norm or by agreeing that the concrete case cannot be seen as an instance of the abstract, then the interpretative task is concluded. If some of the four levels are in contradiction to others, then their respective clarity and importance in the case must be weighed against each other. In this process neither of the canons has an automatic priority over another.[3] The resulting system thus provides a more sophisticated method for understanding norms, by introducing distinct levels of interpretation. While this is a significant step forward in legal interpretation, it also poses two new challenges: 1) How is the quality of validity of an interpretation within a given canon to be judged? And 2) How are the results of the different canons to be weighted against each other? This paper proposes that

---

[2] These "canons of interpretation" in the continental legal tradition are not to be confused with the similarly called '"canons" in the common law tradition. Comp. e.g. Scalia & Garner's (2012) definition of 'canon of construction' as "A principle that guides the interpreter of a text on some phase of the interpretive process" (p. 426). The authors then go on to introduce and discuss 52 canons.

[3] With the exception of the semantic argument under certain conditions. See below for more details.

modern argumentation theory can make a significant contribution to the first of these questions.

The model that will be used for this purpose is Robert Alexy's more contemporary version of the canons of interpretation. In his 1978 *Theorie der juristischen Argumentation* Alexy distinguishes between six canons or arguments: 1. The *semantic argument*, 2. The *genetic argument*, 3. The *historical argument*, 4. The *comparative argument*, 5. The *systematic argument*, and 6. The *teleological argument*. Similar to von Savigny with his four canons, Alexy here offers six distinct aspects of norm interpretation that can lead to compatible or contradictory results. The canons or arguments can be roughly summarized as follows:

1. The *semantic argument* is being used when the concrete case in questions (C) is classified as an instance of the legal norm (N) with reference to linguistic usage (Alexy, 2010, pp. 235f.). This linguistic usage in turn can be backed up by the interpreter with reference to "[...] *his or her own linguistic competence, empirical inquiry, and recourse to the authority of a dictionary.*" (Alexy, 2010, p. 235). This argument differs from the other five in that it can be decisive on its own under certain circumstances. If the semantic interpretation suggest that C *must* be an instance of N or that C *cannot* be an instance of N then no further interpretation is required. If the semantic argument leads to the results that C *can* be an instance of N then it is concluded that N is vague and the other five canons are needed.

2. The *genetic argument* is employed when C is classified as an instance of N based on the intent of the historical legislator of N (Alexy, 2010, pp. 236-239). This can happen in one of two forms: a) the historical legislator has made explicit reference to cases like C and intended them to fall under N, or b) the historical legislator pursued a specific goal with passing N, and this goal can only (or best) be achieved by classifying C as N. Alexy acknowledged the similarity of the second form of the genetic argument with the teleological argument below, and also briefly addresses the notorious problem in trying to determine any specific historical "intent", especially in the case of large legislative bodies that are composed of many individuals with potentially starkly differing personal intentions.

3. The *historical argument* relies on drawing similarities between the present interpretative problem and similar situations (in the same community) in the past (Alexy, 2010, p. 239). Alexy only explicitly addresses cases in which previous interpretations of "C as N" or "C as not N" have led to undesirable results and should therefore be avoided, but it is easy to imagine that the same canon could also be used in a positive manner to learn from the past.

4. The *comparative argument* is the counterpart to the historical argument, but rather than examining the past of the same community the comparison here reaches out to other societies and their experiences in attempting to solve the interpretative problem at hand (Alexy, 2010, pp. 239f.). Since Alexy explains the comparative argument only with reference to the historical argument it is once again unclear if the interpreter here should be limited to learning from the mistakes of others or if this canon is meant to include both analogies of success and of failure.

5. The *systematic argument* relies on the relationship between N and its neighboring norms (Alexy, 2010, p. 240). Since an individual norm (especially in a continental law system) is usually part of a larger code or collection of norms, one can take valuable information about the best way to interpret N, based on where N resides within the code and whether a given interpretation of C as N would contradict established interpretations of other norms in the same system.

6. The *teleological argument* (or *objective-teleological* to distinguish it clearly from the *historical-teleological* argument above) is probably the most complex of the six canons (Alexy, 2010, pp. 240-244). It relies on the overarching aim of the legal system for its interpretation or on *"[...] those aims which decision-makers deciding within the framework of the valid legal order would posit on the basis of rational argumentation, are rational aims or aims objectively prescribed in the framework of the valid legal order."* (Alexy, 2010, p. 241). The interpreter ultimately makes the claim that seeing 'C as N' or 'C as non N' is required to achieve the aims of the legal systems in question. Alexy also calls this *"a kind of argument from principles."* (Alexy, 2010, p. 243)

It is easy to see that dividing the norm interpretation into these six distinct canons has the potential to lead to a significantly more sophisticated interpretation than an all-encompassing holistic approach (or worse: a narrow loyalty to just one canon or its equivalent).[4] It also triggers a number of new questions, not least of which is the problem to distinguish between weak and strong or valid and invalid interpretations within each canon. This is a problem that can be addressed the argument scheme studies and the concept of critical questions.

3. ARGUING FROM AUTHORITY

---

[4] Comp. narrow-minded "literalist" or "originalist" approaches.

Studying modes of reasoning in general and the argument from authority or expert opinion in particular has a very long tradition. Aristotle's *topoi koinoi* are the well-known predecessors of modern argument schemes.[5] Among the valid *topoi* Aristotle lists as number eleven the argument from authority:

> *Another [topic] is from a [previous] judgement [ek kriseos] about the same or a similar or opposite matter, especially if all always [make this judgement] – but if not, at least most people, or the wise (either all of them or most) or the good; or if the judges themselves [have so decided] or those whom the judges approve or those whose judgment cannot be opposed* (Arist. Rhet. 1398b20-23 transl. Kennedy).

He echoes a similar appreciation for valid reasoning based on authority in his definition of the common starting point (endoxon) in a dialectical conversation in the Topic:

> *Generally accepted opinions, on the other hand, are those which commend themselves to all or to the majority or to the wise – that is, to all of the wise or to the majority or to the most famous and distinguished of them.* (Arist. Top. 100b21-23 transl. Forster).

In both books he also hints at the kind of quality a valid authority must possess, but he does not give us a full set of criteria for testing the reasonableness of an argument from authority or expert opinion yet. These clear sets of criteria will only be introduced in modern argumentation scheme studies.

Before the advent of modern argumentation theory in the 20th century, we find a similar relapse for the appreciation of the argument from authority as we did in the case of legal interpretation. In his attempt to establish a rigidity of reasoning after the model of geometry (*more geometrico*) Rene Descartes outright rejects any (non-divine) form of argument from authority. In his *Principles of Philosophy* (1644) he writes "*That we ought to prefer the Divine authority to our perception; but that, apart from things revealed, we ought to assent to nothing that we do not clearly apprehend.*" (Descartes, 1901, I, 76). In other words, arguments from authority must be either based on Christian revelation or be considered as fallacious or invalid. This could be easily written off as an insignificant period in the history of argumentation, were it not for

---

[5] Comp. Hoppmann (2008b) for an overview of Aristotelian modes of reasoning.

the fact that Descartes' shadow still reaches all the way to modern treatments on reasoning which follow the geometric tradition. Copi's early editions of his influential *Introduction to Logic* does not even acknowledge the argument from authority in the index (or anywhere else in the book (Copi, 1956)), whereas more recent editions of the book forward the reader from "argument from authority" to "argumentum ad verecundiam" in the index, thus treating this scheme by default as a fallacy.[6]

In contrast most modern treatment on argument schemes include the argument from authority under a variety of names. Perelman and Olbrechts-Tyteca (1958) list the "Argument from Authority" as one of their "Arguments based on the Structure of Reality", Hastings (1962) includes the "Argument from Authority (Testimony)" in his list of nine modes of reasoning (Hastings, 1962, pp. 126-138.), Kienpointner (1992) dedicates a whole chapter of his *Alltagslogik* to "Autoritätsargumentation" (Kienpointner, 1992, pp. 393-401), and Walton, Reed, and Macagno (2008, pp. 309-313) open their large list of argument schemes with a number of variations of authority based arguments. More recent individual articles on the argument from authority or expert opinion are too numerous to list.[7] The key element for the present purpose, which many of these publications provide is a list of critical questions associated with each argument scheme. These lists provide a crucial advantage over earlier treatments of say, the topoi, as well as the canons of interpretation in that they further distinguish the critical elements of a valid argument. Applying these lists to each of the canons would allow the legal interpreter to make the next significant step in testing the strength of an interpretation.

There is no universal agreement on the precise list of critical questions for each of the schemes. For the argument from authority Hastings lists five questions (Was the source enabled to observe the situation? Is the authority competent in his field? Is the authority motivated to be accurate? What internal evidence is there of the truth of the conclusion? Does the testimony have factors which are highly correlated with accuracy?) (Hastings (1962) pp. 135-137), Kienpointner includes six questions (Is the person of authority cited correctly and completely? Is the person of authority passing its judgment in its area of

---

[6] Copi & Cohen (2010). Modern edition do acknowledge reasonable argumentation from authority under very limited conditions, but these seem like a rather reluctant afterthought. Comp. Copi & Cohen (2010, p. 142)

[7] A good starting point is however the superb collection of essays in the special edition of *Argumentation* 2011 devoted to the topic.

expertise? Is the person of authority in his/her area respected (without reservations), meaning accepted by other authorities as well? Is the person of authority passing its judgment without bias? Is the person of authority immunized against criticism? Are there critical or conflicting judgments by counter authorities?) (Kienpointner (1996) p. 176, my translation), and Walton et al. (2008, p. 310) list a slightly different set of six questions for their argument from expert opinion (How credible is E as an expert source? Is E an expert in the field that A is in? What did E assert that implies A? Is E personally reliable as a source? Is A consistent with what other experts assert?).

While there is a certain variety within these sets of critical questions these do not need to concern us for the present purpose. Yes, a superior set of critical questions is superior for the task at hand, but any set of critical questions is a significant step ahead of the status quo. Among the core questions that can be used on the canons of interpretation when understood as different argument from authority are:

1. Has the invoked authority actually made the statement ascribed to it?
2. Is the invoked source an authority in its field?
3. Does this field match the case at hand?
4. Is the authority making the statement free from bias or similar extraordinary circumstances?[8]

The main challenge that remains then is to translate Alexy's canons into argument schemes.

4. CANONS AS SCHEMES

At first glance the canons of interpretation and the argument schemes seem to be very different concepts; after all they come from quite different historical backgrounds and attempt to solve different academic questions. At second glance it could be argued that the canons of interpretation are a specialized form of argument, supporting a specific claim (C is an instance of N, or C is not an instance of N). Most crucially, each canon corresponds to one specific argument scheme (most frequently a form of the argument from authority) thus allowing a fixed assigning of a specific list of critical question to the canon. Assigning these lists would thus add a second dimension to the six-fold division that is at the disposal of the legal interpreter, extending the set of

---

[8] For a more detailed discussion of the critical questions attached to the argument from authority comp. also Hoppmann (2009).

discreet aspects available from six to about twenty and allowing for a most differentiated approach to norm interpretation.

For the purpose of illustration it will be useful to provide a specific norm for interpretation (N) and a specific case (C).

No single norm will be a perfect example for all canons, but for the purposes of this illustration we can use the following U.S. law, which has recently attracted a certain amount of international attention:

> § 904 Aiding the enemy
> Any person who—
> (1) aids, or attempts to aid, the enemy with arms, ammunition, supplies, money, or other things; or
> (2) without proper authority, knowingly harbors or protects or gives intelligence to, or communicates or corresponds with or holds any intercourse with the enemy, either directly or indirectly;
> shall suffer death or such other punishment as a court-martial or military commission may direct. This section does not apply to a military commission established under chapter 47A of this title.

To give the interpretation of N a focal point we can use the Wikileaks case – more specifically the transmission of data, such as the infamous "Collateral Murder video" from Chelsea Manning to Julian Assange as C. The illustrative (and not purely hypothetical) version of "Is C and instance of N?" for this purpose thus will be "Did Manning's actions qualify as aiding the enemy?"

In an attempt to answer this question with the help of the canons of interpretation a judge or attorney can focus on the six levels outlined by Alexy. Each of them can then in turn be understood to be an instance of a particular argument scheme.

i. The semantic argument

There are a number of key terms in §904 that can lead to more than one possible interpretation. Among these certainly are the words "enemy" and "giving intelligence". According to the results of her trial Manning did indeed transfer data from her computer to Assange without proper authorization. And since Assange in turn made the majority of these data widely available on the WikiLeaks platform (as well as – indirectly – via publications from *The Guardian*, *The New York Times* and *Der Spiegel*) they could have been accessed by groups hostile to the United States, such as the Taliban. But does providing information to a neutral party who in turn shares it with (among others) a hostile party qualify as "giving intelligence", and can the Taliban be classified as "enemies" of the U.S.?

In order to answer these questions on the semantic level the legal interpreter can turn to his or her own linguistic competence or the authority of a dictionary. The conversion of this canon into an argument from authority is certainly the easiest one, and Alexy himself already indicates as much in his phrasing of the canon. In the above case one can then for example turn to the Oxford English Dictionary where under A. I. "Enemy" is defined as "An unfriendly or hostile person." This definition seems to encompass the Taliban. The critical questions that can be connected to this canon to test its strengths are as indicated above:

1. Does the Oxford English Dictionary actually define 'enemy' as 'an unfriendly or hostile person'?
2. Is the Oxford English Dictionary an authoritative dictionary?
3. Does the Oxford English Dictionary's field include legal language?
4. Are there no indicators that the Oxford English Dictionary's entrance on 'enemy' is subject to extraordinary circumstances?

While there is necessarily still a subjective element in judging the answer to each of these questions, using them as an instrument for evaluating the strength of the evidently leads to a more sophisticated toolset and a more differentiated answer than a pure holistic approach. It might for example point the interpreter to a weakness of the argument in critical question 3. Since the OED is not a specialized legal dictionary, one could then turn to Black's Law Dictionary, which defines "enemy" as signifying "[...] either the nation which is at war with another, or a citizen or subject of such nation." Since there has never been a U.S. declaration of war against any nation involved with the Taliban, this might leave the answer to "Is C a case of N?" more uncertain.

ii. The genetic argument

Translating the second canon into an argument from authority also seems to be a rather straightforward endeavor. Appealing to the intent of the historical legislator is an appeal to the authority of a person or a group of people who are in a particularly good position to know how the law should be interpreted. Their authority is based on the double qualification of a) being well enough respected people to be entrusted with the legislation of a society and b) presumably having detailed knowledge of any law they personally vote for.

If one was then to find – for example – the transcript of a speech of a well-respected member of the legislative body addressing §904 and justifying it with reference to the special loyalty that can be expected from a citizen and soldier during the brief periods of an openly declared war, then one could similarly test this interpretation with the respective set of critical questions.

1. Did the member actually include this statement in his speech?
2. Was the member an authority in its field? (e.g. was he a well-respected representative?)
3. Does this field match the case at hand? (e.g. was he involved with this law?)
4. Did he speak free from bias or similar extraordinary circumstances? (e.g. do we have no indicators of vested interests of inappropriate partisanships influencing this speech?)

iii. The historical argument

Translating the third canon into an argument scheme is a little more complicated. It is tempting to interpret the historical canon again as an argument from authority; this time referring to the expertize of previous judges involved in interpreting N. After all, similar to the historical legislators, previous judges have a) been selected to an elevated position in society because they were well respected in some regard and b) presumably have worked up a certain amount of knowledge surrounding the case ("Is C' a case of N?"). This reconstruction might work reasonably well with positive interpretative precedents, but it seems to fail for negative cases. This is particularly relevant since Alexy explicitly only mentions arguments of the form "C' was previously understood as falling under N and the results were undesirable". It seems to be a stretch to interpret this as some form of reverse authority. A more straightforward translation would instead see the historical canon as a case of the argument from analogy: 1) C' has been judged to be an instance of N and the results were (un)satisfying; 2) C' and C are essentially similar; 3) Therefore judging C to be an instance of N is (un)satisfying.

The related set of critical questions for the argument from analogy (and thereby the historical canon) will be a variety of the following:[9]

---

[9] Comp. for a more detailed justification of tis choice of critical questions also Hoppmann (2009)

1. Does A have proposition p? (Has C' indeed been judged to be an instance of N and have the results been X?)
2. Are A and B essentially similar? (Are the cases C' and C essentially similar?)
3. Are the proposition of the initial object and the target object identical? (Is the X in question kept stable?)

In interpreting the Manning case with regards to §904 one might for example refer back to the Pentagon Papers and the treatment of Daniel Ellsberg, thus triggering the critical questions:

1. Was Ellsberg indeed not convicted of aiding the enemy?
2. Were the actions of Ellsberg and Manning essentially similar? (e.g. with reference to their professions as journalist and member of the military respectively)
3. Did the lack of convicting Ellsberg indeed lead to desirable consequences?

iv. The comparative argument

The fourth canon is the exact counterpart of the third, with the only difference that the parallel is not along the time axis, but across societies. Accordingly, translating the comparative canon into an argument from analogy can be done in a similar manner to the translation described above for the historical canon.

v. The systematic argument

The systematic argument is perhaps the most complex canon when it comes to translating it to an established argument scheme. At first glance one might be tempted to once again reconstruct it as an argument from authority – in this case the authority of the compiler of the legal code. After all the norm in question has been placed in a particular position within the code and this placement has been done by a human agent. Unfortunately there are a couple of problems with this interpretation: a) Identifying the agent responsible for compiling the code might be challenging, b) the compiler and the historical legislator may coincide – thus collapsing this canon with the second canon; but most importantly c) if the compiler isn't identical with the historical legislator, what qualifier him/her/them as a significant authority?

A more promising direction of the translation would instead be twofold – depending on the kind of systematic argument that is being made. If reference is being made to the interpretation of neighboring norms it is easy to see that the underlying scheme is that of analogy (with the critical questions discussed above).[10] If reference is made to superior norms and the position of the norm in the entire code then the applicable scheme will be parts and a whole. In this scheme one reasons from the membership of an entity in a group to certain qualities. Typical critical questions for this scheme include:[11]

1. Is A a member of G?
2. Is p a constitutive quality of G?

In the case of §904 there is a wealth of interpretative information that can be taken from the position of the norm in the complete code. §904 is part of Subchapter X, which is in turn part of Chapter 47, which is part of Part II, which is part of Subtitle A, which is part of Title 10, of the U.S. Code. Or to put it into a slightly less nauseating form:

U.S. code → Title 10: Armed Forces → Subtitle B: Army → Part II: Personnel → Chapter 47: Uniform Code of Military Justice (UCMJ) → Subchapter X: Punitive Articles → §904 / Art. 104. Aiding the enemy

If one would thus for example try to interpret the opening words of §904 "Any person who", the systematic canon might guide towards limiting the meaning of "person" to a member of the military in general or the army in particular.[12] The critical questions in this case could be formulated as:

---

[10] This translation would probably also best account for Alexy's standard case, in which the interpreter addresses the consistency of the word usage rule W with regard to neighbouring norms N and N'.

[11] Unfortunately Walton et al. do not list critical questions for their scheme 9.2; The closest scheme in Walton (Argument from Verbal Classification) lists two critical questions: 1) Does a definitely have F, or is there room for doubt?; 2) can the verbal classification (in the second premise) be said to hold strongly, or is it one of those weak classifications that is subject to doubt? (1996, p. 54). Kienpointner's equivalent scheme is the slightly more complex Art_Gattung: Unter- und Überordnung, for which he lists a total of six critical questions, including regarding the membership of G, its general acceptance and alternative hierarchal constructions (Kienpointner, 1996, pp. 97f.).

[12] This would of course be in contrast to the interpretation suggested by the semantic canon where one might make reference to the OED's definition of "person" as "An individual human being; a man, woman, or child." (IIa)

1.
2. Is §904 really part of Title 10?
3. Is it as constitutive quality of Title 10 to only be applicable to members of the armed forces?

vi. The teleological argument

The final canon is the only one, in which a translation into an argument from authority scheme seems to be immediately out of the question. Instead the objective-teleological argument seems to be a clear case of cause-and-effect reasoning, or to be more precise, a manifestation of the argument from effect to cause. I have argued previously that there are four critical questions associated with this scheme, which are[13]:

1. Does the "effect" have the claimed proposition p?
2. Is the relationship between cause and effect indeed as assumed? (i.e. necessary or quasi-necessary)
3. Is the proposition expressed in the "cause" the same as the one expressed in the "effect"?
4. Is the "cause" chronologically prior or synchronous to the "effect"?

However, due to the characteristics of this particular argument and the qualities of the legal system as a whole, three of the four critical questions will practically always return positive results. Question 1. tests if the aims of the entire legal system are indeed desirable (short of extremely tyrannical societies or open civil war, these can probably be taken as a given). Question 3. tests against a naturalistic fallacy, which given that we are dealing with normative interpretations on both sides will not effectively be a problem. Finally, question 4. tests against fallacious correlations in the guise of causal relations. This too, can be ruled out for the relationship between individual norm and entire code. In practice this means that while the teleological canon can indeed be translated to an argument from effect to cause, this translation does not lead to a more detailed testing structure, because the only remaining critical question only inquires whether the interpretation of "C as N" is indeed necessary or quasi-necessary to reach the overarching goals of the legal system.

---

[13] Hoppmann (2011); that paper also includes an extensive review of alternative lists of critical questions for causal schemes.

3. Is interpreting "C as N" / "C as non N" indeed a necessary or quasi-necessary condition for fulfilling the larger purpose of the larger norm or norm system?

This result is largely identical to Alexy's own analysis of the questions inherent in the teleological canon (Alexy, 2010, pp. 241ff.).

5. CONCLUSION

The best way to interpret a legal norm has long been one of the central questions in legal theory and rhetoric. The works of modern legal scholars inspired by von Savigny, most recently Robert Alexy, have allowed legal theorists to make a huge step forward, moving from a complex holistic interpretation of the norm to a more sophisticated six-level analysis. Translating each of these six levels to a well-established and analyzed argument scheme, allows us to use the insights of modern argumentation theory to provide an additional dimension to norm interpretation and the testing of its felicity conditions. Equipped with this extension a person interested in the strength of a particular interpretation of a norm thus has not only the six distinct levels (provided by the canons) at his or her disposal, but also a total set of eighteen critical questions, that are attached to the canons. Breaking up the complex problem of "Is C a case of N?" can thus lead to a significantly more differentiated approach to norm interpretation.

REFERENCES

Aristotle (2007). *On Rhetoric*. 2nd ed. Trans. G. A. Kennedy. Oxford & New York: Oxford University Press.
Aristotle (1960). *Topica*. Trans. E. S. Forster. Cambridge, MA & London: Harvard University Press.
Alexy, R. (2010). *A Theory of Legal Argumentation*. Trans. R. Adler & N. MacCormick. Oxford: Oxford University Press.
Copi, I. (1956). *Introduction to Logic*. London: Macmillan.
Copi, I., Cohen, C., & McMahon, K. D. (2010). *Introduction to Logic*. 14th ed. Upper Saddle River, NJ: Pearson Education.
Descartes, R. (1901). *The method, meditations and philosophy of Descartes*. Transl. J. Veitch. Washington D.C. & London: M. W. Dunne
Engisch, K. (2005). *Einführung in das juristische Denken*. 10th ed. Stuttgart: Kohlhammer.
Feteris, E. (1999). *Fundamentals of Legal Argumentation*. Dordrecht: Kluwer.
Hastings, A. C. (1962). *A Reformulation of the Modes of Reasoning in Argumentation*. Unpublished Dissertation. Northwestern University Evanston, IL.

Hoppmann, M. (2007). Statuslehre. In G. Ueding (Ed.), *Historisches Wörterbuch der Rhetorik*. Vol. 8 (pp. 1327-1358). Tübingen: Max Niemeyer.
Hoppmann, M. (2008a). *Argumentative Verteidigung*. Berlin: Weidler.
Hoppmann, M. (2008b). *Rhetorik des Verstandes (Beweis- und Argumentationslehre)*. In U. Fix, A. Gardt & J. Knape (Eds.), *Handbuch der Sprach- und Kommunikationswissenschaft (HSK 31.1). Rhetorik und Stilistik. Vol. I* (pp. 630-645). Berlin et al: de Gruyter.
Hoppmann, M. (2009). The Rule of Similarity as Intercultural Basis of Defeasible Argumentation. In J. Ritola (Ed.), *Argument Cultures: Proceedings of OSSA 09*. Windsor, ON.
Hoppmann, M. (2011). Correlation and Causality. In F. Zenker (Ed.), *Argumentation: Cognition and Community. Proceedings of OSSA 10*. Windsor, ON.
Kienpointner, M. (1992). *Alltagslogik*. Stuttgart: Frommann-Holzboog.
Kienpointner, M. (1996). *Vernünftig Argumentieren*. Hamburg: Rohwohlt.
Kriele, M. (1976). *Theorie der Rechtsgewinnung*. 2. ed. Berlin: Duncker und Humblot.
Matthes, D. (1958). Hermagoras von Temnos. *Lustrum*, *3*, 58-214.
Larenz, K. (1991). *Methodenlehre der Rechtswissenschaft*. 6. rev. ed. Berlin et al.: Springer.
Perelman, C. (1979). *Juristische Logik als Argumentationslehre*. Trans. J. M. Broekman. Freiburg & Munich: Verlag Karl Alber.
Perelman, Ch. & L. Olbrechts-Tyteca (1969). *The New Rhetoric*. Transl. J. Wilkinson & P. Weaver. Notre Dame: University of Notre Dame Press.
Savigny, F. C. von (1840). *System des heutigen Römischen Rechts*. Bd. 1. Berlin: Veit und Comp.
Scalia, A., & Garner, B. A. (2012). *Reading Law*. St. Paul: Thomson/West.
Thiele, G. (1893). *Hermagoras. Ein Beitrag zur Geschichte der Rhetorik*. Strasbourg: Verlag von Karl J. Trübner.
Walton, D. (1996). *Argumentation Schemes for Presumptive Reasoning*. Mahwah, NJ: Lawrence Erlbaum.
Walton, D., Reed, C., & Macagno, F. (2008). *Argumentation Schemes*. Cambridge et al.: Cambridge University Press.

Argument Schemes, Authority and Legal Interpretation. Commentary on Hoppmann's Canons of Legal Interpretation and the Argument from Authority

MICHAŁ ARASZKIEWICZ
Jagiellonian University
michal.araszkiewicz@uj.edu.pl

MARCIN KOSZOWY
University of Białystok
koszowy@uwb.edu.pl

1. INTRODUCTION

In his paper, Michael Hoppmann discusses the issue of legal interpretation from the viewpoint of argumentation schemes theory. This topic is of crucial importance and very relevant to the ECA conference. The author proposes to treat six well-known canons of legal interpretation, discussed by Alexy, as argumentation schemes for which matching sets of critical questions are provided. We find this task commendable. In addition to Hoppmann's proposal, we think that this discussion could also involve a discussion of some related research strands both in the field of legal interpretation and in the recent study of argumentation schemes.

Our discussion is divided into three parts. In Section 2, we discuss the state of art in the study of legal interpretation and argumentation schemes which, in our opinion, provides an important background for author's analyses. Section 3 deals with certain distinctions related to the concept of authority which may be applied by Hoppmann in the course of his research. Finally, we present some conclusions.

2. CANONS OF LEGAL INTERPRETATION – THE STATE OF THE ART: THE HISTORY, AND THE PRESENT DAY

The author makes two types of references to the state of the art concerning legal interpretation: (1) historical ones, where the main reference is the work of Savigny and (2) references to the contemporary state of the art, mainly to German jurisprudence, where Alexy's contribution is the point of departure of author's original analysis. As a matter of course, the literature on legal interpretation is so immense that it is entirely impossible to refer even to the most important work in the field in one paper. Therefore, any author dealing with the problem of legal interpretation has to make choices and this is also the case of Hoppmann's contribution. However, some of those choices, together with author's comments on the selection of the literature, may provoke some comments.

Let us begin with historical references. When discussing canons of interpretation, the author claims that they were "introduced" by Savigny. Doubtless, the referred work of Savigny is one of the most influential jurisprudential treatises published in the 19th century. But it would be historically inaccurate to claim that the canons of interpretation were not present in the state of the art in earlier writings.

The art of interpretation of different texts (including religious, literary and, last but not least, legal ones) traces its roots back to Antiquity (let us note that the author accurately observes that Ancient Greeks were aware of the problems of legal interpretation) and is commonly known under the name of hermeneutics. Of course, ancient and medieval scholars did not use the conceptual scheme proposed by Savigny, but the canons of interpretation such as textual or teleological arguments were recognized many centuries before the emergence of the German historical school of jurisprudence.

Through the centuries, methods of interpretation of different types of texts evolved in separation, until the 19th century when Schleiermacher proposed a unified method of interpretation (Stelmach & Brożek, 2006). Savigny referred to this philosophical account in his work. Therefore, Savigny's contribution may be praised in particular for grounding legal interpretation in a broader context of philosophy of interpretation, but he should not be indicated as the "inventor" of canons of interpretation. In the "older" legal hermeneutics, similar topics were also a subject of investigations (Stelmach & Brożek, 2006). Extensive analyses concerning the subject of historical development of juristic methods of legal interpretation may be found in Padovani & Stein (eds.) 2007. We agree with the author as regards the significance of Savigny in the history of legal philosophy, but the reader of the commented work may have an impression that too much is attributed to the founder of German historical school of jurisprudence.

As far as the contemporary literature is concerned, the author bases his investigations on the catalogue of six interpretive canons as presented by Alexy in his famous *Theory of Legal Argumentation*. This choice is of course justified due to prominence of Alexy's work in the field of legal theory. However, the presentation would presumably be more complete and convincing if the author referred also to alternative accounts such as those proposed by Tarello (1980), Wróblewski (1992) or Peczenik (2008). It should also be noted that since 1991, the collaborative comparative work edited by MacCormick and Summers (1991) remains the standard point of departure for any investigation accounting for legal interpretation by means of argumentation theories. In summing up the above considerations, we think that presentation of a broader state of the art background could contribute positively to the author's analyses. In particular, presentation of a broader set of "canons" would make the discussion of the paper's main thesis (reconstruction of these canons as argumentation schemes based on authority) more complete and refined.

Having suggested the historical and contemporary literature which may be fruitful for the author in his future work, let us now move on to the discussion of the most important context of his research. Although Hoppmann refers to the general work of Macagno, Reed and Walton on argumentation schemes, he does not compare his results to the paper by Macagno, Walton and Sartor (2012). In the said paper, the three authors reconstruct the interpretive canon based on intention of the actual legislator and *argumentum a contrario* as argumentation schemes. Therefore, the scopes of the commented paper and the work referred here overlap to a significant extent. Interestingly, the sets of critical questions to the argument scheme based on legislative intention, developed, on the one hand, by Hoppmann, and, on the other hand, by Macagno, Walton and Sartor, differ considerably. The set of critical questions elaborated by Hoppmann focuses more on the very person of the parliamentary representative expressing the will of the legislator. The three referred authors formulate more abstract critical questions. A comparative discussion of these sets of critical questions may lead to very interesting results, not only in the domain of legal interpretation, but also to very general ones (how the sets of critical questions to argumentation schemes should be developed?). Let us also note that the argumentation schemes-based approach to legal interpretation is nowadays very common in more formal elaborations of this topic, for instance in the field of AI and Law research (Araszkiewicz, 2013, Araszkiewicz, 2014, Sartor, Walton, Macagno & Rotolo, 2014).

3. ARGUMENTS, AUTHORITY AND NORMATIVITY

Although the paper addresses some problems regarding arguments from authority, its title ("Canons of Legal Interpretation and The Argument from Authority") may be conceived as not fully adequate with respect to the investigated problem. It seems that the majority of interpretative arguments are not arguments based on authority. We are aware of the fact that in footnote 1 Hoppmann explains that the title was based on the former hypothesis that has been modified later, but it seems that the linkage between arguments from authority and canons of legal interpretation could be discussed thoroughly in author's future work.

We also think that it should be explicated in a greater detail why the topic of canons of legal interpretation is linked in the paper to the topic of arguments from authority. As the author points out in footnote 1, tailoring this topic exclusively to arguments from authority does not do full justice to the diversity of argumentative techniques related to the canons of legal interpretation. Hence this sort of justification could be explicitly given in the paper.

An important question there arises: what is the *differentia specifica* between arguments which may rightfully be reconstructed using the notion of authority and other types of interpretive canons (i.e., non-authoritative ones?). Let us add here that while the law as such is often attributed with the feature of authoritativeness (some scholars, for instance Raz (2009), even claim that it is the defining feature of law), it remains doubtful whether such feature may be assigned to canons of legal interpretation. Indeed, interpretive arguments, although they can both guide and justify behaviour of relevant actors (lawyers, judges), remain weak and debatable constraints rather than more rigid rules. They do not have specific "pre-emptive" or "exclusionary" function attributable to at least some legal rules (Gizbert-Studnicki, 2015). These observations lead us to profound questions concerning the sources of (alleged) normativity of interpretive canons in general and of their particular types. Let us take, for instance, an argument based on plain ordinary meaning. What is the relation between descriptive and prescriptive function of dictionaries? How to draw a boundary between semantic knowledge and subject-matter knowledge? And, most importantly, can semantic considerations bring any sufficient grounds for an interpretive decision? (see the discussion of this topic in Gizbert-Studnicki, 2015). Similar questions may be raised with respect to other sources of alleged "authority" of interpretive arguments. Importantly, let us note that Schauer (1987) explicitly juxtaposes an argument from precedent (which is actually authoritative) and argument based on analogy (similarities themselves are not authoritative, but they may be persuasive).

Another remark concerns the interchangeable use of terms "argument from authority" and "argument from expertise". This approach has been explicated in the paper and is of course a legitimate one, but – particularly in the field of legal argumentation where some other sorts of argumentative appeals to authority play an important role – a justification of a broader account could be also considered. For instance, a distinction between (1) expert authority regarding knowledge and (2) administrative or deontic authority regarding what should be done could be here employed in order to distinguish arguments appealing to expertise from those that appeal to the authority of law (see for instance Goodwin (1998) and Walton (1997)).

4. CONCLUSION

Hoppmann's paper is a contribution to the on-going research on application of argumentation schemes theory to the problems of legal interpretation. It offers an interesting, although to some extent controversial, account of reconstruction of six canons of legal interpretation as argumentation schemes with assigned sets of critical questions. In particular, the author attempts to reconstruct some of these canons as (subtypes of) arguments from authority. As we show above, this attempt is debatable for several reasons. As different types of authority are discussed in argumentation schemes theory, these distinctions should be taken into account when applying the said theory to specific domain of law. Doubtless, interpretive canons carry a load of normativity, but this normativity is apparently weaker than in cases of commonly understood arguments based on (deontic) authority. The nature of this normativity should be investigated carefully. The structure of developed argument schemes as well as sets of critical questions should be compared to the relevant literature and methodologically justified. We are of the opinion that the direction taken by the author is very promising, albeit it leads to difficult philosophical issues. The discussion of those issues may however be inspiring for future attempts of elaborating models of legal interpretation in the future.

ACKNOWLEDGEMENTS: We gratefully acknowledge the support of the Polish National Science Center for Koszowy under grant 2011/03/B/HS1/04559.

REFERENCES

Araszkiewicz, M. (2013). Towards systematic research on statutory interpretation in AI and Law. In K. Ashley (Ed.), *JURIX 2013: The Twenty-Sixth Annual Conference* (pp. 14-24). Amsterdam: IOS Press.

Araszkiewicz, M. (2014b). Scientia Juris: A Missing Link in the Modelling of Statutory Reasoning. In R. Hoekstra (Ed.), *JURIX 2014: The Twenty-Seventh Annual Conference* (pp. 1-10), Vol. 271. Amsterdam: IOS Press.

Gizbert-Studnicki, T. (2015). The Normativity of Rules of Interpretation. In M. Araszkiewicz, P. Banaś, T. Gizbert-Studnicki & K. Płeszka (Eds.), *Problems of Normativity, Rules and Rule-Following* (pp. 243-254). Cham: Springer International Publishing.

Goodwin, J. (1998). Forms of authority and the real *ad verecundiam*. *Argumentation, 12*, 267-80.

Macagno, F., Sartor, G. & Walton, D. (2012). Argumentation Schemes for Statutory Interpretation. In M. Araszkiewicz, M. Myška, T. Smejkalová, J. Šavelka, & M. Škop (Eds.), *ARGUMENTATION 2012: International Conference on Alternative Methods of Argumentation in Law* (pp. 31-44). Brno: Masaryk University.

MacCormick, N., & Summers, R. (1991, Eds.). *Interpreting Statutes. A Comparative Study*. Dartmouth: Ashgate.

Padovani, A., & Stein, P. (2007, Eds.). *The Jurists' Philosophy of Law from Rome to the Seventeenth Century. A Treatise on Legal Philosophy and General Jurisprudence Vol. 7*. Dordrecht: Springer.

Peczenik, A. (2008). *On Law and Reason*. 2nd ed. Dordrecht: Springer.

Raz, J. (2009). *Authority of Law*. 2nd ed. Oxford: Oxford University Press.

Sartor, G., Walton, D., Macagno, F. & Rotolo, A. (2014). Argumentation Schemes for Statutory Interpretation: A Logical Analysis. In R. Hoekstra (Ed.), *JURIX 2014: The Twenty-Seventh Annual Conference* (pp. 11-20). Amsterdam: IOS Press.

Schauer, F. (1987). Precedent. *Stanford Law Review, 39*, 571-605.

Stelmach, J., & Brożek, B. (2006). *Methods of Legal Reasoning*. Dordrecht: Springer.

Tarello, G. (1980). *L'intrepretazione della lege*. Milano: Giuffré.

Walton, D. (1997). *Appeal to expert opinion*. University Park, PA: Pennsylvania State University Press.

Wróblewski, J. (1992). *The Judicial Application of Law*. Ed. by Z. Bankowski & N. MacCormick. Dordrecht: Springer.

11

Modeling Argumentative Activity in Mediation with Inference Anchoring Theory: The Case of Impasse

MATHILDE JANIER
Centre for Argument Technology, University of Dundee, UK
m.janier@dundee.ac.uk

MARK AAKHUS
School of Communication and Information, Rutgers, USA
aakhus@rutgers.edu

KATARZYNA BUDZYNSKA
Polish Academy of Sciences, Poland / University of Dundee, UK
budzynska.argdiap@gmail.com

CHRIS REED
Centre for Argument Technology, University of Dundee, UK
c.a.reed@dundee.ac.uk

The goal of this paper is to model the moves mediators and disputants make in mediation activity. What is of interest here is the generic modeling of the moves in the discussion so that the insights of various theories of argumentation can be brought to bear. For this we turn to the Inference Anchoring Theory (IAT) approach. IAT, in particular, allows showing how sequences of utterances work together to form arguments in a dialogical context.

KEYWORDS: inference anchoring theory, argumentative structure, mediation discourse

1. INTRODUCTION

Our interest is in modeling the moves mediators and disputants make to manage impasse in mediation activity. While mediation is conducted to repair a relationship, whether personal or professional, the activity of mediation itself can breakdown. That is, the negotiation or planning that a mediator is helping disputing parties achieve can begin to go wrong or fail (Jacobs & Jackson, 1992). It is under the conditions of impasse when the breakdown of mediation activity becomes apparent. Aakhus (2003) has identified three sources of impasse that undermine the conditions for holding a critical discussion (or the approximation of such a dialogue in real life).

For present purposes we do not comment on whether mediation is a full blown institutionalized activity organized around the resolution of disagreement or whether it is a practice that varies considerably in terms of its formality and values for communication (e.g. Jacobs & Aakhus, 2002; Greco Morasso, 2011, 2008). We do recognize and highlight that mediators play a role in shaping and conditioning the argumentative possibilities and qualities of discussions aimed at managing conflict. What is of interest here is the generic modeling of the moves by disputants and mediators to a relevant discussion so that the insights of various theories of argumentation can be brought to bear, just as Hamblin (1970), Walton and Krabbe (1995) or Prakken (2006) did for other types of dialogues. For that we turn to the Inference Anchoring Theory (IAT) approach (Budzynska & Reed, 2011). Indeed, IAT allows for the exploration of the link between argumentation and dialogical processes. It relies on the assumption that argumentation structures are anchored in the communicative process via illocutionary connectives related to the illocutionary force (Budzynska & Reed, 2011). The notion of illocutionary force, introduced in speech act theory (Austin, 1962; Searle, 1969), refers to communicative functions. In IAT the dialogical act "Bob says p" is linked to the propositional content p via an illocutionary connection (here, *asserting*).

In this paper, we take up the task of accounting for the argumentative moves in dispute mediation by drawing upon and expanding the insights of Inference Anchoring Theory. A feature of this approach is its ability to remain agnostic about particular argumentation theories while incorporating many of the most important insights of pragmatic theories of argumentation. For these reasons, IAT seems well-suited to the challenge of modeling the moves of mediators and disputants. Dialogues analyzed in IAT are represented as graphs that make it possible to describe dialogue dynamics and structure in a very precise way. IAT also allows eliciting dialogical specificities that other theories fail to grasp, in particular how sequences of utterances work together to form arguments in a dialogical context even in the absence of obvious linguistic indicators (Yaskorska, 2014).

Disputants, along with the mediators, will generally face impasses throughout the mediation. Impasses refer to situations in which the discussion leads nowhere and nothing constructive comes out of it. Aakhus (2003) studied the three main sources of impasse that can occur during mediation: irreconcilable facts, negative collateral implications and unwillingness to be reasonable. Irreconcilable facts are discussants' claims concerning their opponent's state of mind, or "unwitnessable events" that cannot be verified and are subject to digressions. Negative collateral implications refer to disputants' claims challenging their opponent's character or competence. Unwillingness to be reasonable refers to moments when a disputant recognizes that the opponent's argument is legitimate, but refuses to take it into account in the pursuit of the argument. Given that impasse threatens the discussion, strategies to manage the sources of impasse need to be developed. In (Aakhus, 2003; Greco Morasso, 2011) three such strategies are identified: redirecting, temporizing, and relativizing. Mediators relativize the assumptions by discounting the party's claims or actions; they temporize the dispute by fostering temporary arrangements when no agreement seems possible on key issues; finally, they redirect the discussion toward more relevant issues when it seems to lead nowhere.

We propose here a method for analyzing the structure of mediation discourse using the IAT framework, and focusing on impasse (when the discussion between parties is blocked) and the strategy that is directly deployed to overcome it. The mediator's role, through the deployment of strategies, exposes aspects of argumentation that are, in other contexts than mediation, usually implicitly managed by discussants. Given that impasse is a typical breakdown in the mediation session, the visibility of what mediators have to do to overcome it, make apparent aspects of dialogical argumentation. Applying IAT to mediation impasse improves the likelihood of capturing the non-obvious markers and indicators of argumentation in dialogue. This method aims at providing a generic modeling of mediation discourse for the comparison of patterns proper to mediation.

The analyses presented here were carried out on one single transcript of a mock mediation, where two of the three sources of impasse defined in (Aakhus, 2003) were found: negative collateral implications and unwillingness to be reasonable; we show that in reaction to these particular cases two different strategies are deployed: redirecting and temporizing. We will focus on these two sources of impasse and the strategies that are employed to deal with them. Our goal indeed is not to make general claims such as "this type of impasse is always/generally overcome by this particular strategy". Rather, we present a method for the analysis of mediation discourse and mediation strategies.

The paper is structured as follows. In Section 2, we will introduce the theoretical background (i.e. Inference Anchoring Theory). In Section 3 we show how IAT can be used to analyze impasse in dispute mediation. In Section 4 we show how IAT helps in modeling the features highlighted in Section 3. Finally we summarize and present the next steps of the research in Section 5.

2. AN INTRODUCTION TO INFERENCE ANCHORING THEORY

Inference Anchoring Theory (IAT) (Budzynska & Reed, 2011) is designed to show and explain how dialogues create arguments. The argumentative aspects of dialogues being rarely obvious, particularly because of the absence of obvious linguistic markers, IAT aims at deriving the arguments through the analysis of a dialogue. The theory explores the dialogical structure of a text to extract its argumentative structure and allows for the representation of the link between the two. An IAT analysis, thus, takes the form of a graph that elicits both dialogical and argumentative structures. All the IAT analyses presented in this paper were produced using OVA+ (Janier, Lawrence, & Reed, 2014), a software tool for the analysis of arguments online, accessible from any web browser[1]. The tool was built as a response to the Argument Interchange Format theory (Chesveñar, McGinnis, Modgil, Rahwan, Reed, Simari, South, Vreeswijk, & Willmott, 2006): it is a tool allowing what the AIF has advocated for, i.e. the standardized representation of arguments which gives the possibility to exchange, share and reuse argument maps. The system uses the framework provided by IAT, what allows for a representation of the argumentative structure of a text, and more interestingly, of dialogues. IAT indeed provides a theoretically well-founded counterpart to AIF. The analyses presented in this paper, plus many more, are available on the AIFdb-Corpora (Lawrence, Janier, & Reed, this issue) webpage[2]. This interface allows gathering and sorting analyses made in OVA+ into corpora. The aim is to provide a framework where analyses can be shared and reused.

For a better understanding of IAT and OVA+, let's consider the example (1) and its IAT analysis in figure 1.

---

[1] http://ova.arg-tech.org

[2] http://corpora.aifdb.org

(1)
Participant1: *Scotland is the best country on Earth.*
Participant2: *Why is Scotland the best country on Earth?*
Participant1: *Sceneries are breath-taking.*
Participant2: *Winters are too cold there.*

The IAT analysis of example (1) illustrates the representation of both the dialogical and the argumentative structure of the dialogue:

- The right-hand side of the graph shows the dialogical structure with:
  - Locution nodes: the reports of the discourse events
  - Transition nodes: the transitions between the locutions or rules of dialogue (TA-nodes)

- The left-hand side of the graph shows the argumentative structure with:
  - Information nodes: the propositional content of each locution (in front of the corresponding locution node)
  - Relations of inference: the relations connecting premises to conclusions
  - Relations of conflict: the relations connecting conflicting information
  - Relations of reframing (two pieces of information which mean the same despite a different linguistic surface)

- The relation between the dialogical and the argumentative structure:
  - Illocutionary connections anchored in the locutions (such as *asserting, challenging, questioning,* etc.)
  - Illocutionary connections anchored in transitions (such as *arguing, explaining, disagreeing,* etc.)

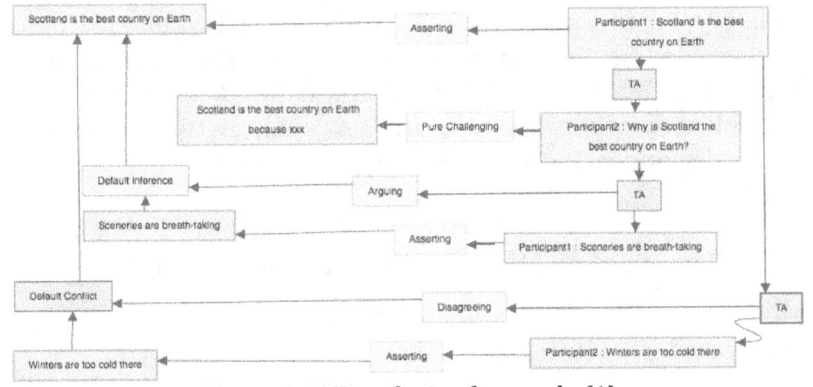

Figure 1. IAT analysis of example (1)

Figure 1 is to be read as follows: Participant 1 asserts that Scotland is the best country on Earth, and Participant2 asks her the ground for stating this. Despite the absence of conventional linguistic indicators such as 'because', the reader (or hearer, or analyst) understands that Participant1's second claim is actually supporting the first one. It is in virtue of the very fact that "Sceneries are breath-taking" was uttered just after Participant2's challenge (shown by the transition node between them) that we know that the latest claim acts as a premise to the first one. This is shown through the illocutionary connection of arguing anchored in the transition node. Participant2 in his turn asserts that Scotland is too cold during winters. Once again, it is in virtue of the very fact that this claim was pronounced after Participant1's argument (shown by the third transition node) that we understand that Participant2 is disagreeing with Participant1, even if no linguistic indicator signals this. This is represented by the illocutionary connection of disagreeing that is anchored in the transition node and takes as a propositional content the conflict node between the participants' claims.

3. ANALYZING IMPASSE IN MEDIATION

3.1. Using IAT to annotate argumentative mediation discourse

Inference Anchoring Theory is designed to capture specific details of argumentation, however, not all of them: some characteristics of dispute mediation cannot be expressed. For example, non-verbal communication cannot be captured here. It is a good start, however, for analyzing dialogical argumentative discourse. This framework has proven particularly stable when used to study real-life argumentation such as debate (Yaskorska, 2014; Budzynska, Janier, Kang, Konat, Reed, Saint-Dizier, Stede, & Yaskorska, 2016; Janier & Yaskorska, 2016) and already revealed useful to analyze many facets of mediation discourse (Janier, Aakhus, Budzynska, & Reed, 2014; Janier & Reed, 2016). For the purpose of this paper, we use IAT to analyze two specificities of mediation discourse: the sources of impasse and the strategies to deal with them.

To begin with, table 1 and 2 summarize the illocutionary connections provided by IAT found in mediation discourse.

Illocutionary connection	Abbreviation
pure question	PQ
assertive question	AQ
rhetorical question	RQ
assertive challenge	ACh
pure challenge	PCh
rhetorical challenge	RCh
assertion	A
ironic assertion	IA
popular concession	PCn

Table 1. List of illocutionary connections anchored in locutions in mediation

Table 1 provides the list of illocutionary connections anchored in locutions that were found in mediation discourse. Assertions and ironic assertions are used to communicate one's opinion, however, with an ironic assertion the speaker says (deliberately) the contrary of what she means and thinks. Popular concessions (PCn) are used to communicate general knowledge (e.g. "Everybody knows that p"). There are three types of question: pure questions (PQ) are used to ask about the hearers' opinion; assertive questions (AQ) and rhetorical questions (RQ) both convey an assertive intention, but when a speaker uses a rhetorical question, she does not expect any reply (contrary to assertive questions). The distinction between pure, assertive and rhetorical holds for challenges as well. Challenges are used to ask about the grounds for the hearer's opinion. The illocutionary connections anchored in transition nodes are presented in table 2.

Illocutionary connections	Functions
arguing	the speaker provides one or more premises to a conclusion
explaining	as above except that this time, all speakers generally know or agree on the conclusion
agreeing	the speaker agrees with another speaker
disagreeing	the speaker disagrees with another speaker
contradicting	the speaker contrasts or concedes

Table 2: List of illocutionary connections anchored in transitions in mediation discourse

For a more detailed explanation of IAT illocutionary connections, the reader can also refer to (Budzynska, Janier, Reed, & Saint Dizier, 2013).

For the purpose of this paper, we will focus on impasses, the trickiest moments of a mediation session. The dialogues analyzed here are taken from the transcript of a mock mediation provided by Dundee's early dispute resolution center[3]. The mediation session involves two parties, Viv and Eric, and two mediators, George and Mildred. The transcript only captures a small part of an entire typical mediation session, but the video it is extracted from is used by training mediators, and we found two of the three sources of impasse presented by Aakhus (2003): negative collateral implications and unwillingness to be reasonable. For these reasons we think that this transcript contains realistic data and thus suits our needs. In the transcript, Viv initiated mediation because she is not happy with the way her boss Eric regards her work and she wants more acknowledgements.

3.2. Negative collateral implications and redirecting.

Negative collateral implications refer to disputants who make claims that challenge their opponent's character or competence (Aakhus, 2003). In the specific example presented below, the source of impasse is followed by the strategy of *redirecting*; mediators redirect the dialogue by shifting the topic of the discussion towards more relevant issues. In our corpus (see example (2)), this source of impasse appears after one of the mediators pointed out the fact that the two parties Viv and Eric have a communication problem when they talk about a particular project they have to deal with. Eric, the boss, does not want to give some tasks to Viv because he is not sure she can deal with them.

>(2)
>a. Eric: *I'm just a bit reluctant to hand over to Viv at this early stage, because of the complexity and if you make a mistake, you waste such a lot of time. But I don't know whether Viv thinks that she's up to it or whether you think you could handle that project.*
>b. Mildred: *What about if we perhaps separate it, had a bit of time and we spoke with each of you to look at the finance project and just see our different expectations and what you would see dealing with that project and then perhaps when we*

---

[3] *http://dundee.ac.uk/academic/edr*

> had a picture from both of you, if both of you came back to discuss your different pictures. Do you think that would work?

In example (2), Eric says that he does not want to hand over one of the projects to Viv because the task is very complicated. The first sentence highlights the complexity of the task and the cost of mistakes that could result from handing the project over to Viv too soon. The second sentence pushes the choice away from Eric to Viv, as though he is not the one to take the responsibility for the decision. Both sentences though seem to carry the implication that Viv is either not qualified or not yet ready or both. The mediator then opens a conversation to avoid this subject and shifts the topic of the discussion from whether Viv is qualified and whether Viv or Mildred should decide whether Viv is qualified, toward discussing the task itself and the expectations around it. Thus Viv's competence is taken out of the discussion. The IAT analysis of this excerpt is given in figure 2.

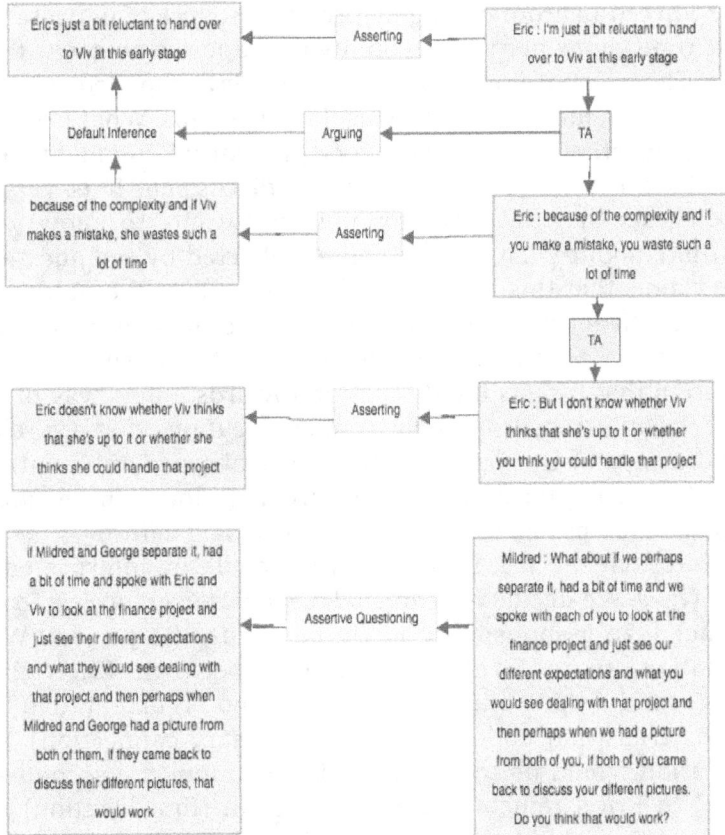

Figure 2. IAT analysis of example (2)

The use of IAT to analyze this extract allows for the detection of the different moves corresponding to the source of impasse and to the mediator's moves to deal with it. Here, Eric casts doubts on Viv's competence and provides an argument for this (see the illocutionary connection of *arguing* between his two first locutions: "I'm just a bit reluctant to hand over to Viv at this early stage" and "because of the complexity and if you make a mistake, you waste such a lot of time"). However, it does not make the discussion move forward since the other party, Viv does not answer to those critiques: this is the impasse. In (Jacobs & Jackson, 1992), the authors describe this frequent situation in dispute mediation i.e. when the parties make claims that have potential argumentative strength but their relevance is lost by the fact that they appear in a moment when they do not serve the argumentative process. Here, Eric's argument is irrelevant considering the current discussion. The mediator is supposed to detect this and to restore the argumentative relevance (van Eemeren, Grootendorst, Jackson & Jacobs, 1993). This is what Mildred does in this extract: her question to shift the topic is redirecting around a highly probable source of impasse while at the same time giving her the possibility to propose a new way to broach the issue. This move is not surprising given it is acknowledged that most of the mediators' moves consist in asking questions. What is interesting is that the question appears as a very procedural comment (or meta-comment) on how to proceed with the discussion. It is very direct, which suggests that the mediator not only wants to know what the parties think about what she proposes (reflected by the question), but she also claims that this is how the discussion should unfold (reflected by the assertiveness of the question). The mediator has actually *redirected* the discussion: the question creates a space for a new conversation that directs the discussion towards a new way of tackling the issue. In IAT, transition nodes connect locutions that are related by rules of dialogue or by logical relations. The absence of transition node between Mildred's question and the previous locutions means, here, that this question has no relation to what was said before.

Note that the Eric's third locution could be interpreted in two different ways. We decide here to analyze it as a way for Eric to say that he will not take responsibility if Viv fails with the project. With this interpretation, there is obviously a link between Eric's first two locutions and the third one. It has however no argumentative function (hence the transition node without illocutionary connection). Another interpretation would be to see this third locution as a second support (premise) to Eric's reluctance: he is reluctant (first locution) because the task is complicated (second locution) and because he does not know if Viv feels she has the ability to handle it (third locution). Both interpretations are possible and correct, and they do not change the following of the analysis: in both cases, Mildred's question redirects the discussion and has no relation with Eric's standpoint.

3.3. Unwillingness to be reasonable and temporizing

In (Aakhus, 2003) and (Jacobs & Aakhus, 2002), unwillingness to be reasonable is defined by parties refusing to reason together and resisting proposals. In our corpus, this happens in example (3), where the source of impasse is followed by the mediation strategy of *temporizing* i.e. temporary arrangements are proposed when no agreement seems possible on key issues.

> (3)
> a. Eric: *I don't know whether Viv could handle that she has the ability.*
> b. Viv: *Well come on, you employed me, surely you thought I had the ability to, you know. But…*
> c. Eric: *Well I did, so there is a way forward then. But I can also check on how she's doing the project and if she's succeeding with it and that will give me a milestone, an indicator of her.*
> d. Viv: *I would quite like to just maybe take time out to look at what my job description was, actually, and from that, given what we've been talking about, it might signal up to me the key points that I want to clarify with you and see what your opinion is. Whether I've read it, whether it's been hieroglyphics to me, or whether I've got it right.*
> e. George: *It's quite possible and again, it's our experience in this sort of situation, it's all about expectations and where your expectations and Viv's expectations match, you have happiness and a smooth life and everything works well. Where they don't, there is conflict, there is uncertainty, there is confusion and those are the sorts of things that contribute to having this sort of discussion. If what we can do today is to help you to get a degree of clarity about the expectations, then if you feel that would be useful…*
> f. Eric: *Well, anything that, as I said at the start, anything that will give me more time back.*

This discussion between Eric, Viv and the second mediator, George, happens sometime after the one in example (2). Here, Eric again casts doubts about Viv's ability to handle the project. However, this time, Viv answers to the critique and claims that if he employed her it is because he knew she was able to deal with it. Eric agrees with her but he does not take it into account; instead, he claims that time has passed since then and he needs to check if she actually can handle the project. Viv does not directly answer to this; rather, she proposes to have a look at her job description to check whether she understood what Eric expected from her. The mediator intervenes only then, by saying that Viv's proposal is a good idea, and Eric eventually agrees as well. The IAT analysis of this excerpt is presented in figure 3. For clarity and space purposes, only the most relevant moves of this dialogue are analyzed.

The analysis shows that Viv disagrees with Eric's first claim and gives an argument. Eric agrees with Viv, however he does not take this argument into account. Note the contradicting node: after agreeing with Viv that he employed her because he thought she had the abilities, he says that he would like her to prove she can do well with the project. He implies that Viv's argument does not hold, although he agrees with her. This is the impasse: Eric is unwilling to be reasonable since he first agrees with his opponent but then refuses to take it into account. Viv then makes a proposal and provides an argument for this proposal (she wants to look at her job description because it may indicate key points she would like to clarify). George agrees with her; more precisely, with the premise of her argument i.e. that it may signal points that need to be clarified. Eric, in his turn, agrees with the conclusion of Viv's argument: he implicitly says that having Viv looking at her job description is a good idea because it will give him some time back. Viv, interestingly, is the one who reacted to the impasse: she made a proposal that concerns a particular issue (here, their expectations concerning Viv's abilities) and not the dispute itself. This is called *temporizing* (Aakhus, 2003). In (greatbatch & dingwall, 1997), the authors show that disputants very often manage to exit arguments without the intervention of the third-party. This is precisely what happens in example (3).

Modeling argumentative activity   257

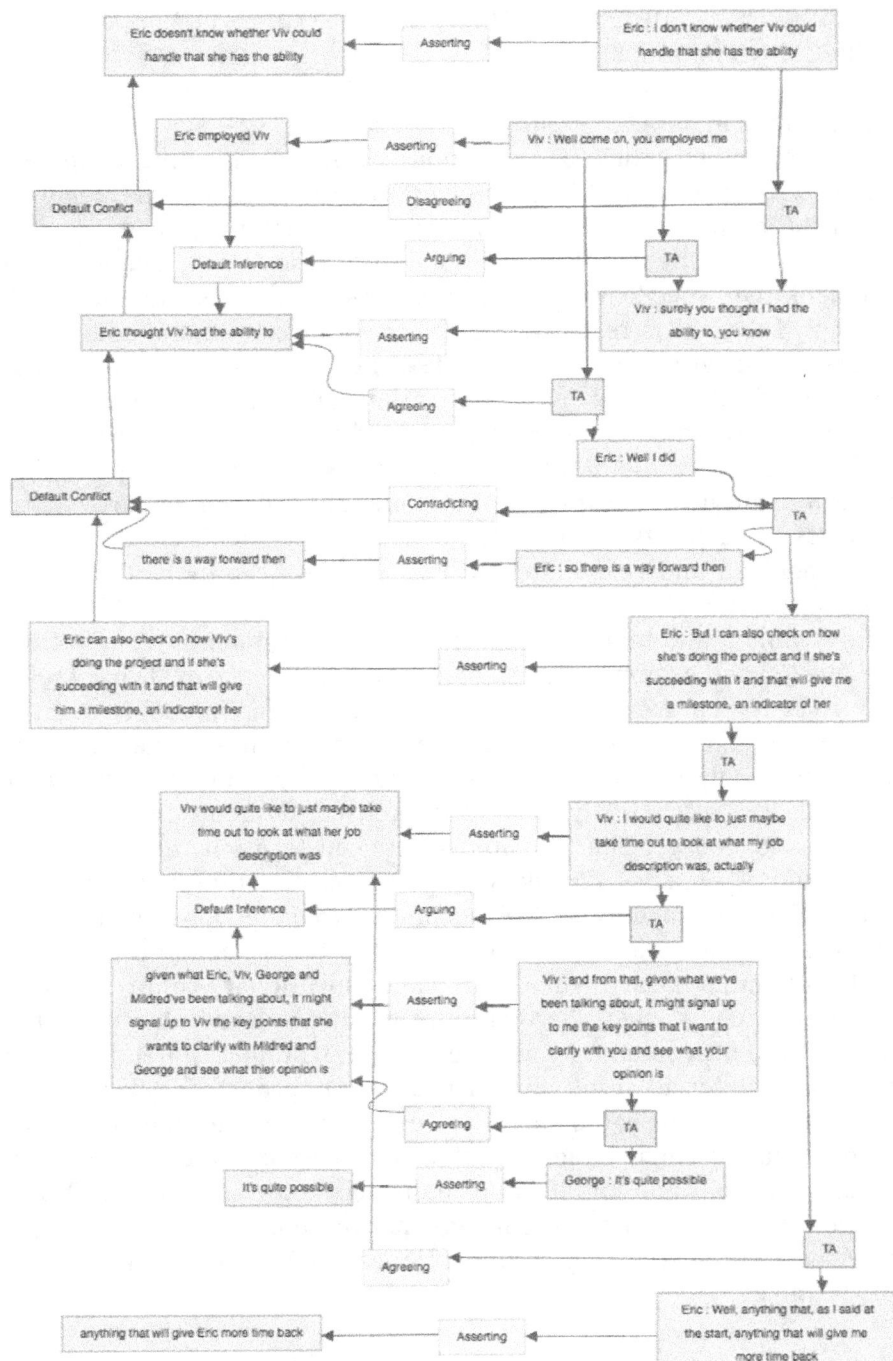

Figure 3. IAT analysis of example (3)

4. TOWARDS A DETAILED ACCOUNT OF THE STRATEGIES FOR OVERCOMING IMPASSSES

The application of IAT to mediation discourse has allowed for the analysis of the structure of mediation discourse with respect to two sources of impasse and the strategies to resolve them. The analyses present the argumentative elements of mediation discourse in a graphical manner; it is then possible to associate a sequence of argumentative specificities to each source of impasse and to each strategy. The goal is to define the patterns proper to the sources of impasse along with the patterns of the strategies. This is a fundamental step towards the formal modeling of the dialogical structure and its recognition. On one hand, modeling mediation impasse presents a challenging task, while on the other hand it shows that IAT is a viable means for capturing specificities of argument in dialogue; in particular it improves the capacity to discover how those specificities and patterns are recognized and responded to. We illustrate this potential in the following sections.

Table 3 and table 4 below present the argumentative moves of example (2) and example (3) revealed by the IAT analyses. Those tables capture every feature highlighted by the graphs. The first column of the tables represents the locutions in order of appearance in the dialogues (and in the analyses). In the second column, *party₁* and *party₂* stand for Eric and Viv respectively; *mediator* is used without distinguishing Mildred and George. The transitions between locutions appear in the third column (e.g. Loc_1; Loc_3 means there is a transition node from the first to the third locution). The illocutionary connections anchored in the locutions and the ones anchored in the transitions appear in the fourth column; ø is used when there is no illocutionary connection to a transition node or when there is no propositional content. Finally, the letters in the fifth column symbolize the propositional contents of each locution (a different letter for each different propositional content). Note that every table is independent from the other: e.g. when the letter *a* appears in one single table, it symbolizes the exact same propositional content; this does not hold if *a* appears in e.g. table 1 and in table 2. The notation *default inference (a,b)* means that there is an inference from *a* to *b* ; similarly, *default conflict (b;[a,c])* means that *b* is in conflict with both *a* and *c*.

4.1. Patterns of example (2): negative collateral implications and redirecting

In Section 2, we have seen that in example (2) the source of impasse was negative collateral implications and that the mediator dealt with it through redirecting. All the argumentative and dialogical features of this passage are presented in table 3.

LOCU-TION	PARTICI-PANT	TRANSITION	ILLOCUTIONARY CONNECTION	PROPOSITION-AL CONTENT
Loc_1	$party_1$		A	a
		$Loc_1;Loc_2$	arguing	default inference ([b,c];a)
Loc_2	$party_1$		A	b
		$Loc_2;Loc_3$	∅	∅
Loc_3	$party_1$		A	c
Loc_4	mediator		AQ	d

Table 3. Negative collateral implications and redirecting

We see that $party_1$ argues (see the *arguing* illocutionary connection in the fourth column) but $party_2$ does not answer i.e. she does not take part in this discussion. This represents the impasse since only one party is actually arguing. Moreover, we see that the mediator uses an assertive question (bottom of the fourth column) that is not connected at all to any of the precedent moves: Loc_4 does not appear in the TRANSITION column. This represents the strategy of redirecting: the fact that there is no relation between her question and the previous moves (e.g. no sequence such as $Loc_3; Loc_4$) shows that she shifted the discussion to another topic. In other words, there is no link between Loc_4 and another locution because the mediator has redirected the discussion. In this particular case, we cannot claim from table 3 solely that the source of impasse presented in this table *is* negative collateral implications: a pragmatic, linguistic and semantical analysis is necessary to see that $party_1$ is challenging his opponent character. The fact that $party_2$ is not taking part in the dialogue however is a strong indicator of impasse in the dialogue.

4.2. Patterns of example (3): unwillingness to be reasonable and temporizing

In example (3) Eric was unwilling to be reasonable and Viv reacted to this source of impasse by proposing temporary arrangements. Let's represent this in table 4.

LOCU-TION	PARTICI-PANT	TRANSITION	ILLOCUTIONARY CONNECTION	PROPOSITIONAL CONTENT
Loc_1	$party_1$		A	a
		$Loc_1;Loc_3$	disagreeing	default conflict (c;a)
Loc_2	$party_2$		A	b
		$Loc_2; Loc_3$	arguing	default inference (b;c)
Loc_3	$party_2$		A	c
		$Loc_3;Loc_4$	agreeing	c
Loc_4	$party_1$		ø	c
		$Loc_4;[Loc_5,Loc_6]$	contradicting	default conflict ([d,e];c)
Loc_5	$party_1$		A	d
Loc_6	$party_1$		A	e
		$Loc_6; Loc_7$	ø	ø
Loc_7	$party_2$		A	f
		$Loc_7;Loc_8$	arguing	default inference (g;f)
Loc_8	$party_2$		A	g
		$Loc_8;Loc_9$	agreeing	g
Loc_9	mediator		A	h
		$Loc_7;Loc_{10}$	agreeing	f
Loc_{10}	$party_1$		A	i

Table 4. Unwillingness to be reasonable and temporizing

Table 4 shows that *party₂* disagrees with *party₁* and that she provides an argument (see the third, fourth and fifth columns); *party₁* agrees with it but discards it immediately after: this is the *unwillingness to be reasonable*, represented in table 4 by the illocutionary connection of *contradicting* that follows the one of *agreeing*. The transition $Loc_6; Loc_7$ only shows the continuity of the dialogue: the transition does not anchor any illocutionary connection (note the symbols ø that follow in the last two columns). *Party₂* argues later on: (see the Transition *Loc₇;Loc₈* that anchors *arguing*). This is the strategy of temporizing: *party₂* continues the dialogue (and argumentation) by proposing temporary arrangements. This is shown by *party₂* introducing an argument (*Loc₇;Loc₈*) that relates to the discussion (*Loc₆;Loc₇*) but that does not attack or support *party₁*'s moves (hence the empty transition: ø). The table also shows that the mediator agrees with her argument (*Loc₈;Loc₉*), and that *party₁* agrees with the proposal (*Loc₇;Loc₁₀*).

5. CONCLUSIONS

Mediation discourse has not been subject to a lot of attention, even less its argumentative facet. Fine-grained analyses of the argumentative structure prove necessary to highlight how argumentation in dispute mediation progresses. This context helps us reveal specificities about arguments that are important for argumentation theory in general. We have shown, for example, a means for modeling the relationship between dialogue and arguments in a context where conventional and obvious indicators of argumentation are not always present.

In this paper, we have illustrated that Inference Anchoring Theory enables the analysis of mediation discourse argumentative structure. The analyses presented in Section 2 make it possible to grasp the subtleties of mediation strategies when sources of impasses occur. This allowed us to relate dialogical features to argumentative strategies. The analyses were then used in Section 3 for the definition of the patterns of the sources of impasse and the strategies to overcome them.
By comparing analyses from different cases of the same source of impasse and the strategies deployed in those cases, it becomes possible to model argumentative sequences of moves and to verify whether such sequences generalize to all mediation sessions. As an example, we could check if all the analyses of *redirecting* present the same features, namely that the mediator interrupts the discussion via an assertive question that has no link with the topic addressed just before (Sections 2 and 3). Applied to several entire mediation sessions, this method will lead us towards the definition of a dialogue protocol in mediation that could be implemented in a tool designed to support mediators during their training.

ACKNOWLEDGEMENTS: The authors would like to acknowledge that the work reported in this paper has been supported in part by the British Leverhulme Trust under grant RPG-2013-076 and in part by the Innovate UK under grant 101777.

REFERENCES

Aakhus, M. (2003). Neither naïve nor critical reconstruction: Dispute mediators, impasse, and the design of argumentation. *Argumentation*, *17*(3), 265–290.

Austin, J. L. (1962). *How to do things with words*. Oxford: Clarendon.

Budzynska, K., Janier, M., Reed, C., & Saint-Dizier, P. (2013). Towards Extraction of Dialogical Arguments. *Proceedings of 13th International Conference on Computational Models of Natural Argument (CMNA13)*.

Budzynska, K., Janier, M., Kang, J., Konat, B., Reed, C., Saint-Dizier, P., Stede, M., & Yaskorska, O. (2016). Automatically identifying transitions between locutions in dialogue. In D. Mohammed & M. Lewiński (eds.), *Argumentation and Reasoned Action: Proceedings of the 1st European Conference on Argumentation, Lisbon, 2015. Vol. II, 311-327*. London: College Publications.

Budzynska, K., & Reed, C. (2011). Whence inference. Technical report, University of Dundee.

Chesveñar, C., McGinnis, J., Modgil, S., Rahwan, I., Reed, C., Simari, G., South, M., Vreeswijk, G., & Willmott, S. (2006). Towards an Argument Interchange Format. *The Knowledge Engineering Review*, *21*(4), 293–316.

Greatbatch, D., & Dingwall, R. (1997). Argumentative talk in divorce mediation sessions. *American Sociological Review*, *62*, 151–170.

Greco Morasso, S. (2008). Argumentative and other communicative strategies of the mediation practice. PhD thesis, Università della Svizzera italiana.

Greco Morasso, S. (2011). *Argumentation in dispute mediation*. Amsterdam: John Benjamins Publishing Company.

Hamblin, C. L. (1970). *Fallacies*. London: Methuen.

Jacobs, S., & Aakhus, M. (2002). What mediators do with words: Implementing three models of rational discussion in dispute mediation. *Conflict Resolution Quarterly*, *20*(2), 177–203.

Jacobs, S., & Jackson, S. (1992). Relevance and digressions in argumentative discussion: A pragmatic approach. *Argumentation*, *6*(2), 161–176.

Janier, M., Aakhus, M., Budzynska, K., & Reed, C. (2014). Games mediators play: Empirical methods for deriving dialogue structure. In not published *1st International Workshop for Methodologies for Research on Legal Argumentation (MET-ARG)*.

Janier, M., Lawrence, J., & Reed, C. (2014). OVA+: An argument analysis interface. In *Computational Models of Argument (COMMA)*, volume 266, (pp. 463–464). Amsterdam: IOS Press.

Lawrence, J., Janier, M., & Reed, C. (2016) Working with open argument corpora. In D. Mohammed & M. Lewiński (eds.), *Argumentation and Reasoned Action: Proceedings of the 1st European Conference on Argumentation, Lisbon, 2015. Vol. I, 367-380*. London: College Publications.

Prakken, H. (2006). Formal systems for persuasion dialogue. *The Knowledge Engineering Review*, *21*(2), 163–188.

Searle, J. (1969). *Speech acts: An essay in the philosophy of language.* Cambridge: Cambridge University Press.
van Eemeren, F. H., Grootendorst, R., Jackson, S., & Jacobs., S. (1993). *Reconstructing argumentative discourse.* Tuscaloosa, Alabama: University of Alabama Press.
Walton, D. N., & Krabbe, E. C. W. (1995). *Commitment in dialogue: Basic concepts of interpersonal reasoning.* New York: State University of New York Press.
Yaskorska, O. (2014). Recognising argumentation in dialogical context. In *Proceedings of the 8th Conference of the International Society for the Study of Argumentation* (ISSA).

Commentary on Janier, Aakhus, Budzynska and Reed's Modeling Argumentative Activity in Mediation with Inference Anchoring Theory: The Case of Impasse

SARA GRECO
Università della Svizzera italiana, Switzerland
sara.greco@usi.ch

1. INTRODUCTION

The aim of this paper is modelling the moves that dispute mediators make in order to exit situations of impasse during the parties' discussion. The paper makes use of Inference Anchoring Theory (IAT, see Budzynska and Reed, 2011) and of a visualization tool for the analysis of arguments, called OVA+ (Online Visualisation of Argument), recently released at the University of Dundee, UK.

The topic considered in this paper is highly relevant to an understanding of mediation as an argumentative practice, as moments of impasse represent crucial situations in mediation. This is easily explained if we consider the rationale beneath this Alternative Dispute Resolution method. As Fisher, Ury and Patton (1991, p. xvii) put it in the opening lines of their very-well known dispute resolution book *Getting to Yes*, "Like it or not, you are a negotiator". This means that we are all confronted with conflict in our everyday professional and personal life. We normally deal with those conflicts via negotiation, namely by directly bargaining with our counterpart; ideally, we do this by means of an argumentative discussion. However, we do need mediation as an institutionalised way to deal with disputes because, sometimes, interpersonal communication fails and we are simply unable to deal with the other party directly. Thus, when two parties enter mediation, the conflictual story behind them makes it impossible for them to solve their problem by themselves. In such a situation, that there are moments in which the discussion gets to an impasse is certainly not surprising. This happens, for example, when parties go back to their original conflictual positions. In this sense, the aspect modelled in this paper is particularly relevant in order to understand a typical reason why mediation might fail; and to analyse possible ways of getting out of impasse.

In the following sections, I will consider some aspects on which the research project outlined by Janier and co-authors could be brought forward.

2. RATIONALE OF THIS RESEARCH PROJECT

A first important question that comes to mind concerns not only this specific paper but also, more in general, any effort of modelling discussions via IAT and with the support of OVA+. One might ask what the purpose of this type of analysis is. As the authors explicitly say, this type of modelling might not add (*stricto sensu*) to the theoretical understanding of dispute mediation as an argumentative institutionalised activity type. For the specific purposes of this paper, in fact, the authors mainly rely on previous literature on dispute mediation and, more in particular, on strategies for exiting situations of impasse as described by Aakhus (2003).

So what is the purpose of modelling? The authors hint to it when, at the end of their work, they say that, by comparing the strategies used in different cases of the same source of impasse, it will be possible to model argumentative sequences of moves and to verify whether such sequences generalize to all mediation sessions. In this paper, they take as an example the analysis of the strategy of *redirecting*. Janier and her co-authors say that, in future research, it could be verified if redirecting always implies that the mediator "interrupts the discussion via an assertive question which has no link with the topic addressed just before". This is, however, what could be done with this method rather than what has been done with this paper. Thus, the construction of a broader dataset seems to be important in order to fully exploit the potential of this type of modelling. However, two types of advantages could be already envisaged.

A first possible advantage goes into the direction of what has been called the *design* of argumentative situations (Aakhus, 2007). If one has a broad data base of successful strategies for exiting cases of impasse, then it could be imagined to use them for the training of new dispute mediators.

Ideally, a second positive fall out might touch upon a theoretical level. In fact, the analysis of a broad set of data could equally show if there are new emerging strategies for exiting cases of impasse, thus making our theoretical understanding of mediation richer.

3. MODELLING SPECIFIC ARGUMENTATIVE MOVES

A second and more specific question to be asked about this paper concerns the relevance of the strategies for exiting impasse in relation to argumentation. Obviously, exiting from situations of impasse is important because otherwise the discussion will be blocked. Anyway, it would be theoretically important to situate this type of mediator's strategies within an argumentative discussion; even though strategies for exiting impasse are not necessarily linked to the mediator (or the parties) advancing standpoints or arguments.

One of the cases discussed by Janier and her co-authors, for example, is the strategy of *redirecting*, i.e. shifting the topic of the discussion from something that is not productive or not adequate to mediation to something else that could be more appropriate and relevant to the resolution of the parties' conflict (cf. Aakhus 2003). From an argumentative viewpoint, this means that one discussion is closed and another discussion is opened. I suggest that a direction for future research might concern the following: how does this type of question serve the institutional purpose of dispute mediation? How is this "change of issue" helping the overall discussion between the parties? These strategies should be evaluated not only at a local level but also for how much they contribute to what I have called a *macro-text of argumentative discussions* (Greco Morasso, 2011); i.e., in other words, how much they help construct an argumentative discussion between the parties that will ideally lead to the settlement of their conflict.

Also, the authors could model more specific argumentative moves. It is well-known (since van Eemeren, Grootendorst, Jackson & Jacobs, 1993) that mediators cannot advance specific issues or standpoints concerning the resolution of the dispute beforehand, because that would betray their institutional role. However, it has also been shown that mediators sometimes advance standpoints and arguments at a meta-level – for example, on the type of issues that the parties need to address. It would be interesting to understand whether Inference Anchoring Theory and the visualization approach supported by OVA+ is helping in modelling and classifying the types of *argument schemes* used by mediators in mediation sessions. In other words, can we give a name to the "default inference" that is mentioned in OVA+ diagrams? Also, it would be interesting to compare whether these argument schemes are similar or different to those used by the parties; and if there are interesting *argumentative patterns* at a micro-level (van Eemeren & Garssen, 2013).

4. CONCLUSION

Overall, the direction of research outlined in this paper looks promising for understanding the communicative and discursive dynamics within dispute mediation. Yet we still need more data and more analyses to understand its full potential. It is worth, however, working in the direction that the authors of this paper have shown, because this could bring both to advantages in our understanding of mediation and to theoretical advances in the study of argumentation within this activity type.

Let me conclude by a local remark, which is philosophical in nature. At the beginning of this paper, the authors say that Inference Anchoring Theory is "agnostic" with regards to argumentation theories. I am sceptical about the possibility of remaining agnostic when proposing a theory; as any theory comes with strings attached, i.e. with its philosophical presuppositions. So it would be interesting to discuss what the presuppositions of Inference Anchoring Theory are. It could be that IAT's agnosticism derives from the fact that IAT is not working at the same level as the other argumentation theories, i.e. it is more concerned with discourse relations and basic relations of inference. This type of interpretation would position the relevance of IAT more clearly within current approaches to argumentation.

REFERENCES

Aakhus, M. (2003). Neither naïve nor critical reconstruction: Dispute mediators, impasse, and the design of argumentation. *Argumentation* 17(3), 265-290.
Aakhus, M. (2007). Communication as design. *Communication Monographs*, 74(1), 112-117.
Budzynska, K., & Reed, C. (2011). *Whence inference.* Technical report, University of Dundee.
Eemeren, F.H. van, & Garssen, B. (2013). Argumentative patterns in discourse. *OSSA Conference Archive.* Paper 42.
Eemeren, F.H. van, Grootendorst, R., Jackson, S., & Jacobs, S. (1993). *Reconstructing argumentative discourse.* Tuscaloosa: The University of Alabama Press.
Fisher, R., Ury, W., & Patton, B. (1991). *Getting to yes: Negotiating agreement without giving in*, 2nd edition. New York: Penguin Books.
Greco Morasso, S. (2011). *Argumentation in dispute mediation: A reasonable way to handle conflict.* Amsterdam/Philadelphia: John Benjamins.

12

A New Approach to Argumentation and Reasoning Based on Mathematical Practice

ANDRZEJ KISIELEWICZ
Institute of Mathematics, University of Wrocław, Poland
<u>Andrzej.Kisielewicz@math.uni.wroc.pl</u>

> The aim of the paper is to propose a new approach to general reasoning and argumentation based on practice of mathematical reasoning. This is in opposition to the classical approach dominating in textbooks on logic and argumentation, which is based on a formal model of mathematics. It is argued that the new approach better fits the practice of argumentation.
>
> KEYWORDS: analysis of possibilities, mathematical reasoning, practical logic, practical reasoning, new approach

1. INTRODUCTION

The view that everyday reasoning has in fact nothing to do with the laws of formal logic, and that there is something wrong in our formal approach to reasoning, is not new. This view was one of the driving forces in creating the field of informal logic. The main argument is based on the facts that almost nobody applies formal rules of inference in practice and that examples of applications in textbooks are either artificial, trivial or inoperable. Many students learning logic in the elementary university course complain that what they are learning may be beautiful and wise in theory, but is utterly useless in practice. Many mathematicians and physicists, who are not interested in philosophy and methodology, claim that the reasoning they apply doing mathematics or physics is highly informal in nature.

However, this does not change the fact that the dominant point of view in philosophy of logic considers formal rules of inference to be an ultimate basis of logical reasoning, and that these formal rules are still presented as the fundamental laws of thought in textbooks on logic, argumentation and critical thinking. This makes an intolerable gap between theory and practice. I think it is high time to recognize the problem and to undertake attempts at a radically different approach. It should be an interdisciplinary effort. The aim of this paper is to make a contribution from mathematical practice.

2. APPLICATIONS OF FORMAL LOGIC

The strongest argument so far that paralyzes any serious criticism of the formal approach was the very successful applications of formal logic in mathematics and computer science. This argument is, however, weaker than it would seem. There are three kinds of these applications:

1) applications in IT and computer science,
2) accomplishments of mathematical logic,
3) applications to the practice of mathematical reasoning.

Applications in computer science are impressive, but mostly they have nothing to do with reasoning. In principle, these are applications of certain calculuses and other mathematical ideas to create computational software and hardware. It is only the attempts to apply logic in Artificial Intelligence to create thinking machines that have referred to the hypothesis that formal rules of inference create a base for logical reasoning. Yet, these attempts, already lasting for more than fifty years, have produced no practical results. It is now fair to say that these attempts have failed. Of course, the people still working in the area of AI do not want to recognize this failure (although there are notable exceptions, like (Minsky, 2011)). Among scientists working in other areas of computer science it is nowadays a rather common view that AI failed: it did not fulfilled its great promise. This failure, after careful analysis, can be viewed as *new and strong evidence that treating formal rules of inference as a base for logical thinking may be wrong.*[1]

The accomplishments of mathematical logic, led by the celebrated Gödel's incompleteness theorem, are impressive and, from a mathematical point of view, beautiful. Nevertheless—it must be emphasized—they pertain not to mathematical theories themselves, as they are, but to their formalizations. Thus, they are subject to interpretation. In fact, in spite of great expectations, no philosophical problem of mathematics has been resolved "with purely mathematical methods"—as some logicians hoped to do within the framework of mathematical logic. All philosophical schools in philosophy of mathematics survived proposing various interpretations of the most important and surprising results (Fraenkel et al., 1973, pp. 331-345).

---

[1] A description and analysis of this failure, as well as of the role of formal logic in scientific research, is given in my book (Kisielewicz, 2011).

Formal theories are mathematical objects and they ignore many aspects of real mathematical theories. It should become clear that mathematical logic deals only with one particular aspect of mathematics, and that *the very fact that mathematics can be formalized does not explain anything, but itself requires explanation.*

This view is supported by the most crucial claim concerning the third point: *formal logic has actually (almost) no applications in the practice of mathematical reasoning.* Let us state facts which logicians have long ignored, or were not even aware of. Most mathematicians nowadays (and for quite a long time since) do not take any interest in the developments of contemporary logic. Mathematicians in their work very rarely refer to the laws of logic. The language they use (in journals, monographs, during lectures) is a natural language, only slightly spiced up with artificial symbols (experience teaches that the overabundance of those only obscures the meaning, making arguments unclear and unreadable). Very rarely a more formal approach is used to clarify a detail. The arguments referred to by mathematicians are neither laws of logic nor series of inferences with hidden premises, as some logicians suggest (Ajdukiewicz, 1974, p. 117). The essence of mathematical reasoning and proofs lies in referring to *imagination*, to a particular type of cognitive models; it is the experience and knowledge of the subject that matters, the intuition of sorts, and not formal rules and schemas constructed by logicians.

Let us emphasize it. I do not question a great influence of the accomplishments of logic on mathematics whatsoever. I claim merely that formal logic is of no essential use in the *practice* of mathematical reasoning.

I may deliver more arguments to support my view, but it will be still debatable—just one of many possible views. Such arguments will not convince people who claim that applying formal logic to AI is well on its way, or that their students do not complain and find rules of formal logic useful in the practice of everyday reasoning. From the very beginning of my research in this area, I was perfectly aware that a decisive argument would be to present another concept of logic or another approach to issues of practical reasoning—one that could have a visible advantage over any other concept. In this article, I present the main ideas of a possible new practical approach.

3. TWO TYPES OF MATHEMATICAL PROOF

Mathematical reasoning is recorded in the form of *proofs*. Mathematical proof may be viewed as a series of elementary inferences. In each step we infer a new conclusion using assumptions and the conclusions established at earlier steps. Thus it can be viewed as a sequence of statements—as formal logic does. Yet, this abstracts from the real structure of proofs, which, in particular, is usually more or less branched. In fact, in the real proof we very often have steps in which we consider two or more cases, and this gives the real proof a *tree-like* structure.

In an extreme form we have to consider a few cases, each case has a few subcases of its own, etc. Another extreme is, when the proof is essentially a series of inferences, and the structure is, in fact, linear. This may be associated with the notion of the *direct proof* and the former with the notion of the *proof by cases* (also called the *proof by exhaustion*). It is true that mathematicians prefer direct proofs, considering them more elegant, more aesthetic, and that such proofs dominate rich and well-developed mathematical theories. Yet, in the newer areas of so called *discrete mathematics*, where theories are less developed and there are fewer interconnections between various claims and theorems, considering cases and subcases is a typical method of reasoning.

Now, if we want to use a mathematical proof as an exemplar of logical reasoning to follow in other less strict areas, it should be rather obvious that it is the first type, reasoning by exhaustion, that should be taken as a more suitable exemplar. In theories and situations where you have less knowledge you need simpler logical analyses. In everyday reasoning we usually have no lemmas and theorems to apply to infer direct consequences (and to form a linear reasoning), while considering various possibilities can be spotted in typical examples of practical logical reasoning.

A series of inferences can be regarded as an extreme case of the analysis of possibilities; such a convention is entirely natural, and certainly more natural than the reverse—treating a proof by cases as a series of inferences (as formal logic does). As a matter of fact, even proofs in the most developed mathematical theories have a repeated element of considering cases. Mostly, to prove a particular statement we assume the contrary and show that it leads to a contradiction. Often, this is done in one sentence: "The fact is so and so, since otherwise... which is impossible, because... "—making the consideration of two possibilities (either *A* or *not-A*) hidden. Yet, the practical idea behind such arguments is the consideration of two complementary cases. While this technique cannot be recommended for everyday logical reasoning (since in reality the complementary case is usually too comprehensive to infer any consequences), the very idea of considering cases is certainly highly recommendable.

Thus, *the basic assumption of formal logic treating the proof as a sequence of statements is an abstraction which ignores all aspects of mathematical reasoning that are not essential to the study of "what can be proved and what not"*. In this abstraction, a spectrum of natural tools is reduced to the iterative application of a few simple steps so that what can be proved at all may be proved—theoretically—with repeated application of these simplest steps. This makes it possible to prove various astonishing theorems about the extent of mathematical method, but is useless in practically proving theorems.

4. LOGIC AS ANALYSIS OF POSSIBILITIES

On the other hand, consideration of cases (possibilities) may be spotted in all instances of logical reasoning in various areas. There is no place here to give extensive examples, which could help to properly understand the theses formulated below. I will give only some indications.

First of all, the best source of purely logical reasoning seems to be solving logical riddles. Here we have all the necessary knowledge presented in a short description, and what remains is to draw conclusions. It can be demonstrated that in the case of logical riddles drawing conclusions is always based on considering various possibilities, in one way or another. Yet, it can be demonstrated that even in the case of puzzles of the same kind there is no universal method to solve them. On the contrary, various people use various methods and approaches. Dividing a problem into proper cases, choosing which cases should be considered, and in what order, is a creative act largely decisive for the efficiency of the whole reasoning. (This alone shows that the general problem of AI is much harder then it seemed to the researchers of fifty years ago.)

Following (Wittgenstein, 1953), it is illuminating to consider what is a common use of the term 'logical reasoning' (which may be treated as the real meaning of the term 'logical'). Let us consider here one simple instance in some detail. I am at home, watching TV, and suddenly the lights go out. The fuse has blown—I conclude. Can the conclusion be called logical? I believe an average language user would have some doubts. For there is a connection between the lights going out, the TV not working, and the fuse blown, so the reasoning follows some imaginable causal link—but more in a manner of conditioned reflex. Drawing such conclusions hardly requires the *art* of logical thinking.

Let us develop the example as follows. The TV stops working and the lights go out. I am asking myself: what happened? Either the fuse has blown or the electricity has been cut off (there have been power cuts at the power plant). But—as I look outside the window—all the other apartments are well lit. Therefore—I conclude—it must be the fuse. Logical?

At this point most of us would probably answer instantly—yes! Someone might observe that not necessarily: that there might have occurred a peculiar type of electricity breakdown which left the fuses intact. There might be other possibilities, even less likely but still impossible to exclude (like "all laws of physics have suddenly changed"—you cannot exclude it with 100 percent certainty!). In practice, however, before considering such possibilities we would probably check the fuse first.

The second argument differs from the first one mainly in that it entails some consideration of possible causes, some *analysis* of possibilities. Thus, I claim that considering various possibilities—various possible states of affairs—is a very important component of all arguments which we are ready to accept as 'logical', in the popular, everyday meaning of the word. The process in question consists of accepting some possibility as a conclusion whenever, as a result of considering other possibilities, we believe that there is no other (reasonable) possibility that might fit better to what we know, that could replace the conclusion we are about to accept.

The analysis of possibilities is involved when a mechanic checks an engine or tries to fix a TV, when a doctor examines a patient and makes a diagnosis, when a detective looks for a perpetrator, a director makes an important decision, a computer programmer designs a program or tries to find a mistake in the code, a journalist analyses the political situation, an everyman faces the challenges of everyday life. Who among them knows formal logic, uses formal rules of inference? They have all learned the necessary skill through practice. They have all learned to conduct such analysis of possibilities, for better or worse. Many of them, having considered various possibilities, draw their conclusions with as strong a conviction as a mathematician discovering his abstract theorems and facts.

5. THE MAIN THESIS

The condition of acceptance of a logical conclusion described above may recall the reader what is called the *argument to the best explanation*. There are however essential differences in our approach. First, in the terminology of 'explanation', we speak of 'the *only* explanation', the only reasonable possibility, not the best one. More importantly, in our approach, this is not the point at all. The point is in the process of analysis: in considering possibilities. Once we start to consider alternative possibilities and discard them on the basis of further observations or other knowledge—we start logical reasoning. If some possibilities cannot be discarded in the "first round" and demand further analysis, considering their sub-cases, we start the complex logical reasoning typically featured in solving logic puzzles and mathematical tasks. On the other hand, practical logical reasoning (that made by mechanics, doctors, detectives, etc.) is often combined it with action: performing various tests, asking questions, checking sources—this is done in order to confirm or exclude some possibility.

My main general thesis is that the essence of logical reasoning lies in the analysis of possibilities. *Logic is considering possibilities.* We accept a conclusion as *logical* if—as a result of the analysis—we are convinced that there is no other (reasonable) possibility.

This gives a completely new perspective (comparing with approach based on formal logic). In particular, *our conclusions in practical logical reasoning are no simple consequence of the premises*, but rather an upshot of the total image of the subject in our mind—an image shaped by conscious thought but also by unclear, semi-conscious convictions and beliefs. While in mathematical reasoning we need to identify premises, in nonmathematical reasoning this is not necessary, and sometimes even pointless or impossible. Practical logic is not a quest for premises.

Such a concise formulation needs to be specified and developed, and simultaneously justified, to acquire its proper meaning. As a new proposal, it in fact needs a whole book to explain what is really meant (and such a book is in preparation). Here I will only try to indicate the most important points.

6. COMPLETE AND INCOMPLETE REASONING

Our thesis (which can also be treated as a stipulative definition) covers both mathematical and nonmathematical reasoning. In referring to conviction the thesis entails that logical reasoning is, generally, not certain. This pertains to practical mathematical reasoning, too, because people may be wrong in their convictions. Mathematicians use here the concept of *obviousness*. Each step in a complete practical mathematical reasoning should be obvious. This does not provide *absolute* certainty, because like every other type of logical reasoning mathematical reasoning depends on imagination-based analysis of possibilities, and here there is room for gaps. Even brilliant mathematicians may happen to err. There are many examples of famous errors in mathematical proofs, which usually have the form of a gap in the proof. A gap is not taking into account some possibility.

On the other hand, even in our investigations about reality we are capable of almost *mathematical certainty*: if there are footprints in the sand, someone must have walked on it; if an object hangs on a rope, and we cut the rope, the object will fall, with so called "hundred percent certainty". As far as validity is concerned, then, there is no clear-cut distinction between demonstrative and non-demonstrative reasoning. What really differentiates these two classes is the scope of possibilities taken intentionally into account.

The thing is, while in mathematics we consciously aim to analyze *all* the possibilities, outside it we confine our scope (more or less consciously) to the "reasonable" ones. To express the distinction we shall refer to mathematical analyses (reasoning, logic used in mathematics) as *complete*, whereas the real life (real world-related) reasoning shall be called *incomplete*. In other words, in mathematics every possibility is reasonable. Only in this sense can mathematical conclusions, explicitly intended to exclude *all the other possibilities*, be considered certain, perfectly valid (as long as they are correct).

At this point we need to realize that *the border between complete and incomplete reasoning is the very border between mathematics and non-mathematics*, between formal logic's area of applicability and the realm in which no formal language is able to express the richness of possibilities. Indeed, for reasoning to be complete the very subject must *be descriptively complete*, that is fully describable in terms of a finite fragment of language. Generally the experience of mathematics proves that every such object can be "mathematicized", can be described as a mathematical entity. A complete argument regarding such objects can always be rendered as a formal mathematical proof. (This pertains, for instance, to the logical puzzles).[2]

When our reasoning concerns the real world, when our knowledge is limited and uncertain, and we have no complete description of the subject—then our logical analysis cannot be complete. Then, we usually choose possibilities which we perceive as worthy, whereas we skip those unlikely, insignificant, not worth considering. There are also possibilities which we are unable even to comprehend, either due to lack of knowledge or to the very nature of the subject under study—these also escape our analysis. Therefore we need to

---

[2] It is worth noticing that even if the subject matter is descriptively complete, it may be more reasonable to apply an incomplete reasoning. The game of chess is a good example.

make our definition of logic more specific: *Logic (in practice) is analysis of* reasonable *possibilities.*

It is more specific, but not more precise. Our guide in the process of selecting reasonable possibilities is, most of all, our common sense, our intuition based on knowledge and experience. To specify some universal criterion here is next to impossible. Whether a possibility is taken into consideration or not depends not only on the possibility itself, but also on our subject matter, purpose, the general level of our knowledge, etc. What is unreasonable today may prove more than reasonable tomorrow. As the borders of our knowledge expand, new possibilities arise: what used to be inconceivable becomes a sure fact. Moreover, what is a matter of serious analysis in one field is in another, more practice-oriented, but idle speculation.

The notion of 'reasonable possibility' introduced here is, thus, relative and imprecise. Yet it is practical and intersubjective—in the sense that people, using their common sense, knowledge and experience can quite easily reach an agreement over what is a reasonable possibility in some case and what is not. In this, it reminds the notion of 'obviousness' used by mathematicians (which will be discussed further in Section 8). Confronting our inferences with reality on a daily basis constantly verifies the idea of reasonable possibility, making our reasoning work despite its fallibility.

It is the ability to differentiate between reasonable, meaningful options and the unworthy ones that is the key to our reasoning's effectiveness. It largely depends on our individual imagination (about both the principles and the details of the field in question). We assume that all reasonable possibilities have been analyzed when we cannot imagine anything else worth considering. And the imagination, in turn, is shaped by our entire knowledge (general and specialist), our experience, practice and so on.

7. REDUCTION

In my forthcoming book (Kisielewicz, 2015) I am trying to explain in detail how it happens that in spite of the fact that mathematical reasoning in practice refers to subjective mental images it can be reduced to the repeated application of certain simple schemas of formal inference of a purely linguistic nature. Because accepting the fact that *this phenomenon is characteristic only for complete reasoning* has important consequences for the whole field of logic, I will indicate very briefly the main point of this explanation.

The objects of consideration of mathematicians are mental worlds of ideal, precisely and unambiguously defined, abstract entities, and the relationships between them. These mathematical *thought worlds* were developed during the history, initially, mainly as abstractions from real world phenomena, later also by means of purely theoretical interest. They are *intersubjective* in as much as different people—all people interested—agree as to various facts and properties of these worlds. (As existing objects they are similar to the world of the Olympian gods, or the universe of Dostoyevsky's novels). What determines and constitutes a given world is expressed in language entirely, precisely and unambiguously, and in this sense these worlds have linguistic nature. Since they are descriptively complete, reasoning about them can be complete, hence: true deduction is possible in this case. The way the knowledge of a given thought world is presented, the definitions, theorems and how the material is arranged in textbooks—the whole complex, multi-level system—this is also a part of the intersubjective side. In this sense mathematical worlds are much richer than their truncated versions in formal logic. (These worlds also have a totally subjective representation: the mental picture a person builds in his or her mind, the individual conceptualization. The more accurate the picture, the better a person's orientation in the given world; the differences can be, and indeed are, tremendous.)

If a step in a mathematical argument is questionable, a mathematician can break it down into smaller, more obvious steps, repeating the process of reduction until it reaches the level of the smallest steps: conclusions following directly from definitions, reformulating definitions, and the like. Since, as mentioned above, what is objective in mathematics is founded solely on its language and definitions, no wonder that ultimately all logical reasoning is potentially reducible to "manipulating" definitions, that it can attain a purely verbal character, lacking any reference to external reality. The fact that the principles according to which definitions can be manipulated (reformulated etc.) can be reduced to a few simple formal rules (concerning only the form of a definition) may be surprising, and it is a very significant achievement of formal logic.

It is important *however to realize the scale and the totality of the reduction*. While we can fairly say that mathematics was successfully reduced to a system of axiomatic theories (or even one axiomatic set theory) so that inferences have been replaced by a system of formal rules (whose appropriateness and obviousness resulted only from the meanings that had been conventionally attributed to logical connectives), we have to add that at the same time it removed from its scope the whole tremendous, multi-layered system of ideas. It has been proved that what remains—the axioms and the rules—is enough, in principle, to recreate the whole of mathematics, that is, to introduce respective definitions, to formulate and prove formally all theorems. What has not been emphasized enough was the fact that *to recreate mathematics as it was we need, in particular, the whole system of ideas as it was*. The degree of this reduction is best characterized by the fact that contemporary computers, though equipped with full knowledge of these axioms and rules, are incapable of proving anything of interest (yet)— because they lack the insight, they cannot choose which direction to pursue since too many possibilities open up at every step.

The reduction in question has very important theoretical implications, because it makes possible the proof of mathematical theorems about the extent of mathematical reasoning. It is true that this reduction was also carried out with a view to practical applications. It turned out however that for practical implications it is too radical, even for computers. While the very idea of such reduction—reducing certain tasks to elementary, strictly defined steps, executable in an entirely automated manner—found a tremendously successful application with respect to computation, and led to the construction of computers and the creation of information technology, in the realm of mathematical reasoning—creative reasoning, neither planned nor guided by algorithms—the reduction exploited in mathematical logic proved very radical, purely theoretical, and (so far) impossible to use in practice.

What must be stressed is the fact that neither Alan Turing nor anyone else ever suggested that reduction of computation to elementary operations on 0s and 1s could be used to improve *human computation*, that they could be of any use to people manipulating numbers. *This fact alone is enough to put in doubt the claim that learning formal laws of logic could in any way help us to make real life inferences.*

What is more, formal rules of inference apply only to complete (mathematical) reasoning. Linguistic completeness and strictness is a necessary condition for formal rules of inference to make sense. So, if they could be really used in real life reasoning, it is only in a mathematical context. But in such a context it is certainly more efficient and natural to use suitable mathematical tools (e.g. arithmetic).

Still one may claim, that single inference rules can be used to make an instance of reasoning more clear. The problem is however that these rules correspond to such simple linguistic transformations that alone, or in small sequences, they yield practically no progress in reasoning. You do not need to know these rules to say things in another way. More than that, you should avoid using any formal linguistic transformations without knowing what you mean. To consider cases efficiently you need to imagine various possibilities, and present them in a possibly clear way—not play with linguistic transformations.

In this situation the introduction to classical logic and the rudiments of predicate calculus in textbooks of practical logic and critical thinking should be adjusted. They need a commentary, explaining that these calculuses are of use only in mathematics, and only as a theoretical device, to analyze the scope of mathematical inferences. *They are of little use for everyday, practical purposes, just like binary calculations are.* But then a natural question can be expected: why teach such a calculus to an average lawyer, or anyone outside mathematics? This is the core of the great misconception we face. Such a calculus can be useful to students in fields like mathematics or computer science, it is certainly a great exercise in abstract thinking, but its status in textbooks of practical logic must certainly change. For practical reasoning does not consist in calculuses; it is based on different mechanics entirely: a mechanics both simpler on the conscious level, and much more complex on the neurobiological one.

Analyzing this matter deeper one needs to come to the conclusion that *treating formal rules of inference as a base for real life reasoning is but a great scientific misconception accumulated over centuries* (and this phenomenon is worth a closer examination all by itself). We face a kind of epistemic failure. *Teaching logic in the way it is done now is a fundamental educational mistake.* I make this bold claim, because I do not want it to be treated as one of many possible points of view. I claim that the dominant views today in this matter are essentially wrong.

8. PRACTICAL METHOD OF LOGICAL ANALYSIS OF REASONING

In accordance with my claims, what I am saying here does not admit any irrefutable proof. I may only collect evidences to try to convince the reader. There is one more strong evidence I need to mention. This is the fact that my approach makes it possible to work out a practical method of verifying logical completeness of reasoning presented in natural language. It is described in my forthcoming book (Kisielewicz, 2015). Here, I can only describe its rudiments.

First, we need to realize how a mathematical proof is actually verified. Contrary to the popular yet misguided belief, *the procedure used in real life is neither mechanical, nor sure, nor even clearly defined* by any fixed rules. Formal logic has indeed developed a strict, mechanical and infallible method, but this method is purely theoretical. Meanwhile, in real life the key notion in verifying proofs is *obviousness*. This is very far from being strict or precise. There is no definition of obviousness, no clearly established conditions (necessary or sufficient) to qualify something as 'obvious'. It is practice that decides what is obvious and what is not, whether the steps of a proof are sufficiently clear or need to be developed further. Besides, the whole notion of obviousness is relative in mathematics. What is obvious in one situation does not have to be obvious in another, or to the contrary—may be deemed too obvious, encumbered with overabundance of trivial explanations and details. (In all that, it reminds the notion of 'reasonable possibility').

If the author of a proof cared more for clarity than for anything else, checking such a proof is virtually "nodding" over subsequent conclusions in admiration for their unquestionable obviousness. (*Understanding* a proof, grasping its essence is an entirely different thing; with too many details in sight it might be much more time-consuming). In serious mathematical works the conclusions are not always that obvious; the reader must sometimes ponder over them a little longer (or even do some routine calculations and proofs himself) before he is able to declare the conclusion correct and indeed obvious.

In such cases, however, there are two possible outcomes. Either the whole checking procedure goes trouble-free (because it really requires only routine, obvious steps), or—if the conclusion is not that obvious—doubts and problems arise. If the person checking the proof, having failed to ascertain its obviousness, starts to doubt whether it is true at all, then he or she tries to disprove it—find a *counterexample*, which would prove the conclusion is not valid (generally). To find a counterexample means to *show another possibility*, to prove that the whole proof has failed to take this possibility into account!

This is the essence of rebuttal in mathematics. You can say: the proof is unsatisfactory, it is unacceptable because this step and that conclusion are unclear, unfounded. But only after you have found a counterexample may you say that the proof is invalid, the reasoning is faulty, for there is the possibility that...

And conversely, when we are unable to find a counterexample we start analyzing why. The analysis usually (that is, when the conclusion was correct but insufficiently substantiated) makes us understand why there cannot be any counterexample, any other possibility. Then we are able to reconstruct the missing steps, and see that the conclusion is indeed valid.

As we can see, the knowledge of formal rules of inference offers little help in checking actual proofs, because their conclusions rarely follow from the premises in a strictly formal manner. If you doubt a proof and wish to verify it, you need most of all a good knowledge of the branch of mathematics you are working with, and imagination—the ability to invent counterexamples.

Searching for counterexamples, for possibilities unaccounted for, is in fact the basis of critical verification of mathematical proofs. As long as they are not caused by sheer incompetence, most errors in proofs result from failing to see some veiled possibility. A counterexample refutes the conclusion; formal logic confirms it in the following way. If there is a counterexample for a thesis, there cannot be any formal proof of that thesis. In other words: what follows logically from some set of premises is true in *every possible world* in which the said premises are true.

With the last thesis, taken from Leibniz, some philosophers tried to construct a universal definition of logical truth. Still, they failed to see that in the light of this thesis practical logic must be simply analysis of possibilities.

At this point many readers, believing in the strict and flawless methods of mathematics, are probably disappointed with the actual method of proof verification. The truth is, it is hardly a method. You could as well call it a commonsense analysis of possibilities, and you would be right, for it consists entirely of accepting the conclusion when it is evidently the only possibility, and looking for other possibilities otherwise. Moreover, the person checking the proof can rely only on his or her own imagination and general knowledge of mathematics.

Yet these are the facts. This is what it actually looks like. In real life such a method proves not only effective, and the most efficient, but also to some extent infallible. And if so, then why not apply it to non-mathematical arguments? Of course we may expect some shifts of focus, some aspects will require a more detailed analysis (for instance distinguishing claims from conclusions), others will have to be modified (the notion of obviousness). Yet such a method, modeled on the verification strategies used in mathematics, is not hard to imagine. Meanwhile it should be clear that *searching for another method, more efficient or more mechanical, is futile since even mathematics does not actually possess any automated strategy.*

The method described in (Kisielewicz, 2015) requires various prerequisites, but one may just try to jump into the deep waters and examine any instance of non-mathematical reasoning, having as the only guideline *seeing that no possibility has been omitted*. For example, if you consider any instance of Sherlock Holmes' reasoning, having this guideline in mind, you probably discover many possibilities Sherlock did not take into account. Many of them you may dismiss as unreasonable or not reasonable enough. But in some instances you will also find examples where quite reasonable possibilities have been overlooked.

In (Kisielewicz, 2015), as one of the examples, I apply this method to (Searle, 1990). It shows that the text presented to American students as an instance of perfect logical reasoning contains a few evident logical gaps. In fact the method discovers gaps in most of the texts you take into consideration, because it is extremely difficult (for the author) not to miss any reasonable possibility in more complex reasoning concerning real world phenomena, especially when confronted with other points of view. What is more, reasonableness of various possibilities can be debatable. As a result, *the proposed method is a tool for rational dialogue rather than a verification tool.*

9. OTHER ISSUES

There are still a lot of issues not mentioned here that would make the whole conception more complete and more convincing. The most important are the concept of *meaning* and the concept of *truth* that must now be viewed in a new light. For these I need to refer the reader to (Kisielewicz, 2015). Still, I would like just to mention four other important issues.

First, let us note that in developing this approach one may show that various types of reasoning considered so far (by abduction, induction or analogy) can be better understood when viewed as more or less general methods of considering possibilities. As such they are not sufficient to infer any conclusion. (They are merely *educated guesses*, in spirit of Peirce.) A phase of considering whether there are no other reasonable possibilities is necessary to complete the reasoning process. This leads to natural solutions of the so-called problem of induction and controversy about abduction.

Second, a consequence of the new approach to logic is a new look at the phenomenon of argumentation. *Argumentation itself may be logical or not.* It is logical if it is essentially a presentation of logical reasoning—if all the arguments and evidences are intended to demonstrate that there is no other reasonable possibility to accept but the proposed conclusion. When arguing, you do not look for premises supporting the conclusion. You look for other reasonable possibilities, and then, for reasons why any other possibility is out of the question.

Third, the fact that I emphasize the informal character of logical reasoning does not mean that I argue against the formal approach in logic and argumentation. On the contrary, computer programs are formal and any attempt to endow computers with capabilities of logical thinking must be formal. The problem is that there are many possible formalizations, on various levels, and we need to decide which can be fruitful, and which will be most efficient.

Finally, a special attention must be given to the fact that practical logical reasoning is usually combined with action. Performing various tests, form simple experiments to asking questions and checking sources, is what makes practical logical reasoning so effective. Logic in practice is *logic in action*. This fact is almost neglected in textbooks on logic and critical thinking.

10. CONCLUSION

We argue that in practice there is no clear distinction between deductive and inductive reasoning. Moreover, in practice, they have a common feature, which is an analysis of possibilities. A natural distinction is whether a subject of discourse is descriptively complete or not. This implies the distinction of whether a conclusion follows necessarily or not. The converse implication does not hold, so the former distinction is more fundamental. Deductive reasoning is rather a very special case of logical reasoning. The phenomenon of deduction is, as a matter of fact, the phenomenon of mathematics.

I believe, that this approach to general reasoning based on *practical experience* in mathematical reasoning—as opposed to one based on a *formal model* of mathematics—is essentially new and original. Also the conclusions mentioned in the text are new. Even if most of them may be traced in works of other scholars, the fact that they follow naturally from my main thesis, that they form a consistent whole, is new, and gives to them a new meaning. This explains the small number of references in the text. Of course, I tried to find out whether these ideas have appeared in earlier works of other authors. And of course, I could find many related thoughts.

The most striking similarities are in Descartes' four postulates of good reasoning proposed in his *Discourse on the Method*. Two of them correspond very strongly to the idea of viewing logic as an analysis of possibilities. Yet, Descartes did not develop these ideas, and the successes of the formal approach prevented other scholars from appreciating the relevance of his postulates. The condition of acceptance of a logical conclusion has much in common with various methods of analysis close to arguments to the best explanation (Lipton, 2004), which was mentioned in Section 5. Also, it turns out that the concept of possibility plays a crucial role in the ontological works of Roman Ingarden. In particular, his *Controversy over the Existence of the World* treats ontology as a *science of pure possibilities* (Ingarden, 1960, pp. 44-45). Yet to see any connection requires deeper philosophical study. The concept of possibility was also analyzed in connection with the semantics for modal logic by Lewis, Armstrong, and many others (see Textor, 2014, Yagisawa, 2014). But this, as a part of a formal approach on elementary level, is rather in opposition to what I propose.

ACKNOWLEDGEMENTS: I thank Andrew Aberdein for numerous suggestions improving my English style. The research of the author is supported by Polish NCN grant 2012/07/B/ST1/03318.

REFERENCES

Ajdukiewicz, K. (1974). *Pragmatic logic.* Dordrecht: Reidel.
Descartes, R. (1960). *Discourse on method and Meditations.* New York: The Liberal Arts Press.
Fraenkel, A. A., Bar-Hillel, Y., & Levy, A. (1973). *Foundations of set theory*, (2nd ed.). Amsterdam & London: North-Holland Publishing Co.
Ingarden, R. (1960). *Controversy over the existence of the world*, vol. 1. Warszawa: PWN (in Polish).
Kisielewicz, A. (2011). *Artificial Intelligence and Logic* (*an account of a scientific venture*), Warszawa WNT, 2011 (in Polish).
Kisielewicz, A. (2015). *Common sense logic,* manuscript.
Lipton, P. (2004). *Inference to the best explanation*, (2nd ed.), London: Routledge.
Minsky, M. (2011), The Golden Age: A Look at the Original Roots of Artificial Intelligence, Cognitive Science, and Neuroscience (Keynote Panel), *MT150 Symposium, Brains, Minds and Machines*, May 3–5, 2011.
Searle, J. R. (1990). Is the brain's mind a computer program? *Scientific American 262*(1), 26–37.
Textor, M. (2014). States of Affairs, *The Stanford Encyclopedia of Philosophy* (Summer 2014 Edition), E. N. Zalta (ed.), http://plato.stanford.edu/archives/sum2014/entries/states-of-affairs/.
Wittgenstein, L. (1953). *Philosophical investigations.* Oxford: Blackwell.
Yagisawa, T. (2014). Possible Objects, *The Stanford Encyclopedia of Philosophy* (Fall 2014 Edition), E. N. Zalta (ed.), http://eplato.stanford.edu/archives/fall2014/entries/possible-objects/.

Commentary on Kisielewicz's A New Approach to Argumentation and Reasoning Based on Mathematical Practice

ANDREW ABERDEIN
Florida Institute of Technology, USA
aberdein@fit.edu

1. INTRODUCTION

Andrzej Kisielewicz maintains that formal logic is largely irrelevant to mathematical practice: the proofs that are the object of the formal logician's attention have little to do with the proofs composed by most working mathematicians (Kisielewicz, 2016). In particular, he tells us that formal proofs are linear sequences of derivations, whereas actual mathematical proofs of any interest characteristically display a branching structure, in which several alternatives are considered in turn.[1] Kisielewicz analyses logical reasoning, understood as a broader enterprise than formal logic, as fundamentally concerned with the analysis of possibility. That is, we accept conclusions as logical if we are convinced that there are no other possibilities. This leads to a demarcation of the border between mathematics and non-mathematics: mathematical reasoning is complete, in the sense that *all* alternatives have been eliminated; in non-mathematical reasoning it suffices to eliminate only the more plausible possibilities. Kisielewicz is sceptical about the potential for artificial intelligence in mathematics. Mathematicians routinely address questionable steps in their reasoning by reducing them to simpler steps. Computers can be adept at this task, but they lack the crucial creative insight to know when it needs to be

---

[1] As a reason for denying that formal logic captures mathematical proof, this overstates the significance of the choice of presentation. It is true that most accounts of formal proof focus on linear derivations, but that is not because formal proofs are necessarily linear. Indeed, there are formal systems in which proofs have exactly the branching structure Kisielewicz proposes, such as analytic tableaux (Smullyan, 1968). However, as we will see in Sec. 2.1, there are many other arguments for rejecting the identification of rigorous proof with formal derivation.

performed.[2] For Kisielewicz, proof checking is checking for obviousness—checking that no possibility has been omitted, and thereby that every alternative possibility besides the conclusion truly has been eliminated. He maintains that the different forms of reasoning which logicians distinguish (such as deduction, induction, abduction, and analogy) may be understood as different ways of considering possibilities. Finally, he briefly draws attention to some historical precursors to his picture, notably René Descartes and Roman Ingarden.

Kisielewicz's paper is the prospectus for a substantial programme, itself the subject of a forthcoming book. As such, the paper is rich in tantalising proposals, many of which are frustratingly lacking in detail. I will focus my remarks on Kisielewicz's larger picture and a couple of his more specific proposals.

2. WHERE WE (DIS)AGREE

The approach that Kisielewicz takes to mathematical reasoning is one which I broadly share. Hence I shall begin by indicating our points of agreement and the difference of emphasis between our approaches.

2.1 Where we agree

The following two theses are often presented as Conventional Wisdom about the relationship of logic to mathematics, at least by philosophers, including philosophers of mathematics:

> (1) Mathematical reasoning is fundamentally different from everyday reasoning;
> (2) Formal logic adequately models the practice of mathematical reasoning.

Kisielewicz rejects both theses, as do I. We are in excellent mathematical company in so doing. For example, the eminent mathematician Michael Harris complains that philosophy of mathematics, "typically pays little attention to what mathematicians actually do, unless they happen to be logicians" (Harris, 2015, p. xiv). Harris acknowledges that a minority position, sometimes referred to as the philosophy of mathematical practice, is more attentive. The even more eminent mathematician Michael Atiyah comments of one collection of essays on mathematical

---

[2] Here Kisielewicz may be unduly pessimistic. See, for example, the work of Simon Colton on computational creativity and its application to automated mathematical theory generation (Colton, 2002).

practice that, "I was pleasantly surprised to find that the book does not treat mathematics as desiccated formal logic" (quoted in Hersh, 2008, p. 96).

Indeed, in recent years, thesis (2) has been the target of a sustained barrage. For example, Yehuda Rav has argued that it overlooks the unaxiomatized nature of much of mathematics and that a central purpose of proof is the development of new techniques, techniques which formalization characteristically obscures (Rav, 1999, 2007). Alan Bundy and colleagues point out that mathematical proofs often contain errors, even though formal proofs are in principle mechanically checkable (Bundy et al., 2005). Marianna Antonutti Marfori argues that, since informal proofs are seldom actually formalized, thesis (2) makes a mystery out of their success and the consensus with which mathematicians accept them (Antonutti Marfori, 2010). And Carlo Cellucci, whose approach is perhaps closest in spirit to Kisielewicz's, states that formal proof disregards the process mathematicians follow. Rather than reason forward from axioms to theorems, he maintains that they habitually reason backwards from open problems to plausible hypotheses (Cellucci, 2008).

2.1 Where we disagree

If I agree with Kisielewicz that mathematical reasoning has more in common with everyday reasoning than it does with formal logic, where (if anywhere) do we disagree? As I understand our respective projects, we each begin from the rejection of the two theses stated above, but we then proceed in opposite directions. I am one of several authors to have argued that techniques that work for everyday reasoning should work for mathematical reasoning too (for a summary, see Aberdein & Dove, 2013). Conversely, Kisielewicz's argument seems to be that techniques that work for mathematical reasoning should work for everyday reasoning too. So we agree that similar techniques should work in both domains and that those techniques are distinct from formal logic, but we arrive at this conclusion by different routes.

This could represent a difference of principle, if we disagreed as to whether the determination of the correct method for mathematical reasoning is conceptually prior to that for everyday reasoning or vice versa. However, it may be no more than a difference of emphasis, or of tactics—if we accept that the same method is used in both domains, we may pick which to analyse; so we might as well pick whichever is easier. Nonetheless, the potential remains for a more substantial disagreement between us: what is the content of our shared destination, an account of reasoning adequate for mathematical and everyday contexts? I shall return to this question in Sec. 4 below, after briefly addressing the pedagogical implications that Kisielewicz finds in his work.

3. A FUNDAMENTAL EDUCATIONAL MISTAKE

Kisielewicz takes a firm line on logic instruction. He asserts that

> treating formal rules of inference as a base for real life reasoning is a great scientific misconception... We face a kind of an epistemic failure. Teaching logic in the way it is done now is a fundamental educational mistake (Kisielewicz, 2016, p. 280).

Plenty of people share this perspective, and they are not without empirical support (see, for example, Cheng *et al.*, 1986). However, empirical studies inevitably focus on a fairly narrow range of pedagogy in formal logic. It is not unreasonable to suspect that their results may not replicate for different approaches to teaching the same material. For example, some logicians have suggested teaching an abbreviated system of formal proof, which forfeits completeness in exchange for a focus on inference rules that more closely reflect everyday reasoning (Sherry, 2006). More conservative approaches to logic instruction may also have real world value. That formal rules of inference are seldom directly applicable to real life reasoning does not entail that their study is without value for real life reasoners. Just as road work can be invaluable for boxers without much resembling anything that happens in the ring, formal logic can be a highly effective form of cross training.

4. LOGIC AS CONSIDERING POSSIBILITIES

The kernel of Kisielewicz's new account of reasoning is a familiar one. As Holmes says to Watson in *The Sign of Four*, "How often have I said to you that when you have eliminated the impossible whatever remains, however improbable, must be the truth?" In very similar terms, Kisielewicz tells us that his

> main general thesis is that the essence of logical reasoning lies in the analysis of possibilities. *Logic is considering possibilities*. We accept a conclusion as *logical* if ... we are convinced that there is no other (reasonable) possibility (Kisielewicz, 2016, p. 275).

This is also the pattern of reasoning attributed in the ancient world to Chrysippus's dog: when it traced a scent to a crossroads, and found that all but one of the exits lacked the scent, it immediately ran down the last remaining exit (Aberdein, 2008). The underlying rule of inference is, of course, the familiar disjunctive syllogism: $p \vee q, \neg p \therefore q$. But is this adequate as a description of mathematical reasoning? We may, in traditional fashion, subdivide this question into two parts. Is the account sound with respect to mathematical reasoning, that is does it only endorse mathematically acceptable arguments? Is the account complete with respect to mathematical reasoning, that is does it endorse all mathematically acceptable arguments?

Soundness would seem to be trivial, since disjunctive syllogism is admissible in classical and, indeed, constructive, systems of formal logic. However, since Kisielewicz has repudiated such systems as analyses of mathematical reasoning, he can't help himself to this result. Moreover, there are other logicians who do not admit disjunctive syllogism as a valid rule, arguing that it does not reflect actual reasoning practice. Relevance logicians observe that, if one comes to believe $p \vee q$ merely because one believes p, which one subsequently learns to be false, without revising the beliefs that were grounded on it, then disjunctive syllogism would license the inference that q, for arbitrary q (Anderson & Belnap, 1975). Luckily for Kisielewicz, the disjunction central to his account may be understood as implicitly intensional. That is, it may be analysed as "if not p, then q", where the sense of "if" is not the material implication of classical logic, but something closer to the "if" of everyday reasoning. Intensional disjunctive syllogism is acceptable even to relevance logicians (see Aberdein, 2008). Understood in these terms, Kisielewicz's account is sound with respect to mathematical reasoning.

Completeness must be a tougher nut to crack, since the methods employed in rigorous mathematical proof are many and various. I have insufficient space in this commentary to assess how well Kisielewicz's account captures them; indeed, *he* has insufficient space in his paper to address this issue. I await his forthcoming book with renewed anticipation.

REFERENCES

Aberdein, A. (2008). Logic for dogs. In S. D. Hales (Ed.), *What philosophy can tell you about your dog* (pp. 167–181). Chicago, IL: Open Court.
Aberdein, A., & Dove, I. J. (2013). Introduction. In A. Aberdein & I. J. Dove (Eds.), *The argument of mathematics* (pp. 1–8). Dordrecht: Springer.

Anderson, A. R., & Belnap, N. (1975). *Entailment: The logic of relevance and necessity*, vol. 1. Princeton, NJ: Princeton University Press.
Antonutti Marfori, M. (2010). Informal proofs and mathematical rigour. *Studia Logica, 96*, 261–272.
Bundy, A., Jamnik, M., & Fugard, A. (2005). What is a proof? *Philosophical Transactions of The Royal Society A, 363*, 2377–2392.
Cellucci, C. (2008). Why proof? What is a proof? In G. Corsi & R. Lupacchini (Eds.), *Deduction, computation, experiment: Exploring the effectiveness of proof* (pp. 1–27). Berlin: Springer.
Cheng, P. W., Holyoak, K. J., Nisbett, R. E., & Oliver, L. M. (1986). Pragmatic versus syntactic approaches to training deductive reasoning. *Cognitive Psychology, 18*, 293–328.
Colton, S. (2002). *Automated theory formation in pure mathematics*. Berlin: Springer.
Harris, M. (2015). *Mathematics without apologies: Portrait of a problematic vocation*. Princeton, NJ: Princeton University Press.
Hersh, R. (2008). Mathematical practice as a scientific problem. In B. Gold & R. A. Simons (Eds.), *Proof and other dilemmas: Mathematics and philosophy* (pp. 95–107). Washington, DC: Mathematical Association of America.
Kisielewicz, A. (2016). A new approach to argumentation and reasoning based on mathematical practice. D. Mohammed & M. Lewiński (eds.) *Argumentation and Reasoned Action: Proceedings of the 1st European Conference on Argumentation, Lisbon, 2015. Vol. I, 269-285*. London: College Publications.
Rav, Y. (1999). Why do we prove theorems? *Philosophia Mathematica, 7*(3), 5–41.
Rav, Y. (2007). A critique of a formalist-mechanist version of the justification of arguments in mathematicians' proof practices. *Philosophia Mathematica, 15*(3), 291–320.
Sherry, D. (2006). Formal logic for informal logicians. *Informal Logic, 26*(1), 199–220.
Smullyan, R. M. (1995 [1968]). *First-order logic*. Mineola, NY: Dover.

13

Symbolic Condensation in Visual and Multimodal Communication

JENS E. KJELDSEN
University of Bergen, Norway
jens.kjeldsen@uib.no

This paper explores the concept of symbolic condensation in pictures in order to explain the possibility of visual argumentation and the benefits of so-called thick representation offered by many pictures. Symbolic condensation makes it possible for pictures to perform argumentation enthymematically and provides an aesthetic plenitude — a thick representation — that adds an epistemological gain to the communication of arguments as premises and conclusions.

KEYWORDS: condensation, condensation symbols, gun control, visual argument, reception, rhetoric

1. INTRODUCTION

As pointed out by Willard, argument "is a form of interaction" (Willard, 1989, p. 92). Visual argumentation is possible because the presentation of an argument is a communicative act performed in interaction, and because the communication of an Argument$_1$ as a cognitive instance of premises and conclusions is not tied to the verbal form of expression. The same argument can be prompted by different manifestations as long as the audience understands it in its proper argumentative context.

Obviously, a picture and a short caption – the use of which is how we usually encounter visual or multimodal argumentation in print – do not explicitly put forward premises and conclusions verbally. So, if pictures are to prompt arguments in the audience, some sort of symbolic condensation must be present. By symbolic condensation I mean the condensing of many different ideas into one, so that the effect and meaning of a picture or other visuals is grasped in one single instant – in a blink of an eye so to speak.

2. TRANSCRIPTIONS AND THICK REPRESENTATIONS

The immediacy and potential instantaneous character in the reception of pictures do not mean that the communicated relationships are any less complex. On the contrary, we may say that their complexity is not limited by the linear structure and discursivity of verbal communication. Langer, for instance, notes that "An idea that contains too many minute yet closely related parts, too many relations within relations, cannot be "projected" into discursive form: it is too subtle for speech" (Langer, 1980, p. 93).

This projection, however, is possible with the non-discursive symbolism we find in pictures because the primary function of the non-discursive is "conceptualizing the flux of sensations, and giving us concrete things in place of kaleidoscopic colors or noises", and this function "is itself an office that no language-born thought can replace" (Langer, 1980, p. 93).

The rhetorical and argumentative value of symbolic condensation is that it allows for a cueing and evoking of a wide range of emotions and more extensive trains of thoughts. This is made possible because of the semiotic richness of pictures. Photographs, notes Roland Barthes, have a feeling of "analogical plenitude" so great that verbal description is literally impossible. There are so many details in a photograph that it would require a lengthy book to try to describe it, and still you would not succeed because to describe a picture is "not simply to be imprecise or incomplete, it is to change structures, to signify something different to what is shown" (Barthes & Heath, 1977, p. 18). The image, in its connotation, writes Barthes, is "constituted by an architecture of signs drawn from a variable depth of lexicons" (Barthes & Heath, 1977, p. 47). Or, to put it another way: the semiotic richness based on the many different semiotic resources working simultaneously in pictures makes symbolic condensation possible.

Think of something as simple and ritualised as a wedding photograph. In order for such an image to make sense, we must call upon our knowledge of certain photographic conventions and formal traits (e.g. perspective, composition, colour), of distance to objects, interpersonal proximity, body movement, gestures and facial expressions, clothing and flowers; this in turn activates knowledge of interpersonal behaviour, cultural norms and values.

Borrowing from Umberto Eco's theory of semiotics, Robert Hariman & John Luis Lucaites in their book *No caption needed* (Hariman & Lucaites, 2007, p. 35) refer to this multiplicity of simultaneous codings as transcriptions. They write, that because the camera:

records the décor of everyday life, the photographic image becomes capable of directing the attention across a field of cultural norms, artistic genres, political styles, ideographs, social types, interaction rituals, poses, gestures, and other signs as they intersect in any event.

Hariman & Lucaites refer here to iconic press photographs, but the theoretical point applies in general. We can have the same kind of multiple transcriptions in paintings, cartoons and other pictures.

The rich representation of pictorial imagery, then, is made possible by the multiplicity of codes, resources, or transcriptions, working simultaneously, and providing pictures with the possibility of plenitude (cf. Barthes & Heath, 1977, p. 18f.).

So, if we seek a rhetorical understanding of pictorial argumentation, we cannot simply extract the verbal lines of reasoning, transform them into propositions and present them in argumentation models. There is a difference between the two modes of representation. Pictures are able to provide vivid presence (evidentia), realism and immediacy in perception (Kjeldsen, 2012a), which are difficult to achieve with words alone. Pictures are rich in visual information because they provide innumerable details for the eye.

Leaning on Clifford Geertz' term "thick description" (Geertz, 1973), we may say that pictorial representation has the ability of performing a sort of "thick representation" that in an instant may provide a full sense of an actual situation and an embedded narrative connected to certain lines of reasoning. This visual richness and semiotic "thickness" disappears if we reduce the pictorial representation to nothing more than "thin" propositions.

However, it is not a case of either/or: In general, theories of visual communication can be divided into two main strands: a phenomenologically influenced tradition regarding pictures as event – a sort of mediated evidentia – and a semiotically influenced view regarding pictures as a codified language system (cf. Kjeldsen, 2015; Kjeldsen, forthcoming). I argue that the power of pictorial rhetoric and argumentation is that it may work both as event and language system. The first perspective helps us see how the analogical plenitude creates presence, realism and immediacy; the second helps us see how pictures can work as a culturally coded language. The former offers emotional condensation, the latter rational condensation.

3. EMOTIONAL AND RATIONAL CONDENSATION

Emotional condensation means that an image, or part of an image, is capable of eliciting an extensive emotional response. As argued by visual communication scholar Paul Messaris (1997, chapter 1), the persuasive power of a pictorial representation consists of its ability to recreate visual cues, which in the real world are connected to specific emotional responses. When we see a small girl in real life, our nurturing instincts may be evoked. A picture of the same subject invites a similar response. Both theoretical and empirical research document that there is "considerable continuity between picture perception and every day, real-life vision" (Messaris, 1994, p. 13).

Emotions are not only evoked by what we see, they are also aroused by the manner in which the subject matter is presented, such as the point of view in a picture. A high-angle shot may infuse us with a sense of power and control, while a low-angle shot may conjure up situations in which we are less powerful. Because we are positioned as viewers in a way that is analogous to positions we recognise from the physical world and our social life, we are invited to react in similar ways.

While the emotional condensation is mostly connected to the phenomenological view of imagery as event, the aspect of rational condensation as a means of evoking argument and reasoning is more connected to the semiotic view of imagery as a codified language. Rational symbolic condensation is, I suggest, the basis for the possibility of visual argumentation. Such condensation is similar to Freud's psychological concept of condensation in the dream work (Freud & Crick, 1999, p. 260), which signifies the condensation of many different ideas into one.

Because pictures have the same ability to hold such condensation, they also have the potential to present arguments. If we understand the enthymeme as a "cooperative interaction", as proposed by Lloyd F. Bitzer (1959), we see that a rhetorical hallmark of pictures is that they offer a rhetorical enthymematic process in which something is condensed or omitted, and, as a consequence, it is up to the spectator to provide the unspoken premises.

Rational condensation in pictures, then, is the visual counterpart of verbal argumentation. In order to be able to reconstruct the implied arguments, the viewer may draw upon knowledge of the context of the picture, such as in the circumstances of the situation at hand (Kjeldsen, 2007). At other times – particularly in advertising – the viewer's reconstruction of arguments may be enabled through visual tropes and figures, which help delimit the possible interpretations (Kjeldsen, 2012b).

Because of their lack of clear syntax and the difficulty in distinguishing between different meaning-making units, understanding the argumentative context is essential for the audience to identify, reconstruct, and interpret argumentation performed by pictures. This is why we should be aware that argumentation is a process, a communicative interaction between people, which means that specific arguments must always be understood in terms of the ongoing debate or discussion they are part of. A rhetorical argument never exists in itself, it is never presented in a vacuum, rather it is always part of human interaction, some kind of communication; if not, then it is simply not a rhetorical argument.

When engaging in visual communication, despite its lack of clear rules of grammar, it is our knowledge of contextual, situational and procedural circumstances that allows us to use images to perform argumentation.

4. FINDING THE ARGUMENTS IN THE SITUATION AND THE AUDIENCE

I have suggested that not everything can be directly materially located in the text itself, but that there are elements present that are decisive for the audience's reconstruction of argumentation prompted by visuals. As pointed out by Pinto, argument may be understood as "an invitation to inference" (Pinto, 2001, pp. 68-69; cf. Groarke, 2015, p. 135). If we do not take this into consideration, we will fail to understand how visual and multimodal rhetorical argumentation actually works. Combining textual studies and empirical reception studies of the interaction between argumentation and audience allows us to create a more nuanced understanding of visual rhetoric and argumentation.

We should not only be concerned with how an argument or any rhetorical appeal is constructed, but also with the degree to which it is audience-oriented, how it is received, interpreted, and processed – that is: how actual audiences actually respond to rhetoric. As proposed by Schiappa: "We need to find out what people are doing with representations rather than being limited to making claims about what we think representations are doing to people" (Schiappa, 2008, p. 26).

So, if we seek an understanding of pictorial and multimodal rhetoric and argumentation, we first have to understand that argumentation is not only text, but rather a cognitive phenomenon, situated in communicative action (cf. Willard, 1989). We may analytically extract premises and argument from texts, but we cannot just locate visual tropes and figures, find ideographs cognitively, or simply extract the verbal lines of reasoning, transform them into propositions and present them in argumentation models. Because of the phenomenological quality of pictures, their visual plenitude and semantic "thickness" disappear if we reduce the pictorial representation to only theoretical concepts, cognitive phenomena or simple verbal propositions. We also have to acknowledge that in visual and multimodal argumentation, the audience is active in the reconstruction of enthymematic arguments. Therefore, the best way to understand how visual rhetoric and argumentation works is to examine the interplay between discourse, context, and reception, applying a combination of close readings of rhetorical utterances, contextual analyses of the situation, and empirical studies of audience reception and response (cf. Kjeldsen, 2007).

I will now further explore the notion of symbolic condensation through theories of condensation symbols. I will then apply this to an example of multimodal argumentation, taking into consideration the interplay between discourse, context, and reception. The example is a public service announcement (a PSA ad) from States United to Prevent Gun Violence, released in April 2013.

5. CONDENSATION SYMBOLS

If it is the case, as I argue, that pictures may hold condensations of emotions and reasoning, an important question arises as to how these condensations are cued and elicited in an audience, and how can we be sure that the same kind of responses are elicited in different people? Or to put it differently: how does such condensation and elicitation of emotions and arguments work?

In examining how pictures may carry a condensation of reasoning and emotions, the notion of condensation symbols may prove helpful.

The concept of condensation symbols probably originates from Edward Sapir, who distinguishes between referential symbolism and condensation symbolism. The former "embraces such forms as oral speech, writing, the telegraph code, national flags, flag signalling and other organizations of symbols which are agreed upon as economical devices for purposes of reference". The latter "is equally economical and may be termed condensation symbolism, for it is a highly condensed form of substitutive behavior for direct expression, allowing for the ready release of emotional tension in conscious or unconscious form" (Sapir, 1934, p. 493).

In terms of actual behaviour, the two types of symbolism are generally blended; however, condensation symbolism is characterised by being of a more primary nature and connected to emotions and the unconscious, diffusing their "emotional quality to types of behavior or situations apparently far removed from the original meaning of the symbol (Sapir, 1934, pp. 493-494).

The concept was further developed by political scientist Doris Graber, who defines "condensation symbols" as "a name, word, phrase, or maxim, which stirs vivid impressions involving the listener's most basic action" (Graber, 1976, p. 289). These stimuli "carry little independent information, but activate a wide variety of similar perceptions and connotations in their listeners" (Graber, 1976, p. 318), and they "evoke multi-faceted deeply involving images" (Graber, 1976, p. 312). In short, Graber assigns three rhetorical functions to condensation symbols: 1. "Evoke rich and vivid images in an audience", 2. "capacity to arouse emotions", and 3. "supply instant categorizations and evaluations" (Graber, 1976, pp. 191-192). This is precisely what pictorial presentations do well.

Neither Sapir nor Graber pay much attention to how condensation is made possible, but Kaufer and Carley introduce the notion of "well-connectedness" as an explanation. Words that have high impact "evince a high degree of connectivity in context" (Kaufer & Carley, 1993, p. 201). By connectivity they mean "the number and strength of associations that link explicit utterances with the insider elaborations of their audience" (Kaufer & Carley, 1993, p. 201). Such symbols, Kaufer and Carley explain, are not simply networked with other concepts, but are ""well-connected" in a network of meaning primed by the context" (Kaufer & Carley, 1993, p. 202). Thus, a concept – or in our case of pictorial communication: a visual sign or symbol – is never well-connected in itself, but only in relation to its use in specific contexts or situations. The expression 'fat cat', the authors explain, has no special connectedness with historically rooted social beliefs when we use it to refer to an obese animal, but when taking about election campaigns the expression "compresses or condenses a network of historical meaning" (Kaufer & Carley, 1993, p. 202). The defining characteristic of condensation symbols is that they are especially "well-connected" in their contexts of meaning.

Kaufer and Carley introduce three parameters for this kind of linguistic "well-connectedness": situational conductivity, situational density, and situational consensus. It is essential to underscore that these parameters are all context dependent. Conductivity, density and consensus of the same concept or form will differ depending on the situation. The rhetorical impact and perceived argumentation of visual or verbal condensation will always rely on the interplay between the symbol, the situation and the audience.

Situational conductivity refers to "the capacity of a linguistic concept both to elaborate and to be elaborated by other concepts in a particular context of use" (Kaufer & Carley, 1993, p. 202). Concepts that have central associations with other situation-specific concepts will be highly conductive and thus easily introduced into the foreground of the situation. In a zoo context, for instance, aardvark is a more conductive concept than ants, but in a picnic context, ants will be the more conductive concept.

Situational density refers to "the frequency with which a linguistic item is used in relation to others, within a delineated context and social group. Density denotes how often a word or expression is likely to recur as part of larger sentences, paragraphs, genres in context" (Kaufer & Carley, 1993, p. 204). In a newspaper house, the authors write, the word "headline" will be more dense that the word "aardvark".

Situational consensus refers to "the extent to which a concept is elaborated in similar ways across a given population in a given context" (Kaufer & Carley, 1993, p. 204). For a concept to function as a condensation symbol it must be high in at least one of these parameters:

1. Network primacy: high in situational conductivity
2. Frequency: high in situational density
3. Agreement: High in situational consensus

The most well-connected concepts, of course, are the ones that entail high degrees of all three parameters. While conductivity, density and consensus in Kaufer and Carley's paper are potential attributes of linguistic concepts used in context, I use these three parameters to examine the rhetorical and argumentative condensation in visual and multimodal representations. I approach not only the elements in a multimodal representation as discrete condensation symbols, I also consider the multimodal representation as a whole as a condensation symbol. The aim is to examine how such rhetorical and argumentative condensation may work in order to evoke emotions and reasoning. As with words, the possibility for visual representation to function as condensation for emotions and reasoning is the degree of their well-connectedness. Thus, I will examine how such well-connectedness is established and this connectedness may lead to specific kinds of reasoning.

6. VISUAL AND MULTIMODAL REPRESENTATIONS HAVE BOTH CONCEPT AND FORM

Compared with pure linguistic communication, visual communication works on the levels of both form and concept. I can write the word "tree" or the words "Jesus Christ" in either Bauhaus font or, for instance, in Bradley hand font, but this change of form wouldn't really make much of a difference for the concepts themselves. However, in a drawing, painting or photograph the way in which a concept is represented does make a difference, and sometimes the form itself, such as visual contrast, represents a concept.

Figure 1: Jacques-Louis Davids painting *The Death of Marat* (1793)

This means that visual and multimodal communication have the potential to be well-connected, both conceptually and formally. We find this in Jacques-Louis David's famous painting The Death of Marat (Fr. La Mort de Marat, 1793; see figure 1). In his argumentation analysis of this painting, Leo Groarke (1996) shows convincingly how David makes the following argument:

> MC: You must strive to emulate Marat in support of the revolution.
> C1: Marat was, like Christ, a great moral martyr
> P1: Marat was a man of great dignity and composure
> P2: Marat's assassin recognised his reputation as a benefactor of the unfortunate;
> P3: Marat gave his last penny to the poor

We will not go into the details of this analysis here, but explore only the claim that Marat was like Christ. The painting of Marat follows the central formal traits in the traditional religious iconography:

> The recumbent pose with the extended, trailing arm recalls, in detail, Renaissance depictions of the Disposition of Christ. (Cf. Girodet-Trioson, Caravaggio, Pontormo, Fiorentino, van der Wyden, Jan van Scorel, Michelangelo, etc.). The gaping wound below the clavicle with the stream of blood parallels the wound in the Saviour's side. (Louis Groarke, 1999, p. 74)

In the painting, the dignity and moral character of Christ is thus evoked by a formal resemblance to a traditional and recognised visual theme. In the context of art and visual representation, this theme is high in conductivity, since it can both elaborate and be elaborated by other themes and formal expressions. It is high in density since it was a dominant theme in its time. Finally, it may be expected to have high situational consensus since we can expect most audiences to recognise the visual similarity to the religious iconography depicting Christ.

So far I have only dealt with the formal visual aspects; however, the visual theme has a conceptual side: the dignity and character of Christ. The visual theme thus connects to many visual representations (as mentioned in the quote above), which are all connected to conceptual aspects of the character and dignity of Christ. Once the audience is connected to this conceptual side, it may tap into other connected conceptual themes, discussions and arguments. Since pictorial representation in this way communicates both concept and form, the situational well-connectedness may be carried out both conceptually and visually.[1]

7. CONDENSATION OF ARGUMENTS AND EMOTIONS IN A PRO GUN-CONTROL AD

Let us try to examine this further through an example of multimodal argumentation in the gun debate in the US. On April 15, 2015 the organisation States United To Prevent Gun Control (States United) released a pro-gun control video-ad (see figure 2). The release came after four months of intense and heated debate on gun control in the US following the Sandy Hook Elementary Shooting where 20-year-old Adam Lanza killed 20 children and 6 adults using two handguns and a Bushmaster AR-15 rifle. The renewed debate especially centred on assault weapons, and automatic and semi-automatic guns – such as the Bushmaster rifle –that allow a shooter to fire many rounds in rapid succession. The debate led to proposals for universal background checks for gun buyers and to federal and state legislation banning the manufacture and sale of certain types of semi-automatic firearms and magazines with more than ten rounds of ammunition. One month after the shooting, Democratic senator Dianne Feinstein introduced the Assault Weapons Ban of 2013. The bill was defeated on April 17, 2013, just two days after States United launched their ad. Incidentally, the ad was released on the same day as the Boston Marathon bombings.

The ad begins with an ordinary but gloomy-looking man in his 60s enters a nondescript office suite at a resolute and quick pace. He passes the receptionist, who shouts after him: "Excuse me, hey, hey". The man ignores her, and moves determinedly forward into the cubicles of the glum office space. He moves past the camera view and we now see him from behind sharing his perspective while we are looking over his shoulder. He makes a sharp right turn, and it becomes evident that he is carrying a rifle. He approaches an office desk, where a man in a white shirt and tie is standing in front of a desk talking quietly to a woman sitting behind the desk.

---

[1] This is in several ways in accordance with the so-called dual coding theory (Paivio, 1979), which argues that communication stimuli can be represented either visually or linguistically. Some stimuli, however, are better learned and remembered because they activate both systems of representation (or codes). Imagery – such as pictures – activates both systems the most, concrete words a little less and abstract concepts the least. In line with dual coding theory, then, we can expect pictures to represent and condense both imagery and concepts.

The intruder now raises his rifle, the man in the shirt and tie instinctively reacts, looks up, raises his open hands beseechingly, while moving backward shouting "Ed, Ed" at the intruder. Ed shoots. Screaming, shouting and panic engulf the room, but the man in the shirt and tie is seemingly unharmed. Ed moves forward, looks at the man he shot at, but then the lowers the rifle. The man, the woman, and the other employees flee in panic.

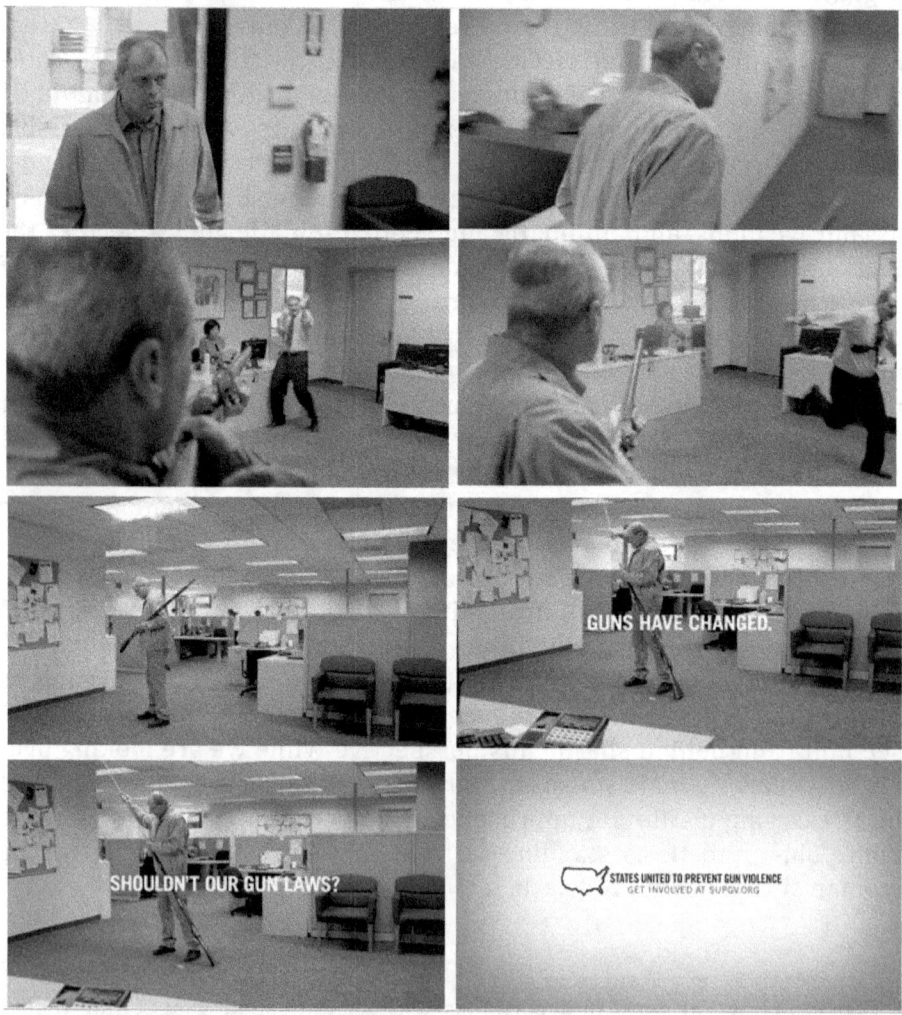

Figure 2: Still shots from a pro-gun control ad from the organisation States United To Prevent Gun Control.

Simultaneously, the perspective has changed so that we now are given a view from behind the desk, where we see Ed placing the rifle on the ground, before he begins the laborious process of cleaning and reloading what we now realise is a musket-style weapon. While he is cleaning the musket, the text "Guns have changed." appears across the screen. The office is quiet, with the exception of the sound of a running photocopy machine, a ringing phone in the background, and the metallic sound of Ed cleaning the rifle. While Ed is still working on reloading the musket a new text appears: "Shouldn't our gun laws?" The ad ends with a line drawing of the continental USA and a text saying: States United to Prevent Gun Violence. Get involved at supgv.org".

8. THE POWER OF VISUALLY DELEGATED EFFICACY AND THE ARGUMENT IT PRODUCES

The ad provides a condensed narrative representing mass shootings. As such, it has a high degree of situational conductivity because mass shootings are connected to many aspects of the gun debate. As a conductive concept – an interconnected rhetorical topos, if you will – mass shootings are readily introduced into debates on gun control. It is a "ready conduit of information" because it has many "ties to many other specific concepts" (Kaufer & Carley, 1993, p. 203). This conductivity goes two ways since the concept of mass shootings can both elaborate dominant aspects of the gun debate and itself be elaborated by such aspects.

The ad is also high in situational density because of the high frequency by which "mass shootings" is used in relation to other concepts and arguments in the gun debate. There is a difference, though, between the gun-control side and the pro-gun side because the concept is more frequent (has higher density) in the former group and is less frequent in the latter. This is because gun-control advocates use it in their probatio argumentation, while pro-gun advocates are forced to use it in their refutatio argumentation.

The concept of mass shootings can be expected to have high consensus within the two groups, but low consensus between the groups. While both groups will agree that the depicted action is a mass shooting (implying high consensus), they will probably differ as to how a mass shooting should be understood and evaluated (implying low consensus). The concept of mass shootings, then, can be expected to be "elaborated in similar ways" (Kaufer & Carley, 1993, p. 204) within the two groups, but in different ways between the groups.

Since the narrative of the mass shooting is high in both conductivity and in density, but low in consensus between the pro-gun advocates and the gun-control advocates, we may expect a shared understanding of which argument is put forward, but a difference of opinion as to how to ascribe meaning and evaluate the concept of mass shooting.

From a reception perspective, we may say that the difference in interpretation between the two groups will not be a matter of polysemy (i.e. not understanding the ad; cf. (Barthes & Heath, 1977, p. 39)), but a matter of polyvalence (i.e. difference of opinion and evaluation; cf. (Condit, 2013)). Both groups will agree that the depicted action is a hypothetical or potential mass shooting (high consensus); however, they will differ in opinion as to how such a mass shooting should be understood and evaluated (low consensus) – thus they will naturally differ in their evaluations of the ad as an argument. This is exactly what we see in the online comments that I examine below. This also means that the added significance of the visual – its thick description – loses its effect on the pro-gun advocates, as is obvious from the comments from people who mock the representation – and especially comments about how Ed acted and used the gun.

The different elements in the ad work as condensation symbols integrated in the visual narrative: Ed is a condensation of the many "mentally unstable shooters" we have heard about in the news. The office suite is a condensation of the many "publicly accessible places" that have been and potentially could be targeted by shooters. The panic and the screams is a condensation of similar situations we have experienced in news and fiction. The musket is a condensation of the kind of weapons that were used when the second amendment was written, and is thus a reference to this law. All these verbal and visual, cognitive and emotional networks are connected in the ad, creating a multimodal syntagm that draws its rational and emotional power from the paradigms (or networks) to which each of the elements also belong. Thus, each of these visual elements represent not just one single thing, but rather function as a condensation of a network of meaning and imagery. Using the word condensation instead of, for instance, representation is meant to emphasise this fact. These interconnections are similar to I.A. Richards' argument that the rhetorical power of a word in use is that it draws upon the missing parts of the contexts to which it is also used: "what a word means is the missing parts of the contexts from which it draws its delegated efficacy" (Richards, 1936, p. 35).

To put it another way: The visual symbols that are present in the ad point to both absent verbal connections (such as positions, arguments, propositions, facts) and to visual connections (such as the terrifying images and sounds we carry from news reporting on mass shootings or from the fictional representations of shootings we remember from movies and television). It is from these contexts that the symbols and the ad as a whole draw their delegated efficacy of emotional appeals and arguments.

As a condensation symbol, the ad directs us to a wide range of connected networks of meaning and imagery, inviting the audience to combine thoughts and images with reasoning and emotion. These connections allow the audience to reconstruct the argumentation of the ad.

Even though the ad is a multimodal instance of argumentation, it is relatively easy to recognise the primary argument since the central elements are represented explicitly and verbally in the caption: "Our guns have changed. Shouldn't our gun laws?". The caption thus provides the main argument:

> 1 We should change our gun laws
> 1.1. Our guns have changed
> (1.1') (When circumstances change, one should also change the law regulating these)

However, even though the caption more or less offers a complete argument$_1$, the argumentation in the ad is more than this. First of all, the verbal text does not say how guns have changed or how this may warrant a change to the gun laws. These reasons are provided by the audio-visual narrative. A more complete account of the argumentation offered by the ad would look something like this:

> 1 We should change our gun laws
> 1.1 Our guns have changed
> (1.1') (When circumstances change, one should also change the law regulating these)
>
> (1.1.1) Modern guns are much more efficient for killing in mass shootings than traditional guns from the time the law was created
> (1.1.1.1) If someone would attempt a mass shooting with a traditional gun of colonial times, this person would not be able to kill many

This is connected to a second line of reasoning:

> 1 We should change our gun laws
> (1.2) (It would save lives)
> (1.2.1) (Restricting or banning the use of modern automatic weapons would limit the ability to kill many people in mass shootings)
> (1.2.1.1) (If someone would attempt a mass shooting with a traditional gun, this person would not be able to kill many

While the first part of the first line of reasoning (1 to 1.1) is primarily evoked by the verbal caption, the second part of this line of reasoning ((1.1')) and the whole second line of reasoning ((1.1.1) to (1.1.1.1) and 1 to (1.1.1)) is primarily evoked by the audio-visual part of the ad.

This second line of reasoning illustrates a general point made by Gilbert, among others, namely that "nondiscursive communications are required in order to clarify discursive communication" (Gilbert, 2001, p. 243). Furthermore, as illustrated, the nondiscursive in this example not only clarifies, but also constitutes the argumentation.

9. THE RECEPTION OF THE AD

It is clear, then, that the ad offers an argument, and that some parts are offered by the caption and some parts are offered through images and sound. Let us examine how audiences perceived the argument and reacted to it?

In order to do this, I have examined several comments threads on webpages showing the ad (such as States United to Prevent Gun Violence and YouTube) and on pages that describe the ad and links to it (such as Mediaite and Democratic Underground).[2]

Studying the reports on the ad, the discussions, debates and online comments, it is clear that people commenting on these pages clearly understand the argument put forward in the ad. Matt Wilstein from the website Mediaite, for instance, summarised it thus:

> Without giving too much away, the 30-second ad uses a terrifying workplace shooting scenario to make the case for updating the country's "antiquated" gun laws. While not explicitly denouncing the Second Amendment, the group is making the point that a lot has changed since the founders declared that Americans have a "right to bear arms." In half a minute, the PSA compellingly sums up what many pro-gun control advocates have been arguing in favor of for decades.

---

[2] The following online comment threads have been examined: 1. The YouTube thread for States United to Prevent Gun Violence (https://www.youtube.com/watch?v=LORVfnFtcH0), 2. Discussion threads on the Facebook page for States United to Prevent Gun Violence on respectively April 15, 2013 and February 15, 2015 (https://www.facebook.com/StatesUnitedToPreventGunViolence?fref=nf), 3. The discussion thread on Mediaite (http://www.mediaite.com/online/watch-new-pro-gun-control-psa-puts-the-second-amendment-in-sharp-perspective/), and a short discussion thread on Democratic underground (http://www.democraticunderground.com/1017112444).

When the user "Inside Guy" made an online comment on this page saying "Brutal, but effective", the user "BillBuckley" gave the following reply:

> I'm not so sure it is all that effective. Granted, it grabs your attention but it's basically arguing that we need more control of guns because of the greater damage they can do, as in this case when a deranged man walks into a workplace. That's fine for the people who already think like that, but how does this persuade anyone on the other side that our present laws are stupid? They'll just say it's not the firepower. Keep the guns away from crazy people and you don't have these problems, or at least not as many.

It is obvious that both Wilstein and "BillBuckley" extract an argument that is quite similar to the one reconstructed above. The same argument (i.e. claiming that effective guns are a problem and should be restricted or banned) is obvious in the comment from the user "The nextinline" on the YouTube channel for States United to Prevent Gun Violence. The user writes:

> So the problem is GUNS? Ok lets say that they ban all guns, lets say 0 civies have 0 guns. When they will want to kill, they will use swords. Ban swords, they'll use knives. Ban knives, they'll use spoons and forks. Ban them as well, they'll use guitar cords and broken glass/dishes. Ban those as well....they will use fuckin rocks tied to a wooden club. Result? Still the same. Those who want to kill, will do it by any means.

The response from user "1missing" is laconic:

> Stupid argument. You can't walk into a mall and kill 50 people with a fucking spoon.

A similar exchange occurs between two other commentators on the site: "Matthew Johnson" argues that:

> The psycho at Newtown had free reign for 20 minutes. Even with a Brown Bess muzzle loading rifle, he could have shot 60 rounds. Tech changes, but psychos are psychos, and laws won't change that.

However, "Panatronic Freud" does not agree, and dismisses the argument:

> Ha. Only a gun nut would go down such rabbit holes. If the kid at Newtown was using a musket, there would not have been 26 dead bodies. This is self-evident.

A similar exchange happened on the website Democratic Underground, where the user "D" commented that it was:

> Kind of funny that you would post this 'video' on the 'internet' using your 'computer'. All things used to express your first amendment rights that were not available in the 1700s.

However, "kurtzapril4" was not impressed:

> Let me know when i can walk into an office or school with my computer and kill a whole bunch of people in seconds, okay? P.S. Guns are not "speech.

These brief excerpts teach us several things. Firstly, there seems to be general agreement on the argument the ad puts forward:

1	We should change our gun laws
1.1	Our guns have changed
(1.1')	(When circumstances change, one should also change the law regulating these)

Secondly, these comments – and the large majority of comments on the issue – come from people arguing either for or against gun control – usually quite passionately. Studying the comments on these sites makes it clear that only the (rather few) pro-control advocates actually mention or reproduce the main argument of the ad, and argue that guns (more specifically automatic guns) should be either banned or better controlled because of their efficiency in harming and killing many people. The pro-gun advocates posting comments tend to leave the main argument of the ad behind and put forward other kinds of arguments – often simply reverting to plain insults.

Thirdly, certain topical moves recur in the online debates, when pro-gun advocates make their arguments. One topical move is to reject the argument analogically by arguing that forms of communication have changed, but we haven't changed the first amendment. Thus, we shouldn't change the second amendment, because guns have changed. Another move is to simply reject the claim of deadly efficiency of contemporary guns. This is either done by arguing that Ed could have been much more efficient if he had only known how to handle the musket, or if he also had had a bayonet, or if he had had more weapons; or it is done by claiming that if the people in the office had efficient weapons, they could stop the intruder before he could harm anyone. This line of argument points to a more general fact of the reception: Overall, while the gun-control advocates tend to accept the visual presentation and base their comments and argumentation on this, the pro-gun advocates tend to reject the visually presented narrative. On the YouTube thread for States United to Prevent Gun Control, for instance, "David Louis" argued:

> actually if ol' Ed was drilled properly with that flintlock he was using, he could have taken out half of that office (especially if the bayonet was attached) because a lot of people don't get up and run when this type of shit happens in the real world...

The user "Frank LeClair" offered a similar argument on the Facebook thread of States United to Prevent Gun Violence, when dismissing the visual narrative:

> First of all, I saw 1, 2, 3, 4 people, that had they been armed at their workplace, saw the shooter walk in with a gun, were ignored by the shooter, and could have stopped the shooter immediately after his intentions were made absolutely clear with his first shot. Second, gun control laws aren't effective at preventing violent, dangerous people from getting a gun, just like drug laws aren't effective at stopping people from getting illicit drugs. All that stricter gun laws do is make GOOD people less able to have a gun for good purposes (such as I'm sure it's a rule in that fictional workplace that everyone in that office was disarmed, including the first 4 that could have stopped the shooting early), and make it relatively easier.

It is clear from this example that "Frank LeClair" has engaged with the thick representation of the depicted office shooting. However, it is also clear that he dismisses the narrative, and argumentatively provides a counter-narrative.

10. CONCLUSION

Symbolic condensation, working through the principle of delegated efficacy, visual and multimodal forms of communication, enables audiences to (re)construct arguments elicited by verbal and nonverbal cues. Symbolic condensation also affords audiences an emotional and aesthetic understanding provided by the thick description offered by visual and multimodal communication.

Generally speaking, the argumentative purpose and power of thick representations is to provide audiences with a deeper understanding of the issue at stake, and thus provide an additional means to comprehend and evaluate the argument. This is also the case with the PSA from States United. While the main argument expressed in words ("We should change our gun laws because our guns have changed) provides an abstract kind of reasoning, which only communicates part of the argument condensed in the ad, the audio-visual thick representation of the shooting provides the audience with more premises (reasons) and a deeper understanding of the debated issue.

This deeper understanding is an important part of the argument since it demonstrates the gravity and urgency of the issue (Kjeldsen, 2015), and thus provides support for the claim that gun laws should be changed.

REFERENCES

Barthes, R., & Heath, S. (1977). *Image, music, text.* New York: Hill and Wang.
Bitzer, L. F. (1959). Aristotle's enthymeme revisited. *Quarterly Journal of Speech, 45*(4), 399-408. doi: 10.1080/00335635909382374
Condit, C. M. (2013). The rhetorical limits of polysemy. In B. L. Ott & G. Dickinson (Eds.), *The Routledge Reader in Rhetorical Criticism.* New York & London: Routledge.
Freud, S., & Crick, J. (1999). *The interpretation of dreams.* Oxford: Oxford University Press.
Geertz, C. (1973). *The interpretation of cultures: selected essays.* New York: Basic Books.
Gilbert, M. (2001). Emotional messages. *Argumentation, 15*(3), 239-250. doi: 10.1023/A:1011156918137
Graber, D. A. (1976). *Verbal behavior and politics.* Urbana: University of Illinois Press.
Groarke, L. (1996). Logic, art and argument. *Informal Logic, 18*(2-3), 105-129.
Groarke, L. (1999). The deceitful artwork: Beautiful falsehood or false beauty? *Humanitas, 12*(2), 64-87.
Groarke, L. (2015). Going multimodal: What is a mode of arguing and why does it matter? *Argumentation, 29*(2), 133-155. doi: 10.1007/s10503-014-9336-0
Hariman, R., & Lucaites, J. L. (2007). *No caption needed: Iconic photographs, public culture, and liberal democracy.* Chicago: University of Chicago Press.
Kaufer, D. S., & Carley, K. M. (1993). Condensation symbols: Their variety and rhetorical function in political discourse. *Philosophy and Rhetoric, 26*(3), 201-226.

Kjeldsen, J. E. (2007). Visual argumentation in Scandinavian political advertising. A cognitive, contextual and reception oriented approach. *Argumentation and Advocacy, 43*(3-4), 124-132.

Kjeldsen, J. E. (2012a). Four rhetorical qualities of pictures. Paper presented at the 15th Biennial Rhetoric Society of America Conference, The Loews Philadelphia Hotel, Philadelphia, PA.

Kjeldsen, J. E. (2012b). Pictorial argumentation in advertising: Visual tropes and figures as a way of creating visual argumentation. In F. H. van Eemeren & B. Garssen (Eds.), *Topical Themes in Argumentation Theory. Twenty Exploratory Studies* (pp. 239-256). Dordrecht: Springer.

Kjeldsen, J. E. (2015). The rhetoric of thick representation: How pictures render the importance and strength of an argument salient. *Argumentation, 29*(2), 197-215. doi: 10.1007/s10503-014-9342-2

Kjeldsen, J. E. (forthcoming). Visual rhetorical argumentation. *Semiotica*.

Langer, S. (1980). *Philosophy in a new key. A study in the symbolism of reason, rite and art. Third edition.* Cambridge, MA: Harvard University Press.

Messaris, P. (1994). *Visual "literacy": Image, mind, and reality.* Boulder, CO: Westview Press.

Messaris, P. (1997). *Visual persuasion: The role of images in advertising.* Thousand Oaks, CA: Sage.

Paivio, A. (1979). *Imagery and verbal processes.* Hillsdale, NJ: Erlbaum.

Pinto, R. C. (2001). *Argument, inference and dialectic: Collected papers on informal logic,* with an Introduction by Hans V. Hansen. Dordrecht: Springer Netherlands.

Richards, I. A. (1936). *The philosophy of rhetoric.* New York.

Sapir, E. (1934). Symbolism. In E. R. A. Seligman (Ed.), *Encyclopedia of the social sciences.* New York: Macmillan.

Schiappa, E. (2008). *Beyond representational correctness: Rethinking criticism of popular media.* Albany: State University of New York Press.

Willard, C. A. (1989). *A theory of argumentation.* Tuscaloosa: University of Alabama Press.

Commentary on Kjeldsen's Symbolic Condensation of Visual and Multimodal argumentation

MICHAEL A. GILBERT
York University, Canada
<u>gilbert@yorku.ca</u>

Prof. Jens E. Kjeldsen is offering us a very rich picture of how visual arguments succeed in being arguments. His basal assumption is that argument is a processual endeavor rather than a linear and/or deductive one, a point with which I heartily agree. Indeed, I am largely in agreement with Kjeldsen's points, so this commentary will not be a critique. Rather, I will use a number of Kjeldsen's points and argue that verbal argumentation is not markedly less complex and less context dependent than visual argumentation, and that the differences are more a matter of degree than kind. Let me be clear: I am not saying that there are no differences, but rather that the differences are not as marked as might be thought.

I take two key points from Kjeldsen's paper. First, that rational and emotional condensation are central to understanding visual argument insofar as they allow "for a cueing and evoking of a wide range of emotions and longer trains of thoughts" (2). Secondly, that condensation symbols need to be strongly connected and embedded in their contexts (9). I believe both these points are correct but that they also apply to discursive argument as well as visual argument. The reason for this is that discursive arguments are not communicated by *propositions*, but by *messages*, and unlike propositions messages are packets of information that rely greatly on context and personal and cultural knowledge.

There is a <u>television beer commercial</u> for Bud Light, a popular beer. In the space of 60 seconds the term "dude" is used in roughly 20 different ways. What is required in each case in order to decipher the meaning is the context. In one case, for example, a co-worker in the adjacent cubicle is tapping a pencil on his desk, and a colleague looks over and says, "Dude." The message is: you are making an irritating noise that can be heard by others close by, and that's rude, so stop it. In another, a couple sits on a sofa cuddling while watching TV. Another fellow is at the end of the sofa, and the guy cuddling looks at him and hisses, "Dude." The meaning here is, we want to go beyond cuddling, you are cramping our style, leave. Let me rush to say that these examples are all expressed visually through the form of the commercial video. However, unlike, say *Marat*, they are representations of common situations we can recognize. Moreover, they can be expressed discursively as I just did above.

What is similar to the visual argumentation that Kjeldsen discusses is that the term has sufficiently rich symbolic condensation that the context is essential to interpreting the argument as is the case, for example, in *fat cat*. By saying that arguments come to us as messages I am saying that a) in order to understand we need context, and b) the context includes not only the discursive content but the means of communication as well. This latter involves everything from voice tone to eye contact to body posture, and so on.

This is quite different from the common understanding of a proposition. A proposition, as I understand it, is the meaning of a discursive unit, typically a sentence, *independent of its context*. Thus the two following sentences express the same proposition.

> (1) The cat is on the mat.
> (2) The damn cat is on my Aunt Peggy's hand-woven mat again!

I take it as obvious that (1) and (2) do not have the same meaning and do not express the same message *even if they are the same proposition.* (2) contains much more information than (1) and includes an emotional condensation packet that will be recognised in the context. Of course, (1) might do the same, but also it might not.

In discussing reactions to the anti-gun PSA ad Kjeldsen describes, he points out how many people understood the argument whether or not they agreed with it. In fact, disagreement with the argument was taken as evidence of comprehension. This is also true with discursive communications. Given certain parameters regarding familiarity a la Willard, (1989, p. 8 ff.) the various uses of "dude" will be recognized and the meaning of the message understood. The degree of familiarity required can vary greatly from strictly cultural to potentially intimate. Thus most English speakers would recognize the word "dear" as used from one partner to another, but might not know the significance of an intimate pet name. Similarly, a casual Spanish speaker might take the word "*amor*" to be highly romantic as it means "love," without appreciating it plays a similar role as the English "dear." Kjeldsen also explains how a visual image can require a great deal of discursive space in order to describe it. One picture, as the saying goes, is worth a thousand words; though in some cases 10,000 might be more accurate. But, as I have argued elsewhere, a great deal of what is communicated in a message is ineffable (Gilbert, 2002, 2003), and it is only the context that one can disambiguate it, though not necessarily explicate it, and almost never reproduce it. Good arguments, contra O'Keefe (1982), are not always linguistically explicable.

There are two points here that apply to visual arguments. The first is that they cannot always be fully explicated. For one thing, it depends on the argument itself. The *Marat* explication provided by Kjeldsen requires a level of sophistication that might beyond many observers. Linking the figure of Marat to Christ might not occur to every observer. Which brings us to the second point. All arguments can be misunderstood. The miracle is that we often do understand them. Kjeldsen has given us a rich an interesting paper with applications not only to visual argument, but to argument in general.

REFERENCES

Gilbert, M. A. (2002). Effing the ineffable: The logocentric fallacy in argumentation. *Argumentation, 16*(1), 21-32.

Gilbert, M. A. (2003). *But why call it an Argument?: In Defense of the Linguistically Inexplicable.* Paper presented at the Informal Logic at 25, Windsor, ON.

O'Keefe, D. J. (1982). The Concepts of Argument & Arguing. In J. R. Cox & C. A. Willard (Eds.), *Advances in argumentation theory and research* (pp. 205-237). Carbondale: Published for the American Forensic Association by Southern Illinois University Press.

Willard, C. A. (1989). *A theory of argumentation.* Tuscaloosa: University of Alabama Press.

14

Fairness, Definition and the Legislator's Intent: Arguments from *Epieikeia* in Aristotle's *Rhetoric*

MIKLÓS KÖNCZÖL
HAS Institute, Pázmány Péter Catholic University, Hungary
miklos.konczol@jak.ppke.hu

The paper first seeks to reconstruct, on the basis of Aristotle's explanation and example in the *Rhetoric* (1374a 26–b 1), how the shortcomings of a legal text, resulting from an omission made by the legislator, can be plausibly argued to provide sufficient ground for not applying the rule contained by the text. Second, it argues that the topics of fairness listed by Aristotle (1374b 2–22) cannot be used to reconstruct Aristotle's views on the functioning of *epieikeia* in judicial decision-making.

KEYWORDS: Aristotle, definition, *Epieikeia*, fairness, justice, legislator's intent, *Rhetoric*

1. INTRODUCTION

In Aristotle's *Rhetoric*, Book I, Chapter 10, a general classification of just and unjust deeds (1373b 1–6) is intended to serve as an outline of the possible topics of arguments useful in judicial speeches, where the goal (*telos*) of rhetoric is persuasion about lawfulness. Some of these deeds, Aristotle states (1374a 20–26), are just or unjust (lawful or unlawful) according to unwritten laws, and these can be divided in two groups: those resulting from a high level of virtue or vice, which are regulated by social norms other than written law (cf. Harris, 2013a, p. 30), and those related to some shortcoming (*elleimma*) of a particular written law. In the second case, however, it *should* be regulated by the written law of a specific political community, but the respective law somehow fails to provide the adequate rules. Fairness (*to epieikes*) is a kind of justice applicable to the latter kind of situation: it is justice beyond written law (*to para ton gegrammenon nomon dikaion*, 1374a 27–28).

The interpretation of Aristotelian fairness has always been a favourite topic of legal philosophers and legal historians alike. For both groups, it is important as the opposite of the strict application of the law. Philosophers therefore mostly study it as an historical example of legal decisions being based on moral considerations rather than positive law. For legal historians, the same problem appears as the question of whether Athenian (or ancient Greek) law did recognise grounds of judicial decisions outside of written law. While both approaches have proven fruitful in providing new insights for legal history and philosophy, the following discussion looks at fairness primarily from a rhetorical perspective, focusing on how arguments from fairness function in legal argumentation.

Starting from a brief reconstruction of the concept of fairness on the basis of Book V, Chapter 10 of the *Nicomachean Ethics*, I shall first compare Aristotelian and Platonic fairness. The analysis of the relevant passages of the *Rhetoric* follows in two parts: first the conceptual summary and the example (1374a 26–b 1) given by Aristotle are examined to map the structure of *epieikeia* arguments, then the links between these arguments and the list of *epieikeia*-related topics (1374b 2–22). I hope to show the close relationship between arguments from fairness and arguments from definition on the one hand, and the distance between the formal description of fairness and the subsequent list of topics on the other.

2. FAIRNESS IN THE *NICOMACHEAN ETHICS*

In the *Nicomachean Ethics*, we find the most detailed discussion of fairness in Book V, Chapter 10 (1137a 31–1138a 3), in an excursus between problems related to the notions of "being treated unjustly" and "acting unjustly". Aristotle approaches the topic through the ambiguity of the usage of the term *epieikes*:

> [S]ometimes we praise what is fair and the corresponding man, in such a way that we transfer the term to other features we are praising, too, in place of 'good' [...] while at other times it appears odd [...] that what is fair should be something praiseworthy when it is something that runs counter to what is just. (1137a 35–b 4)[1]

The solution of the problem comes from another ambiguity: that of the term "just". In one sense, "just" means "legally just", and it is in this sense that *epieikeia* "runs counter to what is just" and "is better than what is just in one sense". In a more general sense, however, what is just

---

[1] Quotations from the *Nicomachean Ethics* follow the text of Rowe's translation (Broadie & Rowe, 2002), with occasional modifications.

comprises what is *epieikes* (1137b 8-11). The reason for the ambiguity is that while laws aim at justice by their nature (see 1129b 14-24), they may still lead to unjust decisions in individual cases and may need rectification through *epieikeia*:

> The cause of this is that all law is universal, and yet there are some things about which it is not possible to make universal pronouncements. So in the sorts of cases in which it necessarily pronounces universally, but cannot do so and achieve correctness, law chooses what holds for the most part, in full knowledge of the error it is making. (1137b 13-16)

It is in those cases, i.e. where a general rule fails to take the particular circumstances of a given case into account, that fairness can play a role in the application of law:

> [O]n these occasions it is correct, where there is an omission by the lawgiver, and he has gone wrong by having made an unqualified pronouncement, to rectify the deficiency by reference to what the lawgiver himself would have said if he had been there and, if he had known about the case, would have laid down in law. (1137b 21-24)

Aristotle emphasises that such cases do not result from intellectual errors made by the legislator, nor do they indicate the technical deficiency of a piece of legislation. Rather, they are inevitable consequences of the tension between the universality of the law and the singularity of human actions (see 1137b 17-19).[2] A related, and equally important, point he makes is that 'rectification' does not mean denying the validity of the law, but has to be made with reference to the legislator's intention.

Thus, although Aristotle is aware that legislation may contain errors (see 1129b 24-25), *epieikeia* is not meant to correct that sort of deficiency by amending the law.[3] Its purpose is to bring about justice in the individual case, thus fulfilling the actual intention of the legislator.

---

[2] He also adds that there is another means of regulation, the decrees, which allow for a greater flexibility on the part of the legislator. As decrees are made for individual cases rather than generalised types of behaviour, they do not have to provide rules for an infinite number of cases. See 1137b 27-32.

[3] *Pace* Hurri (2013, p. 154). Cf. also Saunders (2001, p. 80; ibid. n. 29), mentioning the possibility of "a piecemeal modification", with reference to Brunschwig (1980, pp. 525-526). See further Mirhady (1990, p. 395) on the judges acting as legislators.

Consequently, it works through the interpretation of the law rather than against the law (see Brunschwig 1996, p. 140; Harris, 2013a, p. 28).

3. *EPIEIKEIA* IN PLATO

Aristotle's observation that legal regulation in itself cannot provide adequate grounds for decision in each particular case is strikingly similar to what the Stranger says in Plato's *Statesman*:

> [L]aw could never accurately embrace what is best and most just for all at the same time, and so prescribe what is best; for the dissimilarities between human beings and their actions, and the fact that practically nothing in human affairs ever remains stable, prevent any kind of expertise whatsoever from making any simple decision in any sphere that covers all cases and will last for all time. [...] But we see law bending itself more or less towards this very thing, like some self-willed and ignorant person, who allows no one to do anything contrary to what he orders, nor to ask any questions, not even if after all something new turns out for someone which is better, contrary to the prescription which he himself has laid down. (294a 10–c 4)[4]

In the *Statesman*, the conclusion is not formulated with regard to the judge but to the legislator, i.e. the ruler of the state. Laws, imperfect as they are, serve as general instructions for cases where the ruler cannot make a decision himself. In those cases, however, where he is present, he must be allowed to overrule these general instructions, for his personal expertise and ability to consider all the circumstances will probably lead to better decisions than the legal rule would in itself.

It is only in the *Laws* that Plato addresses the same problem with reference to the application of law, apparently accepting the fact that the legislator cannot be present everywhere to adjudicate legal disputes, and that therefore judges need to be authorised to exercise a certain level of discretion. What exactly this level should be depends on the skills of the judges concerned. The establishment of the facts is necessarily subject to judicial deliberation, for the facts of the individual case cannot be legislated upon in advance. Yet the Athenian speaker seems to be quite confident that in Magnesia, the city to be founded along the lines set by the dialogue, there will be a citizenry that can provide competent judges who can be left to decide some other

---

[4] Translated by Rowe (1995).

questions as well. Concerning penalties, for example, the Athenian says that

> the judge must assist the lawgiver in carrying out this same task, whenever the law entrusts to him the assessment of what the defendant is to suffer or pay, while the lawgiver, like a draughtsman, must give a sketch in outline of cases which illustrate the rules of the written code. (934b 6–c 2) [5]

The sketch that follows is, however, a fairly detailed one: Plato apparently seeks to eliminate from Magnesian legislation much of the ambiguity present in its Athenian counterpart (see Harris, 2013b, pp. 205–209).[6]

But can judges in Magnesia go beyond written law in order to reach a more just verdict? It seems that there is at least one type of affairs where they can. One of the Magnesian laws provides that in case a father dies having a daughter but no (natural or adopted) son, the male relative who comes next in the order defined by the law has to marry the daughter (924e 3–925a 2). The legislator, however, has to take into account the possibility that the prospective heir cannot marry the daughter because of her physical or mental illness (925d 5–e 5, 926b 2–6). Such cases have to be adjudicated by a panel of arbitrators (*diaitētai*) who are allowed to grant exemption from the legal obligation (926a 6–7, b 7–d 2).

Arguably, this is a case of *epieikeia*, albeit Plato does not use the term for it.[7] We see the tension between the law formulated in universal terms and the circumstances of the individual case (925d 8–e 2); the preamble to the law asks for understanding (*syngnōmē*) on behalf of both the legislator (for his inability to consider individual cases) and the persons asking for exemption (for their inability to obey the command of the law) (925e 6–926 a 3); and in the procedure reference has to be made to the legislator's intent (926c 2–4). This makes clear that fairness is not directed against the validity of the law and that it actually serves

---

[5] Translated by Saunders (1970).

[6] Cf. also the remark made by Saunders (2001, p. 87): "Not only will the gaps be fewer, but the actual laws will be far less open-ended conceptually; for the Magnesian citizen is conditioned not merely by an intensive educational process but by the frequent legal preambles: he will have fairly firm ideas about what (say) justice, virtue, heresy, good and bad artistic standards, really *are*".

[7] See Saunders (2001, pp. 84–86), with some qualifications based on certain differences between the typical form of *epieikeia* and Plato's description of the situation.

the good of the political community better than a strict enforcement of the general rule.

What makes this kind of fairness interesting (and characteristic of Plato's Magnesia) is that, as Trevor Saunders put it,

> the need for it is recognised, but only in rare and extreme cases; and its operation is taken clean out of private hands and transferred to senior officials who act on criteria subserving the public interest. [...] To put the point in a lapidary manner, Plato has nationalised a private virtue. (Saunders, 2001, p. 92)

This is essentially the same as what happens to rhetoric in Magnesia: it is taken over by the legislator and subordinated to the interest of the state.[8]

4. FAIRNESS AND DEFINITION IN THE *RHETORIC*: AN EXAMPLE

Coming back to Aristotle, we may now see how the very fact of including the discussion of *epieikeia* signals an important departure from Plato's doctrine.

Unlike in the *Nicomachean Ethics*, where he speaks about the legislator's awareness of the problems caused by the inevitable generality of legislation, in the *Rhetoric* Aristotle mentions two possibilities (cf. Kraut, 2002, p. 108, n. 17): shortcomings in written law may be either due to the ignorance of the legislator, or on the contrary, brought about by him through the deliberate use of general terms and the lack of distinctions. He then gives an example for the collision of a strict interpretation of written law and fairness, resulting from the inevitable lack of complete conceptual determination in normative texts:

> In many cases it is not easy to define the limitless possibilities, for example how long and what sort of iron has to be used to constitute 'wounding', for a lifetime would not suffice to

---

[8] It should be noted that this kind of *epieikeia* does not serve justice directly. The aim of the legislator is, rather, to avoid enforcing the law in cases where its addressees would prefer to suffer punishment, as the law cannot fulfil its function of guiding human actions in those cases, and even if they obeyed would not serve the public interest. It could be argued, however, that the exception serves justice indirectly, as it would be unjust to punish those who decline to obey the law only because obeying it would be worse than suffering whatever punishment (cf. 925e 2–5), which amounts to some kind of a necessity (cf. 926a 2–3).

enumerate the possibilities. If, then, the action is undefinable when a law must be framed, it is necessary to speak in general terms, so that if someone wearing a ring raises his hand and strikes, by the written law he is violating the law and does wrong, when in truth he has not done any harm and this (judgement) is fair. (1374a 31–b 1)[9]

In the introductory chapter of the *Rhetoric*, Aristotle has already pointed out the limits of the legislator's competence in terms of questions of fact (1354b 11–16). Here, however, he goes one step further, asserting that even if the legislator has got a definite intent (in the example it may be that people should refrain from assaulting others with weapons made of iron), its formulation as it appears in the written text is likely to be imperfect. Therefore, the argument can be made before the court that in addition to applying the rule previously given to the particular facts of the case, the judges also have to establish what provisions the text of the law actually contains. The result of their examination of the rule may contradict what is generally understood to be the 'ordinary meaning' of the text. Of course, the speaker need not highlight that this is what happens in the court: rather, he may propose a reading of the text as the one that genuinely reflects the intention of the legislator.

Arguments from the legislator's intent have a twofold character. On the one hand, they exemplify what are often termed 'teleological' arguments.[10] As Jacques Brunschwig puts it in his interpretation,

> there exists a perfectly applicable law, but […] a mechanical or blind application of it would be too severe according to the moral intuitions of the judge and those of the society in which he works. (Brunschwig, 1996, p. 139, following Shiner, 1987)

Consequently, the argument is based on the assertion that the legislator would not have intended the law to lead to such a verdict. On the other hand, the legislator's intent is still something referred to in "rule-based reasoning", where it appears as a means of interpretation, which is intended to help establish the meaning of a normative text, by explaining how the legislator actually meant what he put into words. What is important for us to see here is that this method of reasoning, i.e.

---

[9] Quotations from the *Rhetoric* follow the revised text of Kennedy's translation (Kennedy, 2007), with occasional modifications.

[10] The consequentialist nature of teleological interpretation is highlighted by Cserne (2011, p. 38).

advocating fairness by way of interpreting the text, makes it possible for the orator to avoid questioning the authority of written law.

We have seen that in the *Nicomachean Ethics* Aristotle emphasises the link between the "correction" of the law and the legislator's intention (which also appears among the topics of fairness listed later in the *Rhetoric*). The problem here is, apparently, that whatever one thinks about the legislator's writing skills, the most obvious way of knowing his intention is still to read the text of the law. Thus, arguments for a not-so-ordinary meaning of the text have to face a good deal of scepticism. This kind of scepticism is well illustrated by a quotation from L. L. Fuller's fictitious *Case of the Speluncean Explorers*, in which a judge says that

> [t]he process of judicial reform requires three steps. The first of these is to divine some single "purpose" which the statute serves. This is done although not one statute in a hundred has any such single purpose, and although the objectives of nearly every statute are differently interpreted by the different classes of its sponsors. The second step is to discover that a mythical being called "the legislator," in the pursuit of this imagined "purpose," overlooked something or left some gap or imperfection in his work. Then comes the final and most refreshing part of the task, which is, of course, to fill in the blank thus created. *Quod erat faciendum.* (Fuller, 1949, p. 364)[11]

What, then, remains of *epieikeia* for arguments that can be safely used in a speech without appearing to be seeking "to be wiser than the laws" – to use the words of Aristotle (1375b 23–24)? It may be a good idea to come back to the example Aristotle gives for using fairness in a particular case of judging an offence. "[I]f someone wearing a ring raises his hand or strikes, by the written law he is violating the law and does wrong." Here the discrepancy between the law and the truth is due to the fact that the law does not define "how long and what sort of iron has to be used to constitute 'wounding'." The law, as far as it can be reconstructed from Aristotle's words, forbids and punishes assault with iron. In case someone strikes another person with an iron ring on his hand, the conceptual requirements for applying the law obtain and the

---

[11] It should be noted, however, that in Athenian legal discourse the legislator is never regarded as a "mythical being," although the historical identity of legislators is not examined either. References to the legislator are rather used to attribute a single intention to the law of the polis, cf. Harris (2000, pp. 50–51).

action qualifies as "wounding". In such a case, applying the sanctions of wounding would lead to injustice, as it would mean treating different actions (e.g. deliberately using a sword and wearing a ring) in the same way. This is, in the words of the *Nicomachean Ethics*, an error that results from the lacking qualification (cf. 1137b 22).

In such a case, the defendant can suggest that further qualification has to be added by the judges, saying "what the legislator would have included." For example, further details concerning the characteristics of the object made of iron can be described, in order to make the difference between a ring and a weapon appear in the judgement. Or the intention of the person "raising his hand or striking" can be taken into consideration, in order to distinguish between the deliberate use of a weapon and wearing a ring on one's hand—"looking not to the action but to the deliberate purpose," as Aristotle puts it later (1374b 13–14). These qualifications would then concern the concept of 'wounding' as defined by the law. The defendant would argue that he "raised his hand" or "stroke" but did not "wound," denying not the fact itself but its legal qualification.

This way of reasoning would then be strikingly similar to what is described in Chapter 13 in the paragraphs immediately preceding the discussion of fairness. Arguments from fairness as well as those concerning the *epigramma* focus on the moment of decision, which is essentially about the correspondence between the description of what happened on the one hand, and the abstract case contained by the legal rule on the other.[12] In other words, the question in both cases is if a certain rule is relevant for a certain human action. Looking for the difference between the two kinds of argument, we find Aristotle referring to *epigramma* as "what the laws regulate" (1374a 19–20) and to *epieikeia* as related to unwritten law (20–26). Thus, in the case of the former the speaker concentrates on how the individual action can be best described with the legal terms given in the law. In the case of the latter, in turn, the focus is on how the legal provision should be (re)formulated to express the legislator's (presumable) intention. In light of that, Aristotle's advice about having definitions at hand (1374a 6–9) may equally refer to those arguing from fairness.

5. TOPICS OF FAIRNESS

Interpreters rightly note that Aristotle's discussions of *epieikeia* comprise two different perspectives: one that focuses on the corrective

---

[12] Cf. the distinction between *Sachverhalt* and (*gesetzlicher*) *Tatbestand* in German legal doctrine (*Rechtsdogmatik*), see e.g. Larenz (1969, pp. 230–233).

function of *epieikeia* and one looking at *epieikeia* as a virtue (see e.g. Rapp, 2002, p. 503). This distinction is very important because it is only by keeping these perspectives separate that one can account for the difference between the theoretical reconstruction of *epieikeia* at 1374a 26–1374b 1 and the list of related topics at 1374b 2–22. While it is not very difficult to see how *epieikeia* as a way of statutory interpretation can help the speaker persuade the judges on the one hand, and to contribute to a just decision on the other, the topics of fairness, or at least some of them, seem much more puzzling.

The sentence that introduces the list of topics (1374b 2–3) by establishing a link with the preceding discussion of *to epieikes* makes clear, at any rate, that the following list shows characteristic examples of fair and unfair actions and persons.

5.1 Understanding

The actual list of examples begins with having understanding (*syngnōmē*).[13] *Syngnōmē* appears in Book VI of the *Nicomachean Ethics* as a capacity related to deciding about what is *epieikes* (1143a 19–24). While it is sometimes interpreted as some kind of an extra-legal consideration based on empathy alone (see, however, Grimaldi, 1980, p. 302), in the *Nicomachean Ethics* Aristotle makes it clear that it is directed at truth, which is also emphasised at the end of the example in the *Rhetoric*, where *to alēthes* is opposed to the *gegrammenos nomos* (1374a 36–b 1). The framing[14] of the following distinctions between errors (*hamartēmata*) and wrongs (*adikēmata*), and errors and

---

[13] The opening phrase of the list, *eph' hois te gar dei syngnōmēn echein, epieikē tauta*, raises problems in terms of rendering as well. Kennedy (2007, p. 100) takes *tauta* to refer to *eph' hois*, and *hois* to be the indirect subject of *syngnōmēn echein*, which results in the translation "those actions that [another person] should pardon are fair." The reason why one should pardon anything that is *epieikes* is not quite clear, however. The subsequent phrases (about distinguishing between *hamartēmata*, *atychēmata*, and *adikēmata*, see below) suggest that it is *syngnōmēn echein* that is to be considered as *epieikes*, and its indirect subject *eph' hois* [...] *dei*: "it is fair to pardon what should be [pardoned]." The plural form *tauta*, instead of *touto*, may be explained either by the multiplicity of the situations where *syngnōmē* is needed, or by the instances of *epieikeia* that follow. Grimaldi (1980, p. 302) rightly sees *syngnōmēn echein* as an instance of the *epieikē*, but fails to give a satisfactory explanation for his interpretation.

[14] The section on *syngnōmē* seems to be finished by "to be forgiving of human weakness is fair" (1374b 10–11).

misfortune (*atychēmata*), respectively, suggests that making such distinctions belongs to the domain of *syngnōmē*.

Aristotle gives exact criteria for each of the three cases (1374b 6-10). Misfortune, he says, cannot be anticipated by reason (*paraloga*) and is not caused by an evil moral disposition (*mē apo mochthērias*). Errors, in turn, can be anticipated (*mē paraloga*) but do not stem from moral badness either (*mē apo ponērias*). It is only wrongdoing that can be anticipated by reason and result from an evil moral disposition, from which wrongs committed because of desire (*di' epithymian*) are no exception.

Apparently, then, it is only the wrongs that deserve the full rigour of the law, while errors and misfortune call for a more lenient treatment. While Aristotle gives no examples here, his criteria make it quite clear what cases belong to each of these categories. Wrongdoing, which has been defined at the beginning of Chapter 10 (1368b 6-7 and 9-10), is the case the legislator has in mind when drafting a law about a certain crime. Compared to that, an adequate adjudication of errors and misfortune may require some additions to the legal definition of the crime, just as described under the heading of fairness.

In the case of misfortune, the wrongful intention on the part of the person committing the crime is missing altogether. The paradigmatic case of that is the harm caused by a natural disaster, as e.g. in the case of a storm that prevents a ship from reaching a port.[15]

Errors, on the other hand, belong to the actions done "willingly" (*hekōn*), i.e. "knowingly and unforced" (cf. 1368b 9-11). These cases are usually regarded as the class of human actions covered by "negligence" in modern Western legal terminology (see Hamburger, 1971, p. 102; Harris, 2013a, p. 32).

5.2 Letter and intent

The next topics of *epieikeia* oppose the letter of the law and the intent of the legislator (1374b 11-13). As opposed to *syngnōmē*, where the focus was on the perpetrator's attitude, these topics focus on the desirable way of statutory interpretation (cf. Harris, 2013a, p. 32). Looking at the legislator's intent is, as we have seen, essential for building up an argument from fairness, at least if one wants to avoid making the impression of urging a decision *contra legem*.

---

[15] For examples of *trierarchs* being acquitted, most probably due to their excuses of *force majeure*, see Harris (2013a, pp. 44-45).

5.3 Intention

After the opposition of letter and intent, further topics concerning the perpetrator follow. The first of these regards deliberate choice (*prohairesis*) as opposed to the action itself (1374b 13-14), thus continuing the considerations related to *syngnōmē*. On the other hand, this topic seems to respond to that of definition, where Aristotle says that the question of whether an action qualifies as a certain crime should be decided on the basis of *prohairesis* (1374a 11-13). A further link is to Chapters 10-12, where the probabilities are related to intention rather than an action being actually committed.

5.4 Part and whole

The next two topics oppose the part and the whole, first in the abstract, then in terms of the perpetrator's behaviour. The former is, in itself, sufficiently general to be regarded as another formulation of the essence of *epieikeia*, i.e. the requirement of achieving a decision that is adequate to the individual case. The second one, however, may seem more problematic, as it seems to call for a decision based on past events rather than on the action under dispute.[16] While this possibility cannot be excluded, there are other possible explanations which come closer to what seems to be the basic principle of *epieikeia*. First, past events may be considered, if not for deciding about the lawfulness of an action, then for imposing a penalty.[17] Such a reading would also highlight a possible Platonic influence.[18] Second, the general behaviour of the defendant may be used as indirect proof for his moral character and, consequently, his *prohairesis* in the specific case (cf. Saunders, 1991, p. 113; Johnstone, 1999, pp. 95-97; Lanni, 2006, 60-61). Third, it is also possible that

---

[16] References to past deeds do occur in oratory. An example may be mentioning public service, which is rejected as irrelevant e.g. by Lysias 12.38. Cf. Harris (2013b, pp. 127-128), pointing out also that courts may not have paid attention to such arguments (with examples from Aeschines 3.195, Dinarchus 1.14, Demosthenes 21.143-147, 19.273 and 277, 24.133-134).

[17] See e.g. Dinarchus, *Against Philocles* 11. Cf. Saunders (1991, pp. 113-118) and Lanni (2006, p. 62). Harris (2013b, pp. 131-136) points out that in the *timēsis* the scope of relevant information was broader than in the first part of the trial, where the judges had to decide the question of guilt.

[18] Cf. e.g. *Laws* 862c 6-e 2, where the Athenian speaker explains that punishments should differ according to whether the perpetrator can be "healed" or not.

these topics are not only meant to be used in connection with the judges' decision but also for displaying fairness within a speech.[19]

5.5 Memories

There are two further topics that concentrate explicitly on past deeds (1374b 16-18). The first one opposes good things to bad things experienced by the same person, and the second one good things done by someone to good things done to the same person. Here again, it is hard to see how these could contribute to persuasion concerning the lawfulness of a specific action. Moreover, unlike in the previous topics, the opposition is not between one's general character and an individual action but between (perhaps several) particular actions, and the emphasis is not on the actions themselves but on the act of *mnēmoneuein*. Therefore, the second option of interpretation mentioned above in connection with the topics of "part and whole" is out of question. It seems more likely that it is not the judges who remember something but someone of the other participants of the legal procedure, and that *mnēmoneuein* is used here in the sense of mentioning something.

5.6 Attitudes to wrongdoing and litigation

In the case of the last three topics (1374b 18-22) there is no doubt that they do not regard the judges' attitudes but those of the litigants (or someone who is not directly involved in the case but is characterised in the speech). They say that fairness requires patience, and that it is fair to prefer settling a dispute through words to doing so through deeds. The former may be regarded as an echo of the *Nicomachean Ethics*, 1138a 1-2, although the three topics in the *Rhetoric* follow an order from the most general to the most specific, and being patient does not in itself contain any reference to litigation.

The opposition of words and deeds is widespread in Greek literature and the variety of contexts in which it appears does not allow for attributing one single meaning to it. What seems the most likely here is that, as mentioned above, the three topics start with a general attitude (patience) and finish with the choice between arbitration and judicial decision-making. Hence, one may reconstruct the three steps as three choices between (1) being patient and trying to retaliate; (2) trying to

---

[19] A striking parallel for this usage of *epieikeia* can be found in the treatise *On types of style* (*Peri ideōn*, 2.6) attributed to the 2nd-century (AD) rhetorician Hermogenes of Tarsus. For an English translation see Wooten (1987).

settle the dispute through arguments (which includes the possibility of a legal debate) and physical retaliation; (3) settling the dispute through arbitration and taking the issue to court.

The third topic is accompanied by a brief explanation concerning the nature of arbitration, according to which its *raison d'être* is that unlike judges, arbitrators base their decisions upon fairness rather than the laws. While this opposition may seem to suggest that courts are not allowed to take *to epieikes* into consideration, which would contradict both what Aristotle says in the *Rhetoric* and the *Nicomachean Ethics*,[20] and contemporary judicial practice (see Roebuck, 2001, p. 182; Harris, 2013a, p. 34; *pace* Meyer-Laurin, 1961, p. 41), it is in fact the arbitrators who are in the focus here and Aristotle seems to mean only that they do not have to provide an explanation that is supported (exclusively) by an interpretation of the written law.[21]

6. CONCLUSION

Having accepted an argument from fairness, the judges have to "supplement" the text of the law interpreted, thereby making it irrelevant for judging the action under dispute. The intention of the legislator is thus referred to in order to make it clear that it would be contrary to this intention to punish the defendant for having committed the crime he is charged with (cf. Harris, 2013a, p. 31). In this sense, we may agree with Jacques Brunschwig, who argues that the phrase used by Aristotle in the *Nicomachean Ethics* ("what the lawgiver would himself have said had he been present, and would have included within the law, had he known") refers to two different things. Supplementing the text by adding further qualification of the action in terms of facts or intention is done by reconstructing the abstract and general will of the legislator, while deciding that the rule thus obtained is not relevant for the facts of the case is "what the lawgiver would himself have said had he been present." Yet these are two consecutive steps of the same line of reasoning: the teleological or consequentialist part of the argument, which leads to the decision not to apply the law needs the backing of the

---

[20] See *Nicomachian Ethics* 1132a 4–32, where Aristotle describes the judge as *dichastēs*, i.e. who establishes the just mean (cf. Mirhady, 2006, p. 2; see also Harris, 2013a, p. 32, n. 20).

[21] On the general character of arbitration see Meyer-Laurin (1961, pp. 41–45) and the survey of Roebuck (2001). On the difference between the judge (*dikastēs*) and the arbitrator (*diaitētēs*), both Meyer-Laurin (1961, p. 37, n. 130) and Mirhady (2006, pp. 2–3) quote Aristotle's criticism of Hippodamus' ideas concerning the ideal constitution (*Politics* 1268b 4–13).

interpretive or rule-based part, in order to make the judges feel safe in deciding the case, apparently "according to the laws and decrees of the Athenian people." On the other hand, Aristotle's final clause "had he known" highlights the interdependence of the two steps. It is on the basis of the knowledge of the particular circumstances of the case and pondering the consequences of their judgement that the judges can decide where the text says less than what is necessary for a just decision. Taking into account the particular situation and offering a corresponding interpretation of the general rule of decision, the topic of definition can serve the aims of fairness, so that the speaker will be able, once again, "to make clear what is just."

Unlike the conceptual approach summarised in the first part of the paper, the subsequent list of topics seems to have a much broader scope, which does not in every case fit the interpretive method. The last three topics, in particular, do not say anything about how the judges should decide. Neither the importance of patience, nor the opposition of words and deeds can be used as an argument concerning the merits of the legal case. The remark attached to the last one, where arbitration and adjudication by the court are compared, may appear in an arbitration case as a means of reminding the arbitrators of their duty to make a fair decision, but the assertion that "it is fair to prefer arbitration to adjudication" cannot really contribute to such a decision. Therefore, their place in Aristotle's list is best explained if one does not read them as topics for arguments in the strict sense. Together with some of the other items of the list, they seem to serve as topics of characterisation focusing on the *ethos* aspect of the speech rather than the *logos* (cf. Harris, 2013a, p. 32). What connects them to the other, "legal" topics and the preceding discussion of what is *to epieikes* in law is that they likewise stem from Aristotle's definition of fairness and represent popular beliefs of morality. One should not, however, look in them for principles of legal interpretation, nor can they be used to reconstruct Aristotle's views on the functioning of *epieikeia* in judicial decision-making.

ACKNOWLEDGEMENTS: A previous version of the first part of the paper was presented at the 2nd Central and Eastern European Forum of Young Legal, Political and Social Theorists, Budapest 2010. As for its more recent readers, I am grateful to Edward Harris, George Boys-Stones, and Péter Cserne for their comments and advice. I am also indebted to Serena Tomasi for her stimulating response at the conference.

REFERENCES

Broadie, S., & Rowe, C. J. (2002). *Aristotle, Nicomachean Ethics*. Oxford: Oxford University Press.
Brunschwig, J. (1980). Du mouvement et de l'immobilité de la loi. *Revue internationale de philosophie, 34*(133–134), 525–526.
Brunschwig, J. (1996). Rule and exception: On the Aristotelian theory of equity. In M. Frede & G. Striker (Eds.), *Rationality in Greek thought* (pp. 115–155). Oxford: Oxford University Press.
Cserne, P. (2011). Consequence-based arguments in legal reasoning: A jurisprudential preface to law and economics. In K. Mathis (Ed.), *Efficiency, sustainability, and justice to future generations* (pp. 31–54). New York: Springer.
Fuller, L. L. (1949). The case of the Speluncean Explorers. *Harvard Law Review, 62*(4), 616–645.
Grimaldi, W. M. A. (1980). *Aristotle, Rhetoric I: A commentary*. New York: Fordham University Press.
Hamburger, M. (1971). *Morals and law: The growth of Aristotle's legal theory*. New ed. New York: Biblo & Tannen.
Harris, E. M. (2000). Open Texture in Athenian Law. *Dike, 3*, 27–79.
Harris, E. M. (2013a). How strictly did the Athenian courts apply the law? The role of *epieikeia*. *Bulletin of the Institute of Classical Studies, 56*(1), 25–46.
Harris, E. M. (2013b). *The rule of law in action in democratic Athens*. New York: Oxford University Press.
Hurri, S. (2013). Justice *kata nomos* and justice as *epieikeia* (legality and equity). In L. Huppes-Cluysenaer & N. M. M. S. Coelho (Eds.), *Aristotle and the philosophy of law: Theory, practice, and justice* (pp. 149–161). Dordrecht: Springer.
Johnstone, S. (1999). *Disputes and democracy: The consequences of litigation in ancient Athens*. Austin, TX: University of Texas Press.
Kennedy, G. A. (2007). *Aristotle, On rhetoric: A theory of civic discourse*. 2d. ed. New York & Oxford: Oxford University Press.
Kraut, R. (2002). *Aristotle: Political philosophy*. Oxford: Oxford University Press.
Lanni, A. (2006). *Law and justice in the courts of classical Athens*. New York: Cambridge University Press.
Larenz, K. (1969). *Methodenlehre der Rechtswissenschaft*. 2d ed. Berlin, Heidelberg, New York: Springer.
Meyer-Laurin, H. (1965). *Gesetz und Billigkeit im Attischen Prozess*. Weimar: Böhlau.
Mirhady, D. C. (1990). Aristotle on the rhetoric of law. *Greek, Roman, and Byzantine Studies, 31*(4), 393–409.
Mirhady, D. C. (2006). Aristotle and the Law Courts. *Polis, 23*(2), 1–17.
Rapp, C. (2002). *Aristoteles, Rhetorik*. Berlin: Akademie-Verlag.
Roebuck, D. (2001). *Ancient Greek arbitration*. Oxford: Holo Books.
Rowe, C. (1995). *Plato, Statesman*. Warminster: Aris & Phillips.
Saunders, T. J. (1970). *Plato, The laws*. Harmondsworth: Penguin Books.

Saunders, T. J. (1991). *Plato's penal code: Tradition, controversy, and reform in Greek penology.* Oxford: Oxford University Press.
Saunders, T. J. (2001). Epieikeia: Plato and the controversial virtue of the Greeks. In F. L. Lisi (Ed.), *Plato's Laws and its historical significance: Selected papers of the I International Congress on Ancient Thought, Salamanca 1998* (pp. 65-93). Sankt Augustin: Academia.
Shiner, R. A. (1987). Aristotle's theory of equity. In S. Panagiotou (Ed.), *Justice, law and method in Plato and Aristotle* (pp. 173-191). Edmonton, Alberta: Academic Printing and Publishing.
Wooten, C. W. (1987). *Hermogenes, On types of style.* Chapel Hill, NC: The University of North Carolina Press.

Fairness and Legal Reasoning
Commentary on Könczöl's Fairness, Definition and the Legislator's Intent

SERENA TOMASI
CERMEG, School of International Studies, University of Trento, Italy
serena.tomasi_1@unitn.it

1. INTRODUCTION

Könczöl's essay is a critical study on the concept of fairness from a rhetorical perspective in legal argumentation. In my view, this paper has two main features (which correspond to the author's declared aims): i) a philological interest since the author presents an accurate reconstruction of the concept of fairness on the basis of the classical sources. The analysis focuses on Aristotle, taking into account the most relevant passages in the *Rhetoric* and in the *Nicomachean Ethics*. Then, the Aristotelian version of *epieikeia* is compared to the one proposed by Plato in the *Statesman*. ii) A rhetorical approach: the author lists a possible topic of arguments based on fairness that is useful and used in judicial speeches.

In this comment, I argue that these points should imply further theoretical and methodological insights, which are relevant in the contemporary developments of legal theory and legal reasoning. My goal is to recast the outcomes of Könczöl's analysis in a broader dimension linked to the debate in legal theory, by evaluating the relation between positive law and fairness in light of argumentation theory and by drawing attention to the argumentative process of fairness in legal reasoning.

2. MULTIPLE WORDS FOR *EPIEIKEIA*

The classical concept of *epieikeia* has played a key role in legal systems for ages, since Roman lawyers were aware of the inseparability between *ius* and *aequitas*. There are different ways of language translation and multiple (seemingly) equivalent options, including equity, equality, justness, fairness and reasonableness. Regardless the specific linguistic choice, in principle, each word requires true fairness in opposition to the letter of the law. Equity is what allows the law to be applied in practice, in the different and concrete circumstances, which could have not been embodied in law. The common shaping concept appeals to introducing a creative or corrective element for realizing a fair development of law. The tension between the law and the case is, in fact, inescapable and, therefore, the rules depend on their further interpretive and creative concretizations.

This conception of fairness addresses the following questions: what does it mean creative? Is it reducible to the effort of an affordable ruler? Does it depend on the interactions of the parties in trial?

3. LEGAL REASONING, FAIRNESS AND ARGUMENTATIVE REASONABLENESS

This paper deals with the fundamental issues of legal reasoning and interpretation (Canale & Tuzet, 2008), within the framework of theory of argumentation.

According to rhetorical tradition, the principle of fairness implies a non-formalistic approach to legal reasoning, involving the usage of arguments from fairness in legal argumentation. The decision-making process cannot be reduced to a formalistic way of inference: in a deductivistic framework, a legal provision plays the role of a major premise, a statement of fact is the minor premise and the conclusion is an individualized norm inferred by law. In order for a deductivist model to work, it is necessary that both the relevant legal rules and the facts are undisputed, without problems of interpretation or proofs. The fact is that it is not possible to contain a legal decision in a legal deductive syllogism for several reasons.

First, plural sources. Legal pluralism has become a major theme in legal studies: the existence of legal plurality is not just a fact but has turned to be an institutional value. The presence of multiple legal systems (national and international) and therefore plural values in law has become an *institutional* feature of contemporary legal contexts (Puppo, 2013). The selection of suitable norms and their interpretation according to the performing values is not an automatic procedure but a controversial one, which happens in trial.

Secondly, the rule of law in the State-nation is not based anymore on one leader social class which could generate a dogmatic policy of law interpretation, thanks to shared social interests. Legal organisation of society is congruent with its social organisation: social diversity implies different extra-legal expectations. As to law, it is as plural as social life itself.

Third, the concept of legal entity has turned to be fluid because, accordingly to what in facts happens, there is an explosion of multiple legal acts. To describe the state, it has been used the economical term "inflation" describing a chaotic state of policymaking. This legislative inflation makes the rules conflicting or contradictory. Following the doctrine of principles by Dworkin, in 1992, Gustavo Zagrebelsky, an Italian jurist and a Constitutional Judge, identified a criterion of order in the Constitution Act, considering it not as a normative act but as a fabric of principles (Zagreblsky, 1992). This scholar restored in Italian Jurisprudence a new concept of law, open to values than to the strict positive law, influenced by the constitutional law. To his mind, law has to be mild, in declared opposition to the positivistic idea of strict law (*dura lex sed lex*).

The point is that mildness is not a feature of law itself, but it regards the application of law. Recalling Opocher's legal perspective, the core of legal experience is not positive law but the judicial decision-making process (Manzin, 2014). In a narrow sense, interpretation of law can be understood by reference to the argumentative methodologies employed by judges in judgment: the linguistic interaction with the parties, the topical selection of legal arguments, the role of the institutional contexts, the interpersonal dimension of the process, the uses of rhetoric (Tomasi, 2012, 2015).

In the field of argumentation, argumentation theorists draw insights from logic, rhetoric, communication studies, discourse analysis, stylistics. Taking the pragma-dialectical theory as a possible model for understanding and assessing legal reasoning, to be considered reasonable the exchange of thoughts and moves need to be in accordance with the rules for conducting a critical discussion (van Eemeren, 2011). By the way of these rules, the ideal model provides clear points of orientation for the parties involved: if the interaction proceeds in an adequate fashion, the parties will come to an acceptable settlement of the dispute.

4. FAIR AND REASONABLE TRIAL

Aristotle shaped the concept of fairness to the lesbian rule: it was a particular flexible mason's rule made of lead that could be bent to the curves of a moulding and used to measure irregular curves. The rule is a metaphor for flexibility in practical reasoning: in the Nicomachean Ethics Aristotle discussed the difficulty of legal congruence with reality:

> For when the thing is indefinite the rule also is indefinite, like the leaden rule used in making the Lesbian moulding; the rule adapts itself to the shape of the stone and is not rigid, and so

too the decree is adapted to the facts (Aristotle, Nicomachean Ethics, bk V, ch. 10).

Out of metaphors, fairness consists in a fair trail in which the interpretation of the legal provision is argued by the parties and, finally, by the judge. Before an impartial judge, parties must set their positions, demonstrate the soundness and the coherence, resist to objections, persuade each other's. Claiming trial, conflicting parties make a shared decision about their conflict by communicating about their different standpoints, trying to understand each other's reasons and arguing each other in an institutionalized framework. The third party, without expressing a personal option, would play a mediating role to help the parties to solve the conflict in a reasonable way, favoring the setting (Greco Morasso, 2011). During the communicative interaction, if the parties pursue the same goal towards a reasonable solution of the problem, by the end of the debate the parties should show mutual understanding and respect the final decision.

To my mind, the argumentative account squares perfectly well with the classical concept of fairness reconstructed by the Author, demanding for the application in the legal context of argumentative techniques for analysis and evaluation of arguments (Feteris, 1999).

REFERENCES

Aristotle (Ross, W.D. trans.) (1908). *The Nichomachean Ethics*. Oxford: Claredon Press.
Canale, D., & Tuzet, G. (2008), Interpretation and legal theory: a debate. In *Analisi e diritto* 2007 (pp. 123-207). Torino: Giappichelli.
Eemeren, F.H., van. (2011). *In alle redelijkheid. In reasonableness*. Amsterdam: Rozenberg.
Feteris, E.T. (1999). *Fundamentals of legal argumentation: A survey of theories on the justification of judicial decisions.* Dordrecht: Springer.
Greco Morasso, S. (2011). *Argumentation in dispute mediation: A reasonable way to handle conflict.* Amsterdam/Philadelphia: John Benjamins.
Manzin, M. (2014). *Argomentazione giuridica e retorica forense.* Torino: Giappichelli.
Puppo, F. (2013). *Metodo, pluralismo, diritto. La scienza giuridica tra tendenze conservatrici e innovatrici*. Torino: Aracne.
Tomasi, S. (2015). La struttura argomentativa dell'eccezione. In S. Bonini, L. Busatta & I. Marchi (Eds.), *L'eccezione nel diritto*. Trento: Università degli Studi di Trento.

Tomasi, S. (2012). *Teorie dell'argomentazione e processo penale. Un'analisi comparata delle principali teorie argomentative contemporanee con profili applicativi al processo penale*. Tesi di Dottorato. Trento: Università degli Studi di Trento.

Zagrebelsky, G. (1992). *Il diritto mite*. Torino: Einaudi.

15

Fair and Unfair Strategies in Public Controversies: The Case of Induced Earthquakes

JAN ALBERT VAN LAAR
University of Groningen, The Netherlands
j.a.van.laar@rug.nl

ERIK C. W. KRABBE
University of Groningen, The Netherlands
e.c.w.krabbe@rug.nl

In public controversies, should you always remain reasonable no matter how bad the other side behaves? Or should you retaliate in kind? We discuss some strategies used in the recent controversy about induced earthquakes in the Netherlands. To which extent are these strategies fair, i.e. balanced, transparent, and tolerant? We investigate the constructive and destructive effects of choosing particular strategies for achieving a resolution or compromise and conclude with a number of recommendations.

KEYWORDS: argumentation theory, strategy, fallacy, public controversy, (un)fairness

1. INTRODUCTION

Some public controversies are so complex and unruly that participants, instead of trying to reach a resolution or compromise, are continually putting pressure on one another by threats and moral blackmail. It then seems unfeasible to spot, analyze, and expose every individual fallacy: a reason to go for a more global approach starting from discussion strategies. Another reason for preferring a more global approach is given by interwovenness (typical for public controversies) of on the one hand the search for reasonable resolutions for differences of opinion and on the other hand the search for fair compromises settling conflicts of interest. In this situation it seems less appropriate to concentrate on a particular model of dialogue, for instance a model in which participants focus on reaching a reasonable resolution by means of arguments and criticisms, or contrariwise a model in which they focus on reaching a fair compromise by means of offer and counteroffer. The choice for a more general dialogical perspective that includes both argumentation and negotiation seems more appropriate. In this paper, we therefore do not target separate dialogue moves or dialogues, but rather focus on global strategic patterns (strategies) that manifest themselves in various dialogues and moves. We characterize and evaluate these strategies and will try and check whether they are conducive to the forms of cooperation needed for fruitful and reasonable ways of arguing and negotiating.

We see a public controversy as a complex phenomenon consisting of many dialogues about a number of connected issues, most clearly present in the mass media, and stretching over a considerable period of time. The dialogues themselves are complex and belong to various types (Walton & Krabbe, 1995). Moreover, a public controversy is a polylogue, i.e. more than two parties are involved, which can only with difficulty, and perhaps never in a fully adequate way, be reduced to a complex of two-person dialogues (Lewiński & Aakhus, 2014). For the participants there is often a lot at stake, for instance their income, their way of life, or their identity. The recent controversy about the extraction of natural gas and induced earthquakes in the province of Groningen (Netherlands) exemplifies our concept of public controversy. It will be the subject of our case study in this paper.

We presume that participants in a public controversy are motivated by the desire to safeguard and to strengthen their own position, but that does not imply that reason has no role to play. When a participant offers arguments for a standpoint or proposes components for a compromise, he professes to offer considerations that will bring the parties closer to a resolution or a compromise. For his arguments will convince other participants of the correctness of his standpoint only if these participants can see at least some of the adduced reasons as reasonable and subscribe to them. And in the same way, fairness will have a role to play in every serious negotiation. In this paper, we therefore consider global strategies from this perspective of reasonable discussion and negotiation, being all the same aware of the fact that participants may have also completely different objectives (cf. van Eemeren & Houtlosser, 2002).

By a definite type of strategy employed by a participant of a public controversy, we mean a recognizable purposeful pattern of the dialogical moves of this participant in the various dialogues. Here, we not only include moves that are actually put forward by the participant, but also ways he would react to possible moves of his opponents. In our terms, one may change one's strategy, namely by going from a series of moves according to one pattern to a series according to another pattern. Thus it is possible to adapt one's strategy in the light of information about the opponent's strategy (cf. Dascal, 2008).

Strategies differ in the extent to which they are oriented towards resolutions and compromises or, as we shall say, the extent to which they are cooperative. Opposed to strategies that are to a large extent cooperative stand those that are to a large extent obstructive and hinder the achievement of a resolution or compromise.

How does one recognize a (to a large extent) cooperative strategy? Looking for characteristics one may consider the norms that are stipulated in extant models of dialogue for argumentation or negotiation. But since one has to deal here with strategies that refer to a complex of many dialogues of different kinds, we expect it to be more convenient to evaluate strategies globally as fair or unfair, that is: as more or less balanced, transparent, and tolerant. Our justification for these general criteria of evaluation lies in the ideal of a common argumentative search for resolutions or compromises. Here we assume that fairness is a necessary, but likely not a sufficient, condition for a strategy to count as to a large extent cooperative.

Next we distinguish kinds of (intentional or unintentional) effects of strategies. Some strategies are constructively effective: they are found to increase the degree of cooperation within the public controversy; others are destructively effective and decrease the degree of cooperation. It will surprise no one when a cooperative strategy increases or an obstructive one decreases the degree of cooperation. Sometimes, however, one meets with the inverse effect: a cooperative strategy that decreases the degree of cooperation or an obstructive strategy that increases it.

Our question is: What is the best course of action for those participating in a public controversy in order to achieve a satisfactory outcome? Should one take the line of fairness, cooperation, and reason? And should one stick to that line even when other parties invariably adopt an unfair attitude in the debate? Or should one pay the other back in the same coin? To gain an understanding of these matters, we studied a number of contributions to a public controversy on gas extraction and induced earthquakes in the province of Groningen and, on that basis, established a list of strategies. A complete analysis of this discussion has not been our purpose. Neither did we include other public controversies in this study. Yet we do expect that the strategies we found will surface also in other discussions.

In Section 2 we shall characterize a number of types of strategy used in the discussions about induced earthquakes. In Section 3 we explain how we evaluate global strategies in terms of fairness. Before we, in Section 5, formulate some strategic guidelines and, in Section 6, draw some conclusions about the value of fairness and cooperation, we shall, in Section 4 discuss in what ways strategies can have intended or unintended effects on the degree of cooperation within the public controversy as a whole.

2. STRATEGIES IN THE CONTROVERSY ABOUT NATURAL GAS AND INDUCED EARTHQUAKES IN GRONINGEN

Earthquakes in the provinces of Groningen and Drenthe occur since 1986 (Wikipedia, 2014). They are caused by the extraction of natural gas by the NAM.[1] The Groningen situation was, in 2013, investigated by the "Commission Sustainable Future Northeastern Groningen" (Meijer Commission). They wrote:[2]

> "For many Dutchmen, earthquakes are something from far afield. In northeastern Groningen, earthquakes are a part of daily life. More than hundred times a year a quake occurs. Mostly light: with a force of less than 2 on the Richter scale, sometimes with a force between 2 and 3, and in only a few cases higher. But ever since the earthquake hitting the village of Huizinge (municipality of Loppersum), in the summer of 2012, with a force of 3,6 on the Richter scale, inhabitants must take into account the possibility that more and heavier earthquakes will occur. A force of 4,5 up to 5 is not excluded. Consequently, northeastern Groningen faces a new reality."
> (Meijer Commission, 2013, p. 8)

Also since "Huizinge," the public controversy about the extraction of natural gas and its social and economic consequences gained momentum. The controversy consists of a complex polylogue in which many parties with divergent interests are involved. Yet, the contributions we studied made it apparent that in classifying the strategies used in this controversy one may depart from two groups of participants: (A) those that have an interest in the continuation of the extraction of natural gas; (B) those that have an interest in generous compensations and a safe environment. Most strategies can be characterized as either an A-strategy or a B-strategy.

For each (type of) strategy, there is a corresponding *blaming strategy* based upon blaming the other for misusing the first strategy. In

---

[1] *Nederlandse Aardolie Maatschappij* [Dutch Oil Company].

[2] Quotations in this paper from Dutch sources were translated by the authors.

our list of strategies that are being used in the public controversy about gas extraction and earthquakes, these blaming strategies are not separately mentioned. For each item in the list, the examples we present may illustrate the strategy itself or the corresponding blaming strategy; in both cases the aim is to clarify the first strategy.

Let us first consider two types of strategy that are used in the A-camp as well as in the B-camp.

(1) *Misleading*. By telling falsehoods or by keeping relevant information under one's hat, one leads another party up the garden path. The example illustrates the corresponding blaming strategy.

> **Example 1.** *Sticking plasters*
> "[...] the outrage reached a temporary climax when a research report of the Research Council for Safety revealed this month that, over the many years that natural gas had been extracted, the safety the people in Groningen people had been systematically ignored.
> Lambert de Bont, member of the committee of the Groningen Ground Movement,[3] concludes that this Ministerial order about the gas [stipulating the amount extracted to be somewhat reduced and houses in Groningen to be reinforced] is 'misleading' [...]. They are still pulling our leg. The budget had to be fixed; our safety is a subordinate matter. And in the meantime the Minister is merely applying sticking plasters by issuing reinforcement measures." (Luyendijk, 2015)

(2) *Spinning*. Media are approached or avoided, as one sees fit, and when making public statements, the matters at issue are cleverly formulated and framed to shed a favorable light on one's position and avoid inconvenient criticism.

> **Example 2**. *Vestibule*
> After a working visit to Groningen, MP René Leegte (Liberal) had a telephone conversation with his assistant. This conversation took place in a train on a crowded vestibule. Campaigner Rolf Schuttenhelm, who was among the crowd, put details of the conversation on Twitter. According to him, it was "to be appreciated that René Leegte returning from Groningen traveled by train" but "not very smart how he made a telephone call in public and explained how you may fob off Groningen folks." Thus every one could hear Leegte saying that

---

[3] An action group (*Groninger Bodem Beweging*).

during the working visit that day he had tried and avoided the media:

"Today I avoided the media as much as possible."

According to Schuttenhelm, Leegte also divulged that inconvenient questions from the media about the continuation of the extraction of natural gas might be adroitly parried by pointing out a *local* reduction, ordered by Minister Kamp[4], of the quantity of gas to be extracted. Schuttenhelm:

"And then, publicly, how you may always say 'that Kamp reduced the extraction at Loppersum by 80%'" (RTV Noord, 2015)

Let us now consider a number of A-strategies. Most of them can be wholly or partially characterized as *stonewalling* strategies (cf. Gabbay & Woods, 2001a, 2001b). A stonewalling strategy focuses on yielding as little as possible to other parties, whether they are right or wrong. The following stonewalling strategies occur in the controversy about gas extractions and earthquakes:

(3) *Trivialization.* Depicting problems or drawbacks as slight or insignificant. For instance, the quakes are described as "so light as not to be felt by anyone."

> **Example 3.** *Impalpable*
> "[...] In the northern Netherlands, light earthquakes have been observed since 1986. Experts link them to the extraction of natural gas. [...] Most earthquakes are so light as not to be felt by anyone. [...]
> According to the KNMI,[5] the light earthquakes caused by extraction of natural gas can have a force of maximally 3,9 on the Richter scale. The risk of damage with these light quakes is generally slight. If there is damage done, it will usually be minor (no structural damage to buildings)." (NAM, 2012)

(4) *Fobbing off.* When using this strategy the other is fobbed off with fine talk and left none the wiser; no notice is taken of the real problems.

---

[4] Henk Kamp, Minister for Economic Affairs (Liberal).

[5] *Koninklijk Nederlands Meteorologisch Instituut* [Royal Dutch Meteorological Institute].

(5) *Pushmi-pullyu.* Following this strategy one will once in a while give in a bit to the other's wishes just to take back twice as much later. A stonewalling strategy that is typically applicable when a damage claim needs to be settled.

(6) *Belittlement.* A strategy consisting of not taking the other seriously at all. Belittlement is also part of many other strategies.

(7) *Shelving.* In order to avoid an unfavorable outcome of the public controversy, one tries and shelves matters, postponing any possible outcome.

Not all A-strategies are stonewalling strategies. The next two strategies may not display any stonewalling:

(8) *Rationalization.* A rationalizing strategy uses rational arguments and analyses.[6] For instance in an argument weighing pros and cons:

> **Example 4.** *Hospitals*
> "[...] managing director van de Leemput says that the NAM is indeed concerned about the environment. 'But one has to weigh up pros and cons. We must take into account the impact of natural gas extraction on society. On the other hand, there is also the economic value. For instance, the money that we earn through the gas also serves to build hospitals.'" (Van Sluis, 2012)

However, a rational approach often (rightly or wrongly) leads to a reproach of only presenting a frigid analysis that does not go to the heart of the matter.

> **Example 5.** *Frigid analysis*
> "The eleven studies ordered by Minister Kamp to investigate the risk of earthquakes in Groningen are meant to allow a cost-benefit analysis to be made. To the thousands of inhabitants of the Hoogeland [north Groningen], who filed claims for damages with the NAM and are very worried, such an analysis may come across as frigid and cynical. It is not easy to express safety in euros and make calculations that must comprise the price of a human life. Yet, the government often does so, for instance to determine the appropriate height of a sea dyke and the costs thereof." (Blanken, 2013a)

---

[6] The term "rationalization" is not being used here in a pejorative sense.

(9) *Conciliation*. This strategy is used to convince the others, or to make them believe, that one is prepared to cooperate to find a resolution or a compromise. Thus, conciliation is a strategy by which you encourage the employment of fair strategies (and is thereby itself a higher order strategy, cf. van Eemeren, 2010, p. 35). The corresponding blaming strategy amounts to a charge of hypocrisy or sycophancy.

> **Example 6.** *Good neighbors*
> "The criticism keeps the managing director of the NAM, van de Leemput, occupied, so he avows. On the other hand, he is convinced that 'his' NAM is generous for its neighbors. But, obviously, it is not experienced that way. 'It is really a matter of concern. We intend to make changes. We are going to arrange that inhabitants can themselves call in a second assessor. At our costs. We just want to be good neighbors.'"
> (Van Sluis, 2012)

Let us next consider some B-strategies. These are commonly somewhat more emotional.

(10) *Ad baculum*. The strategy *Ad baculum* frequently makes use of fallacies of the same name, or in any case of threats. Gabbay and Woods (2001a, 2001b) speak in this connection of *threat-games* or *threat-strategies*.

(11) *Daddy-gets-angry*. When mummy threatens that daddy is getting angry, we are dealing with a special kind of *Ad baculum* move or strategy.

> **Example 7.** *They won't swallow it*
> "The NAM would be well advised to dispel that anger, so the members of the Groningen Ground Movement[7] indicate. They don't want themselves to incite violence. Not at all, even. 'But if the NAM continues to ignore the dissatisfaction and the problems and to treat the Groningers this way, then they won't swallow that much longer. Groningers are sober-minded, but there will be a point they have had enough. Installations for extraction of natural gas are indefensible. Next, they will be destroyed or set on fire. You don't want that.'" (Van Sluis, 2012)

---

[7] See Note 3.

(12) *Ad misericordiam*. The strategy *Ad misericordiam* frequently makes use of fallacies of the same name, or in any case of appeals to a kind of sympathy.

> **Example 8.** *Crouched in a corner*
> "The Groningen Ground Movement published short films featuring inhabitants, among them one with a girl crouched in a corner in fear for an earthquake. Wigboldus: 'You get scared. After each quake, people walk around their houses and sit in the crawl space. Did that crack expand? Are there new ones? It's to drive one crazy. It may become an obsession.'" (De Veer, 2013)

(13) *Quid pro quo*. A strategy that allows one to make concessions, but not without getting something in return.

> **Example 9.** *Realistic*
> "The Groningen Ground Movement wants the NAM to improve their procedure for settling damage claims. To make it more transparent. And equal for every one. But the Movement also knows that there is not much chance that Minister Kamp will forgo billions of revenue from the natural gas exploitation. 'As to that, you got to be realistic. But I think the government ought to offer some compensation. This region is confronted by many problems, social, economical, and psychological. Take some action. This should still be a region that attracts firms, that has a future.'" (Interview with Janssen, chairman of the Groningen Ground Movement, Blanken, 2013c)

(14) *Asking too much*. A strategy of stiff demands. Often motivated by the idea that even if those demands are unfeasible, room will be created for obtaining concessions.

> **Example 10.** *One billion*
> When Max van den Berg,[8] at the start of 2013, asked for 1 billion euro as a compensation for direct or indirect damage inflicted by earthquakes, this was, at the time, still regarded as an outrageous amount.
> "[...] the Meijer Committee succeeded in calculating how much, in the next 20 years, the government and the province ought to invest: 1 billion euro.
> Governor Max van den Berg must have smiled as he read that. Disregarding a handful of change, this is exactly the amount he

---

[8] Royal Commissioner for the province of Groningen.

asked for early this year – but then it was called 'compensation' and still had the sound of begging." (Blanken, 2013b)

Most of the listed strategies have a highly obstructive character: They are focused on what is rightly or wrongly supposed to be one's self-interest and not primarily on the attainment of a reasonable resolution or a fair compromise. Three types of strategy offer possible exceptions: Conciliation, Rationalization, and *Quid pro quo*. Conciliation aims at the realization of the necessary initial conditions for finding resolutions or compromises. Rationalization, as we use the word, aims at a resolution of the conflict by means of argumentation, whereas *Quid pro quo* aims at its settlement by means of a compromise.[9] These strategies, then, are apparently basically cooperative. But, anyway, the distinction between obstructive and cooperative strategies is one of degree.

3. FAIR AND UNFAIR STRATEGIES

By going through some simple examples we explore what features one would expect a fair strategy to display, namely that it would be *balanced, transparent,* and *tolerant.* For the purposes of the present paper we shall simply speak of a "fair strategy" whenever a strategy comes up to these expectations.

3.1 Three aspects of fairness

(1) *Fair shares*
When two children play together and one of them has been given a bag of sweets for the two of them, the other may point out that it would be only fair to divide up the sweets. In such a case, "fair" stands for the idea of fair shares. To cooperate well when playing together one should fairly share things; therefore, if you are unfair in this respect, the interaction will become more obstructive.

In a dialogue, too, it may happen that one party is pulling the strings. It would then count as fair if this party does not take all, but leaves something for the other to take. We shall call a strategy in dialogue *balanced* if it does take into account the considerations of other parties. In order to be fair, a strategy must be balanced.

---

[9] Our use of the terms "resolution" and "settlement" follows that of van Eemeren and Grootendorst (2004, p. 58). Cooperation can be oriented towards resolutions as well as towards settlements.

(2) *No cheating*
When A and B are playing together and A knows that his parents left some sweets in a tin for them to take, it would count as unfair if A "forgot" to mention that there are sweets in the tin, in order to take these later for himself to eat at ease. It would also count as unfair if he sneakily first ate half of the sweets and only then announced that there were sweets in the tin and proceeded to share them on a fifty-fifty basis. "Fair" here stands for refraining from lying or misleading, or – positively formulated – for acting transparently. Cooperation requires transparency of action; therefore, if you act opaquely, the situation will become more tinged by obstructivity.

In a dialogue, too, a party always has certain aims of discussion, based on particular information and its private preferences. It will count as fair if a party practices transparency about such matters. It is fair, for instance, to share relevant data. It is also fair to be frank about one's aims in a discussion and one's preferences. We shall call a strategy in dialogue *transparent* if it is sufficiently marked by frankness about itself, that is, about what the party using the strategy is aiming at and about the means the party wants to apply to achieve these aims. In order to be fair, a strategy must be transparent.

(3) *No coercion*
When A got the tin with sweets from the larder, it would count as unfair if B pressured him rudely to give the sweets away, for instance, by snatching them forcefully or by threatening with violence or other sanctions. Sound cooperation presupposes that parties refrain from coercion and rude means of putting pressure on one another: therefore, if you act intolerantly, the interaction will become more obstructive.

In a dialogue, too, a party may have means of power at its disposal and use them strategically. Physical coercion and threats may create good reasons for the opponent to select or forgo a certain option but are unfair from the perspective of resolving or settling the original conflict in a reasonable way. We shall call a strategy in a dialogue *tolerant* if it refrains from using means of coercion and thus leaves enough freedom for an opponent to determine and elaborate her position. In order to be fair, a strategy must be tolerant.

3.2 Fairness and unfairness in strategies

We now discuss, for a selection of the strategies described in Section 2, their most striking features in terms of balance, transparency, and tolerance.

(1) *Misleading*
When you mislead another party by providing incorrect or incomplete information, the effect may be that relevant information fails to be taken into account. Therefore Misleading, besides obviously being an opaque strategy, also counts as an unbalanced strategy. In Example 1 (*Sticking plasters*), Lambert de Bont's criticism seems to amount to the reproach that, in order to be able to continue the extraction of natural gas, incomplete or even incorrect information has been provided about the consequences for the safety of the inhabitants: The measures taken for the reinforcement of houses contribute to a semblance of safety but no more. The fair counterpart of Misleading is a strategy in which one leaves out falsehoods and half-truths and informs the other parties in plain terms.

(2) *Spinning*
It is quite possible to apply a strategy of spinning in a transparent, tolerant, and even balanced way, both when approaching or avoiding journalists and when giving a clever presentation of one's own position or of that of one's opponent. However, as may be seen from Example 2 (*Vestibule*), Spinning can also easily degenerate into an opaque and unbalanced, and therefore unfair, rendering of the situation as one keeps harping on data that are in favor of one's own position in order to screen countervailing considerations. In such situations Spinning "derails" into an unfair strategy: manipulation of the media and thereby of the public controversy.

(3) *Trivialization*
When you trivialize the complaints and worries of your opponent, you are attaching little importance to the considerations that are in favor of her position. Trivialization is an outstanding example of an unbalanced strategy that may be used by authorities to smooth over the detrimental effects of their policies. There is reason to suppose that the NAM trivializes the worries and problems of the inhabitants of the earthquake zone by providing only such information as makes it appear that the earthquakes are not much of a problem (Example 3 *Impalpable*). Thus the NAM fails to do justice to the gravity of the situation. Even though it would, in principle, be possible to apply the strategy in a transparent way, Trivialization, being unbalanced, will all the same be an unfair strategy. Trivialization is an unfair counterpart of a strategy in which a party reduces the interests and objections of another party to their proper proportions.

(8) *Rationalization*
When you rationalize, you present pertinent arguments to support your position, possibly even accompanied by a critical examination of counterarguments. Thus understood, Rationalization constitutes a strategy that is pre-eminently oriented towards balance, transparency, and tolerance. Example 4 (*Hospitals*), however, gives us an instance of a strategy that looks like Rationalization but fails to be truly balanced. The point about the hospitals seems just dragged into the argument. If so, it would be an example of rationalizing in a pejorative sense (substituting sham reasons for true motives). According to Example 5 (*Frigid analysis*) inhabitants may get the impression that a cost-benefit analysis fails to take into account non-economical considerations. So, in this case as well, the strategy (of the government) resembles Rationalization, but would not display balance and transparency, and hence be unfair. As we use the term, however, Rationalization is the fair counterpart of a strategy of rationalization in the pejorative sense.

(9) *Conciliation*
When you apply a strategy of Conciliation, you try and inspire your opponent with confidence. The strategy aims at setting up conditions that make your opponent prepared to trust your assertions and commitments, without an overdose of skepticism (Govier, 1997). Example 6 (*Good neighbors*) shows how the NAM tries to restore trust by proclaiming their good intentions. Conciliation is the fair counterpart of a sycophantic strategy by which one pretends to engage in promoting trust but is actually after a strategic plus-point.

(11) *Daddy-gets-angry*
When you apply the strategy Daddy-gets-angry, you are applying an *Ad baculum*-strategy while disguising the threat by pretending that it is not you who is to apply the sanction. In Example 7 (*They won't swallow it*) the speaker is surprisingly clever at exploiting the anger of the inhabitants so that we may suspect him to apply this strategy. Daddy-gets-angry is a special kind of *Ad baculum*. It is an intolerant, unfair counterpart of a strategy by which you warn for unfavorable consequences of the choice for a particular course of action in cases where those consequences are independent of your choices.

(12) *Ad misericordiam*
When you apply the strategy *Ad misericordiam* and expect the opponent to be fooled, and to succumb, without further reflection, to the emotions you evoke, then your strategy will be very opaque. When you expect your opponent to recognize the appeal to pity as your strategy, then you

will be trying to apply moral blackmail; your strategy will be intolerant. Example 8 (*Crouched in a corner*) could be an example of an opaque use of *Ad misericordiam*. This strategy is also an unfair counterpart of a strategy by which you warn for the unfavorable consequences of the choice for a particular course of action.

(13) *Quid pro quo*
When you apply the strategy *Quid pro quo,* you are indeed putting some pressure on your opponent to move and grant a concession, but in the context of negotiation this would be appropriate as a part of the game; it is part of what it means to cooperate. Consequently, this strategy does not count as intolerant. Neither is imbalance or opaqueness inherent to *Quid pro quo,* as may be clear from Example 9 (*Realistic*) where Janssen attempts to explicitly formulate different interests and overtly works round toward a compromise. We therefore consider this strategy to be fair, with as a possible unfair counterpart the strategy of Asking too much to which we now turn.

(14) *Asking too much*
When you are asking too much, you ask for more than you think you are entitled to, pretending to think that you are. In Example 10 (*One billion*) the Governor was apparently asking too much. His later smile was caused by the unexpected outcome of a fair calculation, which yielded roughly the same amount. At the time, the Governor's strategy was an opaque one of bluffing (though in the meantime the amount was raised to two billion). Asking too much is an unfair counterpart of *Quid pro quo*. In practice, it is hard to determine whether someone is applying a fair *Quid pro quo* strategy or an unfair Asking too much strategy.

Some of the strategies discussed are fair, – at least potentially; most, however, are unfair. All unfair strategies here discussed show more or less directly a kind of imbalance, while some of them clearly suffer also from opaqueness, and others from intolerance. Generally, the use of unfair strategies will let the parties drift apart and dissipate cooperation. For, an unfair strategy undermines trust, and trust is a necessary condition for cooperation. Although unfair strategies are basically obstructive, it is not excluded that in circumstances the application of an unfair strategy would benefit cooperation. Maybe parties sometimes behave unfairly "for the benefit of all." In the next section, we look more closely into this matter.

4. EFFECTS OF STRATEGIES

What kind of strategy can a participant in a public controversy best opt for? We assume that it is of importance to all participants that the discussion leads to a result (resolution or compromise), and therefore that the parties cooperate. The impact of the use of a particular strategy on the *degree of cooperation* in the public controversy as a whole (that is to say: on the extent to which a public controversy welcomes cooperative contributions) takes two different routes: on the one hand through those contributions that are themselves part of the strategy (the *immediate* effect), on the other hand through the reactions that it calls forth (the *mediate* effect). The total effect can be that the use of the strategy generally increases the degree of cooperation in the controversy, in which case we say that it has a *constructive* effect (or that the strategy is *constructively effective*), or that it generally decreases the degree of cooperation, in which case we say that it has a *destructive* effect (or that the strategy is *destructively effective*).

The immediate effect of the use of a strategy corresponds to the extent to which the strategy itself is either obstructive or cooperative. But for the mediate effect the reactions of others have to be taken into account. Perhaps the most natural supposition is that parties copy behavior from one another. When that is always the case, the mediate effect of a strategy will always be in line with the immediate effect, so that cooperative strategies will automatically be constructively effective, and obstructive strategies destructively effective. We think, however, that there are important exceptions.

The basically cooperative strategy of Rationalization, for instance, may rightly or wrongly elicit the reproach that the other presents just a "frigid analysis." (Example 5). Rationalization will then come across as an unfair strategy. The mediate effect could be that other parties, getting annoyed, proceed to use unfair strategies as well. In that way, a cooperative strategy may in circumstances, through a "reversed" mediate effect, become destructively effective. Analogously, the cooperative strategy of Conciliation may come across as sycophancy, and *Quid pro quo* as a kind of Asking too much.

Conversely, basically obstructive strategies may display also constructive mediate effects. For, such a strategy may be used, temporarily, as a sanction against an obstructive party to make it mend its ways. A condition here is that the obstructive strategy so used must still be transparent in order to have the desired mediate effect. For, it must be clear to one's opponents for what behavior they are being "castigated" and what choices they should therefore make to avoid further castigation. Thus, *Ad baculum* and *Ad misericordiam* strategies may, for instance, induce the attacked or addressed parties to abandon their use of obstructive strategies. The strategy of Asking too much can also have such a globally constructive mediate effect. Thus, the demand for one billion euro by Governor Max van den Berg stimulated serious research into the true social costs (Example 10 *One billion*).

5. CHOOSING ONE'S STRATEGY

That basically obstructive strategies can be globally constructively effective does not imply that such strategies are simply to be recommended. Opting for an obstructive or unfair strategy, when others are using cooperative strategies, will yield generally only short-term benefits because the others will after some time probably also switch to such strategies and thereby ruin the process of public controversy as a whole. In this way, one will most likely also fail to achieve one's own aims in the controversy. Therefore, it is generally to be preferred to select a cooperative strategy when others do the same.

Yet there is a problem. What is one to do, if others do not select cooperative but obstructive or unfair strategies? Must one let the others do as they please, or pay back in the same coin? One option is to adopt, for a limited period, also an obstructive or unfair (unbalanced or intolerant, but still transparent) strategy and by this means try and bring the others into line.[10] This must be done for only some time because persisting with an *Ad baculum* or other obstructive strategy after other parties have abandoned obstructive strategies has a destructive, instead of a constructive, effect.

This view on the choice of strategy can be compared with the situation of playing an iterated version of the game of *Prisoner's Dilemma*, which has become widely known through the publications of the political scientist Robert Axelrod.[11] Obviously, there are great differences between choosing one's strategy for playing, with the same opponent, an iterated game of *Prisoner's Dilemma* and choosing one's strategy in a public controversy, be it only because for *Prisoner's Dilemma* the game and the payoffs are exactly defined. But the similarity is striking enough to be inspired by Axelrod's views.[12]

---

[10] Morton Deutsch (2014, pp. 23-27) discusses a number of other options.

[11] Axelrod explains the game on p. 8 of his book (Axelrod,1984).

[12] What we call a "cooperative strategy" may be compared with a *nice rule* in Axelrod. *Nice rules* are strategies in iterated *Prisoner's Dilemma* that never allow you to be the first to defect. Our "obstructive strategies" must then be compared with those rules (strategies) in Axelrod that are not *nice*.

Axelrod offers advice about how to choose one's strategy effectively when playing an iterated game of *Prisoner's Dilemma*. He summarizes his advice as follows:

A1. Don't be envious.
A2. Don't be the first to defect.
A3. Reciprocate both cooperation and defection.
A4. Don't be too clever. (Axelrod, 1984, p. 110)

This advice can be converted and expanded to yield advice about how to choose one's strategy effectively when participating in a public controversy. At the same time, a comparison can be made with advice offered by the social psychologist Morton Deutsch. Deutch recommends that, whether or no the other party wants to cooperate, one should oneself always be "firm, fair, flexible, and friendly" (Deutsch, 2014, p. 27).

1. (See A1 and Deutsch's *be flexible.*) When choosing a strategy in a public controversy, aim at sustainable results (resolutions, compromises) rather than at short-term gains. For, what matters is that the result will be for you as favorable as possible, not whether you gain more or less than another. For that, it is required that you don't cling uncompromisingly to your point of view and that you take the interests of others into account.

2. (See A2 and Deutsch's *be friendly.*) Use as much as possible a cooperative strategy and do not be the first to turn away from cooperation.

3. (See A3 and Deutsch's *be firm.*) If needed, make constructive use of an obstructive strategy to get others that use obstructive strategies to cooperate. Return to cooperation as soon as others do.

4. (See A4 and Deutsch's *be fair*[13]) An obstructive strategy used as explained in the third recommendation need not be fair because it need not be balanced or tolerant. Yet, to be constructive it must be a transparent strategy.

---

[13] By the term *fair,* Deutsch chiefly refers to one's refraining from using what he calls *dirty tricks,* which corresponds to what we denote as "transparency."

Axelrod's A4 is a special case of the last recommendation. For, by being "too clever" the complexity of your strategy increases so as to lose its transparency.

6. CONCLUSION

In many cases the contributions of a participant in a public controversy display a recognizable pattern that can be characterized as a strategy of a particular type. Starting from the public controversy about the induced earthquakes in Groningen, we described fourteen types of strategy, a selection of which we evaluated in terms of fairness. We have further clarified the notion of fairness by discussing three aspects: balance, transparency and tolerance. This was a convenient approach because, in this way, we did not have to analyze the various contributions from the perspective of one determinate type of dialogue and could stay on a global level when showing how fair strategies normally contribute to the degree of cooperation in a public controversy.

We would have liked to conclude that fairness is always the best policy. But to achieve compromises and resolutions, it is sometimes expedient to answer unfairness by unfairness in order to get the opponent to return to fairness and to increase, in this mediate way, the degree of cooperation. In order for this *tit-for-tat* like strategy to be effective, your opponent must have no doubt about what constitutes the unfair strategy that occasioned you to inflict the castigation.

Knowledge about fair and unfair strategies is anyhow essential for a successful proceeding of a public controversy. Only if such knowledge is available, is there an option for the parties to go for fair discussion and negotiation as means to achieve resolutions and compromises. Only then, can unfair behavior be denounced. It is, therefore, an important matter of public interest that the knowledge and skills needed for assessing the fairness of strategies be widely spread.

Such knowledge and skills are important for the stronger as well as for the weaker parties in a public controversy. Those in a position of power have often reasons to show fair behavior because power needs support, and support needs approval by critical people. Those in a position of little or no power may have even better reasons to show fair behavior. For, who will most likely be getting the worst of it when it comes to a showdown would be wise to avail herself of the means of reason and to cooperate with other parties towards a reasonable resolution or a fair compromise.

ACKNOWLEDGEMENTS: For critical comments we thank our commentator and the audience of the presentation at the conference in Lisbon as well as the audience of an earlier presentation (VIOT conference, Louvain, December 2014) and two anonymous referees of the Dutch version of our paper (van Laar and Krabbe, 2016).

REFERENCES

Axelrod, R. (1984). *The evolution of cooperation.* New York: Basic Books.
Blanken, H. (2013a). Het lot van Loppersum [The fate of Loppersum]. *Dagblad van het Noorden* (Weekend, p. 6), 18 May 2013.
Blanken, H. (2013b). We zitten allemaal in een dooie hoek hier [We are all in a blind spot over here]. *Dagblad van het Noorden* (p. 4), 2 November 2013.
Blanken, H. (2013c). 'Ze zuigen het leeg en doen niets terug' ['They're sucking it out and do nothing in return'] *Dagblad van het Noorden* (p. 6), 9 April 2013.
Deutsch, M. (2014). Cooperation, competition, and conflict. In P. T. Coleman, M. Deutsch & E. C. Marcus (Eds.), *The Handbook of Conflict Resolution: Theory and Practice,* 3rd ed. (pp. 3-28). Somerset, NJ: John Wiley. [First edition: San Francisco CA: Jossey Bass, 2000.]
Dascal, M. (2008). Dichotomies and types of debates. In F. H. van Eemeren & B. Garssen (Eds.), *Controversy and Confrontation: Relating Controversy Analysis with Argumentation Theory* (pp. 27–49). Amsterdam: John Benjamins.
de Veer, J. (2013). Mensen worden ongeduldig en steeds bozer [People are getting impatient and more and more angry]. *Dagblad van het Noorden* (p. 2), 5 December 2013.
Fisher, R., Ury, W., & Patton, B. (2011). *Getting to Yes: Negotiating an Agreement without Giving In,* 3rd ed. London: Random House.
Gabbay, D. M., & Woods, J. (2001a). Non-cooperation in dialogue logic. *Synthese, 127,* 161-186.
Gabbay, D. M., & Woods, J. (2001b). More on non-cooperation in dialogue logic. *Logic Journal of the IGPL, 9,* 321-339.
Govier, T. (1997). *Social trust and human communities.* Montreal: McGill-Queen's University Press.
Johnson, R. H. (2000). *Manifest rationality: A pragmatic theory of argument.* Mahwah, NJ: Lawrence Erlbaum.
Lewiński, M., & Aakhus, M. (2014). Argumentative polylogues in a dialectical framework: A methodological inquiry. *Argumentation, 28,* 161–185.
Luyendijk, Wubby (2015). Al een jaar belooft de NAM sommetjes over de risico's [For a year now NAM has been promising to do sums about the risks]. *NRC-Handelsblad,* 27 January 2015.

Meijer Commission (2013). *Vertrouwen in een duurzame toekomst: Een stevig perspectief voor Noord-Oost Groningen: Eindadvies van de Commissie Duurzame Toekomst Noord-Oost Groningen* [Trust in a Sustainable Future: A Solid Perspective For Northeast Groningen: Final Recommendations of the Committee Sustainable Future Northeast Groningen]. <
http://www.www.dialoogtafelgroningen.nl/?attachment_id=%20170>
NAM (2012). *Lichte aardbeving bij Sappemeer* [Light earthquake near Sappemeer]. Press release. <http://www.nam.nl/nl/news/news-archive-2012/2012-08-17-aardbeving-noord-groningen.html>
RTV Noord (2015). 'VVD-kamerlid bespreekt gastactiek telefonisch in de trein.' [Liberal MP discusses gas tactics on telephone in train] *RTV Noord* (website), 17 January 2015. <
http://www.rtvnoord.nl/artikel/artikel.asp?p=143958 >
Sluis, B. van (2012). De grond beeft [The ground quakes]. *Dagblad van het Noorden* (p. 10), 1 September 2012.
van Eemeren, F. H. (2010). *Strategic maneuvering in argumentative discourse: Extending the pragma-dialectical theory of argumentation.* Amsterdam: John Benjamins.
van Eemeren, F. H., & Grootendorst, R. (2004). *A systematic theory of argumentation.* Cambridge: Cambridge University Press.
van Eemeren, F. H., & Houtlosser, P. (2002). Strategic manoeuvring in argumentative discourse: A delicate balance. In: F. H. van Eemeren & P. Houtlosser (Eds.), *Dialectic and rhetoric: The warp and woof of argumentation analysis* (pp. 131-159). Dordrecht: Kluwer.
van Laar, J. A., & Krabbe, E. C. W. (2016). Eerlijke en oneerlijke strategieën in maatschappelijke discussies [Fair and Unfair Strategies in Public Controversies]. In D. Van De Mierop, L. Buysse, R. Coesemans & P. Gillaerts (Eds.), *De macht van de taal: Taalbeheersingsonderzoek in Nederland en Vlaanderen* [The Power of Language: Research in Speech Communication in The Netherlands and Flanders](pp. 131-143). Leuven: Acco.
Walton, D. N. (1992). *The place of emotion in argument.* University Park, PA: Pennsylvania State University Press.
Walton, D. N. (1999). *Appeal to popular opinion.* University Park, PA: Pennsylvania State University Press.
Walton, D. N., & Krabbe, E. C. W. (1995). *Commitment in dialogue: Basic concepts of interpersonal reasoning.* Albany, NY: State University of New York Press.
Wikipedia (2014). *Aardgaswinning in Nederland* [Natural gas extraction in the Netherlands]. Accessed 19 November 2014. <
http://nl.wikipedia.org/wiki/Aardgaswinning_in_Nederland#Bodemdalin g_en_aardbevingen>
Young, I. M. (2001). Activist challenges to deliberation. *Political Theory, 29,* 670-690.

Why Be Fair?
Commentary on van Laar and Krabbe's
Fair and Unfair Strategies in Public Controversies

CHRISTOPHER W. TINDALE
CRRAR, University of Windsor, Canada
ctindale@uwindsor.ca

1. INTRODUCTION

This is a welcome study of how argumentation operates at a more global level, beyond the specifics of any particular model of dialogue. Thus, it brings out the essential dialogical nature of much argumentation, where parties identify, address, and respond to each other's strategies by changing their own strategies. This attention also helps us appreciate how parties would recognize each other's strategies, as they strive to match cooperativeness with the optimum outcome for their own position.

The paper also adds to our appreciation of how 'fairness' operates as a value in argumentation, and of ways in which this concept can be understood and recognized.

In these brief remarks I will look (a) at the way 'fair' is understood and used, and then (b) consider how the *publicness* of public controversies may play a role in their outcome, since this seems one point overlooked so far in the discussion.

2. BEING FAIR

Fairness is a central concept in the paper. It is deemed to play a role in every serious negotiation. So how is this concept understood? Strategies are evaluated globally as fair or unfair, "that is: as more or less *balanced, transparent*, and *tolerant*" (This volume, p. 345). So 'fairness' is a complex concept comprised of these three criteria. These three are chosen because of what they are expected to contribute towards cooperative resolutions or compromises. Each of them is developed in part 3 of the paper. Under a discussion of fair shares, strategies are considered *balanced* if they leave enough room for considerations of the other party. Co-operators provide for others to express themselves.

Under a discussion of not cheating, co-operators will be honest about themselves and what they intend, and how they expect to achieve those intentions. Such strategies are considered *transparent*.

Thirdly, under a discussion of coercion (or no coercion) attention is given to the power dynamics that may be involved. Fairness dictates that physical coercion and threats be avoided. But the principal idea here is that of freedom: parties should leave "enough freedom for the opponent to determine and elaborate" a position (this volume, p. 353). In this sense, adopting the criterion of tolerance seems odd. Tolerance would appear to involve an acceptance of another's point of view as reasonable and a further recognition that they have a right to express it. The principal idea involved is that of equality. Tolerance, on these terms, seems closer to what has already been included under the criterion of *balance*, where others are given the space to express themselves. Here, under this third criterion, 'fairness' actually involves a commitment not to be authoritative or threatening. I would suggest a better term to use for this criterion would be simply *non-coercive*. Or, perhaps, *egalitarian*, in the sense that parties do not see their position as privileged in any way.

3. *PUBLIC* CONTROVERSIES

A *public controversy* is defined in the paper as "a complex phenomenon consisting of many dialogues about a number of connected issues, most clearly present in the mass media, and stretching over a considerable period of time" (this volume, p. 345). And such controversies are polylogues, involving more than two parties. Indeed, the bulk of the discussion here necessarily involves the roles and obligations of these multiple parties, exploring the prospect of cooperation between them. But we should not lose sight of the fact that these controversies are public, since this feature may have a role to play in the fairness of the argumentation and the obligation parties feel to be cooperative.

This becomes important when the discussion turns to the prisoners' dilemma and the kinds of cooperation involved there. Typically, these are two-person dilemmas. In the original scenario, two prisoners are separated and then each told, "If neither of you confesses, you will each get one year in jail. If you confess and your accomplice does not, you will go free and he will get twelve years. If your accomplice confesses and you do not, those sentences will be reversed. If you both confess, you will each get five years."

The rational choice of a self-interested person is to confess (and get five years). But if they consider each other's interests and cooperate, *and know* the other will do the same, then they will stay silent. But, of course, this type of rational cooperation pertains for two-party dilemmas in a single case, and what the authors of this paper have in mind is a far more complex phenomenon with multiple parties over an extended period of time. On their scenario there is the added dimension of parties responding to the other's strategies. Here, they do know what the other does (cooperate or otherwise), and the prospect of tit-for-tat ("I will cooperate with you if you do as well; but if you adopt an unfair strategy, I will do the same") can encourage cooperative behaviour. This is the important dialogical aspect to the case in question. Moreover, there is the further temporal dimension that allows the parties to test things out, experience the other's strategies and learn things about them. That is absent from the prisoners' dilemma.

So we might, on balance, judge the prisoners' dilemma to be of limited value here, because what it tells us about cooperation may not be transferable to the complex contexts of public controversies. Where we might turn instead is to the *publicness* of the controversies and the potential influence of reputation. Because parties involved may be expected to care not just about establishing their position or reaching compromises but also about how they will be perceived and how that perception will impact their ability to function at later stages of the controversy or on other issues altogether. This is the aspect of *ethos* that we can borrow from the rhetorical tradition. As Plato reminded us with his ring of Gyges example in the *Republic,* people are concerned to *appear to be* fair. And the best way to do that is, of course, to be fair. A number of the parties whose contributions are explored in part 2 of the paper are parties who have a public image (or character) which they would want to maintain and about which they would be concerned. This is no guarantee that they will then adopt fair strategies, but it is an added incentive for them to do so. One further thing might be added to this account, and that is the role to be played by trust. As the same parties argue together on a controversy over a period of time, they have the prospect of building trust through familiarity. That is another factor that can, reciprocally, encourage the adoption of fairness.

16

Working with open argument corpora

JOHN LAWRENCE
Centre for Argument Technology, University of Dundee, UK
j.lawrence@dundee.ac.uk

MATHILDE JANIER
Centre for Argument Technology, University of Dundee, UK
m.janier@dundee.ac.uk

CHRIS REED
Centre for Argument Technology, University of Dundee, UK
c.a.reed@dundee.ac.uk

AIFdb Corpora provides a facility to group Argument Interchange Format (AIF) argument maps and search for maps that are related to each other (for example, analyses of related texts.) Users can create and share corpora containing any number of argument maps from within AIFdb. By integrating with the OVA+ analysis tool, AIFdb Corpora allows for the creation of corpora compliant with both AIF and Inference Anchoring Theory, a philosophically and linguistically grounded counterpart to AIF.

KEYWORDS: argument corpora, Online Visualisation of Argument, Argument Interchange Format, Inference Anchoring Theory

1. INTRODUCTION

The number of Argument Interchange Format (AIF) (Chesñevar et al., 2006) argument maps contained within the open and publicly accessible database, AIFdb[1] (Lawrence, Bex, Reed, & Snaith, 2012), now exceeds 4,000 with over 60,000 individual nodes in eleven different languages. These numbers are growing rapidly and, as an increasing number of

---

[1] http://www.aifdb.org/

argumentation tools, such as Arvina[2], the AnalysisWall (Bex, Lawrence, Snaith, & Reed, 2013) and ArguBlogging (Bex, Snaith, Lawrence, & Reed, 2014) begin using AIFdb to store their data, the rate of growth is set to increase.

Although AIFdb offers a search interface to locate both a given node and the maps within which that node occurs, there is no real ability to group argument maps or to search for maps that are related to each other in some way (for example, being analyses of related texts.)

AIFdb Corpora[3] offers such an ability, allowing a user to create and share a corpus containing any number of argument maps from within the database. By integrating closely with the OVA+ (Online Visualisation of Argument) analysis tool, AIFdb Corpora allows for the rapid creation of large corpora compliant with both AIF and *Inference Anchoring Theory* (IAT) (Budzynska & Reed, 2011), a philosophically and linguistically grounded counterpart to the AIF.

2. ARGUMENT DATA

The continuing growth in the volume of data which we produce has driven efforts to unlock the wealth of information this data contains. Automatic techniques such as Opinion Mining and Sentiment Analysis (Liu, 2010) allow us to determine the views expressed in a piece of textual data, for example, whether a product review is positive or negative. Existing techniques struggle, however, to identify more complex structural relationships between concepts.

Argument Mining[4] is the automatic identification of the argumentative structure contained within a piece of natural language text. By automatically identifying this structure and its associated premises and conclusions, we are able to tell not just *what* views are being expressed, but also *why* those particular views are held.

The desire to achieve this deeper understanding of the views which people express has led to the recent rapid growth in the Argument Mining field (2014 saw the first ACL workshop on the topic in Baltimore[5] and meetings dedicated to the topic in both Warsaw[6] and Dundee[7]). One of the challenges faced by current approaches to argument mining

---

[2] *http://arvina.arg-tech.org*

[3] *http://corpora.aifdb.org*

[4] Sometimes also referred to as Argumentation Mining

[5] *http://www.uncg.edu/cmp/ArgMining2014/*

[6] *http://argdiap.pl/argdiap2014*

[7] *http://www.arg-tech.org/swam2014/*

however, is the lack of large quantities of appropriately annotated arguments to serve as training and test data. Several recent efforts have been made to improve this situation by the creation of corpora across a range of different domains.

For example, (Green, 2014) aims to create a freely available corpus of open-access, full-text scientific articles from the biomedical genetics research literature, annotated to support argument mining research. However, there are challenges to creating such corpora, such as the extensive use of biological, chemical, and clinical terminology in the BioNLP domain.

In (Houngbo & Mercer, 2014), a straightforward feature of co-referring text using the word "this" is used to build a self-annotating corpus extracted from a large biomedical research paper dataset. This is achieved by collecting pairs of sequential sentences where the second sentence begins with "This method...", "This result...", or "This conclusion...", and then categorising the first sentence in each pair respectively as Method, Result or Conclusion sentences. The corpus is annotated without involving domain experts and in a 10-fold cross-validation, gives an overall F-score of 0.97 with naïve Bayes and 0.987 with SVM.

Legal texts are the focus of (Walker, Vazirova, & Sanford, 2014), where a type system is developed for marking up successful and unsuccessful patterns of argument in U.S. judicial decisions. Building on a corpus of vaccine-injury compensation cases that report factfinding about causation, based on both scientific and non-scientific evidence and reasoning, patterns of reasoning are identified and used to illustrate the difficulty of developing a type or annotation system for characterising these patterns. A further example of legal material is the ECHR corpus (Mochales & Ieven, 2009), a set of documents extracted from legal texts of the European Court of Human Rights (ECHR). The ECHR, over the years, has developed a standard type of reasoning and structure of argumentation resulting in material which, although not specifically annotated for argumentative content is easily adapted to serve as data for argument mining.

Such efforts add to the volume of currently available data for which at least some elements of the argumentative structure have been identified. The most comprehensive and completely annotated existing collection of such data is the openly accessible database, AIFdb[8] (Lawrence et al., 2012), containing over 4,000 Argument Interchange Format (AIF) argument maps with over 60,000 individual nodes in ten different languages. These numbers are growing rapidly, thanks to both

---

[8] *http://www.aifdb.org*

the increase in analysis tools interacting directly with AIFdb and the ability to import analyses produced with the Rationale and Carneades tools (Bex, Gordon, Lawrence, & Reed, 2012).

Additionally, several online tools such as DebateGraph[9], TruthMapping[10], Debatepedia[11], Agora[12], Argunet[13] and Rationale Online[14] allow users to create and share argument analyses. Although these tools are helping to increase the volume of analysed argumentation, they generally do not offer the ability to access this data and each use their own formats for its annotation and storage. In order to help overcome this challenge, AIFdb currently offers the facility to import and convert Rationale analyses into AIF and development is underway to allow for conversion of the DebateGraph and Argunet formats.

In addition to the previously discussed corpora of structured argument data, there are large corpora of unstructured data available that are rich in argumentative structure, from, for example, Wikipedia, Google Books, meeting data from the AMIDA Meeting Corpus[15] annotated using the Twente Argumentation Scheme (Rienks, Heylen, & Weijden, 2005) and product reviews from websites such as Amazon and epinions.com. Whilst these corpora may be useful for certain argument mining techniques, such as those using unsupervised learning methods, the full utilisation of these resources is limited by their lack of annotation. Despite the lack of marked argument structure, Wikipedia, in particular, represents a considerable amount of data rich in argumentative content. IBM's recently announced Debater project,[16] is an argument construction engine utilising a corpus of unstructured Wikipedia text. Debater can respond to a given topic by automatically constructing a set of relevant pro/con arguments phrased in natural language. For example, when asked for responses to the topic "The sale of violent video games to minors should be banned", Debater scanned approximately 4 million Wikipedia articles and determined the ten most relevant articles,

---

[9] http://debategraph.org

[10] https://www.truthmapping.com

[11] http://www.debatepedia.org

[12] http://agora.gatech.edu/

[13] http://www.argunet.org/

[14] https://www.rationaleonline.com

[15] http://corpus.amidaproject.org/

[16] http://www.kurzweilai.net/introducing-a-new-feature-of-ibms-watson-the-

scanned all 3,000 sentences in those articles, detected sentences which contain candidate claims, assessed their pro and con polarity and then presented three relevant pro and con arguments.

Although Debater is able to extract simple pro and con reasons from Wikipedia articles, it falls short of being able to offer a detailed understanding of the argumentative structure. In (Aharoni et al., 2014), work towards annotating articles from Wikipedia using a meticulously monitored manual annotation process is discussed. The result is 2,683 argument elements, collected in the context of 33 pre-defined controversial topics, and organised under a simple structure detailing a debate claim and its associated supporting evidence.

Another possible avenue for increasing the volume of annotated argument is crowdsourcing, as discussed in (Ghosh, Muresan, Wacholder, Aakhus, & Mitsui, 2014), where a two-tiered approach is proposed to determine which portions of texts are argumentative and what is the nature of argumentation. The first step suggested adopts a coarse-grained annotation scheme based on Pragmatic Argumentation Theory (van Eemeren, Grootendorst, Jackson, & Jacobs, 1993) and asked expert annotators to label entire threads using this scheme. A clustering technique is then used to identify which pieces of text were easier or harder to annotate and it is shown that crowdsourcing is a feasible approach to obtain annotations, particularly on those text segments that were identified as being easier for the Expert Annotators.

The availability of large scale corpora of annotated argument data has a wide range of possible applications, for example, allowing for the comparison of analysis to real world dialogue (Goodwin & Cortes, 2010), determining the validity of argument coding schemes (Pallotta, Seretan, Ailomaa, Ghorbel, & Rajman, 2007), and providing insight into patterns of argument in discourse (O'Halloran, 2011). By building a diverse range of corpora spanning different times and domains, it is possible to perform comparative research into argument usage in a discourse field over time and across discourse fields.

3. AIFDB

The Argument Web (Bex et al., 2013) is a vision for a large-scale Web of inter-connected arguments posted by individuals on the World Wide Web in a structured manner. As such it is necessary to provide a service which not only allows for the storage and retrieval of this structured argument data, but is compatible with the widest possible range of currently existing argumentation software and provides a stable and flexible platform around which future software can be developed. AIFdb is a database implementation of the Argument Interchange Format (AIF)

(Chesñevar et al., 2006), allowing for the storage and retrieval of AIF compliant argument structures. AIFdb offers a wide range of web service interfaces for interacting with stored argument data, as well as offering its own search and argument visualisation features all consistent with the formal ontology of the AIF.

At the lowest level, AIFdb's web services allow for the insertion and querying of the basic components of an AIF argument such as nodes, edges and schemes. These components are represented by tables in the database as seen in Figure 1. Building upon these lower level interactions, AIFdb also offers a "middle layer" which groups these simple queries to allow more complex interactions to be easily performed. For example it is possible, with a single query, to determine all of the statements made by a particular person in support of a given information node (I-node). At the highest level of interaction, AIFdb supports modules handling the import and export of numerous formats such as SVG, DOT, RDF-XML and the formats of the Carneades, Rationale and Araucaria tools.

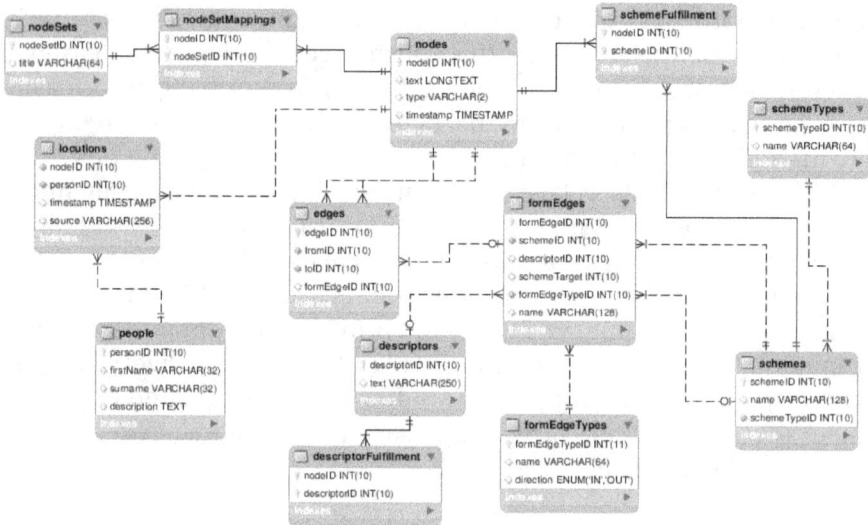

Figure 1 – AIFdb Data Structure Diagram

4. AIFDB CORPORA

AIFdb Corpora extends the functionality offered by AIFdb, allowing a user to create and share a corpus containing any number of argument maps from within the database. Users are able to create their own corpora for specific projects or themes, grouping together argument maps that are related in some way, and enabling these to be viewed together or downloaded in a variety of formats.

4.1 The AIFdb Corpora Interface

In order to create a corpus, the user must specify simple details, including the title of the corpus, a shortname used in the corpus URL, and a brief description, shown in the list of corpora. Once these details are entered, the user is given a unique link to a page where they can edit their corpus. The edit page, as shown in Figure 2, allows the user to update these details, as well as providing text corresponding to the corpus as a whole, for example, if the corpus consists of analyses of different parts of the same text, it can be provided here and will be made available as part of the full corpus download. The edit page also allows the user to manually add individual argument maps to the corpus, and to add sub-corpora; existing corpora that form a part of the corpus being edited. Any subcorpora then act like a part of the parent corpus, for example, when an argument map is added to a sub-corpus, it is added to the parent corpus as well. Additionally, the user may lock the corpus, preventing any other applications from adding maps to it.

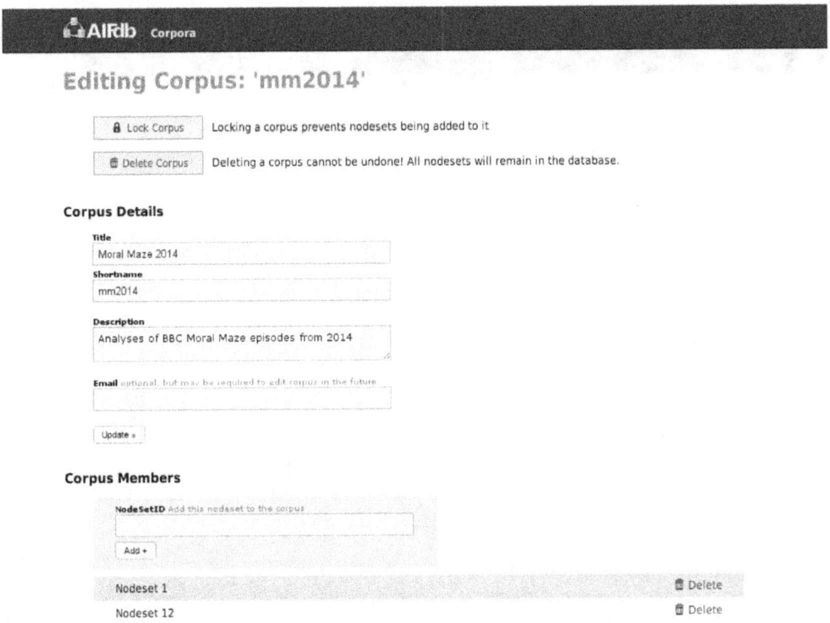

Figure 2 – Corpora Admin Interface

AIFdb Corpora also offers interfaces to list and search corpora as seen in Figure 3, and to display corpora, as seen in Figure 4. This allows a user to share a link to their corpus with others and allows for easy viewing and downloading of the corpus contents, either as individual files in SVG, PNG, DOT, JSON, RDF-XML, Prolog and the formats of the Carneades (Gordon, Prakken, & Walton, 2007) and Rationale (van Gelder, 2007) tools, or as an archive file containing JSON format representations of each argument map as well as the original text of the entire corpus. The display interface also provides links to view, evaluate and edit any of the argument maps contained within the corpus in OVA or OVA+.

Although AIFdb does not allow for the storage of the text corresponding to an argument map, an additional database is available to store these texts and tools such as OVA are able to store the original text for analyses in this database. When the archive file for a particular corpus is generated this database is queried and the text for each individual argument map added to the archive file. If no text has been provided for the corpus as a whole, then the individual texts for each argument map are concatenated and provided together. In cases where a corpus contains subcorpora, again, if no text has been provided for the corpus as a whole, the text for any individual maps is joined with any text that has been given for the subcorpora.

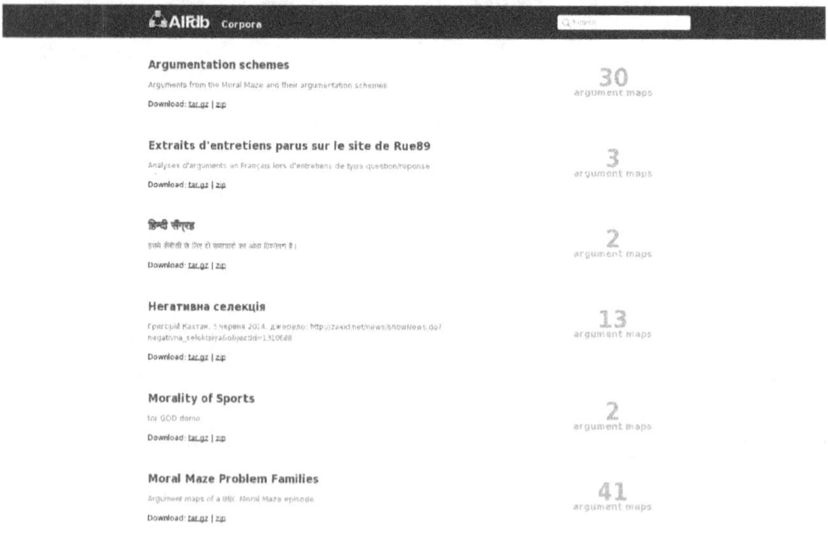

Figure 3 – Corpus Listing

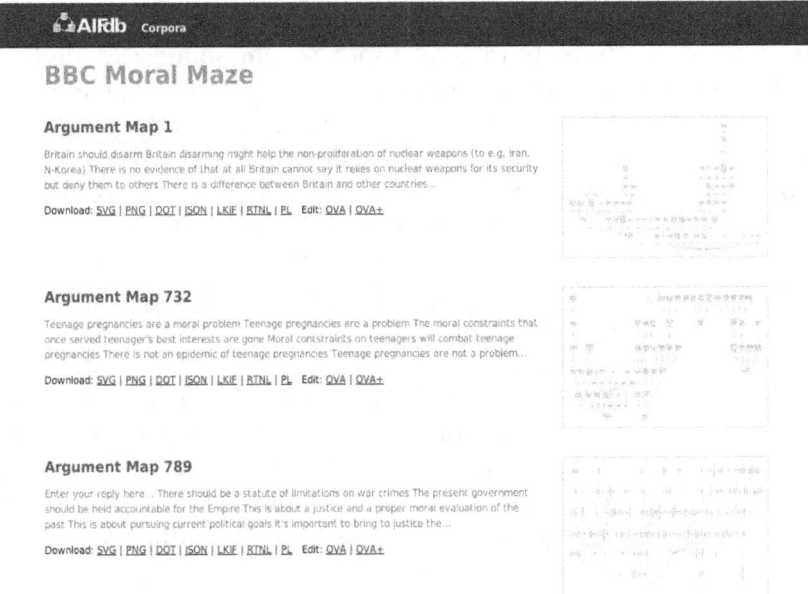

Figure 4 – Corpus Details

4.2 Integration with OVA+

OVA+ offers a web based interface for the analysis of arguments and is accessible from any web browser at http://ova.arg-tech.org/. The tool was built as a response to the AIF theory and allows for the creation and representation of argument structure, combined with the ability to exchange, share and reuse the resultant argument maps. The system relies on the Inference Anchoring Theory (IAT), with the schemes provided allowing for a graphical representation of the argumentative structure of a text and, more interestingly, of dialogues.

OVA+ handles texts of any type and any length. To begin the analysis, the first relevant utterance for argumentation must be extracted in order to create an information node (I-node). Then it is possible to create the locution node associated (L-node) and to specify the name of the speaker (participant); the locution appears, preceded by the name of the participant assigned to it, and arrows link the L-node to the I-node via a YA-node.

YA-nodes are the illocutionary forces of locutions, and can be given a scheme provided by the IAT model. Each following utterance can be annotated accordingly. According to the AIF, it is possible to evidence supports or attacks between arguments. An RA-node (application of a rule of inference) should connect two I-nodes. To elicit an attack between arguments, RA-nodes can be changed into CA-nodes, namely applications of a pattern of conflict. Linked arguments can be established by connecting all the arguments to the proper scheme-node (RA or CA). According to IAT, it is also possible to indicate the transitions (TA-nodes) between locutions by linking two L-nodes. Finally, it is possible for the analyst to assign the illocutionary forces anchored in the transitions. This can be done thanks to the set of IAT schemes which are proposed when a TA-node has been linked to its corresponding scheme-node.

At the end of the analysis, OVA+ permits saving the work on the user's computer as either an image file or as a JSON format file representing the AIF structure. Most interestingly, however, OVA+ offers the possibility of saving an analysis to AIFdb and its further addition to any corpus in AIFdb Corpora. This ability allows for analyses to be reused via AIFdb, and consequently any of the growing number of argument web tools which use AIFdb as their data store; or loaded in OVA+ for consultation or correction, as well as allowing for the rapid and collaborative creation of large AIF and IAT compliant corpora.

4.3 AIFdb Corpora Usage

AIFdb Corpora already collects over 1,000 analyses into a range of corpora, the largest of which are described below:

AraucariaDB An import of 667 argument analyses produced using Araucaria(Reed & Rowe, 2004) and stored in the Araucaria database (Reed, 2006).

AraucariaDBpl A selection of over 50 Polish language analyses created using the Polish version of Araucaria (Budzynska, 2011).

Digging By Debating Argument Study Collection of analyses of 19th century philosophical texts from the Hathi Trust collection, created for the Digging by Debating project[17].

Moral Maze 2012 Analyses of episodes of the BBC Moral Maze Radio 4 programme from 2012. These analyses are split into two sub-corpora **MM2012a** (containing the first four programmes) and **MM2012b** (containing the last four). **MM2012** comprises the full run of 8 programmes from the 2012 summer season. These corpora of analysed arguments have already been used in different projects, in particular the one described in

---

[17] http://diggingbydebating.org/

(Budzynska, Janier, Kang, et al., 2014) and (Budzynska, Janier, Reed, et al., 2014).

Argumentation Schemes Examples of occurrences of Walton's argumentation schemes (Walton, Reed, & Macagno, 2008) found in episodes of the BBC Moral Maze Radio 4 programme

Expert Opinion and Positive Consequences Examples of the Expert Opinion and Positive Consequences argumentation schemes taken from online news sources

Dispute Mediation Argument maps of mediation session transcripts.
This corpus was created to facilitate the retrieving of analysed excerpts of mediation transcripts, as part of the DrEAMS poject (Dialogue-based Exploration of Arguments and Mediation Space) in the University of Dundee. The research project indeed aims at exploring discourse in dispute mediation through analyses of the dialogues between disputants and mediators. A repository of the argument maps (currently 65) revealed necessary in order to keep trace of the phenomena proper to mediation highlighted by these argument maps. This corpus is actually composed of 2 subcorpora: **Dispute mediation: Excerpts taken from publications** and **Mock mediation**. Each corpus of AIFdb Corpora can indeed be integrated to a larger one. In addition to their practical aspect for the project in itself, those corpora offer a valuable data set for the argumentation community, as the analysed mediation dialogues can be freely consulted and downloaded.

Language Opposition Corpus Argument maps of online multi-party interactions
This is currently the largest corpus (1946 argument maps distributed through five sub corpora). It is used in Rutgers for the SALTS project[18], the goals of which are to advance the understanding of how expression of argumentation shapes the ebb and flow of online interactions, and to develop computational models capable of identifying and characterizing the expression of argumentation in multi-party interactions.

ECC - Bank of America and ECC - The Coca Cola Company . Argument maps of Earning Conference Calls.

---

[18] http://salts.rutgers.edu/

The argument maps are the result of a project at Universita della Svizzera Italiana which explores the dialogues between CEOs or managers of big companies and financial analysts during Earning Conference Calls.

Негативна селекція **(Negative Selection)** Analyses of a journal article discussing the Dean's election at Lvov University and the effects of the post soviet system on modern Ukranian academia.

5. CONCLUSION

It is hoped that by making the process of creating and updating a corpus as simple as possible, usage will continue to grow and that AIFdb Corpora will prove to be a useful tool for collecting and sharing AIF argument maps. By close integration with analysis tools such as OVA+, AIFdb Corpora allows for the rapid, collaborative creation of AIF and IAT compliant corpora, and so will offer a valuable resource in areas such as argumentation mining which have a demand for large quantities of such annotated material.

ACKNOWLEDGEMENTS: The authors would like to acknowledge that the work reported in this paper has been supported in part by the RCUK Lifelong Health and Wellbeing Programme grant number EP/K037293/1 - BESiDE: The Built Environment for Social Inclusion in the Digital Economy, and in part by the British Leverhulme Trust under grant RPG-2013-076.

REFERENCES

Aharoni, E., Polnarov, A., Lavee, T., Hershcovich, D., Levy, R., Rinott, R., ... Slonim, N. (2014, June). A benchmark dataset for automatic detection of claims and evidence in the context of controversial topics. In *Proceedings of the first workshop on argumentation mining* (pp. 64-68). Baltimore, Maryland: Association for Computational Linguistics.
Bex, F., Gordon, T. F., Lawrence, J., & Reed, C. (2012). Interchanging arguments between Carneades and AIF - Theory and practice. In *Proceedings of the fourth international conference on computational models of argument (comma 2012)* (pp. 390-397). Vienna: IOS Press.
Bex, F., Lawrence, J., Snaith, M., & Reed, C. (2013, Oct). Implementing the argument web. *Communications of the ACM, 56*(10), 66-73.
Bex, F., Snaith, M., Lawrence, J., & Reed, C. (2014). Argublogging: An application for the argument web. *Web Semantics: Science, Services and Agents on the World Wide Web.*

Budzynska, K. (2011). Araucaria-PL: Software for teaching argumentation theory. In *Proceedings of the third international congress on tools for teaching logic (ticttl 2011)* (pp. 30-37).

Budzynska, K., Janier, M., Kang, J., Reed, C., Saint-Dizier, P., Stede, M., & Yaskorska, O. (2014). Towards argument mining from dialogue. In *Proceedings of the fifth international conference on computational models of argument (comma 2014)* (pp. 185-196). IOS Press.

Budzynska, K., Janier, M., Reed, C., Saint-Dizier, P., Stede, M., & Yaskorska, O. (2014). A model for processing illocutionary structures and argumentation in debates. In *Proceedings of the 9th edition of the language resources and evaluation conference (lrec)* (pp. 917-924).

Budzynska, K., & Reed, C. (2011). *Whence inference* (Tech. Rep.). University of Dundee.

Chesñevar, C., Modgil, S., Rahwan, I., Reed, C., Simari, G., South, M., ... others (2006). Towards an argument interchange format. *The Knowledge Engineering Review*, 21(04), 293-316.

Ghosh, D., Muresan, S., Wacholder, N., Aakhus, M., & Mitsui, M. (2014, June). Analyzing argumentative discourse units in online interactions. I *Proceedings of the first workshop on argumentation mining* (pp. 39-48).

Goodwin, J., & Cortes, V. (2010). Theorists' and practitioners' spatial metaphors for argumentation: A corpus-based approach. *Verbum*, 23, 163-78.

Gordon, T. F., Prakken, H., & Walton, D. (2007). The carneades model of argument and burden of proof. *Artificial Intelligence*, 171(10), 875-896.

Green, N. (2014, June). Towards creation of a corpus for argumentation mining the biomedical genetics research literature. In *Proceedings of the first workshop on argumentation mining* (pp. 11-18). Baltimore, Maryland: Association for Computational Linguistics.

Houngbo, H., & Mercer, R. (2014, June). An automated method to build a corpus of rhetorically-classified sentences in biomedical texts. In *Proceedings of the first workshop on argumentation mining* (pp. 19-23). Baltimore, Maryland: Association for Computational Linguistics.

Lawrence, J., Bex, F., Reed, C., & Snaith, M. (2012). AIFdb: Infrastructure for the argument web. In *Proceedings of the fourth international conference on computational models of argument (comma 2012)* (pp. 515-516).

Liu, B. (2010). Sentiment analysis and subjectivity. *Handbook of natural language processing*, 2, 627-666.

Mochales, R., & Ieven, A. (2009). Creating an argumentation corpus: do theories apply to real arguments?: a case study on the legal argumentation of the echr. In *Proceedings of the 12th international conference on artificial intelligence and law* (pp. 21-30).

O'Halloran, K. (2011). Investigating argumentation in reading groups: Combining manual qualitative coding and automated corpus analysis tools. *Applied linguistics*, 32(2), 172-196.

Pallotta, V., Seretan, V., Ailomaa, M., Ghorbel, H., & Rajman, M. (2007). Towards an argumentative coding scheme for annotating meeting dialogue data. In *Proceedings of the 10th international pragmatics conference, ipra* (pp. 8–13).

Reed, C.(2006). Preliminary results from an argument corpus. In E. M. Bermudez & L. R. Miyares (Eds.), *Linguistics in the twenty-first century* (pp. 185–196). Cambridge Scholars Press.

Reed, C., & Rowe, G. (2004). Araucaria: Software for argument analysis, diagramming and representation. *International Journal on Artificial Intelligence Tools, 13*(4), 961–980.

Rienks, R., Heylen, D., & Weijden, v. d. E. (2005). Argument diagramming of meeting conversations. In *Multimodal multiparty meeting processing, workshop at the 7th intl. conference on multimodal interfaces.* IOS Press.

van Eemeren, F. H., Grootendorst, R., Jackson, S., & Jacobs, S. (1993). *Reconstructing argumentative discourse.* University of Alabama Press.

van Gelder, T. (2007). The rationale for rationale. *Law, probability and risk, 6*(1-4), 23–42.

Walker, V., Vazirova, K., & Sanford, C. (2014, June). Annotating pattern of reasoning about medical theories of causation in vaccine cases: Toward a type system for arguments. In *Proceedings of the first workshop on argumentation mining* (pp. 1–10). Baltimore, Maryland: Association for Computational Linguistics.

Walton, D., Reed, C., & Macagno, F. (2008). *Argumentation schemes.* Cambridge University Press.

17

Towards an Online Social Debating System

JOÃO LEITE
NOVA LINCS, Universidade Nova de Lisboa, Portugal
jleite@fct.unl.pt

JOÃO MARTINS
Computer Science Department, Carnegie Mellon University, USA
jmartins@cs.edu

SINAN EĞILMEZ
NOVA LINCS, Universidade Nova de Lisboa, Portugal
s.egilmez@fct.unl.pt

After the initial boom of the Web 2.0, many people are growing unsatisfied with the depth (or lack thereof) of interactions on social websites. In this paper we discuss some features required by an online debating system aimed at a wide social participation, and present Social Abstract Argumentation, a framework rooted in Dung's Abstract Argumentation that can serve as the underlying backbone of such an online debating tool.

KEYWORDS: debates, ISS algorithm, social abstract argumentation, social web

1. INTRODUCTION

The Web 2.0 (or Social Web) proved extremely successful and its use has become second nature to most of the Internet population. With social networks now widely adopted and their users beating the two billion mark in 2015, the initial boom is over. As social networks become established, the patterns of these new social interactions slowly emerge. It is becoming apparent that many people are growing unsatisfied with the depth (or lack thereof) of interactions on social websites. A growing percentage of users are giving up on the Web 2.0 entirely for lack of intellectually stimulating discussions. Among these we can find experts or enthusiasts who wish to debate some particular topic with people of similar interests, or people who simply want to follow the debate, but who nevertheless have an opinion on what is happening and wish to express it and influence the outcome through some simple interaction mechanisms such as voting.

One of the reasons pointed out is the unstructured, often chaotic, kind of interactions that characterise most of the available Social Networks. Whereas the *trolls* can often be blamed for such chaos, for disrupting interactions with a myriad of intentionally aggressive, racist, uninformative, utterly contradictory or simply irrelevant posts (Torroni, Prandini, Ramilli, Leite & Martins, 2010), existing tools that support interactions should also share part of the blame. Such tools usually force interactions to take the form of a series of comments posted below one another. As the focus of the discussion shifts – not only due to the trolls, but also because current (online) social interactions are no longer dialogues but rather polylogues, with many concurrent topics and lines of reasoning (Lewiński, 2013) – the main topic of discussions and thread of thoughts collected so far is lost.

This prevents a fulfilling experience for those seeking deeper interactions and not just increasing their number of *friends*, *likes*, or similar measures.

There are already some tools that support structured forms of interaction. For example, in debate.org, two users can take opposing sides in debating some issue, following the usual debate rules, while all other users can vote on the winning party according to a set of predefined criteria (e.g. who made more convincing arguments, who had a better conduct, etc.). Despite their merits, websites such as debate.org have several characteristics that limit their adoption in a wide Social Web scale, namely: 1) two (and only two) antagonistic users can engage in a debate, while all remaining users are limited to participate by voting on the winner; 2) the debate structure is very rigid, proceeding in a number of pre-fixed rounds with very strict debate rules that are not known, nor easy to follow, by most; and 3) there are no facilities to reuse arguments and debates. Other existing tools facilitate the representation of structured arguments and argument networks, some even allowing participants to vote on the arguments. Examples include debategraph.org, idebate.org, agora.gatech.edu, and the various tools available at www.arg-tech.org (c.f. Schneider, Groza & Passant, 2013 for a survey).

However, none of these technologies seems adequate to support wide scale social network polylogues, either because they require too much formal knowledge of structured argumentation from its users, or because they fall short of reasoning with the debate data and votes/opinions and only yield very simplistic and naive outcomes.

In order to promote wide scale richer interactions in the form of debates, a Social Network must facilitate:

- More *open participation* where users with different levels of expertise are able to easily express their arguments, even without knowing formal argumentation and any formal rules of debate.

- More *flexible participation* where debates are not restricted to a pair of users arguing for antagonistic sides, but where there may be more than just two sides, more users can propose arguments for each side, and each user is allowed to contribute with arguments for more than one side of the debate.

- More *detailed participation* where users are allowed to express their opinions by voting on individual arguments and on argument relations, instead of just on the overall debate's outcome.

- Appropriate *feedback* to users so that they can easily assess the strength of each argument, taking into account not only the logical consequences of the debate, but also the popular opinion and all its subjectiveness.

1.1 The Envisioned Online Debating System

We envision a self-managing online debating system capable of accommodating two archetypal levels of participation.

On the one hand, experts, or enthusiasts, will be provided with simple mechanisms to specify their arguments and also a way to specify which arguments attack which other arguments. When engaging in a debate, users always propose arguments for specific purposes, like making a claim central to the issue being discussed, or defeating arguments supporting an opposing claim. Thus, the envisioned system should allow users to describe an abstract argument, capable of attacking other arguments, simultaneously with its natural language (or image, video, link, etc.) representation. Therefore, the formal specification of arguments and attacks becomes a natural by-product of the users' intent when proposing new arguments. To make this process as painless and easy as possible, and enable more people to participate, no particularly deep knowledge (such as logics) can be required. It is natural that a new argument might attack a previously proposed argument - indeed, that was likely the object of its creation. However, it is also possible that an older argument attacks the new argument as well. Therefore, the system should allow users to add this new attack relation to the system.

On the other hand, less expert users who prefer to take a more observational role, and do not wish to engage in proposing arguments or attacks, should also be accommodated in the system through a less complex participation scheme. These users may simply read the arguments in natural language (or image, video, link, etc.) and state whether they agree with them. This induces a voting mechanism similar to what is found in current social networks. There are alternatives, such as having argument's social trustworthiness be based on people's opinions of who proposed it. Voting on arguments seems to offer the path of least resistance for being the closest to current social networks. Additionally, it is apparent that not all attacks bear the same weight. Some attacks might have an obvious logical foundation (e.g. undercuts or rebuts), thus gaining trust from the more perceptive users. Other attacks might be less obvious or downright senseless, especially in open online contexts, making users doubt or wish to discard them. Thus, extending the ability to vote to attacks becomes eminently desirable. Not only does voting on attacks more accurately represent a crowd's opinion in a variety of situations, but it also allows the system to self-regulate by letting troll-attacks be "down-voted" to irrelevance.

The system should also be able to autonomously and continuously provide an up to date view of the outcome of the debate e.g. by assigning a value to each argument that somehow represents its social strength, taking the structure of the argumentation framework (arguments and attacks) and the votes into account. A nice GUI e.g. depicting arguments with a size and/or colour proportional to these values would make the debate easier to follow, bringing forward relevant (socially) winning arguments, while downgrading unsound, unfounded (even troll) arguments. So that users may understand and follow a debate, small changes in the underlying argumentation framework and its social feedback (i.e. votes) should result in small changes to the formal outcome of the debate. If a single new vote entirely changes the outcome of a debate, users cannot gauge its evolution and trends, and are likely to lose interest.

Additionally, any debating system as the one envisioned must also ensure that a few crucial properties are satisfied. Online debating systems without the following properties are highly unlikely to be widely adopted by online communities.

- *There should always be at least one solution to a debate.* From a purely logical standpoint, one may consider that some debates simply contain inconsistencies that make it impossible to assign them meaningful semantics. However, we are dealing with the Social Web, where inconsistency is the *norm*. If the system is incapable of providing solutions to every debate, then there is too much risk involved in using it. We believe that most of its users would prefer a system that would, nevertheless, provide them some valuation of the arguments that is somehow justifiable, instead of telling them that the debate is inconsistent.

- *There should always be at most one solution to a debate.* Logicians and mathematicians find it perfectly natural for there to be multiple, or even infinite, solutions to a given problem. However, in a social context as far-reaching as the Internet, it is disingenuous to assume that the general user-base, which likely covers a large portion of the educational spectrum, shares these views with the same ease. It is very hard for someone who has invested personal effort into a debate to accept that all arguments are in fact true (in a multitude of models)!

- *Argument outcomes should thus be represented very flexibly.* In particular, to accurately represent the opinions of thousands of voting users, arguments should be valuated using degrees of acceptability, or gradual acceptability. Two-valued or three-valued semantics risk grossly under-representing much of the user-base.

- *Formal arguments and attacks must be easy to specify.* For example, assuming knowledge of first-order logic for specifying structured arguments would alienate many potential users when the present goal is to include as many as possible. Moreover, simpler frameworks turn implementing and deploying such a system in different contexts (web forums, blogs, social networks, etc.) much easier.

- *Argument strength should be limited by popular opinion, and every vote should count.* In a true social system, there should be no arguments of authority, nor votes without effect. Argument strength can be weaker than its direct support base, since arguments may be attacked by other arguments, but the direct opinion expressed by the votes should act as an upper bound on the strength of the argument. Also, positive (resp. negative) votes should increase (resp. decrease) the strength of the argument/attack on which they are casted (how much can depend on many factors).

- *Computing and updating debate outcomes should be highly efficient.* With the increasing speed of social interactions on the Web 2.0, users would grow impatient if new arguments and votes would not have an almost immediate effect in the debate outcome.

In this paper we discuss Social Abstract Argumentation as a framework which can serve as the underlying formal backbone of an online debating system as the one described above.

Social Abstract Argumentation uses abstract arguments in the sense of Dung (Dung, 1995)[1], but adds the possibility to associate pro and con votes on both arguments and attacks, and a (family of) semantics that goes beyond the classical accepted/defeated valuations assigning each argument a value from an ordered set of values (e.g. the [0,1] real interval, a set of colours, textures, etc.). One particular semantics – based on the popular product T-norm and probabilistic sum T-co-norm and assigning arguments values from the [0,1] real interval – deserves particular attention because of its formal properties, namely guaranteeing the existence and uniqueness of a model, and also because of the existence of an algorithm that can effectively and efficiently compute the debate outcome.

We believe Social Abstract Argumentation provides the theoretical foundations on which to build interaction tools that will provide more robust, flexible, pervasive and interesting social debates than those currently available.

The remainder of the paper is structured as follows: we start by describing the framework and semantics of Social Abstract Argumentation, which we subsequently illustrate with a very simple example, to proceed by discussing some important formal properties and describing the efficient algorithm to compute debate outcomes, before we conclude.

2. SOCIAL ABSTRACT ARGUMENTATION

2.1 Framework and Semantics

We start by describing a Social Abstract Argumentation Framework. First introduced in (Leite & Martins 2011) and later extended in (Egilmez, Martins & Leite, 2014), it is an extension of Dung's AAF, composed of arguments and an attack relation to which we add an assignment of votes to each argument and each attack.

Definition 1 (Social Argumentation Framework). *A social argumentation framework is a triple* $F = \langle \mathcal{A}, \mathcal{R}, \mathcal{V} \rangle$ *where*

---

[1] Abstract Argumentation (Dung, 1995) and Argumentation Theory in general grounds debates in solid logical foundations and has in fact been shown to be applicable in a multitude of real-life situations (c.f. Modgil et al., 2013).

- $\mathcal{A}$ *is a set of arguments,*
- $\mathcal{R} \subseteq \mathcal{A} \times \mathcal{A}$ *is a binary attack relation between arguments,*
- $V : \mathcal{A} \cup \mathcal{R} \to \mathbb{N} \times \mathbb{N}$ *is a total function mapping each argument and each attack to its number of positive and negative votes.*

We use the notion of a semantic framework to aggregate operators representing the several parametrisable components of a semantics:

Definition 2 (Semantic Framework). *A social abstract argumentation semantic framework is a 6-tuple $\langle L, \curlywedge_{\mathcal{A}}, \curlywedge_{\mathcal{R}}, \curlyvee, \neg, \tau \rangle$, where:*
- *L is a totally ordered set with top element $\top$ and bottom element $\bot$, containing all possible valuations of an argument.*
- *$\curlywedge_{\mathcal{A}}, \curlywedge_{\mathcal{R}} : L \times L \to L$ are two binary algebraic operation on argument valuations used to determine the valuation of an argument based on its valuation given by the votes and how weak its attackers are ($\curlywedge_{\mathcal{A}}$), and to determine the strength of an attack given the votes on the attack and the valuation of the attacking argument ($\curlywedge_{\mathcal{R}}$).*
- *$\curlyvee : L \times L \to L$ is a binary algebraic operation on argument valuations used to determine the valuation of a combined attack;*
- *$\neg : L \to L$ is a unary algebraic operation on argument valuations used to determine how weak an attack is.*
- *$\tau : \mathbb{N} \times \mathbb{N} \to L$ is a vote aggregation function which produces a valuation of an argument based on its votes.*

The definition of a semantic framework imposes very little on the behaviour of the operators. As such, many specific semantic frameworks could result in systems whose behaviour would be far from intuitive – a semantic framework where an increase in the strength of the attacking arguments would result in an increase in the strength of the attacked argument would make little sense. There are several basic properties that the operators should obey so that the resulting semantics is adequate for its purpose. For example, $\neg$ should be antimonotonic, continuous, $\neg \bot = \top$, $\neg \top = \bot$ and $\neg \neg a = a$; $\curlywedge_{\mathcal{A}}, \curlywedge_{\mathcal{R}}$ should be continuous, commutative, associative, monotonic w.r.t. both arguments and have $\top$ as their identity element; $\curlyvee$ should be continuous, commutative, associative, monotonic w.r.t. both arguments and have $\bot$ as its identity element; and τ should be monotonic w.r.t. the first argument and antimonotonic w.r.t. the second argument.

Continuity of operators guarantees small changes in the social inputs result in small changes in the models. Were this not the case, outcomes of debates would be very unstable, hard to follow and more easily exploited by trolls. The remaining algebraic properties simply state that the order in which arguments are attacked makes no difference; that an argument's valuation is proportional to its crowd support; that aggregated attacks are proportional to the attacking arguments; and so forth.

Notice also that the valuation set L of arguments is parametrisable. L could be $[0, 1] \subseteq \mathbb{R}$, but it could also be any finite, countable or uncountable set of values such as booleans, colours, textures, or any other set that is deemed appropriate for users of the final application, so long as it is totally ordered.

One particular semantic framework has received great attention because of its properties. It uses a simple vote aggregation function and is based on the well-known product T-norm and probabilistic sum T-conorm. It is dubbed the Simple Product Semantics as is defined as follows:

Definition 3 (Simple Product Semantics). *A simple product semantic framework is $S_\varepsilon = \langle [0,1], \wedge^{\cdot}, \wedge^{\cdot}, \vee^{\cdot}, \neg, \tau_\varepsilon \rangle$, where*
- $x \wedge^{\cdot} y = x \cdot y$, *i.e. the product T-norm*,
- $x \vee^{\cdot} y = x + y - x \cdot y$, *i.e. the T-conorm dual to the product T-norm*,
- $\neg x = 1 - x$,
- $\tau_\varepsilon(v^+, v^-) = \frac{v^+}{v^+ + v^- + \varepsilon}$, *with $\varepsilon > 0$.*[2]

The heart of the semantics is in the definition of a model, which combines the operators of a semantic framework S into a system of equations, one for each argument, which must be satisfied.

Definition 4 (Social Model). *Let $F = \langle \mathcal{A}, \mathcal{R}, \mathcal{V} \rangle$ be a social abstract argumentation framework and $S = \langle L, \wedge_\mathcal{A}, \wedge_\mathcal{R}, \vee, \neg, \tau \rangle$, a semantic framework. A total mapping $M : \mathcal{A} \to L$ is a social model of F under semantics S, or S-model of F, if*

$$M(a) = \tau(a) \wedge_\mathcal{A} \neg \bigvee_{a_i \in \mathcal{R}^-(a)} \left(\tau((a_i, a)) \wedge_\mathcal{R} M(a_i) \right) \qquad \forall a \in \mathcal{A}$$

where $\mathcal{R}^-(a) \triangleq \{a_i : (a_i, a) \in \mathcal{R}\}$, $\tau(x) \triangleq (v^+, v^-)$ whenever $V(x) = (v^+, v^-)$ and $\bigvee\{x_1, x_2, \ldots, x_n\} \triangleq ((x_1 \vee x_2) \vee \ldots \vee x_n)$. We refer to $M(a)$ as the social strength, or value, of a in M, dropping the reference to M whenever unambiguous.

---

[2] The meaning of ε is explained in (Leite & Martins 2011) and, in practice, it should be a sufficiently small value with no significant influence in result of the voting aggregation function.

Each equation encodes the contribution of votes and attacks to the social strength of an argument, for which we now proceed to provide further intuition.

Whenever an argument a_i attacks another argument a, then the strength of the attack is the valuation of the attacking argument a_i reduced by the social support of the attack: no argument's attack is stronger than either its own valuation or the social support of the attack itself. We use $\wedge_{\mathcal{R}}$ to restrict these values.

$$\tau((a_i, a)) \wedge_{\mathcal{R}} M(a_i)$$

As an argument may have multiple attackers, all of their attack strengths must be aggregated to form a stronger combined attack value, using operator $\curlyvee$.

$$\bigvee_{a_i \in \mathcal{R}^-(a)} \left(\tau((a_i, a)) \wedge_{\mathcal{R}} M(a_i) \right)$$

The above equation results in a combined attack strength that must be turned into a restricting value, representing how permissive or weak the attack is, using the $\neg$ operator.

$$\neg \bigvee_{a_i \in \mathcal{R}^-(a)} \left(\tau((a_i, a)) \wedge_{\mathcal{R}} M(a_i) \right)$$

In a social context where the crowd has given its direct opinion on argument a through the votes, it seems clear that a's valuation should never turn out higher than a's social support $\tau(a)$. Thus, an argument's valuation is given by restricting $\tau(a)$ with the value of the aggregated attack using the final operator $\wedge_{\mathcal{A}}$.

$$\tau(a) \wedge_{\mathcal{A}} \neg \bigvee_{a_i \in \mathcal{R}^-(a)} \left(\tau((a_i, a)) \wedge_{\mathcal{R}} M(a_i) \right)$$

Throughout the remainder of the paper, S will stand for the Simple Product Semantics.

2.2 Illustrative Example

Consider a social interaction inspired by (Walton, 2009) where several participants, while arguing about the role of the government in what banning smoking is concerned, set forth the arguments and attack relations depicted in Figure 1a).

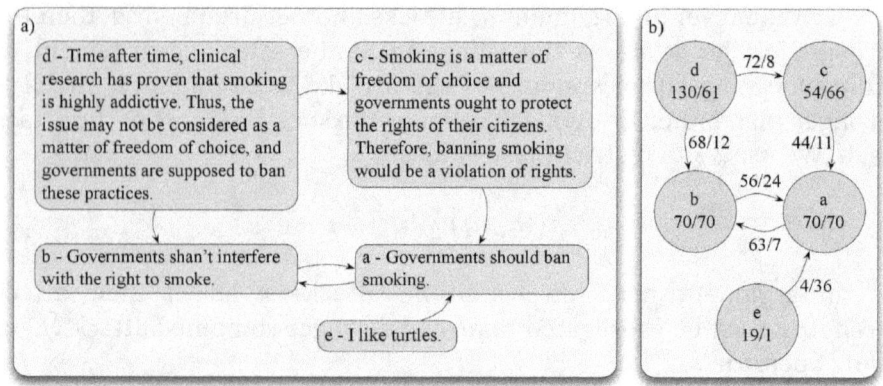

Figure 1: Social Argumentation Framework: a) arguments and attacks; b) votes

Note that these arguments are structurally different: a and b are unsupported claims, c and d contain multiple premises and a conclusion, while e, despite being rather consensual (who doesn't like turtles?), seems to be totally out of context and can hardly be seen as an attack on a (here, the attack by e on a is meant to represent a troll attack). Our goal is to show that SAFs' level of abstraction allows meaningful arguments to be construed out of most participations – in fact, with suitable GUIs, arguments could even be built from videos, pictures, links, etc. – while the participation through voting will help deal with mitigating the disturbing effect of unsound arguments and poorly specified (troll) attacks.

After a while, the arguments and attacks garner the pro/con votes depicted in Figure 1b). Arguments a and b obtain the same direct social support as expressed by the 70 *pro* and *con* votes. Meanwhile, a's attack on b is deemed stronger than its counterpart, judging from their votes. One might speculate that this is a consequence of a delivering a more direct message. Whereas argument c does not get much love from the crowd (a vote ratio of 54/66), its attack on a is still supported by the community (44/11). Perhaps initially there was a better sentiment towards c but the introduction of d, which amassed a decent amount of support itself (130/61), turned the odds against c. Both of d's attacks on b and c materialise to be strong enough, the former being slightly weaker (72/8 versus 68/12). Lastly, argument e received just a mere number of votes, most being positive (19/1). However, there seems to have been a significant effort from the users on discrediting the attack on a by e (4/36). Note that e is a perfectly legitimate argument. Indeed the crowd endorses the fondness for turtles – it's the attack, not the argument, that is not logically well-founded.

With the abstract argumentation framework and the votes on arguments and attacks in hand, we can turn our attention to the valuation of the arguments.

If we consider the social support of each argument, i.e. its value considering only the votes it obtained while ignoring attack relations, we obtain the following values:[3] $\tau(a) = 0.50$, $\tau(b) = 0.50$, $\tau(c) = 0.45$, $\tau(d) = 0.68$ and $\tau(e) = 0.95$, as depicted in *Figure 2*a) (where the size of each node is proportional to its value).

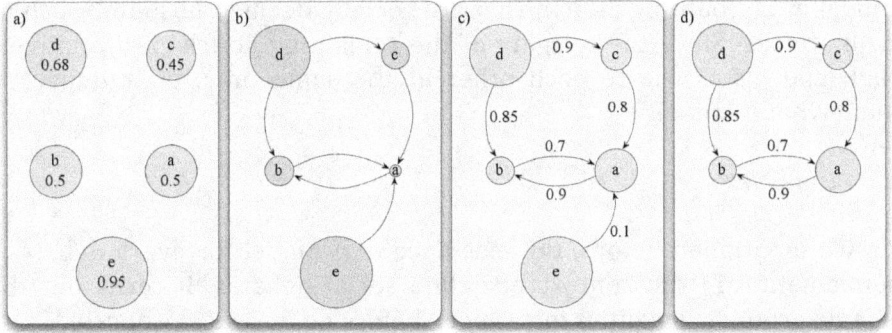

Figure 2: Model of the Social Abstract Argumentation Framework: a) considering social support only; b) considering attacks but not their strength; c) considering attack strength; d) considering attack strength, without argument e.

The original Social Abstract Argumentation semantics (Leite & Martins 2011), which considers attacks between arguments but not the votes on attacks, assigns the following values to arguments: $M(a) = 0.02$, $M(b) = 0.16$, $M(c) = 0.14$, $M(d) = 0.68$ and $M(e) = 0.95$, as depicted in *Figure 2*b). As expected, d and e retain their initial social support values, since they are not attacked, while the remaining arguments see a decrease in their social support value. Argument a decreases the most while b and c maintain a reasonable fraction of their initial strength. Since two of a's attackers – b and c – are attacked by d, which is a non-attacked argument with strong social support, their value is weakened, so their effect on a is lessened. Thus, we can conclude that the main cause for the downfall in a's value is e's attack.

---

[3] We will consider the Product Semantics as in Definition 3, with a neglectable low ε.

We can now turn our attention to the model that also takes votes on attacks into consideration, which assigns the following values to arguments: $M(a) = 0.35$, $M(b) = 0.14$, $M(c) = 0.17$, $M(d) = 0.68$ and $M(e) = 0.95$, as depicted in *Figure 2*c). The value assigned to a by the model increases from 0.02 to the more plausible level of 0.35, mostly due to e's weakened capability to attack a. Indeed, the crowd's overwhelming con votes on the (troll) attack of e on a essentially neutralised it. To confirm, we compare it with the model obtained if argument e was simply removed, depicted in *Figure* 2d), whose valuations of $M(a) = 0.39$, $M(b) = 0.14$, $M(c) = 0.17$, $M(d) = 0.68$ are very similar to those obtained in the presence of e but with a very weakened attack on a, which allows us to conclude for the success of the model in discounting attacks that are socially deemed unsound, such as troll attacks. Since the weights of the remaining attacks are relatively high and also close to each other at the same time, their impact is somewhat minimal.

2.3 Properties

As we mentioned before, the existence of a model for every debate is fundamental in the context of online social debates. It turns out the simple product semantics introduced before enjoys such property.

> **Theorem 1.** *Every social abstract argumentation framework F has at least one S-model.*

The proof of this theorem rests on the use of Brouwer's fixed point theorem and the continuity of the operators used in the simple product semantics.

Also important is that every debate has at most one model. A formal proof of such result has proved harder than initially anticipated, even though we strongly believe in the following conjecture:[4]

> **Conjecture 1.** *Every social abstract argumentation framework F has at most one S-model.*

It turns out that these results hold for a much larger class of semantics, e.g. those based on other Archimedean T-norms and T-conorms. This will permit the development of concrete debating systems tailored to

---

[4] We have a proof, based on Banach's fixed point theorem, for a weaker version with an additional condition, but opted to state the general conjecture given our strong belief in its truth.

specific applications that require, for example, that arguments be more resistant to attacks, or its opposite i.e. that a high social strength be only assigned to those arguments that have no more than very weak attacks.

2.4 Algorithm

The problem of finding a model according to the simple product semantics can be cast to the problem of finding a solution of a nonlinear system where variables represent the arguments and equations encode their attacks, with the following generic form:

Definition 5. *A Social Abstract Argumentation System is a square nonlinear system with n variables $\{x_1, x_2, ..., x_n\}$ and n equations:*

$$x_i = \tau_i \prod_{j \in A_i} (1 - \tau_{ji} x_j) \qquad 1 \leq i \leq n \qquad (1)$$

where $\tau_i, \tau_{ji} \in \,]0, 1[\,$ and $A_i \subseteq \{1, ..., n\}$.[5]

Contrary to the linear case, systems of nonlinear equations cannot be solved exactly using a finite number of elementary operations. Instead, iterative algorithms are usually used to generate a sequence $(x^{(k)})_{k \in \mathbb{N}_0}$ of approximate solutions. These algorithms start with an initial guess $(x^{(0)})$ and, to generate the approximating sequence, follow an iteration scheme of the form $x^{(k+1)} = g(x^{(k)})$ where the fixed-points for g are solutions x^* of the nonlinear system.

The success of iterative algorithms depend on their convergence properties. Given a domain of interest, an iterative method that converges for any arbitrary initial guess is called globally convergent. If convergence is only guaranteed when the initial approximation is already close enough to the solution, the algorithm is called locally convergent. In the case of Social Abstract Argumentation Systems the domain of interest is $]0, 1[^n$ thus the iterative algorithm must converge to a solution $x^* = (x_1^*, ..., x_n^*) \in \,]0, 1[^n$.

Two classical algorithms that can be used to approximate the solution of such a system are the Iterative Fixed-Point Algorithm (IFP)

---

[5] We can exclude $\tau_i, \tau_{ji} \in \{0, 1\}$ because arguments and attacks (x) with $\tau(x) = 0$ have no effect in the system while $\tau(x) = 1$ can never occur, according to the simple product semantics in Def. 3, because $\epsilon > 0$.

where the iteration scheme is directly obtained from the equations (1), and the Iterative Newton-Raphson Algorithm (INR). [6]

Unfortunately, IFP is only locally convergent and often divergent, even for systems with a reduced number of variables, while INR, also only locally convergent, requires the computation of a Jacobian matrix at each iteration, which is prohibitive for large systems.

Based on the Iterative Successive Substitutions Algorithm (ISS) previously proposed for Social Abstract Argumentation frameworks without votes on attacks (Correia, Cruz & Leite, 2014) – itself an adaptation of the Gauss-Seidel method for systems of nonlinear equations –, here we present an adaptation to also admit votes on attacks.

Definition 6 (ISS). *The ISS algorithm uses the iteration rule:*

$$x_i^{(k+1)} = \tau_i \prod_{j<i, j \in A_i} (1 - \tau_{ji} x_j^{(k+1)}) \prod_{j \geq i, j \in A_i} (1 - \tau_{ji} x_j^{(k)}) \qquad (2)$$

From the initial guess $x^{(0)}$, elements of $x^{(k+1)}$ are computed sequentially using forward substitution until the stoping criterion is attained.

Following a similar strategy as (Correia, Cruz & Leite, 2014), we can prove the global convergence of the algorithm.

Preliminary results show that the algorithm performs as well as its original version (Correia, Cruz & Leite, 2014). For example, it is able to approximate the model of debates with 5000 arguments and an attack density of 0.1 (i.e. 10% of all pairs of arguments are related through an attack) in well under 1 second. [7] A thorough analysis of the original ISS algorithm can be found in (Correia, Cruz & Leite, 2014). Additionally, just as with the original ISS, we can exploit the structure of the debate to obtain considerable gains in efficiency.

3. CONCLUSION

We believe that *Social Argumentation Frameworks* lay the theoretical foundations for a deeper, more serious social web. By using abstract arguments and relying on users to specify both arguments and attacks,

---

[6] A comprehensive treatment of methods for solving nonlinear systems of equations with some recent developments on iterative methods can be found in (Ortega & Rheinboldt, 2000) and (Argyros & Szidarovszky, 1993).

[7] The higher the attack density, the slower the convergence of the algorithm. However, an attack density of 0.1 seems to be a rather high value.

it can easily be used by just about anyone, without the requirement for specific knowledge about logic or formal argumentation. By using graphs to structure arguments, it directly supports interactions in the form of polylogues instead of just dialogues. By providing debates with formal, justifiable and yet subjective outcomes, it counteracts the growing trend of superfluous discussion. By providing gradual valuations of arguments, instead of simply label them as accepted/defeated, it widens the potential user base to also include those who could be turned away simply because their minority opinions would be systematically defeated. By providing a voting mechanism both on arguments and attacks, it gives users an additional direct tool to down-vote irrelevant arguments and (troll) attacks, increasing the focus on what is relevant.

A system built on top of *Social Argumentation Frameworks* would maintain a detailed, reusable knowledge-base, and provide the infrastructure for more open, flexible debates than current systems allow. Furthermore, we prove properties that guarantee consistent, understandable feedback, thus facilitating the adoption of the envisioned system by people with a serious interest and experts alike.

Before that can happen, an appropriate GUI needs to be developed. Despite the problems exhibited by existing tools that force interactions to take the form of series of comments posted below one another, it is still the prevailing technology. To replace it with a new different mode of interaction, moving from temporally ordered sequences of comments to graphs of comments related through an attack relation, will require a very intuitive, simple to use GUI – both to specify the arguments and attacks as well as to observe the output and navigate through the graph. In all Social Web success stories, simplicity is the key, and the interface can make or break any novel idea, so its development is not an easy task. Additionally, investigation regarding the appropriate choice of semantic framework is required, which cannot be done in a purely analytical way, and will require the involvement of human users. For example, we can change the vote aggregation function either by employing a different way of aggregating the votes altogether, or by simply adjusting the value of ε in the *Simple Vote Aggregation*. The larger the value of ε, the smaller the effect the initial positive votes will have in the social value of an argument, which can be useful to prevent *trolls* from creating arguments with just one positive vote, but capable of causing great harm to those by them being attacked. We can also change the ∧, ∨, and ¬ operators and, instead of the *Product T-Norm* and the *Probabilistic Sum T-CoNorm*, use other Archimedean operators. For example, using the *Hamacher Product T-Norm* and its corresponding T-coNorm we would obtain a semantics that would diminish the effects of attacks when compared with the *Product T-Norm* while maintaining good theoretical and computational properties.

The framework can also be extended in several ways.

Some authors have advocated the addition of a support relation between arguments (e.g. Amgoud, Cayrol, Lagasquie-Schiex & Livet, 2008; Boella, Gabbay, Van Der Torre & Villata, 2010). Whereas there has been a debate regarding the adequacy of such relation – some argue that since arguments are accepted by default, any support should take the form of an attack on its attackers – its incorporation into Social Abstract Argumentation might prove beneficial, and certainly worth future investigation.

Another possible extension is to consider more general *Vote Aggregation Functions* that take into account other variables to produce the social value of the argument. These additional variables could be, for example, the total number of votes in the debate to include some measure of robustness of each argument (e.g. what should happen with two mutually attacking arguments with 10/10 and 1000/1000 votes?).

Also of interest is to consider attacks on the attack relations themselves, in line of what (Modgil, 2009) has done for acceptability based semantics on top of (Dung, 1995).

There is yet much to be done before we can witness a real change towards more structured, richer, wide scale interactions in social networks. We believe that part of the road to be travelled is in the form of developing appropriate tools that, while very simple to use, support a shift from the old-fashioned dialogues to the new multi-party polylogues. We believe Social Abstract Argumentation to be part of one step in that direction.

ACKNOWLEDGEMENTS: J. Leite and S. Egilmez were partially supported by FCT project ERRO – PTDC/EIA-CCO/121823/2010.

REFERENCES

Amgoud, L., Cayrol, C., Lagasquie-Schiex, M. C., & Livet, P. (2008). On bipolarity in argumentation frameworks. *International Journal of Intelligent Systems*, *23*(10), 1062–1093.

Argyros, I. K., & Szidarovszky, F. (1993). *The theory and applications of iteration methods. Systems engineering.* Boca Raton, FL: CRC Press.

Boella, G., Gabbay, D. M., Van Der Torre, L., & Villata, S. (2010). Support in abstract argumentation. In *Procs. of COMMA'10*, volume 216 of *FAIA*. IOS Press, Amsterdam.

Correia, M., Cruz , J., & Leite, J. (2014). On the efficient implementation of social abstract argumentation. In *Procs. of ECAI'14*, volume 263 of *FAIA*. IOS Press, Amsterdam.

Dung, P. M. (1995). On the acceptability of arguments and its fundamental role in nonmonotonic reasoning, logic programming and n-person games. *Artificial Intelligence, 77*(2), 321–358.

Egilmez, S., Martins, J., & Leite, J. (2014). Extending social abstract argumentation with votes on attacks. In *Procs. of TAFA'13*, volume 8306 of *LNCS*. Springer, Heidelberg.

Leite, J., & Martins, J. (2011). Social abstract argumentation. In *Procs. of IJCAI 2011*. IJCAI/AAAI.

Lewiński, M. (2013). Debating multiple positions in multi-party online deliberation: Sides, positions, and cases. *Journal of Argumentation in Context, 2*(1), 151–177.

Modgil, S. (2009). Reasoning about preferences in argumentation frameworks. *Artificial Intelligence, 173*(9-10), 901–934.

Modgil, S., Toni, F., Bex, F., Bratko, I., Chesnevar, C. I., Dvořák, W., Falappa, M. A., Fan, X., Gaggl, S. A., García, A. J., González, M. P., Gordon, T. F., Leite, J., Možina, M., Reed, C., Simari, G. R., Szeider, S., Torroni, P., & Woltran, S. (2013). The added value of argumentation. In S. Ossowski, (Ed.), *Agreement Technologies*, volume 8 of *Law, Governance and Tech. Series*, (pp. 357–403). Heidelberg: Springer.

Ortega, J. M., & Rheinboldt, W.C. (2000). *Iterative Solution of Nonlinear Equations in Several Variables*. Philadelphia, PA: Society for Industrial and Applied Mathematics.

Schneider, J., Groza, T., & Passant, A. (2013). A review of argumentation for the social semantic web. *Semantic Web, 4*(2), 159-218.

Torroni, P., Prandini, M., Ramilli, M., Leite, J., & Martins, J. (2010). Arguments against the troll. In *Procs. of AI*IA'10* (pp. 232–235). Brescia: Arti Grafiche Apollonio.

Walton, D. (2009). Argumentation theory: A very short introduction. In I. Rahwan & G. R. Simari (Eds.), *Argumentation in Artificial Intelligence* (pp. 1–22). Dordrecht: Springer.

Commentary on Leite, Martins and Eğilmez's Towards an Online Social Debating System

MICHAEL H.G. HOFFMANN
School of Public Policy, Georgia Institute of Technology, USA
m.hoffmann@gatech.edu

1. INTRODUCTION

The paper by João Leite, João Martins, and Sinan Eğilmez addresses an important problem: the low quality of much of online debate. Based on Dung's Abstract Argumentation framework, they develop what they call a Social Abstract Argumentation framework. The main innovation compared to Dung is that they add a formal description for software that would allow the "valuation of arguments." To put it simple: the basic idea is that voting on arguments can improve the quality of online debating.

    I am not a computer scientist, so I leave the debate on the abstract framework that the authors propose to the experts. As a philosopher, I would like to discuss instead four more general problems that I see in the proposal. The first problem concerns the concept of "argument" used here which I find very problematic. The second problem is that there seems to be a trade-off between two goals that Leite, Martins, and Eğilmez want to achieve: the goal to decrease the effects of trolls and the goal to give minority opinions a voice. I would argue that both goals exclude each other in their proposal. The third problem might be the most fundamental. I will raise the question whether it is really a good idea to let people vote on arguments.

2. WHAT IS AN ARGUMENT?

I was worried when I saw statements such as "Governments should ban smoking" and "I like turtles" described as "perfectly legitimate argument(s)." There is indeed a lot of debate among argumentation theorists on how to define exactly what an "argument" is, and if we look at how the concept is used in particular among those who develop argument or debate software, the variety is even larger. But as far as I can see, Leite, Martins, and Eğilmez are the first who are willing to give up completely any reference to the idea of justification or supporting claims by reasons. This, however, leads to the impossibility to distinguish between "arguments" and "opinions." I think it makes a huge difference whether we talk about the quality of arguments in the sense of reason-conclusion complexes or whether we poll people's opinions about certain statements and attack relations among statements.

2. TROLLS VERSUS MINORITIES

Leite, Martins, and Eğilmez put a lot of emphasis on the damaging role of trolls in online debate and how the effects of destructive behaviour can be controlled. Voting is obviously a good way to achieve this goal, because if other users vote down contributions by trolls they will be ranked very low and, thus, disappear from the main focus of a debate. However, exactly the same will most likely happen with the contributions of minorities. By definition, a minority position is a position that is not shared by the majority so that there is nothing in a voting-based system that would prevent that minority positions are also ranked very lowly and disappear, thus, from the main focus of attention. Therefore, there seems to be a trade-off between the goal to decrease the effects of trolls and the goal to give minority opinions a voice. It seems very unlikely—at least in the framework provided by Leite, Martins, and Eğilmez—that you can have both.

3. IS VOTING ON ARGUMENTS A GOOD IDEA?

Opinion polls are certainly a good idea if the goal is to figure out what people think. So, if we do not really distinguish opinions from arguments, the question above is already answered. However, things might look differently if we refer to an understanding of "argument" that at least in philosophy seems to be widely shared. In one of many similar forms it defines an argument as a premise-conclusion sequence so that either one or more premises are intended to support a conclusion or a conclusion is intended to be justified by one or more premises (Hoffmann, 2015; Hitchcock, 2007). Based on this definition it seems to be clearly not an argument if somebody writes "I like turtles" in relation to the statement "Governments should ban smoking." There is obviously no intention to support anything, and no voting is required to figure that out. In cases like these it might be better to have a policy that would allow the administrator of a debate system to remove all contributions of a certain user in case of destructive behaviour. Such an approach would be better than voting because—as I argued above—voting has the tendency to exclude minority positions as well.

My personal opinion on voting is that it depends on the goal you want to achieve. If the goal is to figure out what people think about the quality of arguments, then voting is a good idea. But there are also other goals. Traditionally, arguments are mainly used to persuade other people by good reasons. We want that people change their minds, and if we participate in debate, we should also be ready to change our position based on good arguments. But I would certainly not change my mind based on what other people think. For me, voting makes sense only with regard to the measurement of opinions and everything that is not justified by reasons. If a statement is justified by reasons, then it is much better to let people look at the argument—that is, the relation between reasons and conclusion—and see whether they change their mind based on that argument.

REFERENCES

Hitchcock, D. (2007). Informal logic and the concept of argument. In D. Jaquette (Ed.), *Philosophy of Logic* (pp. 101-129). Amsterdam: Elsevier.

Hoffmann, M. H. G. (2015). Reflective Argumentation: A Cognitive Function of Arguing. Argumentation, 1-33. doi: 10.1007/s10503-015-9388-9.

18

How to Conclude Practical Argument in a Multi-Party Debate: A Speech Act Analysis

MARCIN LEWIŃSKI
ArgLab, Universidade Nova de Lisboa, Portugal
m.lewinski@fcsh.unl.pt

> In this paper I analyse various speech acts which can conclude a practical argument in a multi-party debate (*argumentative polylogue*). To this end, I offer a detailed scheme of practical argument suitable for an external pragmatic account (rather than internal cognitive). Speech acts concluding practical argument – promises, vows, advice, proposals, and others – differ chiefly depending on the agent of action (me, us, you, them) and the conclusion's illocutionary strength.
>
> KEYWORDS: argumentative polylogue, illocution, practical argument, practical reasoning, Searle, speech acts

1. INTRODUCTION

Practical reasoning and theoretical reasoning are typically defined as, respectively, reasoning about what to do and about what to believe. Yet, the innocent "about" might be quite misleading here, and in a dual sense. The distinction does not in fact pertain to the *content* of the *premises* which we are reasoning about, such as when we reason (whether practically or theoretically) "about" the elections in Poland or the corruption in FIFA. Rather, it refers to the *function* of the *conclusion* of reasoning. We thus reason "about" what we conclude we should do about the elections or what we conclude we can believe about FIFA. In this straightforward functionalist approach, the analysis of the function of the conclusion of reasoning is, by definition, crucial. This is my task for this paper. In what follows, I will scrutinise the ways we can conclude our practical arguments in a public debate, especially a multi-party debate (*polylogue*). To this end, I will employ the basic framework of the speech act theory.

2. WHAT CONCLUDES PRACTICAL REASON

2.1 Attitudes or acts of a reasoning agent

Philosophical accounts of practical reasoning (henceforth: PR) are still dominated by a first-person perspective of a single reasoning agent – rather than by a communicative approach grounded in the theory of argument (Habermas, 1984).[1] Accordingly, the philosophical discussion over how to conclude PR revolves around the issue of the nature of the propositional attitude, or intentional state, which properly concludes PR. The conclusion is a result of reasoning from other states (premises) such as desires/intentions and beliefs (Figure 1).

Figure 1: The basic scheme of practical reasoning

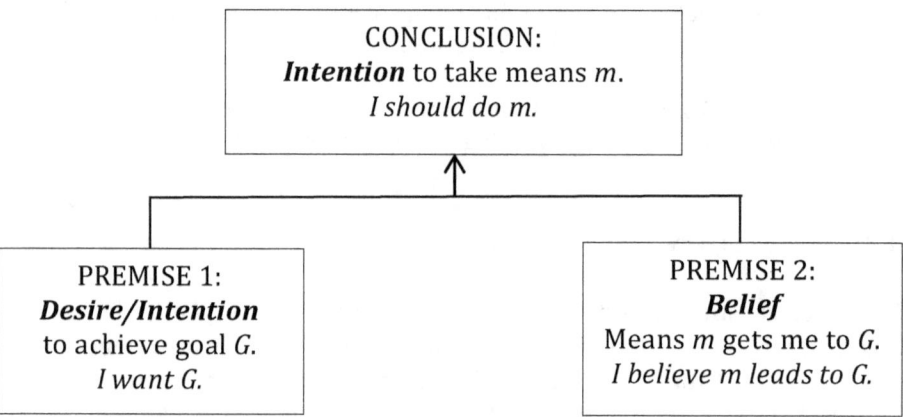

According to Searle, there is "a bewildering variety" of accounts of the elements of PR: they can be "desires, intentions, fiats, imperatives, norms, *noemata*, actions" (2001, p. 242) – and many of these can feature as PR's conclusions. From the weakest to the strongest conclusion, one can recognise the following continuum:

- disposition to act   (*pro-attitude, secondary desire*)
- decision to act   (*prior intention*)
- intention to act   (*intention-in-action*)
- action itself

---

[1] There are, of course, important nuances: internal first-person approaches (e.g., Pollock, 1995), cannot be equated with external first-person approaches (Searle, 2001), which in turn are different from external communicative approaches (Grice, 1957; see Searle, 2014).

The crucial difference of opinion exists between those who think action itself is the proper conclusion of PR (Aristotle, *EN*; Searle, 2001, p. 136) and those who think this is ridiculous, for reasoning is limited to propositional entities and cannot extend beyond them – therefore an intention to act is as far as we can get (e.g., Broome, 2013).

Clearly, we are dealing here with a single reasoning agent pondering over the right course of action for her to take. This would be perfectly fine if the extrapolation from the simplest unit of individual reasoning to various forms of collective reasoning was warranted. But it seems it is not. According to David Hitchcock (2002), an individualistic approach is at risk of producing a "solipsistic, egoistic and unsocial" understanding of PR. Referring to Pollock's (1995) account of PR where the basic scheme of *Beliefs*, *Desires* and *Intentions* is supplemented by a reasoning agent's *Likings*, Hitchcock describes it as solipsistic, since "there is no provision for verbal input from, or verbal output to, other autonomous rational agents, still less for back-and-forth discussion, whether argumentative or non-argumentative" (2002, p. 254). Further, "it is egoistic, in that the function of the entire system is to make the world more to the liking of that system itself" (2002, p. 254). As a result, "nothing [...] permits rational criticism" (2002, p. 255) of an agent's hierarchy of desires and likings. Finally, the "model is unsocial, in that his [Pollock's] rational agent does not (and cannot) belong to any groups of autonomous rational agents with governance structures for making decisions about the actions of the group" (2002, p. 255). As Hitchcock concludes, "[a] comprehensive theory of good practical reasoning would have to remedy all three of these lacks" (2002, p. 255).

2.2 The speech act of proposal

A good starting point towards such a remedy is a shift in the agent of the conclusion from an individual "I" to plural "we". In this case, crucially, the conclusion of PR – "so *I* should do *m*" – would be reformulated to "so let *us* do *m*".[2] This shifts the focus away from *the internal propositional attitude of intention* to *the externalised speech act of proposal*.[3]

---

[2] In Walton's formulation, the conclusion of PR in "multi-agent deliberation" is a "practical ought-statement": *We* ought to do it (2006, p. 204).

[3] Notice, though, that Broome, somewhat inconsistently, also speaks of speech acts which the reasoner performs to herself: "the speech-act you perform is the act of expressing an attitude of yours" (Broome, 2013, p. 253). As we know from the Speech Act Theory, speech acts do much more than just express (propositional) attitudes, such as intentions and desires. Most importantly,

Accordingly, the analyses of proposals have attracted much attention in argumentation theory (Aakhus, 2006; Ihnen Jory, forth.; Kauffeld, 1998; Walton, 2006).

Importantly, proposals straightforwardly connect PR to an argumentative activity of deliberation, a link stressed since Aristotle (see Dascal, 2005). Aakhus analyses proposals in deliberation as speech acts located between Searle's (1969, 1975) *commissives* (such as promises) and *directives* (such as requests) (see Table 1). Commissives are about future acts of the Speaker who, in performing the speech act, commits her/himself to this act ("I will clean the room tomorrow"). Directives are about future acts of the Hearer, whom the speaker wants to get to do something ("Clean the room tomorrow, will you?"). Proposals concern future acts of *both* the Speaker and the Hearer, and their illocutionary point is "to enlist H[earer] in mutually bringing about [act] A" (Aakhus, 2006, p. 406). They would thus be typically expressed by constructions such as "Let's (clean the room tomorrow)!" or "How about we (clean the room tomorrow)?"

According to Aakhus, "[w]hen proposing, a speaker puts forward a future act that requires a joint performance by the speaker and hearer" (2006, p. 405) and, additionally, "the speaker frames the proposed actions as mutually beneficial" (2006, p. 404). In this way, proposing is a speech act through which the conclusion of PR is put forward for consideration in the argumentative activity of deliberation: "A proposer (P) puts forward the proposal in part to get agreement but also to test for doubts and objections [...] that may in turn help P design a more acceptable proposal" (Aakhus, 2006, p. 406). Therefore, proposing belongs to this kind of illocutionary acts in which "speakers necessarily or typically incur probative burdens", that is, "a speaker cannot, other things being equal, responsibly dismiss an addressee's demands for proof" (Kauffeld, 1998, p. 247). What follows is that felicitous proposals concern actions which are: 1) communicated and open for discussion, thus surely not solipsistic; 2) mutually beneficial rather than purely egoistic; 3) jointly performed, and therefore social. In this way, the analysis of proposals addresses all three concerns regarding individualistic approaches to PR identified by Hitchcock. However, as I will argue below – even if a paradigmatic case – a proposal is only one of the possible speech acts which can convey the conclusions of PR.

---

they are communicative, rather than purely mental, acts which therefore always involve at least two parties: the Speaker and the Hearer.

Table 1: The felicity conditions for requesting, proposing, and promising (Aakhus, 2006, p. 406)

Act	Request (Searle, 1969)	**Propose**	Promise (Searle, 1969)
Propositional Content	Future act A of H.	**Future act A of H+S.**	Future act A of S.
Preparatory Condition	H is able to do A. S believes H is able to do A. It is not obvious to both S and H that H will do A in the normal course of events of his own accord.	**H and S are able to contribute to the accomplishment of A. It is not obvious to both S and H that either S or H can do A of their own accord in the normal course of events. That A will leave neither S nor H worse off than not doing A.**	S is able to do A. S believes S is able to do A. It is not obvious to both S and H that S will do A in the normal course of events of his own accord.
Sincerity Condition	S wants H to do A.	**S believes A will mutually benefit H and S or that if it benefits S it will leave H no worse off.**	S intends that in uttering to do A he is under the obligation to do A.
Essential Condition	Counts as an attempt to get H to do A.	**Counts as an attempt to enlist H in mutually bring-ing about A.**	Counts as an attempt to commit S to do A.

2.3 Proviso: Any speech act can conclude PR

All speech acts, including argumentative speech acts, are, well, acts. They are intentionally performed human acts, based on some kind of linguistic, cultural and societal conventions (Austin, 1962; Strawson, 1964; Searle, 1969). As such, while they may be performed without profound deliberation – think of common expressives such as "Ouch!" or

"Sorry!" – speech acts typically result from some prior judgment. That is, we need to practically reason, inside of us, to perform this and no other speech act in this very situation. In any communicative activity, we thus constantly conclude our internal deliberations with a conclusion "I should say X now" or "I should perform speech act of the kind Y (apologise, deny, object to, approve)." In this respect, there is similar PR behind commissive speech acts such as "So I shall catch the 2:30 train to London" and assertions such as "So the cat is on the mat."

In Searle's worlds:

> There is thus a sense in which all reasoning is practical, because it all issues in doing something. In the case of theoretical reason, the doing is typically a matter of accepting a conclusion or hypothesis on the basis of argument or evidence. Theoretical reason is, thus, a special case of practical reason. (Searle, 2001, pp. 90-91)

For Searle, this is the consequence of all speech acts being acts which in their very performance bring about certain commitments on the speaker. And a commitment to defend the truth of a "theoretical" statement is not so much different from a "practical" commitment to fulfil a promise or offer: in both cases, when prompted, one has to produce "argument or evidence." Already for Austin (1962), the fundamental initial distinction between "constatives" and "performatives" eventually collapses in the realm of various illocutionary *acts* defined through their "happiness conditions." Quite tellingly, pragma-dialecticians face a similar complication. While the speech acts of accepting and non-accepting a standpoint at the confrontation stage of a critical discussion are consistently classified as commissives (van Eemeren & Grootendorst, 1984, pp. 101-102; van Eemeren & Grootendorst, 2004, pp. 64-65), "the negative variants of the commissives are themselves strictly speaking to be regarded as assertives rather than commissives" (van Eemeren & Grootendorst, 2004, p. 65, n. 46).

In empirically-oriented studies, the behind-the-scenes working of PR in argumentative activities has been well documented by Hample (2005; Hample, Paglieri, & Na, 2011), Paglieri (2013), and others. In general, forms of instrumental or strategic PR, characterised by a cost-benefit analysis of what and how efficiently one can achieve with a given argumentative contribution, have been identified. Arguers decide to perform and edit their arguments based on considerations such as chance of success, identity and relation management, negative and positive politeness, situational appropriateness, as well as truth and relevance of their arguments (Hample, 2005, Ch. 4). While these results have not been cast in the language of PR about what to do, they clearly can be.

To conclude, there is PR behind performance of any speech act, including assertives and all argumentative speech acts. This, however, is different from analysing the speech acts presented as conclusions of one's practical reasoning – which, when publicly performed, can better be called practical argumentation (henceforth: PA). Concluding that "Aristotle could well have written *Rhetoric* all by himself" requires very different supporting arguments than the conclusion "let's employ Aristotle."

3. DETAILED SCHEME OF PRACTICAL ARGUMENTATION

Before examining the various forms of conclusion of PA, it seems necessary to understand what PA in general consists of. The scheme of PA presented in Figure 2 stems from a rich literature on practical argument in philosophy, Artificial Intelligence, and argumentation theory (see Lewiński, 2014, for a more detailed discussion). In particular, it is derived from a recent comprehensive account of PA by Fairclough and Fairclough (2012). While referring to their work for an in-depth analysis of all the premises constituting the scheme (*Circumstances*, *Goal*, *Values*), I will briefly mention five basic advantages of the scheme, focussing further on the last one: the speech acts which can conclude PA.

First, the scheme shapes the framework of relevance for (multi-party) deliberation. Typically, different parties argue for the contextual betterness of their proposals for action {M, N, O... Z} (see the "M is Best" box). Their deliberation develops then as an *argumentative polylogue* (Lewiński & Aakhus, 2014) along the lines of possible disagreements over the various elements of the structure (basic premises, inference rules and contextual criteria).

Second, the scheme distinguishes between context-independent and context-dependent elements of PA. Its basic general structure (as per Fairclough & Fairclough: all the white boxes in Figure 2) remains constant, while contextual criteria for choosing "the right means" (below the diagram) fluctuate.

Third, the scheme clarifies the notion of the means-goal premise. This premise is grounded in one of the three basic inference licences warranting the choice of "the right means" taking us from the current (unwelcome) *Circumstances* to the (desired) *Goals*. We can thus warrant our conclusions by claiming either that the means are necessary, or that they are satisfactory (good enough), or that they are the best among all the possible alternatives (see Lewiński, 2015, for a detailed analysis).

Figure 2: The scheme of Practical Argumentation

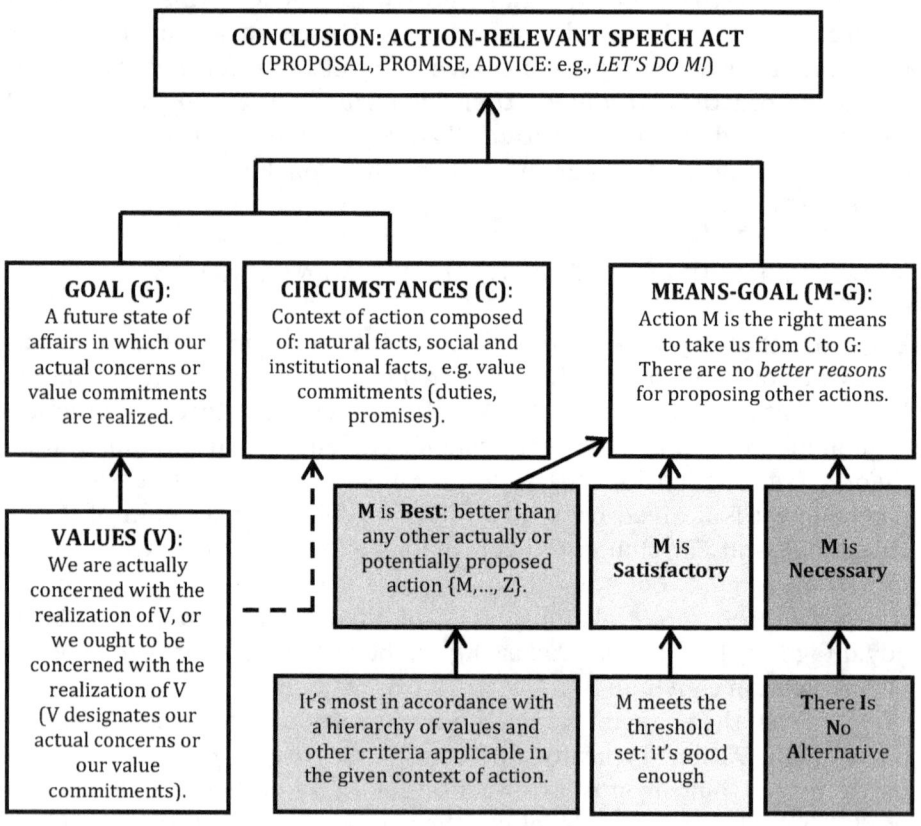

Derived from: Fairclough & Fairclough (2012)

Possible, contextually-determined decision criteria:
direct or indirect costs and benefits (negative and beneficial •
consequences / side effects)
opportunity costs •
practical feasibility •
ethical, moral, or legal implications •
likelihood of realization or of success •
conformance with other goals or strategies, their timing, duration, or location •
(derived from: McBurney, Hitchcock, & Parsons, 2007, p. 99)

Fourth, it provides a new account of how to criticise and evaluate PA. It clearly demarcates the attacks on the main premises (Circumstances, Goals, Values, Means-Goals), from those concerning context-dependent decision criteria, and from criticisms of the inference licences related to the type of inferential step made (necessary, satisfactory, or the best means) (see Lewiński, 2015).

Fifth, the scheme defines its conclusion in terms of a class of action-relevant speech acts, thus pointing to a "unity in diversity" of what we can argue for in PA.

I will now apply this scheme to the analysis of an actual case of complex practical argumentation and then discuss in detail the last point.

4. CASE STUDY: KEEP IT IN THE GROUND

On the 16th of March 2015, the British newspaper *The Guardian* launched a media campaign to divest (dis-invest) from fossil fuels such as coal, oil and gas. The campaign was entitled "Keep it in the ground" and included, at its start, a very short petition, a "note from Alan Rusbridger, [*The Guardian's*] editor-in-chief", a "full story" in which Rusbridger presents an elaborate argument for the petition, as well as other multi-media materials (videos, frequently asked questions, interviews, reportage) related to the dangers of our massive continuing reliance on fossil fuels.[4]

Let me start with a little note on the very slogan of the campaign: "Keep it in the ground." As an utterance in an imperative mood, it seems to be addressed to others, as a directive speech act (request? command? plea?). However, given the nature of such campaigns, and the immediate contextual and co-textual information, it can be better understood as elided "(*We* should) keep it in the ground" or, even better, "(Let *us*) keep it to the ground" – by analogy with "(Let's) give peace a chance" and other such slogans. Here, the addressee – the agent of PA – is the inclusive we.

In what follows, I focus on Rusbridger's "full story" entitled: "The argument for divesting from fossil fuels is becoming overwhelming."[5] This "story" or "argument" – in fact, a well-structured complex argument – is introduced with the following lead:

---

[4] See http://www.theguardian.com/environment/ng-interactive/2015/mar/16/keep-it-in-the-ground-guardian-climate-change-campaign.

[5] See http://www.theguardian.com/environment/2015/mar/16/argument-divesting-fossil-fuels-overwhelming-climate-change?CMP=share_btn_tw. All following quotations are from this source.

> As progressive institutions, the Gates Foundation and Wellcome Trust should commit to taking their money out of the companies that are driving global warming, says the Guardian's editor-in-chief as he launches our climate campaign.

As is appropriate for a journalistic article, it starts with a conclusion. Here, it is clearly a conclusion of a PA – that some agents (the Gates Foundation and Wellcome Trust) "should" undertake certain actions ("should commit to taking their money out of the companies that are driving global warming").

According to Rusbridger, "[t]here are two arguments in favour of moving money out of the biggest and most aggressive fossil fuel companies – one moral, the other financial." In saying this, he explicitly refers to two basic types of PR recognised in philosophy: on the one hand, *moral* or *value-based* reasoning, and on the other hand, *instrumental, prudential, strategic* or *means-end* reasoning (Gauthier, 1963; see Fairclough & Fairclough, 2012; Walton, 2007). The moral argument is analogous to "the push to pull money out of tobacco, arms, apartheid South Africa – or even slavery." Investing big money in the fossil fuel business, even if profitable, is *per se* a bad thing to do – just like making money from arms or slave trade is. The chief value in the fossil fuel argument is generational justice, that is, "concern for future generations": through our current recklessness, we are burdening the future generations with all the negative consequences of climate change. *The Guardian's* financial (or "pragmatic") argument is, interestingly, much more profoundly argued for, likely with the view towards the target agents of change, financial managers. This argument is best summarised in a quote from the Bank of England's deputy head of supervision for banks and insurance companies, Paul Fisher: "As the world increasingly limits carbon emissions, and moves to alternative energy sources, investments in fossil fuels – a growing financial market in recent decades – may take a huge hit." That is to say, those who do not care about the moral implications of climate change *per se*, or even do not believe in it at all, might be driven out of further investment by a prudential risk-assessment.

The "Keep it in the ground" campaign – as is appropriate for any other campaign – has a precise target: managers of endowments, investment and pension funds and, very specifically, the Gates Foundation and Wellcome Trust. It is clear in the text, though, that the problems of climate change policies and energy models based on fossil fuels have a number of other relevant stakeholders. All of them can potentially be the target of the campaign, the addressees of its arguments and, eventually, the agents of the conclusion of the argument. Rusbridger mentions that at first *The Guardian* thought of addressing "governments" and "politicians" in general ("MPs, presidents, prime ministers and members of congress") – but political action on climate change has proven far from satisfying: "the people who represent us around the UN negotiating tables have moved inches, not miles." That's why *The Guardian* decided to address the above-mentioned big institutional investors instead. Another stakeholder in the "story" are the scientists, here endowed with authority and treated with reverence: "If only science were enough […] finance will eventually have to surrender to physics […] the physics is unarguable." Of course, the general public is another crucial stakeholder; here, it is divided into present and future generations, with the responsibilities laid on those who can decide and act now. Finally, there are the fossil-fuel companies' directors, who – by the logic of global capitalistic economy – are compelled to "behaviour that is overwhelmingly driven by short-term returns."

We can pretty straightforwardly reconstruct the complex structure of PA from Rusbridger's argument. The current *Circumstance* is that of a climate change crisis caused by the overreliance on fossil fuels. This premise is briefly stated in the very first sentence of the piece: "The world has much more coal, oil and gas in the ground than it can safely burn." To address, or at least attenuate, this crisis we need to strive for a concrete *Goal*: "80% of the known coal reserves will have to stay underground, along with half the gas and a third of the oil reserves." Attaining this goal is necessary "if we and our children are to have a reasonable chance of living stable and secure lives 30 or so years from now." In general, the main *Values* and duties underlying the entire argument are "concern for future generations," "the protection of the public," "human health and science." But which means can be conducive to reaching this *Goal* and embodying the *Values*? As already mentioned, in *The Guardian's* view the best action to take ("the best" *Means-Goal* premise) is to directly call to divest from major fossil fuel companies – rather than, for instance, "campaigning for a paragraph to be inserted into the negotiating text at the UN climate talks in Paris this December." Appropriately for deliberations, different options are thus considered and one is chosen as *the* option.

Finally, let us look at the *Conclusions* of the argument. These are explicitly indicated with the "so" connective:

> So we ask that the Gates Foundation and Wellcome Trust commit now to divesting from the top 200 fossil fuel

companies within five years. And that they immediately freeze any new investment in the same companies.

This, however, is not the only conclusion of the text: "We will, of course, suggest that the Guardian Media Group does the same, and keeps you informed about its own deliberations and decisions." Finally, the readers are requested to act on the argument too: "Please sign, retweet and generally spread news about the petition."

Shortly, we have here a rather motley assortment of conclusions, at least when compared to a typical first-person-singular conclusion of most PR, as examined in philosophy ("so I shall take the 2:30 train to London"). What can be done about it?

5. CONCLUSIONS AS SPEECH ACTS

I will argue that a speech act approach can bring about the required level of nuance to our understanding of the conclusions of PA. Luckily, one does not need to argue for a theoretical adequacy of using speech act theory in argumentation studies. This is a job done back in the 1980s – and ever since validated and extended (van Eemeren & Grootendorst, 1984; Jacobs, 1989; Jacobs & Jackson, 1983; Snoeck Henkemans, 2014). With this background, we can ask a more detailed question: Which speech acts "convey" or "express" the conclusion of practical reasoning externalised in a text, that is, of practical argumentation?

As a preliminary, consider Searle's claim that "to function in deliberation a reason must be for a type of action, it must be for the agent, and it must be known to the agent" (2001, p. 99). PA, being public, takes care of the last condition all by itself: the reasons are openly presented so – as long as the "normal input and output conditions" for communication obtain (Searle, 1969, p. 57) – they are known to the agent. What remains to be analysed are the type of action and the type of agent that practical argumentation is "for."

Here, Fairclough & Fairclough's work is again very relevant. For them, the conclusion of PA has the form: "Claim for action: Agent (presumably) ought to do A." In their discussion of this point (2012, pp. 45-46), the Faircloughs notice that the first-person-singular analyses of PR opportunistically conceal the fact that the arguer and the agent of action are not always the same. We can, and often do, argue practically with someone else's action in the conclusion ("All things considered, the best thing to do now is for you to go and apologise"). Shortly, the philosophers' *agent-self* (Searle, 2001, p. 99), can be a self different than the arguer-self.

The scope of the claim for action is further complicated by formulations such as "ought to" or "should." These are notoriously ambiguous – if not semantically, then surely pragmatically. In everyday spoken language we manage this ambiguity very well trough skilful, even if unconscious, use of prosody and contextual information. Compare:

> You should *just* let him go – a friendly suggested advice.
> You should let him go *please* – an appeal or entreaty.
> You should let him go *right now* – a command.

> (Note that in spoken discourse, the illocutionary force equivalent to "just," "please," and "right now" can be conveyed solely by prosodic elements, such as rising/falling intonation or accent.)

Even in such simple examples the pragmatic analysis of expressions such as "should" leaves the notion of a "claim for action" largely underspecified. The analysis in the previous section likewise identified that, quite tellingly, *The Guardian* editors conclude their arguments with varied "claims for actions" by agents other than themselves. What exactly can this "claim" be?

In pragma-dialectics, all conclusions of argumentation can be reconstructed in terms of an assertive speech act of advancing a standpoint with its own set of felicity conditions such as the need to defend its propositional content with arguments, when challenged to do so (Houtlosser, 2002). This speech act, however, is defined at the level of "normatively" reconstructed, dialectically relevant moves in a critical discussion. In actual argumentation, at the level of "naïve" interpretations of speakers themselves, standpoints can be expressed through a variety of other conventional speech acts (see van Eemeren et al., 1993; Jacobs, 1989).

As analysed above, in the case of the conclusion of PA, the speech act of *proposal* has been considered a paradigm case (Aakhus, 2006; Ihnen Jory, forth.; Kauffeld, 1998; Walton, 2006). Proposals are speech acts located between *commissives* (such as promises) and *directives* (such as requests). Due to this peculiarity they belong – next to bets, offers and bids – to a class of speech acts inescapably requiring an action of *both* the speaker and the hearer. Taken together, commissives, directives, and their hybrids (such as proposals or offers) form a class of what I will call *action-relevant speech acts* (see Table 2) – in opposition to assertives (representatives), expressives and declarations. Action-relevant speech acts are characterised by their world-to-words direction of fit (Searle, 1975),[6] as their point is to get an agent (whether "I", "you", or "we") to perform an action that will bring the world into a state captured in the intentional content of the speech act. (By contrast, the goal of assertive speech acts with the words-to-world direction of fit is to capture in their content some existing state of the world.)[7] Naturally, so to speak, it is exactly speech acts with an "upward" (world-to-words, world-to-mind) direction of fit that PR is "about." Given their different direction of fit, the similarity between practical and theoretical reason breaks:

> The difference between theoretical and practical reason is in the direction of fit of the conclusion: mind-to-world, in the case of drawing a conclusion from evidence or premises, and world-to-mind, in the case of forming a decision and hence an intention on the basis of considerations. (Searle, 2001, p. 91)

Now, the elements of the class of action-relevant speech acts can be distinguished, as discussed above, along two dimensions: (1) their primary agent: the speaker, the hearer[8], or both; and (2) their illocutionary strength, ranging from the cancellable and nearly off-record to fully endorsed and on-record (see Jacobs & Jackson, 1983). Together, these two axes create an ordered matrix of "claims for action" – or speech acts which can conclude PA (see Table 2).

---

[6] This, of course, has not escaped Searle's attention: "Since the direction of fit is the same for commissives and directives, it would give us a more elegant taxonomy if we could show that they are really members of the same category. I am unable to do this [...] and am left with the inelegant solution of two separate categories with the same direction of fit" (1975, p. 356). See also Searle, 2001, passim.

[7] Note that for Searle (1975) expressives have no direction of fit at all, while declarations have dual direction: both words-to-world and world-to-words.

[8] For the current purposes, it is inconsequential to carefully distinguish between the directly addressed Hearer ("you") and some potential, indirect Hearers, such as third parties ("s/he", "they"). These differences do not affect the analysis proposed here – they do, however, play a significant role in understanding how speech acts function in a multi-party context (Clark & Carlson, 1982; Levinson, 1988; Lewiński, 2013).

Table 2: Action-relevant speech acts as conclusions of practical argumentation

Agent \ Strength	Weak (suggesting)	Neutral (committed)	Strong (solemn)
I / We (exclusive) *commissive*	Announcement (of intent)	Promise	Vow, Oath
You / Them *Directive*	Recommendation, Advice	Request, Appeal	Order, Call, Demand, Plea, Entreaty
We (inclusive) *commissive/directive*	Invitation	Proposal, Offer	Joint Pledge, Joint Guarantee

What is common to these various action-relevant speech acts is, of course, that they can all be complemented with the phrase "to do A": "I promise to do A," "I advise you to do A," "we guarantee to do A." Their strength – here divided for illustrative purposes into three levels – can quite straightforwardly be grasped by our ordinary intuitions: "I've thought about it a lot and: maybe we can go and apologise? / let's go and apologise / we must go and apologise."

Given these distinctions, we can now more precisely characterise the explicit conclusions of *The Guardian's* campaign: First, its very title – "Keep it in the ground" – in the interpretation suggested above, takes as an agent the *inclusive we*. It is, in terms of speech acts, a joint *proposal* or *pledge*: neutral-to-strong combination of a directive (you should) and a commissive (we should too). Second, the main conclusion is addressed to "them" – the Gates Foundation and Wellcome Trust – and explicitly identified as a "call," a strong directive speech act. Third, "[we] the Guardian Media Group" – the *exclusive we* (us) – concludes the argument with an announcement (of intent) or "suggestion," a weak commissive speech act. Finally, "you," "the readers," are requested to spread the news. This is a conclusion expressed through a directive speech act, of a rather neutral force (it is neither only suggested nor strongly required or called for).

6. CONCLUSION ABOUT CONCLUSIONS

The most common formulation of the conclusion of practical argument contains the somewhat enigmatic, if not frightening, expression "I should" / "I ought to" perform a certain action. In this way, PR ends with something like a moral obligation or imperative. This, in the words of Searle, leads to "an unhealthy obsession with something called 'ethics' and 'morality'" and might explain why "the authors are seldom really interested in reasons for action, and are too eager to get to their favorite subject of ethics" (2001, p. 182). For Searle, reasons for action and their conclusions are best understood not through some external ethical systems containing moral obligations we should fulfil, but rather through "the human ability to create, through the use of language, a *public* set of commitments" (2001, p. 183, emphasis original). This set consists of the consequences of various speech acts we perform as a result of our practical argumentation. It is therefore the analysis of such speech acts which can illuminate the conclusions and consequences of our *public* practical argumentation. What I offered here is hardly more than a simple matrix ordered by the agent of the action and the illocutionary force of a speech act. Even this simple matrix, however, can help us better understand what practical argumentation is "about," that is, "why" or "what for" we argue practically. We argue to issue various speech acts: from innocuous private announcements and suggestions to strong commands and solemn joint pledges. This happens in ordinary, typically highly complex communicative situations with multiple speakers and hearers – in *polylogues*, that is (Lewiński, 2014). Further careful analysis of all such speech acts, in their natural context, with their respective felicity conditions and further consequences, can sharpen our understanding of practical argument.

Therefore, we should do it.

ACKNOWLEDGEMENTS: I would like to thank Steve Oswald (University of Fribourg) who provided a number of highly illuminating comments and criticisms on this paper.

REFERENCES

Aakhus, M. (2006). The act and activity of proposing in deliberation. In P. Riley (Ed.), *Engaging Argument: Selected papers from the 2005 NCA/AFA Summer Conference on Argumentation* (pp. 402-408). Washington, DC: National Communication Association.

Aristotle. (1984). *Nicomachean ethics.* In J. Barnes (Ed.), *The Complete Works of Aristotle, Vol. II* (pp. 1729-1867) (transl. by W. D. Ross & J. O. Urmson). Princeton, NJ: Princeton University Press.
Austin, J.L. (1962). *How to do things with words.* Oxford: Clarendon.
Broome, J. (2013). *Rationality through reasoning.* Oxford: Blackwell.
Clark, H. H., & Carlson, T. B. (1982). Hearers and speech acts. *Language, 58*(2), 332-373.
Dascal, M. (2005). Debating with myself and debating with others. In P. Barrotta & M. Dascal (Eds.), *Controversies and Subjectivity* (pp. 33-73). Amsterdam: John Benjamins.
Eemeren, F. H. van, & Grootendorst, R. (1984). *Speech acts in argumentative discussions.* Dordrecht: Floris.
Eemeren, F. H. van, & Grootendorst, R. (2004). *A systematic theory of argumentation.* Cambridge: Cambridge University Press.
Eemeren, F. H. van, Grootendorst, R., Jackson, S., & Jacobs, S. (1993). *Reconstructing argumentative discourse.* Tuscaloosa: University of Alabama Press.
Fairclough, I., & Fairclough, N. (2012). *Political discourse analysis: A method for advanced students.* London: Routledge.
Gauthier, D. P. (1963). *Practical reasoning: The structure and foundations of prudential and moral arguments and their exemplification in discourse.* Oxford: Clarendon Press.
Grice, H.P. (1957). Meaning. *The Philosophical Review, 66*(3), 377-388.
Habermas, J. (1984). *The theory of communicative action.* Vol. 1: *Reason and the rationalization of society* (transl. by T. McCarthy). Boston: Beacon.
Hample, D. (2005). *Arguing: Exchanging reasons face to face.* Mahwah, NJ: Lawrence Erlbaum.
Hample D., Paglieri F., & Na L. (2011). The costs and benefits of arguing: Predicting the decision whether to engage or not. In F. van Eemeren, B. Garssen, J. A. Blair, & G. Mitchell (Eds.), *Proceedings of the 7th Conference on Argumentation of the International Society for the Study of Argumentation* (pp. 718–732). Amsterdam: Sic Sat.
Hitchcock, D. (2002). Pollock on practical reasoning. *Informal Logic, 22*(3), 247-256.
Hitchcock, D. (2011). Instrumental rationality. In P. McBurney, I. Rahwan & S. Parsons (Eds.), *Argumentation in Multi-Agent Systems: Lecture Notes in Computer Science* (pp. 1-11). Dordrecht: Springer.
Houtlosser, P. (2002). Indicators of a point of view. In F. H. van Eemeren (Ed.), *Advances in Pragma-Dialectics* (pp. 169-184). Amsterdam: SicSat.
Ihnen Jory, C. (forth.). Negotiation and deliberation: Grasping the difference. *Argumentation*, online first, doi: 10.1007/s10503-014-9343-1.
Jacobs, S. (1989). Speech acts and arguments. *Argumentation, 3*(4), 345-365.
Jacobs, S., & Jackson, S. (1983). Strategy and structure in conversational influence attempts. *Communication Monographs, 50*(4), 285-304.
Kauffeld, F. J. (1998). Presumptions and the distribution of argumentative burdens in acts of proposing and accusing. *Argumentation, 12*(2), 245-266.

Levinson, S. C. (1988). Putting linguistics on a proper footing: Explorations in Goffman's concepts of participation. In P. Drew & A. Wootton (Eds.), *Erving Goffman: Exploring the Interaction Order* (pp. 161-227). Cambridge, MA: Polity Press.

Lewiński, M. (2013). Debating multiple positions in multi-party online deliberation: Sides, positions, and cases. *Journal of Argumentation in Context, 2*(1), 151-177.

Lewiński, M. (2014). Practical reasoning in argumentative polylogues. *Revista Iberoamericana de Argumentación, 8*, 1-20.

Lewiński, M. (2015). Practical reasoning and multi-party deliberation: The best, the good enough and the necessary. In In B. Garssen, D. Godden, G. Mitchell & A. F. Snoeck Henkemans (Eds.), *The Eighth Conference of the International Society for the Study of Argumentation (ISSA)* (pp. 851-862). Amsterdam: SicSat.

Lewiński, M., & Aakhus, M. (2014). Argumentative polylogues in a dialectical framework: A methodological inquiry. *Argumentation, 28*(2), 161-185.

McBurney, P., Hitchcock, D., & Parsons, S. (2007). The eightfold way of deliberation dialogue. *International Journal of Intelligent Systems, 22*(1), 95-132.

O'Keefe, D.J. (1977). Two concepts of argument. *Journal of the American Forensic Association, 13*(3), 121-128.

Paglieri, F. (2013). Choosing to argue: Towards a theory of argumentative decisions. *Journal of Pragmatics, 59*, 153-163.

Perelman, C., & Olbrechts-Tyteca, L. (1969). *The new rhetoric: A treatise on argumentation* (transl. by J. Wilkinson & P. Weaver). Notre Dame: University of Notre Dame Press. (Original work published 1958.)

Pollock, J.L. (1995). *Cognitive carpentry: A blueprint for how to build a person.* Cambridge, MA: MIT Press.

Searle, J. R. (1969). *Speech acts: An essay in the philosophy of language.* Cambridge: Cambridge University Press.

Searle, J. R. (1975). A taxonomy of illocutionary acts. In K. Günderson (Ed.), *Language, mind, and knowledge, vol. 7* (pp. 344–369). Minneapolis: University of Minnesota Press.

Searle, J. R. (2001). *Rationality in action.* Cambridge, MA: MIT Press.

Searle, J. R. (2014). The structure and functions of language. *Studies in Logic, Grammar and Rhetoric, 36*(1), 27-40.

Snoeck Henkemans, A.F. (2014). Speech act theory and the study of argumentation. *Studies in Logic, Grammar and Rhetoric, 36*(1), 41-58.

Strawson, P.F. (1964). Intention and convention in speech acts. *The Philosophical Review, 73*(4), 439-460.

Walton, D. (2006). How to make and defend a proposal in a deliberation dialogue. *Artificial Intelligence and Law, 14*(3), 177-239.

Walton, D. (2007). Evaluating practical reasoning. *Synthese, 157*(2), 197-240.

What About Perlocution?
Commentary on Lewiński's How to Conclude Practical Argument in a Multi-Party Debate

STEVE OSWALD
English department, University of Fribourg, Switzerland
steve.oswald@unifr.ch

1. INTRODUCTION

There have traditionally been two broad avenues of research in linguistics, which carry over, to a fair extent, to the linguistically-oriented study of verbal communication. One of them encompasses data-driven approaches and builds on meticulous and fine-grained analyses of natural interaction in order to identify context-specific patterns of communicative *behaviour*. Many of these approaches (ethnomethodology, conversation analysis and interactionist approaches more generally, see e.g., Sacks et al., 1974; Schegloff et al., 1977; Hutchby & Woofitt, 1988) emerged in the 1960s and 1970s in reaction to formal and abstract models of language analysis, and have made their scholarly endeavour all about the data: analysis primes over theoretical generalisation, which is often not the chief concern of such accounts. The other direction of research inherits its epistemology from logic and analytical philosophy, which share a concern for formalism, and has traditionally offered moderate to radical reductionist accounts. The idea, in this approach, is to establish, in essentialist terms, descriptively and explanatorily adequate scientific models of language (the Chomskyan tradition in generative syntax is a prime example of this type of research, which also extends to contemporary models in semantics, phonology, morphology and the philosophy of language). Somewhat divided in two, the linguistics research map seemed to offer two quite clearly delineated options, as each tradition was in its own place. And then pragmatics came along.

While Charles Morris is usually credited for the birth of pragmatic research, the study of language in use truly came forward in analytic circles with Austin, Searle and Grice's pioneering work in speech act theory and the study of rational principles of communication (Austin, 1962; Searle, 1969; Grice, 1975, 1989). Speech act theoretic input paved the way for the analysis of communication as action – since *saying* came to be also construed as *doing*. Data-driven approaches in pragmatics saw an opportunity to hop on the P-train (i.e., the performative train), and kept on analysing language from a contextual perspective, now with improved systematicity, taking advantage of developing speech act classifications. As to formal approaches, they hopped on the I-train instead (i.e. implicature train), and were eager to pursue their forefathers' work on meaning by adding yet another layer to it, building on Searle's notion of indirectness and Grice's model of implicature – contemporary pragmatic research on meaning is thus largely devoted to the analysis of the different components of meaning.

Today, any piece of research on any aspect of communication that mentions the word *context* can roughly qualify as pragmatic research. A quick look at the programme of every biennial edition of the International Pragmatics Association (IPrA) conference – which usually features 5 to 6 days of talks with more than 10 parallel sessions – reveals that a large number of researchers investigating extremely different things all operate under the umbrella of pragmatics, since they all address various aspects of communication accounting for the complexity of human verbal exchanges.[1] So 60 years or so after the William James lectures (both Austin's and Grice's), pragmatics still very much looks like the waste-basket Mey (2001, p. 21) and Bar-Hillel (1971, p. 404) mention – at least judging by the internal disparity of all work conducted in the discipline.

However, in the past 10 years or so, increasingly more pragmatic work explores the juncture of both trends and tries to bring both of them together in an attempt to offer at the same time description- and theory-friendly accounts. I believe Lewiński's is an example.

In the remainder of this commentary I discuss, from a methodological and epistemological perspective, how I think Lewiński's analysis contributes to mutually strengthening both trends in pragmatics into one consistent model and how his proposal specifically highlights and rests on the pragmatic notion of perlocution, which pragmaticians have nearly systematically ignored in their accounts.

2. LEWIŃSKI'S PROPOSAL: A STEP CLOSER TO ECOLOGICAL VALIDITY

Lewiński's paper is an inquiry into the nature and function of the speech

---

[1] See for example this year's (2015) programme of the conference at the University of Antwerp:
http://ipra.ua.ac.be/download.aspx?c=.CONFERENCE14&n=1476&ct=1476&e=1 5315.

acts that can count as proper conclusions in patterns of practical argumentation (PA) which draws on speech act theory and pragma-dialectics. This proposal is at the same time descriptively and theoretically fertile. From a descriptive perspective, it offers a very detailed functional typology of the speech acts that may be used (and how they may be used) to verbalise conclusions in practical argumentation. On the theoretical front, analytical categories are functionally justified and Lewiński provides a model with enough generalising power to cover the variety of speech acts potentially involved in the phenomenon he is tackling, i.e., conclusions in PA patterns. In a nutshell, Lewiński offers a theoretically-grounded accurate description of argumentative reality.

This proposal is representative, I believe, of a trend in pragmatic research (construed broadly) that is now gaining momentum and which reaches over to cover experimental research as well. This trend strives for *ecological validity*, in that it purposefully tries to achieve descriptively and explanatorily adequate accounts of communicative phenomena. In this particular case, Lewiński's account of conclusions in PA fulfils in my opinion these goals on several counts.

First, Lewiński adopts a sufficiently broad and unrestricted conception of communication, which by definition is a social phenomenon taking place between at least 2 parties who exchange information and react to each other's messages within a set of specific circumstances (i.e., a context). While this conception seems like a basic minimal requisite for doing research on communication, Lewiński offers more than just that as his framework is tailored to capture communicative complexity. By shifting the perspective from conclusions formulated as "I should do P" to conclusions formulated with the plural "we should do P," the framework puts agency at its core: agents provide reasons, evaluate reasons, but also perform actions based on reasons or are led to reject the performance of actions based on reasons, through argumentative discussion. The incorporation of dialogism (or polylogism, as Lewiński would probably have it) as a feature of the units of analysis is precisely meant to achieve a fuller and more comprehensive account of communication: only by considering as part of the whole communicative process that (i) utterances are designed by people to have an effect on other people (utterances – and many times their effects – are intentional), (ii) utterances trigger a range of different (re)actions, (iii) speakers are free to (re)act by engaging in further communication, and that (iv) actions are relevant in the theorisation of meaning (as they signal the completeness of a verbal exchange), can we pretend to provide an adequate account of communicative exchanges. In this framework, agents, actions and reactions are given proper consideration, and this has implications for more theoretical aspects of how communicative exchanges should ideally be construed in terms of intention recognition and fulfilment. To sum up, the first advantage of this speech act analysis is its attention to the complexity of verbal exchanges, action-wise and agent-wise.

Second, it provides an exhaustive matrix of all the different speech acts that may conclude practical argumentation. In this respect, the account surveys all the speech acts that may actually be used by participants according to two criteria: their primary agent, i.e., who is responsible for performing the action that is predicated in the speech act, and their illocutionary strength, which could also be interpreted as a scale of commitment to the illocutionary force of the speech act. The strength of the framework, on this very issue, is to incorporate both the perspective of the speaker and that of the addressee into the model. As a consequence, descriptive power is increased, since the nature and function of speech acts functioning as conclusions of PA can exhaustively be assessed through consideration of all parties taking part to the communicative exchange. Furthermore, the typology offered here overcomes the difficulties faced by approaches which refrain from generalising and provides an interesting option to systematise the analysis of talk in interaction.

Third, and in connection with the first two advantages, the model introduced here goes beyond the interaction between propositions and offers a hands-on theoretical kit to approach natural data. Many pragmatic approaches are interested in what happens in terms of meaning at the level of propositional (and sometimes non-propositional) content, without taking into account that those propositions are taken up by their addressees – and this does play a role in the communicative process. Of course, you don't necessarily need to consider the entirety of cognitive and behavioural efforts both parties incur in communication to explain how it works;[2] however, only when you do can you aspire to provide a full account of communication, since, as shown by Lewiński, the identification and felicitousness of speech acts may have to take into account their perlocutionary success (see section 3 below).

---

[2] For example, Relevance Theory (Sperber & Wilson, 1995) does not consider cooperation to be a notion that is required to explain how communication works, since mere coordination suffices – the speaker's and the hearer's behaviour happen to dovetail in communication, but that does not mean that both interlocutors cooperate, in the Gricean sense (see Allott, 2007 for a rationale for this).

For those three reasons, I believe that Lewiński's proposal gets us one step closer to a more *ecologically friendly* account of conclusions in PA. His analysis of the "Keep it in the ground" case study convincingly shows that in order to describe and reconstruct argumentative discourse in a way that does justice to the data, such a framework is advisable: it takes into account both (re)actions and agents (as producers, recipients and evaluators of speech acts), both production and reception, both illocution and perlocution. Also, it assesses speech acts and their *consequences*, which, most importantly, opens up new directions for rethinking the very notion of speech act felicity, as we shall see next.

3. SPEECH ACT FELICITY, ILLOCUTION, PERLOCUTION AND PA

In order to assess Lewiński's contribution in light of speech act theory, let us first recall what Austin says about perlocution:

> Saying something will often, or even normally, produce certain consequential effects upon the feelings, thoughts, or actions of the audience, or of the speaker, or of other persons: and it may be done with the design, intention, or purpose of producing them (...). We shall call the performance of an act of this kind the performance of a perlocutionary act or perlocution. (Austin, 1962, p. 101)

Traditionally, perlocution was left out of the study of meaning, as the actual occurrence of such a consequence, in principle, is independent from the success of the illocutionary act. Austin's first example of perlocution was persuasion; one can urge or advise someone (not) to do something (illocutionary act) and fail or succeed in persuading them to comply (perlocutionary act), but whether the latter is the case or not is irrelevant to whether the speaker has effectively urged or advised her interlocutor to (not) do something. One can understand without complying, which is an indication that, crucially, the success of communication requires comprehension but not compliance. Accordingly, Austin sharply distinguishes illocution from perlocution: "[w]e have then to draw the line between an action we do (here an illocution) and its consequence" (1962, p. 110). This is precisely because understanding is distinct from cooperating – in the perlocutionary sense (see also Attardo, 1997). As far as simple speech acts are concerned, perlocution is a non-necessary, optional, consequence of meaning, and its non-satisfaction is no threat to the success of communication – again, construed only as a successful exchange of meaning.

And yet, coming back to Austin's quote, what about cases where the consequence of an illocution is precisely an action (that someone else might be asked to perform)? This is, chiefly, what PA is about, especially if, like Aristotle, Searle and most probably Lewiński, we consider that actions are part of the conclusion of PA and join "those who think action itself is the proper conclusion of PR" (Lewiński, 2016, p. 405). This is where the story becomes complicated, as at least two problems related to the potential role of perlocution in speech act felicity emerge – and these have not been discussed by Lewiński:

- do we consider that the felicity of a speech act (i.e., speech acts akin to proposals in the case of PA) rests on its recognition/identification by the addressee or on its actual ability to trigger the desired effect in the communicative exchange?
- while there is no question that illocution is crucial to characterise the felicity of speech acts, isn't PA THE particular case where perlocution is important, if not necessary, to the success of the speech act?

In other words, is the speech act successful when we understand it, or when we comply with it? In the case of PA, the constraints set by its argumentative nature can be thought to make a case for the latter.

The question, here, is therefore that of speech act felicity,[3] and takes us back to the original Austinian distinction. Even if the idea that illocution is the driving force behind speech act performance in communication remains quite uncontroversial, it seems that PA poses some challenges for speech act theory, and this is mostly due to the fact that these speech acts are used argumentatively and consequently cannot be dealt with exclusively at the propositional and illocutionary levels.

Any speech-act-theoretic account of argumentation needs to consider the speech act itself, but also its *consequences* because of the dialogic (or polylogic) nature of argumentation. It is hard to consider

---

[3] One could also relate this discussion to Vanderveken's distinction between the success and the satisfaction of a speech act (the author thanks Scott Jacobs, personal communication, for pointing this out). Here, speech act felicity is related to the notion of satisfaction: "Elementary illocutionary acts with a propositional content (...) are satisfied *only if* their propositional content represents correctly how things are (...) in the world" (Vanderveken, 1990, p. 132).

that any argumentative speech act is felicitous until it has been appropriately responded to (that is, accepted, called into question or refuted). In the case of PA, Lewiński claims that the conclusion is effectively a claim for action ("we should do X"), and in doing so gives some credit to the idea that the performance of the action expressed in those claims – or at least the performance of an argumentatively-relevant action connected to that action – should be considered at least partly as the natural conclusion of PA. From the perspective of speech act theory, the very point of these claims for action is to go beyond the informative demands of the exchange and, crucially, to make sure that they are acted upon,[4] which are features that need to be accounted for in the theoretical model. Consequently, Lewiński's model seems to be compatible with the idea that in order to analyse PA, we need to consider their perlocutionary consequences, as these encompass the argumentative moves triggered in reaction to the propositional content of the conclusion and, to that extent at least, determine how well the speech act fares in the communicative exchange.

4. CONCLUSION

So where does that leave us? It seems that the nature of PA conclusions, in speech act-theoretic terms, requires the analyst to consider illocution and perlocution together. This warrants a complex analysis that probably needs to extend felicity conditions to capture not only the commitments incurred by the speaker, but also the ones incurred by the addressee in his reaction. That is, while the felicity of the speech acts categorised in table 2 (Lewiński, 2016, p. 417) to a large extent depends on their proper recognition by the addressee, there are grounds to assume that it might also partly depend on the speech act's perlocutionary import. What those perlocutionary aspects amount to precisely remains to be seen; are they restricted to compliance? Or can they take the shape of any other argumentatively relevant move (refutation, request for clarification, requalification, etc.)? This might be a direction of research that would take Lewiński's original proposal farther in a speech act-theoretic account of PA.

---

[4] This is why Lewiński, in line with Fairclough & Fairclough (2012), considers that PA conclusions are "action-relevant speech acts".

REFERENCES

Allot, N. (2007). *Pragmatics and rationality*. PhD Thesis, University College London, ms. available at http://www.phon.ucl.ac.uk/home/nick/content/n_allott_phd_thesis_09_2007.pdf.

Attardo, S. (1997). Locutionary and perlocutionary cooperation: The perlocutionary cooperative principle. *Journal of Pragmatics*, 27(6), 753–779.

Austin, J. L. (1962). *How to do things with words*. London: Clarendon Press.

Bar-Hillel, Y. (1971). Out of the pragmatic wastebasket. *Linguistic Inquiry*, 2(3), 401-407.

Fairclough, I., & Fairclough, N. (2012). *Political discourse analysis: A method for advanced students*. London: Routledge.

Grice, H. P. (1975). Logic and conversation. In P. Cole & J. Morgan (Eds.), *Syntax and semantics 3: Speech acts* (pp. 41-58). New York: Academic Press.

Grice, H. P. (1989). *Studies in the way of words*. Cambridge, MA: Harvard University Press.

Hutchby, I., & Wooffitt, R. (1988). *Conversation analysis*. Cambridge, MA: Polity Press.

Lewiński, M. (2016). How to conclude practical argument in a multi-party debate: a speech act analysis. In D. Mohammed & M. Lewiński (eds.), *Argumentation and Reasoned Action: Proceedings of the 1st European Conference on Argumentation, Lisbon, 2015, Vol I* (pp. 403-420). London: College Publications

Mey, J. (2001) *Pragmatics: An introduction*. 2nd edition. Oxford: Blackwell.

Sacks, H., Schegloff, E., & Jefferson, G. (1974). A simplest systematics for the organization of turn-taking for conversation. *Language*, 50, 696-735.

Schegloff, E., Jefferson, G., & Sacks, H. (1977). The preference for self-correction in the organisation of repair in conversation. *Language*, 53, 361-382.

Searle, J. R. (1969). *Speech Acts: An Essay in the Philosophy of Language*. Cambridge: Cambridge University Press.

Sperber, D., & Wilson, D. (1995). *Relevance: Communication and cognition*. 2nd edition. Oxford: Blackwell.

Vanderveken, D. (1990). *Meaning and speech acts. Volume 1: Principles of language use*. Cambridge: Cambridge University Press.

19

Journalists' Emotionally Colored Standpoints Emerging in Argumentation: A Path Leading to Foster Existing Stereotypes in the Audience?

MARGHERITA LUCIANI
Università della Svizzera italiana, Switzerland
<u>margherita.luciani@usi.ch</u>

> The present research investigates the correlation between journalists' positive/negative evaluative standpoints reflecting their opinion with reference to ingroup/outgroup groups and the degree of specificity/non specificity and/or punctuality/durativity of the arguments supporting the standpoints. I will shed light on this topic analyzing three case studies of newspaper articles. I claim that journalists' standpoints strongly influence the newspaper articles they write and contribute to the maintenance of existing stereotypic beliefs in the audience.
>
> KEYWORDS: abstract nouns, audience, ingroupness, negative evaluation, outgroupness, positive evaluation, stereotypes.

1. INTRODUCTION

Semantic properties of linguistic labels and argumentative structures in which they are embedded convey in a subtle but powerful way information that modifies and alters the objects to which it is referred, playing therefore a crucial function in guiding and shaping cognitive processes. Starting from the certainty that evaluations convey a particular stance (Bednarek, 2009; Bednarek & Caple, 2012; Perrin, 2012) and refer to a principle issue consisting of a value of some sort (Freeman, 2000), the present paper investigates the role of the discursive traces of positive/negative emotions of journalists' evaluative standpoints with reference to ingroup/outgroup conceptualizations emerging in argumentative newspaper articles. I will set up a framework that enables to study this topic starting from the weight given by the presence of concrete or abstract nouns to argumentative structures, observing how this can influence the argumentative force supporting ingroup/outgroup conceptualizations. More specifically, I claim that through the usage of concrete or abstract nouns in argumentative newspaper articles, journalists convey their own individually evaluative reframed standpoint, subsequently contributing to the maintenance -or fostering- of existing stereotyping beliefs in the audience, therefore proving the existence of an intergroup bias factor in mass communication. Nevertheless, I also claim that intergroup bias is only one factor shaping news products, acting together with other factors that lead us to consider journalists as active knowledge mediators and news creators, opening a space for critical review.

2. TRADITIONAL APPROACHES TO THE RELATIONSHIP BETWEEN LANGUAGE AND SOCIAL COGNITION

Even though the conventional nature of language is partially evident, we are often not aware of psychological effects produced by language; in this endeavor, most studies in psycholinguistics and in social psychology mainly focus on categorical perception of terms (Corneille et al., 2002) and on the correlation between linguistic categories and cognition (Maas et al., 1995; Semin, 2000, Stapel & Semin, 2007).

With reference to the research trends investigating linguistic categories able to shape social cognition, one of them reduced its focus of attention on the distinct conceptualizations conveyed by abstractness and concreteness in linguistic categories, in an attempt to establish a correlation between linguistic categories and positive/negative conceptualizations (Maas et al., 1994). The widespread framework that adopted this line of research is known as Linguistic Category Model, henceforth LCM (Stapel & Semin, 2007), which subdivides linguistic structures in concrete and abstract categories; according to these theory, the abstract linguistic category is made up by stative verbs and adjectives, whereas concrete linguistic categories are made up by action depicting verbs and action interpretative verbs. For what concerns the crucial linguistic category of nouns, according to this approach, they have been generally included in the abstract linguistic category (Carnaghi & Arcuri, 2008, pp. 120-121), assuming that the whole linguistic category of nouns can be more abstract than adjectives because they indicate a single property of an object, whereas substantives aim at indicating a systematic whole of properties, also excluding to do a proper distinction in abstract vs. concrete nouns. In this paper I claim that, even though LCM has the merit to investigate linguistic categories, it does not analyze linguistic category in a sufficiently deep and clear way, due to the absence of a semantic analysis, therefore treating terms as mere linguistic labels and establishing fake connections between linguistic choices and cognitive processes. Furthermore, it does not permit to understand how language and communication shape cognition, since it focuses only on meta-semantic properties of terms taken alone, without considering their weight in the whole discourse and without establishing connections between terms and argumentative moves aiming at supporting a certain type of evaluation.

3. CONCRETE VS. ABSTRACT NOUNS EMERGING IN ARGUMENTATIVE STRUCTURES: AN ATTEMPT TO UNDERSTAND THE INFLUENCE OF INTERGROUP BIAS PHENOMENA IN PRINT-JOURNALISM

From the discussion reported in the previous Section, it becomes clear that traditional approaches to the crucial distinction abstractness/concreteness of linguistic categories focus primarily on imprecise and roughly sketched linguistic categories, considering only surface problems of the structures of linguistic categories of words, and felling outside the study of terms as meaningful linguistic units inserted in practices of socialized reasoning and reason-giving, therefore constraining the understanding of the way in which linguistic categories influences positive/negative conceptualizations in real-life practices of interaction.

3.1 Abstract vs. concrete nouns: the importance of a careful semantic analysis

The traditional approaches described in the previous Section underestimate the importance of the linguistic category of the noun in giving a particular perspective to a certain reality as well as the fundamental distinction between concrete and abstract nouns, which in my view represents a focal point in understanding the correlation between abstractness/concreteness and positive/ negative value.

It is therefore important to introduce some notions concerning the linguistic category of the noun as well as some features distinguishing concrete and abstract nouns. For space reasons, I will not include an in-depth literature *excursus* on the category of the noun, which would go far beyond the scope of the paper. Firstly, nouns are also said *substantives* since classical grammar, in which *nomen substantivum*[1] was used to indicate the name of a reality, opposed to

---

[1] See for instance Prisciano "nomen est pars orationis, quae unicuique subiectorum seu rerum commune vel propriam qualitatem distribuit [...]

nomen adiectivum: the former designated the noun as we consider it nowadays, whereas the second designated the adjective. The term *nomen* indicated the semantic function that joins noun and adjective; this has to do with the *qualitas*, a whole of features, or better said, predicates expressed by both the noun as well as by the adjective (Rigotti & Cigada, 2004, p. 203). The difference consists in the fact that the noun refers to this whole of predicates, since it characterizes and subsequently identifies a certain entity, while the adjective expresses predicates as suitable to belong to different entities. The noun can be said to be one of the most relevant and influential linguistic structures for the knowledge and communication of the world, and it has four demands with reference to its *denotatum*: a) the noun expresses the hypothesis that something exists or that some event takes place; b) the noticeability of the *denotatum*, which must be different from something else, and must not be situated in an undifferentiated continuum; c) the relevance of the piece of the world designated by the noun; d) the natural thematic predisposition of the noun to become object of a discourse.

However, in order to reach the main objective of the paper, it is not sufficient to state the importance of the linguistic category of the noun, but rather it is necessary to go one step further and to analyze more in-depth the important and often confused distinction between concrete and abstract nouns. Indeed, the feature of concreteness has often been confused with "physical" and that of abstractness with "non-physical"; nevertheless, it is not a matter of actual existence, but rather it is a matter of semantic structure[2]. The concrete noun designates something that "exists" or "does not exist" (*tree, God..*), whereas the abstract noun indicates something that happens or does not happen, that takes place or does not take place, so that the realities designated by the latter are not entities, but rather properties (*goodness, pallor..*), facts (*fall..*), situations (*damage, help..*), events (*promise..*). The abstract noun is therefore a depredicative noun, which predicates a mode of being: it is in this scenario that I claim that abstract nouns can be said to be a strong component of manipulation played out through language. Indeed, as stated by Rigotti & Cigada (2004, p. 209), the fact that "giving a name" usually presupposes that there is actually something that has this name makes it possible a manipulation of the receiver, by speaking of things that do not exist and through simple name giving: in this way a

---

Nomen quasi notatem, quod hoc notamus uniuscuique substantiae qualitatem" (Inst. Gramm. I, 56-57).

[2] For more details, see Flaux et al. (1996) and Cigada (1999).

consistency of reality to non-consistent entities is subconsciously conveyed, as exemplified by the Gottlob Frege's example *people's will* and in many other ideology-oriented texts (for more details see Cigada, 1999; Greco, 2003; Rigotti & Cigada, 2004, p. 209).

In order to capture the key features distinguishing concrete and abstract nouns, it is necessary to do a proper semantic analysis, which I conduce following Congruity theory, henceforth CT (Rigotti & Rocci, 2006). According to this theory, doing a semantic analysis means to rewrite natural language utterances in terms of predicate-argument structures[3]. In this perspective the semantic contribution of every content word in a language can be represented in terms of a predicate. To analyze the meaning of a lexical item means, first of all, to establish what kinds of predicates it can manifest when it occurs in its different syntactic constructions (Rigotti et al., 2006).

After having carried out a semantic analysis following CT, I will give first evidence that abstract nouns define an intrinsic property of the actor and of the group he belongs and imply a greater likelihood of repetition, therefore enabling us to define positive/negative behaviors as the result of stable characteristics, attributing positivity/negativity to the actor and to its group. By looking at the results of a semantic analysis conducted on predicate-argument structures, it becomes evident that concrete nouns designate substances assuming the role of arguments in predicate-argument structures, whereas abstract nouns, being they depredicative nouns, assume the role of predicates and permit to reconstruct another predicate-argument structure, therefore enabling us to sketch a predicate-argument hierarchy. For instance the concrete noun *door* in the sentence *The door of Cristina's office remained open* is an argument of the predicate *remain open*, and assumes a concrete meaning, therefore designating a singular and determined aspect of reality, namely the *door* of a certain office, which someone forgot to close. On the contrary, the abstract noun *robbery* in the title "Jeweler reacts to a *robbery* and shoots to a bandit" of a newspaper article is at the same time argument of the predicate *to react* and in turn predicate of three suppressed arguments, therefore enabling the reconstruction of a predicate-argument hierarchy. In this case we notice that the identity of X_1 and X_2 remains unspecified.

---

[3] The topic which is at issue in this paper compels me to use the term argument in different passages both in its argumentative and semantic function.

ROBBERY (X_1, X_2)
Presuppositions
X_1 (human being: robber)
X_2 (human being: robbed person)
Implications
X_1 steals with aggressiveness something to X_2

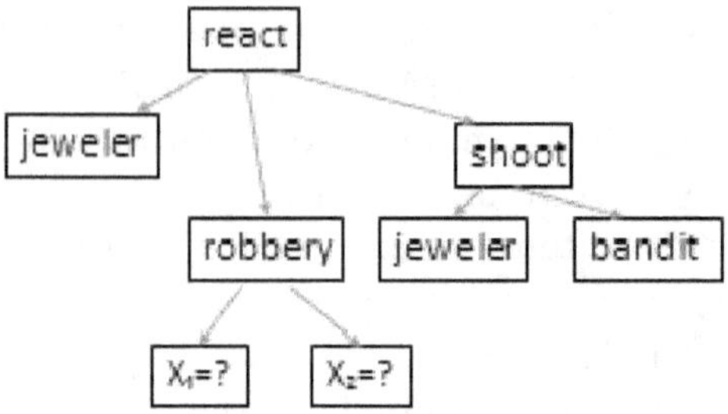

```
                    react
       jeweler       |        shoot
                  robbery  jeweler  bandit
                  /     \
                X₁=?   X₂=?
```

Fig.1 Semantic graph of the sentence containing the abstract noun robbery.

After having shown the semantic analysis of the abstract noun *robbery* and the semantic graph of the sentence in which it is embedded following CT, we have observed that abstract nouns have this feature to be depredicative and to suppress arguments, so that we can state that abstract nouns are: a) manipulatory in the sense that they presuppose the existence of fake realities b) manipulatory in the sense that they induce indefinite and depersonalized general and durable conceptualizations of realities, due to arguments' suppression and to the common suppression of the agent which becomes indefinite like a natural force. These two distinct features (*a* and *b*) characterizing abstract nouns represent the focal point for understanding how their presence in certain arguments conveys a distinct valence (positive or negative) to what is argued for with reference to events performed by the ingroup or the outgroup: an in-depth semantic analysis reveals that there are two distinct reasons leading to the usage of abstract nouns in attributing positivity to ingroup evaluations and in attributing negativity to outgroup evaluations.

3.2 Abstract vs. concrete nouns in argumentative newspaper articles

However, a semantic analysis alone, even the most accurate one, does not appear a sufficient means to grasp the correlation between positive/negative evaluations and stereotype-maintenance induced by abstract or concrete nouns. For this reason, I go one step further by enriching the semantic analysis with the argumentative analysis:

argumentative analysis at the interactional level carried out following pragma-dialectics (van Eemeren & Grootendorst, 2004) offers a valid way to systematically analyze arguments containing abstract and concrete nominal occurrences. Applying the considerations made after the semantic analysis of concrete/abstract nouns to argumentative structures, we can conceive arguments as situated on an ideal continuum shifting from a concrete to an abstract pole, on which arguments are situated depending on the degree of concreteness/abstractness given by the minor/major presence of concrete/abstract nouns. The argumentative perspective permits to overcome the purely linguistic level yet reached via the semantic analysis, and enables a deep comprehension of the reasoning mechanisms connecting the arguments and the standpoint supported by them in a unified and coherent structure, thereby giving us a systematic overview on the real-life practices under investigation.

This exhaustive framework connecting negative/positive valence to abstractness/ concreteness of nouns and arguments is nevertheless not yet sufficient to prove the maintenance or fostering of stereotyping beliefs in mass communication. For this reason, I enrich this work referring to intergroup situations, where the elaboration of information regarding oneself and the others is guided by stereotyping processes and inter-group bias phenomena. According to Maas *et al.* a phenomenon of Linguistic Intergroup Bias (LIB) is at stake when two distinct groups are involved in an evaluation of the events performed by the ingroup or the outgroup; they suggest that a particularly abstract language is used to describe scenes of positive behaviors acted by the ingroup and of negative behavior acted by the outgroup, and in turn participants chose a concrete language to describe negative behaviors acted by the ingroup and positive behaviors acted by the outgroup (Maas et al.,1989).

Once acknowledged that the vague classification of abstract and concrete language is not sufficient and ascertained that a more fine-grained analysis of nouns and of the distinction in concrete vs. abstract nouns is needed, applied to newspaper articles, this suggests that emotionally colored standpoints, conceived as positive/negatively evaluations of ingroup's or outgroup's events, supported by arguments containing abstract/concrete nouns, shifting events' responsibility on the social actor or on the situation, may contribute in a subtle but powerful way to the rising/ maintenance of existing stereotyping beliefs about ingroup/outgroup in the audience. If arguments full of abstract nouns, henceforth abstract arguments, imply that the described event refers to an undefined stable characteristic and is likely to be repeated, then abstract arguments, typically used to describe stereotype –

congruent opinions (for example negatively judged outgroup behaviors or positively judged ingroup behaviors), should foster existing beliefs. Similarly, the usage of arguments full of concrete nouns, henceforth concrete arguments, to support negatively judged ingroup behaviors that do not generalize beyond the specific situation and that do not refer to stability along time, will convey in the audience the idea of an isolated and unrepresentative episode. The same can be hypothesized for cases concerning positive opinions referred to the outgroup; if such opinions are supported by concrete arguments, then such information should not change previous established negative beliefs about the outgroup, denying a generalization of the positivity of the outgroup.

4. CORPUS AND METHODOLOGY

The corpus on which the present investigation is based enables comparative and contrastive studies from a multilingual, multicultural and multimedia perspective, since data are gained from both TV-journalism and print-journalism in the three linguistic areas of Switzerland. Part of the corpus, collected during the Idée Suisse project[4], was collected at the Swiss public service television (SRG SSR) in French and German. A more recent dataset was collected at *Corriere del Ticino* (CdT), the main Italian-language newspaper in the country, within the project "Argumentation in newsmaking process and product"[5]. Both datasets were collected with the same methodology (Progression Analysis; Perrin, 2003, 2013), and comprise audio-visual recordings of various newsroom activities, interviews with journalists, retrospective methodologies, screenshots of writing phases as well as newspaper articles and TV items. For the scope of interest of the present paper, namely for understanding a possible role of journalists as unconscious promoters of existing stereotypes in the audience, caused by certain types of conceptualizations induced by their newspaper articles, I will concentrate on the huge amount of print-journalism data collected at *CdT*, and more specifically on newspaper articles, for two reasons; firstly, I focused on print-journalism and more specifically on newspaper articles, in order to have the possibility to observe a direct

---

[4] "Idée Suisse: Language policy, norms and practice as exemplified by Swiss Radio and Television" (SNF NRP 56, 2005-2007).

[5] The present study is part of the Swiss National Science Foundation research project "Argumentation in newsmaking process and product" (SNF PDFMP1_137181/1, 2012-2015), whose main aim is to investigate the role of argumentative practices in newsmaking processes and products.

feedback of readers' reactions, gained from the section of the newspaper "readers' comments and opinions", where readers comment newspapers' articles; secondly, *CdT* represents an ideal newsroom for my analysis due to the proximity of its seat with the Italian boundary and to the subsequent possibility to find interesting data concerning ingroup/outgroup conceptualizations.

It is in this scenario that I plan to test my hypothesis based on the theoretical considerations proposed in Section 3; I firstly suggest that abstract nouns inducing indefinite general and durable conceptualizations are unconsciously used by journalists within arguments supporting negative standpoints concerning the outgroup, in order to foster outgroup's identification with a negative principle. Similarly, we can expect that abstract nouns subliminally presupposing the existence of fake realities, can be used for conveying indefiniteness in arguments that support positive standpoints concerning the ingroup. On the contrary, I expect that concrete nouns conveying features such as definiteness, specificity as well as temporariness to conceptualizations, are used in arguments supporting positive evaluations of the outgroup and in arguments supporting negative evaluative standpoints referred to the ingroup.

In order to test this hypothesis, I firstly selected newspaper articles of *CdT* by looking at the ones containing explicitly negative or positive or explicitly-judged negative or positive events occurred to Swiss people (ingroup for *CdT*) and to non- Swiss people (outgroup for *CdT*); then, I identified the negative/positive evaluative standpoint of the journalist who wrote the article. Then, I identified the arguments supporting the negative/ positive evaluative standpoints referred to the ingroup/ outgroup, carrying out an argumentative reconstruction following Pragma-dialectics (van Eemeren, 2004). At this stage of the analysis I identified all terms pertaining to the linguistic categories of the noun, and I distinguished them into abstract and concrete nouns following the theoretical principle explained in Section 3. Moreover, I carried out the semantic analysis of some crucial key abstract and concrete nouns, in order to show the weight of this type of linguistic category in determining the argumentative force of the whole abstract or concrete argumentation.

The aim of the proposed methodology unifying semantics and argumentation is to overcome the superficial reductionism of purely atomic linguistic models, that limiting themselves to the analysis of simple linguistic labels, offer only a surface and partial understanding of intergroup bias phenomena in journalism, which instead would benefit from a methodology able to explain the ways in which these intergroup bias phenomena act starting from the investigation of the actual ways in which journalists, consciously or unconsciously, propose and support a certain negative/positive evaluation, such as that described in this paper.

5. CASE STUDIES: THREE DISTINCT TYPES OF EVALUATION

In order to test the validity of the proposed theoretical framework and to elicit an adequate empirical confirmATION, I investigated three case studies concerning three distinct episodes described in *CdT*, and for completeness of analysis in the last case study I compared the event described in *CdT* with the description of the same episode given in a newspaper article of one comparable Italian newspaper, i.e. *Il Corriere della Sera*. I will carry out a semantic and an argumentative analysis of the newspaper articles with the approaches described in Section 3.

5.1 First case study: negative evaluation of events performed by the outgroup

In order to show how a journalist gives a particular frame to a news consisting of a negatively judged event performed by an outgroup member, I introduce the case study consisting of a newspaper article published on the 12th February 2013 on *CdT*. This first case study deals with the violence episodes aimed at causing the resignation of the Egyptian President of the time Mohamed Morsi, occurred in Egypt in February 2013 in the day of the second anniversary of the fall of the dictatorship of the President Mubarak. We can notice a first impression of the negative evaluation of the journalist concerning the situation caused by the violent episodes simply by looking at the title "*In the square without peace*" and at the highlight "*the second anniversary of Mubarak's fall is lived in the symbol place of revolution- a kingdom of violence where first black blocks appear*", in which there is an absolute majority of negative terms such as *kingdom of violence, black block* and in which the only positive term *peace* is negated through the conjunction *without*. The journalist's negative evaluative standpoint is clear: "There is no peace in Tahrir square, which is a kingdom of violence", which is supported by an argumentation focused on the reasons that caused this violence and whose prominent goal is that of attributing the guilt of the whole situation to people causing violence. The journalist reaches his objective by identifying the outgroup with violent people, identifying them with a negative principle that must be eradicated; this process takes place by using the linguistic depersonalization of the members of the outgroup, indeed, when the journalist argues in favor of his negative standpoint, he always uses purely abstract arguments, or at least abstract arguments with some concrete nouns inserted in a categorical phrase[6].

Significant is the lead paragraph, constituting the first argumentative line, which includes abstract terms such as *violence, fall, annihilation, power* and the categorical nouns *dictator, lay opposition, Islamic executive branch*: the only exception is the use of concrete nouns with negative value such as *Molotov, tear-gasses, alarms, stones* and *truncheons*, which are in the plural form, which reinforces the idea of generalization of negativity associated to the outgroup: concrete nouns are used only in plural form and only as specification.

In order to better understand the reasoning processes at play when the journalist deals with the negative evaluations of the outgroup, it is necessary to reconstruct the argumentative structure of the newspaper article under investigation.

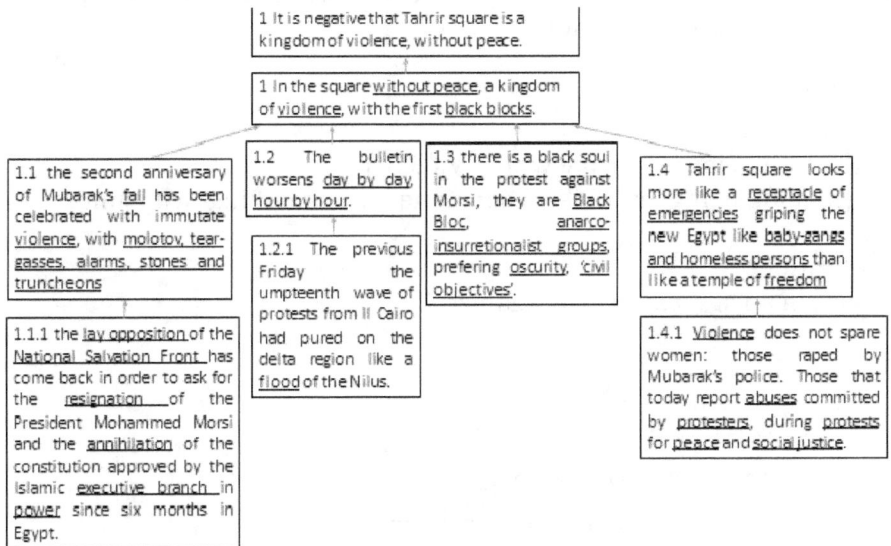

Fig. 2 Argumentative reconstruction of the newspaper article *In the square without peace, a kingdom of violence, with the first Black Blocks*, full of abstract arguments.

The argumentative structure in support of the negative evaluative standpoint is multiple and subordinative, as shown in Figure 1. The journalist argues that the violence is immutate in Tahrir square, even though Mubarak's dictatorship has ended two years before (1.1), which is in turn supported by the argument that the lay opposition of the

---

[6] Categorical phrases are phrases in which we can notice the presence of the designation of individuals or individuated subtypes of a category.

National Salvation Front has come back in order to ask for Morsi's resignation.

The use of abstract arguments leads the audience to conceptualize the protestors as a depersonalized enemy, in which the referent is indefinite, bringing the readers to consider the described events and features of these events as intrinsic properties, not limited to a certain temporary time, bound to the specific context of the present protest, but rather as durable and general characteristics of a situation that never stops though the fall of the dictatorship and though the change of many Presidents.

The semantic analysis following CT of one of the keywords of the article, namely *violence*, an abstract and therefore depredicative noun, reinforces the suggestion that the abstract noun *violence* through the suppression of arguments leads to conceive the act as performed by an indefinite referent, and therefore to generalize the acts of violence well beyond the present situation.

VIOLENCE
Presuppositions
X1 (human being; someone who is violent)
X2 (human being or object; someone or something who undergoes violence)
Implications
X1 acts in an aggressive way against X2

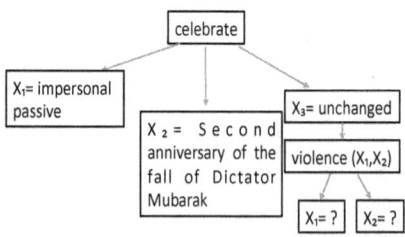

Fig.3 Semantic graph of the first argument containing the abstract noun *violence*.

It follows that we can consider this type of conceptualization as representing a subtle but powerful way to increase pre-existing stereotypes and to foster prejudice in the audience.

Although this first case study provides first evidence for the role of unaware emotionally colored standpoints in mass communication, it is not sufficient because it gives only a partial view of evaluative situations, due to its limitation to a negative evaluation of the outgroup, and a further step needs to be done.

5.2 Second case study: negative evaluation of events performed by the ingroup

Starting from the methodological necessity of analyzing a negative evaluation of the ingroup in order to observe if there is a difference in the usage of concrete or abstract nouns, the second case study selected concerns the analysis of a newspaper article of *CdT* describing an event performed by a Swiss citizen; I found out that the journalist expressing a negative opinion of a member of the ingroup puts forth concrete arguments, therefore attributing negativity to the contextual situation and conceiving the event as an "exception to the rule".

Similarly to the previous case study, in order to introduce how a journalist shapes the audience's conceptualization of a negatively judged event performed by an ingroup member, I introduce a newspaper article published on the 12th January 2013 on *CdT*. This case study deals with the violent episodes performed by a mentally ill person named Peter Kneubühl, who is a Swiss citizen and who shot at policemen a couple of years before the article was published. The issue at stake in the article concerns whether Kneubühl should be considered mentally ill and therefore treated in an asylum or whether he can be considered mentally sane and normally judged for the crimes committed. The journalist acts in an ambiguous manner: by quoting prosecutors' words, still in the title "Kneubühl's trial. The prosecutor: 'treatment in asylum'" he wants on the one hand, to ally with the prosecutor in supporting Kneubühl's mental insanity, and on the other hand, he wants to take the distances from him. The journalist's negative standpoint "Kneubühl must be treated in an asylum" is supported by an argumentation focused on the reasons that motivate this necessity and the prominent goal is that of attributing the guilt of the situation to the mental insanity of the accused, which would soften the idea of guiltiness of the Swiss citizen and unconsciously of the ingroup. The journalist reaches this aim by referring Kneubühl's negative features to the present situation and by leading us to conceive them as bound and limited to the context of occurrence of the fact, namely Kneubühl's shooting at police in a precise and well detailed situation due to his mental insanity; when the journalist argues in favor of his negative standpoint, he always uses almost purely concrete arguments, or at least abstract arguments with some concrete nouns inserted in a categorical phrase, as we can see in Figure 4.

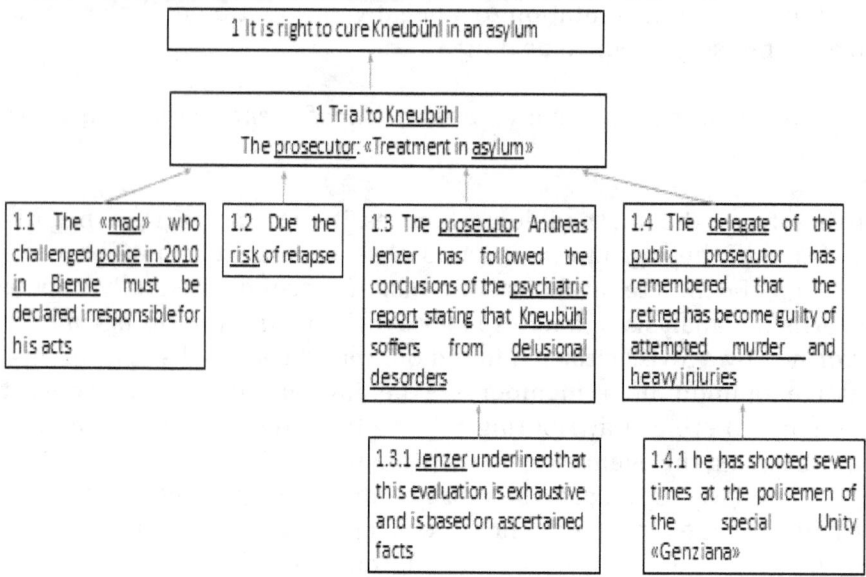

Fig. 4 Argumentative reconstruction of the journalist's argumentation supporting Kneubühl's mental insanity.

The argumentative structure in support of the negative evaluative standpoint is multiple and subordinative. The journalist argues that the Kneubühl must be treated in an asylum, due to the violent acts that he committed and to its mental insanity.

It is worth noting the prevalence of concrete nouns, for the majority explicitly negative, in the whole argumentation, which give force in arguing the limitedness of the negative evaluation; *asylum*, *mad*, *risk*, *psychiatric report*, *attempted murder* and *heavy injuries*.

Interesting is also the precise specification of the year (*in 2010*) and place (*in Bienne*) where the accused committed the crime in the first argumentative line (1.1), which contributes to delimitate the conceptualization of negativity; similarly this happens in the fourth argumentative line with the specification of the number of times that Kneubühl shot at police (*seven times*), instead of using an indefinite adjective (*many times*) and indicating a non-precise quantity.

The third argumentative line, being it an argument from authority reinforces the standpoint by adding the further authorial proof of the psychiatric report, stating that the accused suffers from *delusional disorders*. It is in this scenario that one of the few abstract nouns (*delusional disorders*) is found inserted in a categorical phrase: in this case the insertion in a category, namely that of a specific kind of mental disorders serves to delimitate the negative extension to a precise and definite field. This usage of the abstract noun in a categorical phrase is specular to that done by the journalist in the first case study judging a negative event referred to the outgroup, where it is the usage of concrete nouns – even though they are far and between- within a categorical phrase that hinges at a depersonalization of the outgroup in an anonymous category. This represents a further clue confirming the correlation existing between an ideal continuum shifting from concrete to abstract arguments and valence bound ingroup/outgroup stereotypic conceptualizations.

The sketch of the semantic graph following CT of the focal point of the fourth argumentative line, namely *become guilty of attempted murder and heavy injures* shows that all argument places are expressed, differently from the semantic graph analyzed in the first case study, in which unexpressed argument places are caused by the rules imposed by depredicative abstract nouns to their arguments. This reinforces the suggestion that concrete arguments lead to conceive the act as performed by a definite referent in a precise context at a specific time and in a specific place, and therefore to limit the negativity to the present situation.

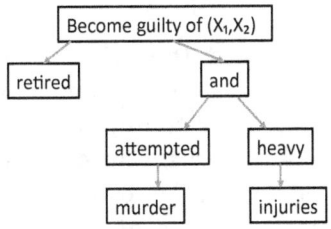

Fig.5 Semantic graph of the focal point of the first argumentative line.

It follows that we can consider the subliminal inducing of this conceptualization as a subtle but powerful way to convey a positive view of the ingroup, attributing negativity to the contextual situation and conceiving the event as an "exception to the rule", even though the judgement of the ingroup member behavior is explicitly considered as negative.

Again, this second case study provides good evidence for the role of unaware emotionally colored standpoints in mass communication; however, it is still incomplete, since I had the possibility to access to articles on Kneubühl only from *CdT*, indeed I did not find the same news in foreign newspapers.

5.3 Third case study: positive vs. negative evaluation of the outgroup and the ingroup in CdT and in a comparable Italian newspaper

Both the first as well as the second case study limit themselves to the study of the evaluation of an event from one group, due to the fact that, for practical reasons, I had not the possibility to access to newspapers from Egypt, and to find consistent amount of articles in foreign newspapers concerning the case study of the Swiss citizen accused of attempted murder.

Therefore, in order to ensure the presence of an adequate methodological setting, I selected a setting that allowed to have access to newspaper articles produced the by two groups (the Swiss group and a foreign group- the Italian group) regarding positive as well as negative behaviors performed by protagonists of each group, comparing *CdT*'s articles with articles from a comparable Italian newspaper, namely *Il Corriere della Sera*; the third case study is methodologically more complete in the sense that it gives us the chance to observe the same event (cross-border commuting) handled in the Swiss-Italian as well as in the Italian written press, even though it is still not exhaustive due to the fact that it does not comprehends a full match of positive/negative evaluations with ingroup/outgroup conceptualization of the same event.

The case study deals with the widespread phenomenon of Italian cross-border commuting at the border Italy-Switzerland: it is well known that Italian cross-border commuters are negatively judged in Switzerland, due to the Swiss accusation of stealing work to Swiss people, and positively judged in Italy, due to their efforts to honestly work at the price of travelling a lot and of being mishandled by many Swiss people.

In order to clearly identify the two groups undergoing evaluation in the articles, I Identified the two frames of cross-border commuters (outgroup for *CdT* and ingroup for *Il Corriere della Sera*) and of residents (ingroup for *CdT* and outgroup for *Il Corriere della Sera*) following frame semantics (Fillmore, 1986, 1992). Then, all negative/positive standpoints referred to each frame were identified, and then the argumentative reconstruction was carried out: results confirmed the predictions made during the previous case studies, so that abstract arguments were used to support the positive evaluative standpoint referred to residents for *CdT* and to cross-border commuters for *Il Corriere della Sera,* namely the ingroup for both groups, generalizing the positivity beyond the specific situation. On the contrary, concrete arguments were used support the positive evaluative standpoint referred to cross-border commuters for *CdT* (ingroup) and to support negative evaluation of cross-border commuters for *Il Corriere della Sera*, restricting the positivity to the contextual situation.

In Fig. 6 and 7 I will show the argumentative structures of the articles of *CdT* and *Il Corriere della Sera* made up by abstract arguments.

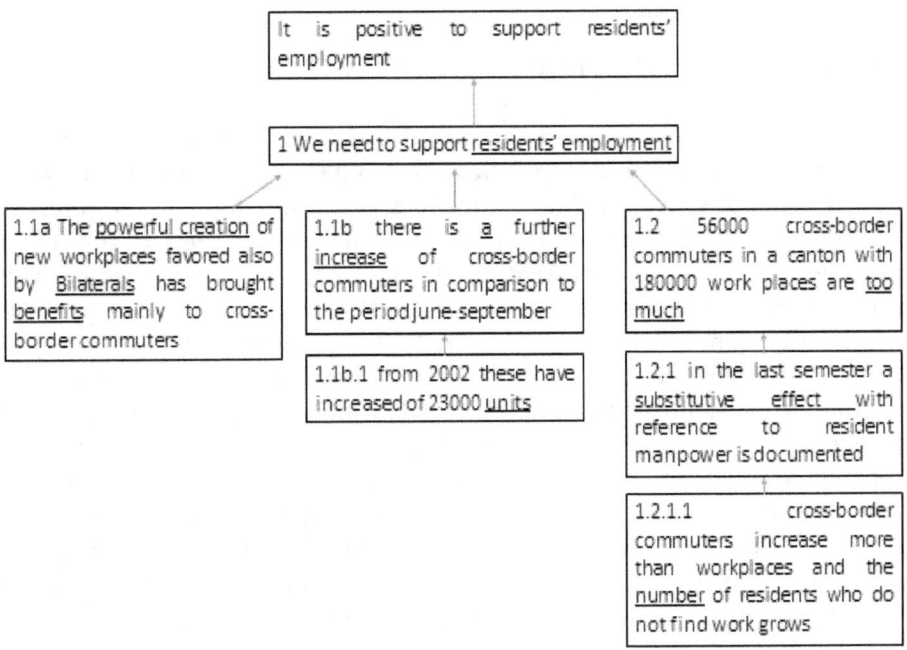

Fig. 6 Argumentative reconstruction of the newspaper article *We need to increase residents' employment* of *CdT*, full of abstract arguments.

In this article the whole argumentation aims at reinforcing a positive view of residents supporting the standpoint that residents' employment should be increased. There is plenty of abstract nouns (*increase, decrease, creation, benefit, substitutive effect*), which together with the indefinite adverb *too much*, as well as with the use of indeterminate articles (*an increase, an effect*) aim at generalizing the idea of negativity referred to cross-border commuters, extending it to all cross-border commuters, classified and depersonalized in *units*, without any exception. In this endeavor it is interesting to highlight the use of *too much*, analyzed according to CT, as shown below:

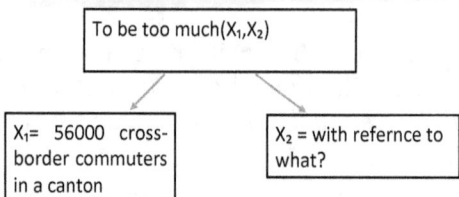

Fig.7 Semantic graph of the focal point of the second argumentative line.

As we can see the second argument of the predicate "to be too much" has remained unfilled: no specification is given with reference to what the number of cross-border commuters is too much, only the physical place (the canton) in which they are too much is specified. In the whole argumentation it is interesting to notice the presence of two hierarchically-ordered evaluations both supported by abstract arguments, generalizing both negativity and positivity: the negative one referred to cross-border commuters is subordinate to (and acts as an argument in favor of) the positive evaluation referred to residents. With reference to this article, I found a further proof confirming the journalists' role of stereotype fosterer, by looking at a retrospective clue, namely a letter of a reader contained in the section letters and opinions: the letter written by the reader, ideally representing the audience, explicitly supports the journalists' standpoint that there is the need to increase residents' employment, and more specifically, it says

> The journalist is right when in his article he writes that there is the need to guarantee the workplace to residents. And I also agree with the reasons that he puts forth in support of this, that can be shared.

This opinion represents a retrospective clue of the hypothesis previously made, being it a proof of the incisiveness of the journalist's arguing in favor of his standpoint, which goes well beyond the newspaper article, and spreads in society through the audience, contributing to knowledge transformation, to the maintenance or even fostering of previous stereotypes.

A similar usage of abstract arguments is made in the following article taken from *Il Corriere della Sera*, in which abstract arguments are a useful means to express the journalist's solidarity with cross border commuters (ingroup), as we can see in Fig. 5;

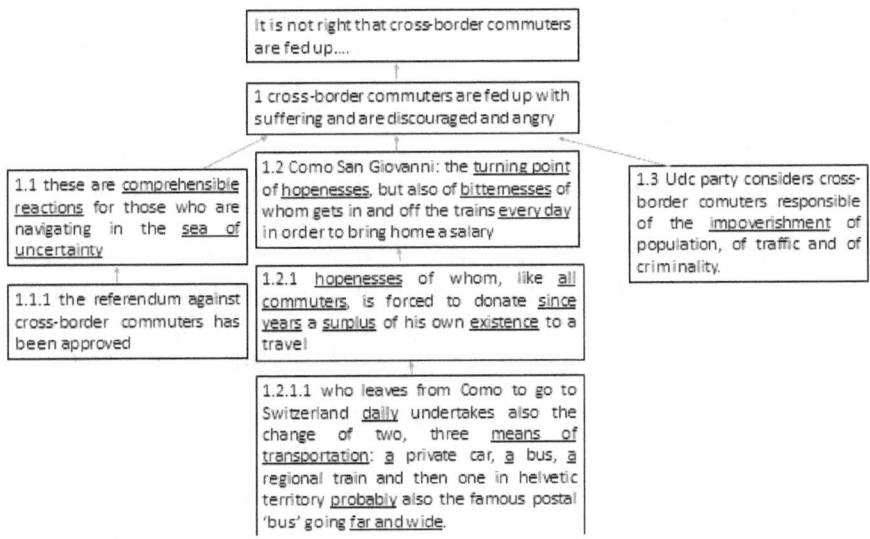

Fig. 8. Argumentative reconstruction of the newspaper article *Cross-border commuters are fed up with suffering* from *Il Corriere della Sera*, full of abstract arguments.

From the observation of the argumentative structure shown in Fig. 5, again, we find *a)* plenty of abstract nouns, most of them used in plural form to accentuate the idea of repetition (*reactions, uncertainty, turning point, hopenesses, bitternesses, years, surplus, existence....*), *b)* undetermined articles used to convey depersonalization, *c)* the preposition *since* conveying the idea of a durable status lasting since long time (*since years*) *d)* adjectives conveying the idea of repetition and collectivity (*every day, all commuters*), *e)* the metaphor of the *sea of uncertainty*, which is used by the journalist in order to stress and to widen the indefinite horizon of cross-border commuters' workplace, as we can see even from the suppression of the second argument after a semantic analysis following CT:

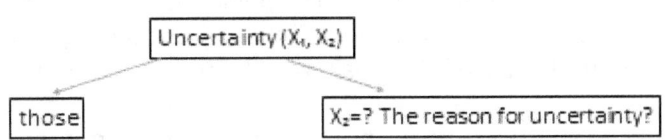

Fig. 9 Semantic graph of the focal point of the first argumentative line.

f) a semantic climax concerning commuters' travels in the second argumentative line (1.2.1.1), starting with a car an ending up in an indefinite location reached by a bus.

All this lexico-grammatical features have a big weight in the argumentative force of the argumentation put forth by the journalist in highlighting cross-border commuters' suffering and in empathizing with them.

Differently, in what follows, I will show two argumentative reconstructions (Fig. 6 and 7) made up by concrete arguments, laying stress on the situational context; they were used to support a positive evaluation referred to cross-border commuters (outgroup) in *CdT* and a negative evaluation referred to cross-border commuters (ingroup) in *Il Corriere della Sera*.

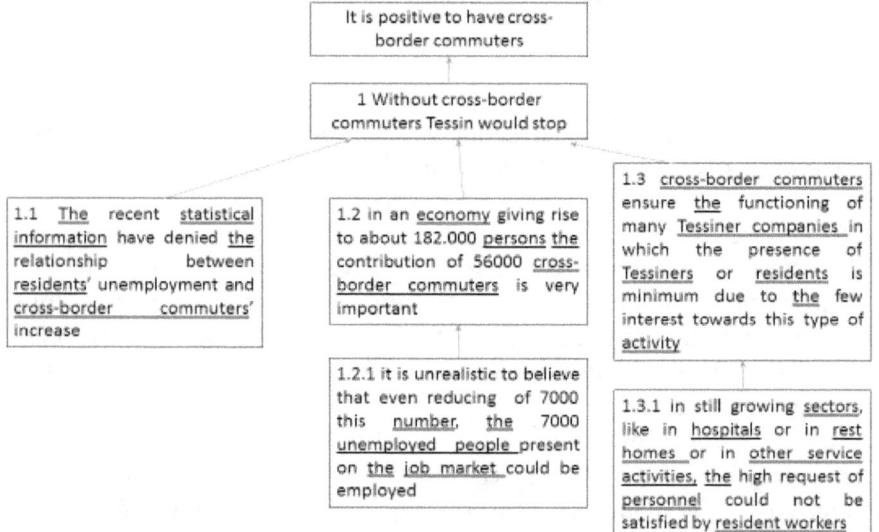

Fig. 10. Argumentative reconstruction of the newspaper article *Without cross-border commuters Tessin would stop* of *CdT*, full of concrete arguments.

In the argumentative structure shown in Fig. 6. it becomes evident the abundance of concrete nouns (*statistical information, economy, persons, job market, companies, hospitals, rest homes, personnel...*), whereas the few abstract nouns are always preceded by a determinative article, ensuring for the delimitation of the positive attribution to the present specific situation.

Again, in a specular way, in the following argumentative reconstruction, supporting the negative opinion of the journalist of *Il Corriere della Sera* concerning the ingroup, there is plenty of concrete arguments;

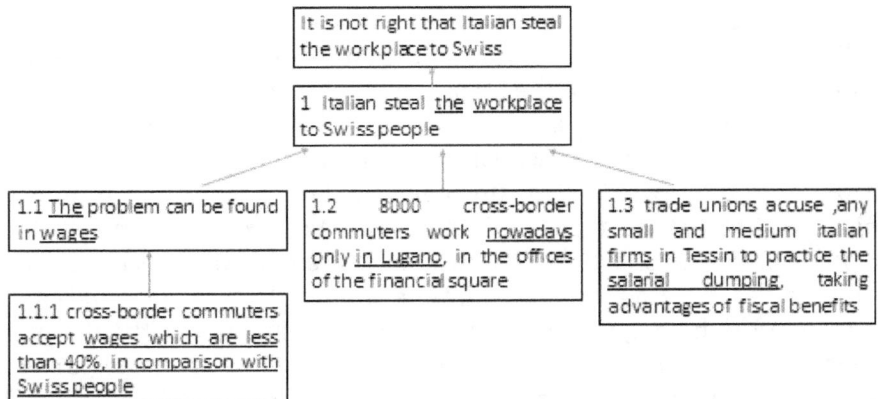

Fig. 11. Argumentative reconstruction of the newspaper article "Italian steal the workplace to Swiss people" of *Il Corriere della Sera*, full of concrete arguments.

As we can see from Fig. 7, in the first argumentative line (1.1.1) the arguments of the predicate *to be less than*, expressing the basis for comparison, are all fully expressed, as I better clarify in the following semantic graph following CT;

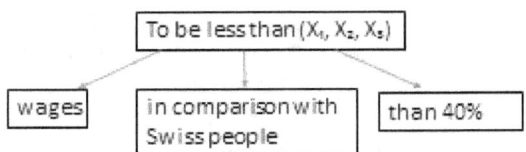

Fig. 12 Semantic graph of the focal point of the first argumentative line (1.1.1).

Furthermore, the time and place references are precisely specified, *i.e. nowadays*, which anchors the argumentation to a specific present moment of time, and **only in** *Lugano*, which delimitates the area in which commuters work to a specific point in space, conceived as a delimited container (for more details on the relationship between prepositions and cognition see Levinson, 2003).

6. CONCLUSIONS

The results drawn from the interweaving of semantic and argumentative analysis shown in the case studies confirmed the predictions and allow us to conclude that argumentation consisting of abstract arguments, generalizing beyond the specific situation, is actually used to support the positive evaluative standpoint of the ingroup as well as the negative evaluative of the outgroup, whereas argumentation made up by concrete arguments, laying stress on the situational context supported a negative evaluation of the ingroup as well as a good evaluation of the outgroup. It follows that results shed light on the importance of the stance taken by the journalist in the news product, which may determine the fostering of existing or newly shaped stereotypes in the audience. The journalist acts as active knowledge mediator and knowledge transformer, imposing a particular framing to the news. In this endeavor I have shown the importance of analyzing in-depth semantic properties of nouns, being them abstract or concrete, in order to reach an adequate comprehension of lexical meaning and to shed light on the major/minor salience of certain parts of the discourse given by the journalist's framing of the news. Moreover, the carefully conducted semantic analysis has laid stress on the weight given by abstract/concrete nouns to the argumentative force of whole argumentative structures. More specifically, in case of evaluation supported by abstract arguments the depersonalization conveyed by abstractness leads to consider non-existent features as existent and intrinsic to an entity, not bound to a specific context, whereas in case of concrete arguments attention is drawn on the situational context and on the exceptionality of the attributed positivity/negativity.

However, much remains to be done, and the analysis poses further challenges. Firstly, in view of these results, shedding light on the influence of intergroup bias phenomena on news products, I suggest that the study would be exhaustive only considering the whole chain of news production comprehending newsmaking decisions in editorial meeting conferences as well as writing processes in news' editing (screenshots of journalists writing, modifying and updating articles), and verifying the implemented strategy in the final product. I have first evidence from previous studies (Rocci & Luciani, forthcoming) and from still ongoing studies that sketching a whole chain of newsmaking processes enables us to conceive the journalist as an active interpreter of the news, able to decide whether to convey positive or negative view of what he is evaluating, being however sometimes victim of possible stereotypic beliefs, which, whether not acknowledged, can dangerously spread in the audience. This changes a lot the deterministic view given by purely product-based analysis of social psychology simply stating the presence of a systematic intergroup bias in mass communication, and enriches the view of the journalist's active role, being able to open a space for critical discussion through argumentation, sometimes constrained by intergroup bias phenomena. Secondly, concerning the improvement of the methodological setting, all possible noise factors undermining the comparability of analyzed newspapers such as style, ideological and institutional norms should be eliminated. More specifically, my future aim is to compare the results of Italian data with those taken from the Italian newspaper *La Padania*, which is a xenophobic newspaper taking opposite positions against immigrates, in order to exclude a correlation of abstract/concrete nouns with the ideology of journalism headlines. Furthermore, starting from the well acknowledged fact that journalists use reported speech to support their own standpoint (White, 2012), the role of journalists' source material should be deepened, for example investigating which type of quotations they select and which role they play to support which type of standpoint. Lastly, it would be necessary to go beyond the interactional analysis and to study abstractness/ concreteness of nouns even at the inferential level, and I intend to do that in future work using *Argumentum Model of Topics* (Rigotti & Greco Morasso, 2010), in order to better understand the missing link represented by implicit premises.

REFERENCES

Bednarek, M. (2009). Dimensions of evaluation: cognitive and linguistic perspectives. *Pragmatics & Cognition, 17*(1), 146-175.

Bednarek, M., & Caple, H. (2012). 'Value added'. Language, image and news values. *Discourse, Context & Media, 1*, 103-113.

Carnaghi, A., & Arcuri L. (2008). *Parole e categorie: La cognizione sociale nei contesti intergruppo.* Milano: Raffaello Cortina Editore.

Cigada, S. (1999). *Nomi e cose. Aspetti semantici e pragmatici delle strutture nominali.* Milano: ISU.

Corneille, O., Klein, O., Lambert, S., & Judd, C. M. (2002). On the role of familiarity with units of measurement in producing categorical accentuation: Tajfel and Wilkes (1963) revisited and replicated. *Psychological Science, 4*, 380-383.

Greco, S. (2003/2). When presupposing becomes dangerous. How the procedure of presuppositional accommodation can be exploited in manipulative discourse. *Studies in Communication Sciences, 3*, 217-234.

Eemeren, F.H. van, & Grootendorst, R. (2004). *A systematic theory of argumentation: The pragma-dialectical approach.* Cambridge: Cambridge University Press.

Fillmore, C. (1968). The case for case. In E. Bach & R. T. Harms (Eds.), *Universals in Linguistic Theory* (pp. 1-88). New York: Holt, Rinehart, and Winston.

Fillmore, C., & Atkins S., (1992). Towards a Frame-based Lexicon: the Semantics of RISK and its Neighbours. In A. Lehrer & E. F. Kittay (Eds.), *Frames, Fields and Contrasts: New Essays in Semantic and Lexical Organization* (pp.75-102). Hillsdale, NJ: Lawrence Erlbaum Associates.

Flaux, N., Glatigny, M., & Samain, D. (1996). *Les noms abstraits. Histoire et Théorie.* Lille: Septentrion PU.

Freeman, J. (2000). What types of statements are there? *Argumentation, 14*(2), 135-157.

Levinson, S. C. (2003). *Space in language and cognition: Explorations in cognitive diversity*. Cambridge: Cambridge University Press.
Maas, A., Salvi, D., Arcuri, L., & Semin, G. (1989). Language use in intergroup contexts: The linguistic intergroup bias. *Journal of Personality and Social Psychology, 57*, 981-993.
Maas, A., Corvino, P., & Arcuri, L. (1994). Linguistic intergroup bias and the mass media. *Revue de Psychologie Sociale, 1*, 31-43.
Maas, A., Milesi A., Zabbini S., & Stahlberg, D. (1995). The linguistic intergroup bias: Differential expectancie or ingroup protection? *Journal of personality and social psychology*, 68, 116-126.
Perrin, D. (2003). Progression analysis (PA): Investigating writing strategies at the workplace. *Journal of Pragmatics, 35*(6), 907–921.
Perrin, D. (2012). Stancing: Strategies of entextualizing stance in newswriting. *Discourse, Context & Media, 1*, 135-147.
Perrin, D. (2013). *The Linguistics of Newswriting*. Amsterdam/Philadelphia: John Benjamins.
Prisciano, *Institutionium Grammaticarum* l. XVIII, H. Keil ed., Teubner, Lipsiae 1855-1859.
Rigotti, E., & Cigada, S. (2004). *La comunicazione verbale*. Milano: Apogeo.
Rigotti, E., & Rocci, A. (2006). Tema-rema e connettivo: la congruità semantico-pragmatica del testo. In G. Gobber, M.C. Gatti & Sara Cigada (Eds.), *Syndesmoi: connettivi nella realtà dei testi* (pp. 3-44). Milano: Vita e Pensiero.
Rigotti, E., Rocci, A., & Greco, S. (2006). The semantics of reasonableness. In P. Houtlosser & A. van Rees (Eds.), *Considering Pragma-Dialectics* (pp. 257-274). Mahwah, NJ/London: Lawrence Erlbaum Associates.
Rigotti, E., & Greco Morasso, S. (2010). Comparing the Argumentum Model of Topics to Other Contemporary Approaches to Argument Schemes: The Procedural and Material Components. *Argumentation, 24*(4), 489–512.
Rocci, A., & Luciani, M. (forth.) Economic-financial journalists as argumentative intermediaries. Special Issue of *Journal of Argumentation in Context*.
Rothbart, M., & Taylor, M. (1992). Category labels and social reality: Do we view social categories as natural kinds? In G.R. Semin & K. Fiedler (Eds.), *Language, Interaction and Social Cognition* (pp.11-36). London: Sage.
Semin, G.R. (2000). Agenda 2000: Communication: Language as an implementational device for cognition. *European Journal of Social Psychology, 30*, 595-612.
Stapel, D.A., & Semin, G. (2007). The magic spell of language: Linguistic categories and their perceptual consequences. *Journal of Personality and Social Psychology, 93*(1), 23-33.
White, Peter R.R. (2012). Exploring the axiological workings of reporter voice news stories. Attribution and attitudinal positioning, *Discourse, Context and Media, 1*(2), 57-67.

Commentary on Luciani's Journalists' Colored Standpoints Emerging in Argumentation

HENRIKE JANSEN
Leiden University Centre for Linguistics, The Netherlands
h.jansen@hum.leidenuniv.nl

1. INTRODUCTION

Luciani's paper deals with the interesting question of how biases that people unconsciously hold can be deduced from the wordings they use. She claims there is an intergroup bias factor at work in mass communication, which is revealed in the kind of nouns that are used in newspaper articles. Her hypothesis is that abstract nouns are used when something positive is reported about the ingroup, and concrete nouns are used when giving a negative report about the ingroup. The explanation for this is that journalists – unconsciously – generalize about something positive relating to the ingroup but, for the same reason, report in a concrete way on something negative relating to the ingroup, because concreteness suggests that this negative thing was a one-off occurrence. Therefore, the way in which journalists give an account of events concerning the outgroup is exactly the other way around, because they have a negative bias towards people outside their own group.

My comments, which are sometimes questions for clarification, will focus on four aspects of Luciani's paper: (1) the way concrete and abstract nouns are distinguished, (2) the demarcation of ingroup and outgroup, (3) the role of the case studies, and (4) the role of the argumentation structures.

2. ABSTRACT AND CONCRETE NOUNS

According to Luciani, the difference between abstract and concrete nouns is a matter of semantic structure. A concrete noun is something that does or does not exist; an abstract noun represents something that either can or cannot happen. A concrete noun can function as an argument; an abstract noun presupposes arguments. The term 'argument' here means a linguistic category – and not the arguments we are talking about when we mean "argumentation". Looking at Luciani's example about the robbery and her explanation of the concept, I get the impression that an abstract noun is often a noun morphologically derived from or related to a verb. As verbs often go together with a subject and one or more objects, these arguments (again, in the linguistic sense) are always evoked in the mind of the hearer, even when they are not explicitly expressed.

Luciani calls this evocation of arguments in the hearer's mind the "manipulative" function of abstract nouns, because "they presuppose the existence of fake realities" (p. 434). Now this, in my view, requires a bit more explanation. I wonder what is meant by "fake" realities – i.e. why would they be fake? Are they fake just because they have been left implicit in someone's utterance? Then the word fake has a connotation that does not seem appropriate here. After all, this word suggests that entities are evoked that should not be evoked. However, looking at the example of the robbery, the entities evoked are a robber and the robbed, and what's wrong with that? Why shouldn't they have been evoked? It would have been interesting if the paper had offered some further instruction on this point.

3. INGROUP AND OUTGROUP

My second comment concerns the fact that I am not entirely sure what Luciani means by the concepts "ingroup" and "outgroup". Her first case study is about a newspaper article (from the *Corriere del Ticino*) on the demonstrations against the former Egyptian President Morsi on Tahrir Square in Cairo. The article contains many abstract nouns and other linguistic features suggesting a generalization; i.e. Concrete nouns are used as well, but they serve as the linguistic arguments for the abstract ones. Moreover, these concrete nouns are plurals, which also contributes to the suggestion of generalizability. Luciani says her semantic analysis of the article's linguistic characteristics reveals that the audience is led to think of the demonstrations as an ongoing thing, not restricted by time limits, and therefore as intrinsic to the group they concern. This case study is contrasted with an example of language use containing many concrete nouns in an article (also from the *Corriere del Ticino*) about a Swiss citizen who shot a policeman, where the shooter is portrayed as being mentally ill. This article also conveys a negative stance on the event, but the use of concrete nouns and the precise indication of place and time etc. suggest that the negativity of this event is limited solely to this event.

I have some doubts about these conclusions. Firstly, I wonder whether the only possible explanation for the concrete or abstract language is bias, or whether it might be explained by simply looking at their respective topics. After all, the article about the Swiss shooter is about one person and it seems difficult to describe a story about a concrete person in an abstract way. In contrast, the Cairo mass demonstrations involve – by definition – many people (demonstrators). Moreover, it was an actual fact that these demonstrations had been going on for a long time and it seems only natural that this should be conveyed in the report. Therefore, in my opinion, both the plurals and the lack of concreteness can be explained by the article's content, which is a more obvious explanation than the respective journalists' biases.

Another reason for my doubt is that I think the concepts of ingroup and outgroup lack definition and demarcation of who their members are. After all, whether the article on the Egyptian demonstrations really reveals a bias held by a Swiss journalist towards the outgroup, also depends on who exactly make up this outgroup. Is it some of the demonstrators that constitute this group, or is it all of them? Is it all Egyptians? Or is it all non-Swiss people? And if the latter is true, does that mean a Swiss journalist sees any Swiss citizen as ingroup? Or could there be categories within one country to which the concepts of ingroup and outgroup also apply?

4. THE ROLE OF THE CASE STUDIES

In her paper, Luciani says that the case studies serve as "evidence for the role of unaware emotionally coloured standpoints in mass communication" (p. 440, p. 443). Presumably she intends these cases as merely an illustration, since three case studies cannot provide this evidence. But it would be interesting to indeed find proof for Luciani's hypotheses. To gather such proof, a systematic search for and language analysis of other articles in the *Corriere del Ticino* is needed. These articles should not only be about the respective topics of the Tahrir demonstrations and the mentally ill Swiss shooter, but also about mass demonstrations in other places (including Switzerland) and other people who have shot another person (including abroad). It would also be interesting to know how the Egyptian newspapers themselves report on the mass demonstrations. Of course, for practical reasons it was not possible for Luciani to examine this, but such research could give some very interesting insights.

The third case study does actually provide the kind of journalistic comparison that is needed for Luciani's research question. In this study, she contrasts the coverage of Italian cross-border commuters in a Swiss newspaper and an Italian newspaper. Luciani's findings on this point could perhaps be generalised if more articles are analysed. In this respect, I would recommend that the comparison of

Swiss and Italian reports on Italian commuters' activities on the Swiss side of the border is supplemented with a similar comparison concerning Swiss commuters' activities on the Italian side of the border.

5. THE ROLE OF THE ARGUMENTATION STRUCTURES

With my last comment, I would like to ask Luciani why she thinks it is necessary for this kind of research to perform an argumentation analysis (an overview of how the standpoint and premises are linked). To my mind, a newspaper article reporting on news events is informative and not argumentative; but indeed, Luciani's argument is that the authors of these articles unconsciously present a standpoint on the situations they describe. However, my point is that the argumentation structures composed by Luciani do not appear to affect the semantic analysis. It seems possible that the latter can be performed without the argumentation structures. The exact role of these structures is therefore not entirely clear to me, and this makes me doubt the benefits of this specific combination of concepts from argumentation theory and linguistics. This does not alter my opinion, however, that insights from both disciplines can offer a very productive way of conducting research.

6. CONCLUSION

Despite my above comments, I think Luciani has offered us a very interesting research question, which certainly warrants investigation. If there are biases at work in journalism, we need to know that. We also need to know how we can recognize them. Luciani has marked the contours of the way such research should be conducted. Her exploration of three case studies is promising; investigation of the use of nouns seems to offer an appropriate method for revealing hidden assumptions. More case studies could indeed deliver the proof that journalists suffer from ingroup and outgroup bias, provided that a workable definition of these concepts is offered.

20

Malleable and Predictable Source Credibility? American Election Candidates as a Case Study

JENS KOED MADSEN
Birkbeck, University of London, UK
j.madsen@bbk.ac.uk

> Source credibility is an important influence in reasoning and persuasion and has been modelled from a Bayesian perspective an amalgamation of epistemic expertise and trustworthiness. Taking point of departure in a domain-dependent conceptualisation of trustworthiness and expertise, four studies conceptualise source credibility in American politics (study 1), party-specific election candidates in the USA (study 2), individual predictions of electoral source credibility (study 3), and priming of trustworthiness facets (study 4). Consequences of findings are discussed.
>
> KEYWORDS: election candidates, epistemic expertise, politics, probabilistic estimations, source credibility, trustworthiness, USA

1. INTRODUCTION

Source credibility has been defined as an amalgamation of trustworthiness and epistemic expertise (Bovens & Hartmann, 2003, chapter 3; Wang et al., 2011), and has proven to have a significant impact on a range of aspects central to political life such as reasoning about evidence (Hahn et al., 2009; 2012; Harris et al., 2015), the estimation of the persuasiveness of a message (e.g. Petty & Cacioppo, 1984; Chaiken & Maheswaran, 1994), generating and intention of whether or not to vote (Householder & LaMarre, 2014), deciding which candidate to vote for (Citirin & Muste, 1999), and complying with public policy (Ayres & Braithwaite, 1992, for a general review see Levi & Stoker, 2000). As such, source credibility is central to humans in at least two ways: it influences how convinced humans become of uncertain information from uncertain sources and it influences both intention generation and decision-making in terms of action. Expertise and trustworthiness, however, are relatively broad and abstract concepts that may take on different operationalizations depending on the context and epistemic domain in which they emerge. That is, both concepts may be domain-dependent and situationally malleable. The paper explores this potential psychological foundation of source credibility.

Epistemic expertise is clearly domain-dependent. For example, an expert podiatrist is not necessarily an expert on 19th century French poetry and vice versa. Perhaps due to the clear domain-dependency, what makes someone appear expert has, to our knowledge, not been studied extensively. Conversely, trustworthiness is frequently thought to be domain-independent and universal and has therefore been subject to a host of studies that aim to define and operationalize the concept. It has been studied in political science (Miller & Listhaug, 1990; Citrin & Muste, 1999), in rhetorical theory (McCroskey & Young, 1981), and in management literature (Colquitt et al., 2007). These studies typically aim to describe trustworthiness a priori, through common-sense definitions, or through factorial analyses. Common to the studies, however, is an aim of uncovering what makes a source trustworthy in general. Comparatively, the current paper argues that conceptualisations of trustworthiness, like epistemic expertise, are domain-dependent and that no universal definition can be offered (aside from general and abstract concepts such as 'integrity', which can take different forms depending on domain and context). Instead, what becomes important is to develop a methodology that elicits the relevant aspects of source credibility in a given domain and socio-cultural setting such that it may be used to define, describe, end predict the influence of that aspect of source credibility on argumentation and decision-making in that particular context for that particular problem.

The paper discusses a series of studies on political source credibility in general and political trustworthiness in particular[1]. These studies explore the operationalisation of political source credibility (study 1), the influence of specific political parties (study 2), predictability (study 3), and whether factors are malleable and may be primed (study 4). Collectively, the studies suggest that trustworthiness, like epistemic expertise, is domain-dependent and that it manifests distinctly in different epistemic domains[2]. We should therefore expect (subtly or less subtly) different factors of trustworthiness to emerge when considering the persuasive potential of a legal source in court

---

[1] Results and discussions are based on three working papers, which are available upon request. The data was collected with Brent Clickard

[2] As the paper only explores one domain, the entailments are indicative rather than conclusive. Results should therefore be read with this limitation in mind.

compared with an election candidate (difference in domains), and we might expect different factors to emerge in the same domain, but given different socio-cultural context (e.g. what makes an election candidate trustworthy in the USA might differ from election candidates in Sweden).

The results presented in the paper point towards the development of a methodology to elicit source credibility. On a theoretical level, the studies suggest that the formulation of trustworthy argumentation and reasoning is domain-dependent and should vary with epistemically different contexts. A malleable foundation of persuasion and argumentation would have implications for the development of normative theories that predict, analyse, and critically evaluate instances of argumentation. For example, it may influence the operationalization of key aspects of subjective probabilistic estimations in a Bayesian framework of argumentation (e.g. Hahn & Oaksford, 2006; 2007; Corner et al., 2011; Harris et al., 2012), it may influence the dialogical rules that constitute fair and honest discourse (van Eemeren & Grootendorst, 2004), and it may influence how to critically evaluate rhetorical artefacts (Foss, 2004). Practically, the development of a method is useful to define and predict the influence of uncertain sources in a specific domain such as politics in the USA. Before presenting the four studies, a brief presentation of the definition and influence of source credibility is useful to set the scene for further discussion.

2. SOURCE CREDIBILITY: DEFINITION AND INFLUENCE

Given the importance of source credibility in argumentation, reasoning, and decision-making, it is hardly surprising that several disciplines have explored the influence of the concept. Amongst these, we find judgment and decision-making (e.g. Birnbaum & Stegner, 1976), advertising (e.g. Braunsberger & Munch, 1998), developmental psychology (e.g. Harris & Corriveau, 2011), the evaluation of legal testimony, both from a normative perspective (e.g. Lagnado et al., 2013; Schum, 1994) and from a descriptive perspective (see Wells & Olson, 2003, for a review).

As mentioned in the introduction, epistemic expertise has not been subject to thorough investigation. However, relevant to the current paper, political science describes trustworthiness as integrity, competence, fairness, flip-flopping, honesty, equitable, and being responsiveness to public needs (Miller & Listhaug, 1990; Citrin & Muste, 1999; Levi & Stoker, 2000), which mirrors definitions offered in rhetorical theory (e.g. McCroskey & Young, 1981; Aristotle, 1995; McCroskey, 1997), cognitive and social psychological findings (e.g.

Bovens & Hartmann, 2003; Hahn et al., 2009; Fiske et al., 2007; Cuddy et al., 2011), and reasoning theory (Walton, 1997).

Most consistently and backed by empirical exploration, trustworthiness has been studied in management literature (Mayer et al., 1995; Mayer & Davis, 1999; Mayer & Gavin, 2005, see Colquitt et al., 2007 for a review). Reminiscent of definitions offered in the previous paragraph, trustworthiness is described as benevolence, integrity, and ability. Benevolence is defined as "...the extent to which a trustee is believed to want to do good to the trustor", ability is "...that group of skills, competencies, and characteristics that allow a party to have influence within some domain", whilst integrity is "...the trustor's perception that the trustee adheres to a set of principles that the trustor finds acceptable" (all Mayer & Davis, 1999, p. 124). These facets might encompass the factors from political science studies (such as "fairness"), and, given the empirical support from management studies, offer a good point departure for testing trustworthiness in the political domain. Study 1 makes use of these factors when exploring political trustworthiness of election candidates in the USA.

The literature yields a notable difference between propensity for trust and trustworthiness, both of which will be discussed in study 1 (see e.g. Colquitt et al., 2007). Propensity for trust can be defined as "...a generalized expectancy that the verbal statements of others can be relied upon" (Rotter, 1967, p. 664) and is measured via a 20-item scale (Kee & Knox, 1970; Couch et al., 1996). Trustworthiness, on the other hand, refers to the qualities that make someone appear to be, literally, worthy of trust. If the trustworthiness factors identified are appropriate and relevant for the particular domain, we predict a correlation between propensity for trust and estimations of trustworthiness. Indeed, "In the ideal case, one trusts someone because she is trustworthy, and one's trustworthiness inspires trust" (Flores & Solomon, 1998, p. 209). For study 1, the propensity of trust provides a check for validity to determine if the factors identified are relevant for the political domain.

Studies have shown the impact of source credibility when concerned with the political domain specifically. For example, Householder and LaMarre (2014) indicate a correlation between level of trust and the intention of voting, studies have shown that lack of trust in government may influence the *choice* of candidate such that low levels of trust yields more votes against incumbent candidates (Hetherington, 1999; Levi & Stoker, 2000), trusting citizens are more likely to comply with government policy (Ayres & Braithwaite, 1992; Pettit, 1995; Braithwaite & Makkai, 1994) as well as engage with the economy and society at large (Fukuyama, 1995; Levi, 1997; 1998).

Distrust in government can further incentivise civic engagement (see Levi & Stoker, 2000 for a discussion of this). However, reviewing voting behaviour of young British citizens in the 2005 election, Dermody et al. (2010) find no significant correlation between level of trust and likelihood of vote (see also Little et al., 2012 for comments on this). Interestingly, voters seem to believe that elected officials should adhere to higher ethical standards than ordinary citizens (Birch & Allen, 2010). As such, trust in election candidates and government has far-reaching, although not deterministic consequences, as it modulates how information is processed, intention of voting and choice in candidate, and compliance with public policies.

Although the above studies might be in disagreement regarding the specific nature of the influence, the conceptualisation, and whether the influence of source credibility is normatively beneficial or detrimental, the studies referenced in this section strongly indicate that source credibility somehow *does* play a major role in how humans make sense of evidence and whether or not they are likely to act in accordance with the information. The following section presents four studies that explore aspects of political source credibility of election candidates in the USA.

3. CASE STUDY: AMERICAN ELECTION CANDIDATES

The following presents the results from separate studies that explore political source credibility of election candidates in the USA. Study 1 reports factorial analyses of data from two surveys (1a and 1b) regarding political trustworthiness and expertise in order to describe the key elements of political source credibility. Although epistemic expertise does not fragment into distinct factors, political trustworthiness of election candidates in the USA departs from the factors identified in management studies and can be described as an amalgamation of capability, consistency & closeness, egotism & opportunism, and communal commitment. In addition to identifying relevant factors, study 1 also indicates that swing voters are trust election candidates significantly less than do staunchly affiliated voters. Study 2 explores party-affiliation in more detail and provides evidence for party-bias, as staunchly affiliated voters rate non-identified election candidates from their preferred party higher than swing voters and voters who are staunchly affiliated with the opposite party. Study 3 uses the items that make up the four trustworthiness factors (amalgamation of capability, consistency & closeness, egotism & opportunism, and communal commitment) to predict political trustworthiness. The items prove to be somewhat predictive and account for 49.5% of the trustworthiness variance. Finally, study 4 explores the malleability of trustworthiness factors by priming participants to consider one factor over the others. Results suggest that when participants were reminded of the importance of a factor, they were subsequently more likely to rate that factor as the most important and that election candidates were less likely to embody it (as priming was negative). This indicates that the emergence of probabilistic estimations of the likelihood of each factor may be malleable and contextually dependent.

3.1 Describing source credibility of American election candidates

Two surveys (survey 1a and 1b) were completed in June 2014. We performed factorial analyses to explore the main factors of source credibility identified in the argumentation literature (trustworthiness and epistemic expertise) for election candidates in the USA.

3.1.1 Political trustworthiness (study 1a)

Materials: Participants completed the 20-item propensity for trust survey (5 items were reverse scored) to elicit how likely they were to trust other people (Couch et al., 1996). Following this, participants responded to 3x20 items relating to the three trustworthiness factors in management studies (benevolence, integrity, and capability, 5 items were reversed scored in each category). Items were phrased as neutrally as possible (e.g. "most election candidates are very qualified"), and participants were asked to rate each item on a 5-point Likert-type scale (1 meaning "disagree strongly" and 5 "agree strongly"). Finally, participants provided demographic information as well as party affiliation. The latter was elicited to determine how favourably disposed participants felt towards the two major parties on a scale from 0-10 (0 = very negative, 10 = very positive). If a participant responded 7 of higher to party affiliation, he or she would be labelled a staunch supporter of that party (e.g. a staunch Republican). Participants who responded lower than 7 to both parties were classified as a swing voter. In sum, propensity for trust (items were randomized), trustworthiness items (randomized), party affiliation, and demographics were measured (presented in that order).

Participants: 600 Participants were recruited from Mechanical Turk (see Paolacci et al., 2010 for analysis and validation of MT as a recruitment platform for social scientific experiments). Participants who completed the survey in less than 10 min or who provided the wrong completion code were discarded from the analyses, leaving 561 participants. All participants were eligible to vote and American citizens.

Hypotheses: Survey 1a was exploratory to probe whether the factors identified in management literature could be replicated in a political domain. We expect that trustworthiness is domain-dependent and should therefore yield different, but distinct factors of political trustworthiness. As a check for validity, we further predict that propensity of trust is positively correlated with subjective probabilistic estimations of trustworthiness. We have no prediction for party affiliation, but we expect swing voters have less trust in election candidates than staunch voters, as level of perceived trust is correlated with choice of candidate and intention of voting (see Levi & Stoker, 2000).

Results: Items scoring < .5 in the factor analysis were not considered strong enough for the analysis. Items from the original trustworthiness categories failed to cluster around the original factors, but instead formed new and distinct factors. With high internal consistency (tested via a Cronbach's alpha), items related to capability provided a distinct and separate category (8 items, α = 0.888). However, items related to integrity and benevolence emerged as three factors that were labelled consistency & closeness (8 items, α = 0.902), egotism & opportunism (9 items, α = 0.898), and communal commitment (6 items, α = 0.909). As a check for validity, a Monte-Carlo analysis was conducted, which indicated that four factors should indeed be identified from the data.

Mean estimations of each factor were calculated as an average of the scores for each item to test for party affiliation and propensity for trust[3]. A Pearson's correlation shows significant correlation between the propensity for trust and participants' estimations of whether or not they trust election candidates (Pearson = .217, $p < 0.001$). This provides a check for the validity of factors, as higher propensity of trust should indeed yield higher estimations of trustworthiness.

Party affiliation was not a "locked question". Consequently 276 participants did not provide this score. These were excluded from the analysis involving respondents' party affiliation, which left 305 staunch voters (221 Democrats and 85 Republicans) and 280 swing voters[4]. A

---

[3] In order to calculate the overall trustworthiness score, the four factors were averaged where egotism and opportunism was reverse scored (such as a low estimate of this factor would yield a higher overall trustworthiness estimation).

[4] There was no statistical difference between staunch Democrats and Republicans and consequently the two groups were merged to one group simply labelled "staunch voters"

paired-sample t-test show significant differences between staunch and swing voters. Staunch voters believe election candidates to be more able (staunch M: 3.34, swing M: 2.96, t = 11.768, df (2239), $p < 0.001$), staunch voters believe election candidates are more consistent and close (staunch M: 2.38, swing M: 1.98, t = 12.534, df (2239), $p < 0.001$), staunch voters believe election candidates are *less* likely to be egotistic and opportunistic (staunch M: 3.57, swing M: 3.81, t = -8.163, df (2519), $p < 0.001$), and staunch voters believe election candidates are more communally committed (staunch M: 3.08, swing M: 2.62, t = 12.147, df (1679), $p < 0.001$). This provides strong support for the prediction that swing voters have less trust in election candidates than staunch voters.

3.1.2 Political epistemic expertise (study 1b)

Materials: The materials for 1a were drawn from previous studies in management trustworthiness. As no such studies were available for the exploratory study on political epistemic expertise, 3x20 items were generated from a common-sense distinction of factors seemingly important in acquiring or having expertise as an election candidate. These were formal education, practical experience, and local knowledge. The items were rated on the same 5-point Likert-type scale as the survey in 1a.

Participants: As in survey 1a, 500 participants were recruited from MT. The same criteria for discarding participants were present and 55 participants were discarded, leaving 445 for the subsequent analyses.

Hypotheses: Study 1b was entirely exploratory and the initial categories from which the 60 items were developed were common sense rather than drawn from previous studies. However, there is evidence from social psychological studies that suggest people attend more carefully to traits of warmth (such as trustworthiness) than of competence (such as epistemic expertise, see Fiske et al., 2007), and therefore we speculate that extracted factors will have less internal consistency (through Cronbach's alpha) compared with factors from 1a. Despite that we have no direct hypothesis for epistemic expertise factors, 1a suggests that swing voters are more sceptical towards election candidates compared with staunch voters. This may be the case also for epistemic expertise.

Results: The factor analysis of items from the three categories failed to replicate the originally intended categories. They further failed to generate any distinct factors, but merely yielded 25 items that clustered around the same factor. This suggests voters did not differentiate between causes of epistemic expertise, but were simply concerned *whether* election candidates were expert. That is, it seems that how the expertise was acquired has less importance for voters, which is in line with the social psychological findings mentioned in the above. In line with findings from 1a, however, a paired-sample t-test shows significant difference between estimations of epistemic expertise when comparing staunch and swing voters (staunch M: 3.50, swing M: 3.22, t = 4.672, df (215), p < 0.001). Staunch voters are more confident that election candidates do have relevant expertise.

3.2 Qualifying support for election candidates: party-specific influences

Survey 1a described political trustworthiness of election candidates in the USA as an amalgamation of capability, consistency & closeness, egotism & opportunism, and communal commitment. It further found no difference between estimations of the likelihood of the existence of each factor between the two groups of staunchly affiliated voters (Democrats and Republicans), but found a significant difference between staunch and swing voters where the latter were significantly more likely to believe that election candidates were less trustworthy overall and were less capable, less consistency and close, more egotistic and opportunistic, and less communally committed when compared with estimates from the staunch group.

Mann & Ornstein (2012) describe politics in the USA as increasingly polarised. This partisanship might conceivably influence how staunchly affiliated voters conceptualise and think of election candidates from their preferred and from opposed political party. If an environment of blind trust of preferred candidates and blind distrust of opposed candidates exists amongst American voters, argumentation studies cited in the above would predict that staunch voters should be less convinced of messages even if these messages are believable (that is, low source credibility with high likelihood of evidence). Such an environment would be potentially disruptive to democratic, open, and honest debate. Study 2 tests this whether the polarisation identified by Mann and Ornstein can be found in voters.

Materials: The four trustworthiness factors from survey 1a yielded 31 items, modified such that "election candidate" became party-specific (e.g. "most Republican/Democrat election candidates..."). Half of the participants saw the Republican condition whilst the other half saw the Democrat condition. To determine staunchly affiliated voters party-affiliation, was elicited in the same was as in study 1.

Participants: 900 participants were recruited from MT with the same criteria for dismissal as 1a and 1b. This left 752 participants for the subsequently analyses.

Hypotheses: Party-affiliation partisanship was expected to influence estimates of the likelihood of the existence of trustworthiness factors for party-specific election candidates. Thus, we predict that staunchly affiliated Republicans will estimate preferred candidates to be significantly more trustworthy overall and for each factor compared with swing voters who will subsequently estimate Republican candidates to be more trustworthy compared with estimates from staunchly affiliated Democrat voters (in other words, Republican voters > swing voters > Democrat voters for Republican candidates). We expect same picture, but reversed for Democrat election candidates (i.e. Democrat voters > swing voters > Republican voters for Democrat candidates).

Results: Findings were strongly in line with the predictions from the partisan hypothesis such that political polarisation can be traced in staunch voters' estimations of their preferred and opposed candidates. All paired-sample t-tests yielded highly significant differences between means in line with predictions (all p's < 0.001, see table 1 for details).

Voter-type	Factor	Mean	Comparison w swing voter	Swing voter w opposed voter
Republican	Capability	3.900	M: 3.033, t = 13.316, df (519), p < 0.001	M: 2.837, t = 3.643, df (1055), p < 0.001
Republican	Consistency & closeness	3.204	M: 2.214, t = 16.463, df (519), p < 0.001	M: 1.941, t= 4.752, df (1055), p < 0.001
Republican	Egotism & opportunism	2.417	M: 3.400, t = -13.091, df (584), p < 0.001	M: 3.835, t = -8.813, df (1187), p < 0.001
Republican	Communal commitment	3.964	M: 2.793, t = 14.816, df (389), p < 0.001	M: 2.366, t = 6.837, df (791), p < 0.001
Democrat	Capability	3.936	M: 3.130, t = 17.723, df (959), p < 0.001	M: 2.717, t = 5.167, df (599), p < 0.001
Democrat	Consistency & closeness	3.356	M: 2.347, t = 20.300, df (959), p < 0.001	M: 2.045, t = 4.678, df (499), p < 0.001
Democrat	Egotism & opportunism	2.385	M: 3.390, t = -17.465, df (1079), p < 0.001	M: 3.797, t = -7400, df (674), p < 0.001
Democrat	Communal commitment	4.146	M: 3.112, t = 18.921, df (719), p < 0.001	M: 2.664, t = 5.661, df (449), p < 0.001

Table 1: Staunchly affiliated voters compared with other voters

The study suggests that political polarization influences how the voters perceive the source credibility of preferred and opposing election candidates.

3.3 Predicting political trustworthiness

The items that make up the four factors identified in survey 1a describe political trustworthiness of election candidates in the USA. Study 3 tests if the items are able to *predict* whether or not a particular voter believes general (non-party specific) election candidates (Madsen, 2015).

It is worth noting that a candidate might be deemed to be trustworthy due to other factors not stemming from her character such as physical and vocal attractiveness (see Little et al., 2012 for a discussion of this) and actions (for example, a candidate is judged to be more trustworthy if she acts *against* her own self-interests, see Cialdini, 2007; Combs & Keller, 2010). The perception of trustworthiness can also be influenced by the actions of other election candidates. For example, insensitive behaviour towards other candidates has been shown to decrease trust (Mutz & Reeves, 2005)[5]. Consequently, we do not predict that the 31 items are able to predict political trustworthiness deterministically, but they should be positively correlated with overall estimates.

Materials: Participants responded twice on a 5-point Likert-type scale for each item identified in survey 1a such that likelihood *and* relative importance of each items was measured. This allowed for relative weighting of each factor[6]. To test the predictability of each factor as well as the overall trustworthiness prediction, ten models were developed (four factors, weighted and unweighted, as well as an overall prediction, weighted and unweighted). Participants were also asked to rate the likelihood that election candidates are trustworthy (the dependent variable against which the models are to be measured).

Participants: 250 participants were recruited from MT (American citizens, eligible to vote). 7 participants failed to deliver the right completion code or did not complete the survey, leaving 243 for the subsequent analyses.

Results: A Pearson's correlation shows that overall trust estimations (weighted and non-weighted) have the highest correlation and outperform individual factors when predicting for the dependent variable (observed trust). To test this further, a stepwise multiple regression analysis was conducted for the ten models against the dependent variable. The overall (non-weighted) scales have the highest

---

[5] Although, contrary to popular belief, negative campaign ads do not seem to significantly alter the perception of trust of either the attacked or the attacker (for a meta-review, see Lau et a., 2007)

[6] Paired-sample t-tests show egotism and opportunism is significantly less important compared with the other factors (egotism and opportunism (M: 3.51), consistency and closeness (M: 3.79), capability (M: 3.89), communal commitment (M: 3.89), ts between 5.809 and 6.959, dfs (242), all ps < 0.001).

level of prediction accounting for 49.5% of the variance in observed trust. The model is highly significant as shown in Table 2.

Predictor	β	R	R²	t	Sig. change
Overall trust prediction	.704	.704	.495	15.373	.000

Table 2: Multiple regression compared with observed trust

3.4 The malleability of trustworthiness: priming political trustworthiness factors

Estimations of items in the above studies are all inherently subjective such that one voter might believe election candidates to be utterly incapable and egotistic whilst another might believe election candidates to be virtual saints. It is by no means clear, however, what factors influence the emergence of such probabilistic estimations. Subjective estimations of the value of products have been shown to fluctuate with context (Stewart et al., 2006), and Kahneman and Tversky famously show that framing influences choice of strategy despite the similarity of the choices between groups (Tversky & Kahneman, 1981, see also McKenzie, 2004). Priming studies suggest that human beings have limited cognitive attention and will take recent stimuli as their frame of reference when considering the meaning of ambiguous words (see e.g. Rodd et al., 2013). To test the influence of framing in Study 4, participants were primed to pay attention to one factor compared with other factors. If priming effects can be shown in the emergence of the probabilistic estimation of the likelihood of the existence of a trustworthiness factor, it suggests that voters' subjective probabilistic estimations are not stable, but are malleable to influences (such as heavy media focus on one particular issue).

Materials: Participants read a short story relating to a woman who lost her house, either due to the incompetence of a local politician or due to a natural disaster. There were four priming conditions involving incompetency, corresponding to each of the four factors identified in survey 1a (the politician was described as either incapable, flip-flopping, egotistic, or too attentive to politics in Washington). Conversely, in the control condition the woman lost her house due to a natural disaster (hurricane). The study was between-subjects, as participants were only presented with *one* priming condition (either one of the four factors or the control condition). All participants subsequently estimated their perceived likelihood of each of the 31 trustworthiness items.

Participants: 950 participants were recruited through MT (American citizens eligible to vote). 64 participants failed to deliver the correct completion code or failed to complete the survey. This left 886 participants distributed across the five conditions for the subsequent analyses.

Hypotheses: As priming was negative (loss of house), we predict that estimates should be lower for primed conditions compared with the control condition (local effect). That is, participants primed with a negative story in which the house was lost due to flip-flopping should estimate the items connected with consistency as significantly lower. More generally, negative priming should have a global effect where participants become increasingly sceptical regarding *all* factors when primed to negatively consider *one* factor. That is, negative priming involving incompetency of an election candidate should yield lower ratings of the likelihood of trustworthiness items compared with the control condition.

Results: In line with predictions, participants who read a negative story that primed their attention towards the incompetence experienced a local and a global decrease in their estimation of election candidates. To test the local prediction, the control condition was compared with the factor of each priming (i.e. comparing capability estimations between control and the capability priming). A paired-sample t-test indicates that the local effect was significant, as participants rated the likelihood that election candidates possessed the quality as significantly lower (or higher in the case of egotism) compared with the control condition (t's between 2.415 and 5.807, p's between 0.016 and < 0.001). The priming also had tentative global effect, as the remaining three factors that were not primed were also estimated to be less likely when participants received the priming story compared with control (all 16 paired-sample t-tests were significant in this regard except for 2; of the significantly tests, t's between 2.096 and 6.377, p's between 0.037 and < 0.001).

Finally, the results nonetheless suggest that the local effect is stronger than the global effect, as the primed factor dipped more than the other factors. Comparing the relevant factor of the primed condition with the same factor for participants who were primed to consider other factors, only 2 out of 12 paired-sample t-tests were not significant (of the significant t-tests, t's between 2.001 and 6.231, p's between 0.049 and < 0.001). This suggests that negative priming has a local *and* a global influence such that participants become increasingly critical in their perception of lection candidates in general and increasingly critical in their perception of the primed factor in particular. As such, it indicates that factors are contextually malleable and subject to cognitive attention.

4. DISCUSSION

The paper explores source credibility in a particular domain, namely political source credibility of election candidates in the USA. In accordance with the hypothesis that trustworthiness, like epistemic expertise, is dependent on the domain in which it manifests, survey 1a identifies domain-specific factors to describe political trustworthiness with high internal consistency (Cronbach's alphas between 0.888 and 0.909). Survey 1b failed to extract specific factors of epistemic expertise. However, as no previous studies had explored this, the results might be a product of flawed onset categories or items. Study 3 made use of the 31 trustworthiness items to predict whether or not participants would find election candidates trustworthy. As step-wise multiple regression shows that the overall trustworthiness prediction outperforms prediction based on the extracted factors and accounts for 49.5% of the variance.

Both surveys in study 1, however, reported that swing voters were significantly more sceptical towards election candidates compared with estimations from staunchly affiliated voters. Study 2 probed deeper into this by assigning party-affiliation to candidates. In line with predictions as well, staunch voters regarded their preferred candidates highly whilst simultaneously rating opposing candidates very low. Where survey 1a reported that staunch voters were more positive for general election candidates, study 2 shows that party-affiliation instigates political polarisation for staunch voters of both parties. Interestingly, this suggests that participants in survey 1a might have had their own preferred candidates in mind when responding to questions regarding election candidates without specific party-affiliation (i.e. "most election candidates..."). Although only tentatively related, study 4 suggests that estimations of the likelihood of the existence of factors *are* influenced by immediate cognitive attention such as priming. Therefore, it is conceivable that the positive estimations from survey 1a were a product of selective consideration, as staunch voters were much more sceptical when being asked to consider politicians from opposing parties.

The current studies point to a potential problem for democratic debate. If a source of information is seen as untrustworthy, it makes normative sense to question the information passed on by that particular source. The results from study 2, however, suggest a systemic bias against candidates from opposite parties such that staunchly affiliated voters hold their preferred candidates in very high regard whilst doubting candidates from opposite parties. This indicates that the political polarisation reported by Mann and Ornstein (2012) can be found at the level of the individual voter. Bayesian argumentation studies have shown that participants are less convinced by arguments presented by an untrustworthy source (see e.g. Harris et al., 2015). If both of these studies are correct, it poses a democratic problem for political debates in the USA given the distrust of opposing political candidates, as this would facilitate an environment in which confirmation biases would dominate. It would be fascinating to apply the Bayesian argumentation paradigm relying on subjective probabilistic estimations of content and source credibility (Hahn et al., 2009) to political debates in the USA. This could be a highly relevant study for future research, as it would bring together in-depth descriptions of a particular socio-cultural and epistemic domain (such as the convincingness of election candidates in the USA) with Bayesian methodologies that have been proven successful in past studies (e.g. Corner et al., 2011; Harris et al., 2012).

Further, if probabilistic estimations are malleable depending on the priming of the epistemic context in which the estimation emerges, the current studies indicate the importance of the media in shaping and maintaining narratives and frames through which issues can be seen. From a strictly normative perspective, it is not yet entirely clear how media coverage influences the subjective probabilistic estimations of content and credibility in elections. This, too, points to possible future studies that integrate real-life democratic issues with probabilistic accounts of reasoning.

5. CONCLUDING REMARKS

Source credibility, defined as an amalgamation of trustworthiness and epistemic expertise, influences how people integrate and approach new information, whether or not they comply with public policies, and who voters are likely to vote for. The paper has explored aspects of domain-specific source credibility of election candidates in the USA. The results suggest that trustworthiness can be conceptualised as being related to the identified factors of capability, consistency & closeness, egotism & opportunism, and communal commitment whilst epistemic expertise could not be conceptualised by distinct factors. Producing an overall trustworthiness prediction, the identified factors account for 49.5% of observed trustworthiness ratings. The results further suggest that political polarisation can be traced to the level of the individual voter such that staunchly affiliated Republican voters are more positive towards generic Republican candidates compared with swing voters who in turn are more positive towards Republican candidates compared with staunch Democrats and vice versa. Finally, the results suggest that trustworthiness is contextually malleable such that the immediate cognitive attention influences the subjective probabilistic estimations of the individual voter. In all, the results suggest that abstract concepts such as source credibility and trustworthiness are grounded in the epistemic domain and socio-cultural context in which they emerge. This points to fascinating potential research that bridges qualitative and quantitative argumentation studies and explores in-depth contextual descriptions together with normative argumentation paradigms.

ACKNOWLEDGEMENTS: Thanks to Brent Clickard for joining me in collecting the data for the studies. The Danish Council of Independent Research supported the research (ID: DFF – 1329-00021B)

REFERENCES

Aristotle (1995). *Rhetoric*. In J. Barnes (Ed.) *The complete works of Aristotle*, vol. 2 (pp. 2152-2270), 6th printing. Princeton, NJ: Princeton University Press.
Ayres, I., & Braithwaite, J. (1992). *Responsive regulation*. Oxford: Oxford University Press.
Birnbaum, M. H., Wong, R., & Wong, L. K. (1976). Combining information from sources that vary in credibility. *Memory and Cognition*, 4, 330-336.
Bovens, L., & Hartmann, S. (2003). *Bayesian epistemology*. Oxford: Oxford University Press.
Braithwaite, J., & Makkai, T. (1994). Trust and compliance. *Policing Society*, 4, 1-12.
Braunsberger, K., & Munch, J. M. (1998). Source expertise versus experience effects in hospital advertising. *Journal of Services Marketing*, 12, 23-38.
Chaiken, S., & Maheswaran, D. (1994). Heuristic processing can bias systematic processing: Effects of source credibility, argument ambiguity, and task importance on attitude judgement. *Journal of Personality and Social Psychology*, 66(3), 460-473.
Cialdini, R. B. (2007). *Influence: The psychology of persuasion*. New York: Collins Business.
Citrin, J., & Muste, C. (1999). Trust in government. In J. P. Robinson, P. R. Shaver, & L. S. Wrightsman (Eds.), M*easures of political attitudes: Measures of social psychological attitudes*, Vol. 2. (pp. 465-532), San Diego, CA: Academic Press.
Colquitt, J. A., Scott, B. A., & LePine, J. A. (2007). Trust, trustworthiness, and trust propensity: A meta-analytic test of their unique relationships

with risk taking and job performance. *Journal of Applied Psychology, 92*(4), 909-927.
Combs, D. J. Y., & Keller, P. S. (2010). Politicians and trustworthiness: Acting contrary to self-interest enhances trustworthiness. *Basic and Applied Social Psychology, 32*, 328-339.
Corner, A., Hahn, U., & Oaksford, M. (2011). The psychological mechanism of the slippery slope argument. *Journal of Memory & Language, 64*, 133-152.
Couch, L. L., Adams, J. M., & Jones, W. H. (1996). The assessment of trust orientation. *Journal of Personality Assessment, 67*(2), 305-323.
Cuddy, A. J. C., Glick, P., & Beninger, A. (2011). The dynamics of warmth and competence judgments, and their outcomes in organizations. *Research in Organizational Behavior, 31*, 73-98.
Eemeren, F. H. van, & Grootendorst, R. (2004). *A systematic theory of argumentation: The pragma-dialectical approach.* Cambridge, UK: Cambridge University Press.
Fiske, Susan T., Cuddy, A. J. C., & Click, P. (2007). Universal dimensions of social cognition: Warmth and competence. *Trends in Cognitive Sciences, 11*(2), 77-83.
Flores, F. & Solomon, R. C. (1998). Creating trust. *Business Ethics Quarterly, 8*, 205–232.
Foss, S. K. (2004). *Rhetorical criticism: Exploration & practice,* 3rd Ed. Long Grove, IL: Waveland Press.
Fukuyama, F. (1995). *Trust,* New York: Basic Books.
Hahn, U., & Oaksford, M. (2006). A normative theory of argument strength. *Informal Logic, 26*, 1-24.
Hahn, U., & Oaksford, M. (2007). The rationality of informal argumentation: A Bayesian approach to reasoning fallacies. *Psychological Review, 114*, 704-732.
Hahn, U., Harris, A. J. L., & Corner, A. (2009). Argument content and argument source: An exploration. *Informal Logic, 29*, 337-336.
Hahn, U., Oaksford, M., & Harris, A. J. L. (2012). Testimony and argument: A Bayesian perspective. In F. Zenker (Ed.), *Bayesian Argumentation* (pp. 15-38). Dordrecht: Springer.
Harris, A., & Corriveau, K. H. (2011). Young children's selective trust in informants. *Philosophical Transactions of the Royal Society B, 366*, 1179-1187.
Harris, A. J. L., Hahn, U., Madsen, J. K., & Hsu, A. S. (2015). The appeal to expert opinion: Quantitative support for a Bayesian network approach. *Cognitive Science 39*(7), 1-38.
Hetherington, M. J. (1999). The effect of political trust on the presidential election, 1968-96. *American Political Science Review, 93*(2), 311-326.
Householder, E. E., & LaMarre, H. L. (2014). Facebook politics: Toward a process model for achieving political source credibility through social media. *Journal of Information Technology & Politics, 11*(4), 368-382.
Kee, H. W., & Knox, R. E. (1970). Conceptual and methodological considerations in the study of trust and suspicion. *Journal of Conflict Resolution, 14*, 357–366.

Lagnado, D.A., Fenton, N., & Neil, M. (2013). Legal idioms: A framework for evidential reasoning. *Argument and Computation*, 4, 46-63.

Lau, R. R., Sigelman, L., & Rovner, I. B. (2007). The effect of negative political campaigns: A meta-analytic assessment. *The Journal of Politics*, 69(4), 1176-1209.

Levi, M. (1997). *Consent, dissent and patriotism*. New York: Cambridge University Press.

Levi, M. (1998). A state of trust. In V. Braithwaite & M. Levi (Eds.) *Trust & Governance* (pp. 77-102), New York: Russell Sage Foundation.

Levi, M., & Stoker, L. (2000). Political trust and trustworthiness. *Annual Review of Political Science*, 3, 475-507.

Madsen, J. K. (2015). Modelling political source credibility of election candidates in the USA. In D. C. Noelle, R. Dale, A. S. Warlaumont, J. Yoshimi, T. Matlock, C. D. Jennings & P. P. Maglio (Eds.). *Proceedings of the 37th Annual Meeting of the Cognitive Science Society* (pp. 1470-1475). Austin, TX: Cognitive Science Society.

Mann, T. E., & Ornstein, N. J. (2012). *It's even worse than it looks: How the American constitutional systems collided with the new politics of extremism*. New York: Basic Books.

Mayer, R. C., Davis, J. H., & Schoorman, F. D. (1995). An integrative model of organizational trust. *Academy of Management Review*, 20, 709–734.

Mayer, R. C., & Davis, J. H. (1999). The effect of the performance appraisal system on trust for management: A field quasi-experiment. *Journal of Applied Psychology*, 84, 123–136.

Mayer, R. C., & Gavin, M. B. (2005). Trust in management and performance: Who minds the shop while the employees watch the boss? *Academy of Management Journal*, 48, 874–888.

McCroskey, J. C. (1997). Ethos: A dominant factor in rhetorical communication. In J. C. McCroskey (Ed.), *An Introduction to Rhetorical Communication*, 7th ed. (pp. 87-107). Boston, MA: Allyn and Bacon Publisher.

McCroskey, J. C., & Young, T. J. (1981). Ethos and credibility: The construct and its measurement after three decades. *The Central Speech Journal*, 32, 24-34.

McKenzie, C. R. M. (2004). Framing effects in inference tasks and why they are normatively defensible. *Memory & Cognition*, 32, 874-885.

Miller, A., & Listhaug, O. (1990). Political performance and institutional trust. In P. Norris (Ed.), *Critical Citizens: Global Confidence in Democratic Government* (pp. 204-216). Oxford, UK: Oxford University Press.

Mutz, D. C., & Reeves, B. (2005). The new videomalaise: Effects of televised incivility on political trust. *American Political Science Review*, 99(1), 1-15.

Pettit, P. (1995). The cunning of trust. *Philosophy & Public Affairs*, 24(3), 202-225.

Petty, R. E., & Cacioppo, J. T. (1984). Source factors and the elaboration likelihood model of persuasion. *Advances in Consumer Research*, 11, 668-672.

Rodd, J. M., Cutrin, B. L., Kirsch, H., Millar, A., & Davis, M. H. (2013). Long-term priming of the meanings of ambiguous words. *Journal of Memory and Language, 68*(2), 180-198.
Rotter, J. R. (1967). A new scale for the measurement of interpersonal trust. *Journal of Personality, 35*(4), 651-665.
Schum, D. A. (1994). *The evidential foundations of probabilistic reasoning.* New York: Wiley.
Stewart, N., Chater, N., & Brown, G. D. A. (2006). Decision by sampling. *Cognitive Psychology, 53*, 1-26.
Tversky, A., & Kahneman, D. (1981). The framing of decisions and the psychology of choice. *Science, 211*, 453-458.
Walton, D. (1997). *Appeal to expert opinion: Arguments from authority.* University Park, PA: Pennsylvania State University Press.
Wang, D., Abdelzaher, T., Ahmadi, H., Pasternack, J., Roth, D., Gupta, M., & Aggarwal, C. (2011). On Bayesian interpretation of fact-finding in information networks. In *14th International Conference on Information Fusion (Fusion 2011).*
Wells, G. L., & Olson, E. A. (2003). Eyewitness testimony. *Annual Review of Psychology, 54*, 277-295.

Commentary on Madsen's Malleability and Predictability of Source Credibility

DALE HAMPLE
University of Maryland, USA
dhample@umd.edu

1. INTRODUCTION

This paper is commendable in several respects. It pursues a central question through a series of studies, accumulating evidence and then pursuing its implications. The question itself is interesting both theoretically and practically: the respects in which credibility (expertise and trustworthiness) is domain-dependent. For example, an expert economist should not be regarded as an expert on sports. This has always been clear, but new evidence on the point is certainly welcome. Trustworthiness – honesty, good will – has more rarely been understood as domain-dependent, and so there is real opportunity for an advance here. Because the methodologies Madsen employed do not require simple yes/no judgments about the applicability of a source's expertise and trustworthiness, the studies yield information about degrees of credibility.

It is worth noting that Madsen discusses credibility independent of particular messages: that is, he asks about people's general trust in candidates rather than how perceived credibility affects reception of a particular speech or advertisement.

Finally, it is interesting to read a paper on credibility that is sensitive to disciplines other than communication and informal logic, such as political science, management, cognitive psychology, and others.

2. CREDIBILITY AND ITS DOMAINS

Traditional practice in studying credibility is to ask respondents to fill out domain-general instruments about their perception of, say, the source's knowledge and integrity. In contrast to that practice, there are two ways to understand the possibility of domain-specificity. The first reflects traditional practice: to have the general items filled out in view of a message from a particular domain, such as politics or health practices. The second, the one toward which Madsen is pointed conceptually, is to ask whether we should actually have domain-specific scales – for instance one set of items for a political candidate and another set for a religious leader.

Either way, we are faced with the problem of defining what a domain is. This has been a conceptual difficulty for our community ever since Toulmin wrote about field dependence. A recent approach to the problem is to provide simple descriptions of domains such as health and law, as has been done in current versions of pragma-dialectics. How detailed our descriptions of a domain need to be remains an issue whenever scholars wish to argue that some process or product is specific to one domain but not expected to assert itself in others. Madsen's paper temporarily avoids this issue by assuming that a particular American campaign season is a domain (or representative of one), which seems reasonable. The larger problem will need to be confronted eventually, however, if this approach is pursued. Madsen speculates, for instance, that American and Bangladesh elections may be different domains.

Of the interesting results in this paper I would like to raise up one or two that might well be important for analysis of arguments that rely on source qualities for their force and quality. Madsen found that expertise in the election domain remained undifferentiated. That is, it was a single thing, as we have been conceptualizing it in our analyses of arguments from authority. However, trustworthiness was empirically dissociated into four discrete concepts. These were the source's capability, consistency/closeness, egotism/opportunism, and communal commitment.

Each of these elements needs distinct evidence if an arguer is to establish trustworthiness. For instance, being able to show that a source is not an opportunist does not necessarily imply that the source also has high communal commitment. We have generally supposed that trustworthiness was a single thing and so Madsen's findings may require that we supply more precise critical questions for argument from authority.

Madsen's findings, of course, will need to be replicated before we make too great an intellectual commitment to them. A larger question – one that cannot be answered by Madsen's study of a single domain – is whether trustworthiness and expertise will have different components in other domains. If that turns out to be the case, we will need to analyze argument from authority differently in each distinguishable domain.

3. CONCLUSION

In sum, this paper constitutes a good and systematic investigation of how authority seems to function in the domain of American elections. Its distinctions and thinking will repay close reading by anyone interested in argument from authority or the place of argumentation in electoral politics.

21

An Epistemological Theory of Argumentation for Adversarial Legal Proceedings

DANNY MARRERO
Institución Universitaria Conocimiento e Innovación para la Justicia, Colombia
danny.marrero@cij.edu.co

The rhetorical view (R) suggests that the goal of factual argumentation in legal proceedings is to persuade the fact-finder about the facts under litigation. However, R does not capture our social expectations: we want fact-finders to know the facts justifying their decisions, and persuasion does not necessarily lead to knowledge. I want to present an epistemic theory of argumentation honoring our expectations. Under my account, factual argumentation aims to transmit knowledge to the fact-finder.

KEYWORDS: adversarial criminal proceedings, epistemological approach to argumentation, extreme adversarialism, factual legal argumentation, knowledge attributions, rhetorical approach to argumentation, strict invariantism

1. INTRODUCTION

Epistemological theories of argumentation (i.e., theories using epistemological concepts and methodologies to understand problems of argumentation) have been prolific accounting for the scope of theories of argumentation, criteria for good arguments, specific argumentative forms and fallacies (e.g., Biro & Siegel, 1992; Feldman, 1994; Goldman, 1999; Johnson, 2000; Lumer, 1990, 1991, 2005; Siegel & Biro 1997). I want to take this approach to another level and show that it also displays some promise of understanding factual argumentation in adversarial legal proceedings. The *rhetorical view* (R) suggests that the goal of factual argumentation is to persuade the fact-finder about the versions of the facts under litigation. However, R does not capture our social expectations for adversarial legal systems because we want fact-finders to know the facts justifying their decisions, and persuasion does not necessarily lead to knowledge. In this paper, I want to present an epistemological theory of legal argumentation that honors our social expectations. First, I will thoughtfully reconstruct R. Shortly, I interpret R is the result of two accounts. On one hand, it is the result of *Strict Invariantism*, which is the view claiming that there is one standard for knowledge attributions, and that that standard is high. Strict Invaraintism denies knowledge in legal contexts because legal inquiries do not reach the cognitive bar set for knowledge attributions. On the other hand, *Extreme Adversarialism*, adopting the perspective of the litigants, claims that the collection and development of evidence is meant to support the parties under litigation. If legal inquiry falls short of knowledge, and legal marshaling of evidence is partial and incomplete, then R follows. Departing from R, I claim that if it is possible to attribute knowledge in legal contexts, and each of the participants in legal procedures has a cognitive role to play, then the goal of factual argumentation is to transmit knowledge to the fact-finder.

2. THE RHETORICAL VIEW

As I take it, R is the account holding that the goal of factual argumentation in legal proceedings is to persuade the fact-finder about the versions of the facts under litigation. A representative example of this view can be found in the book *Legal Argumentation and Evidence*, by Douglas Walton (2002). In this book, Walton suggests a goal-oriented method of evaluation of arguments. In his words, his idea is that "if we can classify the fair trial as a normative model of argumentation that has a definite goal, and that has argumentation structures that are the means of realizing that goal," then "we [can] judge such arguments to be correct or incorrect with respect to how they have been used as part of the trial procedure" (2002, p. 156). If I break down Walton's suggestion in methodological steps, his method would be:[1]

>*Step 1*: Select the argument, A, as a target of analysis.
>*Step 2*: Posit a goal, G, or sets of goals, of the trial.
>*Step 3*: Determine how well the argument, A, would promote the goal, G.
>*Step 4*: If the argument, A, would be ineffective or deficient in promoting the goal, G, the argument, A, would be incorrect.

---

[1] My interpretation of Walton's method is a close paraphrase of Alvin Goldman's (2003) method to evaluate inferences in processes of adjudication (p. 215).

Responding to *Step 2*, Walton claims that the trial is "based on a conflict of opinions" (p. 157). That is, in a trial, two parties defend contradictory propositions such as the defendant being guilty and the defendant not being guilty. The goal of the trial, according to Walton,

> is to provide a setting for dispute resolution through due process. For this purpose, a framework of rationality is imposed on the process by having a trier (judge or jury) who hears all the arguments of both sides, and then comes to an independent decision on who had the stronger, or more persuasive, argument. (p. 158)

If the goal of the trial is to provide a method for conflict resolution via a decision based on the most persuasive argument presented by the parties under litigation, then "the central goal of any argument used in trial is that of persuasion" (p. 156), but not any kind of persuasion. According to Walton, there are two kinds of persuasion, namely, psychological or rhetorical, and rational. Walton identifies *psychological or rhetorical persuasion* with "salesmanship" and defines it as "persuading your audience by any means that will psychologically move them to come to accept the proposition you advocate for them" (p. 157). Alternatively, he defines *rational persuasion* as "using good reasons to persuade your audience by convincing arguments they should accept as rational persons who can weigh evidence on both sides of an issue" (p. 157). Walton claims that the goal of argumentation in the trial is rational persuasion. For one thing, legal argumentation "must be based on what are supposed to be the real facts in a case" (p. 157). In this sense, the parties use expert testimony and eyewitness because their sayings can be verified at trial. For the other, legal argumentation has to meet the standards established by procedural rules such as the rules for collecting evidence, for deciding what counts as evidence, for determining who was to prove what, and the like (p. 159). From Walton's perspective, these sorts of rules specify what counts as evidence, understanding "evidence" as "what is rationally persuasive" (p. 157).

Continuing on with *Step 3*, one wonders how does argumentation, as rational persuasion, promote the goal of a trial? To recall, Walton claims that the goal of a trial is "to provide a setting to dispute resolution through due process," and that such setting is guaranteed "by having a trier (judge or jury) who hears all the arguments of both sides, and then comes to an independent decision on who had the stronger, or more persuasive, argument" (p. 158). Argumentation promotes this goal because once the parties put forward their allegations, the trier of facts in making an independent decision, visualizes the whole picture of the big body of evidence presented at trial. For Walton, "[i]n this judgment, each individual argument used by an advocate, if it was persuasive, has a 'weight,' which is usually a small weight in the large picture" of evidence required for impartial decision-making (p. 158).

3. A PROBLEM FOR *R*: THE COGNITIVE ASPIRATIONS OF LEGAL PROCEEDINGS

From my perspective, R does not capture one of the most important social aspirations of legal proceedings: the aspirations of knowledge. Such aspiration becomes explicit when we realize that for correct decision making, true belief, good reasons, or persuasive arguments are not enough. We do not want judges or juries to make decisions which are coincidentally true, or to decide because of good internal reasons; we want them to decide knowing the facts under litigation. To illustrate this point, let us take a look at the following Gettier-type case proposed by Michael Pardo (2005) in his paper "The Field of Evidence and the Field of Knowledge."

> Two officers plant cocaine in an automobile, and they then give unrebutted testimony at the driver's trial that they found the cocaine after a consensual search of the car. The driver, concerned about his prior record coming out on cross-examination, does not testify and offers no real defense. The fact-finder convicts the driver after finding the officers credible. Now, unbeknownst to everyone except the defendant, he really did have cocaine in the car that never was discovered. (Pardo, 2005, p. 322)

As any other Gettier-type case, Pardo's case shows that the fact-finder does not know that the defendant had cocaine. However, the fact-finder's belief that the defendant had cocaine is true, and the fact-finder is justified in believing that it is true, provided the two officer's "unrebutted" testimony. In other words, this case shows that the fact-finder did not know that the defendant had cocaine; this finding was true, but just as mere coincidence. According to Pardo, in modern legal proceedings, fact-finders are expected to sentence based on the knowledge of the facts under litigation and not on coincidentally truth findings.

If this is right, there is a cognitive aspiration in legal proceedings that theories of legal argumentation should take into account. In other words, theories of legal argumentation are supposed to show how legal argumentation leads to knowledge. In *R*, however, the belief the fact-finder is supposed to be persuaded of is not qualified (Lumer, 2010, pp. 45-46). As a consequence, a fact-finder could be persuaded of something that is not true, or of something that is true by chance; in one case or the other, the fact-finder does not know what we are expecting him/her to know. In Walton's account, a fact-finder is persuaded of a belief about the facts under litigation, when the fact-finder, after weighting the evidence of both sides, selects the heaviest argument for his/her factual reconstructions. Yet, given that Walton does not qualify the belief the fact-finder is persuaded of, he/she can be persuaded of a coincidentally true belief, even with the heaviest argument presented at trial. As in Pardo's case, in Walton's account, a fact-finder could be persuaded that the defendant had cocaine given that the unrebutted testimony is part of a comparatively strong argument

4. STRICT INVARIANTISM AND EXTREME ADVERSARIALISM

My contention is that *R* does not take into account the cognitive aspirations of legal proceedings because it is the consequence of two accounts: Strict Invariantism and Extreme Adversarialism. *Strict Invariantism* is the view that the standard of knowledge is one, and that it is high. For instance, some strict invariantists claim that the standard of knowledge is scientific knowledge. If a putative knower does not satisfy the standards of scientific knowledge, then he/she does not know. *Extreme Adversarialism* is the view claiming that the nature of legal proceedings is adversarial, and therefore, the factual reconstructions in legal proceedings are biased and incomplete. If legal agents cannot meet the standards of scientific knowledge, and their factual reconstructions are biased and incomplete, then the most they can do is to try to persuade the fact-finder of their versions of the facts under litigation.

Let me illustrate these ideas with Susan Haack's distinction between science and law. Before doing so, I need to clarify Haack's concept of *inquiry*. Broadly speaking, "inquiry is an attempt to discover the truth of some question or questions" (Haack, 2004, p. 45). The starting point of inquiry is a question which perturbs a cognitive agent. In solving this issue, he/she formulates one hypothesis and starts looking for evidence which confirms it. Not having confirmation, the agent either modifies or abandons his/her initial conjecture. When the evidence leads to the true answer under consideration, the inquirer's aim is achieved. It is commonly accepted that inquiry is a constitutive part of legal systems because the justice that they want to achieve depends on two sides of the same coin. On one side, justice is conditioned by the application and administration of just laws. On the other side, it is a consequence of the truth determination of legally relevant facts. The latter shows that the law is also one activity whose core is inquiry. However, Haack counter-argues this position juxtaposing science and law. Since the core of science is inquiry, it provides an archetype which law would fulfill if its core were inquiry as well. Yet, law does not conform to science. In Haack's words,

> If the legal system were in the same business as history, geography, or as physics and the other sciences, its way of conducting that business would be peculiar, and inefficient, to say the least. But the law is really not in exactly the same business (2009, p. 13).

For Haack, the main differences between science and law are the following (see Table 1 below). First, the equation of the three main elements of the concept of inquiry (i.e., question, evidence and answer) is different in science and law. A scientific method starts with a question which encourages the search for evidence that could provide an answer for the original issue. Although legal proceedings also start with a question, unlike science, legal agents first provide answers to their initial questions and then look for the evidence which supports their position. Second, the aim of science is to formulate, examine and answer questions which explain how the world works. Alternatively, a legal procedure is a non-violent social means of conflict resolution. To be sure, the legal procedure is aimed to produce a verdict of either guilt or liability, or non-guilt or non-liability according to a body of evidence. This decision ends a dispute between two adversarial parts (e.g., prosecutors vs. attorneys, or petitioners vs. respondents). Third, the interest of science is not only to solve a question, but also to provide explanations for phenomena. Hence, the object of science is general laws which explain particular cases. Legal proceedings, instead, attain their goals through particular cases. Fourth, when scientific results seem to be unsatisfactory under new evidence, scientists wonder about the problems of their partial results and, if necessary, those are modified. In this sense, science is fallible. In contrast, the satisfaction of legal resolutions implies both prompt and definite verdicts. For one thing, extremely slow justice is not justice; for another, constantly modified verdicts conduct legal insecurity. Fifth, science is progressive, whereas law is conservative. To clarify, scientific problem-solving dynamics are reiterative because once a scientific question is solved, a new issue is posited. This leads to progressivism because, normally, the new questions are analyzed and answered using previous results. This contrasts to the importance of precedent in legal decision-making. Tackling the atomism of their case-based orientation, legal systems unify verdicts using previous judicial decisions as patterns for future decisions. Next, whereas science generally has unlimited time in order to solve a problem, law generally has much more rigid time constraints. Finally, scientific investigation is free from formalities, while law establishes rituals for the resolution of social conflicts. In other words, in their investigations, scientists do not use standardized protocols. What is important is the explanatory power of their theories, not the way through which they construct them. Legal resolution of conflicts, on the contrary, homogenizes legal behavior through legal procedures. If legal agents do not follow the procedural itinerary, they cannot achieve their objectives (Hack, 2009, pp. 7-15; 2004, pp. 45-50; 2003, pp. 205-208).

Science ...	Law ...
... formulates a question, looks for evidence, and answers the question.	...formulates a question, answers the question and looks for the evidence which supports this answer.
... searches for the truth.	
... has investigative character.	... determines a defendant's guilt or liability or non-guilt or non-liability.
... searches for general principles.	
... has pervasively fallibilism (i.e., is open to revision in the light of new evidence).	... has adversarial culture.
	... focuses on particular cases.
... pushes for innovation.	... is concerned with prompt and final resolutions.
... has unlimited time for solving a problem.	
	... defers to the precedent.
... has informal and problem-oriented investigation.	... has strong timeline constrictions.
	... relies on formal rules and procedures.

Figure 1 - Juxtaposition between Law and Science

To sum up, Haack's Strict Invariantism fixes the standards for knowledge with the concept of scientific inquiry. If a putative knower does not satisfy the standards of scientific inquiry, then he or she does not know. Haack's Extreme Adversarialism comes from a wrong generalization. She assumes the advocate's perspective and defines all law from this angle:

> The advocacy that is at the core of the adversarial process is a very different matter from inquiry [...] the obligation of an attorney, qua advocate, is to make the best possible case for

his client's side of the dispute—including playing up the evidence that favors his case, and explaining inconvenient evidence away if he can't get it excluded. (2009, p. 13)

However, why should we privilege advocacy and not fact-finding which is the obligation of the triers of facts, investigators and detectives? Haack could replay that "this is not to deny that inquiry plays a role in the legal process [...], but it is to deny that inquiry is quite as central to the law as it is to science" (2009, pp. 12-13). However, if someone asks for a positive account for the role of inquiry in legal procedure, one more time, she does not provide a positive response.

5. AN EPISTEMOLOGICAL THEORY OF ARGUMENTATION FOR ADVERSARIAL LEGAL PROCEEDINGS

If my intuitions are right, a theory of argumentation taking into account the cognitive aspirations of legal proceedings should be able of providing a theory of knowledge attributions for legal contexts and a theory of cognitive division of work for legal proceedings. The reason for this is that if it is correct to attribute knowledge to legal agents, and the parties are part of a cognitive system where each of them has a role to play in the achievement of legal knowledge, then the role of legal argumentation is to transmit knowledge.

5.1. Knowledge Attributions in Legal Contexts

Let me start with a theory of knowledge attributions for legal contexts. As I take it, there is a *knowledge attribution* when an agent, the attributor, asserts that another agent, the putative knower, knows that p. By the same token, there is a knowledge denial when the attributor asserts that the putative knower does not know that p. *The problem of knowledge attributions* is, then, whether the attributor correctly asserts that the putative knower knows (or does not know) that p. Knowledge attributions and denials differ from other knowledge relations such as having or lacking knowledge, getting or not getting knowledge, detecting or not detecting knowledge, and so on. For this paper, it is important to differentiate the problem of knowledge attribution from the *problem of knowledge possession*, which is whether an agent knows that p. While the latter is a first-order knowledge relation, the former is a second-order (or meta-) knowledge relation. That is, the object of a knowledge attribution is a knowledge possession. In one sentence, the problem of knowledge attribution is not whether a putative knower knows that *p*, but under which conditions it is correct to assert that a putative knower knows that *p*.

My working hypothesis is that a knowledge attributor correctly asserts that a putative knower knows that *p* when the putative knower properly closes or advances his/her cognitive agenda. The conceptual background of my account comes from the notions of agent and agenda. Shortly, an *agent* is an entity doing something. *Agendas* are the objectives agents are disposed to achieve (Gabbay & Woods, 2003, pp. 183-185; 195-219; Niño, 2015, p. 39). In this sense, "[a]n agenda is something like a network of tasks or programmes to be discharged" (Gabbay & Woods, 2003, p. 182). Agendas and sub-agendas have *conditions of closure* determining both the actions an agent is expected to perform in order to achieve his/her objective, and the time range in which he/she should do it. An agenda in course is properly *closed* when agents deploy their resources in such a way that its conditions of closure are obtained, but agendas are not closed simpliciter. Instead, the agent's matching of the conditions of resolution comes in degrees. An agenda in course is properly *advanced* when some of its closure conditions have been obtained, but not all of them yet (Gabbay & Woods, 2003, p. 215).

> An agenda may involve things an agent desires to know, or would find it useful to know for the transaction of certain tasks, or the making of certain decisions in some contextually circumscribed circumstances or states of affairs he is disposed to realize (Gabbay & Woods, 2003, p. 183).

I refer to this as *cognitive agendas*. A cognitive agenda is, then, a set of questions that a cognitive agent wants, or needs, to answer for the achievement of his/her objectives. Agents pursue cognitive agendas for the sake of knowledge or they are sub-agendas, which enable them to achieve other purposes. Theories of epistemic risk claim that it is possible to differentiate between two types of cognitive agents in accordance with their attitude toward epistemic risk taking (Fallis, 2007; Levi, 1962; Mathiesen, 2011; Riggs, 2008). Whereas some agents withhold the acceptance of a proposition until all the information has been obtained, other agents act with less caution and accept propositions with incomplete information. Theories of epistemic risk claim that agents accept propositions with incomplete information because of practical reasons. Think of a resident medical officer dealing with clinical emergencies on behalf of admitting consultants in juxtaposition with a professor of biochemistry studying the chemical composition of bacteria resistant to penicillin. Although both, the resident medical officer and the biochemistry researcher, want a true answer for their inquires, the latter, but not the former, can withhold it until all the information has been collected. Cognitive agents adopting cognitive agendas for the sake of the achievement of a practical goal are *practical doxastic agents*. *Theoretical agents*, differently, pursue cognitive agendas when it leads to "the truth and nothing but the truth."

According to Dov Gabbay and John Woods, there are two factors that determine the different types of cognitive agents (2005, p. 11). Firstly, there is the degree of command of resources (time, information and computational capability) an agent needs to advance or close his/her agendas. Secondly is the height of the cognitive bar that the agent has set for him/herself. With this in mind, Gabbay and Woods incorporate a *hierarchical approach to agency*. It postulates a hierarchy in which agents are placed in light of their interests and their capacities. In this model, practical doxastic agents would be placed towards the bottom of the hierarchy and theoretical agents would be higher up (see Table 2). While practical doxastic agents "perform their cognitive tasks on the basis of less information and less time than they might otherwise like to have", theoretical agents "can wait long enough to make a try for total information, and they can run the calculations that close their agendas both powerfully and precisely" (pp. 11-12).

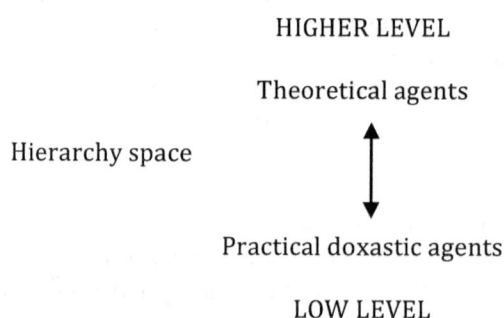

Figure 2 - Hierarchy of Cognitive Agents

My view is that, in terms of the hierarchy, legal agents are practical doxastic agents ranked higher than individuals solving day-by-day-practical problems such as crossing a street, looking for an address or cooking a meal; yet, legal agents are placed at a lower level than are pure theoretical agents such as the researchers of the National Aeronautics and Space Administration (NASA). To be sure, a legal agent commands more cognitive resources than an individual solving day-by-day-practical problems. What is expected in the adversarial system of adjudication is that the litigants, with the incentive of winning the case, look for all the relevant information for the legal inquiry. Additionally, they work in teams of inquirers, witnesses, and experts who seek to make their versions of the case stronger. The idea is that these parallel inquiries exhaust all the relevant information to be known. Another important difference between an individual solving day by day practical problems and a legal agent is the cognitive aim they are disposed to achieve. Legal procedures have a high cognitive aim, namely, to determine the truth of the events under litigation. Legal agents serve this goal in different ways. This is true even for litigants who apparently only serve their respective side's interests. When they take part in a legal inquiry, their vantage point clarifies aspects of the events that are inaccessible by the officials. Individuals, on the contrary, are not always interested in the truth. This explains why individuals are naturally hasty generalizers, or why they do not always use truth-preserving strategies of reasoning (Gabbay & Woods, 2005, pp. 23-25; 2013, p. 212).

Even though legal agents are ranked higher than individuals, they are not perfect inquirers such as NASA's scientific groups are. Since legal inquiries have strong timeline constrictions, they make decisions with incomplete and partial information. Furthermore, in dealing with facts, legal systems expect legal agents to reason as individuals and not as experts. Experts take part in legal discussions as qualified witnesses, but they are not able to make the ultimate decision. The institution of the jury in adversarial systems illustrates this point. *Prima facie*, any citizen can be part of the jury, unless he/she has expert knowledge about the actual issues under consideration. Given this case, the jury is excluded. Alternatively, in legal systems in which the fact-finder should be an educated citizen, what is expected is that such a qualified citizen be educated in the law, but not in factual matters such as forensic science. To sum up, legal agents have the computational capacity of an average person, or a reasonable person (Woods, 2011, p. 226). Therefore, they are ranked lower than theoretical agents, but higher than individuals.

To recall, from my perspective, knowledge attributions have the purpose of stating that a cognitive agenda has been properly closed. Given that the object of knowledge attributions is cognitive agendas, the conditions under which knowledge is properly attributed depends on the nature of the cognitive agenda claimed to have been properly closed or advanced. If my ideas are right, knowledge attributions in legal contexts depend on the nature of the cognitive agenda under account and not on standards belonging to foreign agendas such as the scientific ones. To be sure, legal proceedings impose cognitive agendas on its participants depending upon the activities they have to perform in each procedural stage. These agendas include closure conditions for the required actions. When such standards are not met, knowledge attributions are not justified. For instance, criminal investigations are one of the sub-agendas of the prosecution in criminal cases. The ultimate agenda of a prosecution is to show, beyond a reasonable doubt, that a defendant committed a crime. When the prosecution does not satisfy this burden, the innocence of the defendant is assumed. The goal of criminal investigations is to present a case to the prosecutor. This includes a cognitive agenda of determining whether a crime has been committed, and who did it. Such cognitive agenda is closed or advanced in two ways: either determining that there is good evidence that there was an occurrence of a crime, or determining that there is not good evidence for the occurrence of a crime. Imagine a criminal investigator wondering whether there is good evidence for the occurrence of a crime. Only relevant information that is processed in the right way allows the criminal investigator to come to know if there is good evidence that the crime occurred. There are two main sources of information in criminal investigations: state of affairs, or physical evidence, such as fingerprints, sound recordings, photographs, and so on, and the testimony of eyewitnesses and other collaborators. Let me focus on testimonial evidence. Testimonies are relevant when they provide information useful to advance the agenda of the criminal investigation. That is, they provide information useful to determine that either there is good evidence for the hypothesis that a crime occurred, or there is not good evidence for such hypothesis. When this is the case, criminal investigators attribute knowledge to the eyewitness, victims and other collaborators.

5.2. The Cognitive Division of Work for Legal Proceedings

The cognitive division of work in legal contexts becomes clear if we understand epistemic justification in legal contexts as a type of belief-dependent cognitive process. This understanding calls for some terminology from Alvin Goldman's theory of epistemic justification. To clarify, *theories of justification* are accounts specifying the conditions under which a person is justified in believing (Goldman, 1979, p. 3). Consequently, a theory of justification adopts the next structure:

S is justified in believing that p if and only if:
C_1
C_2
...
C_n

In this structure, S stands for a cognitive agent, p for a fact or proposition, and $C_1 \ldots C_n$ are the conditions for justificatory status.

> As a first approximation, Goldman suggests that:
> S is justified in believing that p if and only if:
> C: p results from a reliable cognitive process.

In this account, *cognitive process* is defined as a function with inputs that have beliefs as outputs (p. 11). Two types of processes are important here. Firstly, the *belief-dependent processes* have other beliefs as inputs. Secondly, the *belief-independent processes* do not have other beliefs as inputs (p. 13). Perception is an example of a belief-independent process. Reasoning, which includes antecedent beliefs within their premises, is an instance of a belief-dependent process. Following this terminology, there are two kinds of beliefs. A *belief-independent belief* is the output of a *belief-independent process*, and a *belief-dependent belief* is the result of a *belief-dependent process*. Finally, "reliability consists of the tendency of a process to produce beliefs that are true rather than false" (p. 10). While in belief-dependent processes reliability depends on the truth of the inputs (i.e., it is conditional), in belief-independent processes, reliability is categorical. From these distinctions, Goldman suggests two forms for evaluating justificatory status.

> The first form is for the belief-independent processes:
> S is justified in believing that p if and only if:
> C_1: p is a belief-independent belief, and
> C_2: p is the result of a categorically reliable process.

> The second form is for the belief-dependent processes:
> S is justified in believing that p if and only if:
> C_1: p is a belief-dependent belief, and
> C_2: p is the result of a conditional reliable process.

If my interpretation of Goldman's ideas is not wrong, legal knowledge is a type of belief-dependent process. This formula captures the core of my interpretation:

> A legal agent is justified in holding an epistemic judgment (j), in a legal procedure, if and only if:
> C_1: j depends on the procedural interventions of other participants in the legal procedure, and
> C_2: j is the result of the truth-conduciveness of the legal procedure.

I will explain these two conditions for legal knowledge below, but a previous distinction is required. Goldman, in *Knowledge in a Social World*, states:

> notice that I am speaking of *judgments* rather than *beliefs* ... The reason for this deviation is that the palpable outputs of legal deliberations are not private beliefs but public judgments of guilt and innocence, liability or non liability" (1999, p. 272).

This is an important distinction between *general epistemology* and *legal epistemology*. While the former studies epistemic justification independently of actual argumentation, the latter is concerned with epistemic justifications publicly justified in legal contexts. Therefore, for Goldman, legal argumentation is a constitutive part of legal epistemology.

To continue, legal agents do not perceive the facts under litigation directly; rather, they form their judgments from different sources of legal knowledge. To mention the most common examples, the presumed fact that "Y was murdered by X" is not perceived by the detective who looks for relevant evidence that establishes whether X murdered Y. Neither the prosecutor who publically accuses X of murder, nor X's attorney perceived the fact under litigation. Instead, they build their respective versions of the case with information provided by their side's detectives, witnesses, material evidence, etc. Finally, the trier of facts—judge or jury—does not perceive the alleged facts. On the contrary, he/she receives the information from the witnesses who are examined and cross-examined at trial. As a consequence, j depends on the procedural interventions of other participants in legal proceedings.

If j depends on the procedural interventions of other participants in legal proceedings, then j is not required to be categorically reliable, but conditionally reliable. In other words, the truth of j depends on the truth of its inputs. Three examples proposed by Goldman illustrate types of cues that juries find very probative (2003, p. 221). First, imagine a witness identification testimony where the witness points the finger at the defendant and states "this is the one." With this information, the trier of facts would probably decide that X murdered Y if the witness pointed at X. However, identifications are not 100% accurate. Witnesses also make mistakes, and this failure is transferred to the fact-finders' decision. Second, when a person confesses that he/she did the crime under inquiry, a judge or a jury tend to believe that the person who confessed actually did the crime. Yet, some confessions are produced by police intimidation or by the possibility of a plea bargain that ends in a negotiation with less serious crime charges for the offender. If X confesses that he murdered Y when he did not do it, all legal judgments drawn from X's confession will not be true. Finally, when a technical clarification is needed to understand the alleged facts, the trier of facts will rely on expert testimonies. Since one of the most important criteria for accepting an expert assertion is the credibility that the expert witness has, some inaccurate expert testimonies are incorporated into legal judgments. Ultimately, the credibility an expert has is not an epistemic criterion (i.e., it is not related to seeking the truth). To conclude, "we cannot expect any [inferential process in the field of law] to make correct (truthful) inferences if its inputs or premises are substantially inaccurate. 'Garbage in, garbage out', as the saying goes" (Goldman, 1999, p. 219).

Under this account, the partisan argumentation might honor the cognitive aspirations of legal proceedings. On one hand, it is correct to attribute knowledge to the parties if they properly close their cognitive agendas. On the hand, the parties' one-sided argumentation has a specific role to play in legal cognitive systems.

6. CONCLUSION

I want to conclude this paper showing how my epistemic account should be understood from the goal-oriented method of evaluation of arguments suggested above. To recall, such method has several methodological steps as follows:

> *Step 1*: Select the argument, A, as a target of analysis.
> *Step 2*: Posit a goal, G, or sets of goals, of the trial.
> *Step 3*: Determine how well the argument, A, would promote the goal, G.
> *Step 4*: If the argument, A, would be ineffective or deficient in promoting the goal, G, the argument, A, would be incorrect.

Regarding *Step 2*, notice neither Walton nor Haack state that the trial has epistemic goals. For Walton the end of legal proceedings is to "to provide a setting to dispute resolution through due process" (2003, p. 158), and for Haack "[t]he core business of a legal system is to resolve disputes; and a trial aims not to find out whether the defendant is guilty or liable, but to arrive at a determination of the defendant's guilt or liability" (2009, p. 12). Why do Walton and Haack exclude the achievement of knowledge as one of the goals of legal proceedings? According to Goldman, theories of legal proceedings fall into two categories: either they are pluralistic or unified (2005, pp. 163-164). *Pluralistic accounts* hold that legal procedures have different aims, no one of which is prior to the other (e.g., fairness, justice, impartiality, allowing pacific coexistence, seeking the truth, protection of civil rights, etc.). *Unified theories*, in contrast, explain procedures with reference to one main end. They do not hold that legal procedures actually achieve the selected goal; better yet, they use it as an explanatory resource to clarify the main activities performed in legal procedures. Within this second alternative, one can find *pure unified theories* and *impure unified theories*. Pure accounts state that the legal practices taken into account are subsumable in one exclusive desideratum. Impure unified alternatives defend that although the aim of legal procedures is such an exclusive aim, it is possible to recognize alternative goals coexisting with the dominant rationale.

Walton's and Haack's accounts are unified pure theories, for they suggest that legal proceedings aim to resolve disputes, or differences of opinion, and they do not prevent alternative goals to be achieved by legal proceedings. My account, on the contrary, is an impure unified theory. Like Goldman, I believe that the aim of legal procedures is to secure "substantively just treatment of individuals." This goal is only achieved if valid laws are correctly applied and the truth of the facts under litigation is achieved (2005, p. 164). Thus, a good argument, from an epistemic account of argumentation, is the one aiming at justified true belief (Lumer, 2005, p. 215). This, in opposition to R, properly responds to the epistemic expectations we have on legal proceedings

REFERENCES

Biro, J., & Siegel, H. (1992). Normativity, argumentation and an epistemic theory of fallacies. In: F. van Eemeren, R. Grootendorst, J. Blair, & C. Willard (Eds.), *Argumentation illuminated* (pp. 85-103). Amsterdam: SicSat.

Fallis, D. (2007). Attitudes towards epistemic risk and the value of experiments. *Studia logica, 86*(2), 215-246.

Feldman, R. (1994). Good arguments. In: F. Schmitt (Ed.), *Socializing epistemology. The social dimensions of knowledge* (pp. 159–188). Lanham: Rowman & Littlefield.

Gabbay, D., & Woods, J. (2010). Logic and the law: Crossing the line of discipline. In D. Gabbay, P. Canivez, S. Rahman & A. Thiercelin (Eds.), *Approaches to legal rationality* (pp. 165-201). Dordrecht: Springer.

Gabbay, D., & Woods, J. (2003). *Agenda relevance*. Amsterdam: Elsevier.

Gabbay, D., & Woods, J. (2005). *The reach of abduction. Insight and trial*. Amsterdam: Elsevier.

Goldman, A. (2005). Legal Evidence. In: M. Golding & W. Edmunson (Eds.), *The Blackwell guide to the philosophy of law and legal theory* (pp. 163-175). Maden: Blackwell Publishing.
Goldman, A. (2003). Simple heuristics and legal evidence. *Law, probability and risk, 2*, 215-226.
Goldman, A. (1999). *Knowledge in a social world.* Oxford: Clarendon Press.
Goldman, A. (1979). What Is Justified Belief? In G. Pappas (Ed.), *Justification and knowledge* (pp. 1-25). Dordrecht: Reidel.
Haack, S. (2009). Irreconcilable differences? The uneasy marriage of science and law. *Law and contemporary problems, 72*, 1-23.
Haack, S. (2004). Epistemology legalized: Or, truth, justice, and the American way. *The American journal of jurisprudence, 49*, 43-61.
Haack, S. (2003). Inquiry and advocacy, fallibilism and finality: Culture and inference in the science and in the law. *Law probability and risk, 2*, 205-214.
Levi, I. (1977). Epistemic utility and the evaluation of experiments. *Philosophy of science, 44*, 368-386.
Lumer, C. (1990). *Praktische Argumentationstheorie. Theoretische grundlagen, praktische begründung und regeln wichtiger argumentionsarten.* Braunscheweig: Viewe.
Lumer, C. (1991). Structure and function of argumentations. An epistemological approach to determining criteria for the validity and adequacy of argumentations. In F. van Eemeren., R. Grootendorst, A. Blair, & C. Willard (Eds.), *Proceedings of the second international conference on argumentation. Organized by the International Society for the Study of Argumentation (ISSA)* (pp. 98-107). Amsterdam: Sicsat.
Lumer, C. (2005). The Epistemological Theory of Argument—How and Why? *Informal logic, 25*(3), 213-243.
Lumer, C. (2010). Pragma-Dialectics and the Function of Argumentation. *Argumentation, 24*, 41-69.
Mathiesen, K. (2011). Can groups be epistemic agents? In H. Schmid, D. Sirtes & M. Weber (Eds.), *Collective epistemology* (pp. 23-44). Ontos Verlag: Frankfurt.
Niño, D. (2015). *Elementos de semiótica agentiva.* Bogotá, Universidad Jorge Tadeo Lozano.
Pardo, M. (2005). The field of evidence and the field of knowledge. *Law and philosophy, 24*, 321-392
Riggs, W. (2008). Epistemic risk and relativism. *Acta analytica, 23*(1), 1-8.
Siegel, H. & J. Biro. (1997). Epistemic normativity, argumentation, and fallacies. *Argumentation, 11*, 277-292.
Walton, A. (2002). *Legal Argumentation and evidence.* Pennsylvania: The Pennsylvania State University.
Woods, J. (2013) *Errors of reasoning: Naturalizing the logic inference.* Milton Keynes: College Publications.
Woods, J. (2011). Abduction and Proof: A Criminal Paradox. In D. Gabbay, P. Canivez, S. Rahman & A. Thiercelin (Eds.), *Approaches to legal rationality* (pp. 217-238). Dordrecht: Springer.

22

Not Just Rational, But Also Reasonable: Critical Testing in the Service of External Purposes of Public Political Arguments

DIMA MOHAMMED
ArgLab, Universidade Nova de Lisboa, Portugal
d.mohammed@fcsh.unl.pt

If a good argument is indeed the "one that fulfills its purpose", then considering the multiple purposes of a (public political) argument becomes indispensable for its assessment. But different purposes may be in conflict, resulting in an inconsistent assessment. In this paper, I argue in favour of considering the distinction between *rationality* and *reasonableness* in order to solve this complication and arrive at a non-fragmented and consistent assessment of the quality of public political arguments.

KEYWORDS: accountability, critical testing, deliberation, European Parliament, political argument, purpose of argument, rationality, reasonableness

1. INTRODUCTION

Assessing the reasonableness of public political arguments is a complicated task. An important aspect of the complication arises from the fact that these arguments are typically multi-purposive. A politician arguing publically is out to achieve several goals by means of argumentation. Public political arguments arise in response to competing demands. This is sometimes the result of the multi-dimensional nature of the responsibility of a politician and other times the result of the multi-purposive nature of political institutions or even simply because public political discourse is open to individuals and groups that have different interests and needs as well different commitments and positions. This is the case in open public debates just as it is in considerably formal institutional contexts, such as the British or the European Parliament. Parliamentary debates, for example, are not only means for deliberating policies and legislations, but also means for holding the executives to account. In examining public political arguments, it is important to consider the multiple goals an arguer pursues in order to understand the strategic function of the argumentative choices made by arguers. Furthermore, if a good argument is indeed the "one that fulfills its purpose" (Johnson, 2000, p. 181), considering the multiple purposes is also indispensable for the assessment of arguments. Taking the multiple purposes into account when assessing arguments saves the assessment from being partial but poses an important challenge: different purposes are often in conflict which may make the assessment inconsistent.

In this paper, I argue in favour of considering the distinction between what is rational and what is reasonable as an important step in arriving at a non-fragmented and consistent assessment of the quality of public political arguments. The distinction runs parallel to another distinction, important when considering the multitude of goals associated with argumentation: the distinction between goals that are *intrinsic* to argumentation and others that are *extrinsic* to it. The distinction is particularly crucial when considering the question of whether or not the norms for assessing arguments can be derived from the functions of an argument. This is in fact an important question, and I would like to start this paper by addressing it (section 2). Following that, I introduce the distinction between what is rational and what is reasonable and explain how it is applied (Section 3). The way the distinction works and the gains we get from it will be illustrated using an example from the European Parliament (EP). I conclude by discussing the implications of the proposal and raising further questions in relation to the relationship between the goals of arguers and the norms of evaluating their arguments.

2. IS A GOOD ARGUMENT INDEED THE ONE THAT FULFILS ITS PURPOSE?

The understanding that a good argument is the one that fulfils its purpose (Johnson, 2000) may seem to be commonplace among argumentation scholars (van Eemeren & Grootendoorst, 1987; van Eemeren & Houtlosser, 2008; Walton & Krabbe, 1995). But it isn't really, at least not without challenges. 'Function claims' that attribute a 'determinable function' to argumentation and warrant deriving norms for assessment from such a function have been challenged by Goodwin (2007), for example. Goodwin's challenge is largely justified by the lack of evidence of one single *a priori* 'determinable function' for argumentation. Goodwin rightly cites many possible functions of arguments, other than the ones referred to by the proponents of 'function claims', and points out to the lack of criteria for deciding which of these functions can be the one that warrants deriving norms (ibid).

Goodwin is right, at least when it comes to the question concerning which of the functions of argumentation is the one once fulfilled the argument can be considered good, and why. The question remains largely unanswered. In order to answer it, more examination needs to be conducted in relation to the different possible goals and functions of arguments, their nature and status.

The goals recognised by argumentation scholars as relevant for the examination of argumentation are of different natures (See Mohammed, 2015, for a detailed survey). While some of these goals can be considered *intrinsic* to argumentation, i.e. goals of argumentation in and of itself and in any context (e.g. the goal to convince or to rationally persuade), there are also goals that are *extrinsic* to argumentation, i.e. goals originating outside of argumentation, usually in the contexts in which arguments occur (e.g. the goal to win votes, the goal to get a customer to buy a product ... etc). Also, within both *intrinsic* and *extrinsic* goals of argumentation, some of the goals identified are goals of the (individual) *act of arguing* (e.g. the goal to convince or the goal to win votes) while other goals are (collective) goals of the *argumentative interactions* in which the act of arguing occurs (e.g. the goal of critically testing standpoints or the goal to exercise accountability).

Looking at the goals identified by argumentation scholars, one cannot but notice that systematisation is needed. Goals, functions, purposes, or aims are often used to refer to more or less the same thing. That in itself is not necessarily a problem given the important common meaning between these terms. Nevertheless, what does create confusion is that the terminology used does not always capture the different natures of the different goals, functions or purposes discussed. As a result, crucial differences get obscured under similar terminology and useful similarities get hidden under apparent differences. Because goals of different natures should play different roles in the analysis and evaluation of arguments, the systematisation of the concept of goal in argumentation scholarship is necessary before we can decide which goals warrant norms for assessing arguments and why (Mohammed, 2015).

Distinguishing between *intrinsic* and *extrinsic* as well as between *individual* and *collective* goals are two crucial steps in the systematisation of the concept of goal (and in deciding which goals warrant norms for assessing arguments and why). These two distinctions can be applied in two successive steps: we first classify goals into two categories along the intrinsic-extrinsic divide and then, within each of the categories, we distinguish between goals of the act of arguing, which represent individual goals, and goals of the argumentative interaction, which represent collective goals.

In order to reflect the particular nature of a certain goal, I have suggested that we refer to a goal that is intrinsic as an *aim* or a *function* and to a goal that is extrinsic as a *use* or a *purpose* (Mohammed. 2015). As a result, we would speak of five different types of goals. On the one hand, we would speak of an *intrinsic constitutive aim* of argumentation (justification or manifest rationality – what makes a certain act count as arguing) and of *an intrinsic function of the act of arguing* (convincing or rational persuasion) as well as *an intrinsic function of the argumentative interaction* (critically testing standpoints). On the other hand, we would speak of *uses of the act of arguing* (when the extrinsic goal is individual) and of purposes of the argumentative interaction (when the extrinsic goal is a collective). Table 1, below, summarises the proposal and situates the different goals identified in the argumentation literature under the categories they would belong to.

Types of Goals / Approaches	Intrinsic			Extrinsic	
	Constitutive aim	Function of the act of arguing	Function of the argumentative interaction	uses of the act of arguing	purposes of the argumentative interaction
Bermejo-Luque	Justify a claim				
Gilbert's coalescent argumentation	Reach a coalescent situation			Face and task goals; Motives	
Johnson's manifest rationality	Manifest rationality	Rational persuasion		Inquiry; Belief-maintenance; Decision-making; … etc	
Toulmin	Justification			Other uses of arguments	
Van Eemeren et al.	Justify an opinion	Convince an opponent of the acceptability of an opinion; rhetorical goals	Critically testing standpoints; critically resolving differences of opinion; dialectical goals	Consecutive perlocutionary consequences of the speech act of arguing	Institutional goals
Walton & Krabbe's dialogue types				Arguers' aims in dialogue types	Goals of dialogue types

Table 1 – Goals in Argumentation Literature, an Overview (Mohammed, 2015.)

To come back to the question raised at the beginning of this section, if a good argument is indeed the "one that fulfills its purpose" (Johnson, 2000, p. 181), and arguments fulfil a variety of functions, uses and purposes, how do we decide which of these to consider when we evaluate (public political) arguments? The different types of goals (functions, uses, or purposes) give rise to different types of norms, all applicable and useful, albeit in different ways. For example, only the norms derived from the intrinsic functions of arguments (i.e. the function of convincing and that of critical testing) can be considered "argumentative" norms, i.e. norms that tell us whether an argument is *good in itself* 'independent' of whether or not it has any positive interpersonal or socio-political consequences. Norms derived from extrinsic uses and purposes are not argumentative in that sense. They are rather context-derived norms that tell us more about the positive or negative interpersonal or socio-political consequences of an argument: does the argument support rational decision-making? does it contribute to the exercise of accountability? ... etc. These context-dependent norms need to be distinguished from the intrinsic argumentative norms. After all, it isn't uncommon that a convincing argument fails in achieving desired context-dependent consequences. The two types of norms need to be distinguished in order to explain what happens in such cases.[1] However, as I will argue in the next section, the two should not be totally independent. Without synchrony between the two, our assessment of arguments may end up being useless and meaningless.

Needless to say that we may derive norms from *individual* functions or uses as well as from *collective* functions or purposes. The choice we make depends more on what we want to assess and why. Are we interested in the extent to which a political campaign is successful in raising support for a certain policy proposal (individual use) or rather in the extent to which the campaign supports good deliberation about this proposal (collective purpose)? In principle, both questions are worthy of investigation. One needs to decide what s/he would like to investigate and ask the question that is relevant. See Mohammed (2015) for a detailed discussion of the norms that are relevant for examining public political arguments, for example.

So, in short, all goals (functions, uses and purposes) warrant norms, it's just important to distinguish the different types of norms

---

[1] No keeping the distinction between the intrinsic functions and extrinsic uses and purposes is probably the most important weakness of Walton and Krabbe's concept of dialogue types. While for some dialogue types, the defining goal is a goal *intrinsic to* argumentation (e.g., persuasion), for others the goal is more of an *extrinsic* goal (e.g., negotiation).

warranted, what they assess and to apply the ones that are useful for the purpose of the analysis and evaluation.

3. SYNCHRONY NOT MERGER: INTRINSIC FUNCTIONS AND EXTRINSIC PURPOSES

Just like I emphasised that it is important to keep the intrinsic argumentative norm distinct from those extrinsic context-dependent norms, I would like to stress that it is actually equally important to maintain some (minimum) synchrony between the two types. Without such a synchrony the assessment of arguments would not be of any use, apart maybe from the mental gymnastics involved in the abstract activity of assessing argument for the sake of assessing them. Arguments are typically employed in the service of some context-dependent uses and purposes. A useful analysis is one that shows us how this happens (i.e. how argumentation contributes to socio-political processes) and a meaningful assessment is one that can explain the success or failure of an argument to fulfil its uses or purposes on the basis of its argumentative quality.

In order to achieve some synchrony between the intrinsic and extrinsic norms, it is necessary to adopt a perspective in which the intrinsic functions of arguments serve its extrinsic uses and purposes. From this perspective, the justification distinctive of argumentation contributes to socio-political processes through convincing and the critical testing of differences of opinion, i.e. through argumentation's intrinsic functions. An arguer who engages in the justification of a standpoint attempts, first, to convince an opponent of the acceptability of the disputed standpoint, and through that, some extrinsic uses of the argument, specific to the context of the argumentative interaction, can be made. For example, by convincing an audience that *the performance of the Government is up to standards,* a Prime Minister arguing in the parliamentary session of Question Time (PMQT) can portray his Government as competent. Furthermore, viewing argumentation as an interaction, arguers' justification fulfils the function of critical testing of the disputed standpoints through which arguers can achieve collective purposes which are specific to the context of the argumentative interaction. For example, by engaging in the critical testing of the standpoints about whether or not *the performance of the Government is up to standards*, arguers in PMQT can achieve the purpose of holding the Government to account.

Here, I'd like to be more specific, and discuss a little bit the functions, uses and purposes that are relevant for evaluating public political arguments in particular, i.e. public argument that contribute to socio-political processes. As do that, I am guided by considerations related to the characteristics of arguments in the public political sphere as well as the benefits that we may get from the different choices we make. The main benefit I have in mind is to be able to conduct an examination that enhances the 'emancipatory potential' (Habermas, 1984) of public political arguments and contribute to empowering members of the general public to become active participants in the political life. In such an examination, the norms derived for the collective function of argumentative interactions and its purposes are more relevant than those derived from the individual function of the act of arguing and its uses. This is mainly because focusing on the goals of argumentative interactions emphasises the agency of members of the general public, which allows us to consider them as active participants in the argument and in the socio-political process to with the argument contributes.

Assessing the quality of the argument in relation to the goals of the act of arguing, namely in relation to convincing and to the individual political uses of it, would tell us something about the rhetorical quality of the arguments and to their political effectiveness as tools to fulfil the individual political aspirations of the arguers. Assessing the quality of the argument in relation to the goals of argumentative interactions would instead tell us something about the dialectical quality of argumentative moves, i.e. about their role in the critical testing of standpoints, and about their contribution to the collective purposes for which arguers engage in argumentation. The latter is definitely more relevant if we intend our examination to help empower members of the general public to be competent participants in democratic life. Keeping the above in mind, assessing the quality of a public political argument will focus on the extent to which the argument fulfils the intrinsic function of critical testing as well as the extrinsic purpose derived from the socio-political process in which the argument occurs. I propose to refer to the former as *rationality* and to the latter as *reasonableness*.

To assess the rationality of an argument on the basis of the norm derived from argumentation's intrinsic function and reasonableness on the basis of the norms derived from its extrinsic purposes is in line with the distinction between rationality and reasonableness in major philosophical works (e.g. Rawls, Perelman; see also Cohen, 2011). Assessing rationality in relation to the intrinsic norm of argumentation is in line with the understanding of rationality as a universal norm that applies at micro/ local/singular levels of assessment. Assessing Reasonableness in relation to the extrinsic socio-political purposes of argumentation is in line with viewing reasonableness as a norm that is more encompassing, and taking more contextual considerations into account.

In view of the above, an argument is rational in as much as it furthers the critical testing of the disputed standpoint, and reasonable in as much as it furthers the socio-political process it is part of (e.g. deliberation, exercising accountability ... etc). Here, it is important to emphasise again the necessity of synchrony between critical testing as an intrinsic function and other extrinsic purposes derived from the socio-political contexts of arguments. Assessing the rationality of an argument makes sense only as long as there is at least one socio-political purpose for which it is beneficial to critically test the standpoints disputed (i.e. to engage in an argumentative interaction). Otherwise, argumentation remains just a (hopefully) enjoyable intellectual exercise. In the good cases where critical testing is instrumental for a broader societal political process, a rational argument would also be a reasonable one (from the perspective of the purpose the argument is supposed to serve). Rationality becomes the basis for reasonableness: the quality of the critical testing procedure offers a good explanation for why a certain argument succeeds or fails in fulfilling its purpose.

But this does not mean that the distinction between rational and reasonable would be totally redundant: (public political) arguments are rarely mono-purposive, which makes the judgment of reasonableness not always a simple and straightforward consequence of the rationality judgment. For example, parliamentary argumentative exchanges that are used for the purpose of deliberation about future courses of action are usually also used in order to exercise accountability and assess the performance of those who are in power. In order to do that, arguers engage in the discussion of two issues: *What is the best course of action to be taken in response to the problematic situation at issue?* As well as *Is the performance of the authorities up to standard?* While the discussion of the first issue serves the deliberation process, the discussion of the second serves the accountability process. The two discussions (about the two issues) can be distinguished analytically but are usually intertwined in reality. Typically, the issues are discussed simultaneously by means of argumentative moves that contribute to the discussion of the two issues at the same time (something like a divergent argument). That means that the same argumentative move can be used in the service of more than just one purpose.

It is not at all unusual that an argument is a rational contribution to the discussion of the future course of action but not so when it comes to the discussion of the performance of those who are in power. Political discourse is full of examples where pro-government supporters employ arguments that are perfectly reasonable deliberation-wise but which obstruct holding governments accountable for the policies and actions that might have caused the problematic situations, i.e. unreasonable accountability-wise. While the rationality is a 'local' judgment that applies within a single discussion, reasonableness is a more encompassing judgment that takes into account the quality of the move in the multiple simultaneous discussion the move contributes to, i.e. its contribution to the multiple extrinsic processes the move is part of. Taking into account the multiple purposes of a public political argument, reasonableness becomes an inter-purpose judgment that synthetises the multiple judgements of rationality across the different discussions, each of which is associated with one of the multiple purposes.

In view of the distinction between rational and reasonable arguments, we are guided to integrate contextual considerations into our assessment of arguments in two necessary ways. The first way is similar to what has been already argued for by van Eemeren (2010, Ch. 5): contextual considerations specify the general universal norm of critical testing into specific criteria that take into account the rules and conventions of the argumentative practices at issue. This is necessary in order to arrive at a judgment of argumentative rationality that is context-sensitive and meaningful. For example, this helps us avoid arriving at meaningless fallacy judgments (those fallacy judgments that do not take constraints of context into account). The second way of integrating contextual considerations, which complements the first one, is meant to take the multiple purposes of arguments into account when formulating judgments of reasonableness. Contextual considerations will determine which external purposes can be attributed to argumentative exchanges and also how these purposes are connected and how they should be considered: should they be ordered in a hierarchy of importance? Or rather balanced as much as possible? ... etc. This is crucial as it enables us to arrive at judgments of reasonableness that consider the contribution of arguments to broader socio-political processes and trace this contribution to specific argumentative moves in their particular contexts.

In the next section, I apply the distinction between rationality and reasonableness in examining a debate in the European Parliament (EP), a relatively young venue for public political arguments, with a fast growing scope of influence. The debate is a good example of argumentative practices that are conventionally multi-purposive and for which the distinction between rational and reasonable arguments is useful. The examination is meant to illustrate the way the distinction works, its merits as well as its limitations.

4. CASE STUDY

The debate I examine (European Parliament, 2013) is about an important recent controversy in the EU, namely the issue of privacy and personal data protection at a time of "security threats". Nevertheless, it is not one of those exceptionally televised argumentative encounters. It is more like an ordinary, even a bit mundane, argumentative event like the majority of public political arguments, which makes it a good case to examine.

The debate came in response to claims by whistle-blower Edward Snowden, published in media, that the US National Security Agency (NSA) monitors bank data in the EU (BBC News, 2013). In it, it was discussed whether or not to suspend the financial data exchange agreement between the EU and the USA, the Terrorist Finance Tracking Programme (TFTP).[2] The debate started by an introductory statement by Commissioner for Home Affairs Cecilia Malmström, who outlined the response of the Commission to the allegations (see Appendix). Her statement was followed by short speeches by representatives of the political groups at the EP, then by speeches by other Members of the EP (MEPs) and ended with a concluding statement by Commissioner Malmström. The Commissioner was not in favour of suspending the agreement, but the general mood among MEPs was in favour of that.[3]

EP debates on Commissioner's statements are characteristically multi-purposive. In them, MEPs attempt to influence the EU policy

---

[2] The TFTP was set up by the U.S. Treasury Department shortly after the attacks of 11 September 2001, and was approved by the EP under the Lisbon Treaty. Supporters of TFTP argue that the agreement has generated significant intelligence data that helped the U.S. and EU States in 'their fight against terrorism'. But problems in the implementation of the agreement have driven many to call it into question. The media reports based on the claims made by Snowden made these sceptical voices even more prominent.

[3] The general mood in the debate was in favour of the suspension, even though the EPP, the largest party in the EP since 1999 was against it. A week after the debate, different political groups proposed motions for action. The vote on the motions took place in the plenary session of 23 October, and a joint motion for a resolution was approved, in which the EP called on the Commission to "suspend the Terrorist Finance Tracking Program (TFTP) agreement with the US in response to the US National Security Agency's alleged tapping of EU citizens' bank data held by the Belgian company SWIFT". The resolution was passed by 280 votes to 254, with 30 abstentions. Although the EP has no formal powers to initiate the suspension or termination of an international deal, "the Commission will have to act if Parliament withdraws its support for a particular agreement", says the approved text.

making by deliberating with Commissioners future courses of action in relation to topical issues. Furthermore, the Commission is held accountable for its past and present conduct in relation to this same issue: the Commissioner defends its conduct in the face of the scrutiny exercised by MEPs. Obviously, argumentation plays a central role in both processes: deliberation is conducted by means of arguments that justify practical claims about future courses of action, and accountability is exercised by means of arguments that justify an evaluative claim about the conduct of the Commission. (see Mohammed, 2015. for more information about EP debates on Commissioner's statements as well as a detailed analysis of this particular debate).

As in other similar debates, the argumentative exchanges in the TFTP debate contribute to these two important political processes. The argumentative exchanges contribute to the deliberation about the measures that need to be taken in response to the alleged NSA spying as well as to the scrutinising of the conduct of the commission in what relates to the protection of the data of EU citizens. This happens by engaging in the discussion of (at least) two issues: whether or not the TFTP should be suspended and whether or not the conduct of the Commission is up to standard. In both cases, the critical testing of standpoints can be assumed to be instrumental for the relevant socio-political process, i.e. there is synchrony between the intrinsic function and the extrinsic purposes of the argument.

Commissioner Malmström opposed the suspension of the TFTP and defended her performance as satisfactory. In her statement, she described in details her efforts to investigate the situation: the numerous meetings she had had with the US officials, the many confirmations she has obtained from them and the many further information she had requested them to provide. In view of the disagreement about the suspension of TFTP, the actions listed can be understood as attempts to justify that *there is no need to suspend the agreement* since *investigations show no evidence of the alleged spying*. But by listing her efforts in detail, Commissioner Malmström was obviously not just concerned with denying the need to suspend the TFTP. The detailed list of actions is clearly also an attempt to justify the claim that *The conduct of the Commission is satisfactory*. In fact, observing the formulations used to represent the several meetings conducted, the requests made and the confirmations obtained by the Commissioner, one cannot miss the Commissioner's effort to defend her own conduct as someone who has done a lot in order to make sure that the private data of EU citizens is protected. The Commissioner emphasises that she "has immediately taken action" following the appearance of the "first allegations in the press"; that she was firm in requesting explanations from US officials; that she called them, sent them letters and met with them in person to obtain more information; that she "insisted" to have written confirmation of the information she received ... etc.

Here we have an example of those argumentative moves that contribute to two discussions and to two political processes at the same time. Taking both processes into account and reconstructing the argumentative exchanges as part of two simultaneous discussions (Mohammed, 2015) sheds better light on some important strategic choices made in the debate. In particular, it helps us understand the Commissioner's strategic choice to detail the description of the efforts she has made in investigating the alleged spying. The detailed descriptions of the Commissioner make more sense once we understand that the Commissioner is not just refuting the opponent's claim that *TFTP should be suspended* but also, maybe even more importantly, defending her conduct. After all, had the Commissioner been only interested in asserting that there is no evidence that the NSA is spying on EU citizens, it would have been enough to cite findings of investigations conducted. There would have been no need to emphasise that she herself has made all this effort to investigate had the Commissioner not been interested in defending her conduct.

Reconstructing the exchange into simultaneous discussions also sets the ground for an assessment of the quality of an argumentative move in a way where rationality and reasonableness can be distinguished. Let us take, for example, the move of listing in detail the actions taken by the Commissioner. The rationality of this move can be evaluated by assessing its contribution to the critical testing of each of the standpoints it defends. We have here two standpoints (in two simultaneous discussions) defended by this move, so we have two judgments of rationality. In the discussion about the conduct of the Commission, the move counts as a relevant defence for the standpoint; in this discussion, it is a rational move. In the discussion about the suspension of the TFTP, the move is not as relevant for the defence of the standpoint; in this discussion, we may say that it is a mild *ignoratio elenchi*, i.e. an argument that does not really justify the standpoint it is supposed to justify. As a result of its being a rational move in the discussion about the conduct of the Commission, the move furthers the exercise of accountability over the Commission. But as a result of it being irrational in the discussion about the suspension of the TFTP, the move does not advance the deliberation about the future course of action.

Is the move then reasonable or not? Well, here contextual consideration would need to guide us in seeking a synthesis of the assessments across-discussions. I would say, in this case, we may think that the two processes (deliberation and accountability) are equally important, both are characteristic of the venue and we have no indication that one is more important than the other. That means that we do not need to make the judgment of rationality in one overrule the other (no hierarchy of norms). Instead, we may consider the gravity of the problem the move causes in the discussion where it is not rational. In this case, the hindrance to the critical testing is mild, and so is the hindrance to deliberation. Consequently, overall, and taking the two processes into account, the move can be considered fairly reasonable. This judgment reflects the overall quality of the move, its contribution to the different socio-political processes it is part of, based on the quality of the critical testing (i.e. argumentative norm) in relation to each.

5. CONCLUSION

I started this paper by highlighting a complication that faces us when assessing public political arguments. On the one hand it is important to take the multiple functions, uses and purposes of an argument into account. But on the other hand, the different goals of argumentation are not always in harmony, and we do not have clear criteria for deciding which ones to consider more important in case of conflict. In order to solve this complication, I proposed to distinguish between the different types of goals of argumentation, especially between argumentation's intrinsic functions, on the one hand, and its extrinsic uses and purposes, on the other hand. Each type of goals warrants different types of norms. All the norms are relevant albeit differently. On the basis of this distinction, I also proposed to distinguish between the judgment of rationality and the judgment of reasonableness. The former based on intrinsic norms, the latter on extrinsic ones. The two are distinguishable (it is important not to merge them) but not totally independent (it is important to adopt a perspective from which there is synchrony between them).

Applying these distinctions needs to be made from a perspective on public political arguments where the critical testing intrinsic to arguments is in the service of socio-political processes. As a result, we can have distinguishable assessments, each related to each of the socio-political processes to which the argument contributes. We are able to explain the positive or negative socio-political consequences of an argument in the argumentative procedure. And we are also able to synthesise the different judgments into one overall assessment if needed.

Despite these merits, it's most important to acknowledge this is work-in-progress that needs to be refined, and which needs to be applied to more cases, more suitable, and more interesting. The point would be to show how exactly the distinction between rationality and reasonableness indeed brings us to non-fragmented and yet consistent and meaningful assessments of public political arguments.

ACKNOWLEDGEMENTS: I acknowledge the financial support of the Portuguese Fundação para a Ciência e a Tecnologia (FCT) through grant no. SFRH/BPD/76149/2011.

REFERENCES

BBC News. (2013). The week ahead at the European Parliament. Retrieved on 26 October 2013 from http://www.bbc.co.uk/news/uk-politics-24419555.
Bermejo-Luque, L. (2010). Intrinsic versus instrumental values of argumentation: The rhetorical dimension of argumentation. *Argumentation*, 24(4), 453-474.
Cohen, D. (2011). Reasonableness, Rationality, and Argumentation Theory. Presentation at the *International Colloquium: Inside Arguments*, Coimbra, Portugal, March 24, 2011.
Eemeren, F.H. van. (2010). *Strategic maneuvering in argumentative discourse: Extending the pragma-dialectical theory of argumentation*. Amsterdam: John Benjamins.
Eemeren, F. H. van, & Grootendorst, R. (1987). Fallacies in pragma-dialectical perspective. *Argumentation*, 1(3), 283-301.
Eemeren, F. H. van, & Grootendorst, R. (2004). *A systematic theory of argumentation: The pragma-dialectical approach*. Cambridge: Cambridge University Press.
Eemeren, F. H. van, & Houtlosser, P. (2008). Rhetoric in a dialectical framework: Fallacies as derailments of strategic manoeuvring. In E. Weigand (Ed.), *Dialogue and Rhetoric* (pp. 133-151). Amsterdam: John Benjamins.
European Parliament. (2013). Suspension of the SWIFT agreement as a result of NSA surveillance (debate). Retrieved on 2 December 2013 from http://www.europarl.europa.eu/sides/getDoc.do?type=CRE&reference=20131009&secondRef=ITEM-019&language=EN.
Goodwin, J. (2007). Argument has no function. *Informal Logic*, 27(1), 69-90.
Habermas, J. (1984). *The theory of communicative action. Vol. 1: Reason and the rationalization of society*. Boston: Beacon Press.
Jacobs, S. (1989). Speech acts and arguments. *Argumentation*, 3(4), 345-365.
Johnson, R. H. (2000). *Manifest rationality: A pragmatic theory of argument*. Mahwah, NJ: Lawrence Erlbaum Associates.

Mohammed, D. (2015). Goals in argumentation: A proposal for the analysis and evaluation of public political arguments. *Argumentation*, online first, doi: 10.1007/s10503-015-9370-6.
Patterson, S. (2011). Functionalism, normativity and the concept of argumentation. *Informal Logic, 31*(1), 1-25.
Toulmin, S. E. (1958). *The uses of argument.* Cambridge: Cambridge University Press.
Walton, D., & Krabbe, E. C. W. (1995). *Commitment in dialogue: Basic concepts of interpersonal reasoning.* Albany: SUNY Press.

APPENDIX

The introductory statement of Commissioner Malmström

Madam President, ladies and gentlemen, I am here tonight to inform you about the actions I have decided to take following the press allegations about the possible access of the US National Security Agency (NSA) to the data exchange through the EU-US Terrorist Finance Tracking Programme (TFTP) Agreement.
On 24 September I met many of you in the LIBE Committee and informed you about the ongoing efforts to follow up on this matter, which is of course of great concern. The discussions in LIBE were helpful and confirmed the need to clarify a number of issues.
Since the first allegations appeared in the press, as I told you then, I have immediately taken action. In July, I sent a first letter to my US counterpart, and on 11 September I called on the Under-Secretary for the Treasury Department, Mr Cohen, and told him that I was waiting for substantial information on the alleged tapping. The next day I also sent him a letter, in which I requested the opening of consultations under Article 19 of the TFTP Agreement. As you know, this is the procedure that is regulated in the agreement in case there are questions or things that need to be clarified.
In reply to my letter – and I shared the letter with the LIBE committee on 23 September – the US Authority provided some explanations. But several important questions remained unanswered. I therefore, this Monday, met with Under-Secretary Cohen in Brussels and I appreciate that he came despite the budgetary constraints. We had open and very long discussions and he clarified a number of points.
During that meeting, Under-Secretary Cohen explicitly confirmed that since the entry into force of the TFTP Agreement the US Government has not collected financial messaging from SWIFT in the EU. He also said that the US Government has not served any subpoenas on SWIFT in the EU during that period. I insisted to have that very important confirmation statement confirmed in writing.
We also discussed in some detail the established channels through which the US does obtain financial information in SWIFT format used by financial institutions worldwide. Also on this, I asked for further explanations in writing, in order to be absolutely sure that these mechanisms do not conflict with the TFTP Agreement.

At this stage, therefore, our contacts with SWIFT and the US Government have not really given any evidence that the TFTP Agreement had been violated. Some further clarifications are, however, needed before we can draw full conclusions. Concluding the consultations with the US remains on the top of my agenda and also for my staff and we intend to do our best to get all information needed in the very near future.

Of course, I will make sure that you are fully informed about future developments at the outcome of these consultations.

How to Be a Better Functionalist. Commentary on Mohammed's Not Just Rational, But Also Reasonable

JEAN GOODWIN
Iowa State University, USA
goodwin@iastate.edu

1. INTRODUCTION

"Argument has no function" (Goodwin, 2007) remains a thesis I stand ready to defend. My aim in nailing it to argumentation theory's front door was in part to encourage increased sophistication in functionalist theorizing. It is easy to say that "an argument is good if it fulfils its purpose," but a cloud of vagueness hovers around that little 'it.' In the paper I here respond to, and in other work within the same project (2015), Dima Mohammed makes significant advances in dispelling that vagueness, through (a) thinking systematically about the kinds of telos-stuff that can be associated with kinds of argument-stuff, and (b) thinking through how to integrate the normative standards that can be derived from them.

2. SORTING THINGS OUT

Function theorists have not infrequently been guilty of loose talk about the aim, goal, end, purpose, function, etc. (in general, "telos-stuff") of argumentation, arguments, arguing, various kinds of argumentative transactions, etc. (in general, "argument-stuff"). Mohammed does the field the great service of clearing the way towards much greater precision. I'm not sure that all her terminology will catch on, since the ordinary meanings of words like "use" and "purpose" will likely win out. Nevertheless, the chart she has developed deserves our attention. She identifies two distinct dimensions to any function claim. First, telos-stuff must be distinguished as intrinsic or extrinsic. Second, argument-stuff must be identified, as either the individual act of arguing or as the collective argumentative interaction.

Every function theorist should have to assign their claims to the boxes Mohammed has devised. In the meantime, while we wait for their responses, the chart enables us to start asking useful questions. For example:

(a) Regarding intrinsic v. extrinsic. Mohammed identifies the goals of Waltonian dialogue types as *extrinsic* purposes of argumentative interactions. That's interesting; my impression is that Walton puts forward these goals as *intrinsic* functions. As Mohammed notes (n. 2), ambiguity on this point is one of the theory's weaknesses. Our differing interpretations of Waltonian theory raise a more interesting question, however. That is: *how do we tell whether some particular telos-stuff (aim, end, purpose, and so on) is indeed intrinsic, i.e. a function?*—as opposed to being merely extrinsic, one among likely many uses or goals? I raised this question in §3 of my original article; Mohammed's chart makes the question all the more pressing.

Mohammed's own discussion suggests that a piece of telos-stuff is intrinsic if it is a "goal [...] of argumentation in and of itself and in any context" (2015). This appears in part to be an empirical test: we need to examine argument-stuff as it appears across a variety of contexts; if we observe telos-stuff T throughout, then we have evidence that that T is indeed *intrinsic*. The problem is that when we look at argument-stuff, we observe lots of variety and little uniformity. I suspect that for any asserted T, an example of actual usage can be found that doesn't contain it. The paper Innocenti and I have submitted to this volume puts forward one such example against a large range of asserted functions. Of course, any example of argument-stuff can probably be *reconstructed* to exhibit T. But what drives the reconstruction is a foundational assumption that that T is indeed intrinsic. That reasoning is circular.

Mohammed's chart shows that a wide range of telos-stuff has been asserted to be intrinsic. Figuring out how to justify any of these assertions is a key task facing function theorists.

(b) Regarding individual v. collective. Being precise about what argument-stuff is being talked about is the single most important thing we could do to reduce confusion and talk-at-cross-purposes among argumentation theorists. I admit I might quibble with Mohammed's two categories; in particular, the "act of arguing" seems to me to be better described as the "activity of making arguments." However, the general thrust of Mohammed's proposal is invaluable. Each of us should take responsibility for being explicit at all times about exactly what aspect of argument-stuff we are discussing.

Mohammed's chart does have one conspicuously odd feature, however. In addition to separate categories for argument-acts and argumentative-interactions, the "intrinsic" side has a third column, the "constitutive aim of argumentation." This strikes me as problematic. The word "argumentation" in English is rare in ordinary usage. To me, it conveys nothing more than a vague sense of what I've here been calling "argument-stuff": anything relevant to the making and exchanging of arguments, including the arguments themselves, activities and inter-activities involving arguments, the arguers (their virtues, traits, cognitive processing and planning), institutions hosting argumentative activities and inter-activities, language registers associated with arguments, and on and on. The only place I'm really comfortable using the term "argumentation" is in referring to argumentation theory or studies, which is a theory about or study of any or all of this argument-stuff. In these two usages, the vagueness is strategic: it helps bring all of us to the same wonderful conference. Everywhere else, it is a disaster.[1]

The vagueness of the column about the "constituent aim of argumentation" stands in contrast with the determinacy of the individual and collective activities Mohammed distinguishes. Some theorists undoubtedly embrace that vagueness. I feel relatively confident, for example, that Ralph Johnson really did mean to include all argument-stuff in his claim that the practice of argumentation had the constituent aim of making rationality manifest in the world. I am less confident, however, about what it would mean for all this argument-stuff to be aimed at justification. This raises the question: Function theorists who assert what Mohammed has designated a "constituent aim of argumentation," *what do you mean?*

3. PUTTING THINGS BACK TOGETHER

Argument-stuff in important contexts is complex. Mohammed shares the interest traditional to rhetorical approaches in "public political arguments," and recognizes that they inevitably "arise in response to competing demands." Citizens' responsibilities have a "multi-dimensional nature," she points out; citizens also have multiple needs, desires and interests, and are acting within civic institutions that are themselves subject to multiple expectations. Any halfway respectable function theory is going to have to respect this complexity while at the same time producing *integrated* or "synchronous" accounts of civic argument-stuff.

Mohammed is confident that the normative standards derived from the extrinsic purposes of argumentative interactions can be

---

[1] I am aware that the word "argumentation" (or similar) in other languages has a more determinate meaning—often, an extended sequence of arguments on one topic, something like what in English would be called a "case." It does not look to me, however, like theorists who speak of the "function of argumentation" are using the word in that sense; I await correction.

reconciled with those derived from their intrinsic function. For example, an argument that makes a good contribution to critically testing a standpoint should, by that very goodness, also make a good contribution to making a decision or holding an official accountable. Mohammed considers that the more serious challenge is integrating the normative standards derived from different and sometimes competing extrinsic purposes—e.g., somehow fitting together the potentially dilemmatic purposes of decision-making and accountability.

Mohammed proposes calling the intrinsic/functional goodness of arguments "rationality," and the extrinsic/purposive goodness of arguments, "reasonableness." Will this proposed vocabulary stick? It has some plausibility: a charge of irrationality seems harsher than a charge of being unreasonable—more connected with the person's basic orientation to reason, more "intrinsic." On the other hand, there also seem to be some drawbacks to the proposal. For one thing, rationality seems to be a binary: an argument is rational or it isn't. But as Ralph Johnson long ago pointed out, arguments are assessed on a scale: they are generally more or less good. Reasonableness does allow degrees, so it doesn't share that problem. But to say a person's conduct is reasonable is, as Rawls points out, to make a specific kind of assessment: it's saying it is making a *fair* contribution to common life. In functionalist theorizing, an argument is good if it makes *a* contribution to achieving an extrinsic purpose; *fairness* would seem to be an additional requirement. And it's not clear that fairness even makes sense for all extrinsic purposes; what, for example, would be a fair contribution to an eristic dialogue? Whether or not the proposed terms stick, however, at least Mohammed's analysis has made clear the different yet integrated roles different telos-stuff can play in argument assessment.

4. CONCLUSION

I remain opposed to functional approaches to constructing normative theories of argument-stuff; I think there are better ways forward. With that understood, I recommend: function theorists need to respond to the challenges Mohammed advances in this project, continuing to develop function theories that are more precise, better defended, and more responsive to real-world complexities.

REFERENCES

Goodwin, J. (2007). Argument has no function. *Informal Logic, 27*, 69–90.
Mohammed, D. (2015). Goals in argumentation: A proposal for the analysis and evaluation of public political arguments. *Argumentation*. Online first (doi: 10.1007/s10503-015-9370-6).

23

Prosodic Constraints on Argumentation: From Individual Utterances to Argumentative Exchanges

FRANÇOIS NEMO
Diasémie (LLL), University of Orléans, France
francois.nemo@univ-orleans.fr

CAMILLE LÉTANG
Diasémie (LLL), University of Orléans, France
camille.letang@univ-orleans.fr

MÉLANIE PETIT
Diasémie (LLL), University of Orléans, France
melanie.petit@univ-orleans.fr

The paper discusses the relationship between prosody and argumentation in the interpretation of what is said, its role in the determination of the argumentative orientation of utterances and the description of argumentative exchanges. It also discusses the wider implications of such a role for argumentation studies as a whole and for specific models such as Argumentation in Language Theory. The automated corpus based methodology used for its study is presented.

KEYWORDS: argumentative contributions, argumentative frames, argumentative orientation of what is said, automated discrimination of prosodic comment, controlled attention, prosodic encapsulation of the argumentative scene, prosodic marking of argumentative orientation

1. INTRODUCTION

In order to understand the precise relationship between argumentation and prosody, our aim will be to discuss successively two interrelated issues, namely:

- the role of prosody in the interpretation of the argumentative orientation of what is said;

- the role of prosody in the structuring and explanation of argumentative exchanges;

In both cases, we shall have to clarify[1]:
- the nature of the "information" associated with prosodic contours or features, and the argumentative significance/importance;
- the way prosody/meaning pairs can be studied in a reliable way.

Because the two requisites for any matching of prosodic features with interpretative ones are the availability of a robust initial analysis of the data at stake, and a "shuttle" process between prosodic and semantic features which allows to improve this analysis, we shall first have to spell out (and recall) the analysis of the source of utterance's argumentativity which we have used as a background for this work:

- the nature of the "information" associated with prosodic contours or features, and the argumentative significance/importance;
- the way prosody/meaning pairs can be studied in a reliable way.

1.1 Accounting for the argumentative orientation of what is said

The idea that *all* utterances are associated with an argumentative orientation has been defended initially within the ALT approach (Argumentation in Language Theory developed by Ducrot & Anscombre[2], Théorie des blocs sémantiques developed by Carel & Ducrot), with the aim of demonstrating that utterances and exchanges do not concern primarily information as such, as advocated for instance by Grice or Sperber & Wilson but the implicit argumentative *conclusions* associated with them. ALT and posterior work have indeed been very successful in demonstrating that describing a sequence of utterances and/or the discourse connectives within such sequences implies describing the relations of co-orientation or anti-orientation between utterances and its linguistic marking, which is why we shall assume here *the existence of a specific layer of interpretation* associated with

---

[1] For earlier work on related issues, see Vincent & Demers (1994), Rittaut-Huttinet (1995), Saunier (1999), Bose & Gutenberg (2002), Fernandez & Picard (2002), Lacheret-Dujour & Beaugendre (2002), Thomsen (2002), Wichmann, (2002). Beyssade & alii (2004), Auchlin & Simon (2004, 2006), Yang, L. (2006), Kral & alii (2007), Sudhoff (2007), Mertens, (2008), Kohler (2009), Mello & alii (2012).

[2] See for instance Anscombre & Ducrot (1976), Ducrot & Anscombre (1983).

argumentative orientation[3] and the necessity to study the role of prosodic constraints in its determination. Among the language-centered approaches to argumentation, ALT has also defended the idea that linguistic signs encoded argumentative orientations and vice-versa that argumentative orientations were obtained through a process in which linguistic signs were playing a leading role.

Following Nemo (1985, 1988, 1992, 1995, 1999, 2001, 2006a, 2007, 2014), we shall also admit that this argumentative orientation (and ALT's notion of *conclusion*) is the consequence of the modal and scalar framing of utterances in general and of the attentional nature of argumentative contributions.

1.1.1. Scalar (and modal) framing of utterances

Discussing the argumentativity of utterances (along with informativity, directivity, performativity, etc.), Nemo (1992, 1995, 1998) introduced a scalar/argumentative constraint on utterance interpretation, whose everyday formulation is "Do not say anything that makes no difference" and showed that ALT's conclusions could precisely be identified for each utterance as *answers* to the question "what difference does it make?". Nemo also showed that this scalarization constraint according to which an utterance U may be used as an argument for an utterance R, if and only if it makes a difference for R that E is the case or not was actually presupposing a modalization constraint (Nemo, 1988) and the description of any utterance, minimally, as the association of a proposition with a modal frame[4], introducing alternatives within a *comparison set*. The comparison process at stake is crucial to the determination of the scalar/argumentative value of the utterance (i.e. the answer to the question "what difference does it make?") which has been called *conclusion* within the ALT framework until then. Other constraints related to the *scalarization constraint* (and important to understand the role of prosody in the determination of argumentative strength) were also described, namely the comparison, the scalar slope and the modal slope constraints.

---

[3] See for instance Raccah (2010).

[4] The description of utterances as embedded in modal frames acting as conditions of relevance is now widely shared, the only difference between models relying on alternative sets (e.g. Levinson, Krifka, Horn, etc.) and the modal and scalar frames introduced in Nemo (1985, 1988, 1992, 1999) is that the latter are not reduced to comparison sets but include the possibilities which have been ignored or taken for granted (i.e. presuppositions) and the scalar value associated with them.

The comparison constraint states that given the fact that the scalar/argumentative value of an utterance U depends on a comparison of the different possibilities introduced by the utterance, the scalar orientation of the utterance depends on the possibilities which are, or which are not, introduced[5].

The scalar slope constraint states that given the scalarization constraint, the argumentative strength of an utterance U depends basically on whether the difference that U makes for R is small or big[6].

The modal slope constraint states that given the fact that it makes a difference for an utterance R if U is the case or not, the more not-U will be possible (likely), the biggest the difference the fact that U is the case will make. And hence, the stronger its argumentative/scalar value.

1.1.2. Attentional nature of the argumentative layer

Another crucial dimension of argumentativity was introduced later (Nemo, 1999) with the observation that because saying something is ultimately a matter of **"attracting someone's attention on something and asking him/her to take it into account"**, and thus belongs to a mechanism called controlled (and joint) attention in psychology, and because there is actually no way to attract someone's attention on something without making manifest one's relation (being worried, happy, surprised, etc.) with that thing – which is what cognitive psychologists call social referencing - all utterances may be considered as associated with an argumentative layer which comprises three components, namely attentional selection (i.e. the fact of choosing to consider certain things and to ignore others), attentional bias (i.e. the fact of looking at something from a certain attitudinal or thymic perspective) and manipulation of attention (i.e. the fact of directing or trying to direct someone else's attention on such or such thing). This results in the fact that all linguistic exchanges will forcefully be embedded in such attentional mechanisms[7], as can be shown in the

---

[5] For instance, the Sapir/Ducrot argumentative contrast between *peu* P (little) and *un peu* P (a little) may be accounted for by the fact that the former introduces a *peu* P vs *beaucoup* P (a lot) alternative which presupposes P, ignores the possibility of not-P and presents the quantity at stake as insufficient whereas the latter introduces a *un peu* P vs *not*-P alternative, in which any small quantity is better than nothing.

[6] As a result, minimizing (sometimes prosodically) the difference something makes is a key aspect of argumentative exchanges.

[7] See Nemo (2006b).

limited space available here by considering strictly uninformative utterances, such as *"I am not three years old"* (uttered by a teenager), *"I am not your dad"* (uttered by a mother) or *"I am your dad"* (uttered by a dad) whose attentional and scalar values illustrate how attentional expectancies ultimately lead to argumentative conclusions in the ALT sense, i.e. as operational consequences (in terms of action or attitude) of taking something into account: a teenager saying *"I am not three years old"* to his parents is saying "*I would like to attract your attention on the fact that I am not three years old and that you should take it into account*", presumably because they have not been fully aware of it (considering their behavior until then) and in order to lead them to the conclusion that they should behave accordingly.

This is why taking into account the attentional dimension of the argumentative interpretation of utterances may prove to be crucial both for the study of argumentation and for a correct understanding of the role prosody is playing in the fine-grained management/manipulation of the interlocutors' attention and in shaping the attentional bias.

1.1.3. Attentional dimension of the argumentative/interlocutive scene

Attentional games however are not limited to individual utterances, and a key level appears to be the level of argumentative contributions. Such contributions:

> are sets of utterances bound by joined attention - which implies that they must be considered together—and are individual or collective definitions of what has to be taken into account at a given moment about a given topic. (Nemo, 2006a, p. 416).

Without entering here in all the details of the question, certain things are worth stressing out. They concern the collective definition of what we propose to call the attentional field and the way interventions and contributions to this shared focus of attention are taking place. Notably, what must be understood is that in conversational exchanges, individual contributions are contributions to a collective definition of what has to be taken into account and how, but that this collective definition is ultimately based on the participants ratification of each of the elements to be considered and of their relative importance.

What is at stake in the description of argumentative contributions is thus initially nothing less than:

- co-defining whether something is worth the attention or not, in other words whether something should be taken into account or not;

- co-defining what should be considered as central or not in the attentional field;
- co-defining how what has to be taken into account must be taken into account (social referencing).

As importantly, and in contradiction with the idea that only common ground is important in the study of conversational exchanges, what must be understood is that even though the accepted purpose of exchanges is supposedly to reach a consensus on those three issues, the reality of exchanges is that disagreements and diverging ground on each issue are the rule and not the exception (Nemo, 2007). This makes it necessary to study in great detail the ratification process of successive contributions and to describe what we propose to call the interlocutive/argumentative scene.

As for the first question, it appears indeed that the contributional value of a contribution will depend on whether the attentional movement and shift associated with an argumentative contribution (or contributional segment) are:

- rejected (as irrelevant or inacceptable);
- ignored (as irrelevant or inacceptable);
- tacitly ratified by a ratifying silence;
- ratified as secondary or marginal, i.e. as deserving little attention;
- ratified as important, i.e. as deserving full attention.

Because only a part of what has been said will be ratified and because the ratification process is not a simple two-step process, a sort of contributional memory which we shall call the interlocutive scene will emerge. While in some cases the interlocutive scene will take the form of a "dialog of the deaf" in which all participants are trying to get the collective attention to focus on competing and disjoined objects or perspectives, resulting in the absence of any shared attentional field, most of the time it will take the form of shared knowledge of both the successive interventions and their collective fate which will result in the coexistence of converging and diverging grounds.

As for the internal structure of individual contributions, it must finally be stressed out that the relationship between the argumentative orientation of individual utterances and the argumentative orientation of an individual contribution as a whole must be approached in terms of contribution modification. In other words, it is important to realize that argumentative discourse is not made of segments but of contributional additions to what has been said, so that for example (Nemo, 2006a, 2007) a discourse connective C between two utterances A and B must not be described as describing/marking the relation between A and B, as is often assumed, but as commenting the relation between saying [A] and saying [AB] and hence as relating two alternative discourses/contributions[8].

2. PROSODY AND THE ARGUMENTATIVE INTERPRETATION OF WHAT IS SAID

As mentioned earlier, it has often been assumed that the argumentative orientation of what is said would be predictable from its semantic content and hence that language-centered studies of argumentation could focus exclusively on this content in order to understand the linguistic dimension of argumentative mechanisms. We shall on the contrary argue here for the necessity to admit that because prosodic contours are crucial to the understanding of "what is said about what is said", but also because such comments about what is said often have the capacity to ultimately alter the content of "what is said", the argumentation orientation of utterances can *never* be predicted without considering it in detail. We shall illustrate this reality by considering the role of prosodic constraints in the determination of the argumentative orientation of utterances including (and sometimes reduced to) linguistic signs such as French *enfin* and *quelques* (i.e. *some*) and by considering the nature of the "information" provided by the prosodic comments associated with various uses of French "*oui*" (*yes*).

We shall first show how the prosodic contours or features associated with the various *use-types* of these signs are indeed largely responsible for the actual argumentative orientation of the utterance, which cannot be said, as assumed in ALT, to be directly predictable from the signs themselves regardless of the way they are prosodically realized.

Because until recently, prosodic studies of interpretation have often reduced the study of intonational meaning to the study of "the use of suprasegmental phonetic features to convey "post-lexical" i.e. sentence-level pragmatic meanings in a linguistically structured way" (Ladd, 2008), it is worth mentioning that things started to change importantly when semantic and pragmatic work on polysemy (in the larger reading of the notion) started to consider prosodic features

---

[8] Typically, adding B will imply that B should also be taken into account that taking it into account might radically transform the global conclusion which can be drawn from considering A and B jointly. In that sense, the contributional value of B is ultimately dependent on the argumentative difference between saying |A| and saying |AB| and thus on the way adding B modifies |A|.

seriously some fifteen years ago (e.g. Bertrand & Chanet, 2005), and with the first PhDs completely devoted (Petit, 2009) to the issue of understanding the relationship between semantic polysemy and prosodic marking.

The first results showed that while identified interpretations of a word could often not be matched with any specific prosodic contour or features, this failure was most of the time caused by the insufficient precision of the semantic description of words' uses. In other words, matching prosody and interpretation appeared possible only through considering fine-grained interpretative and prosodic features and required a methodology of its own.

Moreover, not all prosody being meaningful and meaningful prosody being far from homogenous, we shall distinguish non-structural prosody (NSP), defined (Nemo & Petit, 2015) as:

> any form of prosodic realization of a linguistic sequence of any kind which is free and capable of coercing interpretation (at any interpretative level)

including "free lexicalized prosody" defined as:

> any free form of prosodic realization of a lexical unit which associates one of its uses with both a specific and recurrent interpretation

from non-meaningful "free variation" and non-meaningful "structural" prosody. As we cannot discuss here these distinctions in detail, it must be stressed out that all reference to prosodic features or contours in the next sections will concern exclusively NSP and "free lexical prosody".

2.1 Prosodic constraints on the argumentative orientation of signs

When it comes to measuring the possible role of prosody in the determination of argumentative orientation, it is necessary to define precisely what aspect of argumentative orientation is concerned and how this role can be proven.

For these two issues, our starting point will be to claim that the only way to establish such a role is either to concentrate on signs whose argumentative orientation appears to be variable, and hence which falsify the classical hypothesis according to which the linguistic signs themselves encode argumentative orientations, or to concentrate on exchanges in which the actual orientation of an intervention appears to be at odd with the orientation that could be predicted from the signs themselves.

We shall thus start by considering the uses of French *enfin* and its polysemy, whose first argumentative description can be found in Cadiot *et al* (1995). This morpheme has a wide range of uses, ranging from uses associated with irritation "*Taisez-vous* enfin" ("Shut up will you!") to uses associated with the expression of relief ("*He finally arrived*") and all kinds of other uses, whose interpretations may be glossed by "*at last*", "*forget it*", "*I mean*", etc. Accounting for this variety of uses is possible (Nemo, 1998) by postulating that *enfin* encodes the semantic indication that a problem taking place at a moment t will, somehow, be over in t_{+1}, and by relating all the above mentioned interpretations to various interpretations of this indication, for instance by describing the relief use as corresponding to a situation in which a problem which was taking place has been solved at the moment of speech and the irritation use to a situation in which a problem is taking place at the moment of speech and should come to an end immediately.

Because these uses are associated with clearly distinct prosodic forms, *enfin* was among the first items for which the matching of prosodic form and interpretation was tested on large amount of data (Bertrand & Chanet, 2005). Whereas the initial results (Petit, 2009; Petit, 2010) showed for instance no homogeneity of prosodic form among all the uses testing positive to the "past problem has been solved" test, later results (Petit, 2009, Petit & Nemo, 2009) showed that this absence of homogeneity was caused by diverging prosodic *comments* on this situation. In other words, it turned out to be the case that in a situation in which a problem has been solved, two distinct prosodic comments are possible, one associated with the expression of a feeling of relief, as expected, and one associated with the expression of residual irritation about the fact that it took so long to fix the problem. In other words, within a single interpretation-type (Nemo & Petit, 2012), two opposite argumentative orientations (and attentional biases) could be observed, depending on the specific standpoint taken by the speaker, both of which being prosodically marked and unpredictable without such prosodic marking. In the simplest terms, argumentative orientation was indeed shown to be linguistically marked, as claimed by ALT, but this marking appeared to be a prosodic one rather than a morphemic one[9].

---

[9] We are leaving aside here the question of knowing whether the indications provided by the morpheme are "argumentative orientation"-free, or not (as would be predicted by Galatanu's SPA model). What has been proved in any case is that prosody can over-rule any preexisting orientation.

Moving now to the argumentative orientation(s) associated with the use of French *quelques*, it becomes possible to consider signs whose argumentative orientation appears to be somehow versatile. Despite being presented in Ducrot (1973, p. 7) as oriented towards a positive conclusion in an utterance such as "il a lu *quelques* romans de Balzac" ("He has read some Balzac's novels"), *quelques* appears according to Gaatone (1991, 3; 1991, 9) to be most often paraphrased by *peu* ("little") in dictionaries:

> The seme «peu élevé» is used almost unanimously in the definition of *quelques*, by both dictionnaries and standard French grammars. [10]

with the result of predicting that its orientation should be negative (i.e. anti-oriented with what it applies to). And accordingly, one may observe a constant oscillation in all the existing linguistic work between claims that *quelques* characterizes a limited quantity as positive, as Gondret (1976, p. 149) or Gaatone (1991) and claims of a characterization of a quantity as negligible, with Bacha (1997, p. 54) for instance insisting on the fact that:

> choosing between *quelques* and *plusieurs* for a quantity which may be identical in both cases, is a matter of orienting the interpretation and the conclusion of the interlocutor respectively toward the negative or the positive. [11]

What must thus be acknowledged as a temporary conclusion is that even though any use of *quelques* may be associated with an argumentative orientation, this orientation appears unstable and use-related. This is precisely why testing the role of prosody in the determination of this orientation may become important, as it shows that the prosodic realization of *quelques* appears to be playing an interpretative role at four levels:

- it appears that with the right prosodic realization, it is possible for *quelques* to present a quantity as a *significant* one, deserving full consideration, exactly as it is possible to present it as negligible;
- it also appears that when the first syllable is realized with a variation of intensity, it indicates (Petit, 2009) that the quantity at stake should

---

[10] Our translation. "Peu élevé", when referring to a quantity, means "low amount".

[11] Our translation.

receive full attention, and this regardless of the exact nature of the positive or negative orientation. In other words, this contour is associated with the marking of the scalar slope, i.e. with signaling that the difference something makes is important;
- it similarly appears that the realization of *quelques* is not forcefully associated with any form of prosodic salience, and that in such cases it is the prosodic form of the nominal phrase as a whole that determinates the nature of the argumentative orientation of the utterance (Petit, 2009);
- it finally appears that apart from the marking of the scalar slope on the first syllable, marking the quantity as significant is achieved by adopting a high-pitched melody for the *quelques* segment (contrastively with the rest of the sequence).

Beyond *quelques* or *enfin*, what must be clear is that it is ultimately the whole lexicon which is concerned by the possible lexicalization[12] of such use-types, for it means that describing lexical signs as the association of a phonological form and a lexicalized interpretation, a *signifiant* and a *signifié* in the classical terminology, is not equivalent with reducing the former to a phonematic form and the latter to a prosody-free semantic content as has been assumed since Saussure. It must indeed be clear that all use-types may be described as the association of a *dual phonological form* and a *dual interpretative meaning* and that a lexical entry should in fact (Nemo & Petit, 2012) be represented as:

Use-type:
    Phon : σ   φ : value (phonematic form)
                    π : value (prosodic form)
    Sem : s   ψ : value (interpretation-type)
                    ρ : value (argumentative comment)
    Grammatical status γ :   value

---

[12] Despite being commonsensical, the idea that prosodic clues are important in the study of interpretation has emerged within linguistic semantics only some 20 years ago, and more recently in prosodic studies. Calhoun & Schweitzer (2012), Simon & Grobet (2002), Gibbon (2002) and Elordieta & Romera (2002) introduced the first systematic attempts to match different interpretations with their prosodic realisation, notably about discourse connectives, and provided (Gibbon, 2002) the first *semantic* formulation of the notion of « lexical prosody ». Further work (Auchlin *et al.*, 2004; Grobet & Auchlin, 2006) was focused on the nature of the "information" at stake, notably in terms of reframing of experienciation, transformation of the environment through action, polyphony and marking of attitudes to the discourse (acceptance or rejection, interest).

This explains directly why different linguists, if they fail to consider the π form or relevant π features, can make opposite claims about the nature of the argumentative bias/comment associated with the φ form. Or, as it was claimed in the SPA framework (Galatanu, 2002), that the effective argumentative orientation of signs is discursively profiled.

Moreover, it seems quite clear that because these results can indeed be generalized, for instance to account for the diversity of argumentative orientations of a verb like[13] *minimiser* (*minimize*), the ALT thesis according to which argumentative orientation is linguistically driven can be proven to be correct precisely because of its prosodic marking. Whereas the morphemes themselves are not the ones encoding the argumentative orientation, it is nevertheless the case that prosodic constraints are playing a similar role and are able to *shape* the argumentative orientation or at least to *reshape* it. Argumentation is indeed linguistically marked, and prosodic constraints play a major role in this process.

As for the role that the prosodic marking of a sign such as *quelques* may play in the determination of the argumentative orientation not only of the utterance at stake but of the whole argumentative contribution/intervention, it is important to observe that *quelques* being often used inside a restrictive addition B to what has just been said [A][14], one may observe that the effect of the prosodic marking of a downplaying/minimizing prosodic comment[c] is to considerably weaken this restriction: instead of having an anti-orientation of A and B, as could be expected otherwise, the argumentative sequence is transformed into a AB[c] sequence in which B being marked by the prosodic comment C as negligible, the B[c] segment transforms the restriction B into a restriction that can be neglected, thus modifying the global conclusion which can be drawn from the AB sequence. Moreover, it appears that the prosodic comment [c] on what is said is most of the time considered as reflecting the true position of the speaker, which in such cases clearly backs both A and C (and not the anti-oriented B).

2.2 Lexicalization of argumentative orientations: methodological issues

The first lexicographical description of words to integrate the π prosodic dimension in the description of its polysemy was proposed by

---

[13] See Nemo (2014).

[14] For instance with the B sequence "Sauf quelques personnes" ("But for some/a few people").

Dostie (2004) on discourse connectives. Similar efforts, notably on *enfin*, have led (Nemo & Petit, 2013) to the demonstration that such an integration allowed to obtain much more coherent and exhaustive descriptions of lexical polysemy and opened a new chapter for e-lexicography.

Meeting such an objective imposes however to produce reliable (and thus provable) π/ρ pairs, and thus a complex methodology which we shall only mention here. The study of π/ρ pairs cannot but rely on corpus and oral authentic data taken from the greatest possible diversity of pragmatic contexts. It must then be extracted for each word a data bank of uses, allowing to study the correlation between prosodic and interpretative features. In the *oui* case that we shall discuss here, the size of the data bank has reached 2800 occurrences/uses, with extraction of the whole utterance (or exchange) and of the word form (φ,π) itself. The next step is to obtain a double semantic and prosodic characterization of each use, each characterization being initially made separately. The third and more important step is to cross-check the lexical stability of such or such π form and the prosodic stability of such or such semantic interpretation, and to account for all observed discrepancies, in order to obtain more precise semantic descriptions and to confirm the stability of the prosody/comment pair.

The whole approach has been automated during the on-going DIASEMIE[15] program through the use[16] of automated classification techniques allowing to predict tested interpretative features at stake on prosodic ground, and the automation of the "shuttle" step of analysis of classification errors (in order to enhance the precision of the semantic description) and selection of the most relevant prosodic features[17]. The DIASEMIE approach has demonstrated that automated discriminability of π/ρ pairs was provable and also, through extraction tests using playback of the extracted segment to test the capacity to associate a

---

[15] The DIASEMIE acronym stands for Automated Discrimination of in-Use Word Meanings.

[16] After the sequencing of the signal into 30 ms segments, and the vectorization of their dynamic prosodic features.

[17] The whole process, described in its early form for the first tested feature (convinced versus unconvinced *oui*) in Hacine-Gharbi & al. (2015), allowed machine learning and the use of authentic (but acoustically heterogeneous) data with a classification rate of 80% (prior to the shuttle task) and 97% (after).

meaning to it, that it was possible to cross-check the stability and intersubjective status of the interpretation of a π form[18].

2.3 Prosodic marking of the argumentative/interlocutive scene

At this point, we shall reverse the approach adopted so far, which consisted in testing the prosodic marking of identified argumentative features (e.g. appreciative or depreciative standpoints) and adopt a prosody-based approach in which the main issue to identity the kind of "information" which appears to be associated with a specific prosodic realization of a given sign or sequence.

What has indeed proven to be the case in the study of French *oui* (*yes*) is not only that the prosodic marking of *oui* may be important enough to reverse its meaning, in other words that there are clearly *oui(*s*)* whose interpretations are actually *no*, but also, quite often, that their prosodic realization does not provide only clues about the position/standpoint of the speaker regarding the issue at stake, as could have been expected but also clues about the respective positions of the interlocutor as presented or internalized by the speaker.

In other words, it appears first that the prosodic form π does not only reflect the sole standpoint of the speaker but actually reflects the standpoints of both interlocutors and the degree (and nature) of convergence and divergence between them: the prosodic realizations of "*oui*" appear to mirror the whole interlocutive scene.

Importantly, it can also be shown on the same data, that the prosodic marking of *oui* plays a direct and central role in the dynamics of the argumentative exchanges at stake:

- it is often the case in a dialog that the content of the intervention of the interlocutor which directly follows the utterance of the "*oui*" turns out to be exclusively focused on the prosodic comment^C of the "*oui^C*";
- it is also frequent that the *oui* at stake is only a polite *oui* which reflects no adherence to what is at stake, and may eventually turn out to be a fake ratification, whose main goal is to close the conversation ;
- it is often the case, when the *oui* is only the start of an intervention, that the content of the intervention will prove to be exclusively concerned with elaborating (i.e. developing) on the comment^C;
- it may finally be observed in argumentative exchanges, by considering only the prosodic comments, that a complete prosodic dialog is actually taking place.

---

[18] People are actually able for many (φ, π) forms, to spell out rather robust hypotheses about the situation in which the word was used and the nature of the comment.

Typically, one may notice that π will often mirror the fact that the speaker is choosing to assume saying something which will be rejected by many, or that he handles tactfully his interlocutor, or that what he says is just to please someone, etc.

In any case, it is clear that the fact such a simple sign as *oui* can in argumentative and ordinary exchanges convey such a wide spectrum of fine-grained information about the respective standpoint of the participants, literally integrating viewpoints into structure (Nølke, 1993) and that this commonsensical reality has remained largely clandestine in the study of interpretation and argumentation, is somehow puzzling for anyone analyzing one by one hundreds of its uses.

3. PROSODIC PROFILES AND ARGUMENTATION THEORY

The aim of any argumentation theory is to model argumentative exchanges. As mentioned earlier, considering argumentative exchanges as a special type of conversational exchanges which would deserve a study on their own could prove to be a wrong way to approach the problem, for it seems that such exchanges are ultimately based on an argumentative layer which is somehow consubstantial with language use and that the necessity to exchange arguments and to convince an audience or an interlocutor is inherent to any interlocutive scene[19] and actually can be shown to exist, far from institutional settings, for any attentional move in everyday life.

In the very limited space available here, we have tried to show that it can be considered as established or provable that prosody plays the following roles:

- determining the argumentative orientation of a sign or a sequence, not only as positive or negative but also in the social/collective construction of a shared perspective about what is said;
- determining the scalar slope (in both directions) and attentional/scalar focus;
- determining the attentional hierarchy in the attentional field, in other words determining the ordering in importance of the various elements to be considered (central, secondary or marginal);
- prosodic weakening of counter-arguments or apparent ratification;
- prosodic mirroring of the interlocutive scene, and notably of the ratification scene;

---

[19] It is however clear that the study of argumentation is particularly concerned with ratificational issues and objectives.

- contributing to the manipulation of the attention of others for social ends.

This list being far from limitative, it does seem thus that prosodic constraints play an omnipresent role in argumentative processes, which cannot be underestimated.

Given the attentional and collective nature of contributional games, what we have indeed tried to show is that because what is said cannot be described without considering both how it is said (prosodically) and what this prosodic comment indicates about what is said, the description of interlocutive exchanges – from the prosodic profiling of signs in language use to the prosodic marking of interlocutive relations and ratification – cannot be deprived of its prosodic dimension. Hence, prosodic features should not be called in clandestinely as part of the intuition of the analysts, let alone be considered as marginal features which can be ignored at no cost, but should be an inherent part of both the description of argumentative situations and of the mechanisms which shape them.

In this respect, we have shown first of all that because the most elementary bricks in the argumentative interpretation of utterances and exchanges can be proven to be prosodically profiled, argumentation theory as a whole may not pretend to isolate its object from such prosodic constraints, and secondly that because emerging techniques and methodologies are starting to provide tools which will allow gradually the complete mapping of π forms and features, it will soon become possible to test individual occurrences of the mapped signs and sequences in order to spell out the ρ comments which are associated to it, and to study empirically the way such comments are dealt with in argumentative exchanges.

REFERENCES

Anscombre, J-C., & Ducrot, O. (1976). L'argumentation dans la langue. *Langages, 42*, 5-27.
Auchlin A., & Simon A.C. (2004). Gabarits prosodiques, empathie(s) et attitudes. *Cahiers de l'Institut de Linguistique de Louvain, 30*(1-3), 181-206.
Auchlin A., & Grobet, A. (2006). Polyphonie et prosodie : contraintes et rendement de l'approche modulaire du discours. In L. Perrin (Ed.), *Le sens et ses voix. Dialogisme et polyphonie en langue et en discours* (pp. 77-104). Recherches linguistiques, 28. Metz: Université Paul-Verlaine.
Bacha, J. (1997). Entre le plus et le moins : l'ambivalence du déterminant *plusieurs. Langue française, 116*, 49-60.

Bertrand, R., & Chanet, C. (2005). Fonctions pragmatiques et prosodie de *enfin* en français spontané. *Revue de sémantique et pragmatique, 17*, 41-68.

Beyssade, C., Delais-Roussarie, E., Marandin, J.M., Rialland, A., De Fornel, M. (2004). Le sens des contours intonatifs en français : croyances compatibles ou conflictuelles ?. *Actes de JEP-TALN 2004*. Fès, Maroc

Bose, I., & Gutenberg, N. (2002) Enthymeme and Prosody – A Contribution To Empirical Research In The Analysis Of Intonation As Well As Argumentation. *ISSA Proceedings 2002*.

Cadiot, A. et al. (1985). *Enfin* marqueur métalinguistique. *Journal of Pragmatics, 9*, 199-239.

Calhoun, S., & Schweitzer, A. (2012). Can intonation contours be lexicalised? Implications for discourse meanings. In G. Elordieta & P. Prieto (Eds.), *Prosody and Meaning* (Trends in Linguistics) (pp. 271-327). Berlin: De Gruyter Mouton.

Dostie, G. (2004). *Pragmaticalisation et marqueurs discursifs. Analyse sémantique et traitement lexicographique*. Bruxelles: De Boeck & Duculot.

Ducrot, O. (1973). *Les échelles argumentatives*. Paris: Editions de Minuit.

Ducrot, O., & Anscombre J-C. (1983). *L'argumentation dans la langue*. Bruxelles: Mardaga.

Elordieta G., & Romera M. (2002). Prosody and meaning in interaction: The case of the Spanish discourse functional unit *entonces* 'then'. *Proc. Speech Prosody 2002* (pp. 263-266). Aix-en-Provence.

Fernandez, R., & W. Picard. R. (2002). Dialog act classification from prosodic features using support vector machines. In *Speech Prosody, ISCA International Conference* (pp. 291–294), Aix-en-Provence, France.

Gaatone, D. (1991). Les déterminants de la quantité peu élevée en français. Remarques sur les emplois de *quelques* et *plusieurs*. *Revue Romahe, 26*(1), 3-13.

Galatanu, O. (2002). La dimension axiologique de l'interaction. In M. Carel (Ed.), *Les facettes du dire. Hommage à Oswald Ducrot* (pp. 93-107). Paris: Kimé.

Gibbon, D. (2002). Prosodic information in an integrated lexicon. In *Proc. Speech Prosody 2002* (pp. 335-338). Aix-en-Provence.

Gondret, P. (1976). Quelques, plusieurs, certains, divers: Etude sémantique. *Le Français moderne, 44*(2), 143-152.

Hacine-Gharbi A., Petit M., Ravier P., & Nemo F. (2015). Prosody based Automatic Classification of the Uses of French *'Oui'* as Convinced or Unconvinced Uses. In *Proceedings of the International Conference on Pattern Recognition Applications and Methods* (pp. 349-354).

Kohler, K. (2009). Patterns of prosody in the expression of the speaker and the appeal to the listener. In G Fant, H. Fujisaki & J. Shen (Eds.), *Frontiers in phonetics and speech science* (pp. 287-302). Beijing: The Commercial Press.

Král, P., Cerisara, C., Klečková, J. (2007). *Importance of prosody for dialogue acts recognition*. http://textmining.zcu.cz/publications/SPECOM07_kral.pdf.

Ladd, D. R. (2008). *Intonational phonology*. 2nd edition. Cambridge: Cambridge University Press.
Lacheret-Dujour, A., & Beaugendre F. (2002). *La prosodie du français*. Paris: CNRS Editions.
Mello H., Panunzi A., & Raso T. (2012). *Pragmatics and prosody: Illocution, modality, attitude, information patterning and speech annotation*. Firenze: Firenze University Press.
Mertens, P. (2008). Syntaxe, prosodie et structure informationnelle: Une approche prédictive pour l'analyse de l'intonation dans le discours. *Travaux de Linguistique, 56*(1), 87-124.
Nemo, F. (1985). Contraintes énonciatives et argumentativité. *Semantikos, 9*(2), 21-34.
Nemo, F. (1988). Relevance. Book review. *Journal of Pragmatics, 12*(5-6), 791-795.
Nemo, F. (1992). *Contraintes de pertinence et compétence énonciative : l'image du possible dans l'interlocution*. Thèse de l'EHESS. Paris.
Nemo, F. (1995). A description of argumentative relevance. In Van Eemeren *et al* (Eds.), *Analysis and Evaluation. Proceedings of the Third ISSA Conference on Argumentation* 2 (pp. 226-232). Amsterdam: SicSat.
Nemo, F. (1998). Making a difference or not: Utterances and argumentation. *ISSA Proceedings 1998* (pp. 594-598). Amsterdam: SicSat.
Nemo, F. (1999). The pragmatics of signs, the semantics of relevance, and the semantic/pragmatic interface. *The Semantics-Pragmatics Interface from Different points of View* (pp. 343-417), CRiSPI Series, Amsterdam: Elsevier Science.
Nemo F. (2001). Pour une approche indexicale (et non procédurale) des instructions sémantiques. *Revue de Sémantique et de Pragmatique, 9-10*, 195-218.
Nemo, F. (2006a). Discourse words as morphemes and as constructions. In K. Fischer (Ed.), *Approaches to Discourse Particles* (pp. 415-448). Amsterdam: Elsevier Science.
Nemo, F. (2006b). The pragmatics of common ground: From common knowledge to shared attention and social referencing. In A. Fetzer & K. Fischer (Eds.), *Lexical Markers of Common Grounds* (pp. 143-158). Amsterdam: Elsevier Science.
Nemo, F. (2007). Reconsidering the Discourse Marking Hypothesis. In A. Celle & R. Huart (Eds.), *Connectives as Discourse Landmarks* (pp. 195-210). Amsterdam/Philadelphia: John Benjamins Publishing.
Nemo, F. (2014). Plurisémie et argumentation entre signification morphémique et signification lexicale. In A-M. Cozma, A. Bellachhab & M. Pescheux (Eds.), *Du sens à la signification. De la signification au sens* (pp. 301-312). Bruxelles: Peter Lang.
Nemo, F., & Petit, M. (2009). De la prosodie en discours à la prosodie en langue: Lexicalisation de la forme prosodique des emplois-types. In H-Y Yoo & E. Delais-Roussarie (Eds), *Actes d'IDP (Interface Discours & Prosodie) 09* (pp. 302-312). Paris.

Nemo, F., & Petit, M. (2012). Sémantique des contextes-types. In L. de Saussure & A. Rihs (Eds.), *Etudes de sémantique et pragmatique françaises* (pp. 379-397). Berne: Lang.
Nemo, F., & Petit, M. (2013). Electronic dictionaries and the integration of prosody in the lexicographical treatment of polysemy. In Kwary, Wulan & Musyahda (Eds.), *Lexicography and Dictionaries in the Information Age* (pp. 236-242). Denpassar: *Airlangga University* Press.
Nemo, F., & Petit M. (2015). Prosodie non-structurale et plurisémie. *Revue de Sémantique et Pragmatique, 37*, 87-110.
Nølke, H. (1993). *Le regard du locuteur*. Paris: Kimé.
Petit, M. (2009). *Discrimination prosodique et représentation du lexique: Application aux emplois des connecteurs discursifs.* Thèse de doctorat. Université d'Orléans.
Petit, M. (2010). Discrimination prosodique et représentation du lexique: Les connecteurs discursifs. *Etudes de linguistique appliquée, 157*, 75-93.
Raccah, P.-Y. (2010). Racines lexicales de l'argumentation: La cristallisation des points de vue dans les mots. *Verbum 32:1. L'inscription langagière de l'argumentation.*
Rittaud-Hutinet C. (1995). *La phonopragmatique*. Berne: Peter Lang.
Saunier, E. (1999). A propos de l'interaction entre propriétés des formes intonatives et propriétés de certaines unités morpholexicales. *Faits de langues, 13*, 191-208.
Simon A.-C., & Grobet A. (2002). Intégration ou autonomisation prosodique des connecteurs. In *Proc. Speech Prosody 2002*, (pp.647-650). Aix-en-Provence.
Sudhoff, S., et al (2007). *Methods in empirical prosody research.* De Gruyter.
Thomsen, C. (2002). *Oui* : il y a oui et oui – marqueurs de la syntaxe conversationnelle. In H. L. Andersen & H. Nølke (Eds.), *Macro-syntaxe et macro-sémantique* (pp. 189-206). Berne: Peter Lang.
Vincent, D., & Demers M. (1994). Les problèmes d'arrimage entre les études discursives et prosodiques. Le cas du « là » ponctuant. *Langues et linguistique, 20*, 201-212.
Wichmann, A. (2002). Attitudinal intonation and the inferential process. In *Proc. Speech Prosody 2002* (pp. 11-16). Aix-en-Provence.
Yang, L. (2006). Integrating prosodic and contextual cues in the interpretation of discourse markers. In K. Fischer (Ed.), *Approaches to Discourse Particles* (pp. 265-297). Oxford: Elsevier.

Argumentative Orientation, Prosody and the Coordination of Attention
Commentary on Nemo, Létang and Petit's Prosodic Constraints on Argumentation

ANDREA ROCCI
IALS, University of Lugano, Switzerland
andrea.rocci@usi.ch

1. BIG NEWS ON SMALL WORDS

How many things can a *yes* say in an argumentative exchange? I think that most of us would say "well, very many, and so what?" As if there were not much we can do or should do about that. Many would say that it depends on the context. Some would elaborate that there is not much inside that *yes* that we should look for, it gets most of its meaning from what comes before and what comes after in interaction and we should rather consider *what* is being locally accomplished each time in the argumentative interaction irrespectively that we use *yes*, *no*, hand weaving, whistles or bells to attract our interlocutor's attention. Some would realize that they never paid attention to those signals, but they surely have called them in "clandestinely as part of the intuitions of the analyst". Well, François Nemo, Camille Létang and Mélanie Petit have some very important news for us. They provide this news in a dense, highly technical, paper which tries to pack in a few pages the results of decades of research. I'm not sure that the information density makes it the paper most capable of instantly securing our full attention, not as linguists (for those who are or where linguists), but as argumentation theorists. The unavoidable brevity and omissions – omission of more numerous and more developed examples, and omission of technical details about prosody that would have required to embed a crash course in phonology in the paper – coupled with unusual technical jargon (*scalar slope constraint, modal slope constraint*) do not make this paper the easiest to digest by non-specialists at a conference. That's why I see my duty as a commentator, at the risk of being trivial, to stress what are (in my view) the big news that this article contains for us argumentation theorists.

A first piece of news is that many little words that appear to fit so well in certain argumentative roles and yet occur in completely different roles in other context do carry a precise contribution that can guide the modeling of the argumentative exchange. In order to uncover these linguistic indicators we need a richer model of the linguistic sign recognizing lexicalized use-types of lexemes where a certain prosodic form of the lexeme is associated with a certain argumentative orientation. The analysis of use-types is the object of a new kind of "prosodic lexicography" based on empirical methods and computational tools. So, the authors tell us: the information is there, or, at least, starts to be there and there is no reason for us to continue to fudge those things if we want to be serious about the fine grained analysis of arguments in conversation or spoken monologue. This is, in my view, the first piece of news. Which is also, the authors would say, an argument, because in the life of language, there is no such a thing as news without an argument.

2. A FRESH LOOK TO THEORIES OF ARGUMENTATION IN LANGUAGE

And this brings us to what is in my view the second important piece of news that the authors of this article have for us argumentation scholars: the theory of argumentation in language is back. To be honest, the venerable approach to *argumentation dans la langue* (Anscombre & Ducrot, 1976), initiated in the 1970s, has been with us all this time producing various developments including the theory of *topoï*, the insightful and painstakingly detailed analyses of polyphony of the Scandinavians (Nølke, Fløttum, & Norén, 2004), the radically structuralist theory of argumentative blocks (Carel & Ducrot, 1999), stereotype theory, and several others. Even the most distracted and linguistically insensitive ISSA goer recalls listening one time or another to a French scholar presenting the often insightful, but often paradoxical analyses informed by this approach. But I think that it is a relatively common experience that even the insightful bits were quite hard to integrate in one's work on argumentation. Now, this article brings to our attention again the argumentation in language approach with a strong invitation to reconsider our settled assumptions and hasty generalizations about it. At least, this is what the paper has done *to me*. It brought me to reconsider my views of this approach, which I considered pretty much settled.

Let us briefly review my general reasons of dissatisfaction with some of the "classic" versions of the theory of argumentation in language. Roughly, I'd say I can distinguish reasons of principle or "from above" and reasons of fact, or "from below" – but don't take this distinction as too deep.

From above, I was put off by the perceived insistence of this approach in remaining confined within the abstract system of the *langue* disregarding actual discursive performance and, most importantly, renouncing to say anything about the cognitive and interactional functioning of argumentation. To my discontent, for instance, ALT did not have anything to say about the relationship between argumentation and inference other than a generic skeptical warning that what might seem to us inference based on information is nothing more than the deployment of argumentative orientation hard-wired in the langue. Similarly, the theory had little to say about argumentative exchanges as interactions.

From below, my main problem with the theory was its limited empirical coverage. The analysis covered a limited number of lexical items using select suites of constructed examples as evidence. The feeling that the theory was not really open to empirical disconfirmation was hard to dispel. The review of alternative hypotheses in the literature was also minimal – a characteristic that was initially justified by the novelty of the approach, but became later problematic as critical appraisals and alternative explanations started to emerge. Finally, the idea that argumentative orientation could *exhaust* meaning was hard to swallow and, more specifically, the idea that the meaning of a lexical entry, say, the verb *travailler* (to work) had to be represented as the sum total of the *donc/ pourtant* concatenations it licenses seemed to have dire consequences on the feasibility of practically useful semantic analyses. The biggest units considered were often concatenations of two constructed sentences and little was said about dialogical exchanges and extended discourse.

It is not important here whether the above is a completely fair assessment of each of the other versions of ALT or, at least in part, a misunderstanding, caricature, or hasty generalization. It is more important to stress that few, if any, of these common misgivings about ALT apply to the paper presented by Nemo, Létang and Petit. At the same time some quite radical claims remain and invite us to question our assumption about the essence of argumentation and the proper way to study it. Regarding the "from below" angle: there seems to be a remarkable effort in adopting empirical methods and aiming at broad coverage – even as the semantic and prosodic analysis remains extremely fine-grained, difficult cases and counter-evidence spur theory developments and new analyses. The commentary on argumentative orientation remains a *component* of the meaning of a use-type without exhausting it. At the same time one very strong hypothesis remains:

natural languages are characterized by a generalized layer of argumentative meaning that linguistic semantics has to account for. Items with orientations are the norm, not the oddities. What is interesting for me as an argumentation theorist is that this hypothesis is not only tested from below, but also theoretically motivated from above. Here we encounter the intriguing hypothesis that utterance meanings are to be understood for the difference they make in a set of modal alternatives, and, more importantly the idea that coordination of attention and action requires that the reasons for the saliency of what is uttered to be made available.

3. PRAGMATIC ROOTS OF ARGUMENTATIVE ORIENTATIONS: THE COORDINATION OF ATTENTION REQUIRES REASONS

According to Nemo, Létang and Petit it makes perfect sense to consider argumentative orientation as a generalized level of semantics because all dialogical exchanges can be considered as arguments and give us a reason that is rooted in the nature of joint action. All dialogues are argumentative because the coordination of attention constitutively requires reasons. I quote Nemo and colleagues "in conversational exchanges, individual contributions are contributions to a collective definition of what has to be taken into account and how, but that this collective definition is ultimately based on the participants ratification of each of the elements to be considered and of their relative importance", thus even though "the accepted purpose of exchanges is supposedly to reach a consensus" on the acceptability, centrality and evaluation of each contribution, "the reality of exchanges is that disagreements and diverging ground on each issue are the rule and not the exception".

From the idea that an argumentative dimension is present "in any attentional move in everyday life" Nemo and colleagues draw the invitation to stop considering arguments a special type of conversational exchange. Honestly, I'm not convinced that this should be the case. Even if accept that in every linguistic exchange there is an argumentative dimension consisting as countless micro-issues are debated to reach a common understanding of the joint purpose of conversation, this is not the same thing as having an argument, that is a conversation whose purpose at a macro level is discussing an issue and dealing with disagreement. Compare with pragmatic theories centered on cooperation: saying that every exchange is a case of communicative cooperation to reach common understanding does not mean automatically that each and every exchange is about cooperation and coordination of action at a macro-level.

In every case, Nemo and colleagues have performed a very compelling attentional move obliging us to pay close attention to the empirical work done in the area of prosody by argumentative semanticists and to reconsider the ALT hypothesis of argumentative orientation as a constitutive and generalized level of meaning.

REFERENCES

Anscombre, J.C., & Ducrot, O. (1976). L'argumentation dans la langue. *Langages, 42*, 5-27.

Carel, M., & Ducrot, O. (1999). Le problème du paradoxe dans une sémantique argumentative. *Langue française, 123*(1), 6-26.

Nølke, H., Fløttum, K., & Norén, C. (2004). *ScaPoLine. La théorie scandinave de la polyphonie linguistique*. Paris: Kimé.

24

Repetition as a Context Selection Constraint: A Study in the Cognitive Underpinnings of Persuasion

DAVIS OZOLS
University of Fribourg, Switzerland
davis.ozols@unifr.ch

DIDIER MAILLAT
University of Fribourg, Switzerland
didier.maillat@unifr.ch

STEVE OSWALD
University of Fribourg, Switzerland
steve.oswald@unifr.ch

Repetition of information has been shown to affect the perceived validity of the items repeated with these effects carried over to an inferred assumption. We believe this is highlighted in everyday communication and can result in acceptance of fallacious argumentation. We explain this phenomenon via the notion of Context Selection Constraints and discuss the effectiveness of the *ad populum* fallacy with the help of an experimental design.

KEYWORDS: *ad Populum*, argumentation, context selection constraint, Fallacies, manipulation

1. INTRODUCTION

In this paper we conduct a theoretical and experimental investigation of the influence of repetition and *ad populum* arguments on the perceived validity of conclusions inferred from them in argumentative contexts. Building on previous work in cognitive pragmatics which provides insights on how some argumentative devices, such as fallacies, may cognitively operate (Maillat & Oswald, 2009, 2011, 2013; Oswald, 2010, 2014; Maillat, 2013a, 2013b), and inspired by recent research in cognitive psychology on repetition and familiarity of information (Ozubko & Flugelsang, 2011), we present an experimental framework designed to bridge the gap between both disciplines. Specifically, we investigate whether the repetition of information and the mention of repetition can act as pragmatic mechanisms of contextual constraint on informational selection, as the CSC model predicts (Maillat & Oswald, 2009, 2011).

The *ad populum* argument, i.e., the use of the majority view on a given statement as definite evidence for a conclusion (see e.g. Walton, 2006, pp. 91-96), may be taken to be persuasive by virtue of its propensity to make the proposition it targets familiar: knowing that many people believe X in principle suggests that X is familiar in the community, and consequently that it would make sense for you to believe X too in that community. From a cognitive perspective, it could furthermore be argued that familiarity guarantees a greater degree of accessibility by acting as a constraint on context selection. While Ozubko & Fugelsang (2011) have shown that this can be achieved via repetition of information, we want to test whether mentioning that information has been repeated is likely to yield the same effects. Furthermore, we embed the original design within an argumentative structure as we test the transfer of the familiarity effect from a premise to the conclusion that it supports. Finally, we will hypothesise further that, as Ozubko & Fugelsang (2011) show for their memory retrieval condition, accessing an argument from memory implies that the relevant context set is perceived as more familiar. We posit that the *ad populum* argument combines these two effects (mention of repetition and retrieval from memory) to constrain context selection.

Section 2 of the paper presents the pragmatic framework used, as well as the process of Context Selection Constraint. Section 3 provides the rationale by which the account could be extended to account for argumentative phenomena and fallacies in particular. Section 4 presents the experimental design used to test our assumptions and discusses the results. We conclude by providing some directions for future research that would allow us to overcome the difficulties found in this first round of experimental research on the issue.

2. THE PRAGMATICS OF MANIPULATIVE DISCOURSE: CONTEXT SELECTION CONSTRAINT

In order to understand the framework within which we wish to analyse fallacious argumentative moves, we must first get a sense of the theoretical assumptions that underlie the kind of pragmatic model which will be entertained in the following sections, and more specifically, which grounds the principle of Context Selection Constraint that we take to be responsible for the argumentative efficiency of fallacious arguments such as the *ad populum*.

From a traditional perspective, starting with Grice's original propositions made in the 1950s and 1960s (see the 1989 reprint), pragmatics is founded on an assumption of cooperation between the participants of a talk exchange. Grice's cooperative principle (Make your contribution such as is expected at that moment in the talk exchange) set the ground for an essentially benevolent view of human communication, in which utterances are geared towards working for the joint completion of the communicative event in which participants are involved. While other models of pragmatics have offered different explanations as to what might constitute the driving force(s) behind this initial cooperative, benevolent impetus, all of them have retained at one level or another the view that human communication is a cooperative endeavour towards a common goal.

In Relevance Theory for instance (Sperber & Wilson, 1995; Wilson & Sperber, 2012; Blakemore, 2002; Carston, 2002; Clark, 2013), this same insight is captured in the second principle of relevance: the communicative principle of relevance, which could be defined in the following way (see Clark, 2013 for a good introduction to the relevance-theoretic model):

> Communicative Principle of Relevance:
> Every ostensive stimulus conveys a presumption of its own optimal relevance.

This captures the same idea that there is an expectation bearing on speakers that they would work towards providing an optimal input through their utterance.

For Grice, cooperation – and the four maxims attached to the Cooperative Principle – follows from a general human drive towards rational behaviour (yet relatively vaguely defined). In Relevance theory, on the other hand, optimality is a function of an input to the cognitive environment (henceforth CE) of an individual: the hearer's, in prototypical communicative interactions.

Specifically, an input's relevance is measured in terms of its ability to induce changes in the hearer's cognitive environment – i.e. the sum of the assumptions that are available to the hearer at a given point in time.[1] In that sense, benevolence can be thought of as an effort towards maximising the hearer's cognitive benefit in processing the

---

[1] Following the practice in relevance-theoretic literature, we construct examples in which the speaker is female, and the hearer is male.

utterance. This notion of cognitive relevance has thus been captured by two cognitive principles which govern utterance interpretation processes:

> Relevance of an input to an individual:
> a. Other things being equal, the greater the positive cognitive effects achieved by processing an input, the greater the relevance of the input to the individual at that time.
> b. Other things being equal, the greater the processing effort expended, the lower the relevance of the input to the individual at that time. (Clark, 2013, p. 106)

One of the striking aspects of this approach lies in the fact that interpretation is seen as a process in which the cognitive system follows a path of least effort, combining assumptions conveyed by the input utterance with previously held assumptions available in CE in order to infer new assumptions, maximising the resulting modifications in CE.

In that respect, utterance interpretation is seen as the result of a selection procedure that identifies a set of relevant assumptions which are taken to represent the communicative intention of the speaker when she used the utterance in question in the given interactive situation. The set of assumptions thus constructed stands for the cognitive representation of the context associated with the utterance. That is to say that within Relevance Theory, interpretation relies on a procedure of context selection, where selection is driven towards maximising the cognitive gain and minimising the expenditure of cognitive resources.

In a series of papers, we have argued that a fruitful way of understanding how deceptive discourse works, and in particular of looking at the way people can be manipulated by it, consists in focusing on how it affects the very interpretative process described above (Maillat & Oswald, 2009, 2011, 2013; Maillat, 2013, 2014; Oswald, 2010, 2014). Coming back to the idea of benevolence discussed earlier, while it can be applied in a majority of everyday instances of conversational interactions, it should be clear that in a number of instances the speaker's attitude towards the hearer when communicating with him would best be described as NOT benevolent, or even malevolent. Whereas Grice (1989) does not talk much about such instances, non-benevolent communication has been a preoccupation in Relevance Theory right from the outset. For example, Sperber (1994) referred to standard interpretative procedures as being based on a form of "naive optimism". As it turns out, RT has insisted on the fact that the relevance-based heuristics they propose are susceptible to yield erroneous interpretations, i.e. interpretations that misrepresent the speaker's original communicative intention. In other words, the relevance-driven optimisation procedure may at times miss its target. This is due in great part to the fact that relevance of an input can never be evaluated from an absolute perspective (against all possible sets of contextual assumptions), but in the relative perspective of the selected sub-set of contextual assumptions).

Unsurprisingly, therefore, humans evolved a form of communicative strategy that takes advantage of the limitations of the interpretative system of their interlocutor to deceptively convince them of something. In this sense, manipulation exploits some inherent flaws of the interpretative process in order to induce sub-optimal – from the hearer's perspective – changes in the hearer's cognitive environment. In that respect, the strategy discussed here is similar to the strategy deployed by a computer virus that seeks to enter the system by targeting a technical flaw of that system.

In this context, the flaw targeted in the system is the context selection procedure we described previously. As we saw then, context selection follows a path of least effort to reach the first relevant interpretation of an utterance it can arrive at by combining old, new and inferred assumptions. A deceptive communicator who wants to convince a hearer of U will deliberately try to force context selection down a path – to impose the selection path – in such a way that it ensures the uptake of U in the hearer's cognitive environment.

Thus, whereas a benevolent speaker will try to achieve this goal by selecting an utterance U that is optimally relevant in her own representation of the hearer's cognitive environment, a deceptive – and hence malevolent – speaker will try to induce the same change in the hearer's CE by changing the degree of salience and accessibility of certain assumptions in his CE, thereby ensuring that context selection for U is processed within a constrained sub-section of the hearer's CE. This deceptive manipulative strategy we call Context Selection Constraint and it is defined as follows.

> Context Selection Constraint:
> CSC is a twofold process by which a constraint that limits context selection is combined with a target utterance U in order to force the interpretation of the latter within a limited set of contextual assumptions and to effectively ensure that the interpretation is reached before a known, alternative (contradictory) subset of assumptions is accessed.

On the basis of the two sub-principles of relevance defined above, we predict that there will be two ways to lure context selection towards a sub-optimal set of contextual assumptions. You can either increase the expected cognitive yield of the target utterance in the sub-optimal set of contextual assumptions (in terms of positive cognitive effects); or you can increase the accessibility of the sub-optimal set of contextual assumptions (thereby reducing the cognitive effort required to process the target utterance in that context).

To the extent that such CSC strategies are designed to affect the inferential path followed during the interpretative process, they can be said to convey procedural information in the interpretation of the on-going exchange. Procedural here is intended in the sense defined by Blakemore (2002, p. 89) when she refers to expressions which constitute "means for constraining the inferential tasks involved in utterance interpretation".

In the following sections, we will take the research agenda of CSC one step further by investigating its explanatory power within simple argumentative structures, in which a constraining CSC strategy is used to reinforce a premise in order to convince a hearer of a given conclusion. Having established the theoretical predictions we make based on CSC, we will discuss some on-going experimental evaluations of the model in sections 4 and 5.

3. CSC AND ARGUMENTATION

While the CSC model described in the previous section was originally designed to account for deceptive communication, it can be extended in order to account for some aspects of argumentative communication, notably on the cognitive processing side. The CSC has in the past years indeed been used to offer pragmatic and cognitive insights on a range of argumentative phenomena, among which mostly argumentative moves traditionally defined as fallacies. As part of an effort to demonstrate the usefulness of the model for the analysis of argumentation, this research has tackled, both from a theoretical and an empirical perspective, source-related fallacies such as the *ad verecundiam*, the *ad hominem*, and the *ad populum* – which is experimentally investigated in this paper – in the works of Maillat & Oswald (2009, 2011), Oswald (2010, 2014), Maillat (2013, 2014) and Oswald & Hart (2013), the straw man fallacy (Oswald & Lewiński, 2014; Lewiński & Oswald, 2013), but also extended metaphors and their argumentative potential (Oswald & Rihs, 2013). In what follows we provide the rationale for the incorporation of the CSC model into a genuinely argumentative investigation.

Fundamental developments of argumentation theory have traditionally concerned two core issues related to the practice argumentation. The first has to do with finding ways to distinguish "good" from "bad" argumentation and has therefore concentrated on the identification of reliable normative criteria for argumentative evaluation.[2] The research field originates in Aristotle's development of logic (see Smith, 1995, p. 27), and has been expanded over time by, among others, epistemological, dialectical and informal approaches. These, in turn, include normative (or normative-like) standards, among which functional criteria enforcing the promotion of justified true belief (see e.g. Siegel & Biro, 1997; Goldman, 2003), dialectical rules of critical discussion (see e.g., van Eemeren & Grootendorst, 2004), argumentation schemes and critical questions (see e.g. Walton et al,. 2008), and the criteria of relevance, sufficiency and acceptability offered by informal logicians (see e.g., Blair, 2007; Johnson & Blair, 2006; Johnson, 2000).[3]

The second fundamental area of research in argumentation studies is closer to rhetorical concerns and has to do with the effectiveness of argumentation as an instrument of persuasion. Like those of logic, the roots of rhetoric can be found in Aristotle's work; over time, however, rhetoric failed to develop like her sister disciplines, logic and dialectics, and it is only in the middle of the 20th century that rhetoric started to reclaim scholarly attention, notably through the works of Toulmin (1958) and Perelman & Olbrechts-Tyteca (1958). Contemporary research in rhetoric strives to identify the reasons behind the success of argumentative moves, in terms of their convincingness (see Roque, 2012 for an up-to-date collection of works around the status of persuasion research in argumentation theory), and is nowadays leaning towards interdisciplinarity by recruiting the input of psychological and cognitive science (Herman & Oswald, 2014). To a fair extent, understanding why arguments (be they fallacious or not) might be rhetorically effective and appealing is a psychological question, as they are verbal messages meant to influence people's mental states (thoughts, beliefs, attitudes, etc.). This assumption underlies recent work in cognitive psychology inspired by a Bayesian probabilistic framework which develops a model of inference allowing to predict what kind of arguments people are likely to find strong and weak (see

---

[2] Since we are not addressing evaluative issues in this paper, we deliberately remain vague by mentioning "good" and "bad" arguments so as to avoid a discussion on the relative merits of notions such as validity, soundness, acceptability, etc.

[3] See Zenker (2013) for a thorough review and discussion of these normative criteria.

Hahn & Oaksford, 2006, 2007; Hahn et al., 2012). We claim in what follows that the CSC model can take part in the ongoing effort to bridge the gap between cognitive science and argumentation studies by contributing to shed light on the cognitive underpinnings of argument processing.

Our interest in argumentation theory is psychologically motivated: what happens, on the processing side, when people process arguments and find them convincing? Answering this question requires a focus on the cognitive machinery involved in argument processing; in this paper we will particularly be focusing on what argumentation scholarship has identified as fallacies. The standard treatment of fallacies – identified as such by Hamblin (1970) – considers a fallacious argument to be "one that seems to be valid but is not so" (Hamblin, 1970, p. 12), or, as summarised by Johnson (1987, p. 241), "reasoning which appears to be good but is not". Beyond its obvious shortcomings (see Hamblin, 1970; Johnson, 1987), this definition nevertheless does capture an intuition we can exploit by interpreting it in light of the CSC. The standard claim takes fallacies to be effective (i.e. convincing or persuasive) because they seem valid, or good, which is to say that they resemble good arguments. This last feature can be further discussed by (i) considering a fallacy's propensity to obscure the fact that it manifests a faulty inference and (ii) considering that this propensity plays a role in its success. That is, we hypothesise that fallacies are effective at least partly because we do not notice that they are fallacies. From this informal characterisation emerges the idea that it seems crucial to their persuasive success that they manage to conceal their fallacious nature. The consequence of this, in processing terms, is that successful fallacies manage to keep their addressees from processing specific critical information sets – those the awareness of which would alert the addressee to the fishiness of the argument, such as information about speaker intention, about the inconsistency of the argument, about a discrepancy between its content and the addressee's beliefs and values, etc. For this very reason, fallacies seem to trigger the very processing CSCs trigger: they operate a twofold constraint on the addressee's processing of information, so that chances of representing critical information are weakened and chances of representing "fallacy-friendly" information are increased.

From a discursive perspective, fallacies are verbal manifestation of inferential – specifically argumentative – articulations between pieces of information that are directed at an audience. They are couched in language and as such need to be interpreted before any argumentative processing can take place. Before one can evaluate whether a conclusion follows from a set of premises – even if this is the result of intuitive rather than reflective inference –, one needs to understand their content. This *ipso facto* makes cognitive argumentative processing operate on an input representation that is actually the output of the comprehension process. In other words, what you understand may play a role in what you end up accepting or rejecting after you evaluate a given argument. Assuming that the CSC operates at the level of information selection during the comprehension process, we can therefore assume that argumentative evaluation will be affected by the CSC.

Our claim is thus a claim of processing similarity: what happens in people's minds when they are being duped by a deceptive message is similar to what happens in their minds when they fall for a fallacious argument.[4] We therefore submit – and try to experimentally investigate – that fallacies can be described as devices which induce cognitive constraints on the selection of information which unfold along the two dimensions related to Sperber & Wilson's extent conditions of relevance, namely processing effort and cognitive effect (1995, p. 125, but see also the definition of relevance to an individual given above in section 2). Building on our previous work, we therefore claim that some fallacies will act as devices meant to increase the accessibility and the epistemic strength of information, while others will decrease the accessibility and epistemic strength of antagonistic information. More precisely, we will consider that the fundamental property of fallacious verbal material is its ability to constrain the interpretation of the message in order to prevent a context set C' containing critical information from being considered for the subsequent evaluative stage of processing, most of the times by simultaneously foregrounding a context C which is sufficient to establish interpretative relevance.

To illustrate how known fallacies can be interpreted in this framework, let us take the pair of fallacies known as *ad verecundiam* and *ad hominem*. The first is traditionally assumed to rely on irrelevant expertise (see e.g. Walton, 1995) and the second brings to the fore some personal characteristic of its target in order to undermine their credibility (*ibid.*). Crucially, both fallacies work along the same dimension, but in opposite directions: the *ad verecundiam* emphasises the assumed credibility/reliability/trustworthiness of the source so that the epistemic strength of the source's arguments or claims is increased (you are more likely to believe what your doctor says than what a known pathological liar tells you), and the *ad hominem* does the exact

---

[4] Note that whether the fallacy has deliberately been produced by the speaker is of little importance here, as we are taking the perspective of the addressee. For a discussion see Oswald (2014).

opposite by undermining the credibility/reliability/trustworthiness of the source so that the epistemic strength of the source's arguments or claims is decreased (again, you are less likely to believe what a known pathological liar says than what your doctor tells you). So the propositions conveyed by sources will be differently regarded – and their cognitive status in terms of reliable information will vary accordingly – depending on whether the source is assumed to be trustworthy or not. From the cognitive perspective adopted here, *ad verecundiam* and *ad hominem* are thus interpreted as fallacies which attempt to constrain the message in a way that the epistemic strength of information that will be used for their evaluation by the addressee can be manipulated, thereby influencing the inclusion or exclusion of respectively (perceived as) reliable and (perceived as) unreliable information in the context sets of information which will be used for argumentative evaluation.

We assume that the *ad populum fallacy*, usually defined as the use of the majority view or of a generally accepted statement as definite evidence for a conclusion (see Walton, 2006), works similarly to the *ad verecundiam*, only that instead of relying on the likely epistemic strength we associate with what authorities tell us, it relies on the likely epistemic strength we tend to attribute to widespread beliefs. The next section describes how this effect might be tested in light of the theoretical model introduced here and introduces an experiment meant to test whether *ad populum* arguments do have some appeal.

4. TESTING THE INFLUENCE OF THE AD POPULUM AS A CSC

The effects of repetition on human cognition have already been discussed and studied to some extent (Zajonc, 1969; Hasher & Chormiak, 1977; Cacioppo & Petty, 1979; Bacon, 1979; Arkes et al., 1990; Boehm, 1994; Ozubko & Fugelsang, 2011). It has been shown that repetition of an item or stimulus affects the subjective ratings of both its validity (i.e., the extent to which it is perceived to be true) and its "likeability" (the associated positive affective state). In the literature the two corresponding effects have been termed the validity effect (Boehm 1994), which denotes the increase in perceived validity, and the mere exposure effect (Zajonc, 1969), which denotes an increase in positive affect.

The differences between both of these effects tend to be attributed to the cognitive mechanisms behind them. Zajonc (1969), in discussing the mere exposure effect, argues for an explanation using frequency discrimination, which according to Hasher & Chormiak (1977) is an innate mechanism. At the same time various authors advocate a familiarity-based approach to repetition (Bacon, 1979; Arkes et al., 1990; Boehm, 1994). On their account, the main variable at play in the validity effect is familiarity with information, which can be enhanced through various means, e.g., repetition, source disassociation, reputation of the source, familiarity with the topic, etc. Arkes et al. (1990) provide a complex interaction schema between validity, source disassociation (the idea that the message comes from various different sources) and repetition, arguing that familiarity tends to enhance the validity effect in all cases.

In connection with issues raised in argumentation theory, Ozubko & Fugelsang (2011) have shown that repetition and familiarity not only affect the perceived validity of the repeated item, but that their effects carry over to an inferred assumption. By repeating an item that serves as a premise for a conclusion in an inferential process (e.g., an evidence statement such as "Roses need 20 minutes of sunlight per day to grow properly" and an inferred statement: "Roses can grow even with very little sunlight" Ozubko & Fugelsang, 2011, p. 276) the subjective validity of the conclusion is increased. Further, in what authors call "the memory retrieval condition" – where participants are presented during the evaluation phase with only the conclusion (inferred statement) – the effects are more robust than in the "classical condition" – where both the premise and the conclusion (evidence statement + inferred statement) are jointly presented during the evaluation stage.

We believe that this type of design can lend empirical support for the Context Selection Constraint Model in an argumentative sequence, as was discussed in the previous sections, specifically since Ozubko and Fugelsang have shown that the validity effect persists in an argumentative structure. If repetition and familiarity with information do increase validity ratings towards the repeated and/or familiar statement, this can be exploited in argumentative communication by constraining context selection in order to make targeted assumptions more salient via repetition or induced familiarity. While actual repetition is quite obvious (its use in advertisements could be a prime example), induced familiarity can be achieved in more subtle ways, for example with statements such as "everyone believes X" or "X is accepted by everyone." These instances seem to relate quite straightforwardly to what the literature on argumentation calls the *ad populum* fallacy, defined in the previous section. Now, if familiarity with a given piece of information is sufficient to trigger the validity effect, referring to statements as known by everyone or a majority should make them more salient by assigning familiarity to them within the community. In effect this would then correspond to a CSC strategy that can be exploited to reinforce a claim or conclusion supported by the familiar statement. The next sub-section discusses an on-going experimental study to test these claims.

4.1 Experiment

The current experiment is an attempt to see if the validity effect, triggered by familiarity, can be tested in an argumentative setting. As mentioned before, Ozubko and Fugelsang's results (2011) show that the validity effect can carry over to an inferred assumption or, in argumentative terms, from premise to a claim, which suggests that our assumption has some prospects of being empirically valid. If this is the case, then, based on the claims provided in the CSC model, we would assume that repetition of information would work as a CSC making a premise statement more salient and thus easily accessible in a person's cognitive environment. This in turn should result in a valid strategy for increasing the persuasive strength of the message. Additionally, since the validity effect is mediated by familiarity and not only by direct repetition, we claim that just referring to repetition or to the fact that the present statement is in everyone's cognitive environment (e.g. "everyone knows that X") should tap into similar processing mechanisms and result in an increased validity judgment for an argument.

In order to test these claims we have set up an experiment that manipulates two variables, namely repetition of information and accessibility of information (memory retrieval in Ozubko & Fugelsang, 2011). Repetition of information is manipulated by providing participants with scenarios that have either no repetition of the critical statement, repetition of the critical statement and critical statement in the form of *ad populum*. Accessibility of information is manipulated by providing participants with two types of statements concerning the scenarios of the form: Premise + Claim or Claim. The design is summarised in table 1 below.

Stimuli	Critical Task	Filler task
Scenario consisting of a narrative of approximately 130 words **Conditions** *(Variable: familiarity of a target statement U in the scenario):* 1. U uttered once (frequency 1) 2. U is uttered multiple times (frequency 3) 3. U mentioned as being uttered by everyone (frequency 1)	Assess the degree of agreement with a claim supported by the repeated statement **Conditions** 1. Premise + Claim 2. Claim	Assess the degree of agreement with other claims in connection with the scenario

Table 1 – Experimental design

Based on the previous studies and the theoretical model of the CSC we make two predictions. Firstly, we predict that repetition of critical statements and mention of critical statements in *Ad Populum* form will trigger the validity effect for that statement. Secondly, we argue that the validity effect for critical statements will be increased in cases where only the claim is presented, since memory retrieval is predicted to increase familiarity with the information accessed.

4.2 Participants

Participants in the experiment were 60 individuals from the United States whom we contacted via the Amazon's Mechanical Turk service. The experiment was set as a within participant design in which each individual participated in each of the 6 conditions. As a result 60 participants responded in each condition.

4.3 Material

The experimental material consisted of 18 unique scenarios, 130-words long on average that were pretested on native English speakers for language and argumentative acceptability. We used both scenarios previously used in Evans' (1983) investigation in syllogistic reasoning, together with other scenarios we created ourselves from news bulletins. Each of the 18 scenarios were made to vary on a familiarity scale ranging from no repetition (i.e. one occurrence), to three occurrences, to mention of familiarity via an *ad populum* (see Table 1). For each of the scenarios two critical questions were created (one for the Premise + Claim condition and one for the Claim condition). The Premise + Claim condition consisted of the critical statement found in the scenario previously mentioned and a claim it supported; while in the Claim condition only the claim was mentioned. This second parameter, i.e., the contrast between Claim vs. Premise + claim conditions, echoed the design used by Ozubko and Fugelsang (2011). Finally, each scenario featured two filler questions that, while thematically related to the scenario, ensured that participants would not realise what the purpose of the task was.

4.4 Procedure

Participants completed the experiment in the form of a survey on the Qualtrics survey builder platform. Each participant received all of the 18 scenarios but with various permutations. In general each participant saw 6 items with the No Repetition condition (for 3 of those scenarios, statements with the Premise + Claim condition were presented and for the other 3 scenarios, statements with the Claim condition were presented) together with 6 items for the Repetition condition (the same variation in questions) and 6 items with the Familiarity condition (again, the same variation for question holds). To guarantee this setup, a Latin square design with 6 groups with 18 items in each was constructed. This helped to ensure that each item got represented with all the various permutations – "Repetition", "No Repetition", "Ad Populum" and "Premise + Claim", "Claim." Participants were randomly assigned to one of the 6 groups and the 18 items were randomised.

When accessing the Qualtrics survey the participants were instructed to read each of the scenarios presented and afterwards to rate three statements (1 critical + 2 fillers) based on the scenario presented. Once the scenario had been read and after the participant pressed the "Continue" button, there was no way of returning to the previously read scenario, so the evaluation task was conducted on their recollection of the scenario. At any given moment participants only saw either the scenario or one of the 3 statements related to it. Each of the statements was presented with a 7 point Likert scale from "Strongly disagree" to "Strongly agree" in order to ensure the comparability to previous studies using the same scale.

4.5 Results and discussion

The experiment was designed to yield insights on two measures of interest: (i) whether the validity ratings given by the participants for the statements are affected in the Repetition and Ad Populum conditions and (ii) whether memory retrieval (Ozubko & Fugelsang, 2011) has a noteworthy effect on validity judgments (i.e., whether the Claim condition exhibits higher validity effects than the Premise + Claim condition). The results obtained during the experiment are shown in table 2 and are further illustrated in Figures 1 and 2.

	No repetition		Repetition		Ad Populum	
	Premise + Claim	Claim	Premise + Claim	Claim	Premise + Claim	Claim
M	5.01	5.13	5.22	5.07	5.19	5.28
SD	1.76	1.63	1.57	1.64	1.61	1.39
MSE	0.44	0.41	0.40	0.41	0.40	0.35

Table 2 – Means (M), standard deviation (SD) and mean standard errors (MSE) across all conditions

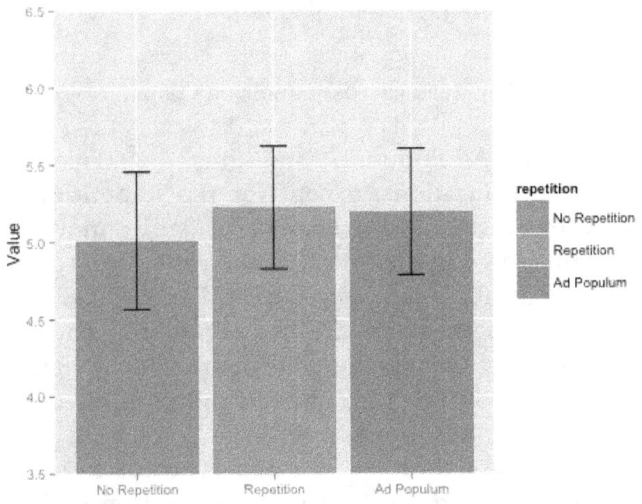

Figure 1 – Means across condition Claim + Premise

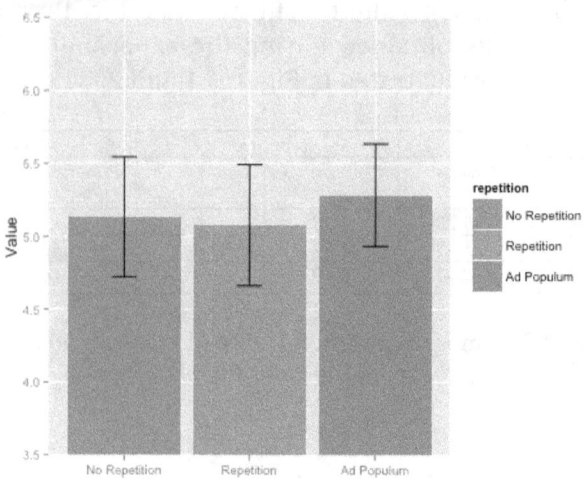

Figure 2 – Means across condition Claim

For the Repetition and Ad Populum conditions, the expected effect was observed for all the conditions, except for the Repetition with Claim discussed below. In general for the Premise + Claim conditions a similar effect to the one observed in the literature on repetition studies (usually 0.30 scale points) was observed (Gigerenzer, 1984). Crucially, the effect of familiarity is replicated in the Ad populum condition. Thus, participants rated the claims in the conditions with Repetition and Ad Populum as more valid than in the control condition (effect size 0.21 for Repetition and 0.18 for Ad Populum, $F(2, 537) = 0.868$, $p=0.4$. For the Claim condition a reversed effect was observed for Repetition (effect size -0.06) and an effect on the Familiarity condition (effect size 0.15, $F(2, 537) = 0.838$, $p =0.4$). Statistical significance at a .05 level was not achieved for the above-mentioned cases. Statistical significance was also not achieved for their combined effects ($F(2, 1074) =0.737$, $p = 0.5$).

Since the effects found here did not reach statistical significance, they should be discussed with caution because the null hypothesis, that there are no differences in the means between our groups cannot be rejected. With that in mind, the familiarity effect seems to be persistent in a similar fashion as discussed in the literature with effect sizes close to those observed (Gigerenzer, 1984). Interestingly, there seems to be very little difference between the Ad Populum condition (induced by mention of repetition) and actual Repetition (0.03), which lends support to the claim that *ad populum* arguments are effective. It seems that the familiarity effect is persistent at the same level in both cases. The question remains if in both cases the same mechanisms are at play, or if we are looking at different mechanisms that just produce similar results. Again, this is consistent with the literature as it highlights the complex relationship between validity, repetition and familiarity (see Bacon, 1979; Arkes et al., 1990; Boehm, 1994). Overall, the Ad Populum, in both Premise + Claim and Claim conditions, exhibited the highest validity ratings of all conditions. It seems that the effect induced by memory retrieval noted by Ozubko & Fugelsang (2011) is best replicated in the *ad populum* condition which can be taken as an indication of the inferential work required to reconstruct both the representation and the mentioned familiarity of the target statement. As a result, the *ad populum* argument paired with an implicit reconstruction of the premise seems to be the most efficient way of triggering the familiarity effect.

Since significance was not achieved in the present experiment, and we can only report a confirmation of the predicted trend, in particular in the Premise + Claim condition, the next steps would be to create a follow-up study, for instance by increasing the number of participants or items, in order to improve statistical power for an effect which is obviously rather small[5].

5. CONCLUSION

In this paper, we have shown how the *ad populum* fallacy can be accounted for within the Context Selection Constraint framework. We have argued that a pragmatic account, based on CSC, of the kind of effect triggered on the hearer by an *ad populum* fallacious argument can be explained in terms of the assumption of familiarity it assigns to the argument: the assumption of familiarity constrains the way we interpret the conclusion supported by that argument.

In doing so we have argued that an *ad populum* conveys procedural information regarding the kind of inferential assumptions that can be derived from the argument embedded in the *ad populum*. Specifically, the *ad populum* licenses an inference on the familiarity of the embedded argument, which results in a greater cognitive effect for the argument. Interestingly, from an argumentative point of view, the trends observed in the experimental results appear to support a model in which the cognitive strength of a premise is reflected in the cognitive strength of an inferred conclusion.

Nevertheless, the on-going experimental evaluation of our predictions has had mixed results. While on the one hand our first set of

---

[5] Additionally employing mixed models for our statistical analysis gives us a better fit of the model and a better analysis of our data indicating a potential for future studies of a similar type.

results appear to confirm the impact of familiarity on the perceived strength of a premise, as well as to confirm the transitivity of familiarity from a premise to the conclusion it supports in an argumentative structure, these trends, although they echo the effect size noted in previous studies, fail to reach statistical significance.

At this stage a number of hypotheses can be entertained to explain these mixed results. Our current best bet is that we need to improve the statistical power of our data by increasing the number of subjects taking the test, or by increasing the number of items. This is the direction currently pursued for our next investigation.

REFERENCES

Arkes, H. R., Boehm, L. E., & Xu, G. (1991). Determinants of judged validity. *Journal of Experimental Social Psychology, 27*, 576 – 605.

Bacon, F. T. (1979). Credibility of repeated statements: Memory for trivia. *Journal of Experimental Psychology: Human Learning and Memory, 5*(3), 241–252.

Blair, A. (2007). Relevance, acceptability, and sufficiency today. *Anthropology and Philosophy, 8*, 33-48.

Blakemore, D. (2002). *Relevance and linguistic meaning: The semantics and pragmatics of discourse markers.* Cambridge: Cambridge University Press.

Boehm, L. E. (1994). The validity effect: A search for mediating variables. *Personality and Social Psychology Bulletin, 20*(3), 285–293.

Cacioppo, J. T., & Petty, R. E. (1979). Effects of message repetition and position on cognitive, response, recall and persuasion. *Journal of Personality and Social Psychology, 37*(1), 97–109.

Carston, R. (2002). *Thoughts and utterances.* Oxford: Blackwell Publishers Oxford.

Clark, B. (2013). *Relevance theory.* Cambridge: Cambridge University Press.

Evans, J. St. B. T., Barston, J. L., & Pollard, P. (1983). On the conflict between logic and belief in syllogistic reasoning. *Memory and Cognition, 11*(3), 295-306.

Gigerenzer, G. (1984). External validity of laboratory experiments: The frequency-validity relationship. *American Journal of Psychology, 97*(2), 185-195.

Goldman, A. (2003). An epistemological approach to argumentation. *Informal Logic, 23*(1), 51-63.

Grice, H. P. (1989). *Studies in the way of words.* Cambridge, MA: Harvard University Press.

Hahn, U., & Oaksford, M. (2006). A normative theory of argument strength. *Informal Logic, 26*(1), 1–24.

Hahn, U., & Oaksford, M. (2007). The rationality of informal argumentation: A Bayesian approach to reasoning fallacies. *Psychological Review, 114*(3), 704-732.

Hahn, U., Harris, A. J. L., & Oaksford, M. (2012). Rational argument, rational inference. *Argument & Computation, 4*(1), 21-35.

Hasher, L., & Chormiak, W. (1977). The processing of frequency information: An automatic mechanism? *Journal of Verbal Learning and Verbal Behavior, 16*, 107-112.

Herman, T., & Oswald, S. (Eds.). (2014). *Rhétorique et cognition: Perspectives théoriques et stratégies persuasives - Rhetoric and cognition: Theoretical Perspectives and persuasive strategies* (Bilingual edition). Bern: Peter Lang.

Johnson, R. H. (1987). The blaze of her splendors: Suggestions about revitalizing fallacy theory. *Argumentation, 1*(3), 239–253.

Johnson, R. H. (2000). *Manifest rationality: A pragmatic theory of argument.* Mahwah, NJ: Routledge.

Johnson, R., & Blair, J. A. (2006). *Logical self-defense.* New York: International Debate Education Association.

Lewiński, M., & Oswald, S. (2013). When and how do we deal with straw men? A normative and cognitive pragmatic account. In D. Maillat & S. Oswald (Eds.), *Biases and constraints in communication: Argumentation, persuasion and manipulation.* Special issue of the *Journal of Pragmatics, 59*(B), 164-177.

Maillat, D. (2013). Constraining context selection: On the pragmatic inevitability of manipulation. In D. Maillat & S. Oswald (Eds.), *Biases and constraints in communication: Argumentation, persuasion and manipulation.* Special issue of the *Journal of Pragmatics, 59*(B), 190-199.

Maillat, D. (2014). Manipulation et cognition: Un modèle pragmatique. In T. Herman & S. Oswald (Eds.), *Rhétorique et cognition: Perspectives théoriques et stratégies persuasives* (pp. 69-88). Bern: Peter Lang.

Maillat, D. & Oswald, S. (2009). Defining manipulative discourse: the pragmatics of cognitive illusions. *International Review of Pragmatics, 1*(2), 348-370.

Maillat, D., & Oswald, S. (2011). Constraining context: A pragmatic account of cognitive manipulation. In C. Hart (Ed.), *Critical discourse studies in context and cognition* (pp. 65-80). Amsterdam: John Benjamins.

Noveck, I. A., & Sperber, D. (2006). *Experimental Pragmatics.* Basingstoke: Palgrave Macmillan.

Oswald, S., & Hart, C. (2013). Trust based on bias: Cognitive constraints on source-related fallacies. In D. Mohammed & M. Lewiński (Eds.), *Virtues of Argumentation. Proceedings of the 10th International Conference of the Ontario Society for the Study of Argumentation (OSSA), 22-26 May 2013.* Windsor, ON: OSSA, 1-13.

Oswald, S., & Lewiński, M. (2014). Pragmatics, cognitive heuristics and the straw man fallacy. In T. Herman & S. Oswald (Eds.), *Rhétorique et cognition: Perspectives théoriques et stratégies persuasives - Rhetoric and cognition: Theoretical Perspectives and persuasive strategies* (pp. 313–343). Bern: Peter Lang.

Oswald, S., & Rihs, A. (2013). Metaphor as argument: Rhetorical and epistemic advantages of extended metaphors. *Argumentation, 28*(2), 133–159.

Oswald, S. (2010). *Pragmatics of uncooperative and manipulative communication.* PhD thesis, Université de Neuchâtel.

Oswald, S. (2011). From interpretation to consent: Arguments, beliefs and meaning. *Discourse Studies, 13*(6), 806-814.

Oswald, S. (2014). It is easy to miss something you are not looking for: A pragmatic account of covert communicative influence for (critical) discourse analysis. In C. Hart & P. Cap (Eds.), *Contemporary Studies in Critical Discourse Analysis* (pp. 97-120). London: Bloomsbury.

Ozubko, J. D., & Fugelsang, J. (2011). Remembering makes evidence compelling: Retrieval from memory can give rise to the illusion of truth. *Journal of Experimental Psychology: Learning, Memory, and Cognition, 37*(1), 270–276.

Perelman, C., & Olbrechts-Tyteca, L. (2008) [1958]. *La nouvelle rhétorique: Traité de l'argumentation.* Bruxelles: Editions de l'Université Libre de Bruxelles.

Roque, G. (2012). (ed). Argumentation and Persuasion, *Special issue of Argumentation, 26*(1), Dordercht: Springer.

Siegel, H., & Biro, J. (1997). Epistemic normativity, argumentation, and fallacies. *Argumentation, 11*(3), 277-292.

Smith, R. (1995). Logic. In J. Barnes (Ed.), *The Cambridge companion to Aristotle* (pp. 27-65). Cambridge: Cambridge University Press.

Sperber, D. (1994). Understanding verbal understanding. In J. Khalfa (Ed.), *What is Intelligence?* (pp. 179-198). Cambridge: Cambridge University Press

Sperber, D., & Wilson, D. (1995). *Relevance: Communication and cognition.* Oxford: Blackwell Publishers.

Sperber, D., Clément, F., Heintz, C., Mascaro, O., Mercier, H., Origgi, G., & Wilson, D. (2010). Epistemic Vigilance. *Mind & Language, 25*(4), 359–393.

Toulmin, S. (2008) [1958]. *The uses of argument.* Cambridge, New York: Cambridge University Press.

Walton, D. (2006). *Fundamentals of critical argumentation.* Cambridge: Cambridge University Press.

Walton, D. N. (1995). *A pragmatic theory of fallacy.* Tuscaloosa: University of Alabama Press.

Walton, D., Reed, C., & Macagno, F. (2008). *Argumentation Schemes.* Cambridge: Cambridge University Press.

Wilson, D., & Sperber, D. (2012). *Meaning and relevance.* Cambridge; New York: Cambridge University Press.

Zajonc, R. B. (1968). Attitudinal effects of mere exposure. *Journal of Personality and Social Psychology, Monograph Supplement, 9,* 1-27.

Zenker, F. (2013). What do normative approaches to argumentation stand to gain from rhetorical insights? *Philosophy & Rhetoric, 46*(4), 415-436.

25

Practical Argumentation and Multiple Audience in Policy Proposals

RUDI PALMIERI
University of Liverpool, UK
rudi.palmieri@liverpool.ac.uk

SABRINA MAZZALI-LURATI
Università della Svizzera italiana, Switzerland
sabrina.lurati@usi.ch

We study the connection between the audience structure and the structure of practical argumentation in financial communication involving multiple stakeholders. Considering corporate stakeholders as text stakeholders, we examine the case of Ryanair's hostile bid for Aer Lingus with the following questions: How multiple stakeholders affect the design of the argumentative strategy supporting the proposal? How corporate leaders frame the different issues entailed by their offer? How these issues are integrated in the practical argumentation structure?

KEYWORDS: Argumentum Model of Topics, audience analysis, corporate argumentation, multiple audiences, practical argumentation, ratified readers, (text) stakeholders, strategic communication, takeovers

1. INTRODUCTION

Between 2006 and 2013, the Irish airline company Ryanair tried for three times to take control of Ireland's flag carrier Aer Lingus by making a public offer for all Aer Lingus shares. All three attempts failed before the Aer Lingus shareholders could decide whether to accept or reject the financial proposal. The reason was that the European Competition Commission and the Irish government, who also held more than 25% of the Aer Lingus shares, had raised serious competition concerns. From a communicative point of view, the Ryanair-Aer Lingus story highlights a crucial aspect that typically characterizes the rhetorical situation (Bitzer, 1968) entailed by public proposals such as takeover bids: in designing their argumentative strategy, corporate leaders (i.e. managers and directors) have to be aware and cope with the presence of multiple audiences, constituted by different groups of stakeholders of the concerned organization. These audiences usually have different and sometimes even partially conflicting demands (van Eemeren, 2010) that arguers have to meet in order to positively modify their exigence and achieve their strategic goals.

The problem of "what a rhetor can do when facing multiple audiences" (Benoit & D'Augustine, 1994, p. 89) has been recognized and discussed by a few scholars (see Perelman & Olbrechts-Tyteca, 1958; Benoit & D'Augustine, 1994; Myers, 1999; van Eemeren, 2010). Much is left unexplored with regard to the impact that a rhetorical situation with multiple audiences has on the structure of the argumentation designed by the arguer, more in particular when a proposal is announced and argumentatively defended in public contexts. In similar situations, simply writing a separate text to each stakeholder seems neither sufficient nor safe, since all other interested parties may have access to and read the message, making it necessary for the writer to account for several audiences within the same text (Myers, 1999) to avoid dangerous communicative side effects.

In a related paper (Palmieri & Mazzali-Lurati, 2016), we imported the stakeholder concept from management theory (Freeman, 1984; Post et al., 2002) into argumentation theory, by introducing the refined notion of text stakeholder (see also Mazzali-Lurati, 2011; Mazzali-Lurati & Pollaroli, 2013). We maintain, in fact, that the role a stakeholder plays in the context of any kind of corporate initiative is inherently argumentative. Any interest (stake) held in relation to an organizational activity becomes an interest in the content of the related organizational messages and it takes the form of an argumentative issue (see Goodwin, 2002). Organizational rhetors (Green, 2004; Hartelius & Browning, 2009) are expected to acknowledge all these issues and have the exigence of responding to them with an effective argumentative strategy (see Jacobs, 2000; van Eemeren & Houtlosser, 2002; Rigotti, 2006; Rocci, 2009).

The notion of text stakeholder represents a starting point to identify and reconstruct multiple audiences, as it is instrumental to characterize not only the addressee of an organizational message but all types of readers/hearers who might raise an argumentative issue (e.g. "Will this business achieve its profit targets?"; "Did this company behave ethically?"; "Can investors trust the new management team?").

By considering the inevitably multi-audience setting of takeover proposals, this paper examines how this kind of contextual factor affects argumentative discourse and, more in particular, the structure of the practical (i.e. action-oriented) argumentation, supporting or criticizing the proposed transaction. How does the presence of multiple text stakeholders affect the design of the practical argumentation supporting the proposal? How are the different issues integrated in the structure of the practical argumentation? How are they framed? To answer these questions we consider, as a case in point, Ryanair's second offer to Aer Lingus and, more specifically, the offer document – i.e. the official text by which an offer is performed – published on December 15, 2008.

Our analysis of practical argumentation is largely based on the theoretical framework of the *Argumentum Model of Topics* (Rigotti & Greco Morasso, 2010), which is discussed in the next section. Section 3 explains the text stakeholder notion and how this notion can support the analysis of multiple audience situations. In section 4, we examine the Ryanair's offer document by (i) identifying the different audiences and the issues they raise and, (ii) reconstructing Ryanair's pragmatic argumentation, focusing on the strategies used to deal with the previously identified multiple audiences. In section 5 we conclude by discussing the main results obtained from our analysis.

2. PRAGMATIC ARGUMENTATION FROM THE AMT PERSPECTIVE

2.1 Practical reasoning in proposal speech acts

Practical reasoning is generally understood as the inference used to justify decisions on possible actions and, in this sense, it is distinguished from epistemic or theoretical reasoning (Hitchcock, 2001). The distinction can be traced back to Aristotle, who in the Topica writes that issues (and standpoints) – named "dialectical problems" – can be oriented either at "choice and avoidance or at truth and knowledge" (I, 104b; see Ross, 1958). While knowledge-oriented reasoning may support different types of standpoints such as descriptive, predictive, evaluative and explanatory standpoints (see Palmieri et al., 2015), pragmatic argumentation is about prescriptive (or policy) standpoints (see Rocci, 2008; van Eemeren, 2010; Fairclough & Fairclough, 2012).

The structure of practical inference has received much attention by scholars in philosophy and informal logic who defined schemes for practical reasoning mainly having in mind the individual decision-maker who has to choose the most prudent course of action to achieve his/her goal (see von Wright, 1963; Walton, 1990; Pollock, 1995; Bratman, 1999; Garssen, 2001; Searle, 2001; Broome, 2002). As such, practical reasoning always presupposes a theory of action, i.e. an ontological account of the constituents of the human action and of their mutual relationships which defines all relevant factors that must be taken into account when deciding what to do (see Rigotti, 2003). These factors include, besides the desires and goals of an agent, also the alternative means and causal chains that can be activated to achieve the goal, the consideration of other possible outcomes and side effects and, so, the comparison of competing ends and values, the quality of the information from which the agent obtains knowledge of actual and possible worlds (see Rigotti, 2008).

More recently, argumentation scholars interested in contextualized argumentative activities have examined practical argumentation not as an individual decision-making process, but rather as a communicated inference (Rocci, 2006). In many communication contexts, indeed, the protagonist/proponent uses pragmatic argumentation to rationally convince the antagonist/opponent of the acceptability of a prescriptive standpoint (e.g. Feteris, 2002; Palmieri, 2008; Ihnen, 2010; van Poppel, 2012). In such cases, the decision-maker is invited to draw an inference (see Pinto, 2001) belonging to the domain of practical reasoning (see section 2.2), for example: inferring the expediency of an action from the desirability of its consequences; or inferring the necessity of choosing one particular action from the exclusion of all potential alternatives (see Rigotti, 2008, p. 566-567; Palmieri, 2014, p. 34).

The structure of the communicated practical inference (and so the criteria to evaluate its soundness) becomes more complex when the decision at issue does not pertain merely an individual action, but more specifically a joint action (Clark, 1996), under the form of interaction, cooperation (see Rigotti, 2003) or collective action (Vega & Olmos, 2007; Lewiński, 2014). Typical is the case of proposals (see Aakhus, 2006), like takeover bids, commercial ads or dinner invitations, where the projected action (trading shares, buying/selling products and services, having a meal together) involves also the proposal-maker with his/her goals, interests and commitments.

When an agent – like the target shareholder in a takeover bid – receives a proposal, the implied argumentative issue takes the form of a yes-no question[1] (should he/she accept the bid or not?). What is at issue

---

[1] This situation is different from the cases of open questions (what should I/we/you achieve X?), where an agent has already defined a precise goal and tries to determine the most prudent course of action that could lead to attain such a goal (Walton, 1990). In fact, practical reasoning can be used both for advocacy and inquiry purposes. A different type of argumentative situation is also at stake with closed-list questions (should I/we/you do A, B or C?), which occur when competing proposals are made (see Lewiński & Aakhus, 2014). However, in all three cases, the standpoint asserting or negating that an action

in the first place is a practical standpoint affirming or negating the desirability, necessity or feasibility[2] of the proposed joint action. The justification of this type of standpoint coincides with a simple or complex reasoning structure in which one or more constitutive factors of an ontology of joint action are presented as an argumentative support. The precise structure of a pragmatic argument depends each time on the particular justification put forward by the arguer, which has to be reconstructed starting from his/her discourse. To this purpose, we adopt the Argumentum Model of Topics procedure, explained in the next sub-section.

2.2 The AMT reconstruction of pragmatic argumentation

The Argumentum Model of Topics (henceforth AMT) is a theory of argument schemes on which basis it is possible to reconstruct the inferential organization of the argument-to-standpoint relationship by keeping together abstract reasoning rules governing arguments and their concrete implementation in context-bound premises (Rigotti & Greco Morasso, 2010). For the sake of illustration, let us consider a very simple example of pragmatic argumentation, which we could find in a business context. Imagine that during a board meeting the CEO seeks to promote a proposal of responsible investment by saying to the Chairman and the other members. "This is a profitable project that will preserve the environment in the long-term. So, please, approve it without hesitation". We can reconstruct the practical standpoint "The Board of directors ought to approve the project proposed by the CEO" and its supporting argument "the project is profitable and will preserve the environment in the long-term". How would the AMT analyse the inferential configuration of this argumentation?

Recovering insights from Aristotle's Topics and subsequent Latin, Medieval and Renaissance authors (see Rigotti, 2006, 2009, 2014), the AMT posits that an ontological relation – the *locus* (*place* in English, *topos* in Greek) – links the premise-argument to the conclusion-standpoint. Loci are very general categories, which broadly coincide with general classes of modern argument schemes, like definition, whole and parts, analogy, authority, cause, and more specifically, efficient cause, final cause, material cause, and many others (see Rigotti,

---

ought to be undertaken is justified by an instance of practical reasoning which connects, in some way or another (i.e. with a specific maxim, see section 2.2), the proposed action to the envisaged goal

[2] For a detailed characterization of modal markers in practical reasoning, see Rocci (2008).

2009). In our example of pragmatic argumentation, the locus at work is that *from final cause (goal) to action* as achieving profit and preserving the environment are taken here as goals that would be attained by the action of undertaking the project figured out by the CEO.

Each locus entails several inferential principles, named *maxims*, which connect the two extremes of the locus by a hypothetical statement. In our case, the maxim is "if an action allows achieving an important goal, the agent ought to undertake it". This general principle can be found in endless concrete reasonings where arguers conclude the expediency of doing something because the obtained outcome would coincide with an important and desirable goal.

Loci and maxims, which represent the context-free component of the inferential configuration, are combined with two material premises, which instead refer to the context where the argument is communicated. One is the *datum*, a fact emerging in the argumentative situation, which usually coincides with the premise made explicit by the arguer in the discussion. In pragmatic argumentation, the datum coincides with means premise (Walton, 1990), in our case the CEO's prediction that "the project will be profitable and preserve the environment in the long-term". The other material premise is the *endoxon*, a concept taken from Aristotle's rhetoric to refer to the background knowledge component bound to the interaction field and the culture shared by the participants to the argumentative interaction. As such, endoxa are normally left implicit, meaning that their acceptability should be taken them for granted by the discussants. In our case, the endoxon is the goal premise of practical reasoning (Walton, 1990): "achieving profit and preserve the environment correspond to important goals of our company". Unlike maxims, the plausibility of endoxa depends on spatial-temporal and cultural variables. Indeed, it is reasonable to imagine that, in other historical periods, the preservation of the environment was not considered to be a priority for a business company as it tends to be claimed nowadays.

In Figure 1, the whole inferential configuration of the CEO's argumentation is reconstructed, indicating the function fulfilled by each element as regard to the practical reasoning. The locus-derived maxim activates a hypothetical syllogism whose conclusion coincides with the CEO's standpoint. The minor premise derives from the material premises, namely from the categorical syllogism conjoining endoxon and datum.

While the locus from final cause should be considered as the prototypical locus generating maxims of practical inference, also other ontological relations can be invoked to justify a prescriptive standpoint: the locus from instrumental cause (e.g. Mauritanians cannot attack us because they do not have weapons", see Rigotti & Greco Morasso, (2010)); the locus from alternatives (e.g. "We do not have food and the supermarket are closed, therefore we must go eating at the restaurant", see van Eemeren & Grootendorst, 1992); the locus from termination (e.g. "You care about your friendship, so you should not destroy it", see Greco Morasso, 2011); and the related locus from setting up which intervenes, for example, when the decision to create an institution is justified by the goodness of having such an institution (see Gobber & Palmieri, 2014).

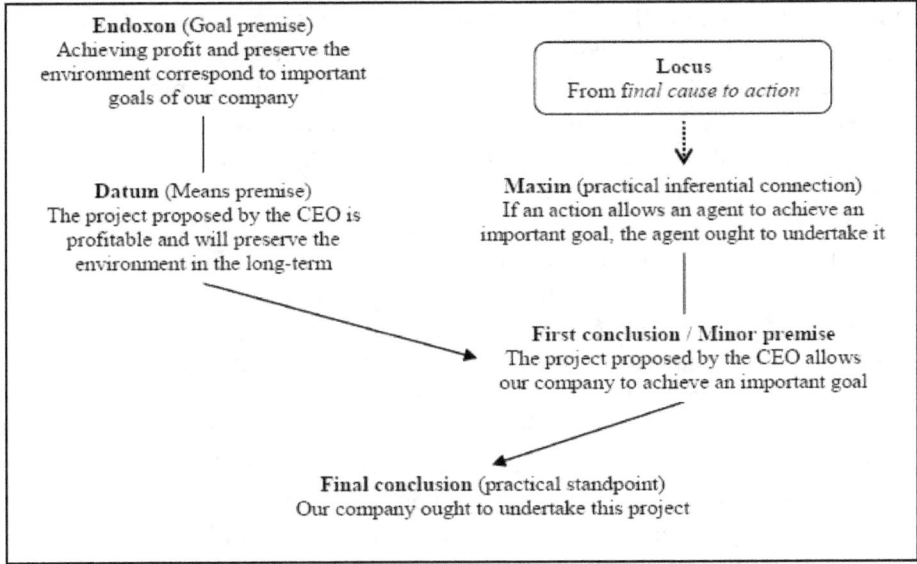

Figure 1 – AMT reconstruction of the inferential configuration of the CEO's argumentation

3. MULTIPLE AUDIENCE: THE NOTION OF TEXT STAKEHOLDERS

When investigating practical reasoning in its real applications, we need to take into account the context where such a practical reasoning is communicated. More specifically, we need to define and characterize the rhetorical situation an arguer finds himself in when using practical argumentation.

According to Bitzer (1968, 1980), a rhetorical situation is "a complex of persons, events, objects, and relations presenting an actual or potential exigence which can be completely or partially removed if discourse, introduced into the situation, can so constrain human decision or action as to bring about the significant modification of the exigence" (Bitzer, 1968, p. 6; 1980, p. 24). The pivotal factor of a rhetorical situation is the exigence, i.e. "an imperfection [present in a situation] marked by urgency" (Bitzer, 1968, p. 6), "which strongly invites utterance" (ibid., p. 5). A set of constraints, "which influence the rhetor and can be brought to bear upon the audience" (ibid., p. 6), and the audience itself, acting as mediator of change (ibid., p. 4), are the two other constitutive factors. The rhetorical audience "consists [only] of those persons, who are capable of being influenced by [that] discourse" (ibid., p. 8) and, therefore, of bringing about the positive modification of the exigence.

Very often, especially in public communication, the rhetorical audience includes different kinds of participants bearing a precise (and, more than often, different) stake in respect to the communicative interaction. By "importing" and elaborating the economic concept of *stakeholder*, which identifies "any group or individual who can affect or is affected by the achievement of the organization's objectives" (Freeman, 1984, p. 46; Post et al., 2002, p.18), we propose the more specific notion of *text stakeholder* (Mazzali-Lurati, 2011; Mazzali-Lurati & Pollaroli, 2014, Palmieri & Mazzali-Lurati, 2016). Text stakeholders participate into a communicative activity type, which according to Rigotti and Rocci (2006) consists of the mapping of an interaction scheme onto a piece of institutional reality named interaction field (p. 172). Within an activity type, the text stakeholder assumes a particular interactional role that is compatible with the interaction scheme and bears a particular stake bound to the interaction field affected by ad constraining the activity. In the case considered by this paper, the offer document implements the interaction scheme of public proposal to a collective decision-maker into the interaction field of the Irish stock market where Aer Lingus is listed.

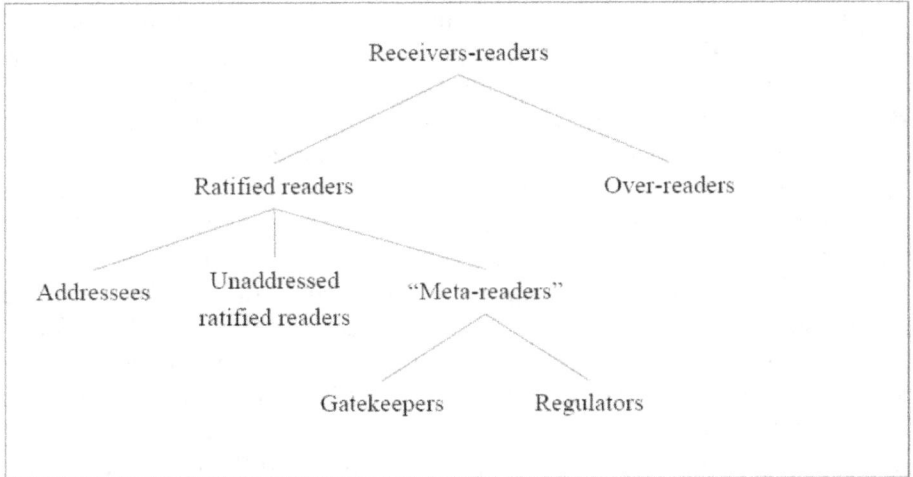

Figure 2 –Types of text stakeholders on the reception side (from Palmieri & Mazzali-Lurati, 2016)

From an argumentative perspective, the stake held by each text stakeholder entails an argumentative issue that the text is expected to deal with strategically. How each issue is handled with by the arguer significantly depends on the interactional role covered by the text stakeholder. Relying in particular on Goffman (1981)'s theory of participation role in face-to-face conversation, we can distinguish different categories of text stakeholders, as schematically illustrated in Fig. 2.

By taking into consideration the physical setting and the social occasion in which an interaction occurs, Goffman observes that behind the words "speaker" and "hearer" more than two different roles are hidden. Focusing on the reception side, addressed recipients (or addressee) are distinguished from unaddressed ratified recipients (see also McCawley, 1999, p. 596) and unratified adventitious participants (or bystanders) – which in written communication become over-readers.

Addressees are those to whom the author directly refers in his/her discourse by designating them with "you". Their stake is always, first of all, to understand the kind of pragmatic change the writer aims at obtaining from them. In each communicative interaction, then, this basic stake is specified according to the shared goals and commitments of the interaction field and according to the arguer's exigence. Unaddressed ratified readers are those being entitled to take part in the communicative interaction and whose "presence" cannot be denied by the author even if they are not directly addressed. Their stake can vary very much from one to the other, because it heavily depends on their role in the given interaction field and on the related shared goals and commitments. On the contrary, over-readers are those who happen to have access to the text even though they would not be expected to take part in the communicative activity type as they do not cover any role in the interaction scheme. Their stake is not considered to be relevant in respect to the arguer's exigence, but to know about and account for their actual presence can be useful to avoid possible communicative side effects and accidents.

We integrate this classification inspired from Goffman with two actors typically involved in public written communication. We refer to them as meta-readers, because they are expected to read the message without, however, entering into the merit of its content. Regulators aim at establishing whether the text can be (legally speaking) published or not, while gatekeepers have to decide whether to diffuse or not the text (or part of it) to a larger public.

4. THE CASE: RYANAIR'S SECOND HOSTILE OFFER FOR AER LINGUS

All three Ryanair's attempts to acquire Aer Lingus were hostile, which means that the offer was made to the Aer Lingus shareholders notwithstanding the opposition from the Board of directors. Hostile bids entail, therefore, an argumentative situation in which the directors of the bidding company (Ryanair) and those of the target company (Aer Lingus) advance and defend two opposite standpoints – that shareholders should accept the offer and that shareholders should reject the offer respectively – thereby playing the dialectical role of protagonist (see van Eemeren & Grootendorst, 2004). At the same time, the target shareholders, who are the final decision-makers, play the dialectical role of antagonists (ibid.), i.e. those who cast doubts on the standpoint and critically scrutinize the protagonist's argumentation to find reasons to decide which standpoint to agree with. This explicit difference of opinion is typically reflected by a public argumentative dispute, in which each side discloses its pro-arguments as well as its refutations of the other side's argumentation (see Palmieri, 2014, pp. 103-116).

Takeover bids, no matter whether friendly or hostile, are subject to precise rules disciplining in particular the conduct of corporate directors around the offer period for the purpose of enabling shareholders to make a properly informed decision (see Haan-Kamminga, 2006). The Ryanair-Aer Lingus case was regulated by the Irish Takeover Rules (*http://irishtakeoverpanel.ie/rules*), which impose on the bidder to publish an offer document containing relevant information about the offer, including not only the financial terms of the offer (Rule 24.2), but also its long-term commercial justification as well as the intentions regarding the future business of the target company, its fixed assets and its employees (Rule 24.1). Takeover rules impose similar duties on the target directors who, after the publication of the offer document, must issue their reasoned opinion about the offer (Rule 25), which in hostile bids take the form of the so-called defence circular (see Brennan et al., 2010; Palmieri, 2014, pp. 100-103). Employees or employee representatives of the target company have the right to receive both the offer document and the defence circular.

The offer document we consider in this paper refers to Ryanair's second takeover attempt and was published on December 15, 2008. Although not mandatory, the document has a cover page featuring as a cantered headline "Creating one strong Irish airline group in Europe". Our analysis shall focus on the first 23 pages which contain two distinct argumentatively-relevant parts: (1) from page 1 to page 16, besides some legal instructions, Ryanair presents arguments supporting the offer and counter-arguments aimed at attacking some of the statements Aer Lingus made at the time of the first offer (2006) or just after the announcement of the second offer (December 1st 2008). These argumentative moves are delivered in an overtly promotional and quasi-advertising style, making use of emphatic titles, visuals, graphs, tables, snapshots of media headlines, etc.; (2) from page 18 to page 23, we find the "Letter from the CEO of Ryanair" addressing the Aer Lingus shareholder, where Ryanair's argumentation is expounded in a plainer and more narrative fashion without any particular use of stylistic devices, except from bold headlines and bullet points.

4.1 Audience analysis

The audience structure of Ryanair's' offer document has been reconstructed by analysing the structure of the corresponding communicative activity type. The complete audience analysis is summarized in the table in Appendix 1, which identifies and describes the different text stakeholders in relation to their institutional stake and interactional role. The mapping of the scheme of public proposal to a collective decision-maker onto the Aer Lingus interaction field assigns the role of "proposer" to the Board of Ryanair and the role of "proposee-decider" to the Aer Lingus' shareholders. As a publicly-made proposal, a further participatory slot made of "third parties" is set by the interaction scheme. It includes other actors of the interaction field, who hold a stake and raise a specific issue in the offer document, although they are not in a position to make a decision about the acceptance/rejection of Ryanair's offer. They belong, therefore, to the category of unaddressed ratified readers. Among them we include, in particular, Aer Lingus' employees, to whom Takeover rules assign precise information rights, financial journalists and financial analysts who might diffuse more or less favourable evaluations and recommendations on the offer, and also customers and the European Commission for reasons that will become evident later. Moreover, the Irish Takeover Panel, which supervises all takeover bids in the Irish market, acts as text regulator, while news media are gatekeepers.

Similarly to the unaddressed ratified readers, also the addressee is in our case a composite audience, as the ownership structure of Aer Lingus does not include only ordinary investors but also other different groups of social actors having partially different stakes. After the 2006 IPO, which transformed the state-owned Aer Lingus into a listed company, the Irish government retained 25 % of the shares, while 21% of the shares were held by the Aer Lingus employees through an Employees Share Ownership Trust. Furthermore, Ryanair possessed a 29% stake, which however does not make it a relevant rhetorical audience since they evidently do not need to be persuaded to accept their own proposal. Indeed, this percentage represents the maximum an investor can buy in the market without being obliged to make an offer for all the shares at exactly the same price. Thus, hostile bidders often try to buy just less than 30% before launching a bid for the remaining shares. What is important to bear in mind is that ordinary investors, the Irish Government and ESOT, one the one side, are all shareholders and thus all of them have a stake entailed by the fact of being shareholder of a listed company, which means to be both owners and capital providers who expect a financial return. On the other side, these two aspects are often unbalanced so that some types of shareholders are largely if not exclusively interested in receiving dividends or making a capital gain while other shareholders are not primarily concerned with profit but with monitoring that the company ensures some non-financial benefits.

Therefore, the different types of Aer Lingus shareholders form a composite audience because they have some peculiar starting points and raise different issues or sub-issues, as indicated on Table 1. As already suggested, composite is also the audience of ratified readers, which include the Aer Lingus directors, who have already argued against by Ryanair and are expected to critically react to the offer document by means of the defence circular and other disclosures; stock market players like non-shareholder investors and financial analysts; the Ryanair shareholders, who expect the deal will be value accruing for the bidder; constituencies of the target corporation such as employees and customers; and the EU commission, which scrutinizes the offer to verify any antitrust issue.

4.1 Argumentative analysis

Dealing with the composite addressee

Given the composite structure of the addressee of the offer document, it is interesting, in the first place, to verify how Ryanair deals with the different types of Aer Lingus shareholders. From several passages of the introductory part, it appears that Ryanair considers the Irish Government as the most important shareholder that has to be persuaded, since many references are made to the advantages that a combination of the two companies would bring to Ireland and its citizens. More specifically, Ryanair argues that Ireland would benefit from an improved service line (e.g. preservation of the Heathrow slot, restoration of the Shannon connectivity, more routes, more punctuality), lower costs for passengers thanks to lower fares and fuel surcharges and a €188 million gain in cash, which is interpreted as "a valuable and timely contribution when departments such as health and education are reducing spending" (p. 19). Instead, the financial attractiveness of this inflow receives much less mention, differently from what usually happens in hostile bids (see Palmieri, 2014, pp. 175-184). Two elements provide further evidence of this adaptation to audience demand (van Eemeren, 2010). First, the title of the cover page is "Creating one strong Irish airline group in Europe" and not something related to any financial gain. Second, in the conclusion of his letter, the Ryanair's CEO does not solicit the reader to accept the offer (as normally bidders do), but rather he invites them to "Support the Ryanair's Offer". So, shareholders are not asked to consider in the first place the premium they would earn by accepting the offer, which would make them wealthier, but the important economic and social project that would be realized thanks to the acquisition. Ryanair, thus, frames its readers-shareholders as joint agents who have the opportunity to cooperate (support) in the realization of a great joint endeavour rather than as counterparts in a financial transaction (see Figure 3):

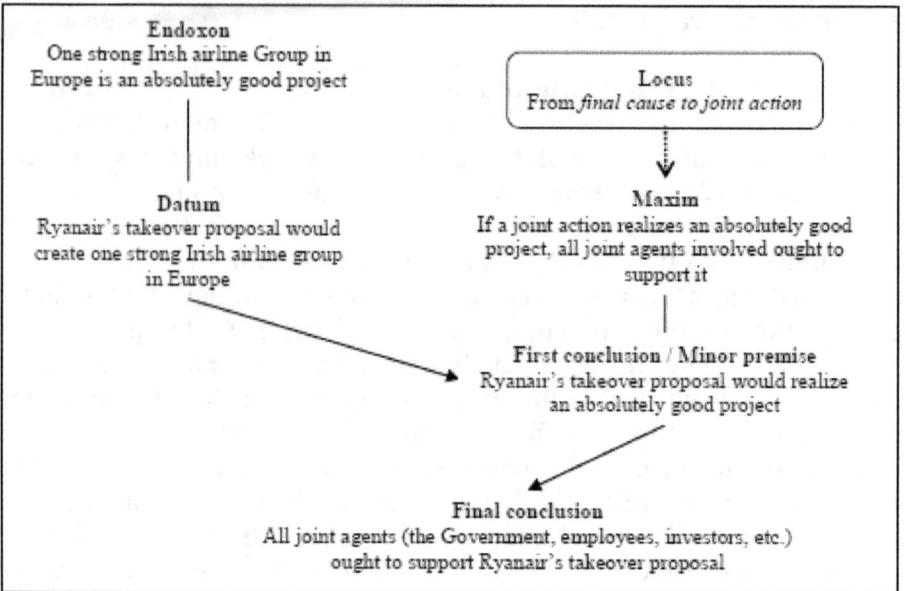

Figure 3 – "Creating one strong Irish airline group in Europe": inferential configuration

The prominence of the Government's stake can be observed very well also on page 15, titled "Ryanair's merger vs. Aer Lingus Independence", which contains a table comparing the benefits that the merger would entail for four classes of stakeholders with the damages that an independent Aer Lingus would bring to the same stakeholders. A clearly separate row of the table is devoted to each of the four stakeholders, who are discussed in the following order: "Ireland", "consumers", "employees", "shareholders". This choice of *dispositio* highlights the social rather than financial dimension of the envisaged deal. The bidder puts on the foreground the decisive role of the Irish government, who is institutionally committed to represent the interests of the country and its citizens. Indeed, the immediate conclusion the reader is invited to draw from this table is: "Create an Irish champion", after which the main practical standpoint of the document is spelled out: "Accept the Ryanair offer".

The benefits and damages for each stakeholder are presented as a list of past or future targets for which the writer does not make explicit any intuitively obvious link. For example, it is reasonable to imagine that the creation of 1,000 new jobs (mentioned under the heading "Employees") would also make the interests of Ireland. This presentational device allows Ryanair to construct a multiple argumentation structure, which would suit the multiplicity of text stakeholders better than any effort to engage with a compound argumentation. Already van Eemeren &

Grootendorst (1992) suggested that multiple argumentation may be subservient to cope with a multiple audience. We follow this hypothesis by suggesting that Ryanair's multiple argumentation could offer the strategic advantage of a sort of reading flexibility by creating the affordance for various interpretations activating particular practical inferences. In our case, we single out the following three audience-dependent readings of Ryanair's table:

> (1) Each independent premise counts as a single argument supporting a specific standpoint in relation to a specific text stakeholder, who holds a particular stake and, accordingly, is mainly or even exclusively interested in the corresponding sub-issue. Imagine, for instance, foreign investors who do not care very much about Ireland. The multiple structure presented in the offer document would allow them to activate the maxim "if a proposed action has desirable consequences for an agent (i.e. consequences coinciding with his/her own goals and interests), the agent should accept it".

> (2) All premises taken together form a compound argumentation structure supporting a "stronger", wider evaluative standpoint such as "the Ryanair-Aer Lingus combination is an absolutely good project", which in turn may lead to the practical standpoint "the offer cannot be rejected". In this case, the maxim comprises all relevant factors defining the reasonableness and prudence of an action and thus has the potential to counter all possible rebuttals.

> (3) The third reading is a mix of the previous two. Imagine that a stakeholder is particularly interested in one of the stakes mentioned by Ryanair (e.g. an employee who is worried about his job or an investor who wants to gain), but such a concern does not make her/him indifferent to other aspects entailed by the deal (e.g. a socially responsible shareholder). This type of reader might construct a compound argumentation, which activates the maxim "if an action allows an agent to achieve his goal while also producing positive side effects for other people, the action should be undertaken".

Accounting for ratified readers

The Aer Lingus board

Let us go back to Ryanair's statement in the cover page "Creating a strong Irish airline group in Europe". Why is it so important to specify that such a group would be "in Europe"? A valid explanation could be that, in hostile offers, the bidder has to account for the argumentative opposition of the target Board, who constitutes an important unaddressed but ratified reader (see Table in Appendix 1). During the first takeover attempt made by Ryanair in 2006, the Aer Lingus directors published a defence document where they expressed their full confidence in the airline standalone prospects. Basically, their pragmatic argumentation activates the *locus from termination* and its maxim "if a present situation is good, it should not be terminated". In the offer document (page 3) we are considering here, an argument belonging to the locus from analogy supports the claim that Aer Lingus' standalone prospects are not credible: Ryanair recalls several mergers and acquisitions in the airline industry, which have occurred or are occurring in most European countries. Therefore, Ryanair's statement in the cover page aims at emphasizing the desirability of the merger project for Ireland and, at the same time, refuting Aer Lingus's pragmatic argument mentioned above.

Besides defending its future value case without Ryanair, Aer Lingus criticized the hostile offer as an anti-competitive move by which Ryanair would have made profit only at the expenses of consumers (i.e. passengers). As explained in section 2, proposals semantically entail a joint action in which the proposal-maker is involved. The actual interest motivating the proposal-maker may become an issue for the decider, who needs to dispel the suspicion that the former is trying to gain only from an unfair redistribution of value. Ryanair's strategy consists in not declaring its possible financial gain from the takeover while highlighting the alignment of interests between the bidder and the target companies:

> Ryanair's Cash Offer delivers excellent value for Aer Lingus' Shareholders whilst ensuring that Ryanair can continue to realise its objective of rapidly growing Aer Lingus and lowering its fares for the benefit of all Aer Lingus stakeholders, including passengers and employees, and Ireland's national aviation policy (p. 23)

Alignment of interests is a crucial factor for the acceptability of a proposal of interaction, as it activates a relevant decision-oriented maxim: "if the interests of my interagent are aligned with my interests, his/her proposal of action is credible". Ryanair manoeuvres within the constraints imposed by the interaction scheme by addressing this important argument to the shareholders while responding to the criticism raised by a relevant ratified reader.

The European Commission

Having prohibited Ryanair's first attempt for antitrust reasons, the European Commission is evidently an unaddressed ratified reader. Its stake and related issue ("would the merger create monopoly?") cannot be ignored by Ryanair especially because it is also a relevant concern for the main decider, i.e. the Irish Government. It is not by chance that this second offer was rejected by the Government mainly on antitrust grounds.

In relation to the role of the EU Commission, we find a very interesting rhetorical strategy devised by Ryanair. In the CEO's letter, at the end of the paragraph titled "Bleak Future without Ryanair", while continuing addressing shareholders (as it is envisaged by the interaction scheme of the offer document), shareholders are assured that:

> Ryanair is confident that this Offer will ultimately be approved by the European Commission for a number of reasons including the following:
> - the Offer will have guaranteed positive competition effects leading to approximately €140 million annual benefits for all consumers as fuel surcharges are eliminated and short haul fares fall by 5% for three years;
> - the unique €200 million bank guarantees and other assurances that Ryanair is making in relation to the Offer; and
> - the Offer would be consistent with the EU's policy of encouraging consolidation in the airline industry (p. 21)

Apparently, Ryanair is only defending a predictive standpoint ("The EU will accept the deal"), which may become an argument from authority for consumer-concerned shareholders (i.e. if the EU commission, which is an expert in antitrust issues, accepts the takeover, the takeover is acceptable from a competition viewpoint). However, we observe that the arguments justifying this prediction coincide with the pragmatic reasons for which the EU should approve the merger. Imagining the EU Commission reading the document, this piece of text would represent a clear pragmatic argumentation. So, the argumentation for the decider-addressee contains the argumentation for a ratified reader. Synthetically, the message to shareholders is "The EU Commission will accept the bid because they should accept the bid", a statement clearly including a pragmatic standpoint for the EU Commission.

5. CONCLUSION

Strategic communication in public contexts is intrinsically directed to a multiple audience that creates constraints and affordances for argumentation. In this paper, we made an attempt to connect the multiple audience analysis with argument analysis, by focusing on pragmatic argumentation supporting a complex business proposal. As a case in point, we consider the takeover proposal made by Ryanair to Aer Lingus in 2008. We tried to show how an understanding of audience as text stakeholders, i.e. as communicative participants who have a precise interaction role and a precise institutional interest, may help to (i) identify contextual constraints and affordances for argumentative strategies:; (ii) reconstruct and explain the strategic rhetorical choices made by the arguer. Within the framework and method of the Argumentum Model of Topics, we went deeper in the inferential structure of the practical inferences (maxims) which Ryanair simultaneously communicates to the different text stakeholders.

Our results contribute in particular to the study of practical reasoning in contexts of public deliberation, to recent argumentation research interested in multi-party debates, discussions and controversy (see Aakhus & Lewiński, 2011) and to the understanding on the connection between audience demand, framing and strategic manoeuvring (van Eemeren, 2010; Greco Morasso, 2011). More in general, we hope our contribution will stimulate further research to go deeper in the analysis and evaluation of strategic communication from an argumentative perspective.

REFERENCES

Aakhus, M. (2006). The act and activity of proposing in deliberation. In P. Riley (Ed.), *Engaging argument. Selected papers from the 2005 National Communication Association/American Forensic Association Summer Conference on Argumentation* (pp. 402-408). Washington, DC: National Communication Association.

Aakhus, M., & Lewiński, M. (2011). Argument analysis in large-scale deliberation. In E. Feteris, B. Garssen, F. Snoeck Henkemans (Eds.), *Keeping in touch with pragma-dialectics* (pp. 165-184). Amsterdam: John Benjamins.

Benoit, W.L., & D'Agostine, J.M. (1994). The Case of the Midnight Judges and Multiple Audience Discourse: Chief Justice Marshall and Marbury V. Madison. *The Southern Communication Journal, 59*(2), 89-96.

Bitzer, L. (1968). The rhetorical situation. *Philosophy and Rhetoric, 1*(1), 1-14.

Bitzer, L. (1980). Functional communication: A situational perspective. In E. White (Ed.), *Rhetoric in transition: Studies in the nature and uses of rhetoric* (pp. 21-38). University Park & London: Pennsylvanian State University Press.

Bratman, M. (1999). *Intentions, plans, and practical reason.* Stanford: CSLI Publications.

Brennan, N. M., Daly, C., & Harrington, C. (2010). Rhetoric, Argument and Impression Management in Hostile Takeover Defence Documents. *British Accounting Review, 42*(4), 253-268.

Broome, J. (2002). Practical reasoning. In J.L. Bermùdez & A. Millar (Eds.), *Reason and nature: essays in the theory of rationality* (pp. 85-111). Oxford: Oxford University Press.

Clark, H.H. (1996). *Using language.* Cambridge University Press.

Fairclough, I., & Fairclough, N. (2012). Values as premises in practical arguments: Conceptions of justice in the public debate over bankers' bonuses. In F.H. van Eemeren & B. Garssen (Eds.), *Exploring Argumentative Contexts* (pp. 23-41). Amsterdam: John Benjamins.

Feteris, E. T. (2002). A pragma-dialectical approach of the analysis and evaluation of pragmatic argumentation in a legal context. *Argumentation, 16*(3), 349-367.

Freeman, R.E. (1984). *Strategic management: a stakeholder approach.* Boston: Pitman.

Garssen, B. (2001). Argument schemes. In F.H. van Eemeren (Ed.), *Crucial concepts in argumentation theory* (pp. 81-99). Amsterdam: Amsterdam University Press.

Gobber, G., & Palmieri, R. (2014). Argumentation in institutional founding documents. The case of Switzerland's Foedus Pactum. In G. Gobber & A. Rocci (Eds), *Language, reason and education. Studies in honor of Eddo Rigotti by his students and colleagues* (pp. 171-191). Bern: Peter Lang.

Goodwin, J. (2002). Designing issues. In F.H. van Eemeren (Ed.), *Dialectic and Rhetoric: The warp and woof of argumentation analysis* (pp. 81-96). Dordrecht: Springer.

Goffman, E. (1981). *Forms of talk.* Philadelphia: University of Pennsylvania Press.

Greco Morasso, S. (2011). *Argumentation in dispute mediation. A reasonable way to handle conflict.* Amsterdam/Philadelphia: John Benjamins.

Green, S. E. (2004). A rhetorical theory of diffusion. *Academy of Management Review, 29*(4), 653-669.

Haan-Kamminga, A. (2006). *Supervision on takeover bids: A comparison of regulatory arrangements.* Deventer: Kluwer.

Hartelius, E.J., & Browning, L.D. (2009). The application of rhetorical theory in managerial research: A literature review. In S.R. Clegg (Ed.), *SAGE Directions in Organization Studies* (pp. 379-404). Los Angeles: SAGE.

Hitchcock, D. (2001). Pollock on practical reasoning. *Informal Logic, 22*(3), 247-256.

Ihnen, C. (2010). The analysis of pragmatic argumentation in law-making debates: Second reading of the terrorism bill in the British House of Commons. *Controversia*, *7*(1), 91-107.
Jacobs, S. (2000). Rhetoric and dialectic from the standpoint of normative pragmatics. *Argumentation*, *14*(3), 261-286.
Lewiński, M. (2014). Practical reasoning in argumentative polylogues. *Revista Iberoamericana de Argumentación*, *8*, 1-20.
Lewiński, M., & Aakhus, M. (2014). Argumentative polylogues in a dialectical framework: A methodological inquiry. *Argumentation*, *28*(2), 161-185.
Mazzali-Lurati, S. (2011). Generi e portatori di interesse: due nozioni-chiave per la scrittura nelle organizzazioni. *Cultura e comunicazione*, *4*, 12-18.
Mazzali-Lurati, S., & Pollaroli, C. (2013). Stakeholders in promotional genres. A rhetorical perspective on marketing communication. In G. Kišiček & I.Ž. Žagar. (Eds.), *What do we know about the world? Rhetorical and argumentative perspectives* (pp. 365-389). Ljubljana: Digital Library of Slovenia & Windsor Studies in Argumentation.
McCawley, J. (1999). Participant roles, frames, and speech acts. *Linguistics and Philosophy*, *22*, 595-619.
Myers, F. (1999). Political argumentation and the composite audience: A case study. *Quarterly Journal of Speech*, *85*, 55-71.
Palmieri, R. (2014). *Corporate argumentation in takeover bids*. Amsterdam/Philadelphia: John Benjamins.
Palmieri, R. (2008). Reconstructing argumentative interactions in M&A offers. *Studies in Communication Sciences*, *8*(2), 279-302.
Palmieri, R., & Mazzali-Lurati, S. (2016). Multiple audiences as text stakeholders. A conceptual framework for analysing complex rhetorical situations. *Argumentation*, online first, DOI 10.1007/s10503-016-9394-6.
Palmieri, R., Rocci, A., & Kudrautsava, N. (2015). Argumentation in earnings conference calls: Corporate standpoints and analysts' challenges. *Studies in Communication Sciences*, *15*(1), 120–132.
Perelman, C., & Olbrechts-Tyteca, L. (1958). *La nouvelle rhétorique. Traité de l'argumentation*. Paris : Presses Universitaires de France.
Pinto, R.C. (2001). *Argument, inference and dialectic*. Dordrecht: Kluwer.
Pollock, J. L. (1995). *Cognitive carpentry: A blueprint for how to build a person*. Cambridge, MA: MIT Press.
Post, J.E., Preston, L.E., & Sachs, S. (2002). *Redefining the corporation: Stakeholders management and organizational wealth*. Stanford: Stanford University Press.
Rigotti, E. (2014). The nature and functions of loci in Agricola's *De inuentione Dialectica*. *Argumentation*, *28*(1), 19-37.
Rigotti, E. (2009). Whether and how classical topics can be revived in the contemporary theory of argumentation. In F.H. van Eemeren & B.J. Garssen (Eds.), *Pondering on problems of argumentation* (pp. 157-178). New York: Springer.

Rigotti, E. (2008). Locus a causa finali. In G. Gobber, S. Cantarini, S. Cigada, M.C. Gatti & S. Gilardoni (Eds), *Word meaning in argumentative dialogue. Special issue of L'analisi linguistica e letteraria, XVI*(2), 559-576.
Rigotti, E. (2006). Relevance of context-bound loci to topical potential in the argumentation stage. *Argumentation, 20*(4), 519-540.
Rigotti, E. (2003). La linguistica tra le scienze della comunicazione. In A. Giacalone-Ramat, E. Rigotti & A. Rocci (Eds), *Linguistica e nuove professioni* (pp. 21-35). Milano: Franco Angeli.
Rigotti, E., & Greco Morasso, S. (2010). Comparing the Argumentum Model of Topics to other contemporary approaches to argument schemes: The procedural and material components. *Argumentation, 24*(4), 489-512.
Rigotti, E., & Rocci, A. (2006). Towards a definition of communication context. Foundations of an interdisciplinary approach to communication. *Studies in Communication Sciences, 6*(2), 155-180.
Rocci, A. (2009). Manoeuvring with voices. In F.H. van Eemeren (Ed.), *Examining Argumentation in Context* (pp. 257-283). Amsterdam/Philadelphia: John Benjamins.
Rocci, A. (2008). Modality and its conversational backgrounds in the reconstruction of argumentation. *Argumentation, 22*, 165-189.
Rocci, A. (2006). Pragmatic inference and argumentation in intercultural communication. *Intercultural Pragmatics, 3*(4), 409-422.
Ross, W.D. (ed.). 1958. *Aristotelis Topica et Sophistici Elenchi.* Oxford: Oxford University Press.
Searle, J.R. (2001). *Rationality in action.* Cambridge, MA: MIT Press.
Van Eemeren, F.H. (2010). *Strategic maneuvering in argumentative discourse.* Amsterdam: John Benjamins.
Van Eemeren, F.H., & Grootendorst, R. (2004). *A systematic theory of argumentation: The pragma-dialectical approach.* Cambridge: Cambridge University Press.
Van Eemeren, F.H., & Grootendorst, R. (1992). *Argumentation, communication, and fallacies: A pragma-dialectical perspective.* Hillsdale, NJ: Lawrence Erlbaum.
Van Eemeren, F.H., & Houtlosser, P. (2002). Strategic maneuvering: Maintaining a delicate balance. In F.H. van Eemeren & P. Houtlosser (Eds), *Dialectic and rhetoric: The warp and woof of argumentation analysis* (pp. 131-159). Dordrecht: Kluwer.
Van Poppel, L. (2012). The strategic function of variants of pragmatic argumentation in health brochures. *Journal of Argumentation in Context, 1*(1), 97-112.
Vega, L., & Olmos, P. (2007). Deliberation: A paradigm in the arena of public argument. In H.V. Hansen et al. (Eds.), *Dissensus and the search for common ground* (pp. 1-11), CD-ROM. Windsor, ON: OSSA.
Von Wright, G. H. (1963). Practical inference. *The Philosophical Review, 72*(2), 159-179.
Walton, D.N. (1990). *Practical reasoning: goal-driven, knowledge-based, action-guiding argumentation.* Savage: Rowman & Littlefield.

APPENDIX 1: Text stakeholders of the Ryanair's offer document (December 15, 2008)

Text participants; role in the interaction scheme	Role in the interaction field affected by the bid	Stake in the interaction field	Issue (textual stake)

ADDRESSEE; decider
2. Aer Lingus' shareholders
   D. *Ordinary investors*
   E. *Irish government*
   F. *ESOT*
	2. Target shareholders: deciding to sell or not their shares to the bidder	10. To improve their financial and/or controlling position
		D. To obtain a financial gain
		E. To guarantee the quality of the AL services for Irish passengers
		F. To increase ESOT value while preserving job rights and conditions
			8. Should we accept RA proposal to buy our shares and takeover AL?
			D. Is the offer financially attractive?
			E. Will RA preserve AL quality services?
			F. Is the offer financially attractive and socially responsible?

UNADRESSED RATIFIED READERS, third parties 9. AerLingus' employees 10. AerLingus' directors 11. RyanAir's shareholders 12. Investors 13. Financial analysts 14. European Competition Commission	11. Target employees 12. Target directors 13. Bidder's shareholders: approve the takeover 14. Possible buyers of RA and AL shares 15. Information intermediaries 16. Antitrust authority: verify monopolistic issues	8. To preserve/improve job conditions 9. To defend the interests of the company following takeover rules 10. To gain from the acquisition 11. To find out gain opportunities 12. Prepare and sell expert opinions on the offer 13. To guarantee fair competition in the airline industry	9. Would the takeover damage job conditions? 10. Should the offer be recommended? 11. Would the takeover increment RA value? 12. Are RA and AL buyworthy? 13. How should we evaluate the offer? 14. Would a RA-AL create monopoly?
GATEKEEPERS, third parties 15. Financial journalists	9. Information intermediaries	9. Report relevant information and evaluations about the bid to all interested readers	17. How should the bid events be reported?
TEXT REGULATORS, third parties 18. Takeover Panel	10. Supervisor of the bid and of the conduct of the directors	10. To verify that the offer document complies with the City Code rules	10. Do the form and the content of the offer document comply with the City Code rules?

Audiences as Normative Roles
Commentary on Palmieri & Mazzali-Lurati's Practical Argumentation and Multiple Audience in Policy Proposals

JEAN GOODWIN
Iowa State University, USA
goodwin@iastate.edu

This paper carries forward the movement in argumentation studies that aims to expand our ability to deal with the complexities of real-world texts. While abstractions like ideal dialogue types can promote careful theorizing and fine distinctions, eventually the theory has to be robust enough to engage with the complex productions of arguers embedded in complex rhetorical situations (Bitzer, 1968). In recent years—and certainly at this conference—there has thus been increasing attention to the presence of multiple goals in argumentation (Mohammed, 2015) or the ways that argumentation functions in non-linguistic media (e.g., the recent and forthcoming special issues on visual and multimodal argumentation). In this essay, Palmieri and Mazzali-Lurati focus on yet another complexity: like Lewiński and Aakhus (2014), they are concerned with understanding argumentation in situations involving multiple participants.

Multiple participants are inevitable in many important real-world contexts. There is no such thing as "the" public; publics are always multiple, permeable and emerging (Asen & Brouwer, 2001). No arguer who enters the public sphere can limit access to her arguments only to the group she desires to address. Political speech and strategic communication must inevitably deal with this fact. The question Palmieri and Mazzali-Lurati raise is what argumentation theory can do to help them.

One obvious approach to distinguishing participants in an argumentative transaction is to identify them by their *standpoints*. This would seem to be a good option when the central issue can be framed in binary terms—e.g., "Ryanair takeover: yes or no?" (Lewiński & Aakhus, 2014). Still, in actual situations there may be two opposed camps who share almost nothing but their position on the central issue; each camp may contain many competing perspectives on how to justify the stance to that issue. Politics often makes strange bedfellows, as when the Catholic Church joins with a feminist faction to try to regulate pornography in a community. Although in cases like this multiple participants may share the same standpoint, it would be vexing to try to squeeze their views into a single argument diagram. The lines of argument would not only fail to overlap, in some cases they might be based on contradictory premises.

Another possibility then would be to pay attention to those differences in argument, and model multiple participants by the different commitment sets they bring to the interaction. That is after all the strategy adopted in the "folk theory" of argumentation evident when people label participants in a controversy as "left," "libertarian," "neo-liberal" or "progressive"; groups are being segmented out from the general mass of citizens by their purported ideological commitments. This also may be a promising approach, especially when the corpus to be analysed includes enough discourse from multiple participants to allow their commitment sets to be reconstructed. But it will be less helpful when the analysis focuses on a single text, as does Palmieri and Mazzali-Lurati's. In such cases, there is little evidence of what any individual audience member is committed too, much less how those commitments cohere or fail to cohere with those of other audience members. Speakers, of course, are always positioned so that they can only guess at their audience's internal segmentation.

Palmieri and Mazzali-Lurati propose an ingenious and, I believe, very promising solution to the problem of modelling multiple participants. Instead of modelling audiences by their standpoints (which may over-simplify) or by their commitment sets (which are difficult to determine), they propose modelling audiences by the normatively established roles they *ought* to play within the argumentative transaction. Audiences in this sense are not defined by their standpoints or commitments—they are *stakeholders*, defined by the specific interest it is their responsibility to maintain on the issue under consideration.

The Ryanair takeover case study shows that such stakeholders are readily identifiable, even in a complex situation. Members of the Board of Directors, for example, have fiduciary duties to pursue the best interests of the company. Members of the European Competition Commission ought to protect the competitiveness of the industry as a whole. The Irish Government (a key shareholder) should be seeking the public good for Irish citizens. And so on. It appears that the institutions supporting this complex argumentative interaction have developed these normative roles specifically to make the participants in the transaction more transparent to everyone involved. To me, this is an important signal that Palmieri and Mazzali-Lurati's approach is correct: we find the world already structuring itself in order to make just the participants they predict, *visible*.

It is interesting to note that the normative underpinnings of these roles are quite diverse: members of the Board and the ECC have legal duties; journalists have professional obligations; employees have a prudential interest in retaining their jobs; the Government has a constitutional obligation. In other words, we do not seem to need a unitary theory of "dialectical obligations" in order to generate participants' roles within an argumentative transaction.

It is also interesting to note that Palmieri and Mazzali-Lurati's approach allows us to identify some individuals as *outside* the transaction. Me, for example, although I may have strong views of how the proposed takeover would be a dreadful case of corporate greed harming ordinary consumers, or a wonderful example of capitalism at its best, my views are irrelevant. I have no *stake* in the matter; I am an un-ratified "over-reader"; it is *none of my business*.

In their analysis, Palmieri and Mazzali-Lurati demonstrate how Ryanair's arguments are structured to respond to the sub-issues that each legitimate stakeholder group is normatively required to consider. Elegantly, they show how at points the same stretch of discourse can be reconstructed as presenting different arguments addressed to the different interests of different stakeholders. This is a nice example in argumentation of what rhetoricians would call *polysemy* (Ceccarelli, 1998). The success of their textual analysis provides additional support for their conception of audiences as stakeholders, since it makes evident that the text constructors were tracking just the distinctions that Palmieri and Mazzali-Lurati have predicted. The one further step I would recommend would be to extend the study to consider the responses of various individuals and groups to the Ryanair proposal—an analysis of the text's *reception* (Ceccarelli, 1998). If these, too, are oriented to the stakeholder interests Palmieri and Mazzali-Lurati have identified, that fact would provide further confirmation for their approach.

In this response, I have focused only on one aspect of this strong paper—the conception of audiences as stakeholders—thus passing over other important contributions, especially the Argumentum Model of Topics (AMT). Let me close by putting Palmieri and Mazzali-Lurati's innovation in a larger context, as a partial explanation of this focus. Within argumentation theory, rhetoric has commonly been assigned the task of elucidating the effectiveness of argumentative discourse in persuading audiences. Actual scholars of rhetoric, by contrast, have uniformly insisted that rhetoric is deeply normative enterprise (e.g.,

Kock, 2009; Leff, 2000). Rhetorical scholars have been tenacious in examining (among other things) the *officia*—the normatively-grounded roles or local responsibilities—of those who undertake to address publics on public business, the *res publica*. The best contributions of rhetorical scholars to the interdisciplinary field of argumentation studies have been detailed reconstructions of how speakers undertake *officia* and the ways those *officia* structure argumentative transactions and the arguments that are exchanged within them (e.g., Innocenti, 2011; Kauffeld, 1998; and, immodestly, Goodwin, 2011). The work of Palmieri and Mazzali-Lurati extends this approach essentially by giving us a conceptualization of the *officia*—the local, normatively grounded responsibilities-of audiences as well. This is something I will be thinking about for a long time.

REFERENCES

Asen, R., & Brouwer, D. C. (2001). Reconfigurations of the public sphere. In *R. Asen & D. C. Brouwer (Eds.), Counterpublics and the state* (pp. 1-32). Albany: State University of New York Press.
Bitzer, L. F. (1968). The rhetorical situation. *Philosophy & Rhetoric, 1*(1), 1-14.
Ceccarelli, L. (1998). Polysemy: Multiple meanings in rhetorical criticism. *Quarterly Journal of Speech, 84*(4), 395-415.
Goodwin, J. (2011). Accounting for the appeal to the authority of experts. *Argumentation, 25*(3), 285-296.
Innocenti, B. (2011). Countering questionable tactics by crying foul. *Argumentation and Advocacy, 47*, 178-188.
Kauffeld, F. (1998). Presumptions and the distribution of argumentative burdens in acts of proposing and accusing. *Argumentation, 12*(2), 245-266.
Kock, C. (2009). Choice is not true or false: The domain of rhetorical argumentation. *Argumentation, 23*(1), 61-80.
Leff, M. (2000). Rhetoric and dialectic in the twenty-first century. *Argumentation, 14*(3), 241-254.
Lewiński, M., & Aakhus, M. (2014). Argumentative polylogues in a dialectical framework: A methodological inquiry. *Argumentation, 28*(2), 161-185.
Mohammed, D. (2015). Goals in argumentation: A proposal for the analysis and evaluation of public political arguments. *Argumentation*. Online first (doi: 10.1007/s10503-015-9370-6).

26

Does Public Deliberation Really Need Normative Constraints? Recovering the Aristotelian Rhetorical Theory[1]

SALVATORE DI PIAZZA
University of Palermo, Italy
salvatore.dipiazza@unipa.it

FRANCESCA PIAZZA
University of Palermo, Italy
francesca.piazza@unipa.it

MAURO SERRA
University of Salerno, Italy
maserra@unisa.it

There was recently an attempt to correlate some sectors of the studies on argumentation and the theory of democracy. The relationship between these two areas concerns the fact that in both cases there is a significant interest in normative models of good argumentation. In these models there is no place for rhetoric. Our work has a twofold aim: (1) to show that, starting from the Aristotelian rhetoric, it is possible to develop a more suitable model of argumentation in the public sphere and (2) to doubt the very need to identify normative constraints for this type of argument.

KEYWORDS: Aristotle, deliberation, democracy, normative constraints, rhetoric

---

[1] Although all the authors collaborated in the conception of the article's general framework, Mauro Serra wrote sections 1, 2, 3 and 4, Salvatore Di Piazza wrote sections 5 and 6 and Francesca Piazza wrote sections 7 and 8. Section 9 was written by all the authors.

1. INTRODUCTION

Since the seventies of the last century an interesting correlation has occurred between *argumentation studies* and some sectors of political theory, in particular those referring to the theoretical paradigm of deliberative democracy. Both paradigms are associated with a marked interest for identification of normative models of good argumentation, even if of course this interest has developed from different perspectives and pays attention to different levels of analysis. Indeed, while theorists of argumentation have primarily focused on *face-to-face* interactions, theorists of deliberative democracy have endeavoured to identify normative models related to social structures and institutions giving shape to political deliberation. Therefore, they have focused their interest on the public sphere, whose characteristics Habermas had started to outline in the sixties. Notwithstanding the inevitable differences, it is possible to identify important common and intersection points between these two theoretical perspectives. On the one hand, indeed, as Habermas himself has many times affirmed, the public sphere comprises different levels and not only social and institutional mechanisms determining deliberation; this means that it includes those communicative interactions, not excluding those *face-to-face*, that somehow contribute to shape public opinion. Thus, recently, John Dryzek, one of the most important theorists of deliberative democracy, has asserted that:

> deliberation itself ought to be dispersed in different locations, such as everyday talk, enclaves of like-minded individuals, social movements, the broader public sphere featuring a plurality of discourses, political parties, legislatures, courts, administrative processes, informal networks, para-governmental actions, and designed forums (Dryzek, 2010, p. 326).

On the other hand, the theory of argumentation, above all in some of its versions, and particularly in the Pragma-Dialectics that we will take into account, has tried to outline a critical rationality model that may be taken as a reference model for the same deliberative theory. After all, Habermas himself has underlined the common points and convergence with the considerations developed by theorists of Pragma-Dialectics.

A complete presentation of the link between these two perspectives, nonetheless, goes beyond the purposes of this paper (but see Rehg, 2005). Starting from such perspectives and some of their points of convergence we will do nothing but lay emphasis on the role to be assigned to normative constraints on public deliberation and we will propose an "Aristotelian" response as a possible alternative to that outlined by these two theoretical paradigms.

2. DELIBERATING IN THE PUBLIC SPHERE

First of all, it will be necessary to define what public deliberation means. In a very general way, it may be defined as "a form of discursive interaction in the public arena" (Vega & Olmos, 2007, p. 1). Therefore, it is an argumentative interaction where information, opinions, preferences are taken into consideration and evaluated to make a practical decision in a critical and reflexive way (Vega & Olmos, 2007, p. 1). In particular, two aspects will be underlined. The topics discussed will be of public interest, but their extent may be remarkably different. They range from *topics* of recognized general interest, such as atmospheric pollution, third world debt, decisions of military intervention in critical situations, to minor interest or even local matters such as those discussed in a board of regents or in a meeting of the city committee members:

> the discursive development of such discussions involves not only a dialectical interaction between several alternatives, but also an interpersonal confrontation of the different participants whose real bodily presence might well result in both power and pressure strategies and, on the other extreme, cautious attitudes which prevent fallacious manipulations under the threat of shame (Vega & Olmos, 2007, p. 2).

Whatever form of discursive interaction, a decision will be taken. This means that although the decision made may be in principle reversed, it may not be indefinitely postponed. At some point the exchange of opinions and argumentations supporting them will be interrupted to allow the decision on the topic being discussed. In the light of such considerations, for our purposes, we propose a rather wide definition of public deliberation: the ensemble of discursive practices aimed at shaping opinions from which comparison a decision on matters of collective interest will be made. Keeping this definition in the background, we will focus our attention, as partially said above, on the one hand on the model of rational argumentation proposed by Pragma-Dialectics, on the other hand, on the concept of Habermasian deliberative democracy, making reference to the model proposed by Joshua Cohen (1997), because:

> the extremity of his position – a direct mapping of Habermas's discourse ethics onto politics – and the abstract way he formulates it make it relatively easy to compare to other views (Shiffman, 2004, p. 90)

Also regarding with the privileged theoretical perspectives, we will not propose a general and complete presentation, rather we will only dwell on some elements from which we will pinpoint the problematic concern of these theoretical suggestions.

3. DELIBERATIVE DEMOCRACY AND PRAGMA-DIALECTICS: A REAPPRAISAL

The essential intuition that is behind deliberative democracy is that the political institutions are legitimate only when representing or embodying will and consensus of people ruled by the same institutions. Concurrently, individuals are considered politically free only to the extent that they have chosen the rules and institutions ruling them. The choice above, nonetheless, comprises an evident normative dimension for it will be guided by reason and will not be submitted to unjustifiable forms of external constraint, such as manipulation, deceit and similar. Therefore:

> the ideal deliberative procedure provides a model for institutions, a model that they should mirror, so far as possible (Cohen, 1997, p. 79).

According to Cohen, in its ideal form, a democratic deliberation will, abide by four criteria: it will be *sincere, reflective, strictly rational, and aiming at consensus.* The outcome of a public deliberation that does not meet these requirements may not be considered legitimate. These criteria correspond in a rather evident manner to presuppositions ruling each communicative interaction according to the model of universal pragmatics outlined by Habermas. According to Habermas the theory of argumentation is not a specific discipline, but coincides entirely with a theory of human rationality (Cantù-Testa, 2006, pp. 122-132). Indeed, the human rationality is communicative and intersubjective and has a specifically argumentative deep structure. This means that the speakers undertaking any communicative interaction, incur the obligations belonging to the argumentation sphere, or rather they undertake, if requested, to give reasons for their statements. The communicative activity which involves every individual daily then is based on a combination of rules whose implicit competence allows him to produce and understand statements. The said rules are not only syntactic or semantic, but also pragmatic standards ruling the development of concrete discursive interactions, making them possible. The pragmatic rules correspond, indeed, to validity claims that every interlocutor may make and that anyway are universal as they are necessarily assumed at any speech act that may be meaningfully produced by a speaker. They are *comprehensibility, truth, truthfulness* and *rightness*. It is from these preconditions that the speakers, according to Habermas, suppose that in any communicative act rationally justified consensus can be achieved. Or rather, it is from these preconditions that it can be affirmed that the consensus is the implicit *telos* in any discursive interaction and that in its ideal form such an interaction cannot but conclude with the triumph of the "unforced force of the better argument" (Habermas, 1996, p. 305). Then there seems to be no doubt that the criteria identified by Cohen are a slightly changed transposition from those presuppositions on which Habermas' theory of deliberative acts is based. Meanwhile, at least two of these elements, those on which we will focus our attention, that is tendency to achieve consensus and truthfulness/sincerity, find a quite clear correspondence in the theory of argumentation proposed by Pragma-Dialectics. As regards the first aspect, it is so evident that it may be enough to mention the following statement by Lumer (2010, p. 43):

> the whole approach of Pragma-Dialectics is constructed starting from one central theorem about the function of argumentative discourse and argumentation in general. The aim of argumentative discourse and of argumentation, as these are seen and constructed by Pragma-Dialectics, is to eliminate or resolve a difference of (expressed) opinion or to resolve a dispute—where "dispute" is understood as: expressed difference of opinion. This resolution has taken place if the participants both agree about the opinion in question (or if the protagonist withdraws his standpoint). The central task of the theory is to develop rules for rational discussions or discourses; and the value of the rules to be developed is regarded as being identical to the extent to which these rules help to attain the goal of resolving disputes.

However, the reference to sincerity is less obvious. One of the four meta-theoretical rules on which the model of Pragma-Dialectics is based is, indeed, the externalization. According to this rule, there is argumentation only when a point of view, a possible opposition to a point of view and the expression of a point of view are supplied in a speech with the patent objective to submit it to a public control. This means that the commitments made by the speakers in a critical discussion will not be identified with their intentions but only with what is entailed by their concrete speech acts. Accordingly, the reference to the clause of sincerity usually comprised in the theory of speech acts – that is you should believe in what you say – fails and is replaced by a more general condition of responsibility, according to which the speaker will bear the conversational consequences of its own speech acts. Meanwhile, van Eemeren and Grotendorst assume that the speakers will be serious, or rather they will say what they intend or believe. Although not explicit, nonetheless, the condition of sincerity may be equally supposed keeping in mind that the normative model outlined by Pragma-Dialectics is based on the idea that argumentation is a critical discussion aiming at the r e s o l u t i o n of a conflict. Above all, the participants contribute to achieve this purpose through a general cooperative aptitude. The dialectic dimension which the theory refers to, indeed, finds its own normative ideal in Socratic dialectics and in its interlinked idea that the attempt to terminate a divergence of opinions will be ruled by the purpose of making prevail the best rational thesis (that of course is considered as identifiable with absolute certainty) even at cost of sacrificing one's personal opinion. But this kind of aptitude makes the idea that the clause, although implicit, of truthfulness/sincerity may fail internally contradictory. Actually, the reference to Socratic dialectics does seem to presume a wider concept of sincerity, analogous to that proposed by Cohen (1997, p. 75), who believes that it entails:

> that discussants are to articulate their view of the common good in a way that they sincerely expect that others might rationally accept (Shiffman, 2004, p. 91).

In whatever way sincerity may be interpreted, it appears anyway, a remarkably problematic element inside a normative model of public deliberation. Of course, this problematic nature does not result from the impossibility to measure in an appropriate way from an empirical point of view the sincerity of participants in a public discussion. After all, just because a thing is not objectively visible or measurable, this does not mean that it is morally or politically irrelevant. The problem, rather, lies in the idea of rationality that is presumed by a similar normative ideal. A kind of rationality that appears quite evident, as Steiner (2012) has proved recently, underlining the bond that the notion of sincerity developed by Habermas and the deliberative democracy maintains with the Kantian concept of *Wahrhaftigkeit*. Of course, this concept means for Kant not telling lies but actually it is much wider, because being *wahrhaftig* means to be true to one's inner self and to find one's innermost identity. In order to make this possible, at least two elements will be presumed: that the inner self is unitary and that it is transparent or rather accessible for an investigation that may be more or less deep. Both assumptions to say the least are complicated. Indeed, on the one hand, it is more reasonable to assume that:

> our most inner self is not something fixed that can we can discover if only we dig deep enough. We rather assume that the inner self is something malleable and elusive that despite all our inner searching we can never quite know (Steiner, 2008, p. 5).

This kind of idea of the inner self seems to make understand the complexity of human psychology better, but it is also congruent with the openness to the possibility to be persuaded by others' opinions that is at the core of deliberative practice. Indeed, this openness could not exist if we were sure of what is our inner self, in such a manner that we do not need to listen to the others' reasons to be *wahrhaftig*. On the other hand, the Kantian idea of a unitary and transparent inner self tends to strengthen a wrong idea of our relationship with speech, as an instrument to be used only in a descriptive way (Markovits, 2008, p. 34). This conviction explains the persistent preconception about rhetoric (O'Neill, 2002), that after all Kant identifies as the practice used to prevent an interlocutor remaining faithful to his/her inner self, depriving him/her in this way of his/her freedom and therefore of his/her dignity as a human being:

> oratory, insofar as by that is understood the art of persuasion [überreden], i.e., of deceiving by means of beautiful illusion (as *ars oratoria*), and not merely skill in speaking [Wohlredenheit] (eloquence and style) is a dialectic, which borrows from the art of poetry only as much as is necessary to win the minds over to the advantage of the speaker before they can judge and *to rob them from their freedom*; thus it cannot be recommended either for the courtroom, or for the pulpit (Kant, CJ, 5, p. 327).

4. CONSENSUS AS A PROBLEMATIC AIM FOR DELIBERATION

Now we move to the issue of consensus. It is tightly linked to the ideal discursive situation, one of the most problematic and discussed notions of the Habermasian model. Before the evident impossibility that such a situation becomes empirically real, Habermas has modified this notion many times. Initially, perceived as a kind of transcendental condition, the ideal discursive situation was later weakened, but in such a way to preserve its idealizing substance. Hence, it retains its role of necessary normative presupposition in order that in the argumentative procedure the unforced force of the better argument may prevail. The main objection that can be made to the normative role ascribed to the connection between ideal discursive situation and consensus concerns, nonetheless, the way it may be connected to the decisions that will necessarily come after a deliberation. The distance between the normative condition and the concrete procedure entails that, as Cohen himself acknowledges, in most, not to say all cases, the rational consensus that can be hoped for is not achieved. However, given that a decision will be anyway made, it is inevitable to resort to the majority vote and consider the vote legitimate up to successive deliberation of the same issue. Actually, its legitimacy is somehow guaranteed by the fact that it is the final outcome of a deliberative process:

> even under ideal conditions there is no promise that consensual reasons will be forthcoming. If they are not, then deliberation concludes with voting, subject to some form of majority rule. The fact that it may so conclude does not, however, eliminate the distinction between deliberative forms of collective choice and forms that aggregate non-deliberative preferences. The institutional consequences are likely to be different in the two cases, and the results of voting among those who are committed to finding reasons that are persuasive to all are likely to differ from the results of an aggregation that proceeds in the absence of this commitment (Cohen, 1997, p. 75).

At this point, nonetheless, we are faced with the following question: how will the individuals that have taken part in the discussion, and have got hammered, be able to explain to themselves the reasons for disagreement persisting after the deliberation? As Gary Shiffman writes:

> if we have what mutually take to be an ideal deliberation, and I end up losing in a vote, what am I to think? Obviously I must

believe that the majority is wrong, otherwise I would have agreed (Shiffmann, 2004, p. 92)

In other words if the participation in the deliberation assumes that individuals will abide by the criteria of sincerity and reflexivity appearing among the presuppositions of every discursive interaction, then its failure cannot but be ascribed, from every speaker's point of view, to one of the following two hypotheses: 1) the interlocutors have not been sincere, or rather they have disappointed one of the conditions of the ideal discursive situation, therefore the disagreement is justifiable in the light of elements such as power, coercion, etc. that usually are excluded, or 2) the interlocutors have not exercised completely their rationality and have in this way prove, although they are the majority, a cognitive limit. In both cases the legitimacy of the decision made, that should have depended on the fact that it was the outcome of a deliberative process, is inevitably compromised. Meanwhile, if you wanted to avoid this unpleasant conclusion, affirming, in a Rawlsian perspective, that the disagreement reflects reasonable comprehensive differences, then in a manner that is not less discouraging we will achieve the conclusion that there is no reason to deliberate because there is no objective common good for us to ascertain. In this way, nonetheless, as Mouffe (2000) has already objected to Rawls, it is the same existence of a political sphere to be called into question.

Although very partial, the considerations made until now allow us to highlight a set of problems that may be, in a wider perspective falling outside the purpose of this paper, extended more in general to a strictly normative approach to public deliberation.

5. THE ARISTOTELIAN FRAMEWORK

On this occasion we will just focus on some aspects, with the objective of making in particular the relation between truth and consensus in the public deliberation emerge. As said above we will resort to the Aristotelian perspective, focused on rhetoric, because we consider that it presents several theoretical advantages. More precisely we claim that the theoretical perspectives, that we are dealing with, lack an appropriate consideration of the specificity of the issues on which we deliberate and the role played by the desiderative component in the deliberative processes, both aspects that, conversely, Aristotle has widely thematised.

Therefore, we will connect to the tradition which tries to recover the role of the Aristotelian rhetoric in public deliberation (Garver, 1994;

Abizadeh, 2002; Yack, 2006; Garsten, 2009), highlighting that in *Rhetoric* Aristotle does not deal with argumentation in general, but with the argumentation specifically related to a deliberation: "its [rhetoric] function is concerned with the sort of things we deliberate" (*Rhet.* 1357a 1, Kennedy, slightly modified); "rhetoric is concerned with making a judgment (people judge what is said in deliberation, and judicial proceedings are also a judgment)" (*Rhet.* 1377b 7-8, Kennedy).
Our first question will be: Which are the consequences of the fact that an argumentation is related to a deliberation?

In order to answer this question we will start from Kock's article (2008). According to Kock, the contemporary theories of argumentation, even those that have recently shown openness towards rhetoric [Johnson (2000), van Eemeren and Houtlosser (1999, 2000, 2001, 2002), and Tindale (1999, 2004)] "define rhetorical argumentation without any reference to a domain of issues" (p. 62). Appealing to the Aristotelian tradition, Kock believes that this domain is that of the "choice of action" or, otherwise called, that of deliberation. Kock thinks that this reference to the domain of the choice is a feature defining rhetorical argumentation that has important consequences. The most relevant is the one heading the article: "choice is not true or false". This means, according to Kock, that "it is a categorical mistake to speak of truth (or probability, for that matter) in regard to a proposal as such" (p. 76). This is a position that, excluding any reference to truth in the rhetorical domain, can be clearly an alternative to the argumentative theories with a normative approach, as Kock himself declares. Nonetheless, we believe that this exclusion of truth is not the most suitable move to face the problems inside that paradigm. If, indeed, on the one side it is true that the choice in itself is not a proposition (and hence may not be judged either true or false), the fact (admitted by Kock only marginally p. 76) that the same choice is supported by propositions of which truth we try to persuade our interlocutor, nonetheless, should not be neglected. This means that rhetorical argumentation has anyway to do with what the interlocutors *believe true*. Hence, we believe that, rather than excluding truth from the rhetorical domain, we have to ask us if, and how, the deliberative purpose has a consequence on the way of understanding the notion of truth and therefore that of consensus.

6. THE TOPIC OF DELIBERATION

This change of mind may represent a starting point also for a reformulation of the opposition between the normative and the descriptive points of view.

In order to do this we will start from the Aristotelian concept of deliberation focusing our attention on the notion of *practical truth*. Indeed, we believe that this notion is useful to weaken the normative constraints and to think in a different way regarding the relationship between rhetorical argumentation and consensus. More precisely, our proposal follows the direction to consider disagreement not as an uneliminable element from an empirical point of view and a marginal element from a theoretical point of view but as an integral feature that a theory of argumentation cannot but consider.

In a necessarily schematic way we may isolate two correlated aspects of the Aristotelian deliberation that seem to us particularly important in comparison with the issues risen above:

a. The topics which we deliberate on are intrinsically questionable;

b. the deliberation necessarily entails both an intellectual component (*logos, dianoia, doxa*) and a desiderative component (*orexis*).

Aristotle restates on several occasions that deliberation concerns objects (*pragmata*) that, not only belong to the domain that can be otherwise than it is, but will necessarily *depend on us*. This means that the field of action is the field of what is *indeterminate*. Let's read the same words of Aristotle:

> no more do we deliberate about the things that involve movement but always happen in the same way (...). But we do not deliberate even about all human affairs. (...) *We deliberate about things that are in our power and can be done* (...). But the things that are brought about by our own efforts, but not always in the same way, are the things about which we deliberate (...). Deliberation is concerned with things that happen in a certain way e for the most part, but in which the event is obscure and with things in which it is indeterminate (EN 1112a 23- b10).

A very similar concept is restated by Aristotle also in *Rhetoric* (1357a 24-27), prior to introducing the features of the preconditions of the enthymeme:

> most of the things about which we make decisions, and into which we inquire, present us with alternative possibilities. For it is about our actions that we deliberate and inquire, and all our actions have a contingent character; hardly any of them

are determined by necessity (Rhet. (1357a 24-27, transl. Roberts).

The main consequence of these features of the field of deliberation is that the subjects of rhetorical argumentation are always intrinsically questionable and this not only for the cognitive limits of the individuals involved but also for the nature itself of the issues involved. In the Aristotelian perspective, indeed, there is a general principle in force according to which every type of object corresponds to an analogous type of knowledge, this being the reason why if the object is indeterminate also the corresponding knowledge will have the same features:

> our discussion will be adequate if it has as much clearness as the subject-matter admits of; for precision is not to be sought for alike in all discussions, any more than in all the products of the crafts. Now fine and just actions, which political science investigates, exhibit much variety and fluctuation (...), and goods also exhibit a similar fluctuation (...). We must be content, than, in speaking of such subjects and with such premises to indicate the truth roughly and in outline, and in speaking about things which are only for the most part to and with premises of the same kind to rich conclusion that are no better (...); for it is the mark of an educated man to look for precision in each class of things just so far as the nature of subject admits (EN 1094 b 12-25).

7. THE PRACTICAL TRUTH

In the Aristotelian perspective, hence, looking for a higher degree of precision than the one the topic allows is a mistake, and it is a mistake that we modern people are used to committing. Being aware of it may become, hence, a first step for the problematizing of the normative perspective and to revise the role of consensus. If, indeed, the topics which we deliberate on have this nature, the possibility that the issue which we discuss admits in principle many solutions – and therefore always entails a margin of disagreement – is not only an empirical fact ascribable to the absence of sincerity or to the cognitive limits of the interlocutors but it is an inner component of deliberative speech due to the nature of the subject.

Continuing to use the Aristotelian terminology, such intrinsically questionable issues, typical of rhetorical argumentation, are the subject of *doxa* and not of *episteme*. Both of them are "powers or habits that we use to judge and we are right or wrong" (DA, 428a 3-4) but, unlike *episteme* that concerns the field of what is necessary (EN 1139b 20-21), *doxa* refers to the field of what is possible, in Aristotelian terms to "what can be otherwise than it is". For the purposes we aim at here, the most interesting feature of the Aristotelian notion of *doxa* is that, on the one hand, it cannot but "appear as true or false" (DA, 428 a18) and, on the other hand, if we look carefully at the way Aristotle analyses *doxa* to discern it from the *imagination* (in Greek *phantasia*) it clearly emerges that truth and falsehood here involved are intrinsically connected to the dimension of *belief* (*pistis*). Actually, here true and false are *believe true* and *believe false*:

> but opinion involves belief (for without belief in what we opine we cannot have an opinion), and in the brutes though we often find imagination (*phantasia*) we never find belief. Further, every opinion (*doxa*) is accompanied by belief, belief by conviction, and conviction by discourse of reason (*logos*), while there are some of the brutes in which we find imagination (*phantasia*), without discourse of reason (DA, 428 a 20-24).

This particular link between *doxa* and the couple truth/false confirms that it would be an error to exclude truth from the rhetorical domain. Indeed, if *doxa* is one of the intellectual components of deliberation we must consider which aspects of truth are relevant with respect to deliberation.

8. DESIRE AND DELIBERATION

In this direction, as said above, the Aristotelian notion of *practical truth* is useful. Firstly, this notion allows us to consider the character of uncertainty of the issues involved (and the consequent problem of the epistemic access to these issues). Furthermore, the notion of practical truth allows to include the other component necessary for deliberation, desire (*orexis*).

Indeed Aristotle believes that no action would be realizable without the intervention of *orexis*, the real driving force of animals (not only humans). Aristotle says that the particularity of human choice consists in the necessary co-presence of desire (*orexis*) and reasoning (*dianoia*):

> what affirmation and negation are in thinking (*dianoia*), pursuit and avoidance are in desire (*orexis*); so that since moral excellence is a state concerned with choice, and choice (*proairesis*) is deliberate desire (*orexis bouletike*), therefore

> both the reasoning must be true (*logos alethe*) *and the desire right* (orexis orthe), *if the choice is to be good, and the latter must pursue just what the former asserts. Now this kind of intellect* (*dianoia*) *and of truth is practical* (*aletheia praktike*); *of the intellect which is contemplative not practical nor productive, the good and the bad state are truth and falsity (for this is the function of everything intellectual); while of the part which is practical and intellectual the good state is truth in agreement with right desire* (*aletheia homologos echousa te orexei te orthe*) (EN 1139 21-30).

The practical truth is hence "truth in agreement with right desire" and it is the type of truth involved in deliberative speech. Of course, also Aristotle thinks that it is a problematic agreement which is hard to be realized also because in Aristotelian anthropology there is no an idea of inner self as something unitary and transparent. Nonetheless, what we would like to underline here is the necessity, for every speech aiming at getting someone to make a decision, to always keep in mind also desire.

The well-known triad of the Aristotelian *pisteis* (proof), *ethos*, *pathos* and *logos*, represents the attempt to keep together the intellectual and the desiderative components of deliberation. Indeed, *ethos* is the *pistis* based on the character of the speaker, *pathos* that based on the emotion of the audience and *logos* is the *pistis* based on the speech itself. Although each of these *pisteis* include in a certain degree both the intellectual (*dianoia*) and the desiderative (*orexis*) component, *ethos* and *pathos* mainly involve the latter instead *logos* mainly refers to the former. In any case, according to Aristotle, the speaker, in order to persuade his audience, must take into account all these *pisteis* together. Actually, this is the only way to enable the audience to make a choice.

9. CONCLUSIONS

In order to summarize synthetically the theoretical suggestion that we are going to propose, it may be useful to go back to the Aristotelian distinction between dialectics and rhetoric (Bentley, 2004). While, as Aristotle underlines at the beginning of *Rhetoric*, between the two disciplines there is a relationship of tight complementarity, nevertheless, there is still a substantial difference between them. Dialectics is a sort of philosophical debate, not by chance with evident common points with Socratic cross-examination, which better suits testing the logic consistency of a specific thesis than ruling a public discussion from which a decision should be made. In the second case we are, indeed, in a situation where uncertainty appears to be the constitutive dimension not only as regards the issues discussed but also the motives of the individuals that are interacting:

> it is a world in which there is uncertainty, tentativeness and ambiguous relations among persons, issues and meanings [...] Uncertainty surrounds the motives and the level of knowledge that a persuasive speaker possesses (Bentley, 2004, p. 131).

Then, in this field, the recourse to rhetoric, as Aristotle had clearly seen, is not something to accept despite everything, but the only real alternative. Therefore, we believe that those models, with a perspective that we could define dialectic, that think they are able to identify a normative dimension of public deliberation, substantially fail in their purpose because they ignore the specificity of both political action and moral agents that, with their unsolvable blending of reason and desire, are employed to make decisions.

REFERENCES

Abizadeh, A. (2002). The Passions of the Wise: *Phronesis*, Rhetoric, and Aristotle's Passionate Practical Deliberation. *Review of Metaphysics*, 56, 267-296.

Bentley, R. (2004). Rhetorical Democracy. In B. Fontana, C.J. Niederman & G. Remer (Eds.), *Talking Democracy* (pp. 115-134). University Park: The Pennsylvania State University Press.

Cantù, P., & Testa, I. (2006). *Teorie dell'argomentazione. Un'introduzione alle logiche del dialogo*. Milano: Bruno Mondadori.

Cohen, J. (1997). Deliberation and Democratic Legitimacy. In J. Bohman & W. Rehg (Eds.), *Deliberative Democracy: Essays on Reason and Politics* (pp. 67-92). Cambridge: The MIT Press.

Elstub, S. (2010). The Third Generation of Deliberative Democracy. *Political Studies Review*, 8(3), 291-307.

van Eemeren, F.H., & Houtlosser, P. (1999). Strategic manoeuvring in argumentative discourse. *Discourse Studies*, 1, 479-497.

van Eemeren, F.H., & Houtlosser, P. (2000). Rhetorical analysis within a pragma-dialectical framework: The case of R. J. Reynolds. *Argumentation*, 14, 293-305.

van Eemeren, F.H., & Houtlosser, P. (2001). Managing disagreement: Rhetorical analysis within a dialectical framework. *Argumentation and Advocacy*, 37, 150-157.

van Eemeren, F.H., & Houtlosser, P. (2002). Strategic manoeuvring: Maintaining a delicate balance. In F.H. van Eemeren & P. Houtlosser (Eds.), *The warp and woof of argumentation analysis (Argumentation Library 6)* (pp. 131-160). Dordrecht: Kluwer Academic Publishers.

Forchtner, B., & Tominc, A. (2012). Critique and argumentation. On the relation between the discourse historical approach and pragma-dialectics. *Journal of Language and Politics, 11*(1), 31-50.

Garsten, B. (2009). *Saving Persuasion: A Defense of Rhetoric and Judgment.* Cambridge, MA: Harvard University Press.

Garver, E. (1994). *Aristotle's Rhetoric: An Art of Character.* Chicago: Chicago University Press.

Habermas, J. (1996). *Between Facts and Norms: Contribution to a Discourse Theory of Law and Democracy.* Cambridge, MA: The MIT Press.

Johnson, R.H. (2000). *Manifest rationality: A pragmatic theory of argument*, Mahwah, NJ: Lawrence Erlbaum Associates.

Kock, C. (2009). Choice is not true or false: the domain of rhetorical argumentation, *Argumentation, 23*, 61-80.

Lumer, C. (2010). Pragma-Dialectics and the Function of Argumentation, *Argumentation, 24*, 41-69.

Markovits, E. (2008). *The Politics of Sincerity. Plato, Frank Speech, and Democratic Judgment.* University Park: The Pennsylvania State University Press.

Mouffe, C. (2000). *The Democratic Paradox.* London: Verso.

O'Neill, J. (2002). The Rhetoric of Deliberation: Some Problems in Kantian Theories of Deliberative Democracy. *Res Publica, 8*, 249-268.

Rehg, W. (2005). Assessing the Cogency of Arguments: Three Kinds of Merits. *Informal Logic, 25*(2), 95-115.

Shiffman, G. (2004). Deliberation versus Decision: Platonism in Contemporary Democratic Theory. In B. Fontana, C.J. Niederman & G. Remer (Eds.), *Talking Democracy* (pp. 87-113). University Park: The Pennsylvania State University Press.

Steiner, J. (2008). Truthfulness (Wahrhaftigkeit) in the deliberative model of democracy.
http://www.yale.edu/polisci/conferences/epistemic_democracy/jSteiner.pdf.

Steiner, J. (2012). *The Foundations of Deliberative Democracy.* Cambridge: Cambridge University Press.

Tindale, C. (1999). *Acts of arguing: A rhetorical model of argument.* Albany, NY: State University of New York Press.

Tindale, C. (2004). *Rhetorical argumentation: Principles of theory and practice.* Thousand Oaks, CA: Sage.

Vega, L., & Olmos, P. (2007). Deliberation: A Paradigm in the Arena of Public Argument. *OSSA Conference Archive*, Paper 147, 1-11.

Yack, B. (2006). Rhetoric and Public Reasoning. An Aristotelian Understanding of Political Deliberation. *Political Theory, 34*(4), 417-438.

Commentary on Di Piazza, Piazza and Serra's Does Public Deliberation Really Need Normative Constraints?

AMNON KNOLL
School of Philosophy, Tel Aviv University, Israel
amnon.knoll@gmail.com

1. INTRODUCTION

Disagreements and conflicts pervade our social and political life. Various theories propose normative reasoned argumentation and deliberation as the proper approach to overcome such disagreements, so as to lead to a better collective reasoned action. Professors Di Piazza, Piazza and Serra's paper raises, in a comprehensive, rich and intriguing manner, important issues that pertain to the core of argumentation and deliberative democracy normative theories and their applicability to political and public deliberation.

In a nutshell, the paper claims that normative dialectics based on structured process, which includes exchange of reasoned arguments between contending sides, is structurally not applicable to the political arena, and proposes instead the use of Aristotle's rhetorical political deliberation. The similarities between argumentation theory (esp. Pragma-Dialectics- PD) and deliberative democracy-DD (esp. Habermas and Cohen) and their problematic reliance on the normative aim of reaching consensus in the public sphere, are well presented and supported in the first part of the paper. I agree with the paper's correct criticism and its proper comparison between the two disciplines; a comparison that is quite scarce in the current argumentation literature and is very much needed. I also think that the core specifications of Aristotle's rhetoric referring to political deliberation, presented systematically in the second part of the paper, are quite relevant to contemporary situations and to the core challenges of the needed integration of dialectics and rhetoric in the public and political spheres.

Together with that, the paper's overall attack on normative approaches to argumentation and deliberation and the authors' proposal for adopting Aristotle's ("non-normative") rhetorical deliberation as a full replacement for normative dialectics are lacking in several perspectives. In the following comments, I shall briefly lay out why although agreeing with a great part of the paper's analysis, there are still some inter-related gaps that might challenge the paper's too all-encompassing conclusions in relation to the need for normative dialectics.

2. NON-CONSENSUS AIMED NORMATIVE APPROACHES

The paper implicitly assumes that all normative deliberative democracy (DD) and argumentation theories are based on the aim of consensus, an assumption that holds a substantial part of the paper detailed (correct) criticism. Nevertheless, although mentioned in a note that there are other DD theories, there is no concrete reference to other relevant DD theories (developed mostly from the 1990s onwards) that are not consensus based (see review in Bohman, 1998; Chambers, 2003), and relate specifically to some of the challenges that the paper raises.

It is important to emphasize that the criticism that exists within the deliberative democracy field on Habermas and Cohen's normative approaches, does not forsake the normative needs for changing the current political discourse and offers other epistemic, moral and prudential normative aims for conducting reasoned deliberation. In addition, some of these normative approaches also relate directly to the issues of integrating rhetoric and dialectics (see Chambers, 2009; Dryzek, 2010a, 2010b). Although also mentioned briefly, these approaches are not explored further and are not compared with the paper's thesis.

This is not to say, that the claims that the paper raises about the applicability of dialectics in the political sphere are fully answered in the DD literature. On the contrary, no definite agreed answer is given in regards to the proper integration between different types of dialectics and rhetoric needed and applicable in the public and political sphere in general, and the place of Aristotle's type of rhetorical political deliberation amongst them in particular. Nevertheless, the proper criticism on consensus-aimed approaches is not necessarily applied - without at least substantial refinements - to non-consensus aimed normative approaches.

The same point applies also to Argumentation. The paper focuses mainly on the Pragma-Dialectics (PD) approach. While PD is prominent and central to the Argumentation field, this selection implies that all argumentation approaches are consensus based. PD, with its critical discussion argumentation process, is not necessarily the only option for a normative argumentation process theory. Although the literature on non-consensus argumentation processes is not yet well enough developed as in DD, it is still explicitly and implicitly mentioned in the argumentation literature and could be developed further. Moreover, some of the normative critical aspects of PD's critical discussion process might still be relevant to political deliberation. However, I will not be able to elaborate on that in this short commentary.

In summary, the challenge that I am trying to raise is whether the paper's general critique on normative dialectics approaches at large should be limited to specific frameworks of consensus based normative dialectics.

3. NORMATIVE VS. NON-NORMATIVE APPROACHES

The paper does not elaborate on its own distinction between normative and non-normative approaches. Although attending to this loaded and complicated question could have diverted the paper from its central path, the inattention to at least the main principles of normativity in relation to the attack and criticism on normative approaches, is lacking. Particularly due to the fact that as opposed to the normative approaches, the paper introduces Aristotle's rhetorical approach as a counter, non-normative approach or at least as missing a significant normative dimension of public deliberation.

The paper, in its closing lines, concludes:

> (I)n this field, the recourse to rhetoric, as Aristotle had clearly seen, is not something to accept despite everything, but the only real alternative. Therefore, we believe that those models, **with a perspective that we could define dialectics, that think they are able to identify a normative dimension of public deliberation**, substantially fail in their purpose. (my emphasis)

In light of these last lines and since clearly Aristotle's approach to political deliberation is not strictly descriptive, what is then the sort of dichotomous significance, in relation to normativity which separates between the criticized approaches and Aristotle's approach?

Moreover, there is a missing clarity in regards to whether or not Aristotle's demand from public deliberation is in and of itself normative (see the following section). It may not be as strong as what he himself demands from dialectics and certainly not as strong as Habermas, Cohen and PD's critical discussion demands, but it nevertheless requires normative demands from those who take part in the process.

In addition, the paper rightly presents the fine line and the needed mixture of reasons, emotions and desires in public deliberation.

This raises the question whether current political propaganda and catchwords exchanges, group polarizations and cognitive and affectual biases could be at least normatively excluded from appropriate public discourse (and in what normative tools besides direct argumentation)?

All that might implies that the claim against normativity should be more limited and more specific. I would like to suggest then, that the basic requirement of normative dialectics and direct reasonable argumentation is based on a facing-each-other process of mutual reference to substantial components of a disagreement, which heavily relies on arguments and attentive consideration of each other's statements (see Dascal & Knoll, 2011). Could this be fully replaced by Aristotle's political deliberation?

4. ARISTOTLE'S POLITICAL DELIBERATION

The paper presents the Aristotelian distinction between dialectics and rhetoric and emphasizes on the one hand that "as Aristotle underlines at the beginning of *Rhetoric*, between the two disciplines there is a relationship of tight complementarity". On the other hand, dialectics according to the paper is a sort of:

> (P)hilosophical debate, not by chance with evident common points with Socratic cross-examination, which better suits testing the logic consistency of a specific thesis than ruling a public discussion from which a decision should be made.

The distinguishing factor is the inherent uncertainty existed in the public discussion composed of both the issues and the motives of the discussants.

The issues of Aristotle's rhetoric vs. dialectics interpretations and their differences from contemporary rhetoric and dialectics are of course very complex and very wide open. However, I would like to point out several possible challenges.

In very general terms, rhetorical political deliberation, according to Aristotle, is focused on oratory to relatively large assemblies in order for the audience to make decisions in specific cases and under certain conditions. Deliberation is not about ends but is rather restricted to the best measures to achieve ends. The minimal normative requirement is that the audience, who votes on the decision, would be exposed to all the different political speeches and opinions. There are no rules for relevance, since according to Aristotle: the audience would immediately observe non-relevant issues. In addition, the assumption is that particularistic interests (e.g. interests of people closer to the boarder when a decision on getting into war is at stake) are excluded.

Under these circumstances and conditions, which at least in principle might be also relevant to contemporary political decision-making, the paper's appropriate analysis of the mixture of reason, emotion, desire and the uncertainty of the issues and motives, applies. In this context of political deliberation it might be that there are no additional concrete rules to how to do the balancing of the pros and cons in all the options offered in order to make the right decisions and certainly there is no structured dialectics process that would lead to consensus.

Nevertheless, even in situations of this type, there is an important place for dialectics - in the sense of facing each other and referring to the opposing arguments involved and even to the way the balancing should be made. This would certainly help the audience (and the speakers as well) in order to set up their minds, not detached from their core identity and their core desires and emotions but in better correlation with the different arguments and reasons involved.

Of course getting into a substantial dialectic process (or debate) in front of the assemblies might not be feasible or efficient, although even in ancient time there were beyond assemblies' potential of face-to-face deliberation. As Chambers put it in reference to Plato:

> By going back to Plato we will see that the strongest objection to rhetoric is not that it appeals to passion over reason. The strongest objection to rhetoric is that it is monological rather than dialogical. (Chambers, 2009, p. 324)

The last point that should be taken into consideration is the substantial changes - on the level of the heterogeneity of political communities, the possibility of agreed upon ends, the role of the public sphere, the place of assemblies and the complexity of the political systems - from ancient Greece onwards till our current networked pluralistic societies. Even without getting into in-depth analysis of these changes, it seems clear that these changes decrease the potential positive role of Aristotle's political deliberation without a substantial direct argumentation and deliberation made in the public sphere.

5. CONCLUSION

In this commentary, I have tried to focus on some of the challenges that the paper raises, in order to question its semi-dichotomist distinction between normative dialectics and Aristotle's rhetorical political deliberation.

To a certain extent, some of the correct criticism of the paper on current consensus normative approaches, together with Aristotle's own (implicit and explicit) assumptions and the very problematic status of our current public and political discourse, promote more strongly the need for a (different) normativity in argumentative deliberation in the public and political spheres.

Aristotle's political deliberation with the right modifications could have an important role in the needed changes, starting with a normative requirement of sincere exposure of decision makers to all potential standpoints. However, in order to improve our troublesome democratic life, aiming towards more desirable reasonable dissensus, this exposure should be complemented with substantial dialectical deliberation (using also rhetoric) done in the public sphere. For this very challenging but potentially, gradually achievable goal – appropriate normative dialectics approaches and normative constraints on public deliberation are still in need.

REFERENCES

Bohman, J. (1998). Survey article: The coming of age of deliberative democracy. *Journal of Political Philosophy*, 6(4), 400-425.
Chambers, S. (2003). Deliberative democratic theory. *Annual Review of Political Science*, 6(1), 307-326.
Chambers, S. (2009). Rhetoric and the public sphere: has deliberative democracy abandoned mass democracy? *Political Theory*, 37(3), 323-350.
Dascal, M., & Knoll, A. (2011). 'Cognitive systemic dichotomization' in public argumentation and controversies. In F. Zenker (Ed.). *Argumentation: Cognition and Community. Proceedings of the 9th International Conference of the Ontario Society for the Study of Argumentation (OSSA).* (pp. 1-35). Windsor, ON (CD ROM).
Dryzek, J. S. (2010a). Rhetoric in democracy: A systemic appreciation. *Political Theory*, 38(3), 319-339.
Dryzek, J. S. (2010b). *Foundations and Frontiers of Deliberative Governance.* Oxford: Oxford University Press.

27

Argumentatively Evil Storytelling

GILBERT PLUMER
Law School Admission Council (retired), USA
plumerge@gmail.com

What can make storytelling "evil" in the sense that the storytelling leads to accepting a view for no good reason, thus allowing ill-reasoned action? I mean the storytelling can be argumentatively evil, not trivially that (e.g.) the overt speeches of characters can include bad arguments. My thesis is that for fictional narratives, the shorter the narrative, the greater the potential for argumentative evil. In other argumentative contexts, length generally appears to make no comparable difference.

KEYWORDS: advertisements, anecdotal arguments, believability, fables, narrative argument, parables, thought experiments, transcendental argument, truth in fiction

1. INTRODUCTION

What can make storytelling "evil" in the sense that the storytelling leads to accepting a view or message for no good reason, thus allowing ill-reasoned action? The general idea that storytelling can have pernicious effects on practical reasoning goes back, of course, at least as far as Plato. My point is that the storytelling can be argumentatively evil, not trivially that (e.g.) the overt speeches of characters can include bad arguments. The storytelling can be argumentatively evil in that it purveys false premises, or purveys reasoning that is formally or informally fallacious. The main thesis of this paper is that there is an aspect involving the very form of fictional narratives, namely, their length, that can distinctively allow a narrative to be evil in the sense indicated. As a rule, the shorter the fictional narrative, the greater the potential for argumentative evil. Here, the notion of length is to be understood such that it is generally a proxy for more abstract features such as how complex and nuanced the piece is. In argumentative contexts other than those involving fictional narrative, length generally appears to make no comparable difference. This feature would put fictional narrative arguments in a special class beyond what is determined by obvious features, such as the definitional fact that they in some way(s) collapse two of the four traditional types of discourse: exposition, description, narration, and argument. The nonobvious features that distinguish this class have been a source of puzzlement and inquiry (e.g., Schultz, 1979; Plumer, 2011; Govier & Ayers, 2012).

2. SHORT FICTIONS

If you place the various major kinds of fictional narratives on a length continuum, on one end you get advertisements and jokes that include a brief fabricated story, as well as short fables and parables; novels lie at the other end, with short stories, films, and plays somewhere more toward the middle. Storytelling poems can lie anywhere on the continuum, but they seem assimilable to other kinds of fictional narratives with respect to argumentative potential. However, narrative "thought experiments" appear to be in a class by themselves, as we will see.

A piece anywhere on the continuum is a "story" in the minimalist sense of being a perspectival or selective depiction of at least two temporally-related events in a further nonlogical (e.g., causal) relationship (adapted from Lamarque, 2004; cf., e.g., Walton, 2012, pp. 191 & 199). A piece anywhere on the continuum is fictional in that at least some of what is depicted is not supposed to be true. A piece anywhere on the continuum can have affective and persuasive force. So, what distinguishes pieces on the short end of the stick, so to speak (other than their word count)?

One feature that such ads, jokes, fables, and parables have in common is that they have a point or message, seemingly by definition, and so are in that (possibly weak) way argumentative. Indeed, it is hard to see what their *raison d'etre* would be without a point or message, in contrast to longer fictional narratives, which instead typically have substantial plot and character development, and fine descriptions ("word paintings") of the natural or artificial world. Ads try to influence you to buy or do something. To "get" a joke is to grasp its point. There is a bit of contention about this regarding fables and parables, but it only concerns whether the point or message has to be implicit. For example, Govier & Ayers appear to be inclined to accept the view that in the western tradition, "a fable comes with 'the moral of the story' stated right there, whereas...a parable must have an *implicit* message" (2012, p. 173). In contrast, Hunt maintains, "both historically and conceptually, that explicitly stating the point of a story is not necessary to a story's being a fable" (2009, p. 381). However, it does seem that longer fictional narratives need not have a point, whether implicit or explicit—consider the recent U.S. television series *Lost* and perhaps James Joyce's *Ulysses*—and even that they are less *literary* if they do have a point or moralize (cf. Hunt, p. 382). "A novel or theater piece need not reach a conclusion or even seem to approach one" (Velleman, 2003, p. 10). Currie (2010, pp. 34-35) offers a kind of explanation of this difference. He says that if you are distinguishing narrative from something on the order of mathematical physics, no doubt parables and the like count as narratives. But if you have in mind something on the order of a short story or novel, you might distinguish narratives from parables and the like, because the latter "have generalizing tendencies that do not fit well with the particularizing, sequential aspirations of narrative."

Before developing some of these ideas further, let us put some illustrations on the table. Here is an example of an ad (from television):

> **Copy and gist from:** *Think small. The story of those Vehicle ads*, by Frank Rowsome, Jr., 1970, pg. 116-7. The company name is changed here. Visual description more or less by: Shazam (Suzanne).
>
> ---
>
> [Dark snowy early morning in country, view is of outdoors through the front windshield of a car. The car's headlights illuminate the falling snow, and the drifts of it, along the untracked, winding, uphill way, and you can see, in passing, snow laden pine and fir branches, bent under the weight of the snow. The only sound throughout: the purring of the car's engine. This trip takes some time.]
>
> [Then the headlights hit and pass a...building, the driver turning the car by it. The car gets parked: the headlights are turned off. A big door of the building soon opens and a powerful snowplow rolls past our view as the ANNOUNCER begins.]
>
> ANNOUNCER
> Have you ever wondered how the man who drives the snowplow drives *to* the snowplow? This one drives a Vehicle. So you can stop wondering.
>
> ---
>
> Note: This commercial was so popular in Florida and Southern California that some stations played it over and over again due to audience requests.

(*http://reocities.com/tvtranscripts/comm/commcar.htm*; accessed on 9 Feb. 2015). Here is an example of a fable (from Aesop):

> The Eagle and the Arrow
> An Eagle was soaring through the air when suddenly it heard the whizz of an Arrow, and felt itself wounded to death. Slowly it fluttered down to the earth, with its life-blood pouring out of it. Looking down upon the Arrow with which it had been pierced, it found that the shaft of the Arrow had been feathered with one of its own plumes. "Alas!" it cried, as it died, "we often give our enemies the means for our own destruction."

(*http://www.aesopfables.com/cgi/aesop1.cgi?1&TheEagleandtheArrow2*; accessed on 15 Feb. 2015). And finally, a Ramakrishna parable:

> WHAT YOU ARE AFTER, IS WITHIN YOURSELF
> A MAN wanted a smoke. He went to a neighbour's house to light his charcoal. It was the dead of night and the household was asleep. After he had knocked a great deal, someone came down to open the door. At sight of the man he asked, "Hello! What's the matter?" The man replied, "Can't you guess? You know how fond I am of smoking. I have come here to light my charcoal." The neighbour said, "Ha! Ha! You are a fine man indeed! You took the trouble to come and do all this knocking at the door! Why, you have a lighted lantern in your hand!"
> What a man seeks is very near him. Still he wanders about from place to place.

(p. 350 of a PDF book, Tales and Parables of Sri Ramakrishna, at: *http://www.archive.org/details/TalesAndParablesOfSriRamakrishna*; accessed on 9 Feb. 2015.)

Velleman (2003) develops the minimalist sense of a story or narrative mentioned above in such a way that it leads to one explanation of how storytelling can be argumentatively evil. As a necessary condition for being a story, he adds: "reliably producing in the audience some emotional resolution" (p. 7, cf. 17). Some examples he gives of such resolution are anxiety relieved, hope dashed, and laughter (for jokes) (p. 7). He uses this theoretical addition to good effect (pp. 3-4) to explain how Aristotle's case of Mitys at Argos can be regarded as a story, even though the relationship between the events is not causal. Mitys was murdered. Later, while attending a "public spectacle," Mitys' murderer was killed when a statue of Mitys happened to fall down on him (*Poetics* 9.1452). Aristotle struggled to elucidate how this case could be a story; Velleman proposes that it is notably because "the sequence of events completes an emotional cadence in the audience" of "indignation gratified." Velleman argues that the trouble is that through experiencing a story's emotional resolution, events become understandable to an audience not through assimilation to "familiar patterns of *how things happen*, but rather to familiar patterns of *how things feel*" (p. 19). The latter, subjective understanding can easily give us a false sense of objective understanding, so skepticism about what a story claims or about its message might be mistakenly dispelled. Hence, "telling a story is often a means to being believed for no good reason" (p. 22), thereby introducing argumentative evil.

Velleman's theory appears to apply nicely to the examples quoted above. Certainly, at least curiosity satisfied plays a role in the vivid "Vehicle" commercial by the time the announcer's voiceover is reached and suggests a generalization. The Aesop fable closes with an explanation/generalization that is a surprise ending to a life-and-death tale. The Ramakrishna parable involves a breathtaking generalization leap, as well as some humor. However, it is not at all clear that the theory applies to longer literary genres where the piece does not have a succinct point, message, moral, or conclusion—for these are what pack the punch or drive the "emotional resolution." One may of course engage emotionally with the meaning of a piece of substantial literature such as a play or novel, but to the extent that its meaning is complex or nuanced, it is unlikely that there will be any definitive—let alone global—emotional resolution (hence there may be a sequel that simply continues the story). About such genres Velleman says "they tend to be described as genres of narrative by extension" (p. 17; cf. 10), but it is more plausible to hold that as a theory of all narrative, his theory overreaches. If anything, it is the shortest genres that are narratives by extension, as Currie suggests above. So, I think we see here one way in which shorter narratives have a greater potential for argumentative evil than longer ones.

In his discussion, Velleman does not distinguish between fictional and nonfictional narration, but on his own theory you would think that the potential for argumentative evil is less for nonfictional narration since by definition it aims at veracity or telling *how things actually happened*. The proper purpose of any nonfictional narrative argument is to be sound in the respect of having true premises, in contrast to the generalizing ad, fable, and parable fictions quoted above, for example.

3. BELIEVABILITY

On the continuum of fictional narrative, if you move in the direction from ads to novels, an interesting feature seems to be that—not immediately but somewhere fairly early on—*believability* becomes a central criterion of assessment. Is the piece successful "make-believe"? This question hardly pertains to shorter fictional narratives; it is not really the "game" in play or an appropriate standard to apply. Rather, such narratives aim at being charming or arresting, and especially at being moving through the emotional resolution packed by their point or message. But whatever they aim at, it seems that the question of believability must be bracketed or suspended for shorter fictional narratives essentially because there is *too little room* provided in such a piece to adequately test out the hypothesis that it is believable.

I don't mean "believability" in the sense that artifices of magic such as talking animals or objects, as are common in fables, would preclude it. For short genres, these are established conventions of expedience and other purposes (see Olmos, 2014); there is no presumption that the author should even acknowledge deviations from accepted science, let alone try to explain or invent any underlying physics. In contrast, there is this presumption for extended science fiction or fantasy narratives, and if they do not conform to our most fundamental shared assumptions about physical reality, their believability in the intended sense is indeed called into question (a possible example is H. P. Lovecraft's novella *The Call of Cthulhu*). More generally, believability seems to be determined mostly by what can be called the "internal" and "external" coherence of the event complex of an extended fictional narrative. I take Schultz (1979, p. 233) to be succinctly explicating internal coherence where he says: "the events must be *motivated* in terms of one another...either one event is a causal (or otherwise probable) consequence of another; or some event's happening provides a character with a reason or motive for making another event happen" (cf., e.g., Cebik, 1971, p. 16). The narrative is not believable if in it things keep happening for no apparent reason or in a way that is inadequately connected with the other events in the narrative. Certainly, this applies to some degree to William Burroughs' *Naked Lunch*, for example. Such "real" connections of efficient, final, and material causes (using Aristotle's terminology), and any probabilistic counterparts, are required. You could not construct a coherent novel or play from a random series of events or even from bolts of cosmic justice, like the *one* we saw in the pithy Mitys story.

But even if the events of a narrative are fully connected, the narrative may still not be believable because those connections do not cohere well with our widely shared basic assumptions about how human psychology and society not only actually, but necessarily work. This is the main component of external coherence. The believability of an extended fictional narrative requires that its plot, characters, and fine description be developed in ways that generally conform to our fundamental shared assumptions about human nature (Max Beerbohm's *Zuleika Dobson* seems to fully recognize this requirement in its intentional violation of it[1]), and secondarily, about physical nature (as noted).

Of course, the believability of an extended fictional narrative does not involve believing that its event complex *is* true; rather, it involves believing that the event complex *could* have been true in a strong sense of "could"—much stronger, for example, than that of mere logical possibility. The possibilities that the narrative evokes, if it is to be believable, must be grounded in "real" event relations and in basic perceived facts of human nature. And as the narrative progresses in developing a theme(s), the possibilities evoked must be salient in that they are thematically relevant. But the shorter the fictional narrative, the closer the possibilities come to being mere logical possibilities. In the shortest, there is almost no plot or character development, or fine description. So there is no way to tell if the narrative is significantly internally or externally coherent.[2]

It seems that generally, believability is experienced by the audience as a simple, unanalyzed datum or measure of the narrative, continuously updated as the audience progresses through the work and imaginatively engages with it. And, as Aristotle said about judging the

---

[1]Consider this description of the novel: "…an ironic fantasy of Oxford undergraduate life a 100 or so years ago. The characters' speech and motives are absurd in about equal measure, but one would be missing the point to hold this against the work. For the author is plainly not seeking psychological verisimilitude…The interest of the work is essentially that of a *tour de force*: how long can the author retain our interest while so consciously eschewing psychological plausibility?" (Currie, 2012, p. 29 & n. 7).

[2]Being believable does not mean that something is on its way to being believed, for that path is never taken for something you know to be fiction. With respect to fictional stories, internal and external coherence constitute more or less all there is to believability; with respect to nonfictional stories, belief may be the only thing there is to believability (possibility is logically implied by actuality). Hence, it is problematic to analyze "believability" ("credibility," "plausibility") indifferently as it pertains to these two story domains, as do Fisher (1987) and Olmos (2013; 2015).

happiness of a person, you do not know for sure about believability until you reach the narrative's end.

Just as we can always ask about an extended fictional narrative—is it successful "make-believe"?—it seems we can ask about any believable plot/character development complex—what principles or generalizations would have to operate in the real world (of human psychology, action, and society), as we conceive it, in order for the fictional complex to be believable? With this question, a *transcendental* argument scheme is generated that is "ambitious" (vs. "modest"—see Stern, 2007) if we presume that our fundamental shared conceptions of human nature are generally true:

(1) This story is believable.
(2) This story is believable only if such and such principles operate in the real world (of human psychology, action, and society).
(3) Therefore, such and such principles operate in the real world.

I have argued elsewhere (2015), on both philosophical and empirical grounds, that our fundamental shared conceptions of human nature are generally true. Let me just summarize the philosophical reasons here. No doubt in certain cases I may find a work of extended fictional narration believable, whereas you do not. But it seems that there is no wholesale relativity of believability because there is such a thing as human nature, which we all share and to which we have significant introspective or "privileged" access, or at least psychological attunement.[3] The believable narrative taps into and relies on these facts, bringing operant principles to the fore—which allows it to function as a perfectly effective psychological "trigger" (cf. Gaiman, 2015, p. xiii). If this general idea were not true, then it would be pretty inexplicable that there is widespread agreement about which novels are good novels, for example. Being believable is a central necessary condition for an extended fictional narrative to be good. So in the transcendental argument, the leap from the inner to outer worlds is limited and facilitated. The leap is from our psychological experience of believability of the narrative to the real world of human psychology, action, and society—which is the primary subject matter of all extended fictional

---

[3] A recent influential article on introspection (Schwitzgebel, 2008) poses little threat to my points here concerning human nature and its operant principles, because the focus of the article is on the untrustworthiness of introspection of immediate conscious experience. Differences among readers in the perceived believability of a novel may be largely attributable to relatively extraneous factors, such as the setting of the novel. For example, if I could get past the fantastic details of Tolkein's trilogy, I think I could better appreciate these novels as implicating truths of human nature.

narratives. This subject matter is basically human nature, I take it. The inner and outer worlds of the narrative argument are significantly the same; it is not as if the worlds are distinct as, for example, thought and a brain in a vat, as in Putnam's memorable transcendental argument (1981, Ch. 1). And, as Nagel (1979, Ch. 12) forcefully argued, because after all we are human, we know *what it is like* to be human in a way we do not know *what it is like* to have a different nature, such as a bat's (and perceive the world primarily through echolocation, be capable of flying, etc.). Such philosophical considerations indicate that the principles evoked in the narrative argument resonate in believability largely because they are *true* of human nature.

As we've seen, storytelling ads and jokes, and short fables and parables, may be charming or arresting. But this affective appeal especially allows them also to be seductive and possibly misleading since they have a point or message. One can be seduced into accepting the message for no good reason and acting on it, for instance, buying a "Vehicle" even though you live in Florida. My key point is that such perniciousness does not apply to longer fictional narratives that are believable, insofar as believability implicates truths of human nature, even though longer fictional narratives in some ways have as much or more affective appeal. Only fictional narratives that are believable exhibit (indirectly, and as wholes) the distinctive narrative argument form outlined above. This form is not only valid but is in a certain way probabilistically sound. (1)-(3) constitute a schematic meta-level representation of the (transcendental) argument of a believable story, which, at the object level, is only indirectly expressed by the story. At the object level, given that premise (1) is true and that our fundamental shared conceptions of human nature are generally true, the conclusion (3) is unlikely to be mistaken. However, at the interpretive meta-level, perhaps especially where the literary critic attempts to directly state which specific truths of human nature are implicated (i.e., flesh out premise (2)), no doubt errors may be committed. Nevertheless, this interpretive enterprise is worth pursuing, for it articulates, insofar as it is successful, the narrative's contribution to human knowledge. Through the transcendental argument and the "work" of progressing through the narrative, true assumptions or conceptions held by the audience about human nature become *justified* true beliefs.

Thus, as compared to shorter fictional narratives, longer ones that are believable have less potential for argumentative evil in the respect that their believability generates a good transcendental argument. This is not to deny that there are other respects in which extended fictional narratives, whether believable or not, may be evil. Some of these respects have nothing to do with argument, and some arise because the overt speeches of characters include bad arguments—which, recall, is not a concern of this paper. Consider the novels of The Marquis de Sade or Ayn Rand. But it is not my intention that my believability *cum* transcendental-argument theory be immune to possible counterexample. Take, for instance, the 1940 Nazi propaganda film *Jud Süß* or even perhaps Sinclair Lewis' *Babbitt*. Both succeeded in turning large numbers against certain classes of people (Jews and small-town businessmen, respectively). A case strong enough to raise questions can be made that these works are believable. Yet there are possible answers. One is that their objectionable stances themselves (anti Semitism, anti_small-town businessmen) are far too specific to be fundamental principles of human nature, and another is that there is no guarantee that fundamental principles of human nature will be pretty (e.g., if we have an innate proclivity to violence). My theory (if correct) would show that a believable narrative must be right about most of the principles it depends on, but it does not preclude that the aim of the narrative nevertheless could be to lead people to false and harmful conclusions about whole classes of people, albeit conclusions that do not rise to the level of principles. Perhaps a more intractable kind of criticism would be constituted by an accumulation of putative empirical counterexamples to my subsidiary thesis that our basic shared conceptions of human nature are generally true, though I have addressed this issue elsewhere (2015).

In any case, some of the preceding ideas have been delightfully, although less formally, expressed by Doody (2009, pp. 155-157). It is worth giving her some room:

> Fiction knows that fable packs the punch, has the charge it wants. At the same time, the prose fiction novel knows that the fable lacks what the Novel always wants to offer—full characterization and length... "This is all you need to know, for my point," says the philosopher, brusquely finishing his fable so he can get on with the job. "Wait, wait," cries the Novel. "This is the job! I want to know more and I don't care so much about your point. For your point might not be true if we knew more. Let us test it by *amplifcatio*."...No parable is safe...We know the story of the Prodigal Son... "But," says the Novel, "that's a great story, but I want to know more. What was the father like? Was there something about him that made the second son want to leave? Suppose the first son's jealousy had existed for a long time, not just at the homecoming. Suppose that son had turned the father against his brother, so the brother lit out and sought affection among the prostitutes?" And so Henry Fielding writes the whole story anew in *Tom*

Jones, the story of the wronged Prodigious Son and the father who must in the end seek forgiveness.

4. OTHER (IL)LOGIC OF SHORT FICTIONS

In contrast to believable fictions, storytelling ads, jokes, fables, and parables, to the extent that they are argumentative, do not exhibit a distinctive narrative argument form, but rather exhibit standard forms such as argument from analogy and inductive generalization. At least partly because of the heavy reliance of such arguments on affective appeal when expressed by such fictional narratives, unsurprisingly, Govier & Ayers (2012, p. 188) found that these "arguments are rarely cogent," and (echoing Velleman) "the form and interest of the story will often distract us from attempting any task of logical assessment." For example, they point out that the parable above, What You are After, is Within Yourself, taken as an argument, involves "hastily generalizing from the highly specific situation of a man wandering about in the dark, with a lighted lantern, to a universal human quest" (p. 178). Not to mention, let us not forget, the single instance on which the generalization is based is *fictional*.

Similarly, the conclusion in The Eagle and the Arrow that "we often give our enemies the means for our own destruction," taking the fable as an argument, is an unjustified leap, although it is more guarded. Understood as an argument, the fable seems best understood as an argument from analogy. Certainly, the source and target domains are distinct but parallel[4]—the fabulous world of talking and reasoning animals, and the human world, respectively. The use of an eagle in particular, might allude to a human type or stereotype (a smart and successful but overly trusting "high-flyer") particularly subject to such a plight. The case seems to fit Hunt's (2009) analysis of fabulous arguments from analogy: they have a "first case/principle/second case" structure, where the principle is in Peircean fashion "abduced" from the first case (the eagle's plight)—the principle "is supported to the extent that it is a good explanation of the first case." The second case, however, is deduced from the principle (p. 373); it is how readers apply the principle "to guide their own moral conduct or persuade others" (p. 379)—as one might think, "I better be careful or there is a real chance that I could inadvertently help my rivals by…"

---

[4]For the importance in drawing an analogy of having two such domains and not merely a similarity relationship, see Perelman & Olbrechts-Tyteca (1958, p. 502), Beardsley (1975, p. 111), and Olmos (2014).

Literally and technically, such analyses indicating the specific illogic of short fictions appear to be correct. However, it must be acknowledged that audiences often take these fictions to be merely suggestive, and not dispositive, of their generalizations or explanations. In other words, audiences often do not take them to be arguments. It nevertheless remains that when they are understood as arguments, their potential for argumentative evil is generally greater than for believable fictions. And this potential is perhaps greatest for children and mentally challenged adults.

5. NARRATIVE THOUGHT EXPERIMENTS

Thought experiments are designed to yield insight. There are many kinds of thought experiments. In perhaps the simplest of taxonomies, Popper (1959, p. 443) identifies three uses of "imaginary experiments"; they may be used to illustrate, support, or undermine a theory (what he calls their "heuristic...apologetic...critical" uses). Thought experiments are all fictional in that a hypothetical or counterfactual situation is visualized or somehow imagined in experience. However, in many thought experiments, storytelling or narrative is not prominent; they are not especially perspectival and "particularizing, sequential" depictions of events. Scientific thought experiments are characteristically in this way non-narrative, for example, Galileo's famous Pisa-type one where he disproves the Aristotelian view that the heavier the object, the faster it falls. On the other hand, narrative thought experiments, like all fictional narratives, are ultimately about human psychology, action, and society.

How do narrative thought experiments otherwise compare to other fictional narratives? Again, length appears to play a critical role in allowing a cogent argument, but in a different way. The most successful narrative thought experiments appear to present an extended and relevant point-by-point analogue of whatever problem is at issue. Thomson's violinist in her paper "A Defense of Abortion" (1971) is paradigmatic. Walton (2012, p. 199) presents a convenient summary of her core source and target "stories":

1. Person finds himself attached to famous violinist.
2. Person had no choice about this arrangement.
3. Having violinist attached is an encumbrance to person.
4. Having violinist attached will hinder person's daily activities.
5. Violinist will die if removed from person.

6. Violinist can only survive if attached to person for nine months.
7. Person can make a choice about removing violinist...

1. Woman who has been raped finds herself pregnant.
2. Woman had no choice about becoming pregnant.
3. Being pregnant is an encumbrance to woman.
4. Being pregnant will hinder woman's daily activities.
5. Fetus will die if removed from woman.
6. Fetus can only survive if carried to term of approximately nine months.
7. Woman can make a choice about removing fetus.

Thomson develops this analogy in different directions during the course of her paper, exhibiting its plasticity and depth. Indeed, the power and cogency of her essay derives from its being a good analogical argument, but not from any embedded fictional narrative being *believable* like a novel, play, or short story. As with other such thought experiments, her violinist story is weak on both external and internal coherence, and it would be astonishing if it were even intended to be believable. As Peijenburg & Atkins say, these are "outlandish stories," even "grotesque"; "ones like Jackson, Searle and Putnam do not eschew the most bizarre accounts of zombies, swapped brains, exact *Doppelgänger*, and famous violinists who are plugged into another body" (2003, p. 305). Walton too, allows that Thomson's violinist's story is only "something that could conceivably happen" (2012, p. 200).

6. ANECDOTAL AND OTHER NONFICTIONAL ARGUMENTS

Finally, rounding out the consideration of argumentatively evil storytelling and bringing the preceding into sharper focus are so-called "anecdotal arguments" and the possibilities they furnish, perhaps notably to politicians. Similarly to Johnson & Blair (2006, p. 70), Govier & Jansen (2011, p. 86) concluded that "anecdotal arguments are bound to be logically and dialectically inadequate if, as is usual, we define them as asking the audience to shift from acceptance of a *particular narrative* to a *general claim* about the world." However, to the extent that the term 'anecdote' connotes that the narrative is nonfictional, such narratives differ from the kinds of narratives considered thus far. Unlike for extended pieces of storytelling such as plays and novels, the actual anecdote in an anecdotal argument cannot itself furnish any argument. This is because, by definition, the point of nonfictional narration (cf. history or biography) involves veracity—sticking to the facts, telling what happened—so there is no theoretical room for the creativity that is needed to invent what happens and thereby construct an argument. Simard-Smith & Moldovan, for example, advance a view of "arguments as abstract objects" that "understands arguments to be objects that can be expressed in different points of space at the same time, and that are creations of human intellectual activity... We often make statements such as 'Searle developed the Chinese room argument'" (2011, pp. 259, 248).

Not surprisingly then, the length of the anecdote embedded in an argument seems to make little or no difference to the cogency potential of the argument. Consider this case presented as an anecdote in Hillary Clinton's speech at the 2008 US Democratic National Convention endorsing Barack Obama (cited by Oldenburg & Leff, 2009, p. 2):

> I will always remember that single mother who had adopted two kids with autism. She didn't have any health insurance; and she discovered that she had cancer. But she greeted me, her bald head painted with my name on it, and asked me to fight for health care for her and her children.

I do not see how the cogency of Clinton's argument for rallying behind Obama could have been significantly affected one way or another if she had presented more or fewer details of the unfortunate woman's situation (or presented more such incidents as anecdotes, which she in fact did). This is mainly because we would still not be told how representative the case(s) cited is, a question that an anecdotal argument in its usual form leaves unanswered.

So it appears that for anecdotal arguments, whatever difference length makes to the potential for argumentative evil, it is not comparable to the difference length makes for fictional narrative arguments. Anecdotal argument seems similar to (nonfictional) induction by enumeration on this score. No number of enumerated black crows identified by ordinary means will get you firmly to the conclusion that all crows are black, though a single perfectly representative one would. I think nonfictional arguments from analogy constitute an exception in that the best present an extended and relevant point-by-point comparison between things in distinct but parallel domains; if you shortchange this, there is no end to the potential for argumentative evil. On the other hand, for deductive arguments, there is simply no case at all to be made that length could make any difference to their validity or soundness.

7. CONCLUSION

In summary, it seems that there are reasons to hold that in fictional narrative the potential for argumentative evil is greatest if the approach taken is "hit and run," so to speak, whereas in other argumentative contexts, length generally appears to make no comparable difference. This is a feature that distinguishes fictional narrative arguments.

ACKNOWLEDGEMENTS: I am grateful to Jason Dickenson, Trudy Govier, Lyra Hostetter, Kenneth Olson, and Teresa Plumer for helpful comments on an earlier draft.

REFERENCES

Beardsley, M. C. (1975). *Thinking straight. Principles of reasoning for readers and writers* (4th ed). Englewood Cliffs, NJ: Prentice-Hall.
Cebik, L. B. (1971). Narratives and arguments. *CLIO, 1*(1), 7-25.
Currie, G. (2010). *Narratives and narrators. A philosophy of stories.* Oxford: Oxford University Press.
Currie, G. (2012). Literature and truthfulness. In J. Maclaurin (Ed.), *Rationis defensor. Essays in honour of Colin Cheyne* (pp. 23-31). Dordrecht: Springer.
Doody, M. (2009). Philosophy of the novel. *Revue Internationale de Philosophie, 63*(2), 153-163.
Fisher, W. R. (1987). *Human communication as narration. Toward a philosophy of reason, value, and action.* Columbia: University of South Carolina Press.
Gaiman, N. (2015). *Trigger warnings. Short fictions and disturbances.* New York: HarperCollins.
Govier, T., & Ayers, L. (2012). Logic, parables, and argument. *Informal Logic, 32*(2), 161-189.
Govier, T., & Jansen, H. (2011). Anecdotes and arguments. In E. T. Feteris, B. Garssen & A. F. Snoeck Henkemans (Eds.), *Keeping in touch with pragma-dialectics. In honor of Frans H. van Eemeren* (pp. 75-88). Amsterdam: John Benjamins.
Hunt, L. H. (2009). Literature as fable, fable as argument. *Philosophy and Literature, 33*(2), 369-385.
Johnson, R. H., & Blair, J. A. (2006). *Logical self-defense.* New York: International Debate Education Association.
Lamarque, P. (2004). On not expecting too much from narrative. *Mind & Language, 19*(4), 393-408.
Nagel, T. (1979). *Mortal questions.* Cambridge: Cambridge University Press.

Oldenburg, C., & Leff, M. (2009). Argument by anecdote. In: J. Ritola (Ed.), *Argument cultures. Proceedings of 8th international conference of the Ontario Society for the Study of Argumentation* (pp. 1-8). Windsor, ON (CD ROM).

Olmos, P. (2013). Narration as argument. In D. Mohammed & M. Lewinski (Eds.), *Virtues of argumentation. Proceedings of the 10th international conference of the Ontario Society for the Study of Argumentation* (pp. 1-14). Windsor, ON (CD ROM).

Olmos, P. (2014). Classical fables as arguments: Narration and analogy. In H. J. Ribeiro (Ed.), *Systematic approaches to argument by analogy* (pp. 189-208). Cham: Springer.

Olmos, P. (2015). Story credibility in narrative arguments. In B. Garssen, D. Godden, G. Mitchell & A. F. Snoeck Henkemans (Eds.), *Proceedings of the eight international conference of the International Society for the Study of Argumentation* (pp. 1058-1069). Amsterdam: Sic Sat (CD ROM).

Peijenburg, J., & Atkins, D. (2003). When are thought experiments poor ones? *Journal for General Philosophy of Science, 34*(2), 305-322.

Perelman, C., & Olbrechts-Tyteca, L. (1958). *Traité de l'argumentation. La nouvelle rhétorique*. Paris: PUF.

Plumer, G. (2011). Novels as arguments. In F. H. van Eemeren, B. Garssen, D. Godden & G. Mitchell (Eds.), *Proceedings of the seventh international conference of the International Society for the Study of Argumentation* (pp. 1547-1558). Amsterdam: Rozenberg / Sic Sat (CD ROM).

Plumer, G. (2015). A defense of taking some novels as arguments. In B. Garssen, D. Godden, G. Mitchell & A. F. Snoeck Henkemans (Eds.), *Proceedings of the eight international conference of the International Society for the Study of Argumentation* (pp. 1169-1177). Amsterdam: Sic Sat (CD ROM).

Popper, K. R. (1959). *The logic of scientific discovery*. New York: Harper & Row.

Putnam, H. (1981). *Reason, truth and history*. Cambridge: Cambridge University Press.

Schultz, R.A. (1979). Analogues of argument in fictional narrative. *Poetics, 8*(1/2), 231-244.

Schwitzgebel, E. (2008). The unreliability of naive introspection. *Philosophical Review, 117*(2), 245-273.

Simard Smith, P. L., & Moldovan, A. (2011). Arguments as abstract objects. *Informal Logic, 31*(3), 230-261.

Stern, R. (2007). Transcendental arguments: A plea for modesty. *Grazer Philosophische Studien, 74*(1), 143-161.

Thomson, J. J. (1971). A defense of abortion. *Philosophy and Public Affairs, 1*(1), 47-66.

Velleman, J. D. (2003). Narrative explanation. *The Philosophical Review, 112*(1), 1-25.

Walton, D. (2012). Story similarity in arguments from analogy. *Informal Logic, 32*(2), 190-221.

Commentary on Plumer's Argumentatively Evil Storytelling

PAULA OLMOS
Universidad Autónoma de Madrid
paula.olmos@uam.es

In this paper, Gilbert Plumer continues, as has been his focus on other recent contributions, to explore certain aspects of narrative arguments. In this case, he is explicitly looking for assessment criteria and claims to have found one that applies to fiction narratives: the shorter the story, the less it will justify certain inferences based on it and therefore the greater the potential for an ill-founded argumentation to be presented through it.

As Plumer uses the term, "short" is, in fact, *short for* schematic, stylized and unrealistic (as in fables, classical and oriental, or ads) and explicitly opposed to the nuanced, complex and rich-in-detail weave and plot of (realistic) novels which are the adequate basis for the kind of argument scheme described by Plumer in his paper which he has presented and used in other contributions. Namely:

> (1) This story is believable.
> (2) This story is believable only if such and such principles operate in the real world (of human psychology, action, and society).
> (3) Therefore, such and such principles operate in the real world.

At the end of the paper, Plumer explores other kinds of narratives, as "thought experiments" and "anecdotes". The latter are non-fictional and are usually advanced as premises for an inductive-like "anecdotal argument", based on the assumed veracity and actuality (not the believability) of the anecdote; and the former, claims Plumer, although fictions, are neither usually embedded in a kind of argument in which the believability of the story would really make any difference for its assessment (this point I think is just suggested and would need further exploration).

So, although one has to finish the paper to locate and put in order all these pieces, Plumer's position is, in my opinion finally clear. It seems he is contemplating only one possible argument scheme (the one he originally developed for novels in his 2011 paper: "Novels as arguments") for which he takes narrative fictions (leaving aside just philosophical thought experiments) would be candidate sources or basis. And he finds that such a scheme is only liable to yield sound arguments only in case the story involved has certain characteristics and actually aims at depicting, in a realistic manner, "the real world". This is not possible if the story is schematic, stylized and unrealistic, and the subsequent reconstructed argument remains ill-founded.

For me, the most valuable contribution of this paper is this attempt to start clarifying a complex panorama by introducing certain distinctions that may be useful for future works. Depending on our analytic aims, I agree that taking in account the fictive or factual character of the stories we use as part of our argumentative efforts might be important (and I therefore assume the criticism he makes to one of my contributions in his paper).

Nevertheless my own view about narrative arguments (which I have presented in other recent contributions: 2014, 2014b, 2015) is that narratives, in general (that is, fictive or factual), may be used to construe very different kinds of arguments and even, that one and the same particular story (a classical fable, for example) might be variously used as basis for construing arguments according to different argument schemes in different contexts (some examples in Olmos, 2014b). So we have to analyze and assess each real case as pertaining to its own argumentative aims in its own context.

Moreover, I find Plumer's kind of "transcendental argumentative scheme for fiction" a somewhat abstract model that probably works better for, let's say, the extraction of very general principles or usable warrants from acknowledged complex fictions than as a genuinely operational form of argument or inference scheme for concrete conclusions. In most real cases, it would be part of a more complex and, at the same time, more concrete argumentation (of a practical, evaluative or theoretical character) and in each case it would support our final conclusions in a somewhat receded way, i.e. as founding the backing of the warrants of our actually operative reasons. That is why, I think, it's called "transcendental": it mostly describes the "conditions of possibility" of the use of certain kinds of arguments than describe those arguments themselves.

In the particular case of the three examples offered by Plumer as support for his thesis (i.e.: the shorter the story, the greater the potential for argumentative evilness), Plumer assumes that they do not offer good enough reasons to support the generalizations they seem to support. My point is that *it depends on what you are going to use those generalizations for* (they might be in need of further support, or not so much), for, in most cases, one's argumentation, most typically a practical one (taking in account that the theme of stories is usually human action), would be much more concrete and referenced to particular circumstances in which those generalizations would either appear as easily applicable or not.

The two fables, I think, do not even aim at "supporting" such generalizations (as even Plumer admits) but only try to "illustrate" and "explain" them (that is, their workings), or just "fix" them in the imagination. Both present and represent, in fact, ideas and warrants that are already assumed as rather *usable* (*prima facie* good enough) in our societies for advising certain behaviour or attitude: "what a man seeks is very near him", "we often give our enemies the means for our own destruction". These fables are just means to teach or recall them. That they are accepted by the interlocutor as supporting a certain conclusion will mainly depend on the circumstances expressed in, and the further objectives of, that particular conclusion. This is so because such warrants are more "(usefully) applicable or inapplicable" to particular cases than "true or false". In any case, the fables do not exactly try to show that they are simply "true".

In the case of the "Vehicle ad", I agree that it certainly creates an atmosphere, emotions, something to remember etc. but what argument it is supposed to support in a direct or receded manner is yet something rather open. So, in my opinion, it cannot be yet assessed as better or worst founded. Most ads merely support the "good name" of a brand, using different reasons and warrants (even their "good taste in publicity"). They might finally aim at advising a purchase, but the steps (argumentative steps) are yet too many. If we assume that this ad (through the story told in it) just shows and conveys the information that the makers of the car have thought about snow conditions and designed their vehicle for better facing them, and this encourages you to visit their store, with the memorable ad in your mind, to ask for more details, I see no evil in it.

ACKNOWLEDGEMENTS: This contribution has been made possible by funds provided by the Spanish Ministry of Economy and Competitiveness: Research Project FFI2011-23125.

REFERENCES

Olmos, P. (2014). Narration as argument. In D. Mohammed & M. Lewiński (Eds.), *Virtues of Argumentation. Proceedings of the 10th International Conference of the Ontario Society for the Study of Argumentation (OSSA), 22-26 May 2013, CD edition*. Windsor: University of Windsor.
Olmos, P. (2014b). Classical Fables as Arguments: Narration and Analogy. In H. Jales Ribeiro (Ed.), *Systematic Approaches to Argument by Analogy* (pp. 189-208). Amsterdam: Springer.
Olmos, P. (2015). Story Credibility in Narrative Arguments. In F.H. van Eemeren & B. Garssen, (Eds.), *Reflections on Theoretical Issues in Argumentation Theory* (pp. 155-167). Amsterdam: Springer.
Plumer, G. (2011). Novels as Arguments. In van F.H. van Eemeren et al. (Eds.), *Proceedings of the 7th Conference of the International Society for the Study of Argumentation (ISSA 2010)* (pp. 1547-1558). Amsterdam: Sic Sat.

28

"Rationality as Use": On the Nature of Rationality in Argumentation

MENASHE SCHWED
Ashkelon Academic College, Israel
m.schwed@outlook.com

The question is how rationality functions in argumentation. The approach is Wittgensteinian in nature as it emphasizes the assumption that argumentation is culture, social and politic laden. If argumentation is understood according to its functions, the approach introduced here argues that functions, such as decision, belief revision or actions, are complex language games. These language games are embedded in the pragmatic and practical uses of language, which constitutes specific forms of life.

KEYWORDS: argumentation, art, epistemology, Joseph Wright, Ludwig Wittgenstein, pragmatic, rationality, science

1. INTRODUCTION

Rationality is one of the philosophers' stones of the Enlightenment and these philosophers seek it in the hope to turn ill society based on irrationality and superstitions into improved, reformed and enlightened one. And as the English painter Joseph Wright of Derby expresses so dramatically in his *The Alchemist*, many other such stones were found elusive again and again (
Figure 3, Appendix below). In spite of its elusiveness, it is assumed here that the concept of rationality is indispensable for the philosophical understanding of argumentation (Biro & Siegel, 2006; Johnson, 2000). However, it is regrettable that we are still not able to arrive to some kind of consensus regarding its meaning and function. This paper is no more than another attempt to get hold of this concept. And this time it is with the particular attention to argumentation. I found that this last consideration functions more as insightful rather than restrictive.

The general idea is to use a Wittgensteinian framework to direct us toward understanding rationality as practices that have differing constitutive rules, norms and values. The Wittgensteinian idea of constitutive rules is extended to constitutive norms and values, which captures both the historically and culturally contingent and changing character of rationality. Furthermore, it captures the sense that there is something at work very much like "essential characteristics" in the sense of something with a determining or *constitutive* force that we need to understand. But these "essential characteristics" are not to be identified with anything like "objective" or "universal" in a sense of necessary and sufficient for some argumentative discourse to be rational. Constitutive rules, norms and values "cause" something to be an instance of rational discourse, but their role is specific to certain practices just as rules are specific to certain games. This last remark is important in order to understand the historically and culturally contingent nature of rational practices.

The importance of these contingencies to the study of argumentation was the subject matter of two of my recent papers. Common to both is the thesis that argumentation essentially involves a *choice* in a constitutive manner and showed how argumentation theory might reflect this feature. In the 2013 OSSA conference, I argued that Humanism, and Enlightenment in particular, distinctively influenced practices of argumentation, as reflected by choices made on political grounds. This thesis was based on historical and philosophical reasons (Schwed, 2013). In the 2014 ISSA conference, I further developed this thesis such that it deals with the problem of rationality in argumentation in a like manner. The general idea was that the demand for rationality is a basic choice, derived from the political and moral ones, which are essential to it (Schwed, 2015).

The purpose of this paper is to understand what place does rationality has in argumentation. However, this purpose is radically different from similar endeavors, such as of Biro & Siegel (2006). Following Putnam sentiments in his "realism with a human face" (1990), the primary issue here is not whether the concept of rationality is definable, posing us with the need to choose between essentialism or universalism and their denials. Rather, following my reading of Wittgenstein, the purpose is to keep *uses* of language in view of their variety and in their broader life contexts. The goal is not to lose hold of the role of contingent contextual factors.

I stress that it is *my* reading of Wittgenstein, since there are no actual remarks of Wittgenstein on argumentation or rationality. It is rather a function of his inspiration on these topics and this primary inspiration is taken from his later approach to various philosophical issues in general, as manifested in the *Philosophical Investigations*

(Wittgenstein, 1967).[1] Furthermore, Wittgenstein offers inclusive approach of integration between uses of language, activities, practices, and the domains or forms, such as social, cultural and historical, in which these take place as one complex matter or "form of life". This multidimensional philosophical approach is obviously only one reading of Wittgenstein and interpretative in nature.[2]

2. PRELIMINARY NOTE: THE WITTGENSTEINIAN APPROACH

A basic Wittgensteinian approach to argumentative discourse and rationality is that there is no Archimedean point in terms of which no one can relevantly criticize whole modes of discourse or, what comes to the same thing, forms of life. The main Wittgensteinian reason is that each mode of argumentative discourse has its own specific criteria of rationality or irrationality, intelligibility or unintelligibility and reality or unreality. This brings to mind Perlman's concept of "audience": "since argumentation aims at securing the adherence of those to whom it is addressed, it is, in its entirety, relative to the audience to be influenced" (Perelman & Olbrechts-Tyteca, 1969, p. 19). However, the possibility for Perlman's 'universal audience' is nothing more than a fiction from the Wittgensteinian point of view. A somewhat cynical response from this point of view would be to say that it transcends the boundary of the particular audience into "the philosophic privilege conferred to reason" (Perelman & Olbrechts-Tyteca, 1969, p. 30).

Instead of the of the more traditional approaches to the analysis of argumentative issues, Wittgenstein's late thought offers some insightful strands that deserve more elaboration. This elaboration is the natural developments of claims made by Wittgenstein. The strands of thought in mind are as follows:

1. To imagine a language is to imagine a form of life (*PI*, §19).
2. What must be accepted – the given – are forms of life (*PI*, §226).
3. Ordinary language is in order as it stands (*PI*, §98, referring to the *Tractatus* 5.5563, Wittgenstein (1961)).
4. Philosophy must in no way interfere with the actual use of language, but can in the end only describe it (*PI*, §124).

---

[1] Hereafter *PI*. All further references to *Philosophical Investigations* will be to section number, as for Part I, or page, as for Part II, of this translation.

[2] The scholarly literature on Wittgenstein's later philosophy is vast. The interpretative approach adopted here has many resemblances to the following works: (Cavell, 1979; Diamond, 1991; McDowell, 1998a; McDowell, 1998b; Putnam, 1990).

The range of application of these four strands of thought is perfectly general, including all aspects of language use as constituting various forms of life. They are part of Wittgenstein's attempt to articulate his sense of what is the sense of ordinary lives with words and eventually what philosophy can and cannot do. A Wittgensteinian approach to argumentation and rationality would be the application of these four strands of thought to a specific area of study.

One important point follows that should be mentioned. If the consequence of applying these claims with respect to rationality in argumentative language games is a species of relativism, then a parallel conclusion must follow from their application to any other mode of rational discourse or game. If a Wittgensteinian approach is relativist about argumentative language games, that will not be because of any view it holds about particular language game, but rather because beliefs or misconceptions to a relativist stance with respect to human language and forms of life in general. What will amount eventually to an entirely universal substitution of objectivity for relativity.

3. THE FAMILY RESEMBLANCE OF THE VARIOUS MANIFESTATION OF RATIONALITY ACROSS ARGUMENTATIVE LANGUAGE GAMES OR DISCOURSES

Although Wittgenstein did not discussed the concept of rationality, he had a lot to say about "concepts" in general. His work is almost a commandment to approach the concept of rationality by considering the diverse practices themselves. If rationality is a pattern of overlapping, norm governed practices with a variety of constitutive rules, norms and values, then the term "rationality" would not have a univocal meaning across cultural eras and historical periods. It does not have a logical structure either. Furthermore, the articulation of rationality in the context of argumentative language games involves criticism of general questions (*PI*, §24). Following Wittgenstein's resemblance discussion starting at *PI* §65, the examination of rationality is the examination of various parts and uses of language, and our assumptions about them and what we picture as their relation to reality. The diversity of the uses of rationality, revealed across the diversity of language games in which we play – evoke the notions of rationality – ranging over personal disputes, political controversies, legal issues, medical deliberations, forming and testing a hypothesis, scientific debates, and commercials, to mention only few. There is no universal concept that can be employed in every such game. This diversity only highlights that "...we see a complicated network of similarities overlapping and crisscrossing:

sometimes overall similarities, sometimes similarities of detail." (*PI*, §66).

What is the Wittgensteinian way to characterize or describe these similarities? The phenomenon is the diversity of the language games, in which rationality is a constitutive component or rather a norm. There are resemblances among the family members underneath this diversity such that the goal is to *characterize* the nature and the diversity among these games (*PI*, §67). However, the study of rationality in argumentative language games can transit from mere descriptive to more explanatory considerations along Wittgenstein's switch from the analogy of family resemblances to the analogy of numbers and the spun thread (*PI*, §67). This transition in analogies is crucial, since it enable us to deal with the highly diverse kind of the uses of rationality and its complexity. How we might be able to use the concept of rationality without clear boundaries, how we could learn this concept, in what sense we can be said to understand it, what "[s]eeing what is common" (*PI*, §72) might be, and more. This kind of study presupposes the understanding of what it is to follow rules.

Since using language involves following rules, the question is how to investigate the use of rationality in argumentative discourses without a definition would involve rules that might not specify all of its applications? Wittgensteinian approach can be summed up in the following methodological instructions:

1. Rationality is not a concept but the uses that we call "rational", in a similar way to the distinction between *language* and *uses of language* (*PI*, §67). Wittgenstein's primary focus is on the practices of language games themselves, the forms of life activities in which those argumentative language-games are an integral part.
2. There is no abstract vantage point of a theory that presupposes to explain and specify the essence of rationality. However, our philosophical stance results from the attitude and commitments we bring to the subject matter (*PI*, §24). Insofar as we are in the business of constructing philosophical definitions, we are committed to the viability of this project in advance of deciding to attend to the actual varied details of the phenomenon of rationality, be it "all that we call" argumentative discourses (*PI*, §67). It is our educated choice to be less inclined to look for general definitions and more to the actual varied details of language uses.[3] It is a self-standing choice regarding the indefinable nature of the "family resemblance" concept of rationality.

---

[3] This general point is developed by Wittgenstein in the preceding sections to *PI*, §67.

3. Furthermore, it is possible to use a single concept of rationality for a variety of phenomena in the absence of a definition in terms of individually necessary and jointly sufficient conditions. In accordance with Wittgenstein's two analogies, it is not only that we discern the sort of crisscrossing resemblances as in families, but also that these resemblances function together as the fibers of a thread or rope. The notion of "games" helps to characterize or describe the patterning of similarities and differences among members of one family, and the analogy of "fibers in a thread" explains how they may form together into one kind, how new and diverse members can be added to a kind, and how our ability to use a concept can encompass such additions.

The last point made above in 3 is crucial from the perspective of debates in epistemology and theories of rationality, where it is said that the *Epistemological* approach (Biro & Siegel, 2006) suggests that the similarities among the various concepts of rationality hold together by means of a prototype or paradigm. That its manifest similarities to paradigms that determine membership in the kind "rationality" and extensions of it. That it is as a paterfamilias sitting in the center of a family portrait. But this will not do according to Wittgenstein, which does not delve into the nature of family resemblances for explanation, but switches analogies to that of a thread or rope in which *no* fibers are more important than any others in constituting a whole through their intertwining relationships.[4]

This switch is important because family membership is not in fact constituted by similarities to paradigms. Accordingly, any sort of crisscrossing resemblances manifest in a family are not the result of a common essence that explains the manifest diversity. The structure of *PI*, §67, shows that Wittgenstein suggests we think of family resemblances to illustrate the possible complexity of a pattern. His second point is that in certain cases, if a structure is large and complex enough, overlapping local relationships among members can determine the whole. But this does not suggest that crisscrossing similarities and differences should and can be explained by underlying essence, prototype or paradigm since it functions more like a thread.

Thus, it is consistent with Wittgenstein's approach to consider manifestations of rationality in relations they stand to one another within practices or theories, as well as in relation with each other. Hence, pragmatic theories of argumentation that focus on the context of complex relationships – theoretical, historical, cultural, or practical – in

---

[4] This point deserves further discussion, especially regarding Wittgenstein remarks on colors in *PI*, §57.

which rationality has nothing to do with essentialism, universalism and so forth.

It begins with seeing that the concept of rationality has a large and complex structure, overlapping many relationships with other key concepts in Western Philosophy. Thus, it will be more effective to analyze it as *multiply intertwining* or simply *fibrous* rather than *family resemblance*. In other words, it is locally overlapping practices with related constitutive norms and values intertwine, with the result that not only similarities but also differences overlap and crisscross. Let us take a simplified example for illustration, consider the legal argumentative style of reasoning before and after the eighteenth century (La Torre, 2002, pp. 377-379).

Consider legal discourses before the French revolution, were the legal reasoning and arguing were understood more as rhetorical in nature and the arguer was seen more as an orator, who relies on his capacity to exercise effective persuasion. These legal practices aimed more at virtuosic performance than formal analysis of the law. It is not that the formal aspects of legal discourse were not important. Still, the constitutive norm of that form of life was that rhetoric was an important subject at the time in the curriculum of the law student.

With the revolution and Enlightenment ideology, there has been a significant shift towards codification of a formal system of laws. With the legal system of the Enlightenment, the argumentative discourse uses legal positivist language. But that emerged as a constitutive norm as the forms of life were also becoming increasingly rationalized. It was a change by different constitutive norms as practices intertwined throughout the early eighteenth century. The theory of law before the revolution centered on rhetorical maneuvers while legal arguments lacked systematization and consistency. However, the theory of law in the time of the Enlightenment aspired to be more cognitive, reasoned, formal, and descriptive as opposed to evaluative, in which the subject, be it the judge or the lawyer, has no substantial place comparing to the *law*. The applying of the law was conceived as a pure cognitive operation, in the model of syllogistic reasoning. What changed in the Enlightenment is that the law became the major premise from which the arguments began. The characteristics of the law has changed and with it its function in arguments. Furthermore, the epistemology of the law has been changed as well, since now reasoning has become a cognitive operation, theoretical rather than rhetorical or practical.

Such local relationships or overlapping between positivistic and rhetorical norms was accounted for both continuity and diversity of the legal discourse over this historical period. The thread of legal discourse is long and thick, beginning in the Roman time and even earlier. Thus, this thread has numerous fibers that might not overlap: the norms of many legal practices along the history of humankind are quite distinct rather than similar. However, this fact would not undermine the integrity of the whole practice of legal discourse insofar as there are local relationships or overlappings between the strands that do not overlap.

The analysis of the concept of rationality from this Wittgensteinian perspective of family resemblance turns the effort to supply a universal and sufficient definition of the concept redundant. What is more important is that according to the Wittgensteinian perspective, we should keep the role of contingency in our explanation of the concept.

4. RATIONALITY AND ARGUMENTATIVE LANGUAGE GAMES AS CONTINGENT PRACTICES

Wittgenstein's *Investigations* offer three explanatory resources that bear on our understanding of rationality in argumentative discourses:

1. First, it highlights the mutually explicatory relationships between activity, language, and context for any rule- or norm-informed endeavor.
2. Second, it emphasizes the role of contingent context both in the sense of practices and customs and in the sense of world conditions.
3. Third, it raises the possibility that some endeavors are informed by constitutive rules, norms or values.

The basic Wittgensteinian idea is that a variety of language games or uses suggests a broader orientation to the life activities or practices to which language is internal. This orientation begins from the very outset of the *Investigations* (*PI*, §2) with the introduction of primitive language games, which emphasizes the contextual relationship between a range of activity and a range of language uses. The idea of language games illustrates what has come to be known as Wittgenstein's maxim: "meaning is use."

The relevant meaning of this maxim to the current issue is that the determining factors of meaning are present in the internal relationships among what we can say, what we can do, and the circumstances in which these take place. In other words, they are internal relationships between human life activities, uses of language and circumstances, and these depend on, enable, and entail one another. It is the reciprocity in forms of life activities and language. Accordingly, activity like rational persuasion comes together with – and is inseparable from – the articulate speech integral to that range of activity, and both can only take shape in external circumstances that allow for arguing with reasons, argumentative schemes and strategic maneuvering and so forth.

One important implication to the Wittgensteinian idea of language games, that seems underemphasized or overlooked, is to focus on the reciprocity among these factors rather than to take any one (action, language, world) of them as primary to the others. This is to my mind the importance of Wittgenstein known statement that "To imagine a language is to imagine a form of life." (*PI*, §19).

Given that using rationality is an activity in the above sense, as any other of our activities, they are all shot through with rules. Thus, the issue of rule following is brought into consideration and more importantly, the role of criteria of correctness, context, and training all stand out. Wittgenstein directs our attention to the basic fact that needs to be countenanced, that "...hence also 'obeying a rule' is a practice." (*PI*, §202). Stressing that it "is a practice" also affirms one of Wittgenstein's controversial suggestion, (*PI*, §199), that "to obey a rule, to make a report, to give an order, to play a game of chess, are *customs*, (uses, institutions)."[5] The bearing on the concept of rationality is clear, since in using a concept we follow rules that specify possible applications. But it is not only the involvement of following rules, but being able to act further on in a specific activity in the right way.

The challenge in understanding the use of rationality in argumentation lays in the nature of its criteria of correctness: Whether the criteria are communal agreement in practice or that communal agreement determines what is correct. McDowell distinguishes between uncovering the nature of criteria of correctness through describing human practices and inventing the criteria through human communal practices (McDowell, 1995, p. 285). This is an important note regarding the scope of Wittgenstein's skepticism in so far as to maintain at least an inter-subjective sense of objectivity of the concept of rationality. Giving that arguing rationally is to do something according to rules, there should be something more than just arguing rationally in a way that a community agrees upon. Given that criteria of correctness are extant in practices, how rationality as a norm-governed activity validates these practices?[6] Following some of Wittgenstein's hints in *PI*, §241-2, the

---

[5] For further discussion of this subject, see McDowell (1998a), which criticizes Saul Kripke's (1982) reading of Wittgenstein.

[6] This question is part of a larger issue regarding the interpretation of Wittgenstein as a sceptics or more specifically that there is nothing at the level of communities that could constitutes the correctness of our practices or of our

validation of a given arguing practice should be part of the mutual relations between our possibilities for action in the world and possibilities afforded by the world. These mutual relations are, above all, historically and culturally contingent and changing.[7]

In order to apprehend the role of practices in the uses of rationality in argumentation and its criteria of correctness one must highlights the issue of rule following. The idea of 'practice' is more of a Wittgensteinian term, which functions more or less as 'audience' and 'discourse' in the argumentation literature. However, Wittgenstein's focus on rule following stresses the notion of practices, and consequently of customs or institutions, as rule- and norm-informed life activities that allow facts and values to become evident and compelling.

5. RATIONALITY, RULE FOLLOWING AND CONSTITUTIVE RULES

This Wittgensteinian focus on rules and their ubiquity in shaping forms of life activities cannot be separated from their being *constitutive*.[8] The idea of rationality in Wittgensteinian terms would be quite expected: the rules that specify what are the uses of rationality in argumentative contexts are constitutive for rationality. To argue in a rational manner *is* to act in specific moves for the sake of rational persuasion, which is the point of "winning an argument". Such rules are *constitutive* as they determine the language games we participate in by specifying the lawful acts and moves and thereby the concept of rationality that figures in an argumentative language game. The idea of rules and following them is powerful, insofar as it indicates that there would be no such concept as rationality in argumentative language games if it were not for the constitutive rules that define what it is lawful in the game for using the concept of rationality. However, participants in an argumentative language game use the idea of rationality without thinking of the rules explicitly and without having the capacity to articulate the rules explicitly. Rather, the point is to recognize the integrated variety in our

---

communal ways of arguing in a specific way rather than another. The starting point in this issue is the skeptical view offered by Kripke's (1982) and Crispin Wright's (2007) arguments.

[7] Wittgenstein exemplifies this important point in his discussion of measurement. See also Richard Eldridge (1987).

[8] See Glüer & Wikforss (2015) for a helpful summary of the idea of constitutive, which emphasizes the affinity between rules and norms. For an important analysis of this idea and its derivations, see also: Haugeland (1998). For an indispensable classical discussion of constitutive rules and practices, see John Rawls (1955).

capacities without prioritizing either perception or action or thought. Argumentative language games exemplify how myriad activities involve rules or norms that are constitutive.

To clarify these last remarks it is important to note that the constitutive rules of an argumentative discourse specify the lawful moves of its components and, thus, give the identity conditions of each component as being "conclusion" or "reason" on one hand or just "sentence" or "utterance" on the other hand. In doing so through these constitutive rules or norms of various kinds, one specifies what it is lawful to do and thereby provides identity conditions for the resulting argumentative discourse. These rules would pertain to the means, techniques, schemes, rhetorical maneuvers, subject matter, and so on, that go into making an argument or arguing.

Thus, the notion of constitutive rules suggests that local or communal rationality practices are informed or determined by normative force and specificity bound to a specific practice. If we take the scientific communities for instance, than this means that 21th century scientists could recognize the constitutive norms of scientific research in the Hellenistic culture in the areas of astronomy, mathematics and geography, for example, without feeling bound by those norms. This approach captures the sense that rational practices diverge significantly as part of cultural and historical phenomena. This cultural and historical contingency is part of our understanding our practices in terms of constitutive rules, norms and values.

Wittgenstein's *Investigations* highlights the culturally and historically contingent contextual details of what we can do and say. It is here where Wittgenstein's family resemblance passage concerning the variety of all that we call language is crucial. The investigation of rationality cannot gain any substantial generality since such generalities are not explanatory by themselves without specification by diversifying detail that belongs to the generalized pattern. Such investigations must keep in view the variety – of the tools of arguments, or language uses – in their context or circumstances, in their lived actuality, rather than form the standpoint of some theory, which aspire to rise above such details as things that obstructs pattern rather than the pattern itself (*PI*, §11-12).

6. THE ARTWORK

How does this approach explain the concept of rationality in argumentation in our time as well as in the past? Given that the discussion here can only be programmatic at best, still examples of constitutive rules and norms integral to particular argumentative discourses and practices is in place. Thus, I will briefly describe how the Wittgensteinian perspective can be used with Wright's dramatic 1768 painting *An Experiment on a Bird in the Air Pump* (

Figure 4 below).[9]

The theme of Wright's scientific candlelight painting *Air Pump* describes a natural philosopher demonstrating the science of pneumatics, the principles of vacuum and that air is vital to life, in a domestic setting for a lay audience. Inside an instrument devoid of air, a cockatoo hovers between life and death while the observers start to grasp that the bird's fate will be determined by the demonstrator, who can open the air passage. The observers display a spectrum of social and emotional reactions to the scientific drama: admiration and hope, tearful distress, scientific absorption, paternal protection, and seeming indifference.

It was not Wright's first "argument" regarding the interplay between science and society of 18th century Europe. The first of the paintings, exhibited in 1766, features an orrery, a kind of planetarium or a model showing the movements of planets in the solar system (

Figure 5   below). However, contrary to the seemingly straightforward interpretation of *The Orrery* that was not intended to evoke sensational drams, the *Air Pump's* visual style creates haunting

---

[9] More on Joseph Wright of Derby, see: Benedict Nicolson (1968). For a general introduction and analysis of Wright's *Air Pump*, see: Paul Duro (2010).

special effects by blending scientific and societal scenes, and it does so to provoke interpretation and further discussion.[10]

Wright was not unusual in his interest in science, as science was becoming popularized in the 18th century.[11] He was part of an ongoing discussion in European society regarding the scientific revolution, technology and industry. Wright's *Air Pump* is an emblematic for the concept of experiment of his time as the vacuum pump experiment is also incorporated in the center of the frontispiece of Diderot and d'Alembert's *Encyclopédie*. He begins with Boyle and Hooke empirical and scientific narrative in experimenting the air pump,[12] but transcends into the cultural, social and moral narratives, rich in metaphors and symbols. He portrayed artfully this drama and sought to bring the ideas of Enlightenment to the general public. However, his *Air Pump* was more ambitious since he wanted *also* to raise important questions about science and the morality of the Enlightenment. These were important questions in an era in which industrial revolution coincided with a period of intellectual revolution of the Enlightenment. It might be that Wright was aiming at the heart of these revolutions when he rose up doubts about the use of rationality.

When using a Wittgensteinian outlook, the painting is contextualized within a broader social history in order to explain how it respond simultaneously to a dual set of circumstances, from its broader social and cultural as well as its artistic context.[13] Obviously, these circumstances do not exhaust the painting's unique meaning. Wright's painting, thus, is part of contextual practice of scholars in the 18th century, in which scholars gathered and discussed the latest developments of science, engineering and industry, the Lunar Society in this case. The painting was part of a wider argumentative discourse referring to seeking to dispel notions of the pre-Enlightenment culture. Notions, such as religious superstition and political and social

---

[10] Laura Baudot (2012) explores the richness of Wright's narratives and metaphors in his *Air Pump*, adding the religious narrative as well, which is not mentioned here.

[11] Terje Brundtland (2011) describes and explains the history of air pumps in the 17-18th centuries and places Wright's *Air Pump* in its scientific and technological context. In doing so, he highlights the cultural triangle of scientific societies at the time, scientific knowledge and research, and the European society.

[12] Simon Schaffern and Steven Shapin (1985) describe the place the air pump had in shaping the new experimental science of the 17th century and the place Robert Boyle and Robert Hooke had in it.

[13] Linda Nochlin (2006) is an example to such an approach in art history.

intolerance, were criticized while promoting a liberal worldview based on rational thought and promoting open and rational society. Wright's painting is part of this ongoing dialog regarding the impact of the new scientific movement on society.

Wright's *Air Pump* is innovatively situated at the intersection of the new scientific theme and the traditional artistic theme of "gathering indoors" scenes, rather than just within the later figurative genre. This is achieved by showing that the scientific apparatus subverts traditional indoors painting along with traditional setting up a gathering. Constitutive norms concerning subject matter, means, and techniques give identity to the painting and these evolve through interplay between the broader social and cultural context and a range of artistic practices and norms. Thus, while Wright's artistic style was customary to his period, his painting illustrates a non-conventional artistic them: the wonder and horror of science as it was known in the 18th century. It challenges us in this context to define new rules and limits for the usage and application of our newly conceived science and technology. In Wright's days as today, the general public is less knowledgeable about science and technology and, thus, have to be more conscious and informed.

The painting illustrates a Wittgensteinian focus on the relationships between a specific artistic practices and broader life circumstances, through which constitutive norms emerge. However, constitutive rules or norms do not only constrain but also liberate, which helps explain innovative works as response to the restrictive effect of rules and norms. The reference to scientific issues is part of the design of a set of constitutive rules, which enable us to see something *new*. But this sort of subject matter came to govern and constrain once it emerged as a norm. There is always a moment in which a rule stops being new and becomes restrictive and vice versa.

Wright was trained with Thomas Hudson, a successful portrait painter in London and his work was typical to the Rococo Era. Wright's artistic style developed into the wide spreading Romanticism of the 18th century. His artistic mark however is due to his befriending scientists, industrialists and engineers besides artists. The combination of influences resulted in a dramatic new argumentative discourse about progress, the ideals of the Enlightenment, and European society in the 18th century. He may not be as well known in art history as his contemporaries, such as Joshua Reynolds or Thomas Gainsborough. But his artistic work undoubtedly gave rise to new constitutive rules and norms regarding the argumentative discourse about the interplay between science and society, a well-established discourse nowadays.

Thus, although Wright's artistic work was very much in keeping with his times, his philosophical challenge was to question preconceived notions and test new ideas. In this point, it is important to remember that when Wittgenstein explains the actual variety and complexity of the notion of language game, he writes: "Here the term "language-game" is meant to bring into prominence the fact that the speaking of language is part of an activity, of a form of life. Review the multiplicity of language-games in the following examples, and in others: ... Constructing an object from a description (a drawing) — ..." (*PI*, §23). Wright was in fact keeping with his age, as part of the Enlightenment practice of revolutionizing traditions. These intellectual practices across the 18th century challenged the constitutive rules and norm of more traditional practices, thus constitute new ones.

A Wittgensteinian approach will highlight that the 18th century humanist artists were challenging more cultural and political ideas than aesthetic norms associated with the artistic world of the time. It might be that Wright found that the norms developed by Enlightenment have ceased to liberate, having rather begun to govern in a way that no longer freed the mind of the people. Since the values of the Enlightenment were identified at the time as liberating, raising arguments in the spirit of Wright's *The Pump* were necessary for checking their validity. This critical dialog between art, science, society and the Humanistic ideas is alive today as in Wright's days. Tamar Schlick (2005), for instance, uses Wright's *Air Pump* as her starting point for a critical discussion of the ramifications of today's Genome Project. Whatever the details of all these important dialogs, it is important to see that the Wittgensteinian framework suggests that arguments, such as *The Pump*, and the broader context of discussion it engender, are part of the process of cultural and intellectual critical dialog, whereby the norms of specific practices evolve. In particular, part of the discussion surrounding the pros and cons of science addresses whether one kind of rationality has a defeasible role in practices.

7. ONE LAST REMARK: THE THERAPEUTIC STANCE

The Wittgensteinian approach is a critical and skeptic one, rejecting any substantial attempt toward essentialism or universalism of any sort. Its promising aspect is in its *therapeutic* conception of philosophizing. It all begins in arguing that philosophy cannot claim general validity over its subject matters or its method, as both traditional epistemology and metaphysics do is merely creating the illusion of their authority over the authority of science or other discourses. Corresponding to Wittgenstein's famous remarks on philosophy (*PI*, §89-133, in particular

§126, §127, and §133), by which the therapeutic stance is inspired, this stance claims that philosophy is neither theory nor method. Rather, the therapeutic stance proceeds in a piecemeal way, adjusting "method", "theory" and "goal" in light of the concerns and problems at hand until satisfaction or clarity is achieved. It is "therapeutic" since once the problem is no longer perceived as a problem, and therefore does not undermine our understanding or challenge our practices, it allows us to proceed without alienate us form the contingency of our practices.

The intellectual puzzle about the rationality and its place and function in argumentation that bothered us is not dealt with the formation of – yet another – philosophical theory and method. Instead, the therapeutic stance aims on understanding of rationality and argumentation through a pragmatic-practical reorientation on or resolution of the puzzle or problem. The goal is to focus on the concrete everyday practices from which our intellectual curiosity and aspiration tend to estrange ourselves through abstractions and metaphysical or epistemological objective validity.

Nevertheless, this contingent way of philosophizing is not uncritically either. It does not imply a conservative uncritical acceptance of argumentative practices that uses rationality as a constitutive rule, norm or value. The therapeutic stance merely claims that we do not want to be cut off from the only actual resources for understanding and criticizing our argumentative practices. We do not want to distance ourselves too far from our concrete rational discourses in our intellectual desire to philosophize or theorize "rationality in general" from an elusive external point of view, be it epistemological, metaphysical or other. Hence, the only source for intellectual or philosophical authority we have available resides in the actual practices or forms of life, in which we actually practice argument-making, claim making, reason giving, rational discourse, understanding, and affirm or refuse our rational criterions in a given argumentative discourse.

Philosophy cannot and should not advance any substantive theory of rationality and argumentations. It cannot be critiqued on their philosophical conception of rationality and argumentations, as it should be read as offering none. Nor should it offer any direction as to how the nature and function of rationality in argumentation ought to be investigated. All that it can do is attempt to explain the actual and particular context from which our puzzles about the place rationality in argumentative discourses must have arisen. And, then, attempt to describe these practices – forms of life – to provide a clarifying perspective. Note again that the terms *rationality* and *argumentative discourses* are here not to be understood in a theoretical way, as substantive objects of inquiry that requires metaphysical, epistemological or ontological grounding, but rather as terms that refer to what we are already familiar with in our various language games.

This does not mean we cannot isolate or explain false belief or misconceptions of arguers about their own practices. However, such explanations and the falsehood of such beliefs only make sense against the background of other argumentative practices and concrete perspectives on argumentative discourses as forms of life. To return to a practical understanding of the *forms of life* within which the understanding of argumentative and rational discourses operate. Indeed, "What has to be accepted, the given, is – one could say – forms of life" (*PI, p.226*). The passage is unique and important for the understanding the role which forms of life are meant to play, since it is the only passage in which Wittgenstein uses the term in the plural and associates it with the "given". Philosophy as therapeutic activity is the investigation of the "argumentative practices" in which our particular intellectual predicaments have their source.

8. CONCLUSION: "RATIONALITY AS USE"

The relevancy of the Wittgensteinian investigation to rationality is that it offers a framework for the reexamination of rationality, as this notion has been demotion in both philosophy and argumentation: First, the cross-disciplinary variability of rationality has seemed too great for it to be a value despite its being one of the core values of the Humanistic Western culture. Second, when it comes to the relation of rationality to argumentation, some sort of essentialism plays a key role in demoting rationality. Descriptive approaches to argumentation argue that the fact that many argumentation discourses does not rely on rationality shows that rationality is not part of the nature of argumentation, regardless of those that seemed to embody the identification of argumentation with the rational. It is good to note that *only if* it is assumed that argumentation has a nature, does the absence of rationality from some argumentation discourses establish that rationality is *inessential* to argumentation.

First, Wittgenstein's work bears on both points. First, Wittgenstein holds that rules, norms and values can only come into view in contingent practices. This outlook provides a framework for a detailed rehabilitation of rationality's specific value and plurality. Part of the task would involve showing how certain form of life makes a certain sense of rationality available. For example, one can examine what is rational in a certain cultural era – the institutionalization of research methods during the Renaissance by Galileo and others – and study the life activities in 16th century humanistic circles in Italy and other intellectual centers to understand that *that* notion of rationality becomes available. However, Wittgenstein ask us to be skeptic as he argues against supposing that because some words figure as an adjective, there is such a property. Rather, it is a word of approbation, and in circumstances of approval the term plays very little role.[14] Thus, the need to understand specific rational argumentative discourses by focusing on the way they are available in specific cultural and historical circumstances.[15]

Second, in displacing any trace of essentialism about rationality, Wittgensteinian approach makes it possible to consider rationality as a constitutive norm or value that inform some argumentative practices. But to the extent that contemporary reaffirmation of the need for rationality in argumentation suggest that rationality's role may be vital and integral to argumentation (Johnson, 2000), they show that rethinking rationality calls for rethinking argumentation and for rethinking argumentation in a way that allows rationality a more integral role. Wittgenstein's investigations offer what is needed by explaining argumentation as overlapping practices with differing constitutive norms and values, so that a certain use of rationality might be a constitutive value for some argumentation practices and a different use of rationality might be constitutive in a range of others. For example, the use of rationality in scientific argumentative discourses and the use of rationality in legal or juridical argumentative discourses. That been said, rationality would not have a constitutive role in other kinds of argumentative discourses, such as some of the political or advertising discourses.

My Wittgensteinian reading of the philosophy of argumentation and rationality is that it regards the various forms of life that have to do with argumentation as founded on some kind of reason rather than faith, emotion, empathy and so forth. This reading's intelligibility depends on criteria of rationality internal to those forms of life, and essentially distinct from those governing other modes of discourse. The relativistic stance of Wittgenstein suggests that one apparently coherent kind of rational evaluation of modes of discourse is really empty. In other words, there is no sense to be made of such evaluation at the level of forms of life taken as a whole. There is no "external" reference frame for critiques of the various forms of life that exercise reason. The rationality of specific language game of exercising reason cannot be

---

[14] These and the following remarks are derived from Wittgenstein's reflection in his "Lectures on Aesthetics" (Wittgenstein, 1966), for example, his remarks on p. 5 (§13).

[15] Wittgenstein (1966, p. 8), passages 25-26.

settled in terms that are determined by other language games and eventually be other forms of life.

Wittgenstein's investigations provide a framework for understanding argumentation, as well as its relationship with rationality, which can be further developed. The main point in this paper is that Wittgenstein directs our focus to the way constitutive rules, norms and values inform our practices and forms of life. This argues that argumentation is interrelated practices that are integral to different forms of life or cultural and historical eras. On this view, argumentation is like a thick rope made of many different intertwining fibers. The strands differ insofar as their constitutive rules, norms and values are different along with divergences in the forms of life in which they figure. This understanding of argumentation is more adequate to the cultural and historical phenomena than the view that argumentation is the sort of phenomenon that has a distinctive essence and is specifiable a-culturally and a-historically in terms of essentialist definition. This perspective offers also an approach to the relationships among the different argumentative discourses, scientific, legal, political, bureaucratic, cultural, and so forth – as interrelations between rules and norms-governed practices within their respective forms of life. In describing these various argumentative discourses, the focus is on the specific ways that the constitutive rules or norms of the various discourses emerge and evolve within forms of life. Thus, we can also ask how specific discourses such as scientific, artistic and political intertwine within any particular form of life, such as liberal states or across several. Rationality as embodied in constitutive rules, norms or values across argumentative discourses might be the philosopher's stone that will enable such understanding.

REFERENCES

Baudot, L. (2012). An Air of History: Joseph Wright's and Robert Boyle's Air Pump Narratives. *Eighteenth-Century Studies, 46*(1), 1–28.
Biro, J., & Siegel, H. (2006). In defense of the objective epistemic approach to argumentation. *Informal Logic, 26*(1), 91-101.
Biro, J., & Siegel, H. (2006). In defense of the objective epistemic approach to argumentation. *Informal Logic, 26*(1), 91-101.
Brundtland, T. (2011). After Boyle and the Leviathan: the Second Generation of British Air Pumps. *Annals of Science, 68*(1), 93-124.
Cavell, S. (1979). *The Claim of Reason: Wittgenstein, Skepticism, Morality, and Tragedy.* Oxford: Oxford University Press,.
Diamond, C. (1991). *The Realistic Spirit: Wittgenstein, Philosophy and the Mind.* Cambridge, Mass., and London: MIT Press.

Duro, P. (2010). 'Great and Noble Ideas of the Moral Kind': Wright of Derby and the Scientific Sublime. *Art History, 33*(4), 660–679. ScienceDirect. doi: 10.1111/j.1467-8365.2010.00779.x.

Eldridge, R. (1987). Problems and Prospects of Wittgensteinian Aesthetics. *The Journal of Aesthetics and Art Criticism, 45*(3), 251–261.

Glüer, K., & Wikforss, Å. (2015, April 27). *The Normativity of Meaning and Content.* (E. N. Zalta, Ed.) Retrieved May 2015, 1, from The Stanford Encyclopedia of Philosophy: http://plato.stanford.edu/entries/meaning-normativity/.

Haugeland, J. (1998). Pattern and Being. In *Having Thought: Essays in the Metaphysics of Mind* (pp. 267–290). Cambridge, Mass.: Harvard University Press.

Johnson, R. (2000). *Manifest Rationality: A Pragmatic Theory of Argument.* Mahwah: Lawrence Erlbaum Associates.

Kripke, S. (1982). *Wittgenstein and Rules and Private Language.* Oxford: Basil Blackwell.

La Torre, M. (2002). Theories of legal argumentation and concepts of law. An Approximation. *Ratio Juris, 15*(4), 377-402.

McDowell, J. (1995). Reply to Gibson, Byrne, and Brandom. *Philosophical Issues, 7*, 283-300.

McDowell, J. (1998a). Wittgenstein on Following a Rule. In *Mind, Value, and Reality* (pp. 221–262). Cambridge, MA: Harvard University Press.

McDowell, J. (1998b). One Strand in the Private Language Argument. In *Mind, Value, and Reality* (pp. 279–296). Cambridge, MA: Harvard University Press.

Nicolson, B. (1968). *Joseph Wright of Derby, Painter of Light: Text and catalogue.* London: Paul Mellon Foundation for British Art.

Nochlin, L. (2006). *Bathers, Bodies, Beauty: The Visceral Eye.* Cambridge, Mass, London: Harvard University Press.

Perelman, C., & Olbrechts-Tyteca, L. (1969). *The New Rhetoric: A Treatise on Argumentation.* (J. Wilkinson, & P. Weaver, Trans.) Notre Dame, Indiana: University of Notre Dame Press.

Putnam, H. (1990). *Realism with a Human Face.* (J. Conant, Ed.) Cambridge, MA: Harvard University Press.

Rawls, J. (1955). Two Concepts of Rules. *The Philosophical Review, 64*(1), 3-32.

Schaffer, S., & Shapin, S. (1985). *Leviathan and the Air-Pump: Hobbes, Boyle, and the Experimental Life.* Princeton, NJ: Princeton Univ. Press.

Schlick, T. (2005). The Critical Collaboration between Art and Science: 'An Experiment on a Bird in the Air Pump' and the Ramifications of Genomics for Society. *LEONARDO, 38*(4), 323–329.

Schwed, M. (2013). Argumentation as an ethical and political choice. In D. Mohammed, & M. Lewiński (Eds.), *Virtues of Argumentation* (pp. 1-14). Windsor, ON: Ontario Society for the Study of Argumentation.

Schwed, M. (2015). Argumentation as a Rational Choice. *Forthcoming in 2014 ISSA Proceedings.*

Wittgenstein, L. (1961). *Tractatus Logico-Philosophicus.* (D. Pears, & B. McGuinness, Trans.) New York: Humanities Press.

Wittgenstein, L. (1966). *Lectures and Conversations on Aesthetics, Psychology, and Religious Belief.* (C. Barrett, Ed.) Oxford: Basil Blackwell.
Wittgenstein, L. (1967). *Philosophical Investigations* (3rd ed.). (G. Anscombe, R. Rhees, Eds., & G. Anscombe, Trans.) Oxford: Blackwell.
Wright, C. (2007). Rule-Following without Reasons: Wittgenstein's Quietism and the Constitutive Question. *Ratio, 20*(4), 481–502.

APPENDIX
Three paintings of Joseph Wright of Derby

Figure 3: Joseph Wright of Derby, The *Alchemist, in Search of the Philosopher's Stone*, 1771.
Joseph Wright of Derby, *The Alchemist, in Search of the Philosopher's Stone, Discovers Phosphorus, and prays for the successful Conclusion of his operation, as was the custom of the Ancient Chemical Astrologers* (1771), oil on canvas, 127 cm × 101.6 cm (50 in × 40.0 in), Derby Museum and Art Gallery, Derby. Source of image: *Wikipedia, Public Domain*:
http://commons.wikimedia.org/wiki/File:JosephWright-Alchemist.jpg.

"Rationality as use" 657

Figure 4: Joseph Wright of Derby, *An Experiment on a Bird in the Air Pump*, 1768.
Joseph Wright of Derby. *An Experiment on a Bird in the Air Pump*, 1768, Oil on canvas, 183 cm × 244 cm (72 in × 94 1/2 in), National Gallery, London, England. Source of image: *Wikipedia, Public Domain*:
http://en.wikipedia.org/wiki/File:An_Experiment_on_a_Bird_in_an_Air_Pump_by_Joseph_Wright_of_Derby,_1768.jpg.

Figure 5: Joseph Wright of Derby, *The Orrery or A Philosopher Giving a Lecture on the Orrery*, 1766.
Joseph Wright of Derby. *The Orrery* or *A Philosopher Giving a Lecture on the Orrery*, 1766, oil on canvas, Derby Museum and Art Gallery. Source of image: *Wikipedia, Public Domain*:
http://commons.wikimedia.org/wiki/File:Wright_of_Derby,_The_Orrery.jpg.

Essentialism, Relativism and Scepticism
Commentary on Schwed's "Rationality as Use"

NUNO VENTURINHA
IFILNOVA, Universidade Nova de Lisboa, Portugal
nventurinha.ifl@fcsh.unl.pt

1. INTRODUCTION

With a few rare exceptions, Wittgenstein's thought has not been consistently applied in argumentation theory. One could perhaps argue that Toulmin's *The Uses of Argument* owes a great deal to the later Wittgenstein's emphasis on the practical dimension of human reasoning, but the fact is that Wittgenstein is only mentioned once in the conclusion to that book (see Toulmin, 2003, p. 233). Henrique Jales Ribeiro, who sees Toulmin's seminal work as contributing not only to an argumentation theory but also to a theory of meaning which should complement that of analytic philosophy, actually claims that both the early and the later Wittgenstein "believed that the concept of meaning did not include argumentation at all" (Ribeiro, 2012, p. 493). It is interesting to consider, however, what Toulmin, in conversation with Sheldon Hackney, said about Wittgenstein:

> One of the things he was struggling with was the question, how communication is possible, how human modes of expression are capable of being meaningful at all. . . .The one thing he was sure about was that any agreement that we could reach would not be a matter of intellectual consensus. It would be a convergence of humane attitudes. (Lifson, 1997)

Schwed's paper does not focus on the issue of Wittgenstein's transition from a formal theory of meaning, as distinctive of the Tractarian period, to an informal one, as paradigmatic of the *Philosophical Investigations*. What Schwed is interested in is the multidimensionality of the meanings we are confronted with in our different uses of language and so it is the concept of rationality, in its fluidity, that is prominent in his approach. According to Schwed, in order to find "the determining factors of meaning", we must look at "the internal relationships" that lie between "what we can say, what we can do, and the circumstances in which these take place" (p. 642), not to any theoretical scheme that aims to determine *in abstracto* what such and such means.

In what follows I shall briefly sketch the main claims of the paper as well as some difficulties I have with specific points of Schwed's view.

2. ESSENTIALISM

As stated at the beginning of his paper, Schwed wants "to use a Wittgensteinian framework" with the purpose of "understanding rationality as practices that have differing constitutive rules, norms and values" (pp. 635-636). What he stresses is not simply that there is more than a single mode of rationality. Schwed wants to indicate that our normativity and valuation are in a process of continuous change in regard to their content, even if the form of rationality that accompanies any concrete application points to what he likes to call "essential characteristics". These, he explains, represent "something with a determining or *constitutive* force that we need to understand" (p. 636).

Schwed immediately tries to make clear that his talk about "essential" and "constitutive" cannot be interpreted as if he were proposing something "'objective' or 'universal' in a sense of necessary and sufficient for some argumentative discourse to be rational" (ibid). It follows from his argument that he could not indeed propose such a thing since, as he also emphasizes, the realm of argumentation is that of contingency as long as it "essentially involves a *choice*" (ibid). However, it seems difficult to maintain that what we call human rationality, seeing it evolving throughout history, is not *the* essence of man, as the Greeks advocated so many centuries ago, as this has specific features. It is really one of the problems with the interpretation of Wittgenstein that, on the one hand, his teaching forces us to abandon Platonism or realism, to use Dummettian jargon, but, on the other hand, his anti-Platonism or anti-realism does not entail an absolutely satisfactory way out from traditional metaphysics.

There is visibly a tension in Schwed's paper regarding the "historically and culturally contingent and changing character of rationality" (ibid) that is highlighted and the recognition of a regularity in what is called rational. What Wittgenstein suggests we do is to concentrate on what is common to all processes falling within that category. We know how much art, philosophy or science have changed since their first manifestations, but the huge developments in our perspectives and techniques are by no means sufficient to exclude the seal of rationality from Prehistoric cave art or Pre-Socratic ontology and physics. It comes as no surprise that Schwed focuses extensively on Wittgenstein's notion of "form" or "forms of life". He calls our attention to §19 of the *Investigations* and to a section from its second part where Wittgenstein uses the singular and the plural forms respectively.[1] For Schwed, this notion captures Wittgenstein's "substitution of objectivity for relativity" (p. 638), but I must say that this reading leaves me uneasy, in particular taking into account Schwed's own claim that there are "essential characteristics". Let us have a closer look at this.

3. RELATIVISM

Schwed's idea is that "a Wittgensteinian approach is relativist about argumentative language games" given that it takes "a relativist stance with respect to human language and forms of life in general" (ibid). This brings into the discussion the problem of communication that Toulmin, in the aforementioned passages, refers to. Of course we cannot pre-establish what kind of arguments in different argumentative situations will do. As Aristotle has shown in the *Rhetoric*, our success in argumentation depends on a wide variety of factors that are not limited to the explicit use of reason as *logos* but also involve his categories of *ethos* and *pathos*. Still, I do not think that an open perspective in argumentation needs to be accompanied by relativism about *human language and forms of life in general*.

3.1 Pragmatism

In laying down the advantages of "pragmatic theories of argumentation that focus on the context of complex relationships" (p. 640), Schwed seems to side with Quine's view about the "indeterminacy of translation", that is, the epistemological limitation to adequately translate what was uttered by someone from another community of speakers or even by someone from our own community. In the eyes of Quine, the best we can attain is pragmatic consistency, not literal translation. I think that Quine's view is quite close to Wittgenstein's but only up to a certain point. Whereas Quine puts the accent on pragmatism, on the need human beings have to establish agreements, Wittgenstein prefers to draw attention to the naturalism that gives shape to our practices. If there were a complete relativism, it would be

---

[1] I discuss this issue in detail in Venturinha, 2010. Wittgenstein also uses the singular form in §§23 and 241 as well as in the first one of Part II, which Peter Hacker and Joachim Schulte have called "Philosophy of psychology – A fragment" (Wittgenstein, 2009).

impossible to understand a poem like Homer's *Iliad*. True, it can be argued that we do not understand it anymore and are lost in translation, as were the first to attempt to translate it. But if it were so, there could never have been a minimum of understanding outside particular spheres, niches of private linguists, so to speak.

What is remarkable in Wittgenstein's later work is that it develops a conception of openness in regard to meaning, amplifying the Fregean context principle already present in the *Tractatus*, but urges us at the same time to resist any temptation to endorse privacy in language. And he does so not for the sake of pragmatism in communication, as if we could decide upon it, but because human life is simply like this and we have to accept it. Thus no matter how far the aesthetical, political or religious discourses of a certain people are from ours in time or space, we are in principle able to grasp them. That we can recognize humanity in other people is the best evidence that we do, and not simply that we want to, understand each other. As Wittgenstein insightfully remarked, "if a lion could talk, we wouldn't be able to understand it" (2009, II, §327), but we can often understand a person even when s/he does not speak, for instance when someone cries or laughs. In those cases we may certainly be under the illusion that we understand the other, and that is why Wittgenstein dedicates so many thoughts to cases of dissimulation in order to identify what is natural for us to assume, to take for true. This is the core of his insistence on "[s]*eeing what is in common*" – the opening phrase of §72 of the *Investigations*, also alluded to by Schwed – which results in following rules that ultimately do not depend on us.

3.2 Variability

Schwed is certainly right in claiming that we can only define "rationality" by pointing to the variety of "uses that we call 'rational'" (p. 639). But I do not agree with him that this implies saying that "[r]ationality is not a concept" (ibid). If it were so, we could only provide definitions of analytical concepts. It is precisely due to the incompleteness of the definitions of our other concepts that Wittgenstein introduces the idea of "family resemblances". For Schwed, Wittgenstein's aim is to eschew that "family membership" involves "common essence that explains the manifest diversity" (p. 640). Schwed wants to escape the positing of an "underlying essence, prototype or paradigm" (ibid), but genomic research has shown exactly the opposite. Schwed tries to remedy the conflict of adopting and at the same time rejecting Wittgenstein's view with the suggestion of examining "the concept of rationality" – something he previously denied exists – "as *multiply intertwining or simply fibrous* rather than *family resemblance*" (p. 641). But this is a simple substitution of words.

Schwed's main concern, to be sure, is that we do not fall into reductionism while attempting to capture what rationality involves. Thus he writes that "the point is to recognize the integrated variety in our capacities without prioritizing either perception or action or thought" (pp. 644-645), with argumentation in the whole appearing as illustrative. Schwed uses the 18th century painter Joseph Wright of Derby to demonstrate that "his artistic work undoubtedly gave rise to new constitutive rules and norms" (p. 648). Schwed attempts to convince us that artists like Wright, who were motivated by science, "were challenging more cultural and political ideas than aesthetic norms associated with the artistic world of the time" (p. 649). I do not dispute that. What I dispute is the relativist status he attributes to what he himself calls "constitutive". It is precisely because human normativity is *constitutive* that we are able to identify variances over time – in the "natural history of human beings" (Wittgenstein, 2009, §415). This brings me to the last point I would like to discuss.

4. SCEPTICISM

Schwed takes Wittgenstein's philosophy as "a critical and skeptic one, rejecting any substantial attempt toward essentialism or universalism of any sort" (p. 649). He links this with the idea of a "*therapeutic* conception of philosophizing" (ibid) that Wittgenstein began to practise at the time of the *Tractatus* and that continues to be central in his later work. Inspired by the so-called resolute reading of Wittgenstein, Schwed is of the opinion that "the therapeutic stance proceeds in a piecemeal way", not in a systematic one, and this means "adjusting 'method', 'theory' and 'goal' in light of the concerns and problems at hand until satisfaction or clarity is achieved" (p. 650). That seems to be in fact what Wittgenstein wants to express when he rejects "*a single* philosophical method" (p. 650) and talks instead about "methods, different therapies, as it were" (Wittgenstein, 2009, §133). However, this does not result in saying, as Schwed believes, that "philosophy is neither theory nor method" (ibid). It is precisely *theory* and *method*, albeit of a different sort.

What Schwed and the self-claimed resolute readers fail to understand is that scepticism is impracticable and the alleged "pragmatic-practical reorientation" that the later Wittgenstein seems to offer is, appearances to the contrary notwithstanding, "the formation of – yet another – philosophical theory and method". That we study it and write about it is the best evidence for this assumption. Schwed claims that philosophy is not to "offer any direction as to how the nature and function of rationality in argumentation ought to be investigated" (ibid). It is this lack of constringency that, for Schwed, leads to a sceptical attitude. But there cannot be any scepticism if we are committed to making sense of other people's practices. It is symptomatic that Schwed states that such a strategy "does not mean we cannot isolate or explain false belief or misconceptions of arguers about their own practices". Why call them *false beliefs or misconceptions* if the framework is a sceptical one? Schwed goes on to say that "such explanations and the falsehood of such beliefs only make sense against the background of other argumentative practices" (ibid), but I imagine these are those of the interpreters. In the end, it is all a question of tolerance and that seems to be the real upshot of Wittgenstein's philosophy.

REFERENCES

Lifson, A. (1997, March/April). A Conversation [of Sheldon Hackney] with Stephen Toulmin. *Humanities*, *18*(2). Retrieved from http://www.neh.gov/humanities/1997/marchapril/conversation/conversation-stephen-toulmin.

Ribeiro, H. J. (2012). On the divorce between philosophy and argumentation theory. *Revista Filosófica de Coimbra*, *42*, 479-498.

Toulmin, S. (2003). *The uses of argument* (updated ed.). Cambridge: Cambridge University Press.

Venturinha, N. (2010). Introduction. In A. Marques & N. Venturinha (Eds.), *Wittgenstein on forms of life and the nature of experience* (pp. 13-19). Bern: Peter Lang.

Wittgenstein, L. (2009). *Philosophical investigations* (4th ed.). Oxford: Wiley-Blackwell.

29

Dialogue grammar induction

MARK SNAITH
Centre for Argument Technology, University of Dundee, UK
m.snaith@dundee.ac.uk

CHRIS REED
Centre for Argument Technology, University of Dundee, UK
c.a.reed@dundee.ac.uk

This paper presents a foundation for inducing formal dialogue games from analysed transcripts of real, inter-human conversations. We describe the *DI-Algorithm* (Dialogue Induction Algorithm), that accepts as input transcripts analysed using the Argument Interchange Format (AIF) enriched with Inference Anchoring Theory (IAT), from which it induces a formal, context-free grammar. This grammar describes the dialogue protocol that was followed in order to generate the original transcript. To illustrate the *DI-Algorithm*'s application, we provide a worked example based on an AIF analysis of a mock conversation.

KEYWORDS: dialogue games, dialogue protocols, grammar induction

1. INTRODUCTION

Philosophical dialogue games, such as those proposed by Hamblin (1970), Walton (1984) and Walton and Krabbe (1995) have been used to influence computational protocols for argumentative inter-agent communication (Reed, 1998; McBurney & Parsons, 2002a, 2002b; Black & Hunter, 2009). These games have traditionally been specified by hand. This, however, is a process that has two significant drawbacks; first, for all but the most trivial games it is time-consuming to generate the rules that define the protocol and account for every possible situation; second, it is rare for dialogue game specifications to accurately reflect the way in which real (human) dialogues progress. This second drawback is particularly relevant to the recent emergence of mixed-initiative argumentation (Snaith et al., 2010), where the participants in a dialogue can be real, virtual or a mixture of both. For users to remain engaged in a mixed-initiative dialogue governed by a formal protocol, it is imperative that permitted moves allow the dialogue to advance in a natural way, especially in domain-specific scenarios. If we are to model such natural dialogue flow in an artificial environment, it is essential that we understand the rules and protocols that are (in many cases subconsciously) followed in real, everyday dialogues.

Types of natural dialogue vary immensely in terms of both structure and the rules that govern their flow; contrast, for instance, a highly strict and formal courtroom setting, with some friends chatting over coffee. In the former, the rules are largely explicit with all parties involved strictly abiding by what they can say and when. In the latter, the structure is much more casual and simply flows based on what has previously been said. Nevertheless, even in formal settings there are certain implicit rules and norms that regulate the conversation (e.g. it is rude to interrupt, you should not suddenly change the topic, questions should typically be answered etc.).

One method of gaining an understanding of the structure of natural dialogues in a given domain or setting is to analyse transcripts of the conversations those dialogues generate. Discourse analysis (or mapping) involves breaking down individual components of an argument or other linguistic structure and re-assembling them in a form that shows the relationship between individual claims and premises (Van Eemeren et al., 2014). The Argument Interchange Format (AIF) (Chesñevar et al., 2006) allows for the analysis of the argumentative structure of a transcript, while Inference Anchoring Theory (IAT) allows this structure be tied back to the illocutionary acts that generated it (Reed & Budzynska, 2011).

In this paper, we present a foundation for inducing dialogue games from analysed transcripts of real inter-human dialogue. We describe the Dialogue Induction Algorithm (*DI-Algorithm*) which extracts sequences of locutions (illocutionary acts and the speakers thereof) from IAT-analysed transcripts and uses those sequences to learn a dialogue grammar. The induced grammar consists of production rules that generate syntactically-valid sequences of dialogue moves, representing the protocol that was followed to produce the original transcript. We illustrate the application of the *DI-Algorithm* by means of a worked example, based on a simple AIF analysis of a mock conversation.

2. BACKGROUND AND MOTIVATION

2.1 Motivation

In real, inter-human dialogues it is rare that an explicit dialogue protocol is followed, and instead it is conventions and norms that dictate the flow of dialogue. Through analysing such dialogues, however, we can expose a protocol encoded within. Speech Act Theory (Searle, 1969) connects propositional reports of dialogue events (e.g. "Bob says it is sunny") via applications of illocutionary force (e.g. "asserting") with their propositional contents (e.g. "it is sunny"). By chaining together the applications of illocutionary force, we can identify an abstract sequence of acts that can subsequently be generalised into a dialogue game representing the encoded protocol (e.g. "asserting" can follow "questioning").

Inducing dialogue games from transcripts of real conversations provides significant benefits both to academia and beyond. As an example of the former, an analysis of a dialogue game induced from transcripts of political debates (e.g. the United States presidential debates) could reveal previously unobserved phenomena that occur in this particular type of discourse.

Beyond academia, application areas range from social computing (where software could support more natural dialogues with mixed-initiative argumentation systems such as Arvina (Lawrence et al., 2012)), to professional training and development (such as successfully inducing a dialogue game that models a courtroom dialogue, providing a training tool for new and existing lawyers).

2.2 Inference Anchoring Theory

Inference Anchoring Theory, IAT (Reed & Budzynska, 2011), provides a set of mechanisms founded in philosophy and linguistics for connecting together two perspectives on argumentation: those that focus on inferential structures and those that focus on inter-agent interactions. IAT unites three pieces of machinery for understanding, representing, manipulating, supporting and creating arguments. First, is a lightweight generalisation of theories of argument macrostructure that identify components of arguments as (more or less) propositions along with relationships between those components focusing on inference and conflict. Second, is an approach to handling discourse structure which focuses on protocol-governed (i.e. rule-governed) transitions between discourse moves. Sitting between these two is the third component, a derivative of speech act theory (Searle, 1969) which connects propositional reports of dialogue events (such as "Bob says it is sunny") via applications of illocutionary force (such as "asserting") with their propositional contents (such as "it is sunny"). The propositional contents are the bread and butter of the first, inferential-oriented, component of IAT; the reports of discourse events are part of the second, interaction-oriented component of IAT; and the third, illocution-oriented component of IAT ties them together. As a result, IAT allows us to explain how an instance of an inference (say from p to q) can be *anchored* in the transition from an agent's challenge of q ("why do you think q is the case?") to the respondent's reply ("well, because p"). Thus IAT captures the intuition that inferences often exist precisely in virtue of the transition from one dialogue move to another – neither challenging q nor asserting p are intrinsically arguing anything. It is only by virtue of asserting p immediately after having q challenged that an inference is established. IAT provides a precise mechanism for showing how inferences are anchored in dialogical behaviour.

2.3 Formal grammars and grammar induction

A *formal grammar* is a set of production rules that allow sentences to be constructed in a formal language, based on two sets of symbols: *terminals*, which cannot be replaced through the application of production rules; and *non-terminals*, which are replaced (by terminals and non-terminals) through applying production rules.

The Chomsky Hierarchy (Chomsky, 1956) is a classification of different types of formal grammar based on the language they generate and the restrictions on the form that rules take. Broadly, there are four levels to the hierarchy, described below where: A and B are non-terminal symbols; α, β and γ are strings of terminal and/or non-terminal symbols; and a is a single terminal symbol.

1. Unrestricted languages (also called recursively enumerable), where production rules have the form $\alpha \rightarrow \beta$; that is, the string α can be replaced by the string β.
2. Context-sensitive languages, where production rules have the form $\alpha A \beta \rightarrow \alpha \gamma \beta$; that is, the non-terminal A can be replaced by the string γ iff A is preceded by the string α and succeeded by the string β.
3. Context-free languages, where production rules have the form $A \rightarrow \gamma$; that is, the non-terminal A can be replaced by the string γ regardless of the context of A.
4. Regular languages, where production rules have the form $A \rightarrow a$ and $A \rightarrow aB$; that is, the non-terminal A can be replaced either by

a single terminal, or a single terminal followed by a single non-terminal.

When written, grammars generally have their rules abbreviated such that those with the same left-hand side have their right-hand sides expressed in a single set. For instance, $A \rightarrow \{\beta, \gamma\}$ represents two context-free rules $A \rightarrow \beta$ and $A \rightarrow \gamma$.

As a simple example to illustrate a formal grammar, consider the following rules which describe valid short phrases in the English language, where "_" is a visible space:

$S \rightarrow \{ARTICLE\_NOUN, S\_V ERB\_PREPOSITION \_ARTICLE\_NOUN\}$
$NOUN \rightarrow \{cat, hat, mat\}$
$ARTICLE \rightarrow \{a, the\}$
$V ERB \rightarrow \{sat, ran\}$
$PREPOSITION \rightarrow \{in, on\}$

The production rule S describes valid phrases in the language: a sentence can consist of an article and a noun, or (recursively) a sentence followed by a verb, a preposition, an article and a noun. Valid phrases in this language include (but are not limited to):

the_cat
a_mat
the_cat_sat_in_the_hat
the_hat_sat_on_the_cat
a_mat_ran_on_the_cat_sat_on_the_mat

The last sentence illustrates an important principle of formal grammars — production rules generate syntactically-valid sentences in the language with no consideration for semantics.

Formal grammars can be used for representing dialogue game specifications. In this paper, we will provide examples of context-sensitive grammars that describe dialogue games; one through a worked example, the others being derived from real data.

2.4 Grammar induction

Grammar induction, sometimes referred to as grammatical inference, is the process of inducing a grammar based on a set of sentences (a language) that grammar can generate. Duda et al. (2001) provide a simple algorithm for grammar induction. This algorithm takes a set of training sentences and returns a grammar that can generate those sentences. Since it is possible for two or more grammars to generate the same language, such algorithms are sometimes expanded to accept as additional input a set of "negative" sentences that are known not to be derivable in the grammar. This limits the potential production rules and thus increases the likelihood of yielding a unique grammar.

In broad terms, given a set of positive (training) examples D^+ and a set of negative examples D^-, the algorithm takes each sentence in D^+ in turn and adds to the grammar the required production rule(s) that allow that sentence to be generated, provided said rule(s) do not allow a sentence in D^- to be generated.

This algorithm also imposes two constraints: 1) that the alphabet (terminal symbols) of the resultant grammar be only those used in the training sentences; and 2) that every production rule in the grammar is necessary in order to regenerate the training sentences.

In our approach to dialogue grammar induction we specify an algorithm that is bound by these constraints, but does not necessarily yield a only a single grammar and thus does not require a set of negative examples. We do not impose a uniqueness requirement because it is possible for some sequences of locutions to be generated by two or more different dialogue protocols, which in turn are described by different dialogue grammars. Our intention is to unearth all possible dialogue grammars from a given training set of sequences.

3. THE *DI-ALGORITHM* FOR DIALOGUE GRAMMAR INDUCTION

In this section, we present the *DI-Algorithm* for inducing a dialogue grammar from transcripts of real conversations. The algorithm takes as input AIF-IAT analyses of transcripts and generates the production rules for a context-free grammar that represents a simplified version of the dialogue protocol.

In common with other grammars, those induced by the *DI-Algorithm* consist of both terminal and non-terminal symbols. The terminals are locutions extracted from the analysed transcripts; the non-terminals are generated by the induction process when creating the production rules.

The *DI-Algorithm* breaks down into a 4-stage process:

1. Locution sequence extraction — from an AIF analysis, extract sequences of locutions in the order in which they were uttered.

2. Minimal valid sequence identification — for each extracted sequence, identify minimal non-atomic sequence(s) that are valid with respect to the transitions in the original sequence.
3. Centre enrichment— find (sub-)sequences in the original sequence that are expansions of the minimal input (i.e. sequences that enrich the minimal expansion with further locutions)
4. Rule generalisation — modifying rules to reduce the size of the grammar by replacing sub-sequences in right-hand sides with the left-hand sides of rules whose right-hand side is that sub-sequence

We now describe each stage of the process in detail with a running example to illustrate each concept. We specify sub-algorithms as logic programs for the sake of brevity.

3.1 Locution sequence extraction

The first stage of inducing a dialogue grammar is extracting sequences of locutions from an IAT analysis. Consider the following mock conversation between Alice and Bob, as they discuss what activity they should do:

>**Alice:** *We should go to the cinema.*
>**Bob:** *Why do you say that?*
>**Alice:** *Because we enjoy watching films.*
>**Bob:** *We should go to the park.*
>**Alice:** *Why do you say that?*
>**Bob:** *Because we enjoy the outdoors.*
>**Alice:** *OK, we should go to the park.*

An analysis of this conversation is shown in Figure 1, with the numbers representing a simplified representation of AIF timestamps (for clarity, we omit the full IAT analysis and anchorings which are not relevant for our purposes). From this analysis, we use the AIF timestamps to extract the following ordered list, S_1, of locutions identified by the analyst, where the subscript denotes the speaker:

$S_1 = (assert_A, challenge_B, assert_A, assert_B, challenge_A, assert_B, concede_A)$

This sequence represents a valid dialogue in the protocol whose grammar we are attempting to induce. To fully illustrate the algorithm, we will use two other valid sequences, derived from other analyses, in a set S of three thus:

$$S = \left\{ \begin{array}{l} S_1: (assert_A, challenge_B, assert_A, assert_B, challenge_A, assert_B, concede_A), \\ S_2: (assert_A, challenge_B, assert_A, challenge_B, assert_A, concede_B), \\ S_3: (assert_A, assert_B, assert_A, concede_B) \end{array} \right\}$$

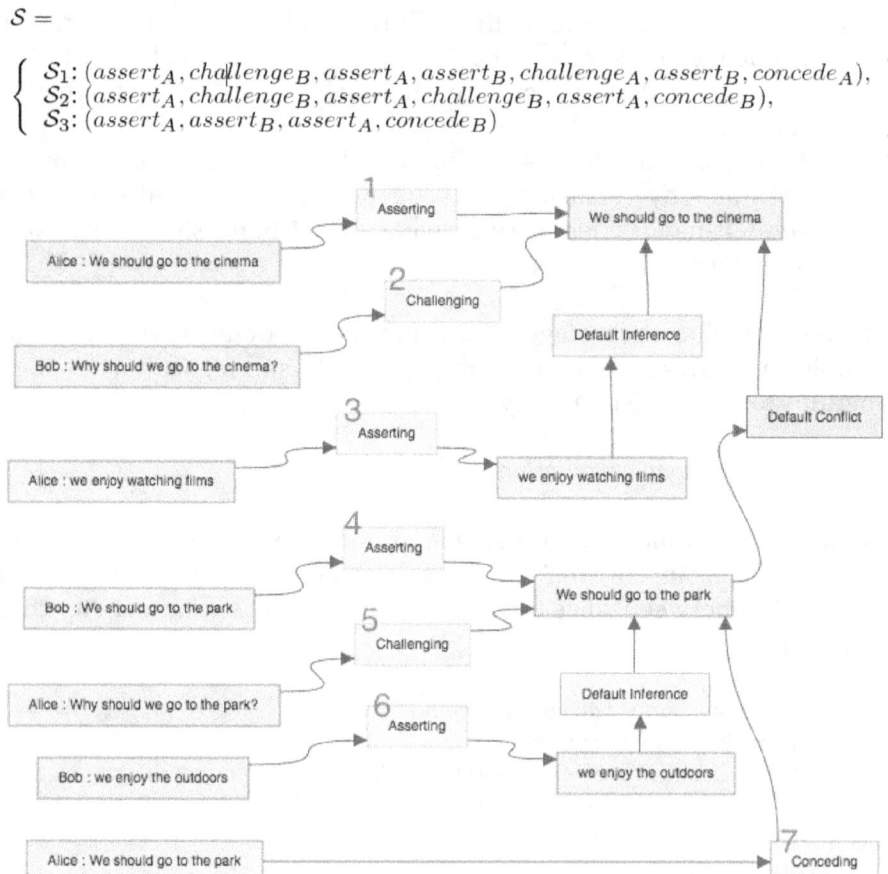

Figure 1 – Example analysis

3.2 Minimal valid sequence recognition

The second stage of the algorithm is to identify minimal valid sequences of locutions, based on the sequences extracted from the AIF analyses. Given a sequence S_i, the minimal valid sequence $S_{i\perp}$ is the shortest subsequence of S_i that is non-atomic (i.e. longer than one element) such that the first and last locutions in $S_{i\perp}$ are, respectively, the first and last locutions in S_i. The non-atomic condition prevents minimal sequences of only one locution; in the simplest case, this can arise from a sequence $S_j = (assert, challenge, assert)$. Without the non-atomic condition,

$Sj \perp = (assert)$ because the first and last locutions are the same.

We look for minimal valid sequences in order to identify the shortest dialogues permitted by the protocol. This is based on the assumption that, for a given sequence $S_i = (s_1, \ldots, s_n)$, s_1 is a valid starting locution in the protocol and s_n is a valid terminating locution. Identifying the shortest permitted dialogues provides the basis for the rules of the dialogue grammar.

Minimal valid sequences are established by either starting at the first locution and working forward through the sequence to find the closest occurrence of the last locution (front-to-back), or starting at the final locution and working back through the sequence to find the closest occurrence of the type and speaker of the first locution (back-to-front).

Once again using $S1$ as an example, we first label each locution in the sequence to provide clarity:

$l_1 : assert_A$

$l_2 : challenge_B$

$l_3 : assert_A$

$l_4 : assert_B$

$l_5 : challenge_A$

$l_6 : assert_B$

$l_7 : concede_A$

The first and last locutions are, respectively, $assert_A$ and $concede_A$. The closet occurrence of $assert_A$ to the end of the sequence is locution $l3$. Thus in this example the minimal valid sequence is that bounded by $l3$ and $l7$ inclusive:

$$S1 \perp = (assert_A, assert_B, challenge_A, assert_B, concede_A)$$

Algorithm 1 shows the sub-algorithm used to determine the minimal valid sequence in a given list using the back-to-front approach.

Algorithm 1 Minimal Valid Sequence determination

```
% mvs(+Sequence,-MinimalValidSequence)
mvs([X | Y], [X | Y]):-
        \+ member(X,Y).
mvs([X | T], [X | Y]):-
        next_sublist(X, T, S),
        mvs([X | S], [X | Y]).


mvs([X | Y], [X | Y]):-
        \+ member(X,Y).
mvs([X | T], [X | Y]):-
        next_sublist(X, T, S),
        mvs([X | S], [X | Y]).

% next_sublist(+Head,+ListStartingWithHead,-ListWithHeadRemoved)
next_sublist(X, [X | S], S).
next_sublist(X, [_ | T], S) :-
        next_sublist(X, T, S).
```

The minimal valid sequences for S_2 and S_3 in the training set are:

$S_2 \perp = (assert_A, concede_B)$ $\qquad\qquad$ $S_3 \perp = (assert_A, concede_B)$

Using the minimal valid sequences, we arrive at the following initial production rules in the induced grammar, where "D" is a non-terminal start symbol:

$$P = \left\{ \begin{array}{l} r_1 : D \rightarrow (assert_A, assert_B, challenge_A, assert_B, concede_A), \\ r_2 : D \rightarrow (assert_A, concede_B) \end{array} \right\}$$

3.3 Centre enrichment

The third stage of the algorithm is to identify ways in which adjacent pairs of locutions can be enriched by inserting sequences between them to create longer valid sequences. When a pair in a minimal valid sequence is enriched, a longer valid sequence of locutions is created, representing a longer dialogue. It is these enrichments that generate the production rules for the dialogue grammar.

Generating production rules based on enrichments is a two-stage process; for every enrichment:

1. generate a rule in which the left-hand side is a new non-terminal and the right-hand side is the (un-enriched) pair
2. for each enrichment of a pair, generate a rule in which the left-hand side is the same non-terminal and the right-hand side is the enriched pair

The first step in centre enrichment is to extract each adjacent pair of locutions from the input sequence. Returning to our example training set, the adjacent pairs in the sequence

S_1: (assert$_A$, challenge$_B$, assert$_A$, assert$_B$, challenge$_A$, assert$_B$, concede$_A$)

are:

p_1 : $(assert_A, challenge_B)$ p_4 : $(assert_B, challenge_A)$

p_2 : $(challenge_B, assert_A)$ p_5 : $(challenge_A, assert_B)$

p_3 : $(assert_A, assert_B)$ p_6 : $(assert_B, concede_A)$

Algorithm 2 shows the sub-algorithm for extracting pairs.
Taking each pair in turn sub-sequences of S_1 are identified which start and end with, respectively, the first and second locution in the pair. Algorithm 3 shows the sub-algorithm for centre enrichment.

Algorithm 2 Pair extraction

```
% pairs(+Sequence,-[Pairs])
pairs([],[]).
pairs([_],[]).

pairs([H, T | S], [[H, T] | Z]) :-
    pairs([T | S], Z).
```

Algorithm 3 Centre enrichment

```
% remove_last(+Sequence,-SequenceWithLastElementRemoved)
remove_last([_], []).
remove_last([X | Y], [X | T]) :-
        remove_last(Y, T).

% enrich(+Pair,+Sequence,-EnrichedPair)
enrich([], [], []).
enrich([X, Y], [X | T], S) :-
        remove_last([X | T], L),
        \+ last(Y, L),
        enrich([X,Y], L, S).
enrich([X , Y], [_ | T], S) :-
        enrich([X , Y], T, S).
enrich([X , Y], [X | T], [X | T]) :-
        last(T, Y).
```

In the case of $S1$, there is only one pair with valid enrichments, $p3$, with two in total. These lead to the production rules:

$r3 : N1 \rightarrow (assert_A, assert_B)$
$r4 : N1 \rightarrow (assert_A, challenge_B, assert_A, assert_B)$
$r5 : N1 \rightarrow (assert_A, challenge_B, assert_A, assert_B, challenge_A, assert_B)$

Applying the same process to sequences $S2$ and $S3$ yields the following production rules:

$r6 : N2 \rightarrow (assert_A, challenge_B)$
$r7 : N2 \rightarrow (assert_A, challenge_B, assert_A, challenge_B)$
$r8 : N3 \rightarrow (assert_A, concede_B)$
$r9 : N3 \rightarrow (assert_A, challenge_B, assert_A, challenge_B, assert_A, concede_B)$
$r10 : N3 \rightarrow (assert_A, assert_B, assert_A, concede_B)$

3.4 Rule generalization

In their current form, the rules can generate only trivial dialogues, because the right-hand side of each rule consists only of terminals. The final stage of the algorithm is to therefore generalise the rules such that, where possible, sub-sequences of terminals are replaced by non-terminals, allowing rules to grow and thereby accurately describe the dialogue protocol.

A rule is generalised if a sub-sequence in its right-hand side can be replaced by the left-hand side of another rule (i.e. a non-terminal). If there is a production rule $r1$ whose right-hand side is a sub-sequence in another rule $r2$'s right-hand side, replace the sub-sequence in $r2$ with the left-hand side of $r1$. For instance, the right-hand side of rule $r3 : N1 \rightarrow$ (**assertA, assertB**) is a sub-sequence of the right-hand side of rule $r10 : N3 \rightarrow$ (**assertA, assertB**, *assertA, concedeB*).

Thus our final set of production rules, including the start rules, is:

$$\mathcal{P} = \begin{cases} r_1 : D \rightarrow (N_1, challenge_A, assert_B, concede_A), \\ r_2 : D \rightarrow (N_3), \\ r_3 : N_1 \rightarrow (assert_A, assert_B), \\ r_4 : N_1 \rightarrow (N_2, N_1), \\ r_5 : N_1 \rightarrow (N_2, N_1, challenge_A, assert_B), \\ r_6 : N_2 \rightarrow (assert_A, challenge_B), \\ r_7 : N_2 \rightarrow (N_2, N_2), \\ r_8 : N_3 \rightarrow (assert_A, concede_B), \\ r_9 : N_3 \rightarrow (N_2, N_3), \\ r_{10} : N_3 \rightarrow (N_1, N_3) \end{cases}$$

Following rule generalisation, it is possible to filter the set of rules in order to eliminate those that can produce the same sequences as other rules, but are less general. This, however, is a step we don't yet wish to implement because we expect propositional content of locutions to have an impact on the filtration. Since it is our intention to build on the *DI-Algorithm* to account for propositional content (see Section 5), we do not want to introduce any processes that will require significant modification.

4. RELATED WORK

Here, we briefly compare and contrast our approach to dialogue game induction with existing work on both dialogue protocol/game induction, and grammar induction in general.

Similar grammar induction techniques are used in Alexandersson & Reithinger, 1997) and (Geertzen, 2009) to extract dialogue structures from corpus data. In both cases, however, the authors examined the problem in the context of dialogue act prediction — given a grammar induced from analysed or marked-up transcripts, how accurately can that grammar predict the next act in a dialogue using the same protocol. This differs from the present work which uses dialogue grammar induction as a step towards inducing an accurate dialogue protocol. It was further noted in (Geertzen, 2009) that algorithms for inducing dialogue grammars could be tested against data that has been annotated by dialogue games. We believe that our work develops a solid foundation for this by using recent developments in automated dialogue game execution (the Dialogue Game Execution Platform (Lawrence et al., 2012)) to generate annotated data against which our induced grammar is tested.

Grammar induction in general is a significant problem in computational linguistics and natural language processing. Approaches to grammar induction are split broadly into the classic machine learning approaches of supervised and unsupervised induction. In supervised induction, the algorithm is provided with sets of annotated sentences from which it learns sentence structure; in unsupervised induction, the algorithm must learn the structure from scratch (Clark & Lappin, 2010).

A variety of different techniques have been used in unsupervised grammar induction. In recent years, much work has built on the generative model of (Klein & Manning, 2002). This model relies on part-of-speech (POS) annotations, where models are trained using sequences of POS tags instead of raw tokens of text. An alternative approach, that parses raw text, is the *common cover links* (CCL) parser (Seginer, 2007). This approach, however, is hard to extend (Ponvert et al., 2011).

Our approach to dialogue grammar induction is set apart from unsupervised natural language grammar induction by the nature of the input to the *DI-Algorithm*. In unsupervised natural language grammar induction, the aim is to induce a grammar from a set of raw, unannotated sentences. Dialogue grammar induction, on the other hand, uses analysed transcripts where the illocutionary acts and speakers thereof are explicitly identified thus removing the need to first identify the type of act an utterance represents; in other words, dialogue grammar induction has already identified the non-terminal symbols (*assert*, *challenge* etc.), with the terminal symbols being the specific locutions complete with propositional content. In (unsupervised) natural language grammar induction, it is first necessary to identify those non-terminal symbols (noun, verb, adjective etc.) based only on the words themselves in the text.

Our approach does, however, have similarities with supervised natural grammar induction that relies on part-of-speech annotations. These POS annotations are similar to the identified illocutionary acts in the transcripts that feed the *DI-Algorithm*. A key difference is that a dialogue has clearly defined start and end rules that are used in the determination of minimal valid sequences that underpin the *DI-*

Algorithm and make dialogue grammars more restrictive than natural language grammars.

5. CONCLUSIONS AND FUTURE WORK

In this paper, we have presented a foundation for inducing dialogue games from analysed transcripts of real, inter-human dialogues, using techniques from formal grammar induction. We described and tested the *DI-Algorithm* for dialogue game induction. Using dialogical AIF, incorporating Inference Anchoring Theory, the *DI-Algorithm* extracts sequences of locutions (illocutionary acts and the speakers thereof) and uses them as a training set for learning a formal grammar. The grammar is a simplified representation of the dialogue protocol, whereby the production rules can verify sequences of locutions that are valid with respect to the original protocol.

In future work, we will design and implement and framework for testing the accuracy of the *DI-Algorithm*. This framework will attempt to induce a dialogue grammar using data generated from executing existing dialogue protocols, with the aim being to arrive at a grammars that accurately reflect those original protocols. Although the test framework is not yet specified, we envisage that accuracy will be measured based on false positives and false negatives in the induced grammars.

Further future work will be to further develop the *DI-Algorithm* so it can induce grammars not just from sequences of locutions, but also their propositional content. Achieving this will allow a DGEP-executable DGDL specification to be generated and, by building on our test framework proposed above, we can compare this new specification to the original one used to generate the test data, thus adding an extra step of verification.

What we have done in this paper is provide a foundation for inducing dialogue games from analysed corpora of real conversations.

ACKNOWLEDGEMENTS: This work was supported by the Leverhulme Trust under grant number RPG-2013-076. The authors are also grateful to colleagues in the Centre for Argument Technology for their comments and feedback on earlier versions of this work.

REFERENCES

Alexandersson, J., & Reithinger, N. (1997). Learning dialogue structures from a corpus. In *Proceedings of the Fifth European Conference on Speech Communication and Technology, EUROSPEECH 1997*, Rhodes, Greece.

Black, E., & Hunter, A. (2009). An inquiry dialogue system. *Autonomous Agents and Multi-Agent Systems, 19*, 173–209.

Chesñevar, C., McGinnis, J., Modgil, S., Rahwan, I., Reed, C., Simari, G., South, M., Vreeswijk, G., & Willmott, S. (2006). Towards an argument interchange format. *The Knowledge Engineering Review, 21(4)*, 293–316.

Chomsky, N. (1956). Three models for the description of language. *IRE Transactions on Information Theory, 2*, 113–124.

Clark, A., & Lappin, S. (2010). Unsupervised learning and grammar induction. In Clark, A., Fox, C., & Lappin, S. (Eds.), *The Handbook of Computational Linguistics and Natural Language Processing*, pp. 197–220. Wiley-Blackwell.

Duda, R., Hart, P., & Stork, D. (2001). *Pattern Classification*. Wiley-Interscience.

Geertzen, J. (2009). Dialogue act prediction using stochastic context-free grammar induction. In *Proceedings of the EACL 2009 Workshop on Computational Linguistic Aspects of Grammatical Inference*, pp. 7–15, Athens, Greece. Association for Computational Linguistics.

Hamblin, C. (1970). *Fallacies*. The Chaucher Press.

Klein, D., & Manning, C. D. (2002). A generative constituent-context model for improved grammar induction. In *Proceedings of the 40th Annual Meeting on Association for Computational Linguistics*, pp. 128–135. Association for Computational Linguistics.

Lawrence, J., Bex, F., & Reed, C. (2012). Dialogues on the Argument Web: Mixed initiative argumentation with Arvina. In Verheij, B., Szeider, S., & Woltran, S. (Eds.), *Proceedings of the Fourth International Conference on Computational Models of Argument (COMMA 2012)*, pp. 513–514, Vienna, Austria. IOS Press.

McBurney, P., & Parsons, S. (2002a). Dialogue games in multi-agent systems. *Informal Logic, 22(3)*, 257–274.

McBurney, P., & Parsons, S. (2002b). Games that agents play: A formal framework for dialogues between autonomous agents. *Journal of Logic, Language and Information, 11*, 315–334.

Ponvert, E., Baldridge, J., & Erk, K. (2011). Simple unsupervised grammar induction from raw text with cascaded finite state models. In *Proceedings of the 49th Annual Meeting of the Association for Computational Linguistics- Volume 1*, HLT '11, pp. 1077–1086, Stroudsburg, PA, USA. Association for Computational Linguistics.

Reed, C. (1998). Dialogue frames in agent communication. In Demazeau, Y. (Ed.), *Proceedings of the Third International Conference on Multi-Agent Systems (ICMAS 1998)*, pp. 246–253, Paris, France. IEEE Press.

Reed, C., & Budzynska, K. (2011). How dialogues create arguments. In van Eemeren, F. H., Garssen, B., Godden, D., & Mitchell, G. (Eds.), *Proceedings of the 7th Conference on Argumentation of the International Society for the Study of Argumentation.*

Searle, J. (1969). *Speech Acts.* Cambridge University Press.

Seginer, Y. (2007). Fast unsupervised incremental parsing. In *Proceedings of the 45th Annual Meeting of the Association for Computational Linguistics,* Vol. 45, p. 384. Association for Computational Linguistics.

Snaith, M., Lawrence, J., & Reed, C. (2010). Mixed initiative argument in public deliberation. In De Cindo, F., Macintosh, A., & Peraboni, C. (Eds.), *From e- Participation to Online Deliberation, Proceedings of the Fourth International Conference on Online Deliberation, OD2010,* Leeds, UK.

Van Eemeren, F. H., Garssen, B., Krabbe, E. C., Henkemans, A. F. S., Verheij, B., & Wagemans, J. H. (2014). *Handbook of argumentation theory.* Springer Berlin. Walton, D. N. (1984). *Logical Dialogue-Games and Fallacies.* University Press of America, Lanham.

Walton, D., & Krabbe, E. (1995). *Commitment in Dialogue: Basic Concepts of Interpersonal Reasoning.* State University of New York Press, New York.

Commentary on Snaith and Reed's Dialogue Grammar Induction

JOÃO LEITE
NOVA LINCS, Universidade NOVA de Lisboa, Portugal
jleite@fct.unl.pt

1. INTRODUCTION

Computational systems that are capable of maintaining intelligent conversations with humans have always taken a very special role when Artificial Intelligence is considered.

Setting aside the recent debate on the threat posed by Artificial Intelligence to Humankind – according to which the role played by humans becomes increasingly irrelevant, possibly until extinction – it is almost always through dialogues with humans that Artificial Intelligence is portrayed, at least outside its scientific community.

Whereas we could blame the Turing Test (Turing 1950) for this (perhaps excessive) focus on dialogue, it seems unquestionable that adequate human-machine interaction is crucial for the development of a new class of artificial intelligent systems that leverages on the new human-human, human-machine, and even machine-machine relationships established over the last 20 years, initially due to the internet and more recently to its online social networks.

2. CONTRIBUTION

Whereas a considerable amount of work has been devoted to developing computational systems that engage in dialogues with humans, the work of Snaith and Reed focuses on a special kind of dialogues, the so-called argumentative dialogues. Usually aimed at trying to reach a consensual conclusion through debate, or just as a record of justification for some assertion, argumentative dialogues can find successful applications both in classical educational systems as well as in a more recent class of systems that employ persuasion as the main driving technique to meet their objectives, such as persuading humans to change their behaviour (e.g., eat healthier) or even their beliefs (e.g., gender equality) (Hunter, 2014).

According to Snaith and Reed, automating the generation of argumentative dialogue moves, in a way that can effectively be used in such new class of AI based systems i.e., that can be simultaneously processed by computers and seemingly accepted by humans, requires not only the use of formal grammars, but that they be induced from real (human) argumentative dialogues.

The input to Snaith and Reed's work is then a corpus of real (human) argumentative dialogues, which have been pre-processed using the Argument Interchange Format (Chesñevar et al., 2006) and the Inference Anchoring Theory (Reed & Budzynska, 2011) in order to overcome some of the traditional bottlenecks of automated natural language processing, and to provide the additional required structure that would otherwise be difficult to obtain automatically.

The contribution of the paper is an algorithm, specified in Prolog, that takes a set of sequences of locutions, each representing an argumentative dialogue, and induces a Context-Free Grammar (CFG). Being preliminary work – as clearly stated by the authors – it is perhaps acceptable that many important details that should accompany the introduction of an algorithm are left out. Nevertheless, since the induction of CFGs is not a novel field – there are even competitions more than 10 years old on learning CFGs – several issues require further attention, namely:

- Is the algorithm correct? Whereas it is known that some languages cannot be identified, in the limit, from positive examples, perhaps the kind of pre-processing through the Argument Interchange Format and the Inference Anchoring Theory provides sufficient restrictions to a) ensure that the language can in fact be learnt and b) use them to prove correctness of the algorithm.
- Is the algorithm efficient? To deal with the amount of information required to develop modern AI applications it is no longer enough to have correct algorithms, but also that they be efficient, both in terms of time and space.
- How does it compare? Given the existence of many algorithms to induce CFGs, one wonders what are the reasons for not adopting an off-the-shelve solution, and develop one from scratch. Unless there are specific features of the input that require a tailored solution – which should be discussed – a comparison with other existing algorithms seems to be in order.

3. FINAL REMARKS

Despite some doubts about the significance of the algorithm proposed, the general line of research where this paper is situated is very promising and should be further pursued, with the ultimate goal of

automating the complete process of participating in argumentative dialogues. To this end, two additional lines of research could be worth pursuing:
- The research on the frontier between Automated Natural Language Processing and Argumentation – e.g., the work by Cabrio and Villata (2012) on identifying arguments from text –could somehow complement the work in this paper, namely by contributing to the automation of the pre-processing of the real (human) argumentative dialogues that the current algorithm requires.
- Given the nature of the argumentative dialogues used, perhaps Probabilistic Context-Free Grammars would be more appropriate, in the long run, for the kind of applications that the authors have in mind.

REFERENCES

Cabrio E., & Villata, S. (2012). Natural language arguments: A combined approach. In L. De Raedt et al. (Eds.), *Proceedings of the 20th European Conference on Artificial Intelligence (ECAI'12), Vol. 424 of Frontiers in Artificial Intelligence and Applications* (pp. 205-210). Amsterdam: IOS Press.

Chesñevar, C., McGinnis, J., Modgil, S., Rahwan, I., Reed, C., Simari, G., South, M., Vreeswijk, G., & Willmott, S. (2006). Towards an argument interchange format. *The Knowledge Engineering Review, 21*(4), 293-316.

Hunter A. (2014). Opportunities for argument-centric persuasion in behaviour change. In E. Fermé & J. Leite (Eds.), *Proceedings of the 14th European Conference on Logics in Artificial Intelligence (JELIA'14). Vol. 8761 of Lecture Notes in Artificial Intelligence* (pp. 48-61). Dordrecht: Springer.

Reed, C., & Budzynska, K. (2011). How dialogues create arguments. In F. H. van Eemeren, B. Garssen, D. Godden & G. Mitchel (Eds.), *Proceedings of the 7th conference on argumentation of the International Society for the Study of Argumentation* (pp. 1633–1645), Amsterdam: SicSat.

Turing, A. (1950). Computing machinery and intelligence. *Mind, 236*, 433–460.

30

On Cognitive Environments

CHRISTOPHER W. TINDALE
CRRAR, University of Windsor, Canada
ctindale@uwindsor.ca

Recent work in argumentation has introduced and explored aspects of the cognitive environment: examining the concept, expanding it, and presenting it as an important idea in our understanding of how audiences operate in argumentative situations. This paper builds on that work in two ways: (i) it extends the discussion of the "cognitive" environment to include values and aspects of the "emotive"; and (ii) it explores the relationship between cognitive environments and universal audiences.

KEYWORDS: Aristotle, cognitive environment, Perelman, universal audience

1. INTRODUCTION

1.1 Picturing the Universal Audience(s)

Conjure in your mind the image of Raphael's "School of Athens". It's so ubiquitous that the task is not difficult. As expected, the eye is drawn to the central figures, strolling together, Plato with his finger pointed heaven-ward, a gesture to the invisible but eternal absolutes, and Aristotle with a measured pressing of the hand toward the ground, with all the down-to-earth realism that that implies, conjuring up the standard of human experience and empirical activity. They make a perfect couple, side-by-side, yet dividing the philosophical hosts down the centuries. Less visible, in fact probably invisible in your version, are their indistinct companions, hidden in the shadows of each central figure. Alongside Plato, stooped beneath his arm, his gaze following the directing finger, is the Cartesian figure, perhaps Descartes himself. And accompanying Aristotle, but hanging back in dutiful recognition of the master's stature is (we're surprised to see) Chaïm Perelman. The

Cartesian and the Perelmanian: Two accompanists, two auditors, two representatives of a kind of auditor, of a universal audience.

Perelman wrote about both of these universal audiences and in his haste, perhaps, created such confusion that both have been associated with his name. But it is only the Aristotelian that is appropriate; the Platonic idea of the absolute that fuels the efforts of philosophers in the Cartesian tradition—that empty universal that strives to be timeless and spaceless—has no relation to what Perelman envisaged.

1.2 The Aristotelian Heritage

The label "universal audience" was one Perelman came to regret (citation). But really for a philosopher of his lineage it was a natural choice. Perelman's actual understanding was very much Aristotelian in spirit. Aristotle broke from the rationalism of his predecessor, who laid claim to a self-evidence that would eliminate all rhetoric that was not based on a knowledge of the truth (Perelman, 1989, pp. 242-243)[1]. So we see him, for example, challenging the Platonic notion of the universal Form of the Good, since it cannot be grasped and known. On Aristotle's terms, we see particulars, not such universals. The physician sees patients rather than "health" itself. And on seeing enough patients, that physician will build up sufficient knowledge as will permit more general claims to be made. And that accumulation of experience is crucial to success, as the physician becomes able to use this "universal" knowledge to act and judge. The knowledge of the particulars is not sufficient; the accumulation of the universal *in the individual's experience* is necessary.

This is how we should understand Perelman's approach to the universal audience.[2] It is a device that breaks from the rationalist objectivity of his philosophical predecessors (like Plato) and relies on the adherence of an audience (Perelman, 1989, p. 242). Accumulating knowledge of audiences through argumentative practices, we gain a proficiency at seeing the universal within the particulars (the actual

---

[1] I would observe here that it does not, as John W. Ray claimed (1978), eliminate all rhetoric from philosophy, since the late Plato clearly saw the philosopher's need for rhetoric when communicating in social contexts.

[2] Perelman writes: "I ask myself, in passing, whether this conception does not underlie the *Topics* of Aristotle, which sets forth the techniques for defending and for attacking a contested thesis with regard to *any interlocutor whatsoever*, while, in his *Rhetoric*, he analyses the means of persuading particular audiences" (1989, p. 244, italics in the original).

audiences we address). This universal captures our understanding of reason as it is alive in the communities within which we live and work and argue.³

1.3 Criticisms of the Universal Audience

But there are problems with Perelman's notion, problems that have been duly identified and addressed in the literature. Not least of these critiques are those provided by Frans van Eemeren and Rob Grootendorst (1995), by Alan Gross and Ray Dearin (2003), and by Scott Aikin (2008). These I take to be representative of the relevant range of criticisms. I will outline several of these problems because they are pertinent to what follows.

Let me begin with the van Eemeren/Grootendorst critique, since it is the earliest (of those I include) and to some degree the most damming, since it is founded on a charge of relativism. They note, that Perelman and Lucie Olbrechts-Tyteca link the soundness of argumentation to an audience that can be either particular or universal (van Eemeren & Grootendorst, 1995, pp. 122-123). But if soundness relies on the success of the argumentation with the intended audience, then argumentation that is sound in one case may not be in another, because there "could be just as many definitions of reasonableness as there are audiences" (124). This makes for an extremely relative standard. They proceed: "Introducing the restriction that argumentation is reasonable only when it is deemed sound by the universal audience does not lead to any necessary limitation. According to Perelman and Olbrechts-Tyteca, each individual is free to choose his or her own universal audience, so this only shifts the source of variation from the listeners to the speakers" (124). In saying this, they have a legitimate concern, which is expressed later in the paper: with every arguer selecting his or her own universal audience, gaps emerge between different groups of people. Argumentation, we might agree, should serve to connect people, not isolate them. As an alternative, the pragma-dialecticians propose a critical-rationalist model where arguments are acceptable to the degree that they withstand a critical interrogation "put forward in a systematic attacking procedure" (128).

So the first serious problem with Perelman's universal audience is that it is a relativistic standard. Much may depend on the real nature

---

³ The contrast at stake is nicely captured by Mary Midgley (1995) who attributes to Plato a *colonial* view of Reason, imposing order on unruly passions and instincts; while Aristotle—"*the* biologist among philosophers"—sees Reason as part of a continuous organic whole (260).

of any "freedom" that arguers have in "choosing" the universal audience. But we will return to this.

Alan Gross and Ray Dearin go to great lengths to make sense of some of the problematic statements associated with the universal audience. But in the process, they present a conception that is much closer to the Platonic ideal than the Aristotelian real. All rhetorical audiences, both universal and particular, are constructions of the arguer. But the constructions aiming at the universal strive to modify or reinforce the real or the true, in the manner of the sciences and philosophy, while those aimed at particular audiences strive to modify or reinforce values, in the manner of public addresses. Here, the sense of the universal is that which is upheld throughout one stream of the philosophical tradition, including Plato and Descartes. And in that light Gross and Dearin commend the critiques of Lisa Ede (1989), John Ray (1978), and Henry Johnstone Jr. (1978), even though Perelman himself identifies the Ede and Ray criticisms as founded on a misreading of the universal audience (1989).[4] Part of Ede's criticism, for example, accuses Perelman and Olbrechts-Tyteca of being caught in a contradiction when they write that reasons addressed to the universal audience "are self-evident, and possess a timeless validity, independent of local or historical contingencies" (Perelman & Olbrechts-Tyteca, 1969, p. 32). While this is indeed what is said in *The New Rhetoric*, the real issue should be what is being described. Still, in separating facts and values by audience, Gross and Dearin assert that a discourse that focuses "on values can never address a universal audience" (Gross & Dearin, 2003, p. 37) because particular values can never be binding on all human beings across time and space. That is, their adoption of Ede's reading of the universal audience influences how they understand argumentation involving both facts and values. They go on to further interpret Ede's reading: she is correct that argumentation addressed to a universal audience has a timeless validity, but not because such a validity exists, but because "speakers arguing for the real in a particular case must assume its existence in the general case" (p. 37). Thus, Perelman and Olbrechts-Tyteca are committed to a paradox of holding that arguers must presuppose a concept of timeless validity even while asserting that each arguer and culture has its own conception of the universal audience. The paradox is explained, or justified, by noting that philosophy addresses audiences that are universal even though there are actual audiences of real people (p. 38).

---

[4] He was more sympathetic to Johnstone, with whom he entered a fruitful dialogue, which in turn led Johnstone to revise his understanding of Perelman's ideas (see Tindale, 2010 for a discussion of Johnstone's views).

So, a second problem to note is that the universal audience is still for Perelman the ideal of the Cartesian tradition, and in that respect cannot include values.

Finally, I turn to Scott Aikin's recent discussion. Like that of Gross and Dearin, Aikin's project is a constructive one that aims to address the confusion that has surrounded the notion of the universal audience. Aikin cites the same remarks on universal validity that were central to Ede's reading and adopted by Gross and Dearin, but then notes in a footnote the controversial nature of this reading (p. 256n1) and Perelman's own later correction of it (p. 240). He draws a distinction between pragmatic and epistemic elements of the universal audience. The pragmatic features "are the sociological facts of cognitive and evaluative overlap between different people, groups, and societies" (p. 240). As I understand his reading, it is this pragmatic aspect that concerns the effectiveness of argumentation. But we must also be concerned about the validity of argumentation, and this is where the epistemic features enter and where the universal audience becomes a rationally regulative notion (p. 241). But the gap between the pragmatic goal of universal adherence (involving numbers) and the epistemic goal of validity (involving competence) cannot be bridged by a single notion of audience. And so Perelman's universal audience must be two different audiences. The pragmatic resolves disagreements and the epistemic produces knowledge.

In his subsequent discussion, Aikin is reluctant to see the epistemic reduced in some way to the pragmatic, and so strives to give credence to the idea of the epistemic universal audience. In doing so, he arrives at a conception that is Perelmanian in sentiment but not what Perelman himself endorsed. The epistemic involves presenting an audience with rationally compelling arguments, so the idea of increasing or eliciting adherence drops by the wayside. In its place stands a unified notion of the universal audience whereby "we address an epistemic UA not for the purpose of changing their minds but for the purpose of winning their approval" (p. 248).

One problem with Aikin's reading, one that he himself acknowledges, is that it does not stay true to Perelman's insistence that individuals, cultures and situations have their own conception of the universal audience. Here, he appeals to the concerns raised by van Eemeren and Grootendorst about relativism. To address this he introduces the distinction between concept and conception:

> The *concept* of epistemic UA is that audience that instantiates all the epistemic virtues and which reflects the ideal position from which to judge the quality of an argument. The criterion

for validity is whether an argument lives up to this standard. Each speaker has his or her own *conception* of what constitutes such an audience both in its virtues and in its commitments, and these conceptions will direct the ways speakers formulate their arguments with an eye to validity (p. 249).

Much then depends on how well an audience's conception matches the concept. The epistemic universal audience is the only one that can fulfill what I take Aikin to take as the primary goal of argumentation—to transmit knowledge. Thus, he concludes: "In order to know fully, believers must be able to give an argument they endorse. In order to put listeners in this situation, they must be addressed in terms with which they are familiar against a backdrop of commitments they share. Arguments that fail in this regard fail to transmit knowledge in the requisite sense and hence fail as arguments" (p. 256).

Aikin's distinction between the pragmatic and the epistemic is an insightful one, and something that should be built upon. His insistence that no unitary theory of the universal audience can accommodate both is the last problem that I will deal with.

It is the position developed in this paper that some, if not all, of these concerns, can be eliminated if we replace the concept of the universal audience with that of the cognitive environment. This latter concept can effectively do the job that Perelman had assigned to the universal audience, and do it without the accompanying difficulties.

2. THE COGNITIVE ENVIRONMENT

A traditional view of communication, favored by such thinkers as Locke (1975) and Grice (1989) saw it as a process by which a speaker's (or writer's) meaning is replicated in the mind of an audience. A more recent model proposed by Robert Brandom (1994) sees the communication of meaning as something negotiated in a social space. Deirdre Wilson and Dan Sperber are closer to the second position when they write: "[communication is] a more or less controlled modification by the communicator of the audience's mental landscape—his *cognitive environment*, as we call it—achieved in an intentional and overt way" (2012, p. 87). In equating this mental landscape with social space (a common physical metaphor, we should note), I understand the cognitive environment as social.[5]

---

[5] In fact, I see it as comparable to the space of reasons noted by Sellars and adopted by Brandom, as well as Habermas' lifeworld (Tindale, 2015, pp. 193-194).

As far as I am aware, outside of computer systems where the term has a specific sense related to interactive environments, "cognitive environments" first arises in argumentation-related literature in Sperber and Wilson's *Relevance: Communication and Cognition* (1986).[6] As they use the concept, it refers to the set of facts, assumptions, and beliefs that are manifest to a person. A belief or assumption is manifest if a cognitive environment provides sufficient evidence for its adoption. A more detailed statement of what is involved is given in the following description:

> To be manifest, then, is to be perceptible or inferable. An individual's total cognitive environment is the set of all the facts that he can perceive or infer: all the facts that are manifest to him. An individual's total cognitive environment is a function of his physical environment and his cognitive abilities. It consists of not only all the facts that he is aware of, but also all the facts that he is capable of becoming aware of, in his physical environment. The individual's actual awareness of facts, i.e. the knowledge that he has acquired, of course contributes to his ability to become further aware of facts. Memorised information is a component of cognitive abilities (1985, p. 39).

The reference to what someone is capable of becoming aware of is crucial here. Cognitive environments involve potential as well as actual knowledge. In this way, they are a more effective replacement for one general sense of "common knowledge." That is, writers and speakers often refer to what is commonly known by an audience when they rarely have grounds for making such a claim. Short of canvassing an audience to access their knowledge and beliefs (and the problems of interpretation associated with such an effort), we do not know what people actually know, we do not know what is commonly known. But given the cognitive environments in which people operate, we can gain a much better idea of what they *should* know. That is, we are aware of what facts, ideas, and beliefs are present within the overlapping cognitive environments of a community. Thus, it is *reasonable* (but not guaranteed) that anyone in that mutual cognitive environment will have access to (will know) what is present there.

An effective analogue for understanding cognitive environments is that of the visual environments that we share. We have access to all kinds of information in our visual field that we do not process at any particular time, but that someone who shares that visual field (or who is

---

[6] It has since been adapted and adopted by others (See Tindale, 1999).

aware of its contents) could expect us to "see." Our attention can be drawn to what is there and in that way our understanding of what is around us is modified.

2.1 Using the Cognitive Environment

In a similar way cognitive environments are modifiable through argumentation. Arguers strive to make present, to actualize, what is there potentially, to make connections between ideas and beliefs that result in further insights, or to move us to action by simply adding to our cognitive environments or weakening some attachment we had to an idea there. Consider as an example a portion of the speech given by Barack Obama during which he announced his candidacy for the Democratic nomination for president:

> But the life of a tall, gangly, self-made Springfield lawyer tells us that a different future is possible.
> He tells us that there is power in words.
> He tells us that there is power in conviction.
> That beneath all the differences of race and region, faith and station, we are one people.
> He tells us that there is power in hope.

In this way, Obama, another "tall, gangly, self-made Springfield lawyer," standing on the steps of the courthouse in Springfield, Illinois where Lincoln launched his campaign, does more than invoke the memory of Lincoln; he summons him to the podium and superimposes his absence over Obama's presence. Or, in our terms here, he draws on his knowledge of his audience (before him and within the nation) and modifies their cognitive environment by taking something he can expect to be present there (an understanding of both Lincoln's appearance and status) and adding to it something new—an association between himself and Lincoln.

As Wilson and Sperber observe, adding a new piece of information to a cognitive environment will modify it, "but it can equally well be modified by a diffuse increase in the saliency or plausibility of a whole range of assumptions, yielding what will be subjectively experiences as an *impression*" (2012, p. 87). Here, of course, a range of rhetorical effects could be adopted. In fact, the notion of cognitive environments provides insight into the cognitive nature of rhetoric and how is can operate in persuasion. Metaphor is an obvious candidate here, drawing as it does on the shared meanings of a community. And allusion can also be effective, as the Obama/Lincoln example illustrates.

3. DEVELOPING THE COGNITIVE ENVIRONMENT

The cognitive environment is, then, a powerful tool for understanding how audiences receive argumentation and adjust their beliefs in relation to it. They are weak in just the right way that supports our appreciation of reasoning on matters of uncertainty, where the conclusions drawn by audiences, and by evaluators about audiences, are characterized by degrees of plausibility.

As it has been developed and then adopted by argumentation theorists like myself, the sense of "cognitive" connected with cognitive environments has been restricted to what falls within the domain of epistemology, principally beliefs and facts. But to think of us in this way harks back to a traditional view of humans as divided beings, split between reason and the rest of the affective system. The path we trace from Aristotle through Perelman challenges that view. As Perelman reminds us: argumentation is aimed at the whole person (Perelman, 1982). Moreover, recent work of relevance to argumentation theory by neuroscientists and cognitive scientists like Paul Thagard (2000, 2006) reinforces the view that we need a more expanded sense of "cognition" to fully capture human experience.

Thagard has written expansively on this, but I will base the discussion here on his theory of emotional coherence (2000). This theory suggests ways in which cognitive interacts with emotion. He calls this "emotional cognition." This idea is premised on the understanding (which is confirmed by work in neuroscience[7]) that all thinking has an emotional component. "[E]motional cognition comprises all of cognition viewed from the perspective that emphasizes the integration of traditional cognitive processes such as reasoning with emotional processes that attach values to mental representations" (Thagard, 2006, p. 237). Thagard's work illustrates the impact emotions have on decision making and the error in thinking we can avoid this influence. Emotional coherence does not just involve what is emotionally desirable, but also connects with kinds of coherence that involve inferences about what is acceptable. Elements in a theory of coherence have an epistemic status of being accepted or rejected. But Thagard also wants to speak of a degree of acceptability. In artificial neural network models of coherence such acceptability "is interpreted as the degree of activation of the unit that represents the element." In addition to this, he suggests that the elements in a coherence system have "an emotional *valence*, which can be positive or negative…For example, the valence of

---

[7] For example, the work of Richard Lazarus (1984) and Antonio Damasio (1994, 1999).

Mother Theresa for most people is highly positive, while the valence of Adolf Hitler is highly negative" (2006, p. 19).

The idea of attaching values to mental representations is important here[8], because it opens the door for expanding our idea of cognition (and hence cognitive environments) to include value as well as belief. In further work (2012, p. 284ff.), Thagard identifies three kinds of values: epistemic, social, and personal.[9] The epistemic covers things like objectivity, bias, evidence and explanation, values that we would share in our mutual cognitive environments. The social covers things like trust, competence and autonomy, which we again share. And the personal covers things like success, fame, and power, which are quite individualized. These values can coincide in an individual's judgment as she frames an explanation for her decision in terms of a personal preference for what she finds just in a particular situation. Given the nature of these values, we can see how those that are epistemic and social would have an objective quality to them. They provide the standards for directing decisions and actions, standards of evidence, for example, for weighing the merits of competing claims.

Once values such as these are included in our discussion, we have a "larger" environment, or certainly one that has a richer sense of cognition. Now, argumentation aims to persuade at the personal level what is available at the epistemic and social levels within one's cognitive environment. This is to reverse the usual relationship assumed between conviction and persuasion. But we are convinced of many things on which we do not act. Our conviction is due to their availability at the epistemic and social, where we are aware of them, but not necessarily committed to them in a way that results in action. Ideas about recycling, for example, have gradually entered our cognitive environments over the last few decades, primarily at the level of social value, to the extent that they are now fixed. We have firm values with respect to recycling evidence, and those values have influenced our personal values to the extent that our behavior has changed in a thoroughly routine way, resulting in personal actions that have become a way of life, but that

---

[8] And, indeed, this work clearing connects values with emotions, insofar as values are viewed as "emotionally valenced mental representations" (Thagard, 2012, p. 284). We may have neutral associations with some concepts, but with most part of the representation will involve an emotionally positive or negative association. Consider the range available for 'house', 'fish', and 'law'.

[9] As I note elsewhere (2015), this triad fits nicely with the three levels of Habermas' lifeworld, with its objective world that is a source of claims about facts, its intersubjective world of norms and values, and its subjective world of private thoughts.

could be justified at the epistemic and social level if necessary. Argumentation brings salience (or, rhetorically, presence) to those relevant features that exist at the epistemic and social levels in order to bring about change at the personal (change that results in action). In one sense, the social mediates between the epistemic and the personal in ways that were suggested by Aristotle's account of the emotions in an account that emphasized both the cognitive and the social (*Rhet.* Bk 2).

This expanded sense of the cognitive environment recognizes the integration of emotion, value and cognition in our deliberations. This is characteristic of all the communities in which we live, whether it be the specialized communities of science and law, or the general communities of an institution, town, state of country. And of course, the new virtual communities in which we operate extended these environments even further, adding to the range of beliefs, facts and values on which arguers might draw. The success of much social media is founded on this assumption.

4. COGNITIVE ENVIRONMENTS AND UNIVERSAL AUDIENCES

As Perelman himself hints (fn2 above), the universal audience was more a correlate of dialectic (as witnessed in the *Topics*), while the particular audience relates to rhetoric (as per the *Rhetoric*). This can suggest several things, but one that is relevant here is between the intersubjective of the dialectical and the personal of the rhetorical. Our discussion of cognitive environments has suggested connections between these, but also that value-based rhetoric in the form of epideictic would be relevant at all levels.[10] If we are now to prefer the concept of the cognitive environment to that of the universal audience then it needs to provide us with all the positive features the latter made available.

4.1. The Role of the Universal Audience

For Perelman and Olbrechts-Tyteca, the universal audience operates as a standard of evaluation. They marked a distinction between efficacy (or effectiveness) and validity (1969, p. 464). Each of these is assessed by an audience. But what can validity mean in this instance? It cannot be

---

[10] Recall here the primacy Perelman gave to this genre of rhetoric, recovering it from the shadows beside deliberation and forensic and placing it as central to discourse because of the role it plays in intensifying values (Perelman, 1984, p. 19).

the formal validity of the tradition, since this concept destroys the need for argumentation. So it needs to be understood on other terms.

In pursuit of such terms, Perelman and Olbrechts-Tyteca approach evaluation in terms of a theory of knowledge that sees, in reasoning that parallels that which Toulmin (1958) was developing around the same time, the criteria of worth as field-dependent. Just as each individual, context and culture has its own universal audience, we could say the same of each field. Science, law, or economics, for example, would each have their own standards of evaluation—their own sense of objective standards within the field and a corresponding audience that represents the standard: "Initiation into a rationally systematized field will not merely furnish knowledge of the facts, truths and special terminology of the branch of learning involved and of the method of using the available tools, it will also provide instruction in assessing the strength of arguments used in these connections" (1969, p. 464). The nature of the instruction involved may be more potential than actual, but the sense is that the ideas and criteria are available "in the field." And we would suggest now that that field (or those fields) amounts to a cognitive environment in which these things are available to and accessible for, but not necessarily known by, all those who operate within them.

But, we may ask, what occurs when disputes arise between fields? And how do we manage argumentation of a general nature that is not field-specific? Here, we operate in a more generalized environment of belief and value, one that may have overlaps with the fields. Here, it is even more appropriate to talk of the cognitive environments. And it may also be appropriate to speak of the specialized fields as being subsumed under this more general environment. After all, it holds within it the objective standards that govern reasoning locally. And as it expands, it has the capacity to capture objective standards that are more global in nature. We can say this particularly now (in a sense in which Perelman could not have said it) because of the virtual world in which we are all involved at some level. This world brings the global into the local; it increases the range of ideas that are accessible in the weak sense that mutual cognitive environments permit. We turn here from talking of audiences to talking of environments in which audiences operate. Those environments are inhabited environments; others with whom we have greater or lesser associations populate them. In these we find and learn the criteria by which we judge what is addressed to us, and the standards by which others would judge them. In these we find the shared values of the social and the epistemic. When we address people in environments we share (thus giving a particular sense to "knowing our audience") we both make our communication relevant to the features of the environment while also assuming that those who operate there know what is relevant. And audiences can understand our approach since relevance is a shared epistemic value. Likewise, we assume people know what is morally acceptable within an environment and we proceed accordingly. Where we get these assumptions right, the burden of proof lies with those we address. Where we get it wrong, it lies with us and would require us to reconsider the cognitive environment and what is accessible there.

4.2. The Three Criticisms

In these ways, we can judge that the concept of the cognitive environment can meet the demands of the role served by the Universal Audience. But it also needs to be able to accommodate or address the problems that have been associated with the Universal Audience and that were canvassed earlier in the paper. Let's review those: (i) it is a relativistic standard, rendering as many universal audiences as there are arguers; (ii) it transmits the timeless validity of the Cartesian model, which does not admit of values; (iii) it is at its heart a fractured standard appealing to both pragmatic and epistemic features that cannot be reconciled.

> (i) By their nature, environments are relative to space and time. As environments in which real people operate, they are relative tothose people. But that they overlap and extend across communities linked by interest and proximity means that while individuals have cognitive environments they are not restricted to those individuals, they overlap and are the means by which people share beliefs and values. Nor are they constructed by individuals; they are the product of cognitive (in the expanded sense) activity in specialized and non-specialized communities. Hence, while we would say that there are as many cognitive environments as there are arguers, since each individual adds something unique to the mutual cognitive environments in which they share, this does not mean we are committed to relativistic standards of the kind that van Eemeren and Grootendorst feared were at stake with universal audiences. Epistemic and social standards of evaluation are shared objective standards sufficient to meet the needs of argumentation.

> (ii) An important thing that cognitive environments lack that universal audiences did not is the label "universal". This is another sense in which they are not relative. Thus, there is no expectation that it would transmit values across time and space except to the degree that there may be mutual cognitive environments shared across time. It is not implausible to think there might be some connections across time, just

as we can connect the events of a person's life by tracing overlaps from day to day, week to week, and so forth, such that we can say that someone is the same person at 50 years of age as she was at 5. But the 5 year old and the 50 year old will have very little in common, the variety of experiences lived over the course of the life will have altered the belief set and values that she possesses. On one level, it does not matter that we share standards with those of the past except to the degree that we might concern ourselves with the authority that those standards have (and I will address that in the last section). It matters more that the standards available to us are standards that have application in our time and space, and that those with whom we argue recognize them. Thus, we avoid the paradox that concerned Gross and Dearin.[11]

Moreover, given the reading they inherit, Gross and Dearin must exclude values from the universal audience, because particular values can never be binding on all human beings across time and space (2003, p. 37). But that requirement was a Red Herring, as we have seen. Even if we stayed with the concept of the universal audience, we could develop it using Thagard's expanded sense of cognition so that it accommodated values. But for other reasons given here we pursue the direction of the cognitive environment. This is an environment in which we see values present at each level. In fact, it is now impossible to detach values from the concept if we want to have an accurate understanding of how audiences experience argumentation, and one that takes advantage of the understanding of cognition available to us from current research.

(iii) Aikin judged that the universal audience was a fractured standard appealing to both pragmatic and epistemic features that cannot be reconciled. This is the criticism that is likely most transferable to the new concept of the cognitive environment and would require more attention than I can give it here. As we will recall, Aikin distinguishes between epistemic and pragmatic senses of the universal audience. The pragmatic features "are the sociological facts of cognitive and evaluative overlap between different people, groups, and societies" (p. 240). And this is the aspect that concerns the effectiveness of argumentation. But we must also be concerned about the validity of argumentation, and this is where the epistemic features enter and where the universal audience becomes a rationally regulative notion

---

[11] In fact, as mentioned above, the view is founded on a misreading of the universal audience that Perelman himself pointed out (1989). When Perelman and Olbrechts-Tyteca speak of absolute and timeless validity (1969, p. 32), they are referring to the universal audience of the philosophers. This is what they proceed to challenge and then replace with a universal audience that is relative to each individual, each culture (p. 33).

(p. 241). He was also reluctant to see the epistemic reduced to the pragmatic, in part I think because of the way he views the epistemic (similar to the sense of validity in the view endorsed by Gross and Dearin). He sees the epistemic to produce knowledge, while the pragmatic deals with disagreements. Thus, he gives his attention to developing an epistemic model that breaks with Perelman's goal of adherence.

My inclination would be to work instead with the pragmatic because clearly, talk of "cognitive and evaluative overlap" echoes something of the cognitive environment. What is at stake here is the nature of the epistemic and the suggestion that the pragmatic cannot produce knowledge. Does this hold true of the cognitive environment as I have described it? Recall Wilson and Sperber's (2012) recognition that rhetorical effects take place within the cognitive environment. These bring us back to consider how things like metaphor and allusion (to name just two devices/effects) operate at the epistemic level. We saw the use of allusion by Obama to modify his audience's cognitive environment. Physical similarities of appearance and background suggest similarities of value. And those for whom Lincoln has positive associations are encouraged to assign a similar value to Obama. Is it "true" that Obama is like Lincoln? Is this knowledge? It is certainly an association that did not likely exist in the mutual cognitive environment of those he is addressing prior to his address. Metaphor also does powerful epistemic work, building on meanings that are shared. Michael Leff (1983), for example, examines the parallels between metaphor and topical invention: "Both involve the organization of belief, the partial structuring of commonly held attitudes and values. Both also function through a process of mediating between extreme elements: in metaphor, the commonplaces associated with various terms provide the ground for connection, and in topical argument, propositions expressing abstract values provide the necessary link. Finally, both require the invention of verbal strategies in circumstances where diverse elements compete for attention" (1983, p. 222). For Leff, metaphors are forms of invention that rely on imaginative rationality. Just as importantly, they depend on social knowledge and the connections that exist within a community.

Admittedly this "social" knowledge may not be sufficient to meet Aikin's epistemic challenge. He seems interested in the kinds of knowledge achieved by science and philosophy. But both the communities of scientists (multiple communities given the distinctions) and philosophers (again a variety of them) have their overlapping cognitive environments, with their epistemic values and standards of evaluation and discovery. Moreover, as argumentation operates across communities and disciplines, it requires tools of broader application. The cognitive environment is one such tool.

5. CONCLUSION: RETAINING OUR ARISTOTELIAN ROOTS

We have now only to ask again more clearly whether the cognitive environment can serve as an objective standard. That is, one that has authority. We might stand back as ask with respect to any objective standard "from whence does it arise?" And, "what authorizes it?" This is a problem that haunts all theories of argumentation, whether it concerns the source of argumentation schemes and their associated critical questions, or the pragma-dialectical rules.[12] This is not the place to enter into an investigation of where *those* schemes, questions, and rules come from and what authority they have.[13] Here, I am simply interested in whether the concept of the cognitive environment can contribute anything to this problem.

Objective standards are judged to be out there, fixed in some way, and removed from the exigencies of any particular circumstance. But there is more than a little vestigial Platonism in such an idea. Working always from the bottom up, in neo-Aristotelian fashion, we look for our standards more readily at hand, in our own immediate environments. The answer to the question can really be none other than human communities themselves. They are the source of our standards and provide the requisite authority for their use. But we need then to explore how to access such communities in a way that allows for assessment. This is where the device of the universal audience has always proven problematic. In representing the standard of reasonableness within communities it has never been clear how it can then be used in assessing the quality of argumentation.

The cognitive environment is not an audience, and so this is a fundamental break from Perelmanian theory. But that the universal audience was an audience was as a much a concern for commentators as that it was universal. The focus on audience brought with it distracting ideas of judge and spectator. When in fact what we want are the standards by which we judge and through which we see. When we

---

[12] J. A. Blair moves in the direction I am advocating when he proposes that the dialectical point of view can provide standards that are independent of individuals but "relative to collective human judgment" (2012, p. 55).

[13] See, for example, Blair's inquiry into the source of argumentation schemes (Blair 2012, chapter 11) and van Eemeren's explanation for the empirical validation of the rules (van Eemeren, 2010, pp. 35-36)

invoked the universal audience, we looked for the standard of reason that was alive in the audiences we addressed. The cognitive environment provides the epistemic and moral standards of judgment. We see what might be seen, what is available to the audiences that share cognitive environments. Thus we can attach notions of reasonableness and plausibility to claims for which evidence exists in those environments. When we reason in situations of uncertainty, the concept of the cognitive environment becomes a fitting tool for the construction and evaluation of argumentation. It thus presents itself as a valuable replacement for the problematic concept of universal audience with which so many have struggled for so long.

REFERENCES

Aikin, S. (2008). Perelmanian universal audience and the epistemic aspirations of argument. *Philosophy and Rhetoric*, *41*(3), 238-259.
Blair, J. A. (2012). *Groundwork in the theory of argumentation: Selected papers of J. Anthony Blair*. Dordrecht, NL: Springer.
Brandom, R. (1994). *Making it explicit: Reasoning, representing, and discursive commitment*. Cambridge, MA: Harvard University Press.
Damasio, A. R. (1994). *Descartes' error: Emotion, reason and the human brain*. London: Picador.
Damasio, A. R. (1999). *The feeling of what happens: Body and emotion in the making of consciousness*. New York: Harcourt Brace & Company.
Ede, L. (1989). Rhetoric Versus philosophy: The role of the universal audience in Chaim Perelman's *The New Rhetoric*." In R.D. Dearin (Ed.), *The new rhetoric of Chaim Perelman* (pp. 141-151). New York: University Press of America.
Eemeren, F. H. van (2010). *Strategic maneuvering in argumentative discourse*. Amsterdam: John Benjamins Publishing Company.
Eemeren, F. H. van, & Grootendorst, R. (1995). Perelman and the fallacies. *Philosophy and Rhetoric*, *28*, 122-133.
Grice, P. (1989). *Studies in the way of words*. Cambridge: Harvard University Press.
Gross, A., & Dearin, R. (2003). *Chaim Perelman*. Albany, NY: State University of New York Press.
Johnstone, Jr. H. (1978). *Validity and rhetoric in philosophical argument: An outlook in transition*. University Park, PA: The Dialogue Press of Man & World.
Lazarus, R. S. (1984). On the primacy of cognition. *American Psychologist*, *39*(2), 124-129.
Leff, M. (1983). Topical invention and metaphoric interaction. *The Southern Speech Communication Journal*, *48*, 214-229.
Locke, J. (1975). *Essay concerning human understanding*. P.H. Nidditch (Ed.). Oxford: Clarendon Press.

Midgley, M. (1995). *Beast and man: The roots of human nature.* Revised edition. New York: Routledge.
Perelman, Ch. (1989). The new rhetoric and the rhetoricians: Remembrances and comments. In R.D. Dearin (Ed.) *The new rhetoric of Chaim Perelman: Statement and response* (pp. 239-251). New York: University Press of America.
Ray, J.W. (1978). Perelman's universal audience. *Quarterly Journal of Speech, 64,* 361-375.
Sperber, D., & D. Wilson (1986). *Relevance: Communication and cognition.* Cambridge, MA: Harvard University Press.
Thagard, P. (2000). *Coherence in thought and action.* Cambridge, MA: MIT Press.
Thagard, P. (2006). *Hot thought: Mechanisms and applications of emotional cognition.* Cambridge, MA: The MIT Press.
Thagard, P. (2012). *The cognitive science of science: Explanation, discovery, and conceptual change.* Cambridge, MA: MIT Press.
Tindale, C.W. (1999). *Acts of Arguing.* New York: State University of New York Press.
Tindale, C.W. (2010). Ways of being reasonable: Perelman and the philosophers. *Philosophy and Rhetoric, 43*(4), 337-361.
Tindale, C.W. (2015). *The philosophy of argument and audience reception.* Cambridge: Cambridge University Press.
Toulmin, S. (1958). *The uses of argument.* Cambridge: Cambridge University Press.
Wilson, D., & D. Sperber (2012). *Meaning and relevance.* Cambridge: Cambridge University Press.

Cognitive Environments, Effectiveness and Dialectical Soundness
Commentary on Tindale's On Cognitive Environments

ANDREA ROCCI
IALS, University of Lugano, Switzerland
andrea.rocci@usi.ch

1. THE NATURALIST ROOTS OF COGNITIVE ENVIRONMENTS

Christopher Tindale's theoretical paper on cognitive environments offers a refreshingly original rapprochement between two strands of research that are quite far apart in the spectrum of the studies that deal with human discourse: Sperber and Wilson's cognitive pragmatics and Chaïm Perelman's rhetorical (and dialectical) theory of argumentation. Neither theory is new, both have been endorsed and criticized in various ways. This reinforces the novelty of Tindale's paper as our – mine at least – habitual framing of these theories makes them strange companions indeed, especially for the task that Tindale assigns to them: finding a viable standard for judging the reasonableness of arguments eschewing the Scylla of Cartesianism and the Charybdis of relativism. That is a standard striving for universality in a manner faithful to the Aristotelian view that "accumulation of the universal *in the individual's experience is necessary*" (Tindale, 2016, p. 688).

It is surprising that Tindale calls the help of Sperber and Wilson's *Relevance Theory* for such a task. Sperber and Wilson's (1986, 1995) theoretical opus stands out as the most unabashedly reductionist, naturalist and mechanistic development of Grice's theory of meaning and understanding. One of the objectives of relevance theory is to replace Grice's maxims – a set of assumptions about cooperative communicative behavior that guide us in understanding what was meant in a conversation – with a principle of relevance based on a cognitive economy of effects and efforts, and to transform Grice's vague schema for working out an implicature, in a deductive computation. This deductive computation is based on a subset of classical deductive logic and takes as input the propositional content of the utterance and a set of assumptions (propositions) that are salient in memory, or, to be more precise, in the *cognitive environment of the hearer*. As a result of the computation the cognitive environment is modified with the derivation of new assumptions – as implicatures – the reinforcement or weakening of existing assumptions and a new organization of memory whereby those assumptions that have been fruitfully used to derive new assumptions, or to strengthen or weaken existing assumptions have come on top of the salience hierarchy and are readily accessible to the hearer in view the processing of new utterance.

Tindale observes that the term *universal audience* "is not as felicitous as perhaps we would like" and carries "considerable conceptual baggage" (Tindale, 2016). In briefly recalling a few of the central tenets of relevance theory, my aim was chiefly to make the point that *cognitive environment* as well carries a heavy conceptual baggage. In relevance theory a cognitive environment is a set of assumptions that is continuously modified in communication (but also in non-communicative contexts of action and perception) through a relevance driven, blindly-operating, computational mechanism of which individuals are not consciously aware. It is not intended as a social space where negotiations of meaning happen. This is the conceptual baggage carried by the construct Tindale turns to in his quest for a better substitute of Perelman's universal audience as a criterion for evaluating arguments.

2. AWARENESS AND INFERABILITY: WHY RHETORIC NEEDS COGNITIVE ENVIRONMENTS

In Sperber and Wilson's cognitive framework, cognitive environments are, first of all, cognitive environments *of an individual*. Yet, in order to allow for communication to succeed to a reasonable degree most of the times we need to admit for *mutual cognitive environments*. These are a development of earlier notions of a *common ground* between the speakers. Vaguely hinted at by Grice, common grounds were given a formal definition in pragmatics by Robert Stalnaker on the basis of David Lewis' definition of *common knowledge*. Common ground is at the foundation of Stalnaker's theory of assertion (1978) and presupposition (1973).

With respect to the notion of common ground based on common knowledge, the move to mutual cognitive environments represents a shift from logic towards psychology and a double weakening both with respect to *common* and with respect to *knowledge*. I leave out the discussion of the common part, which has been, in my opinion, brilliantly solved by another psychological continuator of the common ground tradition: Herbert Clark (1996) – on whose contribution I will have something to say later.

As for the knowledge part, common knowledge inherits the logical omniscience problem from epistemic systems: all the logical consequences of the common ground are already part of the common ground. This feature is somewhat nagging when we deal with argumentation as it turns out that all arguments (or at least all deductive arguments) based on premises that are in the common ground do not add anything to the common ground. This is clearly nagging if you want to use the common ground construct to represent the starting point of an argumentative discussion.

At first blush, cognitive environments are just like common grounds in this respect, if not worse, as they contain all that is *inferable* by the concerned subjects, as Tindale reminds us. Yet, from a psychological point of view we can draw a distinction between *awareness* and *inferability*, which is really a saving grace for argumentation. Awareness, then, can be seen as a matter of degree in terms of cognitive *salience*. Here, as Tindale has shown in his paper, cognitive environments really tell us something about how argumentation happens, what argumentation does and what is the place of rhetoric in argumentation. Tindale shows this perfectly with his beautiful Obama example. An act of argumentation may consist many times in making salient – or rhetorically present – something that we know in order to make us actually infer what is inferable. So that the only outward manifestation of an argument can be a stimulus provoking an attentional modification in the cognitive environment – see Sperber and Wilson's idea of presumption of relevance carried out by ostensive stimuli – so that a certain inferential path is provoked.

Tindale maintains that arguments not only activate parts of the mutual cognitive environment, but also *add* something to them. Interestingly, this distinction between activating and adding can be the basis of a new typology of argument moves: some moves simply activate premises that are in the mutual environment others provide *new* information. For instance the opinion of experts or the testimony of witnesses can lend credibility to new information entering the cognitive environment. In another way promises and proposals create new social facts that reshape the cognitive environment.

3. CAN WE USE COGNITIVE ENVIRONMENTS AS DIALECTICAL STANDARDS?

Having illustrated the potential I see in the notion of cognitive environment for argumentative and rhetorical studies, let me comment briefly on the main thrust of Tindale's paper: the viability of (mutual) cognitive environments as a substitute of the universal audience as a standard of dialectical validity. On the one hand, it is clear how the cognitive environment gives more substance and precision to the construct of the audience, especially a construct of audience that is seen as actively shaping the argumentative discourse in Bakhtinian terms. The quality of the mutual cognitive environment needed for an argument to function effectively as an invitation to inference (see Robert Pinto's definition of argument) tells a lot about the quality of the argument: an argument that can work only in an extreme or disturbed cognitive environment, requiring at the same time very specific factual assumptions, particular interests and values and, simultaneously, a loosening of epistemic criteria is clearly a bad argument. In this sense, the argument of *Mein Kampf* appeals only to a very specific audience, for instance luckily if you are not an "Aryan" there is no way you can appreciate it, while you don't need to be an African American in 1963 to feel the values mobilized by Martin Luther King's *I have a dream speech*. Yet, if we analyze King's speech rhetorically we find that it does exploit every tiny bit of common ground with its various audiences to sway hearts and minds.

Appeal to specific audiences tells us of rhetorical effectiveness while dialectical validity should be measured by the universal audience. The problem that Tindale faces here is how to make cognitive environments universal, albeit in the weak sense, that Perelman attaches to the term. Here Tindale calls in another ever mysterious and ever debated view of validity: Toulminian field dependence. Communities of specialized discourse, let us say *economists*, have their own shared cognitive environments, which include learned "instruction in assessing the quality of argument". Thus an argument can appeal to what is common, what is in principle accessible to all the members of a given discourse community rather than to a specific, individualized audience. This idea of community based cognitive environment is very close to Herbert Clark's (1996) idea of *communal common ground* – common ground that speakers attribute to members because of their belonging to a cultural community and not because of personal histories or shared perceptual experiences. This way of appealing only to the communal common ground seems to inform certain practices of quality control in the academic world, which include firstly and foremostly the practice of the double blind peer review.

But, is the recourse to criteria of validity that are publicly recognized and passed on in a community of discourse the kind of standard we are looking for as argument critics? Here the critical work of economist Deirdre McCloskey (1985) on the rhetoric of economics invites us to caution. Her insightful analyses expose the argumentative poverty of what she calls "economic modernism" and demonstrate how economists' and social scientists' obsession with tactical-level rules of methodology imposes a narrow straightjacket on argumentation to the point of even obscuring the argumentative nature of the scientific enterprise. If her analysis is right, communities of investigators *can* have deeply flawed criteria of argument evaluation and deeply misguided folk theories of argument. It is often the case – and McCloskey would readily concede that for economists – that experts are much better at arguing than they are at producing and disseminating explicit epistemic criteria and rules of argument in their communities. This asymmetry should be taken into consideration, in my view, by Tindale as he develops his view of cognitive environments as standards of validity.

REFERENCES

Clark, H. H. (1996). *Using Language*. Cambridge: Cambridge University Press.
McCloskey, D. N. (1985). *The Rhetoric of Economics*. Brighton: Wheatsheaf.
Sperber, D., & Wilson, D. (1986) *Relevance. Communication and Cognition.* (2nd ed.). Blackwell: Oxford
Stalnaker, R. C. (1973). Presuppositions. *Journal of Philosophical Logic, 2*(4), 447–457.
Stalnaker, R. C. (1978) Assertion. In P. Cole (Ed.), *Pragmatics.* (*Syntax and Semantics 9*) (pp. 315-332). New York: Academic Press.
Tindale, C. (2016). On Cognitive Environments. In D. Mohammed & M. Lewiński (eds.), *Argumentation and Reasoned Action: Proceedings of the 1st European Conference on Argumentation, Lisbon, 2015. Vol. I, 687-704.* London: College Publications.

31

HLA Hart on Logic and Interpretation

COSMIN VĂDUVA
University of Bucharest, Faculty of Philosophy, Romania
cosmin.vaduva@ub-filosofie.ro

Interpretation and *logic* are pervasive concepts in legal adjudication. However, the way these concepts are *used* by legal scholars and judges as well and their *meaning* is not clear at all. Following HLA Hart's insights about logic and interpretation I will provide further conceptual claims about their relationships which might be useful for legal scholars and judges alike.

KEYWORDS: deduction, distinction, factual-normative, HLA Hart, interpretation, logic, mechanical jurisprudence, transcendental argument

1. INTRODUCTION

There is no doubt that *interpretation* and *logic* are pervasive concepts in legal adjudication. Despite their pervasiveness, the way these concepts are *used* by legal scholars and judges as well and thus their *meaning* has not been always fully understood. Thus, according to Hart (1957-8, p. 610), one of the most persistent misunderstandings was expressed in the contention that "the excessive use of deductive logic" leads to the vice of "mechanical jurisprudence". As an inveterate advocate of conceptual clarity, through firmly distinguishing between *interpretation* and *logical deduction* Hart successfully removed this misunderstanding. Therefore, clarifying the use and meaning within legal adjudication of *interpretation* and respectively *deductive logic*, Hart showed why the belief that the "excessive use of logic" is responsible for the mechanical jurisprudence is a mistaken one. The distinction between deductive logic and interpretation appears with great clarity in the following excerpt:

> But logic does not prescribe interpretation of terms; it dictates neither the stupid nor intelligent interpretation of any

expression. Logic only tells you hypothetically that if you give a certain term a certain interpretation then a certain conclusion follows. Logic is silent on how to classify particulars - and this is the heart of a judicial decision (Hart, 1957-8, p. 610).

2. THE PARAPHRASE OF HART'S ARGUMENT ABOUT (DEDUCTIVE) LOGIC AND INTERPRETATION

However, the argumentative framework is far more complex. This is why I find appropriate to *paraphrase* (Copi, Cohen, & McMahon, 2014, p. 36) Hart's argument within which occurs the distinction between the interpretation and deductive logic. It seems to me that, by paraphrasing the argument, the meanings of the two concepts and their applicability within the legal adjudication appear more clearly. HLA Hart's understanding of how deductive logic and respectively interpretation interplay within judicial adjudication and their crucially different roles is apparent in his famous discussion about the "No vehicles in the park" rule (Hart, 1957-8, pp. 607-8 and Hart, 1994, pp. 126-9). Basically, Hart argued that every legal rule has a core of applications and a penumbral area of applications. The argument is complex and this is why I think it is helpful to attempt to provide its paraphrase.

I think that, in fact, Hart provided two distinctive arguments. The first one may be construed as given in the support of the proposition that (1) every legal rule must have a *core* of applications. The second one is meant to justify the proposition that (2) every legal rule *inevitably* has a *penumbral* area of applications. I shall firstly paraphrase what I will call the "core argument" and subsequently what I will call the "penumbral argument". Once these two arguments are made transparent in this manner, one can (1) detect some basic differences between logic and interpretation, (2) see the importance and the limits of the use of deductive logic into legal adjudication, (3) understand the roots of the confusion which lead to the mistaken belief that an excessive use of logic is responsible for the flaws of the mechanical jurisprudence, and finally (4) see other major differences between interpretation and deductive logic which are not to be found expressly within Hart's argument.

2.1 The core argument – a transcendental argument

I think that the core's argument can be illuminatingly captured by seeing it as a *transcendental* argument. The evidence that Hart proposed a transcendental argument is rich. I will consider 3 such occurrences:

we do, as well as recognize the vagueness at the boundary of such notions as discretion, also recognize clear or simple cases, and if we could not do this we should not be able to use the term in communication with each other. (Hart, 2013, p. 653)

If we are to communicate with each other at all, and if, as in the most elementary form of law, we are to express our intentions that a certain type of behavior be regulated by rules, then the general words we use -like "vehicle' in the case I consider -must have some standard instance in which no doubts are felt about its application. There must be a core of settled meaning, but there will be, as well, a penumbra of debatable cases in which words are neither obviously applicable nor obviously ruled out. (Hart, 1957-8, p. 607)

If it were not possible to communicate general standards of conduct, which multitudes of individuals could understand, without further direction, as requiring from them certain conduct when occasion arose, nothing that we now recognize as law could exist. (Hart, 1994, p. 124)

In all those passages can be detected a more or less similar form. The form of a transcendental argument, according to Stanley Paulson is the following:

> A transcendental argument might be set out as having the form: an initial premise, "I am thinking that **X**" or "I have the concept *Y*," followed by a second, transcendental premise to the effect that my thinking that X-or my having the concept Y-is possible only if Z. From the conjunction of these premises, a conclusion follows to the effect that Z is the case. And the existence of Z is incompatible with the doubts raised by the skeptic. (Paulson, 2000, p. 1779)

In what follows I will provide the paraphrase of core's argument.

Premises

P1 There is the concept of law & the concept of communication

P2 The concept of law can be elucidated through the concept of rules (Hart, 1994, p. 123)

P3 "All rules involve recognizing or classifying particular cases as instances of general terms" (Hart, 1994, p. 123).This is a conceptual truth, and not an empirical one. It can also be expressed as follows: "The idea of an individual is the idea of an individual instance *of* something general. There is no such thing as a pure particular" (Strawson, 1971, p. 35).

P4 To apply the rule to the case beforehand involves a *decision* to do that. I should add that I think not only within the penumbra of the rule but even within the core of the rule "No vehicle in the park" one has to *decide* to apply the rule.

P5 To have the concept of rules (and thus of) law and the concept of communication as well is possible only if it is the case that the general terms used in law and communication have core applications. This premise is the transcendental premise of the argument.

P6 What means that general terms have core applications? The general words we use must have some standard instance in which no doubts are felt about its application.

(Given that P1, P5 & P6 it follows that)

Conclusion

P7 There *are* in fact standard applications of rules

The key concept for understanding the core argument it seems to me to be that of doubt. At the root of the differences between the application of a rule within its core and its application within its penumbra area is a correct understanding of what Hart meant by saying that "doubts are felt about the application of the rule". Thus, dealing with the case of an automobile the adjudicator will feel no doubt in applying to it the rule "No vehicles in the park". Apparently for Hart this statement is indistinguishable from the statement there is an *agreement* as to the application of the word to the case at hand. Instead dealing with the case of a bicycle will feel doubts.

2.2 The "we are not gods!" argument

Here it comes the paraphrase of what I will call the "we are not gods!" argument which aims to demonstrate the inevitability of penumbral cases.

P1 It may appear fact situations which cannot be recognized as clear or easy instances of the rule; that is, facts which are not identical with the factual predicate of the standard case – the pending case has only some features in common with the factual predicate of the standard case; furthermore it will lack others or be accompanied by features not present in the standard case. These factual situations are "variants on the familiar".

P2 Given P1, it follows that the general words in which the rule is framed are neither obviously applicable nor obviously ruled out. These are debatable or penumbral cases.

P3 "Human invention and natural processes continually throw up the variants on the familiar". This proposition provides an explanation, i.e. human invention and natural processes, of the variants of familiar.

P4 The variants of the familiar are unanticipated because we are not gods, but men.

P5 If we are to say that these ranges of facts do or do not fall under existing rules...That means *to apply the legal rule within its penumbra. Nota bene!* I take the phrase "To apply the legal rule within its penumbra" to mean not only the "Yes" answer, i.e. the new case falls within the scope of the legal rule, but also the "No" answer', i.e. the new case does not fall within the scope of the legal rule.

P6 The judge *has* to settle the penumbral issue, i.e. to give a "Yes" or "No". That means *inter alia* that the judge cannot refuse to apply the rule on the ground that the new case is a penumbral one.

P7 The facts and phenomena to which we fit our words and apply our rules are as it were dumb.

P8 From P7 it follows that the answer to the issue if these ranges of facts do or do not fall under existing rules cannot be *dictated* by these ranges of facts .

P9 From P8 it follows that to give an answer to the issue if these ranges of facts do or do not fall under existing rules involves taking a decision. What means to take the decision to apply the rules? By contrasting "to dictate the application of the rule" and "to decide the application of the rule" Hart implied that to decide to apply the rule involves necessarily making a *choice*.

P10 From the conjunction of P6 and P9 it follows that the judge who *applies* the rule within its penumbra has to *decide*, that is to *choose*.

P11 The problems of the penumbra are always with us.

P12 When one deductively infer a proposition from another proposition, i.e. use the deductive logic, no choice at all is involved.

P13 From P10 and P12 it follows that to apply the rule within its penumbra is not only a matter of logic. In this area men cannot live by deduction alone.

P14 Legal arguments and legal decisions of penumbral questions are to be rational.

P15 Given P13 & P14, it follows that the rationality of legal arguments and legal decisions of penumbral questions must lie in something other than a logical relation to premises. This rationality is the interpretation of the general terms in which a rule is framed whenever the rule has to be applied within its penumbra.

2.3 Comparisons between the core and penumbra

Now, after I had paraphrased and closely examined the two arguments, I think one can see that to grasp the differences between *logic* and *interpretation* one should notice the differences and resemblances between *the core* of the rules and *the penumbra* of the rules. The deductive logic's use within the legal adjudication as well as its limits cannot be grasped without appreciating firstly the differences between the core and the penumbra of the rule. The mistaken belief that an excessive use of logic is responsible for the vices of mechanical jurisprudence was made possible by the failure to understand what really logic is about conjointly with the neglecting of the penumbral area of the rule. Thus, "Logic only tells you hypothetically that if you give a certain term a certain interpretation then a certain conclusion follows" (Hart, 1957-8, p. 610).

> i. every rule, which means that one and the same rule, has simultaneously a core and a penumbral area of applications. That is, according to Hart, there is no such a thing as rule without any borderline case at all or a rule which has no core of applications at all but only borderline cases.

> ii. Both within the core and the penumbra the rule is *applied*, that is something is seen as a particular of something else, *i.e.* a general term. However, there is a crucial difference. I will bring to light this difference by a linguistic distinction which seemed to have been introduced by Hart with this view in mind when he expressed the view that "All rules involve recognizing *or* (my emphasis) classifying particular cases as instances of general terms" (Hart, 1994, p. 123). Thus while within the core of the rule a particular is *recognized* as being an instance of a general term, within the penumbra of the rule a particular is *classified* as being (or not) an instance of the general term.

iii. One should be careful to distinguish the type of necessity which characterized the truth of the proposition that every legal rule must have a core of applications and the truth of the proposition every legal rule must have a penumbral area of applications. While the former one is, according to Hart, a logical/conceptual/transcendental necessity the latter is only contingently/empirically true. This point is captured into the following passage:

> in the case of *everything* (my emphasis) which we are prepared to call a rule it is possible to distinguish clear central cases, where it certainly applies and others where there are reasons for both asserting and denying that it applies. (Hart, 1994, p. 123).
> Postema too drew attention to this aspect:
> he insisted on the necessity for law of the core of settled application. If there is no core of settled meaning, he argued, legal regulation of human behaviour is impossible (Postema, 2010, p. 263).

iv. Whenever a particular case is recognized by everybody, *without any doubt at all*, as an instance of a general term, that case is a standard case, a clear one. This is why we say that that case is within the core of the rule. On the contrary, whenever there is doubt as to classify a particular case as being or not being an instance of the general term we deal with a borderline case. This is why we say that that case is within the penumbra of the rule.

From the paraphrase of Hart's argument whose conclusion is that every legal rule has a core of applications as well as a penumbra of applications it appears, at least I hope so, more clear the role Hart attributed within legal reasoning to deductive logic and respectively to legal interpretation.

3. CONCEPTUAL DIFFERENCES BETWEEN CORE AND PENUMBRA

Within the second part of the paper, I will make some steps forward beyond Hart's approach on deductive logic and legal interpretation. Thus I will articulate further conceptual differences and similarities between *logic* and *interpretation* which one cannot detect as such in Hart's works. I hope these conceptual articulations will have as a result the improvement of the understanding of the proper place the two key concepts have within judicial adjudication.

3.1 Factual vs normative

When judges are only to draw the correct conclusions from premises, i.e. when they use logic, are faced with *only* normative questions. Instead facing with the charge of bringing the particular cases under general rules, *i.e.* when the judges interpret legal rules, they ought to settle factual and normative issues *as well*.

One illuminating way to capture the point of the distinction between, one the one hand, what means to use deductive logic in legal reasoning and, on the other hand, what means to interpret legal rules in the processes of legal reasoning is to look at the distinction between what *rule to rule relationships* and *rule to world relationships*:

> we must keep separate what might be called 'rule-rule' and 'rule-world', relations; logic and analyticity pertain only to the former, not to the latter kind of relation. The fact that in both cases the criteria for correctness are semantic should not obscure this crucial difference. Suppose someone is pointing at a red object in front of him, saying, 'This is red.' When asked to justify this assertion, one can only appeal to the meaning of 'red'; one would say that this is what 'red' means, thus appealing to a rule about how a word is used in English. Surely, though, it makes no sense to say that we have a logical inference here, or that the ostension expresses an analytical statement. (This is unlike the statement 'Bachelor = unmarried man' which does not concern the application of rules, or expressions, but the semantic relation between them.) ... neither Hart nor any other legal positivist must subscribe to the view that the application of legal rules is a matter of logical inference. (Marmor, 2005, p. 98).

Within the field of logical theory Peter Strawson distinguished between two kinds of grounds of criticism:

> the criticism we offer when we declare a man's remarks to be untrue and the criticism we offer when we declare them to be inconsistent. In the first case we criticize his remarks on the ground that they fail to square with the facts; in the second case we criticize them on the ground that they fail to square with one another (Strawson, 1952, p. 1).

On my view Marmor's and Strawson's distinctions are drawn on the same tracks. The legal theorists and practical jurists as well should keep in mind those distinctions for their ignorance will obliterate the understanding of legal reasoning. Thus, to take just a single example, the common wisdom is that the appeal (and supreme) courts do not adjudicate, as a rule, factual but only normative ones. In Marmor's terms, these courts are called to settle only rule to rule relationships. That is, they are asked to adjudicate only the conformity of the primary legislation under review, i.e. statutes, ordinances of Government, with the supreme law, i.e. Constitution. But as far as both the *object* of the review are legal rules, as are expressed in language, and *the standard* from the perspective of which the review is realized consists also in rules embedded in language, it is clear that all it is about is to settle rule to rule relationships. That is a consistency/conformity relationship, in the case the constitutional court's finding is that the legislation is constitutional or an inconsistency/non-conformity relationship if the constitutional court's finding is that the legislation is unconstitutional.

Or, if we use Strawson's distinction, the ground of criticism in *constitutional* adjudication is not that the object of the review, legislation, fails to square with the facts. This is the case in *ordinary* litigation whenever, for example, the courts are called to settle if factual allegations made by the parties are (or not) supported by the evidence.

3.2 Standards of evaluation: (more or less) correct/incorrect interpretation v. valid/invalid deductive inference

We can ask: "What interpretation is given to that particular term?" and by analogy "How can we establish the logical reasoning where that word is involved?". What means a correct interpretation is quite different form what is meant by a valid logical deduction. The criteria used to assess the correctness of interpretation are quite different from the criteria used to assess the validity of deductive inferences.

Let's suppose that the interpretation given to the particular term *vehicle*, within a penumbral application, is that it *covers* the case of a bicycle. According to HLA Hart in the penumbra the judge has to bring the particular, i.e. bicycle, under the rule, i.e. "No vehicles in the park", taking into consideration social policies and ends. Precisely because, as we have seen, the case of the bicycle is a penumbral one, the interpretation given to the rule, though may be assessed as (more or less) *rational*, it cannot be *conclusive*.

Once the question of interpretation of the term was settled, for reaching the result, to give the judgment, in terms of "yes" or "no", there are three more steps to be reached. Firstly, the court, based on the evidence administered before it, will decide if the case is concernd with a bicycle or not. Secondly, the judge will infer from the conjunction of the propositions "All vehicles are forbbiden to enter in the park", which was interpret as covering the case of bicycles too, and the proposition that in the pending case is about a bicycle X, that the bicycle X has to be fobidden to enter into the park. Thirdly, the judge will decide to infer which *nota bene!* is not the same as the inference itself. As MacCormick said

> Decisions are made, not deduced. To decide is not to deduce. What is entailed by premises is not decided by anybody, though one can indeed decide whether or not to draw the conclusion from the premises one accepts, either in that sense of "drawing" a conclusion by which is meant dwelling conciously upon it in one's thought or that by which is meant acting upon it as a ground of decision (MacCormick, 1992a, p. 183).

Let us suppose that the interpretation given to the term "vehicle" is assessed as a bad one. The evaluation of the (non)validity of the inference from the proposition "No vehicles in the park", as it was already interpreted as covering the case of the bicycles too, has nothing to do with the evaluation of the interpretation of the word in terms of "good" or "bad", "stupid" or "intelligent". Even if, let's suppose for the sake of the argument, there is a (more or less) general agreement concerning the fact that the interpretation given to to the term interpreted is bad/incorrect/mistaken, it is still possible from the proposition which contains the (supposed) mistakanely interpreted term to infer a perfectly valid another proposition. One can perfectly hold *at once* that (1) the interpretation given to a term which occurs into the formulation of a legal rule is *mistaken* and (2) the inference made from the proposition which contains the "defective" term is *valid*. Conversely, one can perfectly hold *at once* that (1) the interpretation given to a term which occurs in the formulation of a legal rule is *correct*/good/rational/intelligent/the best one and (2) the inference made from the proposition which contains the adequately interpreted term is *invalid*.

To learn how to interpret general terms is different from learning how to deduce correctly conclusions from premises. Though both deductive *logic* and *interpretation* are *evaluative* concepts, the standards of evaluation in logical and, respectively, interpretative processes are quite different one from another. That is, to draw *validly* conclusions from premises is quite different from interpreting *correctly* the legal materials. While there is no such a thing as *more or less correct* way to drawing a conclusion from a premise it is intelligible to evaluate an interpretation as being (more or less) stupid or intelligent or (more or less) intelligent interpretations.

Therefore the assessment of what one does when draws a conclusion from a premise involves merely two exclusively solutions. If she *correctly* draws the conclusion from the premise one will say that she *properly* draws the conclusion from the premise. If she does not draw correctly the conclusion from the premise then one will say that she *did not at all* draw the conclusion from the premise and, after all, did not make a deductive argument at all.

The case with the interpretation is strikingly different. Thus, if one evaluates the interpreter's interpretation as being *incorrect* or *bad* or as *less preferable* than alternative interpretations the evaluator will not say, as in the case of drawing the conclusion from the premise, that did not at all interpret the rule. Thus, to recall the most famous hypothetical of legal theory "No vehicles in the park" rule, the adjudicatior/enforcer/interpreter may say that, for the purposes of the rule, the *bicycles* are covered by the rule. She can say that the supposedly *unique* purpose of the rule, i.e. imposing the policy "no accidents at all in the park due to any vehicles at all", does not prohibit covering the case of bicycles.

However, another interpreter of the rule, eventually the upper court, may advance a different view. For example, the view according which the purposes of the rules were much more complex than they were envisaged by the below court. For example, the upper court can maintain that not only reducing the risks of the accidents in parks was the purpose of the rule but also the rule was intended to limit as little as possible the enjoyment of as many as possible people, including of bicycles riders. Given that we are into the penumbra of the rule, no interpretation is conclusive.

3.3 Logic, interpretation and choice

By distinguishing clearly between logic and interpretation the legal actors will understand *the nature of their disagreement* and, as such, it is opened the way for its attenuation which is of a particular importance into the field of law. This is so because while there is no such a thing as a reasonable disagreement about how to deduce validly a conclusion from a premise, there is disagreement, which might be indeed a reasonable one, as to how to interpret legal texts.

3.4 Logic, interpretation and policy/principles

If the judge is to reach *validly* a certain conclusion from a given premise is *not* bound at all to any particular policy. In other words, to make *valid* inferences one needs not to be committed to any moral and/or political policy and/or principles at all. Instead, if a judge, *e.g.* late Associate Justice of US Supreme Court Antonin Scalia, is to endorse what he thinks to be the *correct* or rather *the best* method of interpretation, *i.e.* textualism, he will justify his option by saying that it *is* the appropriate way to further a particular policy/principle, *i.e.* democratic government. Accordingly, disagreement concerning *policy and principles* issues will be reflected only in disagreement concerning the methods of *interpretation* (textualism v. original intent, for example) and has not any implications on the eventual disagreement which might arise as to how the inferences were drawn. In the latter case, it is possible (and it is necessary) to settle the *only* way to infer validly.

3.5 Interpretation, deduction, choice

I think the concept of *choice* is best suited to illuminate and to grasp the different functions deductive logic and respectively interpretation play within legal adjudication. While interpretation of legal texts it implies always choice(s) in the case of deductive logic the choice does not play any role at all. Thus, according to Posner

> Obviously *the choice* (my emphasis) of premises is critical, and that is where public policy comes in. Why enforce only promises supported by consideration, or only promises that are consciously accepted? The reason, if it is a good reason, has to be traceable to some notion of policy rather than just be the result of arbitrary personal preferences or antipathies, or class bias, or some other thoroughly discredited ground of judicial action. It cannot be logic. Logic is used to go from the premises to the conclusion, not to obtain the premises. Of course, a premise may be the result of deduction from some more basic premise, but eventually one is forced back to a premise that cannot be obtained or proved by deduction. The choice of the premises of the argument is an *interpretative* choice which has to be traceable to some notion of policy rather than just be the result of arbitrary personal preferences or antipathies, or class bias, or some other thoroughly discredited ground of judicial action. (Posner, 1986-7, p. 182).

Instead, in the case of logic, one may say, as MacCormick did, that the choice has no role at all, that is that "What is entailed by premises is not decided by anybody" (MacCormick, 1992a, p. 183). Once the interpretive *choice* was made the judge is launched on the deductive mode. That she can take or not the major premise as a major premise shows that it is about an interpretive choice. But once the choice concerning the applicability or not of a general term to a particular situation, i.e. the interpretation, was made, within the process of inferring from one proposition to another no choice at all is anymore involved.

3.6 Logic is always a matter of validity and not of the truth of the propositions

"A deductive argument is one whose conclusion is claimed to follow from its premises with absolute necessity, this necessity not being a matter of degree and not depending in any way on whatever else may be the case" (Copi, Cohen, & McMahon, 2014, p. 27). Indeed "The validity of an argument depends only on the *relation* of the premises to the conclusion" (Copi, Cohen, & McMahon, 2014, p. 29). This point is important because it stresses the concerns of deductive logic. That is, deductive logic is not about the *truth* of the individual propositions which, as premises or conclusion, compound the chain of a reasoning, in particular that of a legal reasoning. Deductive logic is about the assesment of the appropriatness of *the way* one proposition is inferred from another one. Is not concerned directly with the truth of the individual components of the reasoning, i.e. the premises and/or conclusions of the argument. Of course, the truth of a proposition from which is inferred the truth of another one can be settled *through* deductive logic.

3.7 Artificial intelligence issue

One of the most important effects of the understanding of the differences between logic and interpretation is captured by Neil MacCormick:

> Whether or not Artificial Intelligence can, in principle or in practice, yield systems for fact-finding, for fact interpretation, for evaluation or for rule-interpretation or source-interpretation, I do not know, though I rather doubt it. But clearly deductions which take for granted information generated by such processes would be within existing computational capabilities. Creating computer systems for that purpose could be useful in itself, and it could free human beings for the essentially human tasks of investigation, evaluation and interpretation, while democratizing legal information by increasing its accessibility to and user-friendliness for the ordinary citizen (McCormick, 1992a, p. 202).

> There may be many interrelated rules that have to be applied in a fixed order, but where it is difficult for a human reasoned to keep it all in mind. Where this yields long *deductive* (my emphasis) chains, reconstructable as decisions trees, the use of the computer may be thoroughly helpful in alleviating complexity, and in prompting the human questioner to raise the relevant doubts and difficulties at the right point in the reasoning process. The great thing about computer handling of complex concatenations of essentially simple and repetitive processes is that computers appear never to get bored or forgetful in a human way (MacCormick, 1992b, p. 224).

Only by understanding the differences among interpreting legal texts and making valid inferences one can truly understand why MacCormick asserts that the computers might be used to make valid inferences, while they are rather inappropriate to use them in the process of legal interpretation.

3.8 The armchair issue – the a priori issue

Given that logical matters pertain to *rule to rule relationships* the following consequences are drawn. First one, to ground their judgments on *valid* inferences, judges need not go outside of the room or if you prefer another metaphor there is no need to rise from their armchair. On the contrary, to have *true* propositions within their reasoning and to interpret correctly legal terms it is never enough to stay into the armchair. Thus all forms of factual investigations

> share the idea that the investigators go out into the field, ask questions, poke around, interview witnesses, examine physical evidence, and then make the decision themselves. (Schauer, 2009, p. 206).

Given that the conditions to obtain a true proposition within the reasoning means to square it with the facts the judges concerned to issue true propositions have always to rise from their armchairs. Also to obtain a correct or intelligent or wise interpretation judges have to, according to Hart, to take into consideration practical consequences of their decisions, social aims, policies and needs. This means, of course, to rise from the armchair and to investigate facts.

Furthermore, the issue of the review by an upper court of the *validity* of the inference made by a lower court can be settled from the armchair. On the contrary, the issues of the *truth* of the propositions which compound the reasoning of the lower court cannot be at all settle from the armchair.

When reviewing the decisions of lower courts the upper courts, i.e. court of appeals, may find out errors (1) interpretations and/or (2) deductive logic. Given that to draw validly an inference from a given proposition has nothing to do with interpreting correctly the general terms of the legal rule, the judge who reviews the below court's judgment has to disentangle, first of all, the two issues.

4. CONCLUSION

The objective was, departing from what Hart already clarified as to their differences, to advance a little bit the understanding of the distinction between interpretation and deductive logic. Thus I will lay down some conceptual differences between the meanings of deductive logic and interpretation and their roles within legal adjudication which are not to be found expressly within Hart's argument. An investigation into the conceptual differences among deductive logic and interpretation and their mapping is significant for, as it might always be the case not only in philosophy but in law as well, they tend to be forgotten. As Wittgenstein noted:

> Something that one knows when nobody asks one, but no longer knows when one is asked to explain it, is something that has to be *called to mind*. (And it is obviously something which, for some reason, it is difficult to call to mind.) (Wittgenstein, 2009, par. 89).

Even if this map is only a limited one in scope I still maintain that having one will have a *preventive* role towards the recurrence of misunderstandings of legal reasoning. This is the justification of writing this paper.

ACKNOWLEDGEMENTS: I am grateful for invaluable advice and criticism to Lucian Epure, Jaap Hage, Fabrizio Macagno and António Marques.

REFERENCES

Copi, I. M., Cohen, C., & McMahon, K. (2014). *Introduction to Logic*. 14th edition. Edinburgh: Pearson Education Limited.

Hart, H. L. A. (1957-8). Positivism and the Separation of Law and Morals. *Harvard Law Review, 71*(4), 593-629.
Hart, H. L. A. (1994). *The concept of Law.* 2nd edition, with new *Postscript*, written by Hart and edited by Penelope A. Bulloch & Joseph Raz. Oxford: Clarendon Press.
Hart, H. L. A. (2013). Discretion. *Harvard Law Review, 127*(2), 652-65.
Marmor, A. (2005). No Easy Cases? In Marmor, A. (2005). *Interpretation and Legal theory.* 2nd edition. Oxford & Portland, OR: Hart Publishing.
MacCormick, N. (1992a). Legal Deduction, Legal Predicates and Experts Systems. *International Journal for the Semiotics of Law*, V/14, 181-202.
MacCormick, N. (1992b). A Deductivist Rejoinder to a Semiotic Critique. *International Journal for the Semiotics of Law*, V/14, 215-224.
Paulson, S. L. (2000). On Transcendental Arguments, Their Recasting in Terms of Belief, and the Ensuing Transformation of Kelsen's Pure Theory of Law. *Notre Dame Law Review, 75*(5), 1775-96.
Postema, G.J. (2010). Positivism and the Separation of Realists from their Scepticism Normative Guidance, the Rule of Law and Legal Reasoning. In P. Cane (Ed.), *The Hart - Fuller Debate in the Twenty–First Century* (pp. 259-81). Oxford & Portland, OR: Hart Publishing.
Posner, R. (1986-7). Legal Formalism, Legal Realism, and the Interpretation of Statutes and the Constitution. *Case Western Reserve Law Review, 37*(2), 179-217.
Schauer, F. (2009). *Thinking like a Lawyer.* Cambridge, MA: Harvard University Press.
Strawson, P.F. (1952). *Introduction to logical theory.* London: Methuen & Co.
Strawson, P. F. (1971). Particular and General. In Strawson, P. F. (1971). *Logico-linguistic papers* (pp. 28-52). London: Methuen & Co.
Wittgenstein, L. (2009). *Philosophische Untersuchungen. Philosophical investigations.* The German text with and English translation by G.E.M. Anscombe, P.M.S. Hacker, and J. Schulte. Revised 4th edition by P.M.S. Hacker and J. Schulte. Oxford: Wiley Blackwell.

Commentary on Văduva's
HLA Hart on logic and interpretation

ANTÓNIO MARQUES
IFILNOVA, Universidade Nova de Lisboa, Portugal
marquesantoni@gmail.com

Vaduva begins his talk with the claim that reasoning and argumentation in legal matters involves mainly interpretation or interpretive operations and in less degree deductive inferential thought. This is relevant because the shadow of, say, a mechanical judge who works exclusively with deductive logic is a myth that Hart dismisses in the terms already quoted by Voduva. ("Logic only tells you hypotetically that if you give certain term a certain interpretation then certain conclusion follows. Logic is silent on how to classify particulars and this is at the core of a judicial decision.") As I refer to later this myth is perhaps the main pretext for the claims of rule-scepticism as we found in Hart's main work.

Vaduva suggests that the argumentative framework at stake in legal matters presupposes that we focus on the concept of interpretation and Hart's conception that every legal rule has a core of applications and a penumbral area of applications is an excellent inquiry line. At this point I wish to remark that Hart didn't see his philosophical program on law as an interpretive one but as a descriptive stance perhaps under the influence of the late Wittgenstein, although he don't refuse of course explanatory power to descriptivism.

Although the paper of Vaduva has many other important points I want to focus on the issues of interpretation, transcendental argument and the position of Hart concerning the "open texture" of law, because it is the same problem as that of the application of rules on core and penumbral matters.

Regarding legal interpretation, if I understood well, you see it under the form of a transcendental argument. Yet the general form of a transcendental argument requires the presupposition of one or more transcendental principle, which we represent as necessary, at least in the conceptual region where we develop our argumentation. For example, the presupposition of a free will works as a transcendental principle if a judge attributes the culpability of the action B to the person A. In this case free will is a rather general concept, which is not an object of theoretical knowledge but that is to be introduced in the argumentation in favour of the culpability of the accused person. Of course there is space in this argumentation operation to discuss the meaning of the concept, how it is used by the judge, and so on. So it is always possible to start a dialectical scene. That's precisely at this point that we must link the transcendental reasoning with what you call the core's argument. Is the core argument a kind of transcendental argument and must we understand the meaning of the term as the proper, the right region where the transcendental principle applies? For example the principle of the existence of a free will is not well used if we are arguing about acts of persons with some kind of mental illness or acts of children, and so on. In principle these can be core cases, but certain situations of, say, soft mental illness can put the judge or officials of justice in general in the penumbra area you mention. So my question is about the relationship between interpretation and what you designate transcendental argumentation. An even more general question is to know if in your opinion transcendental arguments are a central part of the structure of legal interpretation, and even if there is an overlapping. Also it would be interesting to know from you if there is any link between the concept of core argument and that of transcendental principle that circumscribes the area of legal argumentation.

Another issue related to Hart's philosophy of law. The consideration of core and penumbra areas in the application of legal rules is in his view a distinction that has an important place in the discussion with the sceptical philosopher of law. Rule-scepticism is based on the claim that talk of rules is a myth and laws are simply the decisions of a court and the prediction about it. Hart defines rule-scepticism just in these terms (Hart, 1994, p. 136). So it seems that it is the existence of theses areas of light and shadows by occasion of the application of law that feed the referred to scepticism. The only way, as far as I understand Hart's refutation of scepticism, is thinking on a legal system as a whole of norms. Decisions of the courts are of central importance yet they are part of a system of rules governing a community, which Hart conceives as a complex of primary and secondary rules. I shall not elaborate on this distinction, only remark that the uncertainty that is associated of a persistent penumbra area is solved or better overcome by the unlimited number of secondary rules introduced, applied and obeyed by officials and individuals of a modern state who recognize their validity from their internal point of view. As Hart says "the existence of such a rule of recognition may take any of a huge variety of forms, simple or complex" (1994, p. 94). Such forms are those, which take part in the legal system as concrete form of life. So perhaps interpretation means the same as rules of cognition in a Hartian system. What counts as interpretation is also equivalent to the main operations of practical reason, deliberation and choice.

REFERENCES

Hart, H.L.A. (1994). *The concept of Law*. 2nd edition, with new *Postscript*, written by Hart and edited by P. A. Bulloch & J. Raz. Oxford: Clarendon Press.

32

Speech Acts in a Dialogue Game for Critical Discussion

JACKY VISSER
University of Amsterdam, The Netherlands
j.c.visser@uva.nl

The representation of speech acts is a next step in the formal approximation of the pragma-dialectical model of critical discussion. The project serves two purposes: theoretical investigation of the pragma-dialectical model, and preparation for computerisation. The formal approximation is developed as a dialogue game. To represent the speech act perspective in this dialogue game, the rules for moves and commitments are based on the role of speech acts in critical discussion and their felicity conditions.

KEYWORDS: automated argument analysis, critical discussion, dialogue game, formalisation, pragma-dialectics, speech acts

1. INTRODUCTION

In his 2014 keynote address opening the eight conference of the International Society for the Study of Argumentation (ISSA) Frans van Eemeren mentioned formalisation as one of the most important current topics in argumentation research. Formalisation can serve several purposes in argumentation theory. Most obviously perhaps, it can be seen as a next step in the development of a theory. Through formalisation concepts and relations are defined very precisely, and models can be verified with mathematical means. This function of formal models has been described by Krabbe and others (e.g., Krabbe, 2006, pp. 195-197; Krabbe & Walton, 2011, pp. 256-259) as an abstract theoretical laboratory. This laboratory makes it possible to "experimentally" test the instrumental validity of models, in order to improve the model on theoretical grounds.

Aside from the "laboratory function" formalisation can be a preparation for computerisation. Before an informal theory or model can be implemented computationally it is often necessary to formalise the theory or model to some degree. Through formalisation the concepts and "language" of the theory are brought closer to the formal language inherent to computer programming.

The present paper is a contribution to an overarching research project concerning formal and computational methods within the pragma-dialectical approach to argumentation (van Eemeren & Grootendorst, 1984; van Eemeren et al., 2014, pp. 517-613). The central issue addressed in the paper is how the speech act perspective inherent to the pragma-dialectical discussion model can be accommodated in a so-called formal approximation of the model. The formal approximation is a preparation for the computational application of the pragma-dialectical theory, which I will discuss in section 2. In section 3, I will explain the idea of a formal approximation. In section 4, I will introduce the systematic development of a formal approximation of the pragma-dialectical discussion model by means of a dialogue game. In section 5, I specify the rules of a dialogue game for critical discussion.

2. FORMALISATION IN PREPARATION OF COMPUTERISATION

The computational application of argumentation theory in general has enjoyed an increasing amount of attention over the last decade, not only from argumentation theorists, but also within the research field of Artificial Intelligence. Overviews of the state of the field such as the chapter "Argumentation and Artificial Intelligence" in the new *Handbook of Argumentation theory* (van Eemeren et al., 2014, pp. 615-675), and the volume *Argumentation in Artificial Intelligence* edited by Rahwan and Simari (2009) show a broad variety in computational applications of (insight about) argumentation. For example, there are computer programs to support legal practitioners in constructing and evaluating legal cases (e.g., Verheij, 2005), medical applications where argumentation is used to improve the computerised distribution of donor organs (e.g., Tolchinsky, Cortés, & Grecu, 2008), or educational applications such as interactive teaching methods in which argumentation plays a key role (cf. Scheuer, Loll, Pinkwart, & McLaren, 2010).

Recently another application within "computational argumentation theory" has received a lot of attention: the use of computer programs for *argument mining* (e.g. Budzynska et al., 2014) to search (large numbers of) long texts for argumentative elements and then indicatively reconstruct the arguments. In order to benefit from the work that has already been done in non-computational fields, these computer programs are best based on argumentation analytical methods that have been proven in practice and which are based on solid theoretical grounding. This leads me to investigate how the pragma-dialectical method of analysis (van Eemeren, Grootendorst, & Kruiger, 1983; van Eemeren, Grootendorst, Jackson, & Jacobs, 1993) can serve as a foundation for the computerised analysis of argumentative texts.

It would be overly ambitious at this moment to aim for a computer program that can argumentatively analyse a text in a fully automated way, because this still requires many (technical) preliminary steps. By aiming for small computational tools that support human analysts' sub-tasks, the obstacles in the way of complete automation may be surmounted one by one. An example of a sub-task that a computational tool could support is the construction of an analytic overview.

The analytic overview is the end result of a standard pragma-dialectical analysis of an argumentative text. It provides a systematic overview of all the points (e.g., standpoint(s), argumentation structure, etc.) that are relevant to the resolution of a difference of opinion and to the evaluation of an argumentative text (van Eemeren et al., 1993, p. 86; van Eemeren & Grootendorst, 2004, p. 118). To construct an analytic overview, an analyst follows a two-step method.

In the first step the parts of the original text that are argumentatively relevant are identified by using the *ideal model of critical discussion* as a heuristic (van Eemeren & Grootendorst, 1992, p. 36). In the pragma-dialectical discussion model an ideal procedure is proposed for the resolution of differences of opinion in a reasonable way (van Eemeren & Grootendorst, 2004, pp. 42-68). By using the ideal model as a heuristic, the analyst reconstruct the original text as if it were a discussion aimed the resolution of a difference of opinion. To arrive at this reconstruction in terms of the ideal model, the analyst applies four transformations – deletion, addition, substitution and permutation (van Eemeren et al., 1993, pp. 61-62) – which bring the original text analytically in line with the ideal model.

In the second step an analytic overview is constructed on the basis of the reconstruction in terms of the ideal model that was the outcome of the first step (van Eemeren et al., 1983, p. 290). The content of the analytic overview is fully determined by the reconstruction in terms of the ideal model. Due to this one-on-one matching, the step from reconstruction to analytic overview seems to be very suitable for automation.

To make the automation of this analytical step possible, computational representations are needed of on the one hand the ideal model of critical discussion and on the other hand the analytic overview. Additionally, the computational representations need to include the relations between all the possible (sequences of) discussion moves and the various components of the analytic overview. As a preliminary to the computational representation of the ideal model, a dialogue game is developed as a formal approximation of critical discussion. In the remainder of this paper, I will explain what I mean by "formal approximation", how the dialogue game is construed and what the relation to speech acts is.

3. THE FORMAL APPROXIMATION OF CRITICAL DISCUSSION

The formal approximation of critical discussion is developed as a dialogue game. A dialogue game is a formal rule system defining a set of dialogues that can be "played out". The rules of the game determine which moves can be made when, by which player, and to which effect. Additionally, the rules state the goal of the dialogue and when this is realised. In philosophy, formal dialogue games have been employed to study dialogue and reasoning (e.g., Hamblin, 1971; Lorenzen & Lorenz, 1978; Walton & Krabbe, 1995). Based on the philosophical work, dialogue games are now also commonly used in Artificial Intelligence to model dialogue, and communication between agents (cf. McBurney & Parsons, 2009; Prakken, 2009).

The term "formal approximation" is used to express two considerations. Firstly, the result of a *formal approximation* can be contrasted to the result of a *strict formalisation*. Formalisation in the strict sense can give rise to the impression that the original model is being replaced by the formalised model. This is not the intention with the formal approximation, which is meant to exist next to the original ideal model. The intention is to develop a dialogue game as formal approximation of critical discussion, such that it can be called "formal" in three senses. The formal approximation has to be procedurally regimented (*formal$_3$* in the taxonomy by Barth and Krabbe (1982, pp. 14-19; Krabbe, 1982)) and a priori or normative (*formal$_4$*). Krabbe and others (Krabbe & Walton, 2011, p. 246; Krabbe, 2012, p. 12; van Eemeren et al., 2014, p. 304) have already noted that the existing pragma-dialectical ideal model is also formal in these two senses. By means of the dialogue game a third sense of formal is added: the rigid definition of well-formed linguistic expressions and the way in which these can be combined (*formal$_2$*).[1]

Secondly, the notion "formal approximation" expresses the expectation that not all properties and aspects of the original model can

---

[1] An earlier proposal to formalise the ideal model of critical discussion by Krabbe (2012; 2013), had the same objective: making the system formal2. Although the resulting dialogue systems are different, our approaches and goal are obviously very similar.

be preserved. In this sense the *approximation* is comparable to the conventionalised argumentative activity types as *empirical* approximations of critical discussion developed by van Eemeren and Houtlosser (2005; van Eemeren 2010, pp. 129-162). If the formal approximation yields a result that diverges from the original ideal model, this could be an indication of an imperfection or obscurity in the original model (cf. the laboratory function mentioned in section 1). Conversely, the divergence could also be the result of the streamlining which is inherent to the formalisation of informal models. A reason for this effect can be found in the expressiveness of formalisms, which is (usually) more restricted than that of natural language. Formal models have to be fully explicit and free of ambiguity to define what falls within the model and what is excluded. This restricted expressiveness will mean that the formal approximation is stricter than the original ideal model.

The formal approximation of critical discussion is not developed in its entirety at one time. Instead, a dialogue game is systematically constructed starting from a simplified basis to which additional features are gradually added in extensions and amendments of the dialogue game rules. In this way the scope of features of the dialogue game is brought ever closer to the full extent of the ideal model. This systematic approach has the practical advantage of decomposing a large task into several smaller constitutive tasks, such that these smaller tasks can be carried out at different times and by different people. A second advantage is of a theoretical nature: by gradually increasing the complexity of the dialogue game, the properties of the model can be studied in isolation (without other features of the model complicating the investigation).

A *basic dialogue game for critical discussion* to serve as the starting point for the systematic development of the formal approximation has been proposed before (Visser, in press). In the next sections, I will explain how the role speech acts play in critical discussion can be accommodated in the rules of the dialogue game.

4. ACCOMMODATING A SPEECH ACTS PERSPECTIVE IN A DIALOGUE GAME FOR CRITICAL DISCUSSION

The dialogue game for critical discussion is defined in five types of rules. There are rules for the initial state of the game (subsection 5.1), the available moves (5.2), the effects of moves in terms of commitments (5.3), the possible sequences of moves (5.4), and how the game ends

(5.5).² The rules of the aforementioned basic dialogue game (Visser, in press) are based on the fifteen "technical" rules for critical discussion (van Eemeren & Grootendorst, 2004, pp. 135-157), representing the "dialectical dimension" of critical discussion.³ In the next section, the basic dialogue game is extended with what could be called the "pragmatic dimension" of critical discussion. Van Eemeren and Grootendorst (1984, pp. 98-112) describe the distribution of different (types of) speech acts over the (four) stages of the ideal model and give an overview of the functions the speech acts perform in critical discussion. In this way, every (dialectical) discussion move in the ideal model is associated with a particular (type of) speech act that would prototypically be used to realise it. These associated speech acts form the basis for the extension of the dialogue game rules.

As will become clear in the next section, taking account of the speech act perspective mainly affects two of the five types of dialogue game rules. The rules for the initial state of the game, for the possible sequences of moves, and for ending the game remain largely unchanged in comparison to the existing *basic* dialogue game. The "move rules" reflect the speech acts that are associated with the realisation of discussion moves in the ideal model. This is done by providing paraphrases in natural language for each move and by grouping together some moves that can be realised using the same type of speech act (see subsection 5.2). The "commitment rules" reflect the felicity conditions that go with each of the speech acts (cf. van Eemeren & Grootendorst, 1984, pp. 19-46). The commitments players acquire by making moves consist of propositions reflecting the fulfilment of these felicity conditions (see subsection 5.3).

While the dialogue game in the next section changes the rules of the basic dialogue game to accommodate the speech act perspective, the other simplifications of the basic game remain in place. This means that the "rhetorical dimension" of the pragma-dialectical theory (which is concerned with institutionalised contexts and strategic manoeuvring (van Eemeren, 2010)) is not taken into account yet. As a further simplification, only the argumentation and the concluding stage of

---

² With the exception of the "commencement rules", this way of specifying a dialogue system is similar to Walton and Krabbe's (1995).

³ How precisely the dialogue game rules relate to the fifteen rules of the ideal model is discussed elsewhere (Visser, 2015). These fifteen rules should not be confused with the ten "practical" commandments for reasonable discussants (van Eemeren & Grootendorst, 1992, pp. 208-209), which are intended as a rule of thumb for conducting and evaluating actual argumentative discussions in practice.

critical discussion are part of the dialogue game. The remaining two discussion stages (the confrontation and the opening stage) are only accounted for via the assumptions made in the "commencement rules" for the initial state of the dialogue game.

Regarding the outcome of the confrontation stage, the assumption reflected in the commencement rules is that a single positive standpoint is advanced which meets with doubt. This simplification constraints the dialogue game to single non-mixed differences of opinion about a positive standpoint, thereby excluding differences of opinion about multiple or negative standpoints.

The first assumed outcome of the opening stage is that players agree to only use single argumentation, which may be criticised only by casting doubt, not through contradiction. This means that (potentially mixed) sub-discussions and other complex argumentation are excluded (an extension accommodating complex argumentation is investigated elsewhere (Visser, 2013)).[4] The second assumption is that the justificatory force of the single argumentation may only be based on the inference rules of classical propositional logic (e.g., *modus ponens*). This simplifying assumption with respect to the underlying logic does not preclude the later introduction of more complex or nuanced systems into the dialogue game, such as the pragma-dialectical argument schemes with critical questions (van Eemeren & Kruiger, 1985; Garssen, 1997) or non-monotonic systems for defeasible reasoning (e.g., Pollock, 1987; Dung, 1995).

The assumed simplifying outcomes of the confrontation and the opening stage result in a dialogue game that is a formal approximation of the dialectical and pragmatic dimensions of the argumentation and the concluding stage of consistently non-mixed discussion about one positive standpoint which is defended with a single justificatory argument.

5. THE RULES OF THE DIALOGUE GAME FOR CRITICAL DISCUSSION

Before presenting the dialogue game rules, some remarks are warranted about what the rules do not cover and about a formal language that is used to express the content of dialogue game moves and commitments. The dialogue game rules leave the constitution of the players themselves undefined. In the ideal model discussants are

---

[4] Van Eemeren and Grootendorst (1984, pp. 78-83) follow a similar approach in their introduction of the ideal model of critical discussion, by starting with an elementary discussion from which more complex discussions can be composed.

presumed to be human interlocutors. Because the dialogue game is intended as a preparation for computational application no such assumption is made here. The dialogue game should be such that in principle both human and artificial agents can play it. How the players internally represent the state of the game and how they keep track of their own and their opponent's commitments is also not specified in the rules of the game. In the case of human agents, their internal constitution is a topic for psychologists, and in the case of artificial agents, for software engineers. The rules of the dialogue game do not refer to the internal state or beliefs of the players, but only take into account what the players externalise during the game (cf. van Eemeren and Grootendorst's (2004, pp. 53-55) meta-theoretical principle of "externalisation").

A further factor not accounted for in the rules of the dialogue game is the strategy of players. Players have to make choices about their next move when playing the game. They can use different strategies to try to maximise their chances of winning. Modelling these strategies is considered to be part of the agents playing the game, not of the rules of the game itself – although the rules do obviously pose boundaries to the possible strategies (cf. also the use of game theory for argumentative dialogue games (Rahwan & Larson, 2009), and strategic manoeuvring in argumentative discussions (van Eemeren, 2010)).

A further assumption has to be made in order to facilitate the expression of the content of moves in the dialogue game – i.e. to express what the discussion is about. The rules assume there to be some formal propositional object-language $\mathcal{L}$. In the current study the particular composition of $\mathcal{L}$ is not the main concern. It should be sufficient to think of $\mathcal{L}$ as a propositional language consisting of (an infinite number of) propositions, and one connective: $\Rightarrow$. The propositions can be considered to refer to (atomic or molecular) sentences of classical propositional logic (i.e., $\mathcal{L}$ can be thought of as a meta-language "about" propositional logic). These propositions are used in the dialogue game to express content.

The connective $\Rightarrow$ composes two propositions $\varphi, \psi \in \mathcal{L}$ into $\varphi \Rightarrow \psi$ (to be read as "φ therefore ψ"), and expresses the justificatory force of the first proposition for the second. The justificatory force refers to the underlying reasoning system in which ψ may be inferable on the basis of φ. The current assumption of propositional logic (see section 3) as the underlying system means that any use of $\varphi \Rightarrow \psi$ can be interpreted as denoting the application of some propositional rule of inference to the effect that the acceptability of φ justifies the acceptability of ψ. Without going into detail about specific rules of inference, it is sufficient for this moment to assume there to be some external method which the players can use to determine the acceptability of propositions and inferences in the dialogue game. These external methods are closely related to the intersubjective identification and testing procedures in the ideal model (van Eemeren & Grootendorst, 2004, pp. 145-150). Assumed is finally that the external methods return a positive or negative outcome.

5.1 Beginning the dialogue game

The commencement rules determine the initial state of the game before the first move is made. Because the confrontation and the opening stage of critical discussion are not explicitly modelled, the outcomes of these stages are taken to be part of the initial state of the dialogue game. Based on the assumed outcome of the confrontation stage the dialogue game is played by two players to assess the tenability of one positive standpoint about some proposition $\psi \in \mathcal{L}$.

Based on the assumed outcome of the opening stage the two players are labelled *Prot* and *Ant*, matching the discussion roles of respectively protagonist and antagonist in (the argumentation stage of) critical discussion. *Prot* defends a positive standpoint with respect to ψ, whereas *Ant* critically assesses this defence due to her doubt about ψ. A second outcome of the opening stage is the agreement on a set of mutually acceptable material and procedural starting points. In the dialogue game the material starting points are represented as a (static) set *SP* of propositions that are considered acceptable by both players. Because critical discussants need at least one shared material starting point to have a meaningful discussion (van Eemeren & Grootendorst, 2004, p. 139), *SP* is non-empty: $SP \neq \emptyset$.

The procedural starting points are represented in the dialogue game by three principles. First, the players have to play by the rules (cheating is not possible). Second, the game is turn-based. Players take turns in which they must make one move and then pass the turn to the other player. Third, the players have agreed upon some reasoning system (a logic) and a method to check the acceptability of inferences appealed to.

Finally, it has to be clear from the start what the goal is of the dialogue game: to resolve a difference of opinion about a positive standpoint with respect to a proposition ψ, through *Prot*'s advance of argumentation in defence of ψ, and *Ant*'s critical assessment of the argumentation.

5.2 Moves

The moves players can make are of the form *type(p, "Paraphrase", φ)*, where *type* indicates the function of the move, *p* ∈ *{Prot, Ant}* denotes the player making the move, *Paraphrase* is a description of the move in natural language, and φ ∈ ℒ is the propositional content of the move. Proposition φ can be atomic or in some cases molecular. Every unique instantiation of a move can only occur once per game. This means that no specific combination of the elements that make up a move may be repeated during the course of one game. Moves can also not be retracted, but always stay on the record: the dialogue game is cumulative with respect to the moves. While there are *retract* moves according to rule (M4), the retraction does not concern prior moves, but rather commitments, as will become clear in subsection 5.3.

The different *types* of the moves are based on the speech acts that play a role in critical discussion. For example, the *type* of the move in rule (M1), *argue*, is a representation of the speech act complex "argumentation" in the ideal model, and the *type* of the move in rule (M7), *doubt*, represents the illocutionary negation of an assertive used to indicate non-acceptance in the ideal model. The function of a move in critical discussion terms is indicated in the second part of each of the rules (M1) to (M9).

The paraphrases of the moves are inspired by the performative formulae or standard paraphrases known from speech act theory (e.g., van Eemeren & Grootendorst, 1984, pp. 112-118). The paraphrases are unique to each of the different moves, and can therefore be used to uniquely identify moves without using the fully explicit form. This is useful in computer systems implementing the dialogue game to facilitate a natural language interface.

The dialogue game for critical discussion is asymmetrical where the role of the two players is concerned. This results in two sets of moves, one for each of the players, depending on their role. Player *Prot* has the moves in rules (M1) to (M5) at his disposal to defend his standpoint. Player *Ant* can use the moves in rules (M6) to (M9) to critically test *Prot*'s defence.

>(M1) *argue(Prot, "My reason is that", φ)*: to assert φ as a reason for the ψ at issue.
>(M2) *identify(Prot, "I propose to intersubjectively identify the acceptability of", φ)*: to initiate the agreed upon intersubjective identification procedure in order to assess the acceptability of the propositional content φ of the argumentation.

(M3) *test(Prot, "I propose to intersubjectively test the acceptability of", φ⇒ψ)*: to initiate the agreed upon intersubjective testing procedure in order to assess the acceptability of the justificatory force of φ for ψ of the argument.

(M4) *retract(Prot, "I retract my commitment to", φ)*: to retract commitment to the propositional content or justificatory force of an argument, or to the propositional content of a standpoint.

(M5) *claim_conclusive_defence(Prot, "I have conclusively defended that", ψ)*: to claim the conclusive defence of a positive standpoint with respect to ψ.

(M6) *accept(Ant, "I accept that", φ)*: to accept an argument or standpoint.

(M7) *doubt(Ant, "I doubt whether", A)*: to cast doubt on the acceptability of the propositional content φ or justificatory force $\varphi \Rightarrow \psi$ of an earlier move *argue(Prot, "My reason is that", φ)*.

(M8) *claim_successful_attack(Ant, "I have successfully criticised that", A)*: to claim the successful critical attack of the acceptability of φ or $\varphi \Rightarrow \psi$.

(M9) *claim_conclusive_attack(Ant, "I have conclusively criticised that", ψ)*: to claim the conclusive critical attack of the argumentative defence of ψ by Prot.

5.3 Commitments

By making moves, players change the state of the game. As will become clear in subsection 5.4, every move determines the possible moves the other player can make in the next turn. Some moves additionally affect the commitments of players. Commitments are propositions the acceptability of which a player has to argumentatively defend if prompted to do so. The propositions players become committed to are kept track of in personal commitment stores (cf. Hamblin, 1970, p. 257). The commitment stores of the players are represented in the dialogue game by two sets, $CS_{p,t}$, of propositions $\varphi, \psi, \ldots \in \mathcal{L}$, for player p ∈ {*Prot, Ant*}, and with an index $t \in \mathbb{N}$ as a turn counter. Every turn the t counters of the commitment stores are increased by one.

At the start of the game, at $t=0$, the commitment stores of the players contain some propositions. Based on the requirements at the beginning of the dialogue game, $CS_{Prot,0}$ contains the shared starting points and the propositional content of the standpoint, say some $\psi \in \mathcal{L}$. The shared starting points are also contained in $CS_{Ant,0}$, but ψ explicitly is not – otherwise *Ant* would already be committed to the proposition under discussion when starting the game, preventing a difference of opinion from arising in the first place. Rules (C1) and (C2) specify the content of the commitment stores at $t=0$.

During the game, some of the moves made in a turn t modify the players' commitments. Starting from a speech act perspective makes clear that commitments cannot be equated with moves, but should be treated separately. The moves in the dialogue game are based on the speech acts that are the prototypical realisations of discussion moves, while the commitments are based on the felicity conditions of those speech acts. In contrast to moves, commitments can be retracted (in accordance with rules (M4) and (C4)), meaning that while the dialogue game is cumulative with respect to moves, this is not the case with respect to commitments.

Every speech act comes with a number of conditions that should be fulfilled for the speech act to be performed felicitously. Based on the communication principle (van Eemeren & Grootendorst, 1992, pp. 49-52) the addressee of the speech act can presume the producer of the act to be committed to the fulfilment of these felicity conditions. If the addressee has doubt regarding the fulfilment of a condition, this can give rise to a difference of opinion (i.e., felicity conditions contribute to the "disagreement space" (cf. van Eemeren et al., 1993, pp. 102-104)). The propositional content of the standpoint at issue in the difference of opinion then consists of a proposition expressing the fulfilment of the felicity condition under consideration (cf. van Eemeren & Grootendorst, 1984, pp. 95-98; Houtlosser, 1995, pp. 65-91). In the dialogue game such propositions expressing the fulfilment of felicity conditions can be employed to represent the commitments that players acquire as the result of making a move.

Although it is possible to let all dialogue game moves result in a number of commitments covering all the felicity conditions of the relevant speech act, currently only a small selection is incorporated into the commitment rules. This has to do with the conciseness of the rules. Most of the commitments based on the fulfilment of felicity conditions would not play any part in the remainder of the dialogue game, because they are never referred to in the dialogue game rules, while they would obscure the proceedings. Relevant then in this sense are only those felicity conditions that create or discharge dialectical obligations within the restrictions posed to the full extent of the ideal model in section 4.

Rules (C3) to (C5) define the relevant effects of three moves on the players' commitments. Moves of the type *retract* delete a proposition from a commitment store, thereby removing the dialectical obligation to defend it, but potentially also cutting short a line of argumentative defence. Moves of the type *accept* add commitments, thereby removing the possibility of doubt and criticism. Moves of the type *argue* are based on the speech act complex "argumentation", and cause the addition to the commitment store of two propositions expressing respectively the propositional content and the justificatory force of the argumentation (cf. the different identity and correctness conditions that van Eemeren and Grootendorst (1992, p. 31) describe for the speech act complex). For all the other moves holds that $CS_{p,t} = CS_{p,t-1}$.

(C1) $CS_{Prot,0} = SP \cup \{\psi\}$.
(C2) $CS_{Ant,0} = SP$.
(C3) argue(Prot, "...", φ)[5]: $CS_{Prot,t} = CS_{Prot,t-1} \cup \{\varphi, \varphi \Rightarrow \psi\}$.[6]
(C4) retract(Prot, "...", A): $CS_{Prot,t} = CS_{Prot,t-1} - \{A\}$.
(C5) accept(Ant, "...", φ): if $\varphi \Rightarrow \psi \in CS_{Prot,t}$, then $CS_{Ant,t} = CS_{Ant,t-1} \cup \{\varphi, \varphi \Rightarrow \psi\}$, otherwise $CS_{Ant,t} = CS_{Ant,t-1} \cup \{\varphi\}$.

5.4 Sequences

The dialogue game is always started with a move *argue(Prot, "My reason is that", φ)* in which *Prot* advances a proposition φ as a reason for the propositional content ψ of the standpoint at issue. When other moves can be made, depends on the state of the game at the point when the move is considered. Three properties of the state of the game can be relevant. The move made in the preceding turn is always of importance. In some cases it is also important what the content of the players' commitment stores is, or what the outcome of an external procedure is. Rules (S1) to (S13) determine for each move under which circumstances it can be made (within each rule it holds that $\varphi \neq \psi$). Some moves are just one of the legal continuations of possibly various preceding moves (indicated as "may follow"), while other moves are a mandatory follow-up to a particular preceding move ("follows").

(S1) argue(Prot,"...", φ): starting move.
(S2) identify(Prot, "...", φ): may follow doubt(Ant, "...", φ).
(S3) test(Prot, "...", $\varphi \Rightarrow \psi$): may follow doubt(Ant, "...", $\varphi \Rightarrow \psi$).
(S4) retract(Prot, "...", φ): may follow doubt(Ant, "...", φ) or claim_successful_attack(Ant, "...", φ).[7]

---

[5] To improve the readability of the commitment and sequence rules, the paraphrases are omitted in the moves.

[6] Because these sets are not multisets, any proposition can only be a member of the same set once. Any subsequent attempt to add the proposition will have no effect on the commitment store.

[7] Rules (S4) and (S5) imply that a negative result of an intersubjective procedure cannot be rectified by Prot. This is a temporary effect of the

(S5) retract(Prot, "...", φ⇒ψ): may follow doubt(Ant, "...", φ⇒ψ) or claim_successful_attack(Ant, "...", φ⇒ψ).
(S6) retract(Prot, "...",ψ): follows claim_conclusive_attack(Ant, "...", ψ).
(S7) claim_conclusive_defence(Prot, "... ", ψ): follows accept(Ant, "...", φ).
(S8) accept(Ant, "...", φ): may follow identify(Prot, "...", φ) if φ ∈ SP, or test(Prot, "...", φ⇒ψ) if φ⇒ψ is valid, or argue(Prot,"...", φ), or follows claim_conclusive_defence(Prot, "... ",φ).
(S9) doubt(Ant, "...", φ): may follow argue(Prot,"...", φ), or test(Prot, "...", φ⇒ψ) if φ⇒ψ is valid.
(S10) doubt(Ant, "...", φ⇒ψ): may follow argue(Prot,"...", φ), or identify(Prot, "...", φ) if φ ∈ SP.
(S11) claim_successful_attack(Ant, "...", φ): follows identify(Prot, "...", φ) if φ ∉ SP.
(S12) claim_successful_attack(Ant, "...", φ⇒ψ): follows test(Prot, "...", φ⇒ψ) if φ⇒ψ is not valid.
(S13) claim_conclusive_attack(Ant, "...", ψ): follows retract(Prot, "...", φ), or retract(Prot, "...", φ⇒ψ).

The sequential structure of the dialogue game is visualised in figure 1.[8] The nodes of the graph – the text boxes – represent the dialogue game moves. The edges – the arrows – represent the transitions between moves from one turn of the game to the next. The topmost node of figure 1 is the first move of the game. The route straight through the middle is the shortest, where *Ant* immediately accepts the argumentative defence of the standpoint.[9] The routes on the left and on the right side relate to criticism of respectively the propositional content and the justificatory force of the argumentation. The two nodes at the bottom of figure 1 are the two terminating moves of the game.

---

simplification of the dialogue game by restricting it to single argumentation. An extension of the rules to facilitate complex argumentation would allow Prot ways to continue his defence (Visser, 2013).

[8] The visualisation is meant to elucidate the dynamics of the dialogue game. The relation to the pragma-dialectical notion of "dialectical profile", and a further clarification of the dialogue game with an example of a natural language dialogue are provided elsewhere (Visser, 2015).

[9] The possibility for Ant to accept the argumentation straight away may appear counterintuitive. Commitment rule (C6) makes sure though that as a result of this move Ant becomes committed to both the propositional content and the justificatory force of the argumentation. Moreover, there does not appear to be an obligation to criticise in the ideal model of critical discussion (see e.g., van Eemeren & Grootendorst, 2004, p. 151).

5.5 Ending the dialogue game

The dialogue game ends when neither of the players has a way to continue the game in accordance with the rules. This situation occurs after a player makes one of the moves *retract(Prot, "I retract my commitment to", ψ)* or *accept(Ant, "I accept that", ψ)*, where ψ refers to the propositional content of the standpoint at issue. In principle there are several ways of determining who is the winner of the game. With future extensions to the dialogue game in mind, such as the inclusion of mixed differences of opinion (where either party can advance argumentation in favour of conflicting standpoints), winning and losing are defined on the basis of the players' commitments. If after making one of the aforementioned finishing moves in turn t, it is the case that $\psi \in CS_{Ant,t}$, then *Prot* wins. This means that *Ant* accepted the standpoint about ψ after a conclusive argumentative defence by *Prot*. In all other cases *Ant* wins, because *Prot* had to retract his commitment to ψ after successful criticism of his single argumentative defence.

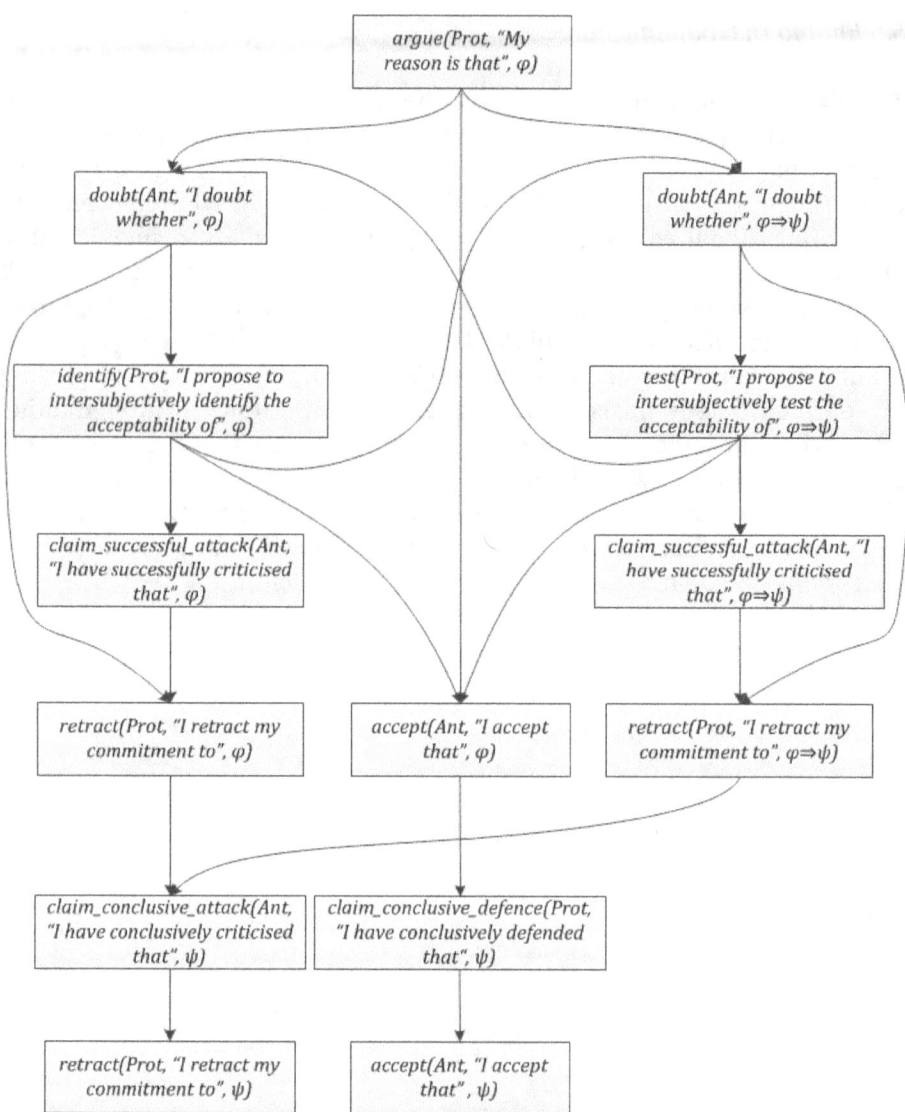

Figure 1 – The sequential structure of the dialogue game for critical discussion

6. CONCLUSION

In this paper, I have specified the rules of a dialogue game for critical discussion as part of a series of dialogue games systematically geared towards the formal approximation of the pragma-dialectical ideal model of critical discussion. The systematic approach starts from a basic dialogue game only covering a very restrictive part of critical discussion, to which more elaborate or complex features of the ideal model can be gradually added. Characteristic of the current instantiation of the dialogue game is the accommodation of the speech act perspective inherent to critical discussion. The speech act perspective is an essential part of the *pragma*-dialectical ideal model. To accommodate it in the dialogue game, I proposed to base the dialogue game moves on the speech acts that prototypically realise discussion moves in the ideal model, and to use the fulfilment of felicity conditions as the basis for commitments in the dialogue game.

By accounting for the speech act perspective of critical discussion, the dialogue game is brought one step closer to being a formal approximation of the entirety of the ideal model. Further extensions to the dialogue game are obviously still required. Several have been mentioned in the preceding sections. At some point, the rules of the (systematically composed) dialogue game have to be computationally implemented to test whether the dialogue game indeed forms a useful intermediary between the pragma-dialectical ideal model (with its practical uses in analysis and evaluation) and computational applications thereof.

ACKNOWLEDGEMENTS: I am grateful to my ECA commentator, Alice Toniolo, for her valuable comments and suggestions, and to Frans van Eemeren and Francisca Snoeck Henkemans for the ongoing fruitful discussions about the research reported in this paper.

REFERENCES

Barth, E. M., & Krabbe, E. C. W. (1982). *From axiom to dialogue: A philosophical study of logics and argumentation*. Berlin: Walter de Gruyter.
Budzynska, K., Janier, M., Kang, J., Reed, C., Saint-Dizier, P., Stede, M., & Yaskorska, O. (2014). Towards argument mining from dialogue. In S. Parsons, N. Oren, C. Reed & F. Cerutti (Eds.), *Computational models of argument, Proceedings of COMMA 2014* (pp. 185-196). Amsterdam: IOS Press.
Dung, P. M. (1995). On the acceptability of arguments and its fundamental role in nonmonotonic reasoning, logic programming and n-person games. *Artificial Intelligence, 77*(2), 321-357.
van Eemeren, F. H. (2010). *Strategic maneuvering in argumentative discourse: Extending the pragma-dialectical theory of argumentation*. Amsterdam: John Benjamins.

van Eemeren, F. H., Garssen, B., Krabbe, E. C. W., Snoeck Henkemans, A. F., Verheij, B., & Wagemans, J. H. M. (2014). *Handbook of argumentation theory*. Amsterdam, etc.: Springer.
van Eemeren, F. H., & Grootendorst, R. (1984). *Speech acts in argumentative discussions: A theoretical model for the analysis of discussions directed towards solving conflicts of opinion*. Dordrecht: Foris.
van Eemeren, F. H., & Grootendorst, R. (1992). *Argumentation, communication, and fallacies: A pragma-dialectical perspective*. Hillsdale, NJ: Lawrence Erlbaum Associates.
van Eemeren, F. H., & Grootendorst, R. (2004). *A systematic theory of argumentation: The pragma-dialectical approach*. Cambridge: Cambridge University Press.
van Eemeren, F. H., Grootendorst, R., Jackson, S., & Jacobs, S. (1993). *Reconstructing argumentative discourse*. Tuscaloosa, AL: The University of Alabama Press.
van Eemeren, F. H., Grootendorst, R., & Kruiger, T. (1983). *Argumentatieleer 1: Het analyseren van een betoog [Argumentation studies 1: The analysis of an argument]*. Groningen: Wolters-Noordhoff.
van Eemeren, F. H., & Houtlosser, P. (2005). Theoretical construction and argumentative reality: An analytic model of critical discussion and conventionalised types of argumentative activity. In D. Hitchcock & D. Farr (Eds.), *The uses of argument. Proceedings of a conference at McMaster University, 18-21 May 2005* (pp. 75-84). Hamilton, ON: OSSA
van Eemeren, F. H., & Kruiger, T. (1985). Het identificeren van argumentatieschema's [The identification of argument schemes]. *Forum der letteren, 26*(4), 298-308.
Garssen, B. (1997). *Argumentatieschema's in pragma-dialectisch perspectief [Argument schemes in a pragma-dialectical perspective]*. Amsterdam: IFOTT.
Hamblin, C. L. (1970). *Fallacies*. 1970. London: Methuen.
Hamblin, C. L. (1971). Mathematical models of dialogue. *Theoria, 37*, 130-155.
Houtlosser, P. (1995). *Standpunten in een kritische discussie [Standpoints in a critical discussion]*. Amsterdam: IFOTT.
Krabbe, E. C. W. (1982). *Studies in dialogical logic*. (Doctoral dissertation). Groningen University.
Krabbe, E. C. W. (2006). Logic and games. In P. Houtlosser & A. van Rees (Eds.), *Considering Pragma-Dialectics: A festschrift for Frans H. van Eemeren on the occasion of his 60th birthday* (pp. 185-198). Mahwah, NJ: Lawrence Erlbaum.
Krabbe, E. C. W. (2012). Formal dialectic: From Aristotle to pragma-dialectics, and beyond. In B. Verheij, S. Szeider& S. Woltran (Eds.), *Computational models of argument. Proceedings of COMMA 2012*. Amsterdam, etc.: IOS Press
Krabbe, E. C. W. (2013). De formalisering van kritische discussie [The formalisation of critical discussion]. In R. Boogaart & H. Jansen (Eds.), *Studies in taalbeheersing 4* (pp. 233-243). Assen: Van Gorcum.

Krabbe, E. C. W., & Walton, D. N. (2011). Formal dialectical systems and their uses in the study of argumentation. In E. Feteris, B. Garssen & F. Snoeck Henkemans (Eds.), *Keeping in touch with Pragma-Dialectics* (pp. 245-263). Amsterdam: John Benjamins.
Lorenzen, P., & Lorenz, K. (1978). *Dialogische Logik [Dialogical logic]*. Darmstadt: Wissenschaftliche Buchgesellschaft.
McBurney, P., & Parsons, S. (2009). Dialogue games for agent argumentation. In I. Rahwan & G. Simari (Eds.), *Argumentation in Artificial Intelligence* (pp. 261-280). Dordrecht: Springer.
Pollock, J. (1987). Defeasible Reasoning. *Cognitive Science, 11*, 481-518.
Prakken, H. (2009). Models of persuasion dialogue. In I. Rahwan & G. Simari (Eds.), *Argumentation in Artificial Intelligence* (pp. 281-300). Dordrecht: Springer.
Rahwan, I., & Larson, K. (2009). Argumentation and game theory. In I. Rahwan & G. Simari (Eds.), *Argumentation in Artificial Intelligence* (pp. 321-339). Dordrecht: Springer.
Rahwan, I., & Simari, G. R. (Eds.). (2009). *Argumentation in artificial intelligence*. Dordrecht: Springer.
Scheuer, O., Loll, F., Pinkwart, N., & McLaren, B. M. (2010). Computer-supported argumentation: A review of the state of the art. *International Journal of Computer-Supported Collaborative Learning, 5*(1), 43-102.
Tolchinsky, P., Cortés, U., & Grecu, D. (2008). Argumentation-based agents to increase human organ availability for transplant. In R. Annicchiarico, U. Cortés & C. Urdiales (Eds.), *Agent Technology and e-Health* (pp. 65-94). Basel: Birkhäuser.
Verheij, B. (2005). *Virtual arguments: On the design of argument assistants for lawyers and other arguers*. Den Haag: Asser.
Visser, J. C. (2013). A formal account of complex argumentation in a critical discussion. In D. Mohammed & M. Lewiński (Eds.), *Virtues of Argumentation. Proceedings of the 10th International Conference of the Ontario Society for the Study of Argumentation (OSSA), 22-26 May 2013* (pp. 1-14). Windsor, ON: OSSA.
Visser, J. C. (in press). A formal perspective on the pragma-dialectical discussion model. In A. F. Snoeck Henkemans, B. Garssen, D. Godden & G. Mitchell (Eds.), *Proceedings of the 8th Conference of the International Society for the Study of Argumentation (ISSA), 1-4 July 2014*. Amsterdam: Sic Sat.
Visser, J. C. (2015). *Kritische discussie als basis voor de geautomatiseerde analyse van argumentatieve teksten [Critical discussion as a basis for the computerised analysis of argumentative texts]*. Manuscript submitted for publication.
Walton, D. N., & Krabbe, E. C. W. (1995). *Commitment in dialogue: Basic concepts of interpersonal reasoning*. Albany, NY: SUNY Press.

Commentary on Visser's Speech Acts in a Dialogue Game for Critical Discussion

ALICE TONIOLO
Department of Computing Science, University of Aberdeen, UK
a.toniolo@abdn.ac.uk

1. INTRODUCTION

I am delighted to be a commentator for Visser's paper "Speech acts in a dialogue game for critical discussion". In this paper, Visser continues his effort in establishing a link between computational models of argumentation-based dialogue and the influential pragma-dialectical theory of critical discussion. Visser elaborates this connection by means of a "formal approximation" of a critical discussion that can be represented via a two-party dialogue game, where participants may be a combination of humans or software agents. He builds upon his previous dialogue game for critical discussion (Visser, 2013), taking a deeper perspective where speech acts are now considered, and directly linked to phases of the argumentation stage.

Visser argues that this dialogue game is a step towards pragma-dialectical theory being ready for computerisation. Indeed, dialogue games and speech act protocols have largely been adopted in computational models of argumentation-based dialogue between software agents. The importance of Visser's work lies in providing a computational perspective of the pragma-dialectical theory in order to bridge the gap between human argumentative practice and practical software applications. In an invited talk at COMMA 2012 (Krabbe, 2012), Professor Krabbe discusses the influential role of formal dialectics (Hamblin, 1970) in the development of computational argumentation-based dialogue models in Artificial Intelligence (AI). This is especially due to the seminal work of Walton and Krabbe (1995), followed by later work of Reed (1998) and McBurney and Parsons (2002). Parallel to the development of formal dialectics, the pragma-dialectics foundations have been formed by the seminal work of van Eemeren et al. (1996). Visser's idea of a formal account of pragma-dialectics may allow such theory to make its way towards computational applications. Interestingly, Professor Krabbe argues that this formalisation would also reach beyond practical applications showing that formal dialectics and pragma-dialectics have very strong connections.

In his paper, Visser enriches the dialogue game for critical discussion with a speech act protocol forming a dialogue system. Henceforth, for convenience I will refer to this system as the cd-dialogue. The focus of this commentary is to highlight how Visser's approach can be applied in some argument-mediated human-computer interaction contexts, approaching Visser's work from the perspective of AI computational models of dialogue (Section 2). Section 3 presents some similarities with existing persuasion dialogue protocols, in particular with the protocol of Prakken (2006). The aim in this section is to show that some features of the cd-dialogue are already present in AI computational dialogues. In Section 4, I discuss some additional features that may be considered to further develop the formal approximation towards a full computational model of the cd-dialogue.

2. APPLICATIONS OF FORMAL CD-DIALOGUES

As Visser argues, a computational model of cd-dialogue would be beneficial for both argumentation in AI and pragma-dialectical theory. Some interest has been developed recently within the AI computational argumentation community in complex problems of automated argument analysis and in applications oriented at capturing the richness that characterises natural argumentation. Some notable examples are the development of tools to support human-argumentation (e.g., the Argument Web (Bex et al., 2013)), to automatically analyse argumentative text (e.g., argument mining (Green et al., 2014)), together with a wider range of applications in law and medicine, engineering, education, intelligence analysis and so on. Similarly, the pragma-dialectical theory aims at analysing and reconstructing human argumentations (van Eemeren et al., 1996) and, hence, a closer integration would help enrich tools and techniques on both sides.

The development of a computational cd-dialogue may be particularly beneficial in investigating richer interaction between human and agents. On the one hand, two or more agents may perform an autonomous reconstruction of a critical discussion, playing the dialogue game while a human would act as an observer. The dialogue rules would support an easy scrutiny of the process, where the observer may ask for explanations of a particular branch of dialogue. The agents may report back an explanation of why and how a standpoint is successfully defended alongside with the possibility for the human to intervene and support the intersubjective procedure. On the other hand, a cd-dialogue may be used to mediate a human-agent discussion, in which the agent employs the normative formalism to check the validity of the steps and signal the user if any of these is fallacious.

In order to develop these applications, the work of Visser plays an important role. In the next section, I will show some characteristics of the cd-protocol that already exists in agent-based dialogue systems and would be useful in further developments of computational cd-dialogues.

3. EXISTING COMPUTATIONAL MODELS OF DIALOGUE

Here I wish to challenge Visser's work to seek a stronger connection with existing computational models of dialogue (e.g. McBurney & Parsons, 2002; Prakken, 2006). Those dialogue systems often refer to the types of dialogue proposed by Walton and Krabbe (1995). In particular, models for persuasion dialogue seem to be appropriate for formalising critical discussion (Krabbe, 2001). Some of these existing models would be helpful in terms of formal representation and to analyse computational properties of the dialogue. Perhaps critical discussion is much richer than what agent-based persuasion dialogue may express, but it would be useful for the reader to discuss what new features are required. Here I would like to suggest some initial points of connections between the dialogue system for persuasion of Prakken (2006) and the cd-dialogue presented by Visser in this paper.

In Table 1, a summary of the speech acts of Visser's cd-dialogue and Prakken's persuasion dialogue (ps-dialogue) are presented. Note that in Prakken's system a dialogue tree can be constructed from the moves, and a dialogical status is assigned: IN if the move is justified, OUT otherwise. Proposing a claim in the cd-dialogue is assumed to happen in a previous initial phase. The second block of speech acts is related to the argumentation phase. In ps-dialogue arguments support or attack the claim, depending on the type of challenge previously moved; arguing in cd-dialogues is moving a support for the claim. In both dialogues, the conclusions of arguments are added in the commitment store, in cd-dialogues the inference rule is also added. This leads to a difference in the *retract* speech act in cd-dialogue where the rule forming the argument may be retracted in addition to retracting a statement. I would suggest the author to further consider the meaning of adding and removing rules from the commitment store. Additional arguments could be formed or retracted at the same time given this revision of the knowledge-base.

Casting a doubt seems to retain a different meaning regarding the acceptability of the moves and of the claim. Note that there is no change in the commitment store. In ps-dialogue, *why* is used to challenge a statement and if not counterattacked, it leads the claim to being not supported (claim OUT). This indicates a shift of burden of proof in the dialogue.

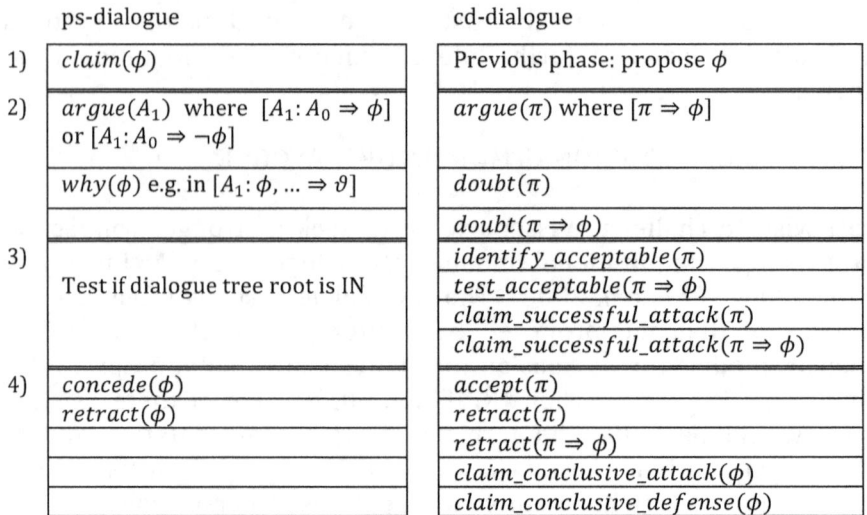

Table 1: Speech act protocol in Visser's and Prakken's dialogue

In cd-dialogue, however, a speech act *doubt* does not seem to have a direct effect in the acceptability of the claim. In particular, it can be used for doubting a statement or an inference rule. (The latter could only be represented in ps-dialogue as an undercutting argument.) Moves of the type *identify_acceptable* or *test_acceptable* guide additional procedures to accept or deny the claim. Some considerations on the difference between *doubt* and *why* would be very useful for developing computational cd-dialogue models.

Finally, although Visser's work is moving towards a formalisation of the critical discussion, I believe there is a challenge yet to be addressed: the lack of formalism to test the acceptability of claims. In cd-dialogues following pragma-dialectical theory the test/identify functions are meant to be performed externally, based on the justificatory force of the inference. The definition of this force of inference in computational models of argumentation may lead to different research approaches investigating simple attacks, values, uncertainty, preference and so on. In order to employ such dialogue in an autonomous system, these procedures must be fully formalised.

4. CONCLUSION AND CHALLENGES

In conclusion, I believe that Visser's work presents a promising approach towards the understanding of computational aspects of the pragma-dialectical theory. There are, however, some open challenges for further developments of computational cd-dialogues. One very challenging line of research is, as previously mentioned, a formal account of the acceptability function. The formalisation of this procedure would require to understand the links between the intersubjective procedures used to establish whether an argument is successfully defended and the current computational theory of abstract argumentation where semantics are used to evaluate the acceptability status of arguments (Dung, 1995).

Visser's formal approximation of the pragma-dialectical theory may lead to insights for enhancing and applying computational models of argumentation-based dialogue. An example is the intersubjective procedure of identification (van Eemeren & Grootendorst, 2004, p. 145), in which participants must agree on what the shared knowledge is, what can be challenged, and what type of information can be introduced in the discussion. This procedure may be useful for interpreting enthymemes and especially for dealing with the introduction of new information within autonomous dialogue. This is a challenge that is yet to be addressed thoroughly in AI dialogue systems. Visser's model provides ways to enrich the research within these open challenges in computational argumentation-based dialogue systems.

ACKNOWLEDGEMENTS: This research is supported by the award made by the RCUK Digital Economy program to the dot.rural Digital Economy Hub; award ref.: EP/G066051/1.

REFERENCES

Bex, F., Lawrence, J., Snaith, M., & Reed, C. (2013). Implementing the argument web. *Communications of the ACM*, 56(10), 66–73.
Dung, P. M. (1995). On the acceptability of arguments and its fundamental role in nonmonotonic reasoning, logic programming and n-person games. *Artificial Intelligence*, 77(2), 321–357.
Green, N., Ashley, K., Litman, D., Reed, C., & Walker, V. (2014). First workshop on argumentation mining. Available at http://www.uncg.edu/cmp/ArgMining2014/.
Hamblin, C. L. (1970). *Fallacies*. London: Methuen.
Krabbe, E. C. W. (2001). The problem of retraction in critical discussion. *Synthese*, 127(1-2), 141–159.

Krabbe, E. C. W. (2012). Formal dialectic: From Aristotle to pragma-dialectics, and beyond. In B. Verheij, S. Szeider, & S. Woltran (Eds.), *Computational Models of Argument*, volume 245, (pp. 11-13). Amsterdam: IOS Press.

McBurney, P., & Parsons, S. (2002). Games that agents play: A formal framework for dialogues between autonomous agents. *Journal of Logic, Language and Information, 11*(3), 315-334.

Prakken, H. (2006). Formal systems for persuasion dialogue. *The Knowledge Engineering Review, 21*(2), 163-188.

Reed, C. (1998). Dialogue frames in agent communication. In *Proceedings of the Third International Conference on Multi Agent Systems*, 246-253.

van Eemeren, F. H., & Grootendorst, R. (2004). *A systematic theory of argumentation: The pragma-dialectical approach*. Cambridge: Cambridge University Press.

van Eemeren, F. H., Grootendorst, R., & Snoeck Henkemans, A. F. (1996). *Fundamentals of argumentation theory: A handbook of historical backgrounds and contemporary applications*. Mahwah, NJ: Lawrence Erlbaum Associates.

33

Speech Acts and Burden of Proof in Computational Models of Deliberation Dialogue

DOUGLAS WALTON
CRRAR, University of Windsor, Canada
dwalton@uwindsor.ca

ALICE TONIOLO
Department of Computing Science, University of Aberdeen, UK
a.toniolo@abdn.ac.uk

TIMOTHY J. NORMAN
Department of Computing Science, University of Aberdeen, UK
t.j.norman@abdn.ac.uk

> We argue that burden of proof (BoP) of the kind present in persuasion does not apply to deliberation. We analyze existing computational models showing that in deliberation agents may answer a critique but there is no violation of the protocol if they choose not to. We propose a norm-governed dialogue where BoP in persuasion is modeled as an obligation to respond, and permissions capture the different types of constraints observed in deliberation.
>
> KEYWORDS: burden of proof, deliberation dialogue, dialogue protocol, multiagent systems, norms

1. INTRODUCTION

Recent work on deliberation dialogue in AI raised awareness of the need to develop formal models that capture the richness of deliberations that typically occur in human dialogue (Walton et al., 2014). Existing models of dialogue in multiagent systems use argumentation schemes, argument frameworks and formal dialogue structures to analyze and evaluate argumentation in deliberation, where the goal of the dialogue is to enable the participants to make an intelligent choice on what to do in a given set of circumstances. Existing research has generally focused on the problem of what to say; i.e. what speech acts can be performed or what critical questions can be advanced when making a proposal. An open problem is that of determining why certain speech acts should be performed and what the implications are for the participants.

In the formal deliberation system of Kok et al. (2011), when the speaker makes a proposal the hearer is required at its next move to either reject the proposal or to challenge it by asking the why-question "Why propose P?". In reply to this question, the hearer is obliged to give an argument supporting P. A similar requirement is considered in a persuasion dialogue, where there is a burden to defend an assertion that is questioned by the other party. In the argumentation literature, this requirement is called the burden of proof. However, there is an issue about whether making a proposal in a deliberation dialogue carries with it a burden of proof (Gordon et al., 2007), which suggests that the obligation on the hearer to support P may not be realistic in human deliberation dialogue. This problem is taken up in this paper where we analyze the problem of burden of proof in deliberation dialogue.

According to the account of the speech act of making a proposal of Kauffeld (1998), a proposal must present a statement of resolve that expresses a determination or conclusion that the speaker has reached. Kauffeld holds that the speaker makes a proposal with the intention of answering objections against it, and therefore incurs a burden of proof to defend the proposal. But it has been argued that making a proposal in a deliberation dialogue is different from making a claim in a persuasion dialogue, which does incur a burden of proof (Gordon et al., 2007). In persuasion dialogue, the participant that holds the burden of proof must answer to critiques to the claim because of the need to satisfy this burden. In contrast, recently Walton (2014) shows that there is no comparable burden of proof attached to the speech act of making a proposal in a deliberation dialogue. In deliberation, the goal is for the group of agents to arrive collectively at a decision on what to do. If the proposing agent fails to defend a proposal by immediately presenting an argument in support of the proposal, it should not have to retract the proposal. The problem is to understand the reasons for which the respondent has a problem with the proposal. Accordingly, the speech act protocol needs to allow participants to postpone answers to critiques to the proposal, to account for other proposals or explanations of circumstances. However, in existing dialogue protocols for multiagent deliberation that are often derived from persuasion dialogue games (McBurney et al., 2007; Kok et al., 2011; Medellin-Gasque et al., 2011; Walton et al., 2014), this flexibility is not permitted, and failing to answer to a critique would require a proposal to be withdrawn.

The absence of burden of proof in deliberation leads to a quest for a different model of protocols that considers what constraints arise between participants when a speech act is moved. In this paper, we propose a formal model of norms used to define protocol rules for deliberative dialogue. In multiagent systems, norms describe the ideal behavior of the agents in a society (Kollingbaum & Norman, 2004). We explore the use of norms to define what speech acts an agent is obliged or permitted to perform, and which are prohibited. Our dialogue protocol is said to be *norm-governed* as the norms regulate when an agent is allowed to make a move in the dialogue.

The core idea of norm-governed dialogues is the association of the burden of proof to an obligation for the agents to defend a claim. In persuasion, an obligation after an opponent's critique requires a proponent to defend or to withdraw the claim. Failing to do so leads to a violation that forces the parties to terminate the dialogue. Here, we argue that in deliberative dialogue such an obligation must be substituted with a permission to answer a critique, where agents may reply with a supportive argument for a proposed action or may exchange explanations, but they are not obliged to. Therefore, there is no violation of the protocol if they choose not to put forward a counterargument.

In this paper, we analyze the problem of burden of proof in deliberation. We survey recent work on formal models of deliberation dialogue in AI in Section 2. We discuss how the rules for performing speech acts need to be reconfigured to account for the absence of burden of proof in deliberation (Section 3). We then present the characteristics of our norm-governed deliberation dialogues in Section 4. Our deliberative protocol permits some degree of flexibility in providing justifications for proposals that were not formally allowed for in previous models.

2. COMPUTATIONAL MODELS OF DELIBERATION DIALOGUE

In this section, we briefly survey recent work on computational models of deliberation dialogue. We show that the configuration of these speech act protocols raises issues about the burden of proof.

Models of deliberation dialogue proposed so far in the AI literature consider a group of agents, with each member pursuing its own goals. At the same time, there is the overall goal of the deliberation dialogue itself: the agents have to decide what course of action to take in a particular set of circumstances requiring a choice between alternative courses of actions (Medellin-Gasque et al., 2011). To achieve their collective goal, the agents have to collaborate with each other. At the same time, each will have its own interests or plans. Proposals from different agents may conflict with each other. Agents may even disagree about what the circumstances of the decision are that constrain the choices. The best way forward is for the agents to present arguments, to critically examine the arguments put forward by others, and to collect evidence drawn from the circumstances of the case to produce other arguments that support or attack the arguments previously put forward.

Current formal argumentation systems developed in AI use argument mapping tools that can be applied to modeling the structure of arguments. This structure may be represented using argumentation schemes, or using defeasible *modus ponens* (Walton, 1995) as rules of inferences in computational argumentation systems. Here we have chosen the Carneades Argumentation System (CAS) to make an argument map of the sequence of reasoning in the examples below (Gordon et al., 2007). In this system, the ultimate conclusion of the sequence of argumentation is shown at the left of the page and the arguments supporting or attacking it are displayed as a tree structure leading to the ultimate conclusion, which is the root of the tree. A pro argument is shown with the plus sign in its node, a con argument is shown with a minus sign.

In this paper we will use Example 1 presented in Toniolo et al. (2012) where the deliberation process sees two agents involved in the operations for responding to a natural disaster.

> (1) Two agents x, a local authority, and y, a humanitarian organization, are concerned with the repair of the water supply in a location that has suffered catastrophic damage. Agent x proposes to stop the water supply to the location in question. Agent y argues that there is a need for water in that location to run a field hospital, which is required to aid disaster victims. Agent x proposes that the supply of water to the location must be stopped because the water is contaminated. Agent x argues that supplying water to the location is not safe because x has scheduled the use of excavators during that time. To solve the problem, x and y should modify their individual plans. Such changes are constrained by the goals of agents x and y, the circumstances of the case (or by what is known of them), and by values such as public safety.

There are several important features of the argumentation in this example. First, the speech act of making a proposal is centrally important. Second, arguments are used to support or attack proposals.

The features can be shown using an argument map. The pro argument used to support x's proposal is shown at the top of Figure 1, while the con argument attacking it is shown at the bottom.

In the disaster example the two parties have conflicting proposals. In this case, agents must continue the dialogue to see whether some compromise can be made. Perhaps the field hospital could be located in a different area or the excavations could be carried out at a different time. The best way for the deliberation to move ahead is for the two parties to engage in a collaborative discussion in which they inform each other of their goals and means that could be devised in order to enable fulfillment of both their goals.

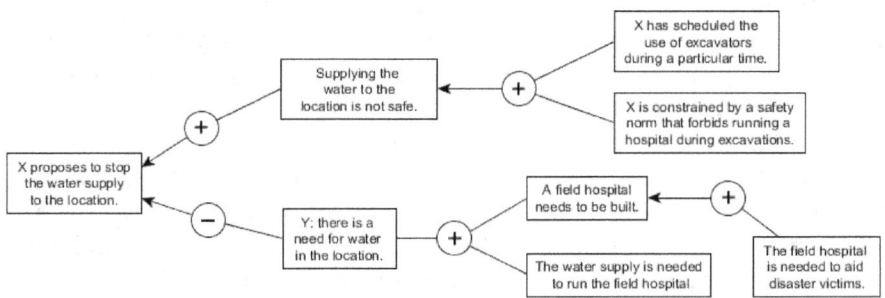

Figure 1 – Argumentation map of the disaster example

2.1 Speech acts in deliberation models

The first model of deliberation dialogue developed in the AI literature was the McBurney, Hitchcock and Parsons (MHP) model, which identified eight stages in a deliberation dialogue (McBurney et al., 2007).

1. *Opening Stage*: the collective goal of the dialogue is an issue or "governing question" that applies to the whole dialogue. The issue is to decide what to do in a given set of circumstances.
2. *Inform Stage*: there is a discussion of goals, any constraints on the actions being considered, and any external facts relevant to the discussion.
3. *Propose Stage*: proposals are put forward by any of the parties.
4. *Consider Stage*: comments are made on the proposals that have been brought forward, and arguments for and against proposals are considered.
5. *Revise Stage*: the goals, the actions that have been proposed, and the relevant facts may be revised.

6. *Recommend Stage*: participants recommend a particular action which others can accept or reject.
7. *Confirm Stage*: participants together confirm their acceptance of one selected option.
8. *Close Stage*: participants arrive at a good decision on what to do.

In a dialogue model, the speakers have to take turns making their proposals and commenting on alternative proposals. A *communication protocol* is a set of rules agent uses to communicate with each other by determining which part of the conversation comes at which point in the exchange and the permissible speech acts that can be made at each move. There are several kinds of distinctive speech acts recognized in the MHP model of deliberation dialogue. These include speech acts for making a proposal, asserting a statement, stating a preference for an action, asking the other party to justify an assertion, saying whether a proposal should be accepted or rejected, retracting a previous assertion, and withdrawing from the dialogue. Permitted speech acts are represented in Figure 2 using a simplified finite state machine diagram. More complex diagrams may report the subject of speech acts. Here, we only represent the type of speech acts (outgoing edges) that can be used to reply to speech acts previously performed (incoming edges).

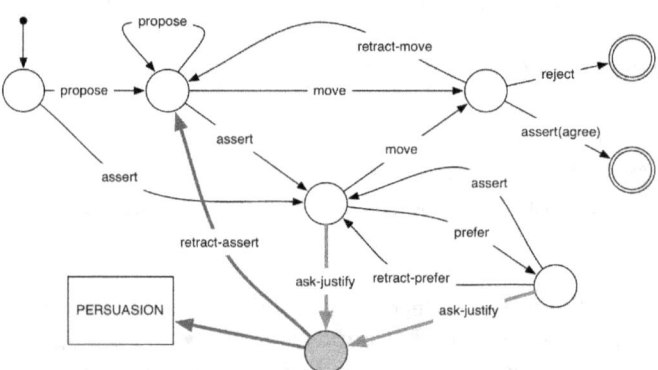

Figure 2 – Speech acts of the MHP model

In this research we argue that the speech acts employed as moves between the agents in a deliberation dialogue need to be refined. In the MHP model, when the speaker puts forward a proposal the hearer is required, in its next move, to either reject the proposal or to challenge it by asking *"Why propose P?"*. In reply to this question, the hearer is then obliged to give an argument supporting *P*. This is due to the burden of proof which is assigned to the participant who makes a suggestion, even after critiques by other participants. The consequence of this BoP allocation is explicitly represented via the protocol rules that regulate the dialogue. In the MHP dialogue, the use of the act *ask-justify* challenges previous statements and at the next turn, requires agents to: either shift to a persuasion dialogue to persuade the opponents that it is worth considering the current action proposal; or retract the statement that was challenged, weakening the support for the proposed action or removing that action from the commitment store. Therefore, we can observe that agents have an obligation to defend their proposal or it is likely that the proposal will be dropped.

Many approaches to deliberative dialogue for collaborative agents consider extensions to the MHP model of dialogue for interaction between agents to decide upon the best action (e.g., Kok et al., 2011) or plan (e.g., Medellin-Gasque et al., 2011) to perform.

The aim of the argumentation-based dialogue in Medellin-Gasque et al.'s work is to evaluate plans according to preferences among values based on the argumentation scheme for practical reasoning where challenges follow critical questions (CQs). The proposal is a partial plan corresponding to a path of actions for achieving intermediate goals; those must be agreed upon before continuing to the next part of the plan.

The dialogue protocol uses speech acts defined via preconditions and effects, and dialogue phases inspired by the MHP protocol. The use of both acts *challenge(CQ)* and *question(CQ)* correspond to asking for a justification. The protocol is, however, more restrictive than the MHP protocol: the answer to challenges must defend the proposal or the proposal itself is retracted. A representation of the speech acts is shown in Figure 3.

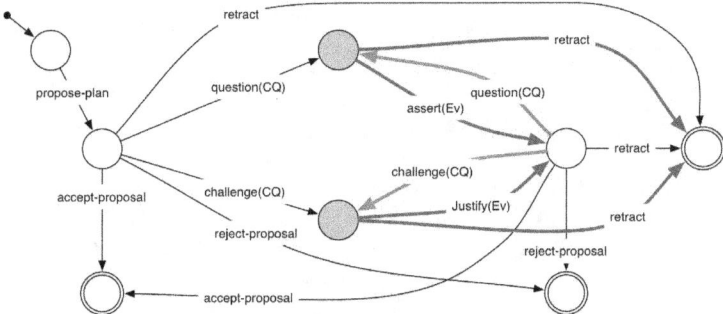

Figure 3 – Speech acts of Medellin-Gasque et al. (2011)

In the approach of Kok et al. (2011) a team of agents aims to choose the best action to achieve a goal. The protocol is based upon the protocol for persuasion of Prakken (2006) and inherits its characteristics of responses to challenges. When one of the *why* questions is asked to proponents, they must reply with *argue* or retract the proposal. If *why* is asked to opponents, they must argue or accept the proposal. Toniolo et al. (2012) developed a further model of deliberation dialogue based upon the protocol of Kok et al. (2011) by using the argumentation scheme for practical reasoning to help agents carry out a collective task by identifying conflicts between their plans. The plans of individual agents are interdependent, driven by different objectives and social constraints, leading to asymmetric knowledge of how the state of the world is modified by individual agents. The use of practical reasoning leads agents to form better collaborative plans to move forward in the deliberation sequence. The speech acts permitted are shown in Figures 4 and 5.

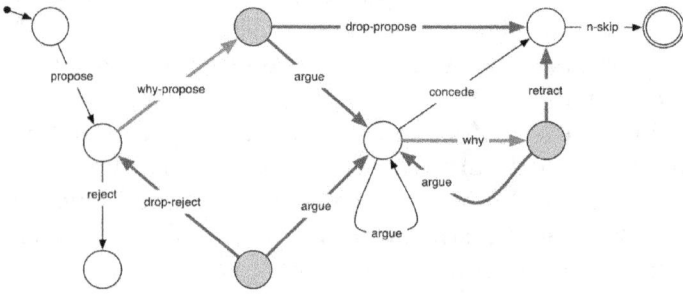

Figure 4 – Speech acts of Kok et al. (2011)

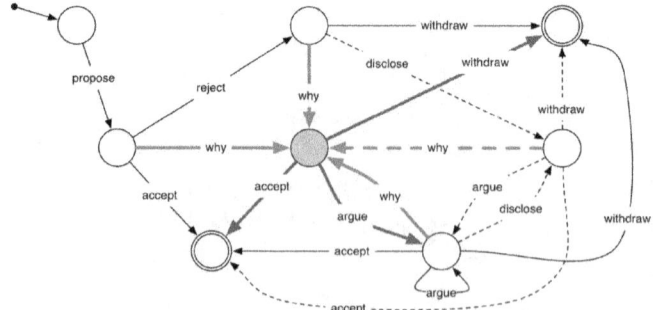

Figure 5 – Speech acts of Toniolo et al. (2012) – Dotted lines represent the Walton et al. (2014) extension

Walton et al. (2014) argued that a computational model of deliberation useful for dynamic multiagent systems is best based on a natural concept of deliberation, meaning that it should share certain important features with real examples of natural language deliberation. One important feature of natural deliberation is that it needs to remain open to collecting new information and considering arguments as long as this procedure continues to be fruitful, but at the same time it needs to be closed off once the circumstances require arriving at a timely decision on what to do. An open knowledge model was designed to capture the capability of a deliberating agent to update its practical reasoning based on information regarding the circumstances of the practical problem acquired as the dialogue proceeds. This feature enhances the adaptability of a plan in relation to circumstances that may be rapidly changing. This way of setting up speech acts in deliberation protocols was followed in Walton et al. (2014). The additional moves in the protocol are represented with dotted lines in Figure 5.

In existing frameworks, the burden of proof is held by the participants that propose an action in the dialogue, and this guides the type of answers that are allowed after challenging a statement. For example, we have observed that to a question "*Why?*", the proponent must present a defending argument, withdraw the proposal or withdraw an argument for the proposal previously presented. This approach is derived from persuasion dialogue, but the problem is that it raises issues about burden of proof in deliberation. This problem is taken up by the present paper.

3. BURDEN OF PROOF IN DELIBERATION DIALOGUE

In this section we analyze the problem of burden of proof in deliberation dialogue, and introduce the reconfiguration of existing dialogue protocols in AI to model this problem.

The participants in a deliberation dialogue take turns making moves during the argumentation stage. These include the speech act of making a proposal, retracting a proposal, making an assertion, retracting an assertion, putting forward an argument, defending a goal, and attacking a goal.

In a persuasion dialogue, there is a burden to defend an assertion that is questioned by the other party. This means that if the proponent fails to defend the assertion it has made, it must immediately retract that assertion. This notion of burden of proof, albeit important, is a slippery one that is hard to define (Walton, 2014). One needs to begin by drawing a distinction between global and local burdens of proof. A global burden of proof is set at the opening stage of a dialogue. Once set in place, it applies through the whole argumentation stage. At the closing stage it is used to determine the outcome of the dialogue. The local burden of proof in a dialogue applies to speech acts made in moves during the argumentation stage. For example, in a persuasion dialogue the global burden of proof, called the burden of persuasion in law, is set by law at the opening stage of a trial. But requirements to supply evidence stemming from this global burden of proof are brought into place during the pro-contra argumentation by both sides during the trial. If one participant makes an assertion of proposition P during the argumentation stage, and the other party challenges that assertion, it can be assumed that there is a burden on the party who made the claim to provide evidential support for it. In law this is called the evidential burden, as contrasted to the global burden of proof set at the opening stage (Walton, 2014).

There is no comparable burden of proof attached to the speech act of making a proposal in a deliberation dialogue. First, the goal of the deliberation dialogue is not to resolve a conflict of opinions by putting forward pro and con arguments on both sides. Instead the goal set at the opening stage is for the group of agents to arrive collectively at a decision on what to do. Suppose an agent puts forward a proposal in a deliberation dialogue and another agent questions it. If the proposing agent fails to defend its proposal by immediately presenting an argument in support of the proposal, the proposal does not need to be retracted. When a proposal is put forward by an agent it is assumed that this proposal is part of a feasible plan of action that this agent has thought about. The problem is that another agent in the dialogue might have a different plan. If so, this other agent can object to the proposal put forward by the first agent. But an appropriate response to the objection is that the plans are interdependent and need to be integrated, as indicated in the disaster response example.

In Walton et al. (2014) it was argued that feasibility can be determined by the increase in the number of conflicts resolved between the two plans. A speech act protocol must enable the timely asking of questions about another party's plan in a deliberation without this questioning being perceived as an attack of the kind one might have in a persuasion dialogue where fulfillment of burden of proof for an assertion is immediately demanded and required. There needs to be room for a discussion in which one party can ask the other about its plans and goals without being perceived to be pressing an argumentative attack against them. While deliberation can have its adversarial aspects, the speech act protocols need to be set up in such a way that an agent may ask constructively about the plans and goals of another agent, and offer an explanation about its own plans and goals. A framework is needed within which an agent can explain its plans and goals, as well as its knowledge of the present circumstances, to another agent who has put forward a different proposal as a means for answering the governing question.

A brief caution also needs to be inserted here. Another important aspect that is often overlooked is that there can be shifts to other types of dialogue, for example a shift from a deliberation dialogue to a persuasion dialogue or to an information-seeking dialogue (Walton & Krabbe, 1995). There are burdens of proof in these two latter types of dialogue, but if one is unaware of the shift, it may seem that the burden of proof resides in the deliberation dialogue itself.

The ambiguity of the asking of a why question in a deliberation dialogue was already acknowledged in Walton et al. (2014) when it was noted that when an agent asks a why question it is asking the question *"Why do you want to perform this action?"*. When the other agent poses con arguments it can do so by explaining some circumstances, using the speech act (Arg_{sup}). This speech act could be described as a counterargument but it has the function of explaining circumstances. The capability to allow for explanation to be embedded in arguments (Bex & Walton, 2011) is very important in deliberation because an agent may need to take the initiative to exchange new information about the circumstances at any point in the dialogue. Accordingly, the speech act protocol needs to be set up to allow that an appropriate response can be either an argument, an introduction of new information about the circumstances, or an explanation of, say, an agent's plans or goals. Tolerance has to be made for the permitting of such speech acts at any point in a deliberation dialogue. For these reasons, a narrow framing of the burden of proof in the way that is characteristic of persuasion dialogue is inappropriate in deliberation dialogue.

3.1 Deliberation protocols with no BoP

To set up a speech act protocol that conforms to the approach described above, we see the sequence as proceeding along these general lines. When an agent puts forward a proposal, arguments supporting or attacking it are naturally advanced, leading to pro and con argumentation in which many arguments support or attack other arguments connected into a network of argumentation. But suppose an agent fails to defend its proposal in the face of an attack against it by another agent. This will mean that the proposing agent will lose ground for support of its proposal by the other participants in the deliberation or unless somehow during the subsequent dialogue some moves are made that overcome the deficiency in the proposal that has been pointed out, unless the proposal is modified to meet the objection. However, the protocol should be that if an agent fails to defend a proposal it has brought forward, the dialogue should move on to other considerations. For example, agents may need to modify their plans to take into account new information acquired during the dialogue. The dialogue terminates when an agreement is found on how to move forward with a practical reasoning sequence needed to solve the problem set at the opening stage. Even an undefended proposal may in the end be adopted if no other alternatives exist that are any better are found, or if new information has come to light during the dialogue that strongly supports the undefended proposal or shows that it is not really open to an objection that was earlier posed.

For these reasons we define a range of reactions by a hearer agent when a speaker agent puts forward a proposal in a deliberation dialogue. The responding agent can, of course, accept or reject the proposal, but this does not matter very much at this stage. The reason is that after the proposal is discussed and explained there will be a point just before the closing stage where agreements or disagreements with the proposal can be voiced. Often a vote is taken, for example. What matters at this earlier point where the proposal has just been introduced is whether the respondent has a problem with the proposal. There can be a range of such problems. One of the leading ones is that the respondent might have a different plan of action, and therefore might want to question the proposal with regard to differences between different plans of action. Another problem is that the respondent might not understand the proposal, because understanding it presupposes knowledge of the complex plan that the proposer has in mind. The proposer in this case needs to explain the plan in a way that responds to the questioner's problem. Another problem is that the respondent, even though it is generally agreeable with the proposal, might think that it needs to be modified in certain respects before it will work. Another option is for the responding agent to present a different proposal. The sequence of dialogue may then take the route of exploring the differences between the two proposals.

The protocol for speech acts in a deliberation dialogue should be flexible enough to admit all these options, even if the participants failed to support the proposals at some point in the dialogue. The aim is to direct the dialogue down a sequence of argumentation in which various proposals are articulated and refined, leading either to some general agreement on a compromise proposal, or at least the formulation of a set of clear proposals so that the participants are well-informed enough that they can have good grounds to either agree with or disagree with any particular proposal. Simply to say that there is a burden of proof to either prove or refute a proposal at the next point in the dialogue where the respondent has to reply to it, would restrict these possibilities.

In conclusion, although it seems like a very controversial thesis to uphold, there is no burden of proof in a deliberation dialogue. There is only a burden of responding appropriately to a proposal by answering it with a range of replies that moves the deliberation dialogue forward. We refer to this as the *burden of responding constructively (BrC)* and in the next section, we will define a norm-governed protocol that accounts for the BrC and the absence of BoP in deliberation.

4. NORM-GOVERNED DIALOGUES

In multiagent systems, norms describe the ideal behavior of the agents in a society (Kollingbaum & Norman, 2004; Gasparini et al., 2015). Norms generally represent permissions, obligations and prohibitions. In this research, we exploit this diversity of constraints to model the requirements in responding to challenges during dialogue imposed by the burden of proof. We claim that the burden of proof in deliberation in existing research corresponds to an obligation to respond to a challenge. Such a burden, however, does not exist in deliberation. In contrast, agents are permitted to respond to challenges, but answers to such challenges can be postponed or omitted. For simplicity, we refer to a deliberative dialogue with burden of proof as a BoP dialogue and to our proposed dialogue with burden of responding constructively as a BrC dialogue. In this section, we define a norm-governed dialogue in a multiagent environment based on the protocol of Walton et al. (2014).

Following Walton and Krabbe (1995), the elements of a communication protocol include: *locution rules* (i.e., moves possible); *structural rules* (i.e., moves allowed); *commitment rules* (i.e., players' commitments); and *termination rules*. The novelty of a norm-governed dialogue lies in the introduction of norms to define structural rules and termination rules.

4.1 Locution rules

In our dialogue, a speech act is $perf(\vartheta)$, where $perf$ is a performative and ϑ is a subject defined among plans of agents. Plans are formed by goals, states of the world and actions, expressed via sentences ℓ_i in a logic language L_t as defined in Toniolo et al. (2012). The dialogue is among a set of agents $Agt = \{x, y, z, ...\}$ and proposals are actions $\mathcal{A} = \{A_k, A_l, ...\}$. Arguments $Args$, defined with sentences ℓ_i, are built as argumentation schemes from negative consequences (Walton, 1995), stating that an action A_k should not be brought about because it conflicts with other goals or actions. We refer to instances of this type of argument as Arg_{def}. Information about circumstances and explanations for proposed actions can be exchanged by offering support to previously stated claims via supporting arguments Arg_{sup}.

The speech acts that we consider are:
- $propose(A_k)$: agent proposes an action;
- $reject(A_k)$: agent rejects the proposal;
- $accept(A_k)$: agent accepts the proposal;
- $withdraw(A_k)$: agent withdraws the proposal;
- $why(\ell_i)$: agent asks a question "why?" to challenge an argument $\ell_i \in Arg_{def}/Arg_{sup}$ or an action $\ell_i = A_k/\neg A_k$;
- $argue(Arg_{def})$: agent presents an attacking argument against a sentence ℓ_i where $\neg \ell_i \in Arg_{def}$;
- $disclose(Arg_{sup})$: agent presents a supporting argument for a sentence ℓ_i where $\ell_i \in Arg_{sup}$.

A dialogue d is formed by a sequence of moves m_i, executed in a turn-taking fashion, where no repetitions are allowed. Each move is identified by the player of the move in Agt, and a speech act $perf(\vartheta)$.

4.2 Structural rules

Structural rules state when agents are allowed to speak and what they are permitted to say. In a norm-governed dialogue, these rules are represented with norms using the CÒIR language (Gasparini et al., 2015). Norms in CÒIR are defined as obligations, permissions or prohibitions to achieve certain states. In our formalism these are dialogue states. Norms are activated according to a condition over previously achieved states. Each norm expires in a new dialogue state. Let DKB represent a knowledge base of a dialogue state.

Definition 1. *A norm n_i is defined as a tuple:*

$$\langle idn_i, mod_i, act_i, goal_i \rangle$$

where: idn_i a unique identifier; mod_i is chosen among $\{O, P, F\}$ where P is a permission, O an obligation and F a prohibition ($\neg P \equiv F$); act_i is the activation condition that, when matched in DKB, causes a norm to activate; and $goal_i$ is the state to be achieved or avoided.

Conditions act_i and $goal_i$ are formulated in terms of predicates q(?v) with variables represented as ?v, and constants as c. Note that a prohibition corresponds to the absence of permission. The following elements mark certain properties of the dialogue:

- agent(?agt): an agent in Agt;
- prop(?agt,?act): a proponent ?agt $\in Agt$ of ?act $\in \mathcal{A}$;
- spoke(?agt): a player ?agt has spoken;
- done(?perf,?subj,?agt): m_i has been moved by agent ?agt containing a speech act $perf(\vartheta)$ composed by ?perf= $perf$ and ?subj = ϑ
- others(?agtx): a function that returns a group of agents ?agty $\in Agt$ excluding ?agtx;
- memberOf(?agt,?group): an agent ?agt $\in Agt$ member of a group ?group $\subseteq Agt$.

In norm-governed dialogues, norm compliance is tested at each dialogue state. Following Gasparini et al. (2015), we use a Kripke structure DKS to represent the possible evolutions of the world as a directed graph with states as nodes. A sequence of moves in a dialogue is a path in the graph where arcs correspond to moves, represented as $\rho_h = s_0 \to^{m_1} s_1 \to^{m_2} \ldots \to^{m_h} s_h$. Intuitively, DKS represents a formalism for the graphs of Section 2. We refer to a formula entailed by the structure as a formula that is evaluated to true in a particular state, $(DKS, s) \vDash \varphi$. The knowledge base DKB in a state s is the set of propositions entailed by DKS such that $q \in DKB(s)$ if and only if $(DKS, s) \vDash q$. Further, we assume that at each transition $s_i \to^{m_{i+1}} s_{i+1}$, spoke(agt) and done(perf,subj,agt) hold in s_{i+1}. After a move $propose(A_k)$, a predicate prop(agt,Ak) holds and the move $withdraw(A_k)$ removes prop(agt,Ak). This maintains a record of the active proposals during a dialogue.

In order to evaluate norm-compliance, we match the activation condition act_i of each norm n_i against $DKB(s')$ in a state s'. If it holds, we refer to the instantiated condition as $act_i(\theta_j)$ where $(DKS, s') \vDash act_i(\theta_j)$. The instantiated goal of the norm $goal_i(\theta_j)$ is tested against the next state s'' to determine the norm-compliance of move m'', where $s' \to^{m''} s''$.

Consider a path ρ_h for dialogue d_h. We define a *norm store* $NS(s)$ for gathering active norms and an evaluation function $\mathcal{V}(s)$ to label violations for each state $s \in \rho_h$.

Definition 2. *The norm store $NS(s)$ is defined for a dialogue state $s \in \rho_h$ as a set of tuples $\langle mod_i, goal_i(\theta_j) \rangle$ of a norm n_i activated in state s.*

$NS(s)$ is updated at each transition, and emptied when a new transition is executed in path ρ_h.

Definition 3. *An evaluation function* $\mathcal{V}(s): S \rightarrow \{$`compliant`, `non-compliant`$\}$ *labels each dialogue state* $s \in \rho_h$. *Given* $s' \rightarrow^{m''} s''$, *state* s'' *is* `non-compliant` *when:*

- *A forbidden goal is achieved:* $(DKS, s'') \models goal_i(\theta_j)$ *and* $\langle F, goal_i(\theta_j) \rangle \in NS(s')$,
- *An obliged goal is not achieved:* $\langle O, goal_i(\theta_j) \rangle \in NS(s')$ *and* $(DKS, s'') \not\models goal_i(\theta_j)$, *or*
- *None of the goals is permitted:* $\nexists goal_i(\theta_j)$ *s.t.* $\langle P, goal_i(\theta_j) \rangle \in NS(s')$ *and* $(DKS, s'') \models goal_i(\theta_j)$.

Otherwise, s'' *is* `compliant`.

The norms that model the protocol of Walton et al. (2014) as a norm-governed dialogue with BoP are:

1. Norms to regulate turn-taking $N_T = \{n_{t1}\}$: The last agent who spoke is forbidden to speak.

2. Norms N_P to define what an agent is permitted to say:
   - n_{p1}: A proposal is always permitted.
   - n_{p2}, n_{p3}: Every agent, except the proponent, is permitted to reject or accept a proposal.
   - n_{p4}: Only the proponent is permitted to withdraw the proposal.
   - n_{p5}: Every agent is permitted to disclose information after a proposal has been moved.
   - n_{p6}, n_{p7}: A why or argue move (both with the intent to challenge a statement) is permitted after accept, propose, disclose, argue, reject and why.

3. Norms N_O that impose obligations on agents to respond in a certain way:
   - n_{o1}: Agents are obliged to move reject, why or accept after *propose*.
   - n_{o2}: Agents are obliged to move disclose, why or withdraw after *reject*.
   - n_{o3}: Any response to *withdraw* is forbidden.
   - n_{o4}: Any response to *accept* is forbidden.
   - n_{o5}: Agents are obliged to move argue, why, withdraw or accept after *disclose*.
   - n_{o6}: Agents are obliged to move why, argue, disclose, withdraw or accept after *argue*.
   - n_{o7}: Agents are obliged to move argue, withdraw or accept after *why*.

These norms are formalized in Figure 6.

Observation 1. *The structural rules of a norm-governed protocol for deliberation dialogue with BoP are formed by $N_{BoP} = N_T \cup N_P \cup N_O$. The obligations n_{o6} and n_{o7} impose a burden of proof on the proponent requiring it to immediately respond to a challenge.*

Here, we argue for the need of a protocol that accounts for the absence of burden of proof in deliberation — BrC. This is represented by a new set of obligations N_O' that includes $n_{o1}, n_{o2}, n_{o3}, n_{o4}, n_{o5}$, but excludes n_{o6} and n_{o7} as they are responsible for imposing the burden of proof. We also remove n_{o3} permitting agents to continue the dialogue after withdrawing a proposal.

Observation 2. *The structural rules of a norm-governed protocol for deliberation dialogue with BrC are formed by $N_{BrC} = N_T \cup N_P \cup N_O'$ that include obligations for propose, reject, accept and disclose. Replies to argue and why challenges are permitted but not obliged.*

N_T:
$n_{t1} = \langle idn_{t2}, F, act_{t2} = \text{agent(?agtx)} \land \text{spoke(?agtx)}, goal_{t2} = \text{spoke(?agtx)} \rangle$

N_P:
$n_{p1} = \langle idn_{p1}, P, act_{p1} = \text{true}, goal_{p1} = \text{done(propose, ?actk, ?agtx)} \rangle$
$n_{p2} = \langle idn_{p2}, P, act_{p2} = \text{IN\{agent(?agtx)} \land \text{prop(?agtx, ?actk)\}FILTER NOT EXISTS} \{\text{prop(?agty, ?actk)}\}, goal_{p2} = \text{done(reject, ?actk, ?agtx)} \rangle$
$n_{p3} = \langle idn_{p3}, P, act_{p3} = \text{IN\{agent(?agtx)} \land \text{prop(?agtx, ?actk)\}FILTER NOT EXISTS} \{\text{prop(?agty, ?actk)}\}, goal_{p3} = \text{done(accept, ?actk, ?agtx)} \rangle$
$n_{p4} = \langle idn_{p4}, P, act_{p4} = \text{prop(?agtx, ?actk)}, goal_{p4} = \text{done(withdraw, ?actk, ?agtx)} \rangle$
$n_{p5} = \langle idn_{p5}, P, act_{p5} = \text{prop(?agtx, ?actk)}, goal_{p5} = \text{done(disclose, ?subj, ?agty)} \rangle$
$n_{p6} = \langle idn_{p6}, P, act_{p6} = \text{done(propose, ?subj, ?agtx)} \lor \text{done(disclose, ?subj, ?agtx)} \lor \text{done(argue, ?subj, ?agtx)} \lor \text{done(reject, ?subj, ?agtx)} \lor \text{done(why, ?subj, ?agtx)} \lor \text{done(accept, ?subj, ?agtx)}, goal_{p6} = \text{done(why, ?subl, ?agty)} \rangle$
$n_{p7} = \langle idn_{p7}, P, act_{p7} = \text{done(propose, ?subj, ?agtx)} \lor \text{done(disclose, ?subj, ?agtx)} \lor \text{done(argue, ?subj, ?agtx)} \lor \text{done(reject, ?subj, ?agtx)} \lor \text{done(why, ?subj, ?agtx)} \lor \text{done(accept, ?subj, ?agtx)}, goal_{p7} = \text{done(argue, ?subl, ?agty)} \rangle$

N_O:
$n_{o1} = \langle idn_{o1}, O, act_{o1} = \text{done(propose, ?subl, ?agtx)}, goal_{o1} = \text{EXISTS\{memberOf(?agty, others(?agtx))} \land [\text{done(reject, ?subj, ?agty)} \lor \text{done(why, ?subj, ?agty)} \lor \text{done(accept, ?subj, ?agty)}]\} \rangle$
$n_{o2} = \langle idn_{o2}, O, act_{o2} = \text{done(reject, ?subl, ?agtx)}, goal_{o2} = \text{EXISTS\{memberOf(?agty, others(?agtx))} \land [\text{done(disclose, ?subj, ?agty)} \lor \text{done(why, ?subj, ?agty)} \lor \text{done(withdraw, ?subj, ?agty)}]\} \rangle$
$n_{o3} = \langle idn_{o3}, F, act_{o3} = \text{done(withdraw, ?subl, ?agtx)}, goal_{o3} = \text{done(?perf, ?subj, ?agty)} \rangle$
$n_{o4} = \langle idn_{o4}, F, act_{o4} = \text{done(accept, ?subl, ?agtx)}, goal_{o4} = \text{done(?perf, ?subj, ?agty)} \rangle$
$n_{o5} = \langle idn_{o5}, O, act_{o5} = \text{done(disclose, ?subl, ?agtx)}, goal_{o5} = \text{EXISTS\{memberOf(?agty, others(?agtx))} \land [\text{done(argue, ?subj, ?agty)} \lor \text{done(why, ?subj, ?agty)} \lor \text{done(withdraw, ?subj, ?agty)} \lor \text{done(accept, ?subj, ?agty)}]\} \rangle$
$n_{o6} = \langle idn_{o6}, O, act_{o6} = \text{done(argue, ?subl, ?agtx)}, goal_{o6} = \text{EXISTS\{memberOf(?agty, others(?agtx))} \land [\text{done(argue, ?subj, ?agty)} \lor \text{done(withdraw, ?subj, ?agty)} \lor \text{done(accept, ?subj, ?agty)} \lor \text{done(disclose, ?subj, ?agty)} \lor \text{done(why, ?subj, ?agty)}]\} \rangle$
$n_{o7} = \langle idn_{o7}, O, act_{o7} = \text{done(why, ?subl, ?agtx)}, goal_{o7} = \text{EXISTS\{memberOf(?agty, others(?agtx))} \land [\text{done(argue, ?subj, ?agty)} \lor \text{done(withdraw, ?subj, ?agty)} \lor \text{done(accept, ?subj, ?agty)}]\} \rangle$

Figure 6 – Norms for a norm-governed dialogue

4.3 Termination rules

A dialogue d terminates if: no further possible state exists after a last state s_h or agents enter an illegal state of the dialogue such that $\mathcal{V}(s_h)$ =non-compliant. In the case of deliberation with BoP, the agents must reply to a speech act argue, or why with a counterargument, otherwise s_h is a non-compliant state that leads to an early termination of the dialogue. In a BrC dialogue, these obligations are removed and this early termination is avoided.

4.4 Norm-governed BoP and BrC dialogues: an example

Let us expand Example 1 to consider norm stores for BoP and BrC norm-governed dialogues in Example 2.

> (2) Agent x proposes action A_1: stop the water supply (move m_1). This is rejected by y (m_2). Agent x explains that the water is contaminated (m_3) but y is not satisfied with this explanation
> and asks for further reasons with $why(A_1)$ in a move m_4. The dialogue path is $\rho_4 = s_0 \to^{m_1} s_1 \to^{m_2} s_2 \to^{m_3} s_3 \to^{m_4} s_4$. The norm stores for BoP and BrC at state s_4 are:
> - DKB:
> $(DKS, s_4) \models$ prop(x,A1) $\wedge$ spoke(y) $\wedge$ done(why,A1,x)
> - Deliberation BrC:
> $NS_{BrC}(s_4) = \{ \langle P,\text{done(propose,?actk,x)} \rangle,$
> $\langle P,\text{done(withdraw,A1,x)} \rangle, \langle P,\text{spoke(x)} \rangle,$
> $\langle P,\text{done(disclose,?subj,x)} \rangle, \langle F,\text{spoke(y)} \rangle,$
> $\langle P,\text{done(why,?subj,x)} \rangle, \langle P,\text{done(argue,?subj,x)} \rangle \}$
> - Deliberation BoP:
> $NS_{BoP}(s_4) = NS_{BrC}(s_4) \cup \{$
> $\langle O,\text{memberOf(x,others(y))} \wedge [\text{done(argue,?subj,x)}$
> $\vee \text{done(withdraw,?subj,x)}] \rangle \}$

In a BoP dialogue, a move after s_4 must adhere to the obligation of replying to why with an argument. For example, x may say that the location is not safe as there are other interventions scheduled. The other option is for x to withdraw A_1 otherwise there is a violation of the protocol. We may observe that $why(.)$ in a BoP dialogue corresponds to $ask\text{-}justify(.)$ in a MHP dialogue. Counterarguments should be immediately moved, otherwise A_1 must be retracted.

In a BrC dialogue, the speech act argue is permitted, but additional options (as those listed in Section 3) are also permitted as per instantiations of N_P. For example, agent x may propose to postpone blocking off the water supply. Agent x may ask why to understand the reasons why agent y is reluctant to agree with the proposal. Agent x may disclose a part of its plan stating that there will be free access to water cisterns. In this way, agent x prevents y committing to refuse the proposal. Therefore, we can see that our new model of dialogue permits agents to postpone or omit answers to a challenge.

5. CONCLUSION

The focus of this research is to improve and enrich existing autonomous systems in the light of new findings on the theory of natural deliberation. Recently, Walton (2014) argued that there is no burden of proof in deliberation. In this paper, we showed that in AI even the most influential protocols for deliberation inherited the characteristics of burden of proof from persuasion dialogues. We proposed to overcome this problem by employing norms to define protocol rules. In norm-governed dialogues with BoP, the proponent is obliged to argue against a challenge. In dialogues without BoP, the proponent is simply permitted to do so. This allows time for the participants to consider different proposals.

Other features of computational models of deliberative dialogue may be affected by the absence of BoP; e.g., how to decide which action to adopt, as agents may be committed to an action that has no support. The shifts to other dialogues with BoP should be modeled to ensure that the BoP requirements are reinstated. Norm-governed protocols may also, permit the definition of other dialogue features; e.g., the "right" to make a move. This involves a permission for an agent to move a speech act, while all the other agents would be prohibited to prevent the agent to speak.

We believe that the research proposed in this paper may give useful insights on how develop AI systems that more adequately capture the richness of natural deliberation.

ACKNOWLEDGEMENTS: Douglas Walton would like to thank the Social Sciences and Humanities Research Council of Canada for Insight Grant 435-2012-0104 that supported work in this paper. This research is also partially supported by the award made by the RCUK Digital Economy program to the dot.rural Digital Economy Hub; award ref.: EP/G066051/1.

REFERENCES

Bex, F., & Walton, D. (2011). Combining explanation and argumentation in dialogue. In *Proceedings of the 12th International Workshop on Computational Models of Natural Argument*.

Gasparini, L., Norman, T. J., Kollingbaum, M. J., Chen, L., & Meyer, J. J. C. (2015). CÒIR: Verifying normative specifications of complex systems. In *Proceedings of the 18th International Workshop on Coordination, Organisations, Institutions and Norms*.

Gordon, T. F., Prakken, H., & Walton, D. (2007). The Carneades model of argument and burden of proof. *Artificial Intelligence, 171*(10–15), 875–896.

Kauffeld, F. J. (1998). Presumptions and the distribution of argumentative burdens in acts of proposing and accusing. *Argumentation, 12*(2), 245–266.

Kok, E. M., Meyer, J. J. C., Prakken, H., & Vreeswijk, G. A. W. (2011). A formal argumentation framework for deliberation dialogues. In *Argumentation in Multi-Agent Systems*, volume 6614 of *Lecture Notes in Computer Science*. Berlin Heidelberg: Springer.

Kollingbaum, M. J., & Norman, T. J. (2004). Norm adoption and consistency in the NoA agent architecture. In *Programming Multi-Agent Systems*, volume 3067 of *Lecture Notes in Computer Science*. Berlin Heidelberg: Springer.

McBurney, P., Hitchcock, D., & Parsons, S. (2007). The eightfold way of deliberation dialogue. *International Journal of Intelligent Systems, 22*(1), 95–132.

Medellin-Gasque, R., Atkinson, K., McBurney, P., & Bench-Capon, T. (2011). Arguments over cooperative plans. In *Theory and Applications of Formal Argumentation*, volume 7132 of *Lecture Notes in Computer Science*. Berlin Heidelberg: Springer.

Prakken, H. (2006). Formal systems for persuasion dialogue. *The Knowledge Engineering Review, 21*(2), 163–188.

Toniolo, A., Norman, T. J., & Sycara, K. (2012). An empirical study of argumentation schemes for deliberative dialogue. In *Proceedings of the 20th European Conference on Artificial Intelligence*, volume 242, 756–761.

Walton, D. (1995). *A Pragmatic Theory of Fallacy*. Tuscaloosa and London: University of Alabama Press.

Walton, D. (2014). *Burden of Proof, Presumption and Argumentation*. Cambridge: Cambridge University Press.

Walton, D., Toniolo, A., & Norman, T. J. (2014). Missing phases of deliberation dialogue for real applications. In *Proceedings of the 11th International Workshop on Argumentation in Multi-Agent Systems*.

Walton, D., & Krabbe, E. C. W. (1995). *Commitment in Dialogue: Basic Concepts of Interpersonal Reasoning*. New York: State University of New York Press.

Commentary on Walton, Toniolo and Norman's Speech Acts and Burden of Proof in Computational Models of Deliberation Dialogue

JAN ALBERT VAN LAAR
University of Groningen, The Netherlands
j.a.van.laar@rug.nl

1. INTRODUCTION

I am very much attracted by the project of developing *rigorous* normative protocols for deliberation dialogue that are *flexible* and allow for strategies that are close to those in natural deliberations. My one comment amounts to the idea that we can subscribe to the basic insights and results of the paper without needing to disparage the role of the burden of proof in deliberation dialogue.

2. THE IMPORTANCE OF THE BURDEN OF PROOF RULE IN DELIBERATION DIALOGUE

The authors argue that there is no burden of proof in deliberation dialogue. I label the rule according to which a challenged statement must either be withdrawn or defended by means of argument, the "Burden of Proof Rule". The authors propose a set of norms for deliberation dialogue without this burden of proof rule. They hold that the proposer, after having her proposal challenged, should not only be allowed to postpone her defense, but may also omit it altogether, without having to withdraw the proposal. As a reason, the authors advance that the goal of a deliberation dialogue is to arrive collectively at a practical decision. In more detail, the idea seems to be the following: Suppose party A has a plan *A* and makes, accordingly, a proposal **A**. Suppose further that there is a party B, who has a plan *B* that runs counter to plan *A*, and that party B objects to proposal **A**. Possibly, he even makes a counterproposal **B**. Party A, then, need not be forced to choose between withdrawing her proposal **A** and defending it. Instead, the two parties can collectively discover that their plans *A* and *B* allow of some integration, and they may proceed to explore an optimal integrative solution introduced by means of a new proposal, say proposal **C**. According to Walton, Toniolo and Norman, norms for deliberation dialogue should be flexible enough for allowing such a dialogue in which proposal **A** does not get defended at all.

I agree that this seems like a fully legitimate scenario, and a quite typical one for deliberation. Different from the authors, however, I do not conclude from this that there is no burden of proof in deliberation dialogue. Even in the minimally argumentative scenario just sketched, I see the notion of burden of proof as being of crucial importance. How is it important?

A deliberation dialogue is a highly collaborative kind of dialogue that starts from a practical problem. It can but does not need to start from a difference of opinion of difference of interests. Its main goal is to find a stable solution for the problem at hand that is feasible within the current social setting by being, in the end, acceptable to the parties present to the deliberation (thus providing a ground for political legitimacy).

The concept of deliberation implies that the final agreement results from a careful consideration of the issue, that is, from a weighing of the pros and cons in a conscientious way (cf. Fishkin & Luskin, 2005). Consequently, a deliberation can be seen as a kind of practically oriented inquiry into a solution that is optimal by being defensible to all parties in the dialogue, albeit possibly in different ways to different parties. In other words, a deliberation is an argumentative examination into an action proposal that best withstands mutual criticism.

The typical way of proceeding in deliberation is that participants move proposals. These proposals can be tentative and explorative, especially at the earlier stages, so that the kind of commitments they incur need not be of a very fixed nature. At more mature stages, the proposals typically become more definite and committal. The parties need to assess the comparative argumentative merits of all serious proposals, so as to find the optimal choice. This argumentative examination can be highly collaborative, and it's not required that tasks are rigidly distributed over a defending proponent and an attacking opponent. But all the same, they need to evaluate each of these proposals as either allowing of a successful defense towards the parties present, or not. Deliberation dialogue, thus conceived, requires that persuasion dialogues are embedded, whether these are developed in a more collaborative fashion - leaving ample room for mutual explanation of stances so as to assist one another to progress swiftly (cf. Van Laar & Krabbe, 2013), or in a moderately antagonistic fashion.

With each proposal, a party can be expected to try to influence the course of events in a direction where she expects to find the optimal and mutually acceptable solution. In this way, each proposal conveys information about where the proposer locates the final outcome, and where not. Any next proposal must accommodate this information, in order to stand a chance at getting accepted by all parties. In this way, they try to step-wise develop an integrative solution that they all consider to be optimal, given their diverging plans and views. In other words, they collaboratively are working towards a proposal such that, if challenged,[1] its burden of proof can be successfully discharged vis-à-vis the parties present.

I can conceive of two deliberative scenario's where the parties shake off a proposal without any argument being advanced. *First*, party B may just "see" the argumentative merits and especially demerits of party A's proposal **A**, and refrain from challenging proposal **A**, moving quickly his counterproposal **B**. In that case, the argumentative message by party B is something like that party A will not be capable of discharging her burden of proof for proposal **A** in a way that could convince him. Party A may take the implicit advice to heart, and abandon her proposal in favor of party B's new proposal or move a new on herself. In this situation, party A must be seen as withdrawing her initial proposal. The role of the burden of proof is indispensible in this course of moves: It is because the burden of proof for this proposal **A** is expected not to be dischargeable that it is abandoned. The *second* scenario is the one we started with. For it's also possible that party B does challenge proposal **A**, but that party A comes to understand that her proposal is indefensible, i.e. that the incurred burden of proof is non-dischargeable, and that on that ground she abandons and withdraws her proposal **A**. Instead of proposal **A**, she may table a new proposal that hopefully improves things from party B's point of view. Again, what drives the deliberation forward is the discovery that the earlier proposal cannot be successfully defended to all parties, that the burden of proof that one incurs after being challenged could not be successfully discharged, so that the earlier proposal cannot possibly be the proper outcome of the deliberation dialogue. In these two scenarios no burden of proof is discharged. But the proposal at issue has been withdrawn, and so the parties have acted in line with the Burden of Proof Rule. I guess my point is that by changing from an earlier proposal to a proposal that accommodates an interest or opinion of the other side, one explicitly or implicitly withdraws the earlier, inferior proposal.

---

[1] Following Ihnen Jory, one preparatory condition for the felicitous performance of the speech act of a proposal is that the proposal serves an interest that is shared by the speaker and the addressee (2014). This common interest can become the target of a critical response, and the addressee may in particular object to the supposition that the proposal serves her interests.

Consequently, the parties can move on and explore this supposedly superior proposal without violating the Burden of Proof Rule.

3. CONCLUSION

From this perspective, I agree with the idea that a normative protocol for deliberation dialogue should be flexible, and leave the participants the option to postpone the discharge of the burden of proof. They even could be given the option to refrain from discharging the burden of proof entirely, so that indeed there is a permission to discharge the burden of proof, rather than an obligation. But we can acknowledge the appropriateness of such a norm and yet retain the Burden of Proof Rule, and stress the importance of the burden of proof as indispensable for a genuinely deliberative dialogue.

REFERENCES

Fishkin, J. S., & Luskin, R. C. (2005). Experimenting with a democratic ideal: deliberative polling and public opinion. *Acta Politica, 40*, 284-298.
Ihnen Jory, C. (2014). Negotiation and deliberation: Grasping the difference. *Argumentation: An International Journal on Reasoning.* Online first (doi: 10.1007/s10503-014-9343-1).
van Laar, J. A., & Krabbe, E. C. W. (2013). The burden of criticism: Consequences of taking a critical stance. *Argumentation: An International Journal on Reasoning, 27*(2), 201-224.

34

Using Argumentation within Sustainable Transport Communication

SIMON WELLS
Edinburgh Napier University, UK
s.wells@napier.ac.uk

KATE PANGBOURNE
University of Aberdeen, UK
k.pangbourne@abdn.ac.uk

In this paper we present the preliminary results of a survey of persuasive communication within the sustainable transport domain. This survey is underpinned by a reconstruction of the arguments used, a scheme oriented analysis of the corpus of reconstructed arguments, and elements of a theoretical and applied framework for using the corpus to effect lasting behaviour change using argumentative techniques within the self-same domain.

KEYWORDS: argumentation schemes, behaviour change, corpus building, motivation, persuasion, sustainable transport

1. INTRODUCTION

In this paper we present the preliminary results of an ongoing survey of persuasive communication within the sustainable transport domain.

This survey is underpinned by a reconstruction of the arguments used, a scheme-oriented analysis of the corpus of reconstructed arguments, and elements of a theoretical and applied framework for using the corpus to effect lasting behaviour change using argumentative techniques within the self-same domain.

Our aim is three-fold; primarily to develop and deploy computational argumentation within a, from the perspective of argumentation theory, novel domain where such technologies have the potential for real societal benefit. A key issue for computational argumentation technologies is how we smoothly move from theoretic issues of argument to useful tools for the domain practitioner. We aim for our experiences to inform this process and for this paper to outline and report on two preparatory steps in such a process, the gathering of knowledge that an intelligent argumentation process can consume and the identification of a gross architectural framework. Secondly, we aim to explore the linkages and overlaps between the behaviour change theory and argumentation theory research areas. Behaviour change theory incorporates rich psychological models of how individuals form, break, and re-form new or adjusted habitual behaviours whereas argumentation theory focusses on how rational and justifiable decisions are made. By aligning both approaches we propose that argumentation performs an important ethical role in ensuring that individuals who are exposed to behaviour change techniques do so in an informed way. Additionally, argumentative interaction can perform a significant role in building motivation, a critical prerequisite in achieving behaviour change; argumentation theory can thus improve the likelihood of behaviour change being successful and sustainable. Finally, we aim to exemplify current best-practice in the construction, initial release and ongoing development of a flagship dataset for sustainable transport and to make clear a range of attributes that should be considered, and ideally satisfied, when building and sharing datasets in order for them to be sustainable resources.

In section 5 we lay out a program of future work that seeks to provide a solid quantitative foundation for the current approach though evaluation of both our corpus of sustainable transport arguments and our interaction mechanisms to ensure that they are both effective and appropriate. In the longer term we aim to identify effective, scalable, and reproducible communicative and argumentative techniques that can be used to help people to make informed and justifiable choices about their behaviours.

The major contribution of this research is to underpin existing motivational and behaviour change communications within the sustainable transport domain with solid argumentation theoretic foundations and to provide an extended corpus of analysed and reusable arguments. This approach brings together two important and complementary research areas, one of which has focussed on psychological models at the expense of practical techniques, and the other which has focussed more heavily on ideal reasoners and normative models, almost to the exclusion of consideration of the messy thinking that characterises human action in the real world.

2. BACKGROUND

Transport, particularly that relating to personal mobility is a huge source of environmentally damaging emissions and pollutants.

Additionally the transport sector alone accounts for 40% of final energy consumption in the European Union. We focus on the problem domain of reducing unsustainable travel behaviours, which is a normative policy goal in many developed world contexts.

In order to tackle climate change, there is an imperative to reduce greenhouse gas (GHG) emissions from all sectors. Whilst emissions in most sectors are falling, those from transport-related emissions have risen by 36% since 1990. Cars alone account for 12% of the total EU CO2 emissions, with similar figures for CO, NO, Ozone, particulate matter, and other toxic and volatile chemicals (figures gathered from the European Commission[1]). In aggregate individual travel habits therefore have a large impact on the quality of the environment, particularly in urban environments in which, as of 2014, 54% of the world's population now live[2]. Transport is a particular 'offender' in relation to GHG emissions and considerable effort has been devoted to addressing the issue. For example, the recent UN Climate Summit in New York announced four global transport-related initiatives to progress the goal of low carbon mobility.

9. The Urban Electric Mobility Initiative (UEMI) aims to increase the number of electric vehicles (e-vehicles) in cities to at least 30% of all new vehicles sold annually by 2030, and to make cities e-vehicle friendly.
10. The International Union of Railways (UIC) Low-Carbon Sustainable Rail Transport Challenge will promote the use of railways for both freight and passenger transport.
11. The International Association of Public Transport (UITP) Declaration of Climate Leadership brings together 1300 member organisations across 92 countries that are providing clean public transport (PT) for city populations.
12. A new commitment by the International Civil Aviation Organisation (ICAO) to develop more sustainable alternative fuels for aviation, to develop a global CO2 standard for new aircraft and design and implement a measure for international aviation from 2020.

---

[1] Climate Action Policies:
http://ec.europa.eu/clima/policies/transport/index_en.htm

[2] World Health Organisation Global Health Observatory Data:
http://www.who.int/gho/urban_health/situation_trends/urban_population_growth_text/en/

The UN regards these initiatives as crucial to reducing GHG emissions. However, it is clear that all these initiatives will require attention to the demand side for success. In essence this translates to achieving behavioural changes. In the main this requires persuading individuals to switch away from private cars fuelled by the internal combustion engine either to public transport and to active travel, i.e. walking and cycling or to low-carbon vehicles. However, whilst it is important to develop low-carbon vehicles, in many urban transport contexts the widespread adoption of low-carbon cars will fail to solve another severe issue related with car use: congestion. Therefore, urban transport authorities have two imperatives in relation to changing citizens' behaviour: reducing GHG emissions (which addresses both climate change targets and local air quality issues) and traffic congestion problems. Indeed, this is an issue that has already received quite a lot of attention from urban authorities, involving many different approaches to influencing travel behaviour, often founded on psychological understandings of behaviour (Howarth & Riley, 2012).

The European Union (EU) has been particularly active in the sustainable transport domain for a number of years, and has explicitly linked achieving greater sustainability in the transport domain to information technology, branded as Smart Mobility. Furthermore, the EU is committed to Intelligent Transport Systems (ITS), deploying the affordances of information technology and big data to improve the efficiency of transport networks across Europe[3], and a number of projects that deliver different aspects of personalised transport information to consumers (either through technology, such as Multi-modal Journey Planners or other means such as those used in SEGMENT) have been funded (e.g. SUPERHUB, MyWay, SEGMENT) by the European Commission, as traffic and travel information is an early priority under directive 2010/40/EU which requires standards for interoperability, compatibility and continuity of systems across Europe by 2017.

Therefore, in this paper we focus primarily on arguments for and against sustainable travel in urban mobility contexts. In such contexts the widest variety of alternative travel modes are most likely to exist: walking, cycling and public transport use as sustainable, and increasingly Mobility 2.0 (Lanzendorff, 2014) innovations which combine new models of vehicle use (such as pooling, sharing or leasing) with new types of vehicle, fuels and powertrains, particularly electric

---

[3] *http://ec.europa.eu/transport/themes/its/road/action_plan/index_en.htm*

vehicles). Where more alternatives to the private car exist, the arguments for travel behaviour change are both more varied, and in theory more persuasive. However, in our preliminary investigations we have ascertained that whilst there is a great deal of effort being put into behaviour change campaigns, much less is known about the effectiveness of such campaigns, particularly in relation to the messages (arguments) that are being promoted to the public (Pangbourne & Masthoff, 2015; Davies, 2012). Nevertheless, small-scale voluntary travel behaviour change programmes appear to have achieved some success in demonstrating that personalised "encouragement, motivation and information" (Rocci, 2012) does result in an increase in the use of more sustainable transport alternatives.

Therefore, if people can be persuaded to modify their habitual behaviours, to choose to use more sustainable transport modes where those are available, then real improvements in the environment can be achieved. To achieve this requires three important factors to be taken into account. Firstly, how to effect lasting behaviour change, in this case we see argumentation-based interaction as a key factor that will enhance the effect of existing digital behaviour change approaches. In section 4.1 we survey current approaches to behaviour change and make the case for argumentative interaction as a key enhancement to those approaches. Secondly, how to effect such behaviour change at scale. We propose that behaviour change supported by digital technology, particularly mobile digital devices, is a way to scale behaviour change to societal scope. Mobile devices such as phone and tablets have achieved high usage within much of western society. These devices usually contain high fidelity sensors which can, with the owner's consent, track behaviour, as well as provide increasingly powerful computational resources to underpin intelligent decision-making software. Such devices therefore offer an unparalleled opportunity to place a behaviour change assistant in the pockets of everyone. We discuss some of the issues associated with this approach in Section 4.2. Finally we must collate sufficient resources, in this case arguments, to populate an autonomous argumentation system and to enable such a digital behaviour change support system to act sustainably within the problem domain.

A preparatory step in developing an argumentation system that targets a specific problem domain is the acquisition, structuring and preparation of useful domain-knowledge. It is this knowledge acquisition step upon which we focus in the core of this paper and which is detailed in Section 3. There have been many public awareness campaigns and communications over the last few decades which have aimed to change personal transportation habits, for example, encouraging cycle use or discouraging car use. These campaigns often reduce the message to the level of a slogan, removing nuance that might otherwise make a developed argument more persuasive. We have collated motivational messages and argumentative communication from a large number of existing sustainable transport communication campaigns deployed around the world. These messages have, where necessary, been treated as enthymemes and have been reconstructed to instantiate unexpressed premises and conclusions. The resulting arguments have been subjected to a scheme oriented analysis to yield more complete argument resources that include consideration of critical questions and the ways in which the argument might be responded to. By reconstructing these arguments and storing them in a reusable way using the Argument Markup Language (AML) and Argument Interchange Format (AIF), we have been able to construct a corpus, that we name the Sustainable Transport Communication (STC) Dataset, which can be explored using appropriate interaction techniques, and whose elements can be framed and presented in the most strategically appropriate way, given consideration of the specific person and the behaviour that is being targeted. This builds on the approach identified in (Reed & Wells, 2007) in which the knowledge bases of agents within a Multi-Agent System (MAS) were populated using AML structures to provide both knowledge and argumentative relationships between items of knowledge. In (Reed & Wells, 2007) it was demonstrated that dialogues generated from such a knowledge base can appear to be realistic. Interaction with argument structured knowledge bases, mediated by dialogue games such as those described in (Wells & Reed, 2012) and collated in (Wells, 2012), thus provide a way to approach natural-seeming mixed initiative interaction even in the absence of more complex strategic capabilities.

We propose that to effect lasting behaviour change, the recipient must make an informed decision about their behaviour and the habits that they wish to change. Non-permanent behaviour change could occur through happenstance, the participant tries something different for no apparent reason, or trickery, the participant does something different because it is made easier to perform the new behaviour, or bribery, the participant is offered some incentive to alter their behaviour. However lasting and persistent habit formation will occur when a person understands the context in which their behaviour must change and can fall back upon their personal reasons for doing so, especially if the old habits are difficult to break or the new habits are difficult to form. Additionally it is also important to recognise that for a person to change their established habits is difficult and that they need to be supported in forming new and different habitual behaviours; behaviour change, especially in difficult problem domains does not easily occur in a vacuum but can require external support. Whilst behaviour change theory can provide relatively rich psychological models, particularly of the process that underpins how new behaviours are formed, it is argumentation that can provide well developed models of (1) dialogical interaction, (2) reasoning, and (3) supporting knowledge representation. Together these will enable effective and repeatable behaviour change in targeted problem domains like sustainable transport. Whilst behaviour change provides mechanisms for supporting the formation of new habits, for example, through the use of targeted interventions and challenges, argumentation provides the mechanisms for ensuring that a person is making an informed choice and has established a personal justification for why they are performing such a difficult task. It is the person deciding to make an informed and justifiable choice, to change their behaviour, that is a key aspect to effecting long lasting behaviour change. Additionally, to achieve this kind of behaviour change at scale requires the adoption of digital technologies and the use of personalised and appropriate interaction techniques to ensure that arguments are both selected and framed so as to be as effective as possible for the given person. Our eventual aim is to use arguments to increase motivation, to use dialogue to interact with users, and to adapt the rich range of argumentation schemes and dialogue models to work with behaviour change theories.

3. THE STC DATASET

In this section we describe the STC dataset.

In subsection 3.1 we describe the data collection process which is available as a Git repository[4] or as a citeable snapshot via DOI[5]. In 3.2 we describe how the data and associated metadata that comprise the dataset are stored, in subsection 3.4 we summarise the contents of the data set, and in subsection 3.3 we describe how we have aimed to construct a reusable data resource by adhering to current best practices for data release.

The dataset has been released in two modes, firstly as a Git repository shared publically using the GitHub site[6] and secondly as an archived release that has been uploaded to the Figshare site[7].

---

[4] *https://github.com/siwells/STCD*

[5] *http://dx.doi.org/10.6084/m9.figshare.1386856*

[6] *https://github.com/*

[7] *http://figshare.com*

Git is a version control system for software development that has become very popular and is used for many large software development projects including the Linux Kernel. Because Git primarily tracks changes to plain text files it is equally well suited to maintaining text-based datasets as software source code. Each change to the dataset, for example, addition of a new resource or updated metadata, is known as a commit and a cryptographic hash is made of each commit so that every change is tracked in a reliable way and so that every earlier commit can be retrieved. The author of each commit is also tracked within that commit so that a history of the dataset is available, viewable, and navigable with respect to what changed, who changed it, and when this occurred. This system also supports trivial branching to create alternative versions of the dataset for experimentation, distribution so that many people can work concurrently on the dataset, and good tooling for merging the results back into a single canonical repository for the dataset. Starting from the initial commit, all subsequent branches and commits form a distributed and cryptographically verifiable connected directed graph of commits which can be recombined in a variety of ways to suit the needs of the user. For our purposes, Git provides history and versioning for the contents of the dataset, enabling the data to be mutated and reused whilst tracking every change that occurs and allowing subsequent changes and updates, as well as the provenance of the changes and updates, to be tracked.

Figshare is a web site that enables research outputs to be allocated a Digital Object Identifier (DOI) and shared in a freely available manner. Subsequent downloads and citations are tracked and metrics can be retrieved from the site. As data reporting is handled by a third party this makes the provenance of metrics associated with dataset uptake more objective and reliable than if the data were hosted by the originators of the dataset, subject to the usual caveats related to the potential for gaming online metrics and so-called alt-metrics.

By adopting both approaches, using Git and Figshare, we are able to provide a versioned and manageable working dataset alongside defined releases which are allocated a DOI and are subject to citation tracking. This enables us to satisfy some of the best practices laid out in section 3.3. The strategy for deciding on a release is that, minimally, any publication based upon the dataset should define a specific release that can be cited by DOI. Any subsequent work that builds upon that publication can thus confidently reuse the data associated with that publication in a form that is both reproducible and replicable. Between publications, the dataset will continue to grow and potentially bugs or incorrect data will be fixed. However this will occur in a manner that allows all previous versions of the dataset to be recovered, and for the dataset itself to be mutated to suit the needs of specific experiments.

3.1 Data Collection

The raw data that forms the basis for the STC dataset was collected during 2014 and 2015 from publically accessible UK based websites associated either with public transport authorities, news organisations, green travel, or environmental awareness groups.

A key question in corpus building concerns the representativeness of the data. Whilst some of the data related to the STC dataset is presented within this paper and associated dataset release, the dataset is far from complete and currently merely serves to collate a variety of arguments that can be found in the wild. One measure of whether an argument corpus can be said to be tending towards completeness could be when the number of new arguments falls toward zero as new resources are added, e.g. new resources incorporated within the dataset merely tend to repeat or rephrase arguments that already exist within the dataset. We are not at that stage yet. An alternative approach might be to exhaustively analyse some of the collections of case studies that now exist. With the EU's interest in sustainable and smart mobility a number of resources have been built that collate English language case-studies of various sustainable transport initiatives that have occurred within the European context. For example, the ELTIS urban mobility observatory[8] maintains a large collection of urban transport and traffic management case studies that could be argued are representative of the European-wide sustainable urban transport context, but unfortunately these are not publically available and the general public, in the role of travellers, are not the targetted audience. For the moment however the STC dataset should be treated as a living resource that will grow over time.

3.2 Data Formatting & Handling Procedures

For each resource that is identified for inclusion a new resource folder, is created in the dataset repository under the resources folder.

Each resource folder is named according to a URI[9] for the resource. All current resources have been sourced online and thus their

---

[8] *http://www.eltis.org/*

[9] For more information about URI syntax see RFC-2986: *http://www.rfcbase.org/rfc-3986.html*

URLs have been used. Into this folder are placed, minimally, the non-optional artifacts from the following list:

1. A single UTF8 plain text file representing the argumentative text of the entire resource prepared for argument analysis using the Araucaria software.

2. Araucaria Analysis saved into both AML and AIF files named analysis.aml and analysis.aif respectively.

3. (Optional) Annotated/extended text to show placement of graphics, pictures or animations/videos for subsequent contextual analysis of non-textual aspects of the presented arguments.

4. (Optional) Additional notes regarding the resources stored in a single UTF8 plain text file.

5. (Optional) Screen shot of the resource in situ (so that the analysis can be revisited to see the resource in the way that it was presented). Screenshots must be archived as PDF or PNG files, named resource.pdf or resource.png, and at a resolution that means text is easily legible.

6. Metadata: Stored in a single UTF8 plain text file named metadata.txt the contents of which are quoted key:value pairs which can easily be converted by script into JSON format by converting the pairs into a comma separated list and enclosing everything in curly brackets. This means that it is a simple matter to load all metadata into a JSON-based, document-oriented, NoSQL database for further processing.
    a. Global Unique ID (GUID) generated using a standard tool[10] e.g. **"guid":"580e0d27-3cc2-41f7-97cb-98631a3832e2"**
    b. (b) Date and time of collection in ISO-8601 format, e.g. **"datetime":"2015-04-12T13:00:28+00:00"**[11].
    c. (c) Location of original resource (URL, URI, DOI as suits the resource). e.g. **"url":"** http://www.transportscotland.gov.uk" NB. If a URL is used then it is assumed that the URL is navigable so that updated resources can be retrieved using automated tools and scripts. A DOI should include only the identifier portion, scripts should

---

[10] The following Python one-liner will generate a new globally unique ID (GUID) each time it is run:

$ python -c "import uuid; print uuid.uuid4()"

[11] The following Python one-liner will generate the current time in ISO 8601-format:

$ python -c "import datetime; print datetime.datetime.now().isoformat()"

add the DOI resolution part to form a complete URL. If a URI is used then no assumption is made that the resource indicated is necessarily navigable.
It should be noted that because Araucaria has been effectively abandoned by its original developers and no longer compiles for modern versions of Java, but is available under the GPL, a new version was bug-fixed and built and the fixed version made available from GitHub[12].

3.3 Best Practises for Working With Datasets

In the past it has been sufficient for a dataset to be merely zipped up and deposited on an institutional website, if it has been made available at all, and often on the personal page of a researcher, and subject, in the longer term, to so-called linkrot.

More recently there have been efforts to make such data more accessible, for example, by providing web interfaces to databases that can be accessed online as exemplified by the AraucariaDB which ran from circa 2001 until it was subsumed into the ArgDB[13]. More recently the idea of data as a citable academic output has gained traction, with impetus particularly from the Biological and Physical Science community that has given rise to the concept of the data journal[14] and increased requirements from funding agencies for open and reusable data from experimental research. In response, efforts have been directed at ensuring that data is appropriately managed in order to support optimal reuse. In (de Waard, 2014) a hierarchy of aspects associated with optimal data reuse is identified which are summarised, along with how we addressed them, as follows:

Preserved *Existing in some format* - We selected the Argument Markup Language (AML)(Reed and Rowe, 2001) in the first instance which can be trivially converted to the Argument Interchange Format (AIF)(Chesnevar *et al.*, 2006). The choice of AML over AIF was due to the simplicity of AML and the option to 'upconvert' the data to AIF if necessary. We adopted the principle of only using a format that is as complicated as necessary for the primary representation of data. Subsequent reuse may, of course, require additional processing in order to address specific questions but that might be from AML into any

---

[12] *https://github.com/siwells/monkeypuzzle*

[13] *http://www.arg.dundee.ac.uk/AIFdb/search*

[14] *http://www.nature.com/sdata/*

number of other suitable formats so beginning with the simplest machine-readable format for data representation was a pragmatic choice. Whilst plain text would be simpler, the goal of eventually reusing this dataset as knowledge within intelligent agents means that the starting format should at least be in a markup language that is easy for both machines and humans to read and use.

Archived *Existing in a long-term durable format* – Each resource is stored as plain-text files using UTF-8 unicode character encoding[15]. By taking this approach we obviate the need to provide virtual machines or similar in order for basic data to be read or otherwise manipulated. The AML format itself has also existed for around fifteen years and is well documented. Because AML is an XML language, the DTD for the format itself is also included within the Git repository. Additionally, by using Git, any source code that manipulates the basic data can also be included within any distribution of the dataset, thus leading towards increasingly reproducible research.

Accessible *Available to others, other than the researcher* - The dataset has been publically released under permissive licensing in two primary locations. Consumers of the dataset who use the Git repository, particularly those using the Github releaase, can optionally request for their alterations or branches to be included in the core branch.

Comprehensible *Understandable by others* - All formats used to represent data within the repository are either international standards, e.g. UTF8 plain text, or else based on well documented and stable data formats, e.g. Portable Document Format (PDF), AML. Where possible information for recreating readers for the data has been included.

Discoverable *Can be indexed by a search engine* - Both the GitHub and Figshare sites are indexed by Google and other search engines which means that they can be serendipitously discovered.

Reproducible *Others can reproduce the experiment* - The STC dataset is not related to an experiment but where possible sufficient metadata for each resource that comprises the dataset has been recorded to enable the dataset to be rebuilt from first principles. In addition screenshots of original resources are included to provide additional contextual information about how the data was originally presented.

---

[15] *http://www.unicode.org/versions/Unicode6.0.0/*

Trusted *Provenance known* - Whilst the data has been collated by researchers who are publically active within both the argumentation and transport fields and employed by UK higher education institutions, this argument from expert opinion is not alone sufficient to ensure the provenance of the data. In addition the data that comprises the STC Dataset has been collated from public resources, which can all be verified by visiting the original URLs and the metadata is structured in such a way that retrieving the original data and reconstructing the dataset from first principles could be automated. However, because much data that is publicly available on the web is transient, is often never archived and may disappear, screenshots of the original data resources in context are included within the dataset.

Citable *Able to link to dataset and track citations* - The public instances of the dataset, at Github and at Figshare can be linked to, for example, using hyperlinks. Furthermore, objects stored at Figshare are allocated a DOI which makes the citation of specific snapshots a trivial and straightforward task.

Usable *Allow tools to run over the data* - The dataset is amenable to processing by any number of text-based, and other, tools, for example Python scripts. Furthermore, the licensing used specifically allows the dataset to be mined. The adoption of Git makes it a straightforward task for a researcher to gain access to their own complete copy of the dataset

Integrated *Upstream and downstream align* - The authors believe that their approach as outlined in the preceding steps means that all aspects of the best practices proposed in (de Waard, 2014) have been satisfied as well as can be expected.

In producing and publishing the STC dataset we have aimed to satisfy as many of these requirements as are practicable.

3.4 Summary & Preliminary Analysis of the Data

At time of writing the STC Dataset contains greater than sixty resources gathered from a range of public-facing websites.
Whilst building this collection a number of features were noted that recur across the sources of data; the use of population segmentation by transport type, the use of testimonials, the use of devil's advocate questions, use of non-argumentative behaviour change interventions, and the use of blog posts.

Population segmentation by transport type is when the messages are split into groups that are directed towards users, or potential users, of particular transport modes, for example, grouping together messages about cycling on one pages and messages about car-sharing on another. Frequently such approaches occur with a matching of both transport modes and life-style preferences. For example, messages about cycling are frequently associated with health benefits. Segmentation occurs frequently and is a feature of campaigns that (1) focus on increasing sustainable transport, and (2) target multiple transport modes.

Testimonials are frequently used to convey a more personable face associated with sustainable transport rather than just the plain facts. Usually these are either aspirational, of the form "I am fed up with x and desire to do y because z" or else are from people who have successfully altered their behaviour and are now advocating the change with messages that typically have the form "Since I started doing x I have seen benefits y because z".

Slightly less popular are the use of devil's advocate questions. A good example of this approach comes from the Walkit urban walking route planner site[16] which poses some tougher questions associated with walking in which the answers are more argumentative and presumptive, for example, "The tube or bus will run anyway, so does walking really save carbon?". It can be assumed that the authors of such communications wanted to aim for honesty, they don't have the answers, but also that in asking for the recipients position, they are prompting people to think, and it is self-reflection that is a critical but often overlooked aspect of persuasive communication.

Many communications also incorporate challenges that are typical of behaviour change theory, e.g. try out walking to work during the "Walk to Work Week"[17] but interleaving the challenge with reasons and argumentation for why the target of the communication should try it.

Finally, on platforms that are generally not associated directly with transport providers there are often blog posts that contain more in depth argumentation with respect to topics in sustainable transport. These provide more reasoned and better framed prose associated with transport behaviours that underpin the, more brief, advertising messages of many campaigns.

---

[16] *http://walkit.com/going- green/*

[17] *http://walkit.com/walking- to- work/walk- to- work- week/*

In addition to the aforementioned features, early analysis of individual statements within the corpus has also given rise to a number of preliminary findings. For example, as a rule the statements used are either positive or neutral in tone rather than negative. In addition nearly all statements are couched in terms of a shift of transport behaviour to a suggested mode. The only occasions when a mode is suggested as one that should be moved away from is when the mode is the car. This is of course excepting car shares, car-pooling, and taxis.

Of course the real test is to determine which presentational approaches and which arguments work best for a given individual. In the remainder of this paper we focus on technical preliminaries necessary to enable us to usefully tackle this question.

4. TOWARD SUSTAINABLE BEHAVIOUR CHANGE AT SCALE

The work thus far reported forms only our attempts to skirt the traditional AI knowledge acquisition bottleneck (Forsythe & Buchanan, 1993) by building a foundational knowledge base.

In this section we explore the roles of arguments and dialogue within models of behaviour change, before sketching a scenario of using mobile devices, behaviour management techniques, and arguments to effect large scale behaviour change.

4.1 Behaviour Change Models

There are two main approaches to behaviour change based upon the psychological models devised by Michie (Michie et al., 2011) and the digital behaviour change, or captology, models due to Fogg (Fogg, 2003).

Michie's approach may be applied in the presence of absence of technology and deals with habit forming behaviours whereas Fogg's approach is predicated on the mediating role of digital technology in providing a trigger for behaviour change.

Fogg's Captology is an approach to building persuasive technology that is based in a presumptive model of behaviour change. In the presence of *Motivation,* the desire to achieve something, *Ability,* the capacity to perform the behaviour, and a *Trigger* or facilitator, then changes in behaviour may occur, such that $M + A + T = $ *behaviour change is more likely to occur.* Additionally, the intersection of motivation and ability form an action line along which the behaviour occurs. The curve of the action line defines how a highly motivated person is able to perform hard tasks whilst a person with low motivation can perform easier tasks. There is a tension between motivation and ability such that in order to give a targeted behaviour change an increased chance of success, either the task must be made easier or else the participants motivation must be increased. Michie's approach use a similar but slightly different presumptive model in which behaviour change occurs in the presence of the correct levels of *Capability, Opportunity* and *Motivation* such that C+O+M → B. This approach is known as the or COM-B model of behaviour change (Michie et al., 2011).

There are many common elements between Michie and Fogg's approaches. Capabilities, and abilities are obviously similar, and opportunities and triggers are related, although in Michie's approach, opportunity relates more to the circumstances in which the behaviour occurs whereas Fogg's triggers relate more to a technological action, the trigger, which causes the behaviour to occur. Both models also recognise motivation as a significant attribute of the behaviour change process. Motivation should however be viewed as critical when addressing the question of how to effect lasting behaviour change rather than mere incidental or "triggered" behaviour change. In the presence of both capability and opportunity, or ability and a trigger if you are that way inclined, a person who is not motivated will likely not perform the behaviour anyway, and even if they do then they are unlikely to do so enough for the behaviour to become habitual. However a motivated person is both more likely to overcome questions of ability, but is also more likely to effect behaviour change that is sustained over the long term.

The problem with both the Fogg and the Michie approaches is that they are predicated on a sufficient level of motivation existing in order to facilitate the behaviour change but neither model provides practical and replicable mechanisms for increasing motivation and therefore increasing the likelihood of a resultant behaviour change. This suggests a role for argumentation, and particularly strategic argumentative dialogue, within behaviour change. At this point we should also note that a person may not even be sufficiently motivated enough to listen to the arguments or engage in dialogue. Behaviour change, particularly societal behaviour change that seeks to change the habits of large numbers of people does not occur in isolation but as a part of a multitude of both independent and complementary practices. For example, governmental policy and public information campaigns often raise awareness of issues, laws may be introduced to curb the worst excesses of specific behaviours, and people often recognise either individually or within their social groups that particular behaviours are problematic. It is therefore pragmatic to assume a form of social diffusion; social norms can and do shift as a result of both local and national political will, the actions of opinion formers, and issue awareness amongst affected communities. Against this background, whilst there will always be some recalcitrant groups, awareness is raised over time and people will either self-select to find out more and tackle problem behaviours or else will be forced, through personal crisis, to engage with some form of behaviour change. Whilst this process has historically been *ad hoc* behaviour change theories and particularly digital technology supported approaches seek to make behaviour change directed, predictable and repeatable.

However, it is unethical to seek to change a person's behaviour without their informed and active consent. Informed choices are made in the presence of sufficient knowledge. Dialogue is a good mechanism for increasing a person's knowledge about the context of their behaviour and argument is a good way to structure information related to the justification of positions. Thus the assumption can be made that for behaviour change to be sustained then a person must be able to make informed choice about their behaviour. Behaviour change must be a conscious and deliberative process, not a side-effect of rote, Skinner-box style mechanisms. We conjecture that informed consent, based upon increased knowledge and capacity for decision making, is likely a contributing factor to longer term behaviour change. By engaging in motivation building processes, which increase both a person's knowledge and their abilty to reason with that knowledge, a person is more likely to decide to change their behaviour, to do so for identifiable, enumerable, and justifiable reasons, and for the change to be lasting.

To achieve this we extend the Michie and Fogg approaches by incorporating argumentative interaction as both an important, motivation building, early step in behaviour change, but also as an important process sustaining activity. This step should occur early in the behaviour change process, before such techniques as goal setting and review, monitoring and feedback, comparison and ranking, or prompts and personalisation are applied. In summation, current behaviour change theories incorporate well developed models for managing behaviour but practical techniques for achieving behaviour change are less well developed and usually involve rudimentary forms of information-seeking and persuasive dialogue type interactions that are more highly developed within argumentation theory. Instead, the rich psychological models of behaviour change should be augmented with arguments aimed at increasing motivation, and dialogue to increase engagement and investment.

4.2 Scaling Behaviour Change using Mobile Devices

It should be noted that behaviour change deals with people in the real world whose decisions can often be characterised as messy, wrong, unjustifiable, or unreasonable; often behaviour exhibits all of these characteristics simultaneously.

Consequently, no single approach can be guaranteed to be successful in altering the behaviour of an individual. As a result multiple techniques must often be combined, for example, within the EU FP7 funded SUPERHUB project a variety of behaviour change techniques were deployed (Forbes et al., 2012), (Gabrielli et al., 2014), (Gabrielli et al., 2013a), (Gabrielli et al., 2013b) with the aim of effecting behaviour change in large overlapping groups of users and encouraging them towards more sustainable travel habits. One to one support has also been shown to vastly improve the success of behaviour change interventions, for example, the smoking cessation nurses used by NHS Scotland and the personal sponsor systems used by twelve step programs, but at great expense in terms of time and/or money. However when the scope of a behaviour change intervention encompasses a city or nation, then such levels of support become untenable and scaling up successful behaviour change interventions becomes a massively expensive proposition.

The advent of, and increasing penetration within society of mobile digital devices suggests one way to target behaviours and support behaviour change at scale. By incorporating intelligent software digital agents into mobile apps for popular phones, behaviour change support can be brought to the masses. However, if a person does not wish to install such software then they risk being missed. A solution to this was also proposed in the SUPERHUB project (Forbes et al., 2012), instead of getting people to opt directly for behaviour change software, they were offered journey management software that was of standalone benefit and which could better solve their personal journey planning problems. However this software also incorporated functionality that provided personalised travel recommendations, it learnt about the individual user and their travel habits, as well as incorporating some behaviour change techniques to attempt to influence the uptake of more sustainable travel choices. Unfortunately this system did not incorporate argumentative capability aimed at increasing the motivation of users to engage with the behaviour change functionality so an opportunity was missed. Whilst SUPERHUB had missed opportunities it did demonstrate how personalisation, behaviour change interventions, and gamified interactions (Wells et al., 2014) could be combined with more mundane journey planning functionality to form a compelling if incomplete behaviour management system.

5. FURTHER WORK & DISCUSSION

There are a number of directions for further work.

One of the drawbacks of analysing existing persuasive communication from sustainable transport campaigns is that there is no data about how the campaigns performed and there is no data that sheds light on whether a given communication campaign had objective effects on the behaviour of individuals who were exposed to it. Similarly there is no data about how individuals react to the individual messages and arguments that are communicated. Additionally there is no data about how the persuasive weight of individual communications is altered by the way in which they are presented. We plan therefore, subject to the vagaries of research funding, to engage in user centred evaluations of the arguments within the STC dataset. It is assumed that individuals will respond differently and that categorising individual, for example using Annable-style user segmentation techniques (Annable, 2005) which build on an expand theory of planned behaviour (Ajzen, 1991) to identify and group different kinds of user and correlating those segments against more and less successful communications from the dataset will provide a mechanism for strategic interaction, enabling more successful arguments to be deployed as required to maximise the motivation building phase of behaviour change.

6. CONCLUSION

We have presented some preliminary results of a survey of persuasive communication from public websites within the sustainable transport domain alongside an approach to storing, sharing and reusing argumentation corpora in a manner that is current best practise.

This dataset however is only a preliminary step in a more ambitious plan for applying argumentation concepts within sustainable transport communication. To this end the major contribution of this research has been to underpin existing motivational and behaviour change communications within the sustainable transport domain with elements of a solid argumentation theoretic foundation.

REFERENCES

Ajzen, I. (1991). The theory of planned behavior. *Organizational Behavior and Human Decision Processes*, *50*(2), 179–211.

Annable, J. (2005). 'Complacent car addicts' or 'aspiring environmentalists'? Identifying travel behaviour segments using attitude theory. *Transport Policy*, *12*, 65–78.

Chesnevar, C., McGinnis, J., Modgil, S., Rahwan, I., Reed, C., Simari, G., South, M., Vreeswijk, G., & Willmott, S. (2006) Towards an argument interchange format. *Knowledge Engineering Review*, *21*(4), 293–316.

Davies, N. (2012). What are the ingredients of successful travel behaviour campaigns? *Transport Policy*, *24*, 19–29.

de Waard, A. (2014). Ten habits of highly effective data. In *Proceedings of the AAAI Discovery Informatics Workshop*.

Fogg, B. J. (2003). *Persuasive Technology: Using Computers to Change What We Think and Do* (Interactive Technologies). Morgan Kaufmann Publishers.

Forbes, P. J., Wells, S., Masthoff, J., & Nguyen, H. (2012). Superhub: Integrating behaviour change theories into a sustainable urban-mobility platform. In *BCS HCI Workshop on Using Technology to Facilitate Behaviour Change and Support Healthy, Sustainable Living*.

Forsythe, D. E., & Buchanan, B. G. (1993). Knowledge acquisition for expert systems: some pitfalls and suggestions. Chapter in *Readings in knowledge acquisition and learning: automating the construction and improvement of expert systems* (pp. 117–124). Morgan Kaufmann Publishers Inc.

Gabrielli, S., Forbes, P., Jylha, A., Wells, S., Siiren, M., Hemminki, S., Nurmi, P., Maimone, R., Masthoff, J., & Jaccuci, G. (2014) Design challenges in motivating change for sustainable urban mobility. *Computers in Human Behavior*, *41*, 416–423.

Gabrielli, S., Maimone, R., Forbes, P., Masthoff, J., Wells, S., Primerano, L., Pompa, M., & Haverinen, L. (2013). Co-designing motivational features for sustainable urban mobility. In *Proceedings of the ACM SIG-CHI Conference on Human Factors in Computing Systems (CHI 2013)*.

Gabrielli, S., Maimone, R., Forbes, P., & Wells, S. (2013). Exploring change strategies for sustainable urban mobility. In *Proceedings of the Designing Social Media for Change Workshop*.

Howarth, C., & Riley, T. (2012). A behavioural perspective on the relationship between transport and climate change. Chapter in *Transport and Climate Change, volume 2*, (pp. 261–286). Emerald Group Publishing Limited.

Lanzendorff, M. (2014). The black spot of policy evaluations: Social change and mobility 2.0. In *Proceedings of the Royal Geographic Society Annual International Conference*.

Michie, S., van Stralen, M. M., & West, R. (2011). The behaviour change wheel. *Implementation Science*, *6*(42).

Pangbourne, K., & Masthoff, J. (2015). The message is the medium - influencing travel behaviour change using persuasive techologies. In *Universities Transport Studies Group annual conference*.

Reed, C., & Rowe, G. (2001). Araucaria: Software for puzzles in argument diagramming and xml. *Technical report, University Of Dundee*.

Reed, C., & Wells, S. (2007) Dialogical argument as an interface to complex debates. *IEEE Intelligent Systems Journal: Special Issue on Argumentation Technology*, *22*(6), 60–65.

Rocci, A. (2012). Encouraging changes in travel behaviour towards more sustainable mobility patterns: the role of information. Chapter in *Transition towards sustainable mobility: the role of instruments, individuals and institutions*. Ashgate.

Wells, S. (2012). Collation of formal dialectical games from the literature. *Technical Report UOD-SOC-2012-001*, University of Dundee.

Wells, S., Kotkanen, H., Schlafli, M., Gabrielli, S., Masthoff, J., Jylha, A., & Forbes, P. (2014). Towards an applied gamification model for tracking, managing, & encouraging sustainable travel behaviours. *EAI Endorsed Transactions on Ambient Systems*, *14*(4).

Wells, S., & Reed, C. (2012). A domain specific language for describing diverse systems of dialogue. *Journal of Applied Logic*, *10*(4), 309–329.

Commentary on Wells and Pangbourne's Using Argumentation within Sustainable Transport Communication

MARK AAKHUS

School of Communication & Information, Rutgers University, USA
aakhus@rutgers.edu

1. INTRODUCTION

The project reported by Wells and Pangbourne offers an important case for reflecting on the relationship between design and argumentation. Their project seeks to integrate computation and argumentation to develop means to cultivate behavioral change in transportation behavior that in turn fosters sustainability. It is a project with big aspirations, which is appropriate because transportation and sustainability are big, wicked problems that require commensurate ambition. Their report focuses on the project aims and plans, and thus offers insights into some kinds of contributions a design stance toward argumentation research could deliver. A design stance seeks knowledge through interventions and inventions, which is different from but complementary to empirical and critical investigations more common in argumentation theory and research (Jackson, 2015). These comments first highlight key design elements of the project and then reflect on design for argumentation and ways in which design projects, such as the one proposed by Wells and Pangbourne, could deliver for argumentation theory and practice in big ways.

2. ARGUMENTATION AS MEANS FOR CHANGE

The project describes a plan for inventing a computational means to intervene on the way people think and talk about their transportation behaviors. The aim was to build a system that could support how individuals develop their reasoning about transportation choices and actions so as to decrease carbon emissions and traffic congestion.

One key design element of the project was the use of argumentation theory to assess the context of intervention for constraints and opportunities. Within the larger context of transportation and sustainability, the project locates a problem and opportunity in supporting behaviour change. Prior approaches focused solely on the development of persuasive messages, while argumentation theory pointed to the need for external support of change. The project planners drew upon the insights about dialogue, reasoning, and knowledge representations as basis for designing behavioural change support. From argumentation theory, the planners also developed a design hypothesis that the intervention should support behaviour change as informed consent and grounding in justifications.

A second key design element was the use of argumentation theory to develop and shape the communicative materials. On one hand this involved collecting examples of how transportation choices and behaviour are reasoned about to identify the variety of justifications people use. That corpus required coding to differentiate the messages and labelling to make the examples searchable and retrievable by a computer. This resulted in an initial corpus that provided a knowledge base about how people reason about transportation choices and behaviour.

A third key design element was the proposed personalization of the intervention. The intervention sought to make the rationales for sustainable transportation choice part of the journey management software people use on the computers and smartphone applications. Rather than requiring people to choose to engage the behavioural change intervention, the proposal was to more subtly incorporate it into an ongoing, more routine technology use.

3. DESIGN AS A WAY OF KNOWING

The sustainable transport communication project reveals a puzzling dimension of designing communication: the object of design is not simply the system, technology, or language-use, but communication. As Lyytinen (1985, p. 61) put it long ago: "The very idea of an information system, however, is to provide a means and an environment for human communication." The aim of interventions and inventions is to make communication possible that was previously difficult, impossible, or unimagined (Aakhus, 2007). The three key design elements described above point to the relationship between the design of digital and computational artifacts and the construction of communicative contexts. An essential aim of the project is for users to experience interaction in making transportation choices.

Attempts to invent and intervene require generating hypothesis about how the arrangement of interaction and language-use can foster some qualities of communication while downplaying others. The transport project had developed particular ideas about how communication works and how it ought to work by using argumentation theory to critique given theories of change and propose new ways forward. The artifacts and systems developed embodied hypotheses about communication that could be tested against the realities of the setting and whether the context could be shaped or disciplined in a particular way. Presumably the artifacts and systems will be revised and developed based on how they work. It will be in the iterative development of artifacts/system and procedures of implementation that knowledge about realizing contexts that support new habit formation will be built.

What is noteworthy about design is that it is a way of knowing. Knowledge is built in, and expressed through, interventions and inventions. While the transportation communication project report focused on its direct design task, it also suggests some further ways in which design can contribute to building argumentation theory and practice.

First, the project suggests a bottom up method for changing practice by changing argumentation. Argumentation research is typically interested in describing the broad discourses and micro-discourse practices people engage in or criticizing those uses. The proposed collection, storage, and re-use of transportation argumentation illustrates a both-and strategy regarding broad discourses and micro-discourse practices as the corpus reflects both. By making the corpus a resource for individuals to invent rationales for new behaviour, the project suggests how information infrastructures are generative of action and communication and not merely repositories. This generativity is made possible because both macro and micro aspects of discourse are available and descriptive and critical uses of that discourse are made possible.

Second, the project suggests how theories of reasoning and persuasion are built into the information and communication technologies and infrastructures. The proposed expansion of the rational model of behavioral change by incorporating common insights from argumentation theory changes the technology from persuasive to reflective. The approach highlights the possibility for participants to engage the very communicative conditions that make their current behaviour rational. The building of the corpus and its means of use scaffolds participation that could become reflective of the argumentative patterns and structures in the corpus to the point that they participate in constructing new patterns and structures for reasoning about the domain represented by the corpus.

Third, the project suggests that argumentation research could be in for a big change if it can further embrace architectures for collecting, storing, retrieving, and sharing that enable large research projects. The current state of the art in scholarly exchange are conferences where participants report on work they have completed elsewhere at another time. The format facilitates a kind of interaction and exchange that is very important but that has not leveraged the collective intelligence of the broad field. By and large, current scholarly exchange preserves the various camps that have developed around particular theories and modes of analysis. Architectures, such as illustrated in the sustainable transport project, can enable entirely new ways of scholarly engagement. For instance, by building large repositories of arguments it becomes possible to test the implications of the perspectives of various camps against each other relative to grand challenges. The perspectives and methodological strategies for description, analysis, and invention of the various camps could be compared relative to common data and common goals. Moreover, it would be possible to discover the actual unique contribution of each perspective relative to the common contributions across perspectives.

REFERENCES

Aakhus, M. (2007). Communication as design. *Communication Monographs*, *74*(1), 112–117. doi:10.1080/03637750701196383.

Jackson, S. (2015). Design thinking in argumentation theory and practice. *Argumentation*, *29*(3), 243–263. doi:10.1007/s10503-015-9353-7.

Lyytinen, K. J. (1985). Implications of theories of language for information systems. *MIS Quartely*, *9*(1), 61–74.

35

Giving Reasons *Pro Et Contra* as a Debiasing Technique in Legal Decision Making

FRANK ZENKER
Department of Philosophy & Cognitive Science, Lund University, Sweden
frank.zenker@fil.lu.se

CHRISTIAN DAHLMAN
Law Faculty, Lund University, Sweden
christian.dahlman@jur.lu.se

RASMUS BÅÅTH
Department of Philosophy & Cognitive Science, Lund University, Sweden
rasmus.baath@lucs.lu.se

FARHAN SARWAR
Department of Psychology, Lund University, Sweden
farhan.sarwar@psy.lu.se

We report on the results of deploying the debiasing technique "giving reasons pro et contra" among professional judges at Swedish municipal courts (n=239). Experimental participants assessed the relevance of an eyewitness's previous conviction to his credibility in the present case. Results are compared to data from lay judges (n=372). The technique produced a small positive debiasing effect in the sample of Swedish judges, while the effect was negative among lay judges.

KEYWORDS: debiasing technique, heuristics and biases, legal decision-making, prior conviction, witness scenario

1. INTRODUCTION

How to improve decisions is a pertinent question whenever judgments are unavoidable. The decisions that judges and juries must reach virtually every day provide a case in point, a fortiori when these bear strongly on the fates of individual and collective agents. Since biased reasoning and decision making is (rightly) thought to occur also in legal contexts (see, e.g., Langevoort, 1998 for a review; cf. Mitchell, 2002), no argument seems required that it ought to be reduced. Rather, empirical knowledge is wanted how reliable reductions may be achieved.

Professional judges tend to assume of themselves, firstly, that non-jurist decision makers regularly err in assessing the relevance of legal evidence; and, secondly, that judges reason in ways that reliably avoid such error. For some five decades, however, empirical research in the heuristics and biases tradition has supported the first assumption also for judges. Relevance-assessments may therefore be assumed to differ widely between intuitive and deliberative modes of reasoning and decision making, both between and within (groups of) agents.

Our research focuses on the second assumption, above. It addresses four related questions through controlled experimentation and interpretative analysis:

(1) What is the accuracy-difference between judges' and laypersons' assessments of the relevance of legal evidence (or: how much better are judges at activating 'system 2' in such assessments)?
(2) Do relevance-assessments improve across both groups subsequent to being instructed to deploy a debiasing technique?
(3) What is an optimal allocation between debiasing techniques and the bias(es) thus mitigated?
(4) How can debiasing techniques be improved?

This paper reports empirical results regarding the first two questions forthcoming from a pilot-study with a sample of professional Swedish judges and a sample of Swedish lay-judges (*nämdeman*). Experimental participants were asked to assess aspects of a mock legal-case that had been manipulated to contain bias-triggering information. In the experimental subgroup, the mock case was followed by explicit instructions "to give reasons pro/con"; in the control group it was not.

The purpose of this experimental set-up is to assess the positive, negative, or neutral effect(-size) of instructions to deploy a debiasing technique in a hypothetical legal decision making scenario vis-à-vis an established cognitive bias, on one hand, and a debiasing method, on the other—where the latter may count as far less established. The relevant bias is a "devil effect", insofar as an information-item about a person is of exaggerated importance in gauging her general credibility (see below). Such research contributes to assessing the average effectiveness of a debiasing technique, itself an instance of prescriptive ameliorative intervention, if and insofar as a technique constitutes an efficient cause whose effect shows as an improved, or perfect, alignment between a normative standard and the decision outcome.

Following a brief introduction to biases and debiasing (Sect. 2), we present the method (3) and main results (4), offer a discussion (5), and close with brief conclusions (6).

2. BIASES AND DEBIASING

Biases are generally considered *latent*, that is, subjects tend to be unaware of them. By definition, a technique does debias when its deployment brings forth a decision that (i) differs markedly from one brought forth by deploying a heuristics, and (ii) also complies with a normative standard, e.g., as set forth by the law.

Broadly speaking, what authors such as Kahneman & Tversky (1982; 1996), or Kahneman (2011) call *biases*, philosophers and scholars of law associate with the *fallacies*. The latter fields share a tradition in Aristotelian scholarship, specifically the critiques of the Sophistic mode of audience persuasion. The 16th century Francis Bacon's delivering his *idolatry* or the 17th century John Locke naming of a range of fallacies fronted by "ad" (e.g., *ad hominem*) have continued this tradition into the modern age. Since Hamblin (1970), fallacies are standard fare in speech communication, rhetoric, and argumentation studies, among others. Around that time, moreover, the interpretation of fallacies as *reasoning errors* became separated from viewing fallacies as *problematic arguments* (e.g., van Eemeren & Grootendorst, 1984). Most psychologist and cognitive scientists, however, continue to strictly endorse the first interpretation.

Despite a vast number of empirical studies confirming the assumed operation of such biases for various groups of subjects, few studies pertain to contexts of legal decision making. Exceptions are, among others, Guthrie et al.'s (2007) study of anchoring, hindsight bias and base rate neglect, and English et al.'s (2006) study of the anchoring effect. Both particularly support that biases also influence legal decision making (for further references see Zenker et al., 2015).

Extant research moreover strongly suggests that humans are especially challenged in the application of debiasing methods, and more so in self-application (Pronin & Kugler, 2007; Pronin, Lin & Ross, 2002; Willingham, 2007; Kahneman, 2011; Kenyon, 2014). Self-assessment for biased thinking generally counts as a difficult cognitive ability to master; the primary challenge is the suspension of latency. But extant research (e.g., Guthrie et al., 2007; Irwin et al., 2010) also identifies debiasing techniques for legal decision making contexts, including the following.

Some of their underlying principles are already incorporated into procedural and substantial law. Debiasing effects thus brought should hence produce decisions that fall within the law.

- *Accountability*: legal decisions are subject to review by higher courts (Arkes, 1991).
- *Devil's Advocate*: Reminding subjects of the hypothetical possibility of the opposite standpoint (Lord et al., 1984; Mussweiler et al., 2000).
- *Giving Reasons* (Larrick, 2004, p. 323; Hodgkinson, 1999; Mumma & Wilson, 1995; Koriat et al., 1980).
- *Censorship*: When evidence counts as inadmissible, this may avoid biases triggered by such evidence.
- *Reducing Discretion*: Formulating legal norms that leave less room for a judge's interpretation (e.g., explicit checklists, or a pre-set damage amount).

An overview of extant research on debiasing in legal contexts including key methodological issues and additional references is provided in Zenker, Dahlman, and Sarwar (2015). As is argued there, successful debiasing techniques must simultaneously address aspects of cognition, motivation, and technology. They need to raise the agent's awareness of the bias (cognition) in ways that sustain or increase her impetus to avoid biased reasoning (motivation), while providing information that agents can in fact deploy to correct extant reasoning (technology).

Empirically testing a debiasing technique vis-à-vis a bias-triggering mock case serves to (i) empirically assess the extent to which a hypothetical (yet realistic) legal decision can be subject to biases, if and insofar as judges' and laypersons' hypothetical decisions "in the lab" are representative of those "outside the lab." Research further serves to (ii) estimate the potential of such instructions at mitigating biases, if and insofar as mitigation in the lab indicates that the same succeeds outside the lab. Finally, research eventually yields (iii) information on the optimal point at, and the optimal manner in, which decision makers would reasonably want to deploy a debiasing technique.

3. METHOD

By regular mail, *all* 667 professional judges at municipal courts in Sweden were asked to answer a pen-and-paper questionnaire that sought to assess whether, and if so to what extent, a previous conviction affects a witness's credibility. By way of a court's chief judge, moreover, 738 lay judges were asked to assess what one may generally call the "prior conviction relevance" in the following mock case.

Sebastian P is charged for assault. According to the prosecutor's charge, Sebastian P assaulted Victor A, on July 20, 2012 at 23:30 outside a cinema in central Malmö, by repeated blows to the head. Sebastian P testifies that he acted in self-defense and denies the charges. One of the witnesses in the trial is Tony T, who was at the site on that particular evening. During the examination of the witness Tony T, it emerges that he had recently served a two-year prison sentence for illegal possession of weapons and arms trafficking.

Which of the following best describes your assessment? (Tick one option only)

- Tony T's previous conviction for illegal possession of weapons and arms trafficking affects the assessment of his credibility as a witness in the current trial. When various factors are weighed, the fact that he had previously been convicted of illegal possession of weapons and arms trafficking is *strongly* to his disadvantage.

- Tony T's previous conviction for illegal possession of weapons and arms trafficking affects the assessment of his credibility as a witness in the current trial. When various factors are weighed, the fact that he had previously been convicted of illegal possession of weapons and arms trafficking is *clearly* to his disadvantage.

- Tony T's previous conviction for illegal possession of weapons and arms trafficking affects the assessment of his credibility as a witness in the current trial. When various factors are weighed, the fact that he had previously been convicted of illegal possession of weapons and arms trafficking is *somewhat* to his disadvantage.

- Tony T's previous conviction for illegal possession of weapons and arms trafficking *does not affect* the assessment of his credibility as a witness in the current trial.

In the experimental groups of both samples (professional and lay judges)—after the scenario, but before the central question and the four alternative answers were presented—participants were asked to state reasons why Tony T's convictions would affect his credibility as a witness in the present trial *and* to state reasons why his convictions would not affect his credibility in the present trial. No such instructions were included in the questionnaire given to control group-participants.

Of the professional judges, 40% returned the questionnaire (*n*=239), where 143 participants, i.e., 59.8% of the sample, had *not* received instruction to deploy any debiasing technique before answering the case (control group), while 96 participants, i.e., 40.2% of the sample, were instructed to state reasons for their assessment (experimental group; later referred to as "debias group"). Among lay judges, 52% returned the questionnaire (*n*=372), of which 171, i.e., 45.9%, belonged to the experimental group and 201, i.e., 54,1%, to the control group. In both samples, the response rate is unbalanced since participants were at liberty to return the questionnaire; they did not receive financial or other compensation for participating in this study.

Typical responses in both samples included the following *pro/con* reasons:

Prior conviction is relevant (pro)
- Tony T. has no barrier to breaking the law
- Tony T. may have an interest (e.g., revenge)
- Tony T. has reduced "citizenship-capital"
- Tony T. has a pro-attitude to violence

Not relevant (con)
- Unrelated event/circumstances
- No evidence that prior conviction matters
- Prior conviction *should* be irrelevant
- Current testimony occurs under oath

Prior to deploying the questionnaire, we did not formulate a point-hypothesis to code a normatively correct response. Rather, we assumed that obtaining differences between the experimental and the control group suggests that "giving reasons *pro et contra*" has a debiasing effect provided that participants in this group do on average display a lower assessment of the prior conviction relevance.

4. RESULTS

The effect of deploying the debiasing technique "giving reasons *pro et contra*" was *prima facie* miniscule. First looking at professional judges, fewer participants in the debias group than in the control group took the witness's previous conviction to be *clearly* or *strongly* to his disadvantage in the present case. Expressed in numbers, these were six and respectively one *vs.* zero participants (4.2% and 0.7% of the sample). This can provide at best *some* reason to believe that the debiasing technique had an ameliorating effect on judges. Moreover, 28 judges in the control group (19.6% of the judges in the control group) register as finding the witness's prior conviction to be *somewhat* negatively relevant. Finally, 20 judges in the experimental group (12.8% of the judges in the control group) so register *despite* a debiasing technique being deployed.

Turning now to lay judges, by contrast, hardly any noteworthy differences arose between the control and the experimental group: 7% and 8% of the total number of lay judges found the prior conviction to be *clearly* or, respectively, *strongly relevant*; 30% in each group found the conviction to be *somewhat relevant*; 61% and 63%, respectively, found the prior conviction to be *not relevant*. Table 1 and Fig. 1 give the full results of the questionnaire.

Responses were coded on a four point ordinal scale (as *not relevant, somewhat, clearly* and *strongly to the witness's disadvantage*; see Table 1). In order to investigate the difference between the four groups, that is, the control and experimental groups, each consisting of either professional or lay judges data was then subjected to an ordered probit analysis.[1]

		not relevant	somewhat relevant	clearly relevant	strongly relevant	N
Judges	Control	108 (76%)	28 (20%)	6 (4%)	1 (1%)	141
	Debias	76 (79%)	20 (21%)	0 (0%)	0 (0%)	96
	Total	*184 (77%)*	*48 (20%)*	*6 (3 %)*	*1 (0 %)*	*239*
Lay Judges	Control	126 (63%)	60 (30%)	12 (6%)	3 (1%)	201
	Debias	105 (61%)	52 (30%)	11 (6%)	2 (2%)	171
	Total	*231 (62%)*	*112 (30%)*	*23 (6%)*	*5 (2%)*	*372*

Table 1. Responses from Swedish judges and lay judges (*n*=number of subjects)

---

[1] See Daykin and Moffat (2002) for paradigmatic applications of ordered probit analysis and its advantages over far-better known, but also less well-suited, linear regression analyses. For instance, ordered probit analysis is not open to the objection that the distances between any two ordinal data points are implicitly treated as being equal. The probit analysis was done in the R statistical environment using the polr function in the MASS package (Venables & Ripley, 2002).

This analysis assumes that underlying the ordinal scale, on which participants' responses are measured, is a continuous random variable representing participants' assessment of prior-conviction relevance (PCR). The value of this latent variable has no direct interpretation but is a relative measure of PCR, where a higher value implies that a prior conviction is deemed more relevant. Crucial for the following statistical analysis, the expected value of the latent variable can be taken as a measure of the general sentiment of the group and can thus be used in comparing the groups.

The distributional parameters of the latent variable were gauged through maximum likelihood estimation, yielding the parameter estimates under which the ordered probit model is most likely to generate the observed data in Table 1.

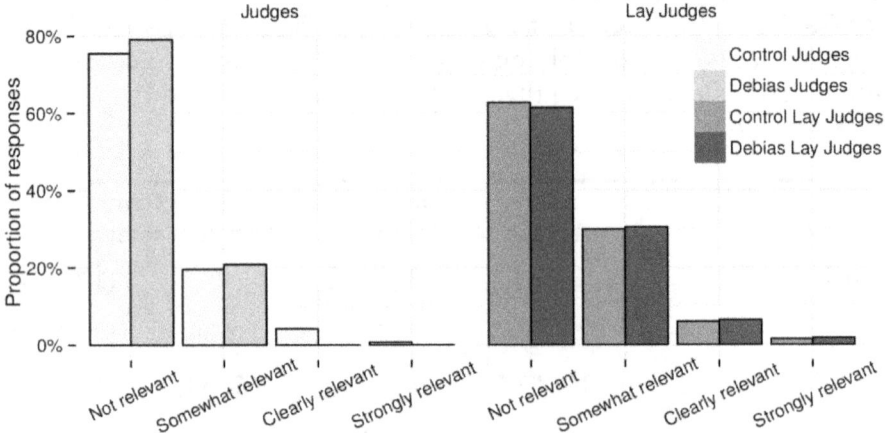

Fig. 1 Proportion of responses from Swedish judges and lay judges

In virtue of being maximally consistent with the original data, the hypothetical model may be interpreted as the *most probable continuous distribution of the latent PCR-variable* among respondents. In this sense, the hypothetical model can be viewed to have probably generated the original data. The shaded curves in Figure 2 show the maximum likelihood estimates of the latent PCR-variable among judges and lay judges in the control and the experimental group. The figure is divided into four regions corresponding to the four possible responses in the survey. The percentage of the area under the curve within each region corresponds to the model's estimate of the probability that a member of these groups produces the corresponding survey response. The dashed vertical lines mark the expected values of the latent PCR-variables, here taken as a measure of the general sentiment of groups.

Comparing panels A-B and C-D in Fig. 2, the displacement of the expected values of the PCR-variables indicates the impact of the debiasing intervention. While there is a visible difference in the general assessment of PCR between judges in the experimental and judges in the control group, there is hardly any difference between the lay judges in the experimental group and lay judges in the control group. But there was nevertheless a substantial overall difference between judges and lay judges: the former judged the prior conviction to be less relevant than the latter.

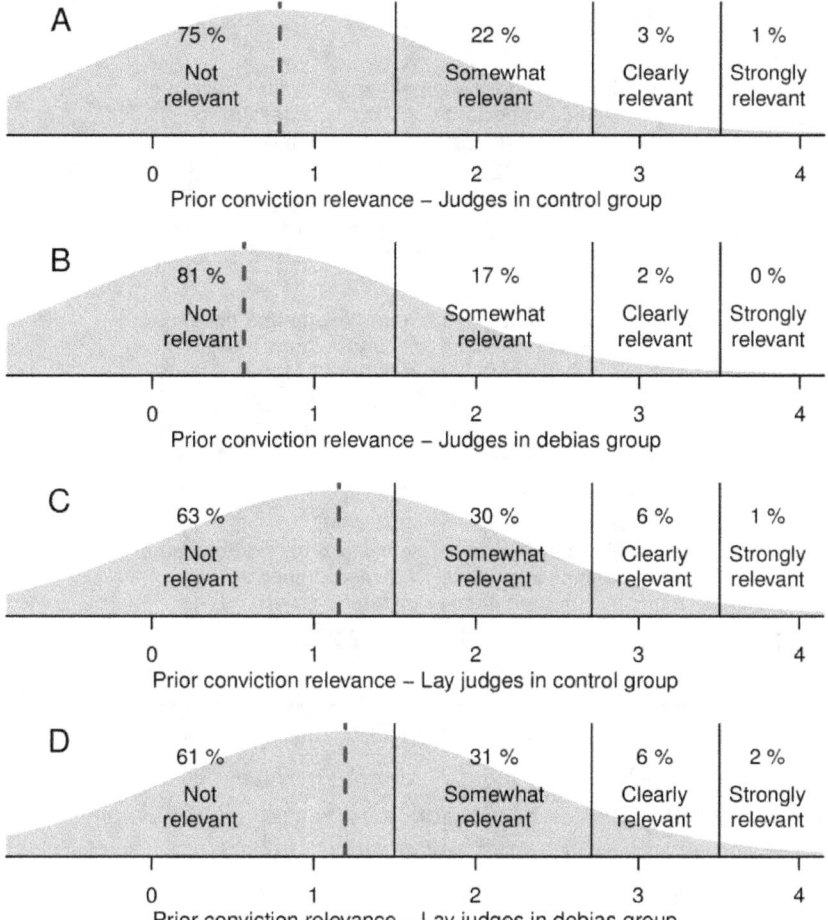

Figure 2. Probability distribution of the latent "prior conviction relevance"-variable for judges and lay judges in the debias and the control groups.

A Bayesian analysis was performed to gauge the uncertainty in the estimates from the ordered probit analysis, and to quantify whether the joint data from judges and lay judges in the experimental and the control group support, or undermine, the hypothesis that the debiasing technique "giving reasons pro/con" had an ameliorating effect.[2]

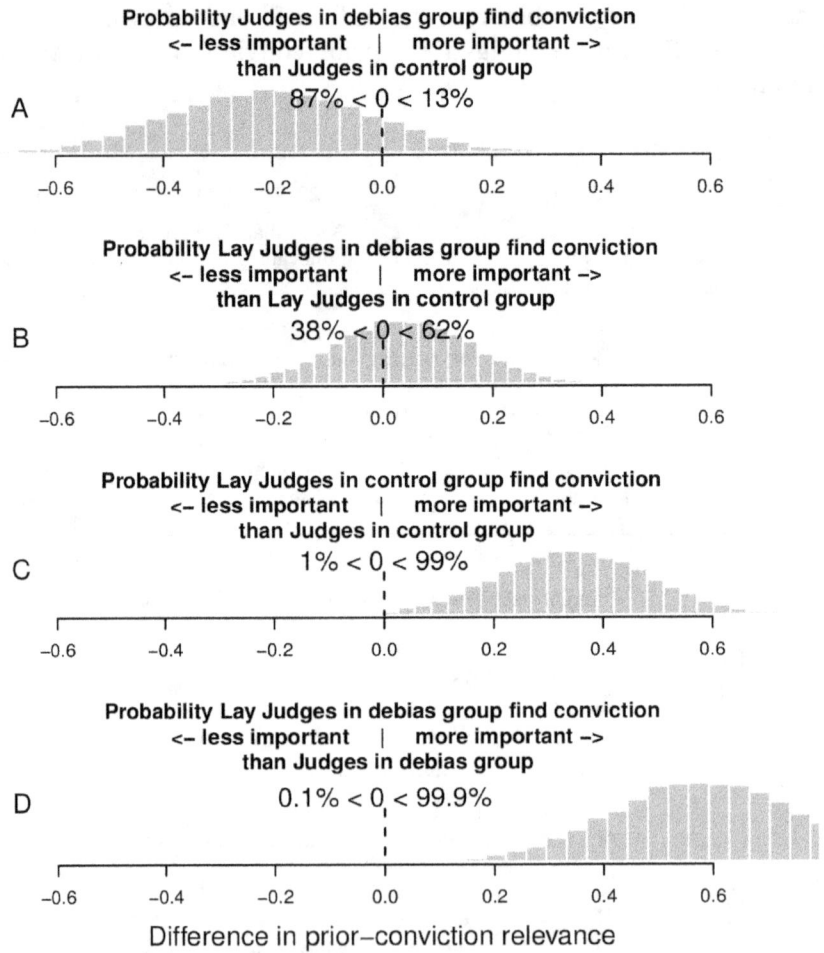

Fig. 3. Distribution of probabilities given model and evidence from professional and lay judges

---

[2] The analysis was performed in the R statistical environment using the MCMCoprobit function in the MCMCpack package (Andrew et al., 2011). The default priors of the MCMCoprobit function was used, which were non-informative uniform priors over all parameters.

Figure 3 shows the probable difference in the expected values of the PCR-variable (marked by a dashed line; Fig. 2) between all four groups. Given model and data, there is a 87% probability that judges in the experimental find the prior conviction less relevant (Fig. 3, panel A), compared to a 38% probability that lay judges in the experimental group find the prior conviction less relevant (Fig. 3, panel B).

This may be interpreted as *rather weak positive* evidence that deploying the relevant debiasing technique has a debiasing effect among judges, but not among lay judges. Moreover, comparing judges and lay judges in the control group (Fig. 3, panel C) and the debias group (Fig. 3, panel D) shows a probability larger than 99%—which may be interpreted as *very strong* evidence—that in both the control and in the experimental condition lay-judges assign a higher prior conviction relevance than judges, with evidence from the experimental condition registering slightly stronger yet. This, in fact, amounts to having observed an interaction of the professional status with the assessment of prior conviction relevance.

4. DISCUSSION

In the experimental data, strong evidence for a mitigating effect of the debiasing method "stating reasons pro/con" onto participants' responses has *not* been forthcoming. Rather, the study found an 87.1% probability for a mitigating effect. This can at best count as weak evidence. In the mock case, lay judges did overall assign a *greater* weight to the previous conviction of the witness than professional judges. Moreover—and perhaps disturbingly—compared to the relevant control group lay judges in the experimental group displayed an *increased* mean score.

Results are broadly negative in the sense that the "Tony T" mock case failed to trigger a *strong* bias among professional or lay judges. By and large, professional judges merely assigned *some* weight to the previous conviction, while lay judges assigned a greater weight. The debiasing technique "stating reasons *pro et contra*" in other words failed to meet with a strongly biased sample of judges and lay judges. The technique nevertheless appears to succeed in "taking the edge off," as it were. After all, compared to the relevant control group, the number of extreme judgements in the experimental group of professional judges is reduced. It stands to reason, of course, that "removing" but one extreme judgement through a debiasing intervention does already constitute an important and desirable outcome. This nonetheless remains a very small effect. And as the debiasing technique met with a comparatively more biased sample of lay judges, its deployment not only failed to mitigate the bias; rather, it slightly worsened the judgement compared to the control group of lay judges. But also this result remains statistically insignificant, and so cannot easily be accounted for as an effect of deploying the technique.

To address the objection that additional data should have been collected in order to assess whether a statistically significant debiasing-effect would after all have been observed, consider that the sample of Swedish judges in the present study ($n=239$) represents no less than 40% of the relevant population(!). To increase this number would no doubt present greater practical difficulties. It remains correct, of course, that small experimental effects must always be confronted with large data-samples. But for the small effect here reported to *potentially* register as statistically significant does necessarily require a sample-size that exceeds the size of the relevant population! This fact hence entails that there might be biases whose presence, and debiasing techniques whose effect, can principally *not* be demonstrated by obtaining strong evidence for a difference between the control and the experimental group whenever the effect is too small to register as significant even against the size of the relevant population. For this reason, the "need more data"-objection is particularly weak in the present context.

Demonstrating the effectiveness of a debiasing technique at conventionally accepted levels of significance could instead be served by maximizing the difference between participants' ratings in the control and the experimental group. In the present study, as we saw, both groups displayed rather low degrees of biasedness. It therefore remains a challenge for future research to create experimental set-ups that induce stronger biases. In view of the idea that already a small ameliorating effect, if it is real, should be viewed as a desirable outcome of deploying a debiasing technique, we suggest that it can be reasonable to accept weaker forms of evidential support, rather than inferring that the debiasing technique was probably ineffective. Since this stance is unlikely to meet with wide acceptance, however, the key-task would remain to induce a stronger bias.

5. CONCLUSION

Among Swedish judges at municipal courts, the "Tony T" mock case failed to meet with "sufficiently biased" respondents, since few assigned a great(er) weight to the witness's prior conviction regarding his credibility in the present case. The debiasing technique "giving reasons *pro et contra*" could thus at best produce a small effect—too small to count as strong evidence relative to the sample or even the relevant population. Rather than inferring that the technique probably had no effect, however, we submit these results as weak positive evidence in favor of the effectiveness of this debiasing technique.

As we also saw, results differed—yet in the normatively "wrong" direction—when the same technique was deployed vis-à-vis the same mock case among lay judges, who seem to have constituted a comparatively more biased sample than the professional judges. The debiasing technique had a weak *adverse* effect on lay judges; subsequent to deploying it, the latter assigned a slightly increased weight to the relevance of previous conviction. As we have stressed, however, this interpretation is subject to caveats as the effect remained too small.

Among all measures taken, we obtained very strong evidence merely for a relation between the profession and the level of biasedness, there being a probability greater than 99% that lay judges were more biased than professional judges. To test the effectiveness of debiasing methods against standard statistical assumptions, future studies seeking to produce strong(er) positive evidence are challenged to find ways of triggering strong(er) biases.

ACKNOWLEDGEMENTS: We thank audience members at the First European Conference on Argumentation, 9-12 June 2015, Lisbon, Portugal, for discussion and Fabrizio Macagno for his commentary. Research was funded by the Ragnar Söderberg Foundation. Rasmus Bååth acknowledges funding through Swedish Research Council grant number 349-2007-8695.

REFERENCES

Arkes, H.R. (1991). Costs and benefits of judgement errors: Implications for debiasing. *Psychological Bulletin, 110*, 486–498.
Daykin, A.R., & Moffat, P.G. (2002). Analyzing ordered responses: a review of the ordered probit model. *Understanding Statistics, 1*(3), 157–166.
Eemeren, F. H. van, & Grootendorst, R. (1984). *Speech acts in argumentative discussions: A theoretical model for the analysis of discussions directed towards solving conflicts of opinion*. Amsterdam: Walter de Gruyter.
English, B., Mussweiler, T., & Strack, F. (2006). Playing dice with criminal sentences: The influence of irrelevant anchors on experts' judicial decision making. *Personality and Social Psychology Bulletin, 32*(2), 188–200.
Guthrie, C., Rachlinski, J. J., & Wistrich, A. J. (2007). Blinking on the bench: How judges decide cases. *Cornell Law Review, 1*, 1–44.
Hamblin, C. (1970). *Fallacies*. London: Methuen.
Irwin, J., & Daniel, L.R. (2010). Unconscious influences on judicial decision-making. *McGeorge Law Review, 43*, 1–20.

Kahneman, D., & Tversky, A. (1982). On the study of cognitive illusions. *Cognition, 11*, 1123–1141.

Kahneman, D. & Tversky, A. (1996). On the reality of cognitive illusions: A reply to Gigerenzer's critique. *Psychological Review, 103*, 582–591.

Kahneman, D. (2011). *Thinking, Fast and Slow*. New York, NY: Farrar, Strauss and Giroux.

Kenyon, T. (2014). False polarization: Debiasing as applied social epistemology. *Synthese, 191*(11), 2529–2547.

Koriat, A., Lichtenstein, S., & Fischhoff, B. (1980). Reasons for confidence. *Journal of Experimental Psychology: Human learning and memory, 6*(2), 107–118.

Langevoort, D. C. (1998). Behavioral theories of judgment and decision making in legal scholarship: A literature review. *Vanderbilt Law Review, 51*, 1499–1540.

Lord, C.G., Lepper, M.R., & Preston, E. (1984). Considering the opposite: A corrective strategy for social judgment. *Journal of Personality and Social Judgment, 47*(6), 1231–1243.

Martin, Andrew D., Quinn, Kevin M., & Park, Jong Hee (2011). MCMCpack: Markov Chain Monte Carlo in R. *Journal of Statistical Software, 42*(9), 1-21.

Mitchell, G. (2002). Why law and economics' perfect rationality should not be traded for behavioral law and economics' equal incompetence. *Georgetown Law Journal, 91*, 67–167.

Mumma, G.H., & Wilson, S.B. (1995). Procedural debiasing of primacy/anchoring effects in clinical-like judgments. *Journal of Clinical Psychology, 51*(6), 841–853.

Mussweiler, T., Strack, F., & Pfeifer, T. (2000). Overcoming the inevitable anchoring effect: Considering the opposite compensates for selective accessibility. *Personality and Social Psychology Bulletin, 26*(9), 1142–1150.

Pronin, E., Lin, D., & Ross, L. (2002). The bias blind spot: Perceptions of bias in self versus others. *Personality and Social Psychology Bulletin, 28*, 369–381.

Pronin, E., & Kugler, M. (2007). Valuing thoughts, ignoring behavior: The introspection illusion as a source of the bias blind spot. *Journal of Experimental Social Psychology, 434*, 565–578.

Venables, W. N. & Ripley, B. D. (2002) *Modern Applied Statistics with S*. Fourth Edition. Springer: New York.

Willingham, D. (2007). Critical thinking: Why is it so hard to teach? *American Educator, 31*(2), 8–19. (reprinted as: Willingham, D.T. (2008). Critical thinking: Why is it so hard to teach? *Arts Education Policy Review, 109*(4), 21–32.)

Zenker, F., Dahlman, C. (2015). Reliable debiasing techniques in legal contexts? Weak signals from a darker corner of the social science universe. In F. Paglieri (Ed.). *The psychology of argument: Cognitive approaches to argumentation and persuasion* (pp. xx-yy). London: College Publications *(forthcoming)*.

Commentary on Zenker, Dahlman, Bååth and Sarwar's Giving Reasons *Pro Et Contra* as a Debiasing Technique in Legal Decision Making

FABRIZIO MACAGNO
ArgLab, Universidade Nova de Lisboa, Portugal
fabrizio.macagno@fcsh.unl.pt

1. INTRODUCTION

The paper presents an insightful and groundbreaking approach to legal reasoning and argumentation. The fundamental assumption of this work is that biases, or rather latent fallacious reasoning (Zenker, Dahlman, Bååth & Sarwar, 2016, p. 811-812) affect legal reasoning as well, and can result in unwarranted conclusions to be reached. Such biases can be un-triggered by specific techniques, and in this paper the authors assess the effect of a strategy of debiasing character attacks in witness testimony. To this purpose, the authors run a mock test in which in a hypothetical scenario in which a witness is shown to have been convicted for previous crimes. The decision-makers (judges and perspective jurors) are divided in two groups (the control and the experimental group), in which the experimental group is subjected to a debiasing technique (to give reasons for their decision). The authors use quantitative methods to assess the effectiveness of this strategy, but as they report, the results are statistically weak, even though relevant for the purpose of discussing the relationship between the technique and its effects. Despite the efforts of the authors to undermine the relevance of the paper for legal argumentation, this work sheds light on relevant theoretical and practical issues that need to be taken into account, and introduces an extremely interesting method of investigation.

2. THE BIASED REASONING: CHARACTER ASSASSINATION

The authors took into account a specific type of biased reasoning, the commonly called "character assassination" in law that can be "devastatingly effective" (Cantrell, 2003, p. 534; Solomon, 2003, pp. 7–8). By showing that a witness (or a defendant) committed a previous crime, the decision-maker (the judge or the jury) is led to conclude that his testimony is less reliable (or that he committed also the crime he is accused of). This type of attack is commonly analyzed in argumentation theory as the *ad hominem* argument (Walton, 1998, pp. 198–199, p. 217; 2002, p. 51). *Ad hominem* arguments consist in showing that the interlocutor's argument should not be accepted based on a negative judgment on different aspects of his or her character, such as logical reasoning, perception, veracity, or cognitive skills (Macagno, 2013). Clearly, the reasonableness of type of argument depends on the type of argument it is aimed at undermining. *Ad hominem* attacks are often reasonable when they undermine arguments based on the expertise or the position to know of a source, namely authoritative arguments. In these cases, if the truth, or rather acceptability, of the testimony depends on some of the character features attacked, the argument can be reasonable. Otherwise, would be simply irrelevant to the conclusion.

Despite their unreasonableness, often irrelevant *ad hominem* arguments have great impact on the evaluator of an argument, and more specifically in law on the judge or especially the jury. Attacking character does not simply amount to showing a flaw in an argument. It means showing that a person's character is somehow negative, which can lead to negative emotions. Emotions such as indignation, fear, contempt, or hate divert the interlocutor's attention from the rational and systematic assessment of the attack, leading to a conclusion based on fast associations between the emotion and a possible immediate reaction (Blanchette & Richards, 2004; Blanchette, 2006; Macagno, 2014).

Attacking a witness's character is allowed by the rules of evidence at common law. According to rule 609 of the Federal Rules of Evidence, it is possible to introduce evidence of the witness's past convictions in order to impeach his character for truthfulness: "One way of discrediting the witness is to introduce evidence of a prior criminal conviction of the witness, which affords the jury a basis to infer that the witness's character is such that he would be less likely than the average trustworthy citizen to be truthful in his testimony" (*State v. Nash*, 475 So. 2d 752, at 754, 1985). In this sense, evidence of prior misconduct can be a rational ground for assessing the trustworthiness of a witness, one of the possible dimensions that need to be taken into account when judging his testimony. However, this evidence often risks becoming a trigger of a fallacious conclusion reached by means of "fast" reasoning (Kahneman, Slovic, & Tversky, 1982; Tversky & Kahneman, 1974). As pointed out by McCormick (McCormick, 1972, p. 104) "a slashing cross-examination may carry strong accusations of misconduct and bad character, which the witness's denial will not remove from the jury's mind." One of the clearest and most famous examples of the force of this type of character attack is *People v. Simpson* (No. BA 097211, 1995), in which the previous misconduct of the witnessing detective (Fuhrman), combined with proof of racial behavior, led the jury to believe that he was not reliable. This character assassination led to the acquittal of O.J. Simpson.

3. DEBIASING CHARACTER ASSASSINATION

Biased reasoning in law can occur both when the judgment is made by a professional judge, and when it is rendered by a jury of laypeople. In this latter case, in particular, the possibility of the jurors being unaware of the possible fallacious or weak reasoning becomes even higher, as they are not trained to make legal decisions and evaluate objectively the various factors of the case. To this purpose, the authors have first analyzed the so-called debiasing techniques, namely strategy used to un-trigger the latent mechanisms leading to a fallacious conclusion. Such strategies have different foci, depending on the dimension of the automatic reasoning that they intend to address. They can be aimed at turning a latent (implicit) mechanism into an explicit one, or leading the decision-maker to a careful assessment of the force of his conclusion, or simply preventing the decision-making from making some inferences. We can classify the techniques in Table 1:

Making the reasoning explicit	Assessing the conclusion carefully	Preventing inferences
Giving reasons	Accountability	Devil's advocate
	Censorship	Reducing discretion

Table 1: Debiasing techniques

These techniques clearly are some of the possible ones that can be used to reduce the possibility that the decision-maker comes to a conclusion based on problematic implicit arguments. Other possible techniques used in law are confronting the interlocutor with a biased but contrary conclusion or argument, so that the reasoning process becomes subject to careful assessment (Macagno & Walton, 2012).

The authors chose to test the debiasing technique of giving reasons, and they proved that its effects, even though not statistically significant, however indicate that there is a high probability that the difference between the control group and the one subjected to the debiasing intervention is due to the intervention itself (5.6 times more probable than not). However, as the authors point out, this outcome does not meet the statistical requirements for significance. Moreover, when they took into account only the lay judges, they noticed that the debiasing technique worsened the judgment compared to the control group.

4. POSSIBLE PROBLEMS

One of the possible criticisms on the experiment can be addressed to the very mock case that the judges had to assess. Perhaps one of the causes of a lower variability is due to external factors and variables that the authors could not control, due to the vagueness of the case. The case reads as follows:

> Sebastian P is charged for assault. According to the prosecutor's charge, Sebastian P assaulted Victor A, on July 20, 2012 at 23:30 outside a cinema in central Malmö, by repeated blows to the head. Sebastian P testifies that he acted in self-defense and denies the charges. **One of the witnesses in the trial** is Tony T, who was **at the site** on that particular evening. During the examination of the witness Tony T, it emerges that he had recently served **a two-year prison sentence for illegal possession of weapons and arms trafficking**.

The judges had to assess whether the previous conviction of Tony T affects the credibility of his testimony (in a strong, clear, some, or no way). However, the case is too generic and leaves too much room to narratives and possible reconstructions of background information. Considering that a very low percentage of judges indicated a strong or clear effect on the credibility of the witness, a low one cannot be excluded if the judge is allowed to reconstruct information that the case does not specify. Did the witness know the defendant? Was the witness involved in other criminal activities after his conviction? Was the witness somehow related to with the defendant or the victim? All such factors cannot be excluded, and are likely to be reconstructed or taken into account as possible reasons of a biased testimony. At common law, a real case usually involves a cross-examination of the key witnesses, especially when their credibility can be undermined by character issues. While professional judges are trained to evaluate the various factors of the case without making additional hypotheses, this could be not the case for laypeople who simply relate the story with the most accessible narratives (the witness may know the defendant, since they are allegedly both violent, and the witness may want to cover for his friend). If the debiasing technique consists in giving reason, some factors that can be used in such reasons need to be taken into account and controlled.

A second possible problem is the language of the variables. The authors indicate a four-point scale, but they fail to define clearly what "somehow" means when referred to "affecting the witness's credibility." As pointed out by common law cases, evidence of prior convictions is allowed because it is an element that the jury may want to take into account when assessing a witness's trustworthiness compared to an average citizen. Clearly, when no other elements are present, this piece of evidence can be irrelevant for evaluating character. However, when combined with an extremely succinct narration of the circumstances, it is not unreasonable to think that an untrained judge may find possible additional reasons making the testimony somehow less reliable. Perhaps the authors could refine their tests introducing more variability. They could formulate clearer hypotheses, less subject to personal interpretations, including or excluding some circumstantial factor, and then use a Likert scale with more or more definite levels (strongly disagree... strongly agree) to assess them. For example, the test could read as follows:

> •"Tony T's previous conviction for illegal possession of weapons and arms trafficking strongly affects the assessment of his credibility as a witness in the current trial." (strongly agree ... strongly disagree).
> •...
> •"Considering that Tony T had never met the defendant or the victim before, Tony T's previous conviction for illegal possession of weapons and arms trafficking strongly affects the assessment of his credibility as a witness in the current trial." (strongly agree ... strongly disagree).

In this fashion, they could measure how much the interpretation of the event can affect judgment and more importantly the reasons underlying it.

5. CONCLUSION

The authors have focused their self-criticisms on the scarce significance of the overall results. However, this study shows clearly how lay judges and professional ones differ concerning the assessment of a case. This difference becomes even more relevant when we consider the fact that the debiasing technique has opposite effects on laypeople, who provided even worse results when they had to give reasons. This effect should be analyzed in depth, and related to the problem of prejudices and background knowledge. How do prejudices affect the reconstruction of a state of affairs? Perhaps it would be interesting to investigate how a layperson can reconstruct the narrative underlying the whole case, including the relationship between the witness and the defendant, and compare them with the reconstruction of professional judges.

To conclude, the authors perhaps failed to make a statistical point, but opened a very broad range of fundamental questions that should be addressed with the method that they used and that is revolutionary in the field of legal argumentation.

ACKNOWLEDGEMENTS: I would like to thank the Fundação para a Ciência e a Tecnologia for the research grant no. IF/00945/2013.

REFERENCES

Blanchette, I. (2006). The effect of emotion on interpretation of and logic in a conditional reasoning task. *Memory and Cognition, 34*, 1112–1125.
Blanchette, I., & Richards, A. (2004). Reasoning about emotional and neutral materials - Is logic affected by emotion? *Psychological Science, 15*(11), 745–752. http://doi.org/10.1111/j.0956-7976.2004.00751.x
Cantrell, C. (2003). Prosecutorial Misconduct: Recognizing Errors In Closing Argument. *American Journal of Trial Advocacy, 26*, 535–562.
Kahneman, D., Slovic, P., & Tversky, A. (Eds.). (1982). *Judgment under Uncertainty: Heuristics & Biases*. Cambridge: Cambridge University Press.
Macagno, F. (2013). Strategies of Character Attack. *Argumentation, 27*(4), 369–401. http://doi.org/10.1007/s10503-013-9291-1
Macagno, F. (2014). Manipulating Emotions. Value-Based Reasoning And Emotive Language. *Argumentation & Advocacy, 51*(2), 103–122.
Macagno, F., & Walton, D. (2012). Character Attacks as Complex Strategies of Legal Argumentation. *International Journal of Law, Language & Discourse, 2*(3), 1–58.
McCormick, C. (1972). *McCormick's handbook of the law of evidence*. St. Paul, Minnesota: West Publishing Company.
Solomon, R. (2003). *Not Passion's Slave*. New York: Oxford University Press.
Tversky, A., & Kahneman, D. (1974). Judgment under Uncertainty: Heuristics and Biases. *Science (New York, N.Y.), 185*(4157), 1124–1131. http://doi.org/10.1126/science.185.4157.1124

Walton, D. (1998). *Ad Hominem Arguments*. Tuscaloosa: University of Alabama Press.
Walton, D. (2002). *Legal argumentation and Evidence*. University Park: The Pennsylvania State University Press.
Zenker, F. Dahlman, C. Bååth, R. & Sarwar, F. 2016. Giving *Reasons Pro et Contra* as a Debiasing Technique in Legal Decision Making. In D. Mohammed & M. Lewiński (eds.), *Argumentation and Reasoned Action: Proceedings of the 1st European Conference on Argumentation, Lisbon, 2015.* Vol. I, 809-822. London: College Publications.

36

Against Visual Argumentation: Multimodality as Composite Meaning and Composite Utterances

IGOR Ž. ŽAGAR
Educational Research Institute & U. of Primorska, Slovenia
igor.zzagar@gmail.com

> This paper concentrates on the (so-called) visual argumentation, more precisely, *on the impossibility of (pure) visual argumentation, its very vague methodology and epistemology*. Following N. J. Enfield's groundbreaking work *The Anatomy of Meaning* (2009), I will try to show that: every meaning is composite and context-grounded; every meaning is multimodal; any analysis of meaning should be conducted in terms of *enchronic analysis* and reconstructed as *composite utterances*.
>
> KEYWORDS: composite meaning, composite utterances, enchrony, framing, mental spaces, multimodality, polyphony, reasoning, seeing, visual argumentation

1. INTRODUCTION

Journal *Argumentation and Advocacy* is celebrating

> the groundbreaking work on visual argument that appeared in the journal's 1996 (double) issue on visual argument. Since that time, visual argument has become a central topic in argumentation theory and been featured in presented papers and published articles that explore case studies and investigate the possibility of a theory of visual argumentation (published on *Argthry*, 28. 8. 2014).

As an interested bystander who was not (and is not) a partisan of VA nor an active participant in more or less heated debates around VA (at least not so far), I would like to start with a very short overview of these passed twenty years, and then - extensively commenting on Leo Groarke's paper "Six Steps to a Thick Theory" - concentrate on some (basic/necessary) concepts AV is (in my view) lacking, but should be (in my view) incorporated in their conceptual framework in order to better explain how visuals (visual argumentation/persuasion included) function, i.e. how they get/catch the viewers, how the viewers break down the presented visuals, and how they reconstruct their meaning. What I will be concerned with, and what I consider as indispensable concepts for the analysis of visuals is the following: frames (Goffman's experiential as well as Fillmore's semantic frames), polyphony (Ducrot), enchrony and composite meaning (Enfield), mental spaces (Fauconnier), maybe even superdiversity (Vertovec, Blommaert) and rhizome theory (Deleuze).

Multimodality, a very handy, more and more popular and fashionable term these days does (potentially and implicitly) embrace all these concepts in an oblique and undifferentiated way, but these concepts (processes and mechanisms), should be addressed and highlighted separately and explicitly, not just tacitly presupposed under the fancy umbrella of multimodality.

2. TWENTY YEARS IN A NUTSHELL

The way I say these twenty years of development of visual argumentation could be expressed contrastively, almost like an antithesis. On the one hand, the introduction to this double issue of A&A on VA, written by D. Birdsell and L. Groarke twenty years ago, is (understandably) still pretty cautious as to what visuals can do (all emphases are mine):

> - "... the first step toward a theory of visual argument must be a better appreciation of both the *possibility* (!) of visual meaning and the *limits* of verbal meaning" (Birdsell & Groarke, 1996, p. 2);
> - "... "we often *clarify* the latter (i.e., spoken or written words) with visual *cues* ..." (*ibid.*);
> - "*Words* can establish a context of meaning into which *images* can enter with a high degree of specificity while achieving a meaning different from the words alone" (*ibid*, p. 6);
> - "... diagrams can *forward* arguments" (*ibid*);
> - "The implicit verbal backdrop that allows us to *derive* arguments from images is clearly different from the immediate context created by the placement of a caption beside an image" (*ibid.*).

If we sum up: visuals may have some argumentative or persuasive potential (there is a *possibility* of visual meaning, visuals can *forward* arguments, and arguments can be *derived* from visuals) but they are usually (always?) still coupled with the verbal, and can achieve these argumentative effects only (?) in combination with the verbal. And the *pièce de resistance* Birdsell and Groarke are offering to illustrate the claims above (i.e. the possibility of visual argumentation) is an anti-smoking poster, published by the U.S. Department of Health, Education and Welfare in 1976 (I'll be commenting on it later on). Here it is:

Figure 1 (Smoking fish)

On the other hand, in the last five years or so years, visuals are more and more often presented by the proponents of VA as directly and unambiguously offering arguments by themselves, without any intervention or help from the verbal (or any other code), and not being conditioned or in any other way dependent on the verbal at all. Here are two reconstructed examples (I say reconstructed because I was unable to get the original materials from the authors).

The first one is a square ball, used as an example by one of the presenters at the 2014 ISSA conference. It was a small drawing of a square ball (unfortunately, the presenter wouldn't send me the exact drawing) with "China" written on it, obviously cut from some newspaper or magazine, but presented without any immediate context: it wasn't made obvious to which section of the newspaper the visual belonged to (and the presenter would not explain it), nor could we see the neighbouring articles (and the presenter wouldn't explain that either). But he was very explicit in claiming that the argument offered by the visual itself was more than obvious: "The Chinese football sucks!".

The counter-argument came up in the discussion. A colleague in the audience understood the square ball with the "China" inscription on it as a metaphor of corruption in the PRC. I, myself, understood it as a metaphor of a hybrid socio-political system: turbo-capitalism under the leadership of the Central Committee of the CPC. Obviously, the argument was not evident from the drawing itself, otherwise so different interpretations could not have been possible. But, if the drawing would have been framed appropriately (so that we were able to see where in the paper the drawing was published, in which section, or what were the neighbouring articles), such an appropriate and sufficient framing would disambiguate the interpretation(s).

Here is another example of insufficient framing:

Figure 2 (Notre-Dame Gargoyles I)

Figure 3 (Notre-Dame Gargoyles II)

Figure 4 (Notre-Dame Gargoyles III)

A photo resembling the three above (unfortunately, this presenter wouldn't send me the exact photo either) was presented at IPrA conference in New Delhi in 2013, with almost the same words as the square ball at the ISSA 2014 conference: "What the argument is, is obvious from the photo itself".

3. VISUAL ARGUMENTATION AND THE NECESSITY OF FRAMING

But is it, really? Maybe we should recall what already Ch. S. Peirce had pointed out more than eighty years ago (Peirce, 1931-58, 2.172): "Nothing is a sign unless it is interpreted as a sign". In other words, nothing is interpreted as a sign (i.e. e. representing or referring to something else) unless there is *intention* to see it, to understand it as a sign.

And these signs (Figures 2, 3, and 4) can have many different interpretations (if not framed appropriately and sufficiently):

- view of Paris (or one of the views of Paris);
- view of Paris from Notre-Dame;
- Notre-Dame on the background of Paris;
- Postcard greetings from Paris;
- some memorial photos from/of Paris;
- details of Notre-Dame architecture;
- examples of sacral architecture;
- motives from the Notre-Dame outer walls;
- mythological motives from the Notre-Dame architecture;
even
- excerpt from a book on plumbing (these Gargoyles were often used as gutters).

What is my point in enumerating all these? Simply, that *we should first know what the (immediate) context of a visual is*, and *only then proceed with the interpretation and meaning construction*. Or, in Wittgenstein's words (1953, I-#663): "Only when one knows the *story* does one know the *significance* of picture". Which is, if we ponder a bit about this problem, just a corollary of a much more famous 7th thesis from his *Tractatus Logico-Philosophicus*: "Whereof one cannot speak, thereof one must be silent." Applied to visuals, we could paraphrase it as: until we know what the visual is (all) about, we cannot talk about it.

Or, if I may (finally) put it in the terms of what I will be proposing: we have to frame the visual (or the verbal, for that matter), and perform a frame analysis first (i.e. before proceeding to any kind of meaning construction).

3.1 Goffman's frames

Frames I am talking about here are not semantic frames as developed and defined by Charles Fillmore in 1977 (though even semantic frames (may) have a role in potentially argumentative interpretation of visuals as I will try to point out at least fragmentary), but frames that help us organize our everyday experience, frames as developed by sociologist Erving Goffman in his influential book *Frame Analysis: An Essay on the Organization of Experience* (London: Harper and Row, 1974).

What are Goffman's frames? In his own words:

> When the individual in our Western society recognizes a particular event, he tends, whatever else he does, to imply in this response (and in effect employ) one or more frameworks or schemata of interpretation of a kind that can be called primary. I say primary because application of such a framework or perspective is seen by those who apply it as not depending on or harking back to some prior or "original" interpretation; indeed a primary framework is one that is seen as rendering what would otherwise be a meaningless aspect of the scene into something that is meaningful (Goffman, 1974, p. 21).

Goffman distinguishes between natural and social frameworks. Natural frameworks "identify occurrences seen as undirected, unoriented, unanimated, unguided, purely physical" (*ibid.*, p. 22). Social frameworks, on the other hand,

> provide background understanding for events that incorporate the will, aim, and controlling effort of an

intelligence. [...] *Motive and intent are involved, and their imputation helps select which of the various social frameworks of understandings is to be applied* (ibid., p. 24).

So, there are different frames one can apply to a single event/entity, as in our two reconstructed examples with a square ball and the Notre-Dame Gargoyles, but "we tend to perceive events in terms of primary frameworks, and *the type of framework we employ provides a way of describing the event to which it is applied*"(ibid., p. 24).

For a contextualized illustration, let us go back to the smoking fish advertisement (Figure 1). The authors (Birdsell and Groarke) first admit that "visual images can, of course, be vague and ambiguous. But this alone does not distinguish them from words and sentences, which can also be vague and ambiguous"(Birdsell & Groarke, 1996, p. 2). And I agree with that. Than they qualify this poster as "an amalgam of the verbal and the visual" (*ibid*.), which, again, sounds quite acceptable. But then they conclude: "Here the argument that you should be wary of cigarettes because they can hook you and endanger your health *is forwarded by means of visual images...*"(*ibid.* p.3). Which is obviously not the case. Without the verbal part, "don't you get hooked!", the poster could be understood (framed) as a joke, as a cartoon, where, for example, smoking is presented as such a ubiquitous activity that even anglers use cigarettes to catch fish. Only when we add the verbal part, "don't you get hooked!" - where "hooked" activates a (this time semantic) frame of (semantic) knowledge relating to this specific concept, which includes "get addicted", and is, at the same time, coupled with a visual representation of a hook with a cigarette on it - is the appropriate (intended) frame set: the poster is now understood as an anti-smoking add, belonging to an anti-smoking campaign.

3.2 Mental spaces

Equally problematic and ambiguous is the UvA poster Leo Groarke is using in his "Logic, Art and Arguing" (*Informal Logic*, Vol. 18, Nos. 2&3, 1996, p. 112):

Figure 5 (UvA chief administrators)

Groarke's argument goes as follows:

> The black and white photograph [...] presents the university's three chief administrators in front of the official entrance to the university. Especially in poster size, the photograph makes a stark impression, placing all this confident maleness in front of (visually blocking) the university's main entrance. **According to the committee,** which commissioned the poster, it is a "statement" which effectively makes the point that "we want more women at our university" and "still have a long way to go in this regard.

But, if we are not acquainted with the committee's "statement" that they want more women at their university (as, I guess, an "average" Amsterdamer is not), and we just, walking the streets of Amsterdam, bump into this poster with three corpulent males, "stating" "UvA for Women", it is not at all clear how the poster was intended to be framed (by its authors). Is it (simply) a bad joke? Should it be taken ironically, maybe cynically, as a meta-statement from somebody who knows and objects the fact that UvA is all male? There is even a (at least implicitly) sexist interpretation that all these males at UvA need more women.

In other words, because of the insufficiently unambiguous framing it is not at all clear that we (the observers) can (and even should) reconstruct the argument(ation) in question the way Groarke does:

The poster thus presents the argument:

P
↓
C

where the premise P is *the* (visual) statement that "The University of Amsterdam's three chief administrators are all men" and C is the conclusion that "The University needs more women (Groarke, 1996, p. 111).

Even if we take P as rather unambiguous (which it is not; for one thing, the fact that the University of Amsterdam's three chief administrators are all men is not a matter of general knowledge), the arrow, leading to C, is in no way so linear, unidirectional, or monotonic (if you want) as to lead exclusively to C, interpreted as "The University needs more women". C could have had many other interpretations (and P many other formulations, for that matter), for example: "UvA doesn't need women!", "UvA is a sexist institution", "UvA needs some women to change appearances".

Much more appropriate representation of how we can read the UvA poster, and how we should interpret it, could be formulated in terms of mental spaces (nowadays more popularly called blending theory). Like this:

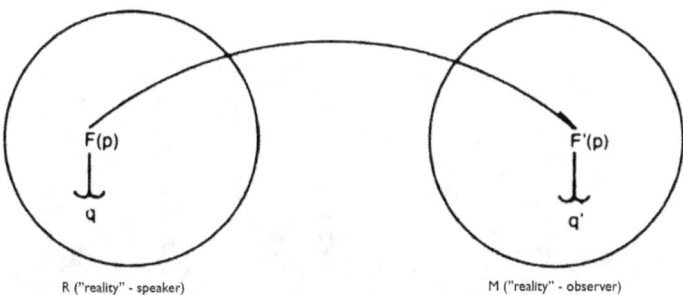

Figure 6 (Construction of meaning in mental spaces)

Figure 6 should be read (interpreted) as follows. R stands for the "reality" of the speaker (speaker's mental space), M for the "reality" of the observer (observer's mental space). *p* represents the poster in question, *F(p)* its (intended) premise, and *q* its (intended) conclusion in R. In M, on the other hand, *p* still represents the same poster in question (hence the long arrow connecting the two spaces), but *F'(p)*, the observer's premise, and *q'*, the observer' s conclusion, may be quite different from speaker's premise and speaker's conclusion (depending on the observer's experience, social and cultural background, education, gender, and many other, even bio-neurological and cognitive factors). On top of that, M spaces may be multiplied in relation to R space, precisely because of observers' different (social, cultural, etc.) background, education, gender, and many other factors.

3.3 Polyphony

A bit different mechanism seems to be at work in Marlboro advertisements Asimakis Tseronis used at the Brač Argumentation Conference in 2012. Actually, these were not advertisements but "subvertisements", produced by a group called Adbusters (a name that is rather indicative as to what they are doing to advertisements).

Chronologically, the original advertisements come first, of course. The background is always the American (Wild?) West, represented in warm, yellowish and brownish colors, and in the foreground there is always one or several cowboys. They may be smoking or not, but a pack of Marlboro cigarettes together with the company's logo is always highly visible and sets the frame (= we are talking cigarettes advertisement here, not, for example, westerns, or horse breeding). Just like in this advertisement:

Figure 7 (Marlboro cowboys - original)

What do Adbusters do to original ads? They can't use the company's logo and packs of cigarettes, of course, so they use the standardized Marlboro background (warm, yellowish and brownish colors in the background, several cowboys in the foreground) to activate the appropriate frame with the observers (= this is (about) Marlboro). And the text within this familiar "Marlboro country", implicitly and indirectly, alludes to the missing packs of cigarettes.
Like in Figure 8:

Figure 8 (Marlboro cowboys - original)

Maybe even more efficient is the following parody (Figures 9 and 10). On the original advertisement we see cowboys on horses in a winter landscape, with Marlboro packs in the lower right corner:

Figure 9 (Marlboro Country - original)

On the "busted" version, we just see horses in an empty graveyard, covered with snow, while the tombstones symbolically replace packs of cigarettes (in the original version, presented by Tseronis, the caption 'Marlboro Country' is missing).

Figure 10 (Marlboro Country - busted)

What is the mechanism at work here? It appears that a kind of "gestalt" (warm, yellowish/brownish colors in the background, cowboys in the foreground ...) sets the frame (= Marlboro advertisements), while the text or the setting in the photo activates a (kind of) polyphonic reading (not a semantic frame): we can only make sense of and understand the busted advertisement if we connect it to the original advertisement, i.e. we can only understand it on the background of the original ad, i.e. as a kind of meta-ad.

And when I am mentioning polyphony, I am referring to Bakhtin, of course, but even more explicitly to Ducrot's theory of polyphony, informed by Bakhtin, but much more elaborated. You may recall that Ducrot (2009, pp. 32-44) is distinguishing between a producer, a locutor and several enunciators/utterers or uttering positions. A producer is the person/organization ... that is the "material" author of a given piece of text (or visual). In our case, the producer(s) would be the Adbusters (and their collaborators), the people who produced the anti-ad in question, those who had the idea, set the scenery, took the photo, developed it, and so on ...

A locutor is the person/organization ... that is (symbolically) responsible for the message of the ad. In our case the message could be reconstructed as something like: "Smoking kills". But this (meta)message is obviously only possible because there is an interplay of (at least two) enunciators or uttering positions; the first one declaring that smoking is cool/attractive/adult (the original Marlboro ads) ..., and the second one subverting, criticizing such a position (the Adbuster ads). And the criticism prevails as the main message (in the Adbuster ads).

4. AN INTERLUDE: RHIZOME AND SUPERDIVERSITY IN VISUAL ARGUMENTATION

At this point, it may be worth mentioning that in dealing with visuals, with construction of meaning and interpretation in visuals, we are obviously dealing with the so-called rhizomatic structure and rhizomatic reading.

Rhizome is (philosophical) concept developed in 1980 by G. Deleuze and F. Guattari (*A Thousand Plateaus*, London and New York, Continuum, 2004), and defined as theoretical approach allowing for multiple, non-hierarchical entry and exit points in data representation and interpretation. In a nutshell, any point of rhizomatic structure can be connected to any other, and ceaselessly establishes connections between (different) semiotic chains, organizations of power, and circumstances relative to the arts, sciences, and social struggles. Something we tried to show in section 3.

Rhizome and rhizomatic structures become conceptually especially interesting if coupled and supplemented with a (rather) new sociological concept that is rapidly gaining importance, the concept of superdiversity. Superdiversity is a concept coined by sociologist Steven Vertovec, and he defines it as (Vertovec, 2007, p. 1025):

> [...] a dynamic interplay of variables among an increased number of new, small and scattered, multiple-origin, transnationally connected, socio-economically differentiated and legally stratified immigrants who have arrived over the last decade.

And what could be the significance of this new concept for the analysis and interpretation of visuals? Exactly the possibility that increasingly different cultural, educational, and ideological background of potential readers/interpreters (not necessarily immigrants, of course), may imply even more different access points and interpretational paths in reading and interpreting visuals. In other words, the allegedly unidirectional and unproblematic arrow connecting P and C in Leo Groarke's interpretation of the UvA poster may not just be multiplied in different ways, pointing in different directions, but may also change its shape, from straight to wavy or curved or even broken, depending on how complex the meaning and possibilities of its interpretation may be. Which also implies that possible C's may come in different forms and formulations.

5. "THE REASONING IS THE SEEING"?

This is the reason why visual argumentation should *concentrate more on different possible entry and exit points in data representation and interpretation of hypothetical visual arguments*. As a kind of a case study - exposing possible *caveats* as well as *cul-de-sacs* of visual argumentation - we will concentrate on Leo Groarke's recent proposal of reconstructing visual arguments as presented and conceptualized in his 2013 article "The Elements of Argument: Six Steps to a Thick Theory", published in the e-book *What do we know about the world? – Rhetorical and Argumentative perspectives*.

Here is the photo Groarke is taking as a starting point:

Figure 11 (Fruit found on the Detroit river I)

If we just take the photo in Figure 11 *per se*, as it is (as we see it *prima facie*), without or before any verbal explanation, and not knowing anything about possible context(s), the photo could be framed in many ways. As, for example:

1) introducing/showing a peculiarly looking fruit;
2) preparing a snack (or some other kind of meal);
3) showing/presenting a new knife;
4) showing/presenting an efficient/robust/... knife;
5) showing the protective gloves, or how do protective gloves look like/how we use them;
6) warning that one should wear protective gloves when using a knife (demonstrating safety procedures),
and there are many other way.

Against visual argumentation

But Groarke does disambiguate the photo rather quickly with the following explanation (all emphases throughout the text that will follow are mine):

> Consider a debate spurred by an unusual fruit I discovered during a kayak ride *on the Detroit River*. When my description ("nothing I recognize; a bumpy, yellow skin") initiated a debate and competing hypotheses on the identity of the fruit, I went back and *took the photographs reproduced below. On the basis of these photographs, the fruit was quickly identified as breadfruit.*

So the frame in question is the first one mentioned: introducing/showing a peculiarly looking fruit. And here is how Groarke reconstructs the argument (actually the process of arriving from argument(s) to conclusion) in question:

> The argument that established this conclusion *compared my photographs to similar photographs found in encyclopaedia accounts of breadfruit.* One might summarize the reasoning as: "The fruit is breadfruit, for these photographs are like standard photographs of breadfruit." But this is just a verbal paraphrase. *The actual reasoning – what convinces one of the conclusion - is the seeing of the sets of photographs in question.* Using a variant of standard diagram techniques for argument analysis, we might map the structure of the argument as:

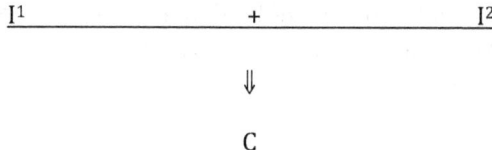

> where C is the conclusion that the fruit is a piece of breadfruit, I^1 is the set of photographs I took, and I^2 is the iconic photographs of breadfruit to which they were compared.

5.1 Comparing the visuals as an argumentative procedure...

But should (and does) the reasoning really consist just of "the seeing of the sets of photographs in question"? Is just seeing and visually comparing photographs *from different sources* really enough for a reasoned, justified conclusion (in question)? And last but not least, let us not neglect Groarke's remark that "on the basis of these photographs, the fruit was *quickly* identified as breadfruit". Is the velocity of (visual?)

reasoning to be considered a necessary and sufficient criterion for good argumentation?

Let us try to replicate Groarke's procedure. Here are some photos of breadfruit found in different encyclopaedias:

Figure 12 (Breadfruit at Tortuguero; Wikipedia, https://en.wikipedia.org/wiki/Breadfruit)

Figure 13 (The fruit of the breadfruit tree - whole, sliced lengthwise and in cross-section; Wikipedia, https://en.wikipedia.org/wiki/Breadfruit)

Figure 14 (Breadfruit; Healthy Benefits, http://healthybenefits.info/the-health-benefits-of-consuming-bread-fruit%E2%80%8F/)

And here, again, are Groarke's two photos (from the point of view of perception, processing and meaning construction, it is important for the viewer that they are incorporated between new photos (of breadfruit), and not just referred to by numbers (e.g. Figure 11): the one we have already seen:

Figure 11 (Fruit found on the Detroit river I)

and the one we haven't seen yet:

Figure 15 (Fruit found on the Detroit river II)

Now, please inspect these photos carefully. Is there really such a resemblance between the two represented fruits that we can *quickly* identify the fruit from the Detroit river as breadfruit? I don't think so, or to put it in Groarke's words, I don't see that resemblance.

Breadfruit, as we have seen, has a kind of knobbly skin with spines or hard hairs, patterned with irregular, 4- to 6-sided face, while in the center there is a cylindrical core. On the other hand, the skin of the fruit found in the Detroit River seems smooth, without spines or hairs, covered with smooth irregular bumps, no 4- to 6-sided face, and there seems to be no cylindrical core in the center (though that may be due to the lightning, the angle or some other disturbing factor).

5.2 ... and disambiguating it with verbal information

In such a case (where some items/entities look alike, but don't exactly the same; though it seems that Groarke was quite satisfied with the comparison, and has even *quickly* arrived at the conclusion that the fruit from the Detroit river was in fact a breadfruit), just "seeing" is not enough, and it is wise if not necessary to consult other reliable sources, like verbal description. All encyclopaedias usually have them (and that is one of the reasons they are called encyclopaedias).

Why verbal descriptions? Simply, because in such a case there is not much else one can consult. On the other hand, language is still the only communicative "medium" that is (rather) linear, straightforward, and unambiguous enough; in combination with pertinent visuals almost error-proof. And if, when consulting encyclopaedias or other relevant sources, *we don't just check the photos, but the text as well*, we find the following description of breadfruit (please, pay special attention to emphases in italics):

> Breadfruit (*Artocarpus altilis*) is one of the highest-yielding food plants, with a single tree producing up to 200 or more fruits per season. *In the South Pacific*, the trees yield 50 to 150 fruits per year. *In southern India*, normal production is 150 to 200 fruits annually. Productivity varies between wet and dry areas. *In the Caribbean*, a conservative estimate is 25 fruits per tree. Studies *in Barbados* indicate a reasonable potential of 6.7 to 13.4 tons per acre (16-32 tons/ha).
> /.../
> Breadfruit, *an equatorial lowland species*, grows best below elevations of 650 metres (2,130 ft), but is found at elevations of 1,550 metres (5,090 ft). Its preferred rainfall is 1,500–3,000 millimetres (59–118 in) per year.
> /.../
> Breadfruit is a staple food *in many tropical regions*. The trees were propagated far outside their native range by Polynesian voyagers who transported root cuttings and air-layered plants over long ocean distances. (From Wikipedia: http://en.wikipedia.org/wiki/Breadfruit)

If we sum up, breadfruit is a tropical plant, usually found (and used) in tropical areas. It is, therefore, not very probable to find it in Ontario, in the Detroit river, though it is not completely impossible, of course, that a specimen of a breadfruit found its way into the Detroit river from one of the (not so many) local Caribbean restaurants or stores.

But if relevant sources (encyclopaedias...) were indeed amply consulted (i.e. browsed through), and the point of departure in investigating the nature of the found fruit was not based on some kind of preconceived idea or a hunch that the Detroit river fruit looked very much like breadfruit, a neutral, objective and interested investigator should have easily found the following photos as well:

Figure 16 (*Maclura pomifera*; Wikimedia Commons, https://commons.wikimedia.org/wiki/File:Maclura_pomifera_Inermis_BotGardBln1105Fruits.jpg)

Figure 17 (*Maclura pomifera*; Plants for a Future, http://www.pfaf.org/user/Plant.aspx?LatinName=Maclura+pomifera)

Figure 18 (*Maclura pomifera*; Acta Plantarum, http://www.actaplantarum.org/acta/galleria1.php?aid=463)

Figure 19 (*Maclura pomifera;* Wikimedia Commons, <https://commons.wikimedia.org/wiki/File:Maclura_pomifera_FrJPG.jpg>)

And once more, here are the two photos of a fruit found in the Detroit River:

Figure 11 (Fruit found on the Detroit river I)

Figure 15 (Fruit found on the Detroit river II)

A close comparative observation between encyclopedic photos of this second fruit and the photos of breadfruit reveals that this second fruit looks much more like the fruit found in the Detroit river: its skin seems smooth, without spines or hairs, and it is covered with smooth irregular bumps, not 4- to 6-sided face as in the bread fruit.

And if we consult the verbal part of the encyclopaedia, connected to this fruit, we find the following (once more, please, pay attention to emphases in italics):

> *Macula pomifera*, commonly called *Osage orange, hedge apple, horse apple, bois d'arc, bodark, or bodock* is a small deciduous tree or large shrub, typically growing to 8–15 meters (26–49 ft) tall. It is dioecious, with male and female flowers on different plants. The fruit, a multiple fruit, is *roughly spherical, but bumpy*, and 7.6–15 centimeters (3–6 in) in diameter. It is filled with sticky white latex. *In fall, its color turns a bright yellow-green.*
> /.../
> Osage orange occurred historically in the Red River drainage of Oklahoma, Texas and Arkansas and in the Blackland Prairies, Post Oak Savannas, and Chisos Mountains of Texas. *It has been widely naturalized in the United States and Ontario.*
> (from Wikipedia:
> http://en.wikipedia.org/wiki/Maclura_pomifera)

As you can see for yourself, the verbal description of *Macula pomifera* actually fits the Detroit river fruit much more accurately than the description of breadfruit. And since we learn that the Osage orange "has been widely naturalized in the United States *and Ontario*" it is much more probable that it fell in the water someplace along the Ontario river than that it found its way into the river from one of the Caribbean facilities in Ontario. (Another way of starting the argumentative search and arriving at (the same) conclusion would be using the framework of (different) mental spaces again. But we don't have time for this here and now.)

6. DO PICTURES TELL A THOUSAND WORDS?

What can we learn from this? Above all that sayings like: "A picture tells a thousands words" should be indeed taken seriously. *But*, to be (absolutely) sure *which of these thousands words* refer to that particular picture we have in front of us in these particular circumstances, *we have to cut down (on) those word considerably*. On the other hand, *without any words at all, we can hardly identify the exact meaning of the picture!*

In other words, there seem to be no pure visual arguments (as there are, probably, very few purely verbal arguments; if any at all), and instead of visual argumentation (or purely verbal argumentation, for that matter) we should (always) talk about multimodal argumentation and multimodal meaning (combining, in our case, at least visual and verbal, but other semiotic modes are usually involved as well, such as gesture and gaze). But multimodal meaning and multimodal argumentation require different (expanded, at least) analytical framework, let us simply call it multimodal analysis. And in relation to that, I would like to emphasize a few points.

In cases where just "seeing" is not enough, and we have to consult verbal (or other) sources (and incorporate other types of signs, like gestures, gazes...), we should be talking of *enchronic analysis* (Enfield, 2009). What is enchronic analysis?

> Enchronic analysis is concerned with *relations between data from neighbouring moments*, adjacent units of behaviour in locally coherent communicative sequences (Enfield, 2009, p. 10).

Enchronic analysis is therefore looking at *sequences of social interaction in which the moves that constitute social actions occur as responses to other such moves, and in turn these give rise to further moves*. The Detroit river fruit is exactly a case in point: from observation of the photos of the fruit taken on the river, we have to move to the observation of the photos in encyclopedias. And to get more complete and accurate information we have to switch from photos to text, and incorporate the textual information as well. And to fine-tune our findings (understanding), we have to switch to yet other photos (if necessary), and from them to yet another text(s) (if necessary), and finally compare all these again with the initial photo (of the fruit taken on the river).

If, when consulting encyclopaedias, we really do that, i.e. *we don't just check the photos, but the text as well*, and then go and (re)check other texts and photos, and compare them with the initial photo(s), the final result we arrive at should be described as *composite meaning*, resulting in *composite utterances*, conceptualized as: " [...] we may define the composite utterance as a *communicative move that incorporates multiple signs of multiple types*" (Enfield, 2009, p. 15).

Here is a visual example of a composite sign (with composite meaning), Enfield is using himself:

Figure 20 (Willy Brandt in Warsaw Ghetto)

And this is his analysis (*ibid.*, pp. 3-4):

> While the kneeling posture may have an intrinsic, ethological basis for interpretation, this particular token of the behaviour has had a *deeply enriched meaning* for many who have seen it, because it was performed by this particular man, at this time and place. The man is Willy Brandt, chancellor of West Germany. *Once you know this, the act already begins to take on enriched meaning.* It is not just a man kneeling, but a man whose actions will be taken to stand for those of a nation's people. It is 7 December 1970, a state visit to Warsaw, Poland. These *new layers of information should yet further enrich your interpretation*. To add another layer: the occasion is a commemoration of Jewish victims of the Warsaw Ghetto uprising of 1943. /.../ The body posture [...] is a *composite sign in so far as its meaning is partly a function of its co-occurrence with other signs*: in particular, the role being played by its producer, given the circumstances of its time and place of production. The behaviour *derives its meaning as much from its position on these coordinates as from its intrinsic significance.*

7. A SHORT CONCLUSION

If after checking and re-checking different photos, different texts, and the strange fruit that was found in Detroit river, we finally point (and probably gaze) at it, declaring: "This fruit is (not) a bread fruit!", we have produced a composite utterance, (enchronically) embracing several ((at least) seven) layers of meaning, belonging to three types of signs (conventional signs: words/text; non-conventional signs: photos, gesture, gaze; symbolic indexicals: demonstrative pronoun "this"). Therefore, to gain analytic credibility and interpretive force, visual argumentation should consider incorporating into its framework all these gradual steps, as well as all these mutually dependent concepts.

REFERENCES

Birdsell, D. S., & Groarke, L. (1996). Toward a theory of visual argumentation. *Argumentation and Advocacy, 33*(1), 1-10.
Blommaert, J., & Rampton, B. (2011). Language and superdiversity. *Diversities, 13*(2), 1-21.
Deleuze, G., & Guattari, F. (1980/2004). *A Thousand Plateaus*. London and New York: Continuum.
Ducrot, O. (1996/2009). *Slovenian Lectures*. Ljubljana: Pedagoški inštitut/Digitalna knjižnica.
Enfield, N. J. (2009). *The Anatomy of Meaning*. Cambridge and New York: Cambridge University Press.
Fauconnier, G. (1984). *Espaces mentaux*. Paris: Minuit.
Goffman, E. (1974). *Frame Analysis: An Essay on the Organization of Experience*. London: Harper and Row.
Groarke, L. (1996). Logic, art and arguing. *Informal Logic, 18*(2&3), 105-129.
Fillmore, Ch. J. (1977). Scenes-and-frames semantics. In A. Zampolli (Ed.), *Linguistic Structures Processing* (pp. 55-81). Amsterdam: North-Holland.
Peirce, Ch. S. (1931-58). *Collected Writings* (8 Vols.). (Ed. C. Hartshorne, P. Weiss & A. W. Burks). Cambridge, MA: Harvard University Press.
Tseronis, A. (2012). Refuting claims visually: the case of subvertisements. PPT presentation from Dani Iva Škarića conference, Postira, Brač, Croatia.
Vertovec, S. (2007). Super-diversity and its implications. *Ethnic and Racial Studies, 30*(6), 1024-1054.
Wittgenstein, L. (1953/1986). *Philosophical Investigations* (Transl. by G. E. M. Anscombe). Oxford: Basil Blackwell.

www.ingramcontent.com/pod-product-compliance
Lightning Source LLC
Chambersburg PA
CBHW071147230426
43668CB00009B/866